McGraw-Hill Yearbook of Science and Technology 1970 REVIEW

1971 PREVIEW

McGraw-Hill **Yearbook of**

McGRAW-HILL BOOK COMPANY

NEW YORK   ST. LOUIS
                SAN FRANCISCO
DUSSELDORF   NEW DELHI
JOHANNESBURG   PANAMA
KUALA LUMPUR   RIO DE JANEIRO
LONDON   SINGAPORE
MEXICO   SYDNEY
MONTREAL   TORONTO

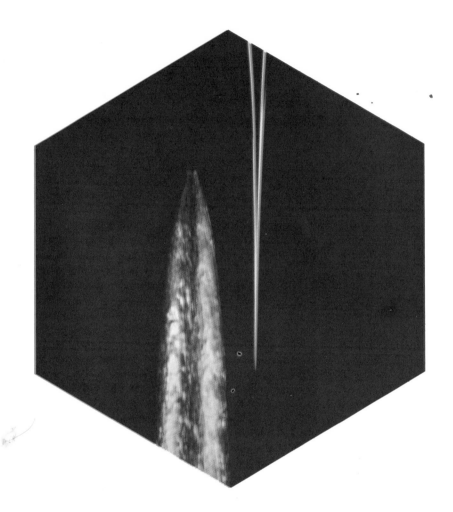

Science and Technology

COMPREHENSIVE COVERAGE OF

THE IMPORTANT EVENTS OF THE YEAR

1970

AS COMPILED BY THE STAFF OF THE

McGRAW-HILL ENCYCLOPEDIA OF SCIENCE AND TECHNOLOGY

On preceding pages:

Left. A mercury drop under the influence of vibration. (Photograph by J. C. Stuten, in H. Jenny, Cymatics, Basilius Presse, 1967)

Right. The micropipette electrode, dwarfed by the point of a common household pin, measures intracellular electrical potentials or changes. (Corning Glass Works)

Library of Congress Catalog Card Number: 62-12028

International Standard Book Number: 07-045225-3

Table of Contents

Editorial Advisory Boards

1960—1970

Dr. Roger Adams
Professor Emeritus of Chemistry
Formerly Head, Department of Chemistry and
Chemical Engineering
University of Illinois

Dr. Detlev Bronk
President (retired), The Rockefeller University

Dr. Sidney D. Kirkpatrick
Formerly Editorial Director,
Chemical Engineering and *Chemical Week*
McGraw-Hill Publishing Company

Dr. Joseph W. Barker
Consulting Engineer
President and Chairman of the Board (retired)
Research Corporation, New York

Dr. George R. Harrison
Dean Emeritus, School of Science
Massachusetts Institute of Technology

Dr. William W. Rubey
Department of Geology and
Institute of Geophysics
University of California, Los Angeles

1968—1970

Dr. Neil Bartlett
Professor of Chemistry
University of California, Berkeley

Dr. Richard H. Dalitz
Department of Theoretical Physics
University of Oxford

Dr. Freeman J. Dyson
The Institute for Advanced Study
Princeton, N.J.

Dr. Leon Knopoff
Professor of Physics and Geophysics
University of California, Los Angeles

Dr. H. C. Longuet-Higgins
Department of Machine Intelligence and Perception
University of Edinburgh

Dr. A. E. Siegman
Professor of Electrical Engineering
Stanford University

Dr. Hugo Theorell
The Nobel Institute
Stockholm

Lord Todd of Trumpington
University Chemical Laboratory
Cambridge University

Dr. E. O. Wilson
Professor of Zoology
Harvard University

Editorial Staff

Daniel N. Lapedes, *Editor in Chief*

David I. Eggenberger, *Executive Editor,
Professional and Reference Books Division*

Gerard G. Mayer, *Director of Design and Production*

Marvin Yelles, *Senior editor*

Edward J. Fox, *Art Director*

Joe Faulk, *Copy manager*

Richard A. Roth, *Art editor*

George Ryan, *Copy editor*

Ann D. Bonardi, *Art coordinator*

Patricia Walsh, *Editing assistant*

Kathleen Skultety, *Editing assistant*

Consulting Editors

Consulting Editors (continued)

Dr. Edward A. Martell. *National Center for Atmospheric Research, Boulder, Colorado.* GEOCHEMISTRY.

Dr. Harold B. Maynard. *President, Maynard Research Council, Inc.* INDUSTRIAL AND PRODUCTION ENGINEERING.

Dr. Bernard S. Meyer. *Professor and Chairman, Department of Botany and Plant Pathology, The Ohio State University.* PLANT PHYSIOLOGY.

Dr. Jacob Millman. *Department of Electrical Engineering, Columbia University.* ELECTRONIC CIRCUITS.

Dr. William Mosher. *Chairman, Department of Chemistry, University of Delaware.* ORGANIC CHEMISTRY.

Dr. N. Karle Mottet. *Professor of Pathology and Director of Hospital Pathology, University of Washington.* ANIMAL PATHOLOGY.

Dr. Royce W. Murray. *Assistant Professor of Chemistry, University of North Carolina.* ANALYTICAL CHEMISTRY.

Dr. Robert H. Noble. *Optical Science Center, University of Arizona.* ELECTROMAGNETIC RADIATION AND OPTICS.

Dr. Harry F. Olson. *Staff Vice President, Acoustical and Electromechanical Laboratory, RCA Laboratories.* ACOUSTICS.

Dr. Jerry Olson. *Oak Ridge National Laboratory.* CONSERVATION; PLANT ECOLOGY.

Dr. Harold S. Osborne. *Consultant.* TELECOMMUNICATIONS.

Dr. Guido Pontecorvo. *Imperial Cancer Research Fund, London.* GENETICS AND EVOLUTION.

Prof. K. R. Porter. *Chairman, Department of Molecular, Cellular and Developmental Biology, University of Colorado.* CYTOLOGY.

Prof. C. Ladd Prosser. *Head, Department of Physiology, University of Illinois.* COMPARATIVE PHYSIOLOGY.

Brig. Gen. Peter C. Sandretto. *Director, Engineering Management, International Telephone and Telegraph Corporation.* NAVIGATION.

W. C. Schall. *Goldwater, Valente, Fitzpatrick & Schall, Members of the New York Stock Exchange.* COMPUTERS.

Dr. Bradley T. Scheer. *Head, Department of Biology, University of Oregon.* GENERAL PHYSIOLOGY.

Prof. Frederick Seitz. *President, The Rockefeller University.* SOLID-STATE PHYSICS.

Dr. Raymond Siever. *Department of Geological Sciences, Harvard University.* GEOLOGY (SURFICIAL AND HISTORICAL); PHYSICAL GEOGRAPHY.

C. Dewitt Smith. *Mining consultant, Dewitt Smith and Co., Inc., Salt Lake City, Utah.* MINING ENGINEERING.

Dr. Fred M. Snell. *Dean, Graduate School, State University of New York at Buffalo.* BIOPHYSICS.

Dr. Mott Souders. *Formerly Director of Oil Development, Shell Development Company.* PETROLEUM CHEMISTRY.

Prof. William D. Stevenson, Jr. *Department of Electrical Engineering, North Carolina State of the University of North Carolina at Raleigh.* ELECTRICAL POWER ENGINEERING.

Dr. Horace W. Stunkard. *Research Associate, Invertebrate Zoology, American Museum of Natural History.* INVERTEBRATE ZOOLOGY.

Dr. E. L. Tatum. *The Rockefeller University.* BIOCHEMISTRY.

Dr. Aaron J. Teller. *Teller Environmental Systems, New York, N.Y.* CHEMICAL ENGINEERING.

Dr. Garth Thomas. *Center for Brain Research, University of Rochester.* PHYSIOLOGICAL AND EXPERIMENTAL PSYCHOLOGY.

C. N. Touart. *Research Physicist, Air Force Cambridge Research Laboratory.* GEOPHYSICS.

Dr. Henry P. Treffers. *Professor of Microbiology, Yale University School of Medicine.* MEDICAL MICROBIOLOGY.

Prof. H. H. Uhlig. *Department of Metallurgy, Massachusetts Institute of Technology.* METALLURGICAL ENGINEERING.

Prof. William W. Watson. *Department of Physics, Yale University.* ATOMIC, MOLECULAR, AND NUCLEAR PHYSICS.

Contributors

A list of contributors, their affiliations, and the articles they wrote will be found on page 447.

Preface

The 1971 *McGraw-Hill Yearbook of Science and Technology* ushers in our coverage of scientific and technical advances of the 1970s. This Yearbook, the tenth volume in the series, continues to analyze and report in detail significant scientific and technical achievements of the previous year. As an individual book it provides a comprehensive record of progress and accomplishment. As a supplement it updates and enriches the basic material in the *McGraw-Hill Encyclopedia of Science and Technology*, third edition (1971), and previous editions. The 1971 Yearbook's format, size, and color usage are patterned after those of the third edition of the Encyclopedia. Following the organization established in previous Yearbooks, the 1971 edition begins with major articles on six subjects selected for their broad interest and growing significance. The second part, a selective pictorial section, features a number of the outstanding scientific photographs of the past 12 months. The third part consists of alphabetically arranged articles on the advances, discoveries, and developments in science and technology during the past year.

The choice of the subject matter in another year of accelerating scientific achievement was the work of 67 consulting editors and the editorial staff of the *McGraw-Hill Encyclopedia of Science and Technology*. But most of the credit should go to the 194 eminent specialists who contributed to the present volume. Their interest, knowledge, and writing ability make them the real creators of the 1971 Yearbook.

DANIEL N. LAPEDES
Editor in Chief

Eugene D. McCarthy carried out research studies relating to the origin of life with Melvin Calvin at the University of California, Berkeley. At Harvard University, he was a Helen Hay Whitney Research Fellow. He is now professor of biochemistry at the Universidad Católica, Valparaiso, Chile.

Origin of Life

THE SUCCESSFUL COMPLETION OF THE *Apollo 11* mission in July, 1969, was the appropriate culmination of a decade that has produced the most definitive evidence for the existence of life at the earliest periods of geological history. The mission provided the first extraterrestrial samples uncontaminated by biological detritus from this planet. For the scientist, therefore, the analysis of the lunar samples constituted a unique landmark. In the first place, they offered the first opportunity of applying a series of analytical techniques, of remarkable sophistication and precision, to determine the organic carbon content of an extraterrestrial sample that had not been exposed to terrestrial contamination on its journey to Earth. In the second place, in the event of finding organic matter, the very criteria that established the biological origin of such material in the Earth's most ancient sediments could now be applied to the lunar sample to determine the origin of its organic matter. Thus it seemed for a while that one was on the threshold of answering the question that has intrigued scientists and philosophers for centuries: Is life as we know it unique to this planet?

The results that have emerged from a rigorous and systematic analysis of the organic content of the first lunar samples have inevitably caused something of an anticlimax. The most optimistic expectations have not been fulfilled, since there has, as yet, been no unambiguous characterization of indigenous organic material at the lowest levels of detection. Moreover, the micropaleontological examination of the lunar sample has produced no evidence of microfossil organisms.

Such a negative picture must be placed in proper perspective. The chemical analyses from the Surveyor lunar landings in 1967 indicated that the maria surfaces were strikingly devoid of carbon, and in this respect showed no resemblance to the carbonaceous meteorites. It was not surprising, therefore, that the returned lunar surface samples should contain no indigenous organic matter. Ideally, cores taken from considerable depth and from a series of differing locations would be preferable for analysis. Thus the results so far represent at best nothing more than a preliminary examination.

POSSIBILITY OF EXTRATERRESTRIAL LIFE

Estimates have been made which indicate that the number of planets suitable for life in the known universe is $\sim 10^{17}$. Moreover, there is every reason to believe that organic carbon compounds could be present somewhere in an extraterrestrial environment. Carbon is considered to be the fifth most abundant element in the solar system.

Meteorites. Indigenous carbon is found in many meteorites, usually in a reduced form other than graphite or inorganic carbides. Because all meteorites have been exposed to terrestrial contamination at some period in their history, conclusions drawn from any meteoritic organic analysis are questionable. This has been strikingly demonstrated by the recent analysis of the Pueblito de Allende meteorite, which fell in Mexico in February, 1969. It contained 0.35% carbon. The sample was analyzed within a month of falling and yet was shown to contain considerable contamination by organic material of terrestrial origin. Such a finding seriously undermines any interpretations regarding the origin of organic compounds found in meteorites of much longer terrestrial history.

Mars. While the meteoritic controversy has remained at an impasse, others have speculated on the existence of life in supposedly more favorable environments. N. H. Horowitz has given some consideration to the possibility of life on Mars. In spite of an average temperature of −55°C, an atmosphere of carbon dioxide, and only a small amount of water vapor, he claims that none of these conditions preclude the surface of Mars as a possible abode of life. Martian conditions are similar to those encountered around frozen lakes in Antarctica, where microbial life has been detected. However, the scarcity of water on the planet is the most serious objection to the existence of life. Such speculations, though not without intellectual value, contribute little toward a definitive answer to the question of whether extraterrestrial life exists.

Abiogenic-biogenic controversy. Most recently, one specific and extremely crucial problem has perhaps received more attention than any other. It concerns the attempt to find criteria which will determine whether organic material, isolated from the geological environment, has an abiogenic or biogenic origin. Directly related to this problem has been the postulate that it should be possible to recognize, at a molecular level, the transition from chemical evolution to biological evolution that must have occurred in Precambrian times. What molecular criteria can we establish that would detect this boundary in the geological environment? And, as a corollary to this question, if organic material is unambiguously detected in an extraterrestrial environment, is it possible to determine the origin of such material? Before attempting to answer such questions, it is necessary to reexamine the philosophy and recent history of the molecular approach to evolution, which received its real impetus at the beginning of the 1960s.

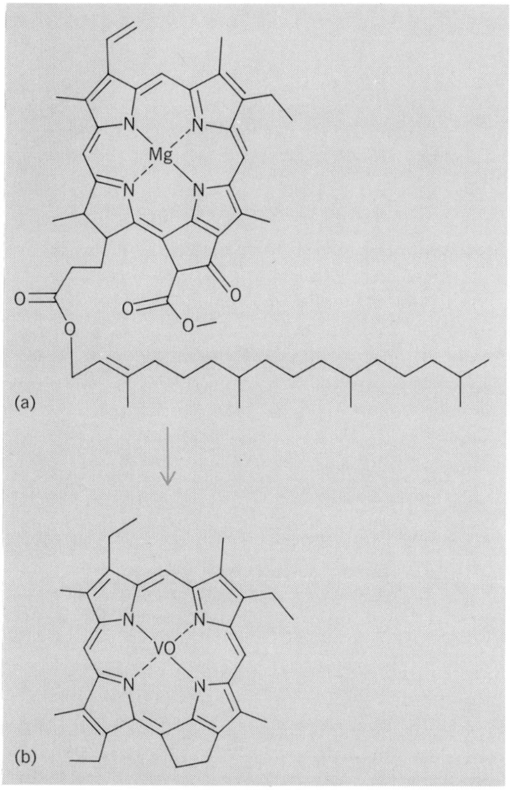

MOLECULAR FOSSILS

There are two possible approaches to gaining an understanding of what the evolutionary history of the Earth might have been. The chemical evolutionary approach has endeavored to reconstruct the early history of the Earth by trying to simulate, in the laboratory, the types of chemical reactions that could have taken place and given rise to the simple organic molecules that constitute living organisms today. The other approach has been to extend the classical approach adopted, in part, by Charles Darwin.

Chemical evolutionary approach. Before a certain period of geological time, approximately 6×10^8 years ago (at the start of the Cambrian), the morphological record becomes virtually extinct. It has become possible to gain insights into the Earth's evolutionary development prior to Cambrian time as a result of two major scientific breakthroughs: the increase in our understanding of biosynthetic pathways and the advent of analytical techniques, specifically gas chromatography and mass spectrometry, which are capable of describing the architecture of individual molecules in acute detail. The search for evidence of life has been continued not only at the morphological level, but more significantly at the molecular level. This approach, now adopted by the organic geochemist, has been used to provide evidence for life in Precambrian times as long ago as 3×10^9 years.

The organic geochemist's approach rests on the following premises: that certain molecules, possessing a characteristic structural morphology, show a reasonable stability to degradation over long periods of geological time; that the formation of this structure should be understood in terms of

Fig. 1. Structural representation of (a) chlorophyll a and (b) vanadyl deoxyphylloerythroetioporphyrin. Chlorophyll, the green pigment of plants, can give rise to two kinds of molecular fossil in the geological environment: a porphyrin and an isoprenoid hydrocarbon.

known biosynthetic sequences; and that such a structure should be formed uniquely by biological sequences. Such compounds are commonly referred to as biological markers or molecular fossils. The various compounds which have been looked for are the ones that are typical constituents of living organisms today, namely, nucleic acids, proteins, carbohydrates, organic pigments, and lipids. The basic structural units of the last two groups of compounds are shown in Fig. 1. All these compounds fulfill the structural specificity requirement stated above, but only the lipids, and to a lesser extent the class of organic pigments known as the porphyrins, survive from the earliest period of geological time still recognizable in their original form.

Lipids as biological markers. Among the lipids, one particular group of structures, based on the C_5H_8 isoprene unit, has been sought after and characterized in Precambrian sediments. These polymers of isoprene, referred to as isoprenoid hydrocarbons, have a methyl branch on every fourth carbon atom of a linear hydrocarbon chain and are extremely stable, so that there would be little degradation of these compounds even over periods of billions of years. Two specific hydrocarbons, a C_{20} hydrocarbon, phytane, thought to be derived from a hydrocarbon side chain of the green pigment of plants (chlorophyll), and a C_{19} hydrocarbon, pristane, which is present in marine organisms, seem to occur ubiquitously in ancient sediments (Fig. 2).

Though one or two functional groups may have been removed from the original hydrocarbon chain of the biological system, the characteristic structure by which the molecule is recognized remains unaltered. In this respect there is a direct analogy with the morphological approach, with the important distinction that the analytical techniques of the geochemist are now capable of detecting much smaller quantities of material.

Two more groups of lipids, also structural polymers of the basic isoprene unit but biosynthesized in a more complex fashion, are the steranes and triterpanes. These tetracyclic and pentacyclic structures also have survived from Precambrian times. The tetracyclic structure, cholestane, is shown in Fig. 3, with other hydrocarbon molecular fossils that have been found in sediments ranging back to the Archean (3×10^9 years).

Thermal stability of molecular fossils. All organic compounds in the geological environment are unstable with respect to temperature. The as-

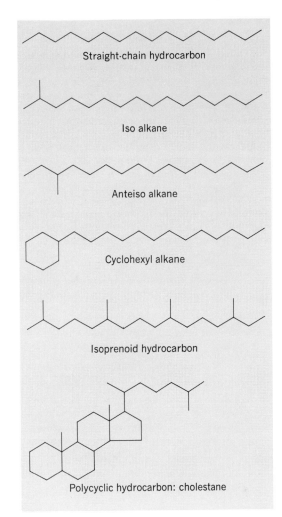

Fig. 3. Structural representation of a series of hydrocarbon molecular fossils that have been characterized in sediments of varying ages.

signment of great geological age to certain hydrocarbon structures and even to amino acids in Precambrian sediments can be justified by invoking thermal cracking as a mechanism in geological degradation processes. Simple calculations indicate that hydrocarbons should be stable for more than 10^{22} years at room temperature and for 10^{12} years at 100°C. Geothermal temperatures are often quite low and may not have exceeded 30°C. Although the presence of catalysts in the geological environment will accelerate the rate of these processes, it is not unreasonable to find intact isoprenoid hydrocarbons in sediments whose age is 10^9 years. In the case of amino acids, D. Conway and W. F. Libby studied the rate of decarboxylation of the amino acid alanine. They showed that, at room temperature, half of the alanine molecules would be expected to have undergone decarboxylation in 10^9 years, but at 100°C the corresponding time would be only 100 years. This experiment suggested that, in environments whose geothermal history had been mild, some amino acids derived from the hydrolysis of proteins could be stable for a billion years. In keeping with this prediction, some amino acids, including alanine, have been found in a trilobite fossil believed to be 45×10^7 years old.

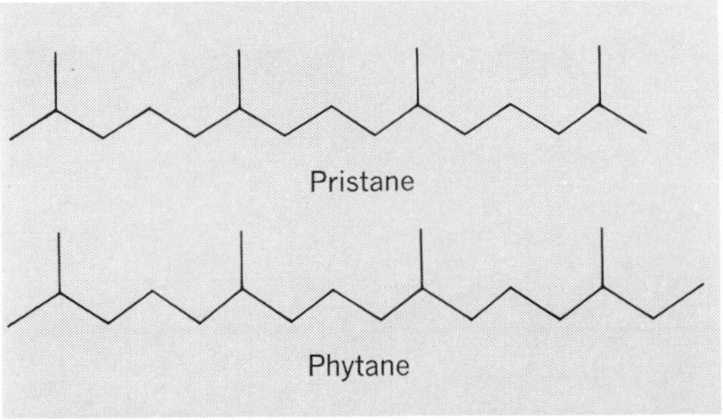

Fig. 2. Structural representation of the hydrocarbons pristane and phytane.

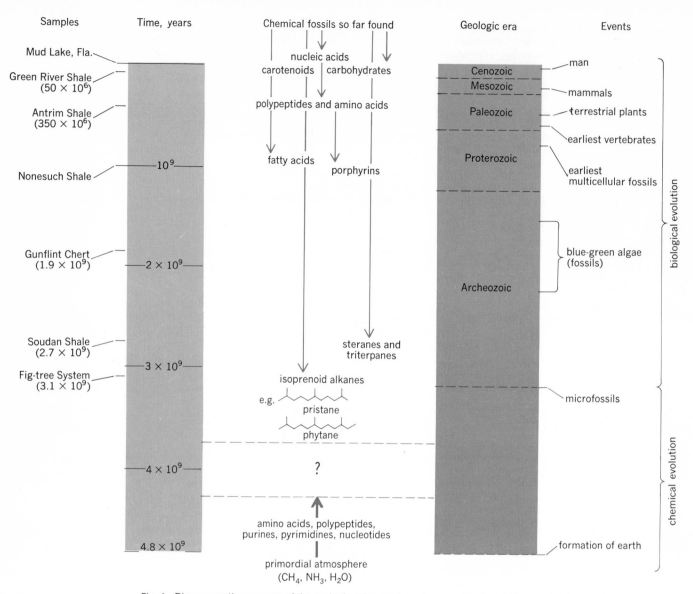

Fig. 4. Diagrammatic summary of the geological time scale and concomitant evolutionary developments.

Although the presence of indigenous Precambrian amino acids receives some degree of plausibility from thermal kinetic studies, the case is seriously weakened by the recent racemization studies of Jeffrey Bada. For example, these racemization kinetics have brought into serious question the indigenous nature of small quantities of amino acids found in the Fig-tree chert from Swaziland (3.1×10^9 years old). These amino acids are optically active, have the L configuration, and are therefore of undoubted biological origin. The rates of racemization of amino acids are sufficiently fast that complete racemization would be expected to occur within 10^6 years even under mild geological temperatures. Racemization of amino acids was found to be complete in a deep-sea marine sediment after only about 5×10^7 years. These studies suggest that a more likely explanation for the presence of amino acids in this Precambrian sediment would be later migration of these organic compounds into the sediment from another source. This kind of kinetic study is of critical importance

when it comes to considering the complex problem of the indigenous nature of molecular fossils. The implications of such findings for the indigenous nature of the hydrocarbons in the Fig-tree chert is an intriguing speculation.

Chronological survey of molecular fossils. A geological time scale giving the important evolutionary developments is shown in Fig. 4. Structurally intact proteins, carbohydrates, and nucleic acids do not survive under geological conditions for any significant period of geological time. When they have been characterized at all, they are found only in very recent sediments. In addition, the absence of analytical techniques capable of handling very small quantities of material has directed attention to those classes of compounds which are amenable to analysis in larger amounts with modern analytical instruments. These compounds are the pigments and the lipids.

During the last decade, many research groups have conducted extensive analyses on the hydrocarbon extracts from Precambrian sediments located in different parts of the world. The chronological picture, at least for the isoprenoid hydrocarbons, seems consistent thus far. The saturated isoprenoid hydrocarbons, pristane and phytane, have been characterized in the oldest Precambrian sediment, the Onverwacht sediment (3.4×10^9 years old) from South Africa, as well as in later Precambrian sediments such as the Nonesuch formation (1.0×10^9 years old), the Gun Flint iron formation (1.9×10^9 years old), and the Soudan shale (2.7×10^9 years old), all from North America. It should be emphasized that in none of the sediments has the characterization of the pristane and phytane involved the designations of their optical centers, and no absolutely unequivocal evidence has been provided for their indigenous nature. Nevertheless, it is thought that they probably are indigenous. This latter aspect, namely, the problem of migration, has been a subject of great controversy, with the geologists so far having most to contribute. Notwithstanding these uncertainties, the evolutionary picture does appear consistent and would seem to establish the existence of life at least 3×10^9 years ago.

Porphyrins and steranes have also been found in Precambrian sediments, although generally less frequently and without the same degree of spectroscopic rigor. Wherever they have been found, however, they have tended to enhance rather than diminish the evidence for life in Precambrian times. Fatty acids, which occur ubiquitously in the plant kingdom, have also been detected in Precambrian sediments. Because the linear fatty acids lack a characteristic structure, there has been a reluctance on the part of organic geochemists to use them as unequivocal biological markers. Nevertheless, these compounds do survive into Precambrian times and provide additional evidence for Precambrian life. The paleontological work of, first, E. Barghoorn and S. Tyler, and later, Barghoorn and W. Schopf, have complemented the finding of molecular fossils in Precambrian sediments. Barghoorn and associates have characterized unicellular bacterium- and alga-like microfossils in several Precambrian sediments, including the Fig-tree chert, more than 3×10^9 years old.

ASSESSMENT OF ISOPRENOID HYDROCARBONS AS BIOLOGICAL MARKERS

Some scientists have surmised that the molecular approach may have more to contribute than any morphological approach, not only because of the relative evanescence of most plant tissues in geological deposits, but also because a systematic examination of the geological record at the molecular level may provide insights into the biochemical development of evolutionary processes. The interface from prebiological to biological evolution is more likely to be recognized in terms of molecular structural changes rather than morphological remains. It should be emphasized, however, that even molecular fossils give no clue about the path of evolutionary development between 3.5×10^9 and 4.5×10^9 years ago, since we have no geological record for that period. Moreover, there are objections to the molecular, or organic geochemical, approach which must be taken into account.

Objections. The organic geochemical approach has been challenged from three very different standpoints: contamination, migration, and abiogenicity. Each of these criticisms questions the validity of the conclusion that the characterization of Precambrian molecular fossils, and in particular isoprenoid hydrocarbons, demands a biological origin and is therefore indicative of life at the time of deposition of the rock.

Contamination. The problem of contamination is one which is constant in any organic analysis. Laboratory contamination, which can arise simply from handling procedures, is a form of contamination which cannot always be completely avoided, but is one which can be guarded against if rigorous controls are established. In early studies on the Orgueil meteorite, the hydrocarbon analysis suggested the presence of compounds of biological origin. Subsequent studies argued very strongly that such organic material originated on this planet. When a meteorite plunges through the Earth's atmosphere and is later stored on a museum shelf, it is exposed to many kinds of terrestrial contamination. Such uncertainties threaten to undermine any conclusions drawn from almost all meteorite analyses.

Migration. The possibility always exists that organic material has seeped into the shale at a period later than its formation. The migration question cannot yet be answered unequivocally because of a lack of knowledge of the factors involved, namely, oil seepage and the origin of oil in general. Study of the geological environment of the sediment under consideration may reveal seepage passages along which organic material could have migrated. Another approach has been to examine the carbon isotope ratios of the organic extractable material and the nonextractable material (known as kerogen). Photosynthetic organisms are known to discriminate against carbon-13 in preference to carbon-12. Both land and marine plants have lower C^{13}/C^{12} ratios than their respective carbon sources, atmospheric CO_2 and ocean carbonates. Several investigators have examined the C^{13}/C^{12} ratios for various carbon reservoirs in nature and a summary of these findings is given in Fig. 5, where δ is defined as in Eq. (1). The isotope data are re-

$$\delta = \frac{C^{13}/C^{12}{}_{sample} - C^{13}/C^{12}{}_{standard}}{C^{13}/C^{12}{}_{standard}} \times 10^3 \quad (1)$$

ported in terms of change in δ per mil of the ratio relative to the standard; for example, $\delta = +10$ means that the C^{13}/C^{12} ratio of the sample is 10

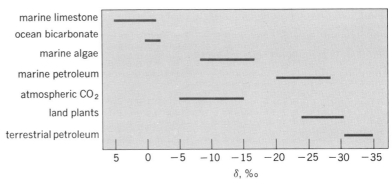

Fig. 5. Distribution of C^{13}/C^{12} ratios of organic material from different geological sources. This is a summary of the findings.

C¹³/C¹² ratios in coexisting organic extractable (that is, soluble) material and insoluble kerogen

Sample	Location	Age	δC^{13} soluble	δC^{13} kerogen
Green River shale	Colorado	Eocene	−27.57	−28.80
Chattanooga shale	Tennessee	Devonian	−27.48	−27.46
Kolm shale	Sweden	Cambrian	−26.19	−27.43
Nonesuch shale	Michigan	Precambrian	−28.14	−28.15
Soudan shale	Minnesota	Precambrian	−25.00	−34.81
Fig-tree shale	Swaziland, South Africa	Precambrian	−27.55	−26.94
Ventersdorp shale	Transvaal, South Africa	Precambrian	−25.78	−36.86

units per mil, or 1%, greater than that of the standard. It can be seen that both marine and land petroleum are enriched in C^{12} with respect to all other carbon reservoirs. It has been shown that the lipid fraction extracted from plants is enriched in C^{12} compared to the whole plant, and since lipids are thought to be primary precursors of petroleum, such an enrichment should be, and is, reflected in the C^{13}/C^{12} isotope ratios for petroleum.

In his approach to the problem of migration, T. Hoering examined the C^{13}/C^{12} ratios of the organic extractable material and the insoluble kerogen material. Crude oils and their marine kerogens exhibit very similar C^{13}/C^{12} ratios. Significant differences occur, however, between the soluble and insoluble material of some Precambrian sediments (see table). This discrepancy could reflect the fact that the two materials came from different places, suggesting that the extractable organic material may have migrated in from some other location. The problem is aggravated by a lack of knowledge concerning the mechanism of formation of the insoluble organic matter, kerogen. Alternatively, the discrepancies might merely reflect different biological source materials which are subsequently degraded to give differing C^{13}/C^{12} ratios. Whereas such ratios provide a most useful and reliable guide as a criterion of the biological origin of organic extracts, such values do not, as yet, offer a definitive answer to the problem of migration.

Abiogenicity. The third and most crucial objection to the organic geochemical approach is that molecular fossils, such as isoprenoid hydrocarbons, may have originated from an abiogenic source and therefore are not necessarily indicative of biological origin. Many attempts have been made to synthesize isoprenoid hydrocarbons abiogenically and to characterize such compounds in synthetic preparations of petroleumlike material. A. Wilson demonstrated experimentally how straight-chain hydrocarbons, which are present in petroleum in significant quantities, might be formed by nonbiological processes. D. R. Davis and Libby have shown that high yields of heavy hydrocarbons result from the polymerization of solid methane by cobalt-60 gamma rays. The Fischer-Tropsch reaction has been considered by the protagonists of the abiogenic theory for hydrocarbon formation to be a significant process in the formation of petroleum (Fig. 6). This process involves the reaction of carbon monoxide and hydrogen at temperatures between 200 and 300°C in the

presence of a catalyst, as in Eq. (2). Using a mix-

$$CO + H_2 \xrightarrow{\text{Catalyst}} \text{Hydrocarbons} \qquad (2)$$

ture of carbon monoxide and deuterium to eliminate the contamination danger, Ryoichi Hayatsu and E. Anders characterized a series of deuterated isoprenoid hydrocarbons ranging in carbon number from C_9 to C_{14}. An analogous series of hydrocarbons has been found in an Italian Cretaceous shale. The Fischer-Tropsch process suffers from an inability to account for the higher molecular weight isoprenoid hydrocarbons, particularly pristane and phytane, which are major components of petroleum hydrocarbons. Furthermore, in the Fischer-Tropsch products there is no evidence for the presence of members from the sterane and triterpane series; these compounds are also significant components of petroleum.

Nonbiological polymerization of isoprene. Recent polymerization studies of G. Natta have shown that the nonbiological polymerization of small organic molecules such as propylene and butadiene proceeds with considerable stereospecificity. The stereospecificity in the polymerization of isoprene had hitherto been considered a characteristic feature of the biological system. The polymerization of the 5-carbon iosprene unit (present in the biosynthetic pathway as isopentenyl pyrophosphate) involves a head-to-tail linkage and takes place in a stereospecific manner (Fig. 7). Using a series of highly specific transition-metal catalysts, Natta showed that isoprene can be converted to a polymer containing 99% of linear 1,4 trans structures by means of an aluminum-vanadium catalyst. The infrared spectrum of the resulting polymer resembles that of natural rubber, where isoprene units are linked together through 1,4 linkages (Fig. 8). By changing the catalyst to $TiCl_3$, a stereospecific 1,4 cis linked polymer is produced. This polymerization process exactly mimics the stereospecific features of the biosynthetic pathway. These findings call for a critical reappraisal of isoprenoid hydrocarbons as biological markers. Certainly the structure of the isoprenoid hydrocarbons, taken in isolation, can no longer be considered to be derived unequivocally from a biological precursor.

Criteria for biogenic origin. Nonbiological polymerization processes of the Natta type do not, however, automatically invalidate the conclusions of extensive studies in organic geochemistry in recent years. What they demand is a reevaluation

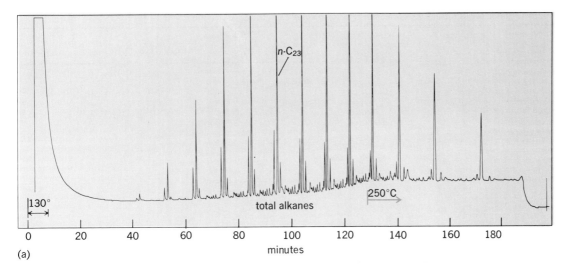

(a)

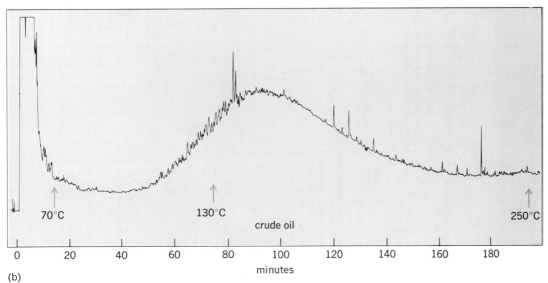

(b)

Fig. 6. The gas-chromatographic patterns of two hydro-carbon mixtures of known abiogenic origin. (a) The Fischer-Tropsch gas chromatograph has features resembling hydrocarbon distributions of biogenic origin, whereas (b) the methane spark product, a complex mixture of unresolved components, shows no such resemblance.

of the concept of the molecular fossil. The evidence still points overwhelmingly to a biogenic origin for almost all the organic extracts of crude oils and shales. The important point about such polymerization processes is that in those areas where the relative abundance of different biological markers is small and where there is no evidence beforehand that the organic extract is derived from a biological source, characterization of the architectural skeleton of the biological marker may be inadequate evidence for biogenicity. It is in precisely the area of greatest evolutionary significance—namely, the transition from chemical evolution to biological evolution which must have occurred some time between 4.5×10^9 years, the age of the Earth, and 3×10^9 years, the age of the oldest microfossils—that it is necessary to establish criteria which will distinguish between those hydrocarbons derived from an abiogenic source and those which have a biogenic origin. These are the same uncertainties which will arise in the organic analysis of any extraterrestrial sample.

Stereochemistry. The designation of the precise stereochemistry of an individual molecule would be a powerful technique which could determine its origin. In addition, such an approach involves designating the configuration of any optical center. Very few attempts have been made to determine the stereochemistry of the isoprenoid hydrocarbons isolated from crude oils and shales, and for this there are very good reasons. Until recently it had been felt that the very specific architectural skeleton of compounds such as pristane and phytane was in itself sufficient evidence for biogenic origin. Furthermore, there are some serious technical difficulties involved in obtaining reliable optical measurements. The optical rotations of such compounds are so small that large quantities of the pure material are required.

Despite these experimental difficulties, there have been successful endeavors to distinguish

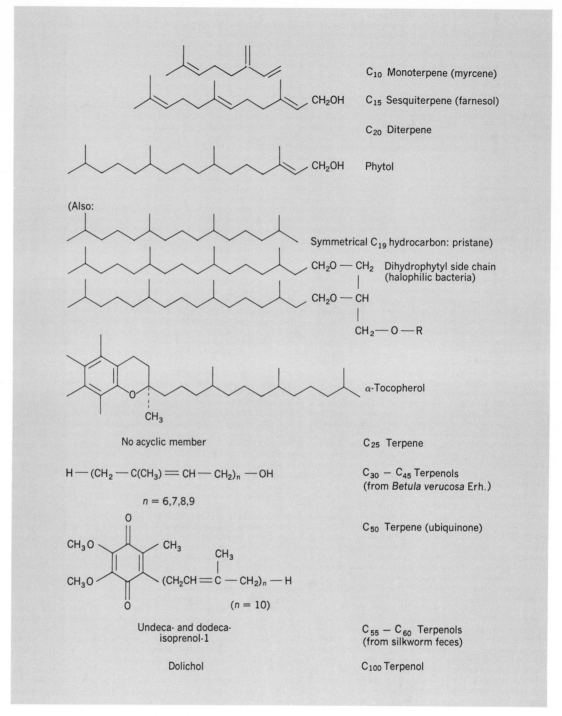

Fig. 7. Head-to-tail linkage in biological systems.

between the diastereoisomers of a given compound. Using gas chromatography, the diastereoisomers of the isoprenoid acid phytanic acid (Fig. 9) have been separated as their methyl esters. The isoprenoid hydrocarbon pristane can exist in two diastereoisomeric forms: (1) the meso form (the configuration isolated from marine sources is shown in Fig. 9), and (2) a mixture of d,l isomers. Preliminary attempts to bring about a separation of these two diastereoisomers have not succeeded. Such a separation would be an important breakthrough in organic geochemistry because it would establish a criterion by which to distinguish pris-

tane derived from an abiogenic source and pristane derived from a biological source. One would not expect beforehand the meso configuration of the naturally occurring material to be produced exclusively in an abiogenic synthesis.

Optical activity. Optical activity has often been invoked as a criterion for biological origin. Such a phenomenon was observed in the early analysis of the Orgueil meteorite, though later investigations suggested that artifacts may have given rise to it. For optical activity measurements to be of value, it is important to relate the origin of the optical activity to a specific component of a complex mixture, and to correlate this optical rotation, in magnitude and direction, with that of a standard compound. Some scientists have speculated on the spontaneous formation of optically active substances from an inactive substance without interference from a directing asymmetric agency. E. Havinga showed that methylethylallyl anilinium iodide may crystallize into either enantiomer under certain experimental conditions. Such an experiment suggests a mechanism for the formation of the first optically active compound and indicates that biological activity may not necessarily be an absolute criterion for biogenicity.

Biosynthetic criteria. If one considers the biosynthetic pathway for isoprenoid hydrocarbons, one can establish further criteria for biological origin. The enzymic polymerization of the isoprene unit, isopentenyl pyrophosphate, involves a 1,4 head-to-tail linkage and takes place in a stereospecific manner. At two specific places in this biosynthetic pathway, the head-to-tail linkage is replaced by a tail-to-tail linkage (Fig. 10). (1) The C_{15} compound, farnesyl pyrophosphate, reacts with another molecule of farnesyl pyrophosphate to give the C_{30} hydrocarbon squalene. (2) In the biosynthetic pathway to the carotenoids, which are C_{40} terpenoid compounds, an analogous tail-to-tail linkage is formed. Thus at the C_{30} and C_{40} level in the biosynthetic pathway, there are formed compounds where the spacing between the branched methyl groups is now four methylene units instead of five. Such a carbon network appears to be uniquely produced by the biological system and can therefore be used as a criterion for biological origin. Several, as yet unsuccessful, attempts have been made to identify this carbon network in organic geochemical material.

Positional isomerism. The relative abundance of positional isomers of an individual compound may be used as a criterion of biogenicity. When isoprene is polymerized abiogenically, one would expect that three compounds should be formed: the head-to-tail (H-T) linkage, the tail-to-tail (T-T) linkage, and the head-to-head (H-H) linkage (Fig. 11). When one considers the addition of another molecule of isoprene to produce a C_{15} compound, and a further molecule to give a C_{20} compound, then several products should result whose structural skeletons are also shown in Fig. 11. All these compounds might be expected to be present in an abiogenic polymerization process. Only the 1,4 head-to-tail compounds (that is, 2,6-dimethyloctane, farnesane, and phytane) should be present in organic extracts derived from biological precursors. It is the presence of these additional positional isomers which may serve as a criterion for abiogenic material. In other words, the search for what is not present must be added to the search for what is present.

Although no extensive study has been carried out in search of other positional isomers, the evidence in most crude oils and shales seems to indicate that the isoprenoid hydrocarbons are the predominant isomers for a given carbon number. It is in those organic materials whose origin is much less clear, that is, in thucholite samples and hydrothermal deposits, that the presence of positional isomers is being looked for with renewed interest. It would be important, in this context, to analyze

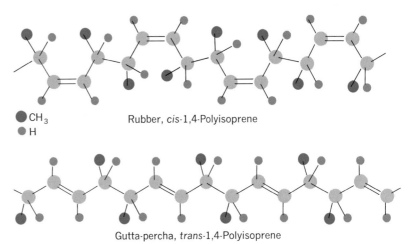

● CH₃
● H

Rubber, *cis*-1,4-Polyisoprene

Gutta-percha, *trans*-1,4-Polyisoprene

Fig. 8. Stereochemistry of natural rubber.

the hydrocarbons resulting from the irradiation of isoprene, with cobalt-60 x-rays in the presence of mineral vermiculite followed by reduction, in which farnesane and phytane have been characterized as components.

Normal alkane ratios. Criteria for biogenic origin such as the odd-to-even distribution of normal alkanes in sediment extracts are only reliable for the most recent sediments. This feature disappears in the old sediments which have been subjected to temperature, pressure, and reducing conditions in the geological environment. The identification of biological markers other than the saturated isoprenoid hydrocarbons is most useful in complementing the evidence for biological origin. However, except for the porphyrins and the

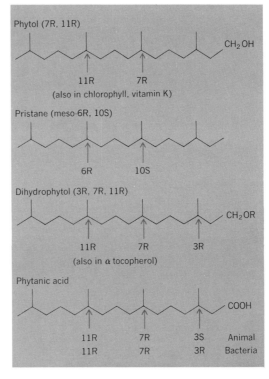

Phytol (7R, 11R)

CH₂OH

11R   7R

(also in chlorophyll, vitamin K)

Pristane (meso-6R, 10S)

6R   10S

Dihydrophytol (3R, 7R, 11R)

CH₂OR

11R   7R   3R

(also in α tocopherol)

Phytanic acid

COOH

11R   7R   3S   Animal
11R   7R   3R   Bacteria

Fig. 9. Determination of the absolute configuration of optical centers may be a necessary criterion to establish the biogenic origin of isoprenoid hydrocarbons.

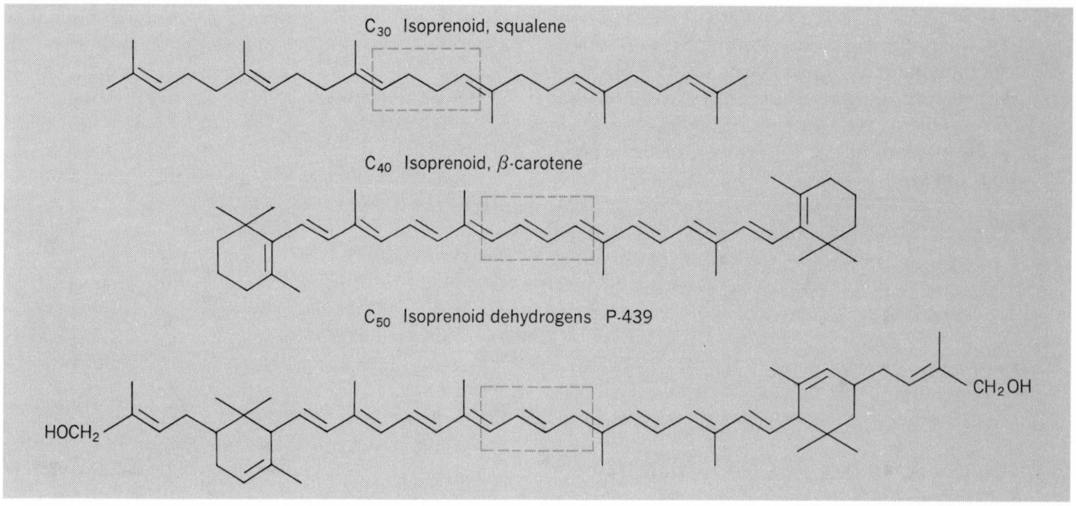

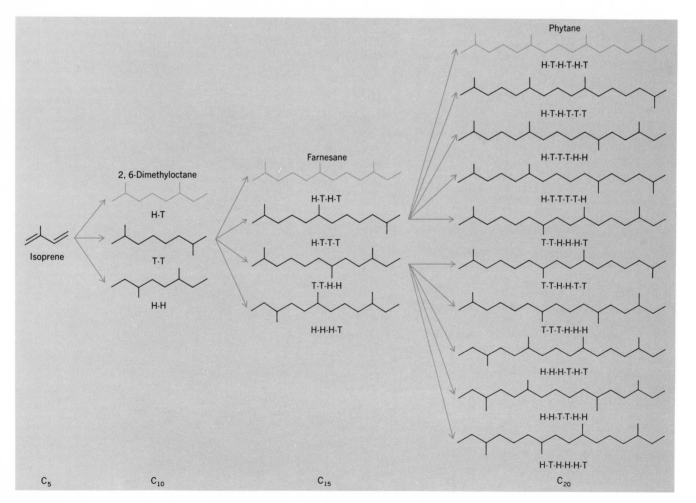

Fig. 10. Tail-to-tail linkage in biological systems.

steranes and triterpanes, other classes of biological markers do not survive into the region of greatest interest, the Precambrian.

Carbon isotope ratios. The measurement of C^{13}/C^{12} isotope ratios may provide the diagnostic marker we are looking for with which to ascertain the origin of hydrocarbons in geological environ-ments. Such measurements already reliably distin-guish between abiogenic and biogenic carbon sources. Although these measurements have been made on total organic extracts, in which many different structural classes of compounds are pres-ent, they have already cast some doubt on pro-posals for a biogenic origin of hydrocarbons in

Fig. 11. A hypothetical, nonbiogenic polymerization of isoprene, showing all the positional isomers that would be expected to result from such a process. Certain iso-mers may be indicative of biogenicity.

hydrothermal deposits. Furthermore, it may be possible to study the carbon isotope ratios of an individual molecule in spite of the practical problems involved in this approach. Isolation procedures involving crystallization and chromatographic techniques themselves bring about isotope fractionation of individual molecules, a difficulty which can cause uncertainty in the conclusions drawn from such measurements. The great value of this approach lies in the fact that because each biological molecule has a different synthetic pathway, the carbon isotope ratios should be different for each of these molecules. Using high-resolution mass spectrometry, it should be possible to determine the carbon isotope distribution at each of the individual carbon atoms of, for example, the pristane molecule.

This approach is a most powerful one in ascertaining the origin of the isoprenoid hydrocarbons in either Precambrian sediments or extraterrestrial materials.

CONCLUSION

The abiogenic controversy has emerged as a most critical problem in an attempt to determine the transition from chemical evolution to biological evolution. In the last decade, a completely new approach to examining the fossil record, at the molecular level as well as at the morphological level, has immeasurably enhanced our understanding of the Earth's evolutionary development. Both the organic geochemical approach and the micropaleontological approach have identified remnants of living organisms either in the form of molecular fossils or microfossils, providing evidence for life forms around 3×10^9 years ago. In spite of this measurable progress, however, no unequivocal criterion has been established by which to determine the indigenous nature of a molecular fossil in a given sediment. Furthermore, only a very small number of samples have yet been analyzed, since few rocks 3×10^9 to 3.5×10^9 years old have been found. What remains is a monumental and tedious task. In regard to the whole question of the origin of life, there are those who feel that the most important steps have yet to be taken. In the words of L. Orgel, ". . . A number of firm strides have been taken . . . only a few of them clearly lead in the right direction. . . ."

[EUGENE D. MC CARTHY]

Pieter Korringa studied biology at the University of Amsterdam. In 1937 he was appointed biologist in the Netherlands Institute for Fishery Investigations and started research on oyster and mussel farming in the field laboratory at Bergen op Zoom. In 1957 he moved to the IJmuiden headquarters of the Institute to become its director. Since 1961 he has taught hydrobiology on a part-time basis at the University of Amsterdam.

Mariculture

IN THIS ERA, WITH ITS ALMOST INCREDIBLY rapid technical development, many of us fail to realize that we have not made much headway in the production of our daily food. In its primary form, our food does not yet come from a factory. We are still completely dependent on a biochemical reaction of green plants which produces organic matter, with the aid of solar energy, from carbon dioxide, water, and various inorganic salts. Everything we eat can be traced back to this biochemical reaction, known as carbon dioxide assimilation. Our forefathers learned long ago how to take advantage of this natural phenomenon, but only in our time have we acquired enough knowledge to measure the influence of all the factors involved and to control the production of organic matter by green plants to a certain extent.

On land, chlorophyll-bearing plants such as grasses, herbs, shrubs, and trees are the main producers of organic matter. Sunlight, moderate temperature, water, nutrients, and sufficient amounts of atmospheric carbon dioxide are necessities for this biochemical reaction. Virtually the same thing happens in the sea. The seaweed attached to the rocks and the eelgrass growing in some shallow areas are not, however, the main producers of organic matter; rather, a multitude of minute plants, invisible to the naked eye, are responsible for this production in the sea. The same factors are involved as on the land: sunlight, carbon dioxide, water, and various inorganic salts; the last-named are the so-called nutrients.

It is difficult to estimate how much organic matter is produced from inorganic constituents by chlorophyll-bearing plants per unit of surface and per unit of time. Agriculturists were the first to measure how much organic matter is produced by a given crop per square meter of arable land. On an oat field in Europe, it was estimated as 380 g of dry organic matter per square meter per year. This figure gives a good general picture, since data published for peas, wheat, hay, and rice do not deviate appreciably from it. We should, however, consider that all these crops are grown in intense culture on well-tilled, well-fertilized land. It is certainly not a good measure of production on an average square

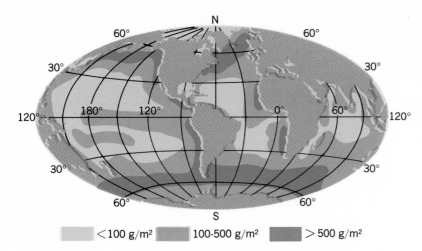

Fig. 1. Primary production in the sea, expressed in grams of dry organic matter per square meter per year. (*Adapted from Kesteven and Laevastu, FAO, 1958*)

meter for all the continents combined; desert, steppes, mountains, and ice-covered areas produce very much less organic matter, if any at all.

It is less easy to measure primary production in the sea, but modern techniques, making use of radioactive tracers, have made it possible to assess how much organic matter is produced by marine microscopic plants (Fig. 1). Sunlight does not penetrate very deep into the water; hence production of organic matter is limited to the superficial layer of the seas, down to some 200 m when the water is very clear. The best way to express the production of organic matter in the sea is to calculate how much organic matter is produced under a square meter of sea surface. It is not surprising that in the sea, as on land, the primary production is found to vary considerably from place to place. Water is not the limiting factor here, as is so often the case on the land, rather it is the amount of nutrients available in the surface layer. Roughly one-third of the surface of all the seas should be considered as marine deserts, producing less than 100 g of dry organic matter per square meter per year. Under about one-fifth of the oceans' acreage, on the other hand, the annual production exceeds 500 g of organic matter per square meter, with maxima sometimes exceeding 1000 g. The production of the remaining areas lies between these extreme values. For all the seas combined, 250 g of organic matter per square meter seems a reasonable average. This is a surprising figure, since it falls only slightly short of the production of 380 g mentioned above for the oat field, which was said to be a figure well above the average for all the land combined. If we then take into consideration the fact that the seas cover about 71% of the Earth's surface, we are forced to conclude that the greater share of all primary production is located in the sea.

MARINE VEGETATION

All of man's food can be traced to primary production, the first link of the food chain. Does the sea yield the greater share of the food, as should be expected if man were to take an equal toll of all organic matter produced by green plants, either directly as vegetable matter or indirectly after incorporation in the body of some animal? The an-

swer must be that only about 1% of the food consumed by man comes from the sea!

Accessibility. Why do we take so little from the sea in spite of the fact that its primary production yields more organic matter than it does on land? An important reason could be the lesser degree of accessibility of marine products. Not only is it easier to harvest crops of various description on the land, our own habitat, but, on land, crops occur in a more concentrated way, both in space and in time. On land, most of the vegetable crops grow from 25 cm below the Earth's surface to at most a few meters above it, and are therefore well concentrated and easily accessible; on the other hand, the organisms responsible for primary production in the sea are distributed throughout a water layer about 200 m thick. This means that they are about 100 times as dispersed as are the plants on the land. In addition, vegetable crops on the land require several months, or up to a year, to grow and ripen, and are ready for harvesting in a brief, well-defined space of time; therefore harvesting is usually well concentrated in time too. The minute plants in the sea, however, grow all the year around, although with marked ups and downs, especially at higher latitudes; there could be no brief period of harvesting in the sea.

Size. Another reason why the percentage of man's food which comes from the sea is very small is the considerable size difference between plants on land and those in the sea. If we think in terms of rice, wheat, beans, potatoes, cabbage, tomatoes, and bananas, to mention just a few examples, we realize that such crops are easy to harvest, either by hand or with the aid of a machine, and, further, that a considerable part, often well over 50%, of the organic matter produced by these plants in the course of one season is directly edible. Some marine plants, for example, the brown, green, and red algae attached to the rocks in and slightly below the intertidal zone, are of similar sizes to land plants. Some of these are edible, a few species of red algae even delectable and locally much sought after, such as the Japanese amanori (*Porphyra*), but the quantities harvested are insignificant and will have no appreciable impact on future food statistics.

Over 99.9% of marine plants are of microscopic size and constitute the phytoplankton. Most phytoplankton, expressed in numbers but sometimes also in volume, are 10 μ or less in size and cannot be caught with nets made of the finest woven material. Even for the larger species, nets of the finest bolting silk (with meshes of 50 μ) are needed to catch them. Such fine-meshed nets are inefficient as catching gear and clog very easily. Harvesting phytoplankton on a realistic scale therefore meets considerable difficulties. Even if it were possible to do so, they would not be an acceptable food item for man. Some of the smallest species of flagellates are thin-skinned and readily digestible, and since there are many species of flagellates, no doubt some are palatable. In the sea, however, there is always a mixed crop and palatable species rarely, if ever, predominate. There are other species, especially among the somewhat larger armored dinoflagellates, which are poisonous, for example, the various species of *Gonyaulax* responsible for the phenomenon called red tide, which may kill fish and render shellfish toxic for predators, including man. Random harvesting of a wild crop of flagellates, even if it were technically feasible, would thus never be a good system whereby to collect food from the sea.

Among the larger species which can be caught with fine-meshed nets, diatoms predominate; these are small plants with siliceous frustules, or cell walls. When ingested intact, a diatom often leaves the intestinal tract of its predator alive and therefore has not properly served as food. Diatom feeders must break open the frustule to reach the edible part. Moreover, many species of diatoms bear needle-sharp siliceous spines which might perforate the more delicate tissues of those who ingest them. Since diatoms, like flagellates, occur in nature in mixed populations, they are as such unacceptable as food for man.

Second link of food chain. Man usually does not prefer to live on a purely vegetarian diet, though this would be biologically possible. A variety of animal products, such as meat, cheese, and eggs, are much sought after, since they are palatable and easy to digest. Moreover, in the process of digestion, animal products yield a combination of amino acids such that the basic material for the construction of the consumer's own body is readily available, whereas amino acids from vegetables are usually present in proportions that differ rather widely from those required to build and restore the human body. It is usually not possible to collect sufficient quantities of animal food merely by hunting or fishing. Practically everywhere in the world some kind of animal husbandry has been developed to meet the demand, just as agriculture was developed long ago to take over from collecting wild vegetable matter.

Animal products are necessarily more "expensive" than vegetable food and therefore must be considered as a kind of luxury, taken on a daily basis by relatively affluent groups only. This is true because conversion of vegetable matter into animal matter is not a very efficient process: Animals digest only part of the vegetable matter ingested. Also, they cannot make use of all the digested food for the purpose of building up their bodies or for producing milk or eggs; a considerable part of the digested food must be used for energy demands, for maintaining the required body temperature, for moving about to search for food or to escape from enemies, and for catching, chewing, and digesting vegetable food. The turnover rate differs from species to species. Evidence suggests that, as a rule, only 10% of the food taken in is converted into the body tissues of the consumer. In animal husbandry this percentage is often somewhat higher; on the other hand, some species living in the wild never reach the 10% level. Thus animal food is more "expensive" and it is advisable to switch over to the production of vegetable food in times of scarcity, since an acre of arable soil can yield about 10 times more vegetable food than animal products.

About 85% of the food taken by man all over the world is of vegetable nature. To produce the remaining 15% of animal foodstuffs, it takes a large quantity of vegetable matter produced by primary production, roughly the same volume as the 85% consumed directly. However, a large proportion of this vegetable food is grass or other vegetable matter not directly edible by man.

When we realize that the marine organisms responsible for primary production are, with the exception of some seaweeds, not suitable as human food, we are forced to conclude that, at most, 10% of the primary production of the sea will be available in the form of animal food. This animal food belongs to the second, third, fourth, and occasionally even the fifth links in the food chain. The animal food we are used to on land virtually all belongs to the second link, such as beef, eggs, milk, cheese, and butter. Some of the species used in animal husbandry belong only partly to the second link and partly to the third, since a certain percentage of their food is not of vegetable nature. The daily diet of pigs and chickens usually contains a fair portion of fish meal or other cheap animal products. If 100% of their diet consisted of animal products, then they would belong to the third link in the food chain, on the same level as lions, eagles, foxes, weasels, and insectivorous birds. This would mean that a still smaller percentage of the primary production would ultimately be available as human food. Following the 10% rule, it is clear that animals of the third link represent, at most, 1% in weight of the original primary production by plants.

Man utilizes the second link of the food chain to only a small extent when he harvests crops from the sea. He incorporates in his diet items such as clams, oysters, scallops, and mussels. These mollusks are filter feeders that live predominantly on microscopic plants or their remains. They therefore belong to the second link of the food chain, and can be produced on a very large scale where the phytoplankton is rich enough. If all other conditions were favorable, one could expect that, in an area where the primary production yields 200 g of organic matter per square meter per year, up to 20 g of mollusk body could be produced.

Over 99% of our seafood belongs, however, to the third or fourth link of the food chain, which means a much smaller harvest per surface unit.

How many people could live exclusively on a diet of third-link animals occurring naturally on land? Very few, indeed. The human population would be sparse if this were its obligatory mode of feeding.

FISHERIES

Of all the food taken from the sea, about 50% comes from under one-tenth of 1% of the surface of the sea. Although the fleets of many nations congregate on traditional fishing grounds and one rarely sees fishing boats in the central parts of the oceans, this is still a surprising figure.

Distribution of fish. Ships were once less sturdy than they are today and, because they lacked reliable charts and navigation instruments, it is understandable that fishermen preferred to work in inshore waters. These conditions may still hold for some developing countries, but certainly not for the experienced and well-equipped fishermen of the countries which are technically developed. Distance cannot be the reason that fishermen still cling to the traditional fishing grounds. In reality, the unequal distribution of the fish compels fishermen to do so. Rich fishing grounds are found in the shallow waters above the continental shelves and in some limited areas where upwelling water causes increased fertility and great primary production. Most of the fish species caught on a well-known fishing ground such as the North Sea are, at least during part of the year, dependent on shallow areas either for their food or for their reproduction. Therefore such fish stocks have a limited range and do not occur in the oceans proper. Some species of fish, such as the mackerels, tunas, and sauries, are truly pelagic and can permanently thrive in the open sea, far away from the continents, but they, too, like to sojourn at least temporarily in the nutrient-rich shallow seas.

Nutrient-poor ocean layers. In tropical and subtropical zones the offshore waters usually have few fish, so few that it is generally not remunerative to fish there. This can be explained in terms of low primary production. The upper layers of these waters are rather warm and do not mix during any season with the colder deeper layers. The upper layers are gradually deprived of their fertility, because wastes and dead bodies of all types of organisms tend to sink and to take with them elements such as phosphorus and nitrogen, which are often the limiting factors in primary production by phytoplankton. These droppings and dead bodies are eventually decomposed by microbes, which thereby set free soluble inorganic components of phosphorus and nitrogen; but if this process occurs at depths of over 200 m, where very little light penetrates, these elements cannot be used again in the cycle of organic production. This explains the existence of a hugh "dead" capital, an enormous water mass rich in phosphates and nitrates that extends to the bottom of the ocean but which is out of reach of sunlight.

Improving yields. Rapid technical development has exerted a noticeable influence on fisheries in many parts of the world. Bigger ships have been built, both for catching fish and as factory ships to process the catches. Stronger engines have come

into use, enabling a rapid switchover from the old passive fishery with drift nets and long lines to active fishing with trawl nets and purse seines, even for rapid swimmers such as mackerel and herring. Electronic apparatus for finding fish schools and for precision navigation has been developed to improve the efficiency of the fishery.

These improvements involve more expensive units, but such units can catch considerably greater quantities of fish per unit of time. Where the fish stocks are large enough, such a development leads to harvesting greater crops of fish and to a higher production of human food from the sea.

On most of the traditional fishing grounds, however, there are clear signs of overfishing: Gradual decreases in the average size of the fish caught and reductions in the number of age groups in the catches indicate that too many fish are being taken from the sea too soon and that the fish hardly get a chance to grow to reasonable size and to take part in reproduction. Another sign of overfishing is a reduction of the catch per unit of effort, that is, per hour of fishing with a standard ship.

Nature produces a given amount of fish, and fishery biologists can predict what is called the optimum sustainable yield, indicating how the cropping rate can be maximized without depleting the stock. There is no objection to the use of modern equipment in the fishing industry, but the number of units used to harvest the fish should be adapted to the productivity of the stocks. When this is not done, more and more money is invested in the fishing industry to catch the same total quantity of fish as before, which means a decline in remuneration. If this continues, there comes a day when reproduction cannot keep pace with the losses suffered and a serious decline of the fish stock occurs. Sometimes this decline is irreversible, as has happened in whaling and in exploiting the Californian sardine. Recently, alarming data have been collected on the influence of too intense fishing of North Sea herring and of cod and haddock in the Barents Sea.

Improvement of the catches of the traditional fishery is, however, certainly possible. In underfished areas, which still can be found here and there, although often at quite a distance from the traditional fishing ports, improvement of catches can be achieved by modernizing equipment. In overfished areas the catches can be improved by regulating the fishery in such a way that the optimum sustainable yield is approached. Fishery biologists have the necessary knowledge for setting up a truly rational fishery, but it is difficult to achieve international agreement on practical ways to reduce fishing intensity to such a degree that the best advantage can be taken from local primary production. Therefore it seems realistic to forecast that the production of fish for human consumption from traditional fishing grounds will show a further decline in the coming decade, and that some valuable stocks will decrease to such a low level that recovery will become doubtful.

Fish meal production, preferably from species unfit for human consumption, may increase still further. The lower the species is in the food chain, the greater the increase to be expected in production. As such, krill (euphausiids), no longer preyed upon as once by nature's own factory ships, the

Fig. 2. Oyster farming in Zeeland, Netherlands. Lime-coated tiles are excellent spat collectors, but their use demands a considerable amount of manual labor.

whales, presents a potential resource of considerable importance.

FARMING THE SEA

Fishing is actually a very primitive way to collect food. It involves harvesting the crops which nature yields freely. On land, only an insignificant share of the vegetable and animal food harvested for human consumption can be ascribed to crops that nature produces freely. Long ago man realized that only agriculture and animal husbandry could yield the large quantities of food required to feed his increasing populations.

Status of present-day efforts. Sea farming has been going on in several countries in Europe and Asia for many years. However, although the yields of food have been increased, the conclusion must be drawn that farming the sea as practiced in these countries is not fundamentally comparable with the traditional systems of agriculture, horticulture, and animal husbandry. An analogy may be drawn between a man keeping a dozen chickens in his farmyard and a modern chicken farm whose main business is creating eggs and chickens. In the modern chicken farm the chickens are kept in a limited space under optimum conditions of food, light, temperature, sanitation, and protection from disease and predators. Because natural phenomena are controlled, the modern chicken farm is true farming, while the keeping of chickens in the farmyard is semifarming. In a similar way, sea farming is a case of semifarming.

Netherlands. Exactly 100 years ago the Dutch government put the natural oyster beds in the Oosterschelde in the hands of prospective oyster farmers. Appropriate farming techniques, including control of predators, competitors, and diseases, all guided by scientific research provided by the government, led to a spectacular rise in the production of oysters (Figs. 2 and 3). The old natural beds had yielded only about 500,000 oysters per year, but, by the adoption of farming techniques, up to 30,000,000 marketable oysters were produced year after year in the same area.

The production on the old natural beds was limited by a scarcity of objects (collectors) on which the planktonic oyster larvae could settle. Hence the great success when man offered the larvae suitable collectors at the appropriate moment. Settlement of oyster spat can be expected in places where current velocities are low for hours in succession, but growth and weight increase are invariably much better in places where somewhat stronger currents, richer in phytoplankton, prevail. This accounts for the success obtained with transplantation of seed oysters to such places.

The oystermen put out their collectors, but have to wait and see whether water temperature, availability of food, and absence of predators will favor the development of the oyster larvae; hence they still depend completely on nature in their spat-producing venture. When the oystermen do manage to produce a great quantity of oyster spat and then try to step up the production of marketable oysters, they repeatedly find that the Oosterschelde waters will not produce more than 30,000,000 oysters. As soon as the local oyster population surpasses a certain level, invariably growth slows down and the mortality rate increases strik-

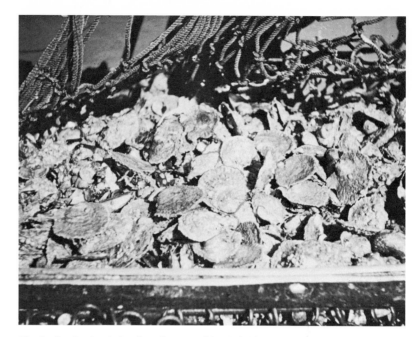

Fig. 3. Dredge load of oysters from a well-kept plot in Zeeland.

ingly. Evidently the amount of phytoplankton in the Oosterschelde sets a limit on the production of oysters. Thus natural phenomena control productivity and Dutch oyster farming must be considered semifarming.

Spain. About 20 years ago, efforts were begun to make use of the rich phytoplankton in the rias of Galicia, in northwestern Spain, by growing mussels in hanging culture. Subsequently special rafts (Figs. 4–6) were designed that had ropes coated with mussels suspended from them. Within 8 months the tiny mussel seed, collected from the rocks in the intertidal zone, grow to large-sized mussels suitable for consumption. Every rope carries about 50 kg of mussels and hundreds are hung from each raft. Today this area produces almost 100,000,000 kg of consumption mussels per year.

Fig. 4. Construction of a mussel raft in Galicia, Spain.

Fig. 5. Mussel rafts moored in a well-sheltered site in sea water rich in phytoplankton near Vigo, Galicia.

The Spanish mussel production equals that of the Netherlands, where mussels are grown in bottom culture, with good plots yielding about 10 kg of mussels per square meter. This is a very high concentration of animals, even though only about 25% of the weight of the mussel constitutes the edible parts. The tides bring phytoplankton to the mussels from a vast area, and calculations on the production of animal matter should be related to that area and not to the acreage on which the mussels are actually grown.

Fig. 6. Each Spanish mussel raft carries about 800 ropes. The ropes are thickly coated with mussels, but since mussel seed often settles profusely among the marketable mussels, grading by hand is a necessity.

The same argument holds for the mussels grown in a suspended culture in Galicia. Food comes from a far greater volume of water than that in which the mussels are bathed at a given moment. Though the rias of Galicia are rich in phytoplankton and though the water temperatures are ideal for mussels all year, there cannot be unlimited mussel production.

Another factor, at least partly beyond control by man, is the procurement of mussel seed. Seed is usually taken from the intertidal rocks, where it settles naturally in varying quantities; sometimes it is collected from ropes suspended for that purpose; and sometimes it is taken from small mussels that settle among the more mature mussels on the rope. The first two methods give varying quantities of seed, while the last method needs a lot of manual labor to sort out the young mussels, and labor is becoming a limiting factor in the Spanish mussel industry. Mechanization of the Spanish mussel industry seems hardly possible. Furthermore, when it is considered that it has been decided to limit the number of mussel rafts in the Spanish rias to avoid adverse effects of the mussel parasite *Mytilicola intestinalis*, one is forced to conclude that here too man cannot break the chains that nature has set to limit production. It is perhaps wise to speak in this case too of semi-farming.

Indonesia. Hundreds of years ago construction was begun of shallow ponds on the northern coast of the island of Java (Indonesia). The ponds were filled with a mixture of sea water and fresh water, and juvenile specimens of the milkfish (*Chanos chanos*) were placed in them. The milkfish is a marine species which grows and thrives best in brackish waters that have a rich vegetation of blue-green algae, on which it feeds. The milkfish therefore is one of the few fishes belonging to the second link of the food chain; as such, a remarkably large production of edible fish per unit of surface can be expected. Nowadays thousands of milkfish ponds are in use on Java and Madura (Figs. 7 and 8), as well as on the Philippines, on Taiwan, and in some other countries in southeastern Asia.

Indonesian fishery biologists know how to stimulate the growth of blue-green algae with a mixture of fertilizers, and therefore it seems as if milkfish farming is dependent to a lesser degree on primary production provided by nature itself than shellfish farming is. This may be true, but the real problem in milkfish farming is not the quantity of food in the ponds but an increasing scarcity of young fry. The milkfish spawns at some distance from the coast, and the young fry is caught with a special type of net when it approaches the coast in search of brackish water. The factors governing the extent of the production of milkfish fry are poorly known. It does not seem impossible that the number of adult fish participating in reproduction has gradually become too small. Therefore the author advised the Indonesian government, concerned about the decline in the catches of milkfish fry, to reserve certain ponds for the production of a spawning stock and to give those fishes the chance to migrate to their spawning grounds in the sea. Pond-reared milkfish do not reproduce in brackish waters, and efforts to speed up ripening of the ovaries with the aid of hormone injections have not yet led to practical results.

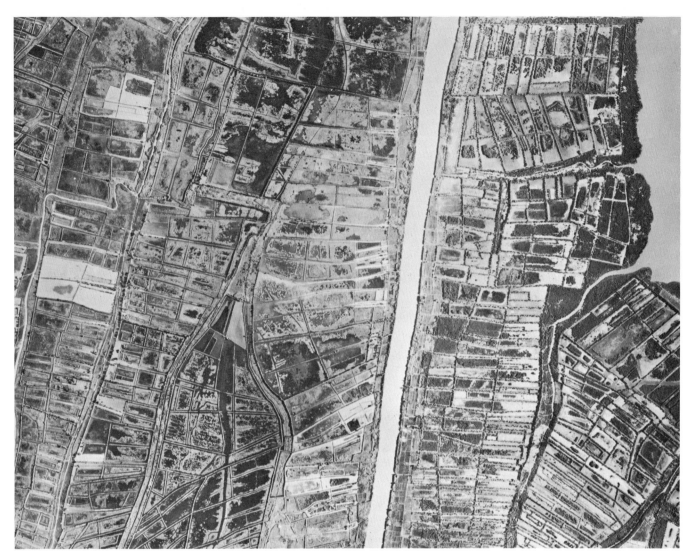

Fig. 7. Milkfish ponds along the Bengawan Solo River, Java, Indonesia. (*Courtesy of W. H. Schuster*)

Only when artificial maturation, spawning, and fertilization become a matter of routine will it be possible to call milkfish farming a case of true farming. Then both the extent of reproduction and the production of food can be guided, which means that a good system to control competitors, predators, parasites, and diseases will ensure that it will be possible to aim at a production of a given magnitude practically independent of the whims of nature. But even then it will be rather arbitrary to call cultivation of milkfish in brackish ponds a true case of farming the sea.

Japan. The same doubt can be expressed concerning eel farming in Japan; though the elvers are born in the ocean, farming is practiced in ponds filled with fresh water (Fig. 9). The elvers travel to inshore waters and try to reach low-salinity waters rich in food. When approaching maturity, they start the return journey to their spawning grounds in the depths of the ocean. The eel ponds do not contain enough natural food to allow a great number of eels—which belong to the third and fourth links of the food chain—to grow and fatten, but Japanese eel farmers, guided by scientific advice, have in recent years offered the eels formula food; now the production of consumption eels has

Fig. 8. Milkfish farming in Sembajat in eastern Java. Pond-reared milkfish are packed in baskets lined with banana leaves, and ice is added to keep the fish in perfect condition during transportation.

Fig. 9. Eel farming in Japan. Mechanical control of the oxygen level is a prerequisite in well-stocked eel ponds.

Fig. 11. Oyster harvesting time in the Bassin d'Arcachon, France. Within a few hours the trays filled with oysters are loaded on a boat.

reached an almost incredibly high level per unit of surface.

But just as in milkfish farming, the stumbling block in eel farming is procuring the fry. Given sufficient pond acreage, all the equipment required, and sufficient capital for buying formula food, electricity, and labor, one still cannot hope for a commercial success when elvers are not available in sufficient numbers to start the produc-

tion of a new crop. In years with a poor run of elvers, Japanese eel farmers try to buy elvers from other countries, recently even in Europe! Therefore Japanese eel farming is in reality only semi-farming, for it remains dependent for at least part of the process on the caprices of nature. A closed cycle, so characteristic of a true culture, is not yet possible.

It would not be difficult to list many more examples of farming the sea, of which the oyster industries in France (Figs. 10 and 11), Australia, Japan (Fig. 12), and the United States are among the economically more important.

True farming. It is clear from the foregoing discussion that development of true cultivation, or full-cycle farming, systems for various fish and shellfish requires foolproof control of two important factors: procuring the fry and providing the food. In the case of mollusks, biologists worked for many years on a system of rearing larvae, obtained by induced spawning, until settlement. Finally success was obtained in several species of bivalves and gastropods: A very high percentage of the larvae can be reared until the benthic phase and then to full size and maturity. The first phase, larval development, requires meticulous care and the utmost cleanliness. Consistent results are obtained only if food organisms grown in pure culture are offered in the correct dose. Even slight deviations from the scientific prescriptions lead to inconsistent results and repeated failure. It is still a specialist's job to produce the desired quantities of young spat, and an inexpensive production of millions of specimens cannot yet be realized. But even the specialist in the rearing of shellfish larvae cannot guarantee prompt delivery of the ordered number of spat.

Hundreds of thousands of mollusks, belonging to various species, have already been cultivated from egg to egg, that is, in a full cycle, especially in

Fig. 10. Oyster farming in France. At low tide the oyster beds in Charente Maritime are tended like gardens. Numerous small sticks are planted in the beds to keep off shellfish-devouring rays.

the United States, Great Britain, and Japan (Figs. 13–15), but a commercially sound and independent industry based on these achievements has still to be developed. There seems to be a fair chance that cultivation of the hard-shelled clam *Mercenaria mercenaria* will soon lead the way. Its larvae will be reared under semilaboratory conditions and, after a sojourn in trays filled with running sea water, the young benthic specimens will be grown on tidal flats on the southern coast of England, protected against predatory crabs under fine-meshed netting. The very high price to be met on the market in France has to provide the key to remunerative farming.

In addition to rearing of larvae, one should also be able to stimulate primary production in order to exercise sufficient control over the quantity of food required in rearing the young specimens to marketable size. Nutrients are often the problem in primary production. During World War II, fertilization experiments were carried out in two Scottish lochs. Though it appeared to be possible to simulate primary production in the lochs, it was not possible to guide this process toward those species of plants and animals which are the links in the food chain that leads to man via consumption fish. Almost all the fertilizer was ultimately used by various plants and animals that are considered "weed" organisms, including birds. If "weed" organisms cannot be controlled in every link of the food chain, as is done in agriculture and animal husbandry, there is little hope that fertilization of sea water will pay for itself or that consistent results will always be obtained.

Experiments being carried out at St. Croix in the Caribbean should be seen in the same light. Nutrient-rich water from a depth of 1000 m is being pumped into a system of man-made ponds. A rich development of plant life in the ponds can certainly be expected, but it will be difficult, if not impossible, to exert an influence on the species composition of these sessile and planktonic plants and to avoid development of "weed" species, including poisonous dinoflagellates such as *Gonyaulax*.

Examples of full-cycle farming. Though we are forced to classify most examples of farming the sea as cases of semifarming, there are a few instances of full-cycle farming, and some others in which the cycle is almost closed.

Algae. An interesting example of full-cycle farming is the cultivation of the red alga amanori (*Porphyra*) in Japan. Really complete gardens have been created in well-selected sites in the sea; these gardens offer various types of collectors for *Porphyra* to settle and grow on. *Porphyra* spends part of its life cycle inside the superficial layer of mollusk shells to produce a spore, and therefore empty mollusk shells are placed in the same area. The spore settles again on branches and netting put out by the farmers and produces the edible fronds so sought after by Japanese gourmets. In addition, since fertilizer is spread in the gardens on a regular basis, the cultivation of amanori can be called a case of true farming.

Flatfish. Recent British efforts to rear flatfish is another example of full-cycle mariculture. Plaice and sole can be induced to spawn normally in captivity, and a high percentage of the larvae can be

Fig. 12. Oyster farming in Japan. Growth and fattening of oysters are successful in hanging culture widely used in well-sheltered bays.

reared until they are mature enough to be called fish. It is possible then to make the fish grow rapidly. This concept could be tested by putting millions of postage-stamp-sized plaice into the North Sea in areas where plaice food is abundant and plaice is scarce. Since fishermen of several nations fish the North Sea, the question has arisen as to how much of the cost will eventually return a profit to England.

Fish and shrimp. Fish and shrimp farming is carried out in Japan. Since every generation starts anew with material caught in nature, this has to be classified as semifarming. In the case of the prawn *Penaeus japonicus*, however, egg-bearing females are taken from the sea. The rearing, to the point at which the prawns are marketable, is done under

Fig. 13. The first step in laboratory rearing of shellfish, as carried out in Japan, is growing various types of food organisms for the larvae in pure culture.

Fig. 14. The European flat oyster (*Ostrea edulis*) is grown from egg to egg in Japan. However, the high costs of rearing make large-scale commercial application impossible.

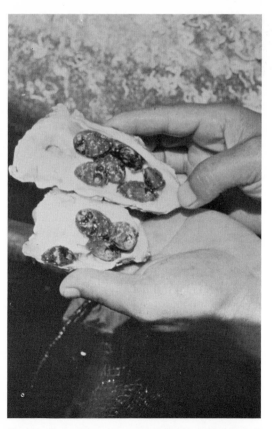

Fig. 15. Young laboratory reared abalones (*Haliotus*). It will take a long time and considerable labor and expense to grow them to marketable size.

carefully controlled conditions, with specially reared food organisms administered in the first delicate phase of the prawn's life. This is almost a case of true farming. In Japan, fishes such as yellowtail and red sea bream are already grown on a large scale. Yellowtails are caught at sea as small fish and then put out in enclosures where they grow rapidly when adequately fed. Sea bream larvae can be grown on a large scale from eggs under artificially controlled conditions.

Prawns, yellowtails, and sea bream are fed with fresh or frozen fish and shellfish. Since the conversion factor is about 8:1, the price of the fish used as food must be considerably less than one-tenth the price to be made for the farmed fish, because food constitutes only part of the expenses. Netting to enclose the fish in spacious ponds, manual labor, electricity to operate aeration machines, and the usual overhead expenses require a certain percentage of the gross profit. Since the food to be administered should meet high quality standards and since formula food used in eel farming is not cheap, one must expect that, even in a country with rather low wages such as Japan, cultivated fish must necessarily command high prices on the market. This is the case. In countries with high wages, the prices that must be asked for farm-raised prawns will inevitably be higher still, and thus the Japanese style of farming is prohibitive. Experiments are now in progress in the United States in rearing young prawns from eggs under semi-laboratory conditions and distributing the young prawns over large acreage of ponds in which they can obtain sufficient natural food. This is an extensive method of farming in the second phase of cultivation, following an intensive cultivation in the first, more delicate phase. It remains to be seen, however, whether this system of farming will be remunerative.

PROSPECTS

Industries based on a system of semifarming, such as mussel and oyster farming, certainly are of considerable commercial importance and their products are very welcome on the market, commanding a rather good price. Farming of mollusks is confined to coastal waters that offer sufficient protection against gales and other extreme weather conditions. When special equipment such as racks and rafts is used, shelter is even more important than it is in cultivation on the sea bottom.

Water pollution. Since some phases of farming oysters, mussels, and clams require considerable manual labor and since a high degree of mechanization is not always applicable, this type of farming can best be carried out in well-populated areas. Since the cultivated product is rather perishable, especially when it is to be eaten raw, a densely populated area in the immediate vicinity or at least excellent transportation facilities are welcome factors. On the other hand, a densely populated area inevitably means increasing pollution of the inshore water, especially where industry prevails. Many formerly productive oyster beds are already so badly polluted that larval development and settling have become impossible, and conditions for

growth of oysters have deteriorated so much that the oysters have disappeared completely. In areas of lower levels of pollution, the shellfish produced are often tainted to such an extent that they are not accepted on the market or are refused by public health authorities because of intolerable contamination with domestic sewage. Not only in Europe and the United States but also in Japan, oyster farming has been ousted bit by bit from well-protected bays and inlets by ever-increasing water pollution, and has been pushed to less sheltered sites where the risks of damage by severe weather and by a variety of predators are greater. Similar results must be expected where engineering works cut off bays and estuaries and lead to deterioration of the biological environment, which means destruction of existing and potential resources in the coastal waters. Therefore the prospect is that many of the traditional shellfish farming areas will gradually lose their suitability unless really serious efforts are made to control water pollution.

World food supply. Since farming the sea is necessarily limited to well-sheltered coastal areas and unpolluted water, the impact of its total production on food statistics cannot possibly be very great. If we are very optimistic and assume for a moment that the yield of the traditional fishery will not change, mariculture could perhaps lead to an increment in the crop produced from the sea from the present 1% to 1.1%. This is hardly the prospect the hungry millions are looking forward to!

[PIETER KORRINGA]

Jay W. Forrester is professor of management at the Massachusetts Institute of Technology. He directs a research and teaching program on dynamics of corporate, urban, and social systems and is author of Industrial Dynamics, Urban Dynamics, *and* Principles of Systems.

Social-System
Dynamics

THE DEVELOPMENT OF CIVILIZATION has been characterized by a sequence of frontiers in the understanding of man and his environment. There have been periods when the focus was on developing new forms of government. At other times the challenge has been the exploration of geographical frontiers. Some eras have been characterized by the development of great literature. Another stage in human progress has been the development of religions and man's relationship to the unknown. More recently, the challenge has been science and technology and the use of nature to raise the material standard of living. But science and technology are receding into the everyday fabric of human affairs; they are no longer frontiers. Now the focus is changing to exploration of our social systems, understanding their behavior, and learning to design social systems that better serve humanity. This will be the intellectual frontier of the next three decades.

A new science and methodology for dealing with complex systems is emerging. A theory of complex systems is being developed. Now the perceptive strengths of a person can be combined with the computational strength of electronic computers to generate new insights into the behavior of our social systems. We are entering a new frontier in human endeavor—the frontier of understanding the dynamic behavior of our social systems.

BACKGROUND

Throughout the world, the social systems of mankind are in crisis. Local communities no longer serve their traditional function of focusing diverse individuals on common objectives or of providing continuity and communication from one generation to the next. The industrial corporation, in the past so important in raising the standard of living, is under attack; it has served its purpose of organizing human effort for efficient production but may be failing to meet human aspirations for independence, individual importance, and self-fulfillment. The larger, older cities are falling into distress with costs rising and revenues declining, while all corrective actions seem to accelerate the downward spiral.

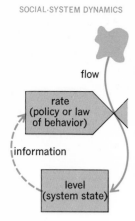

Fig. 1. Simplest possible system structure: one level and one rate in a single feedback loop.

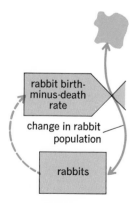

Fig. 2. Simple loop, either positive or negative feedback depending on how an increase in rabbits changes the net birth-minus-death rate.

On a national scale, inflation, economic stagnation, and an increasingly visible gap between rich and poor all suggest that we do not understand the interrelationships between fiscal and monetary policy, unemployment, economic growth, and equal opportunities for all citizens. Between nations, military conflict, international trade imbalance, international monetary crises, and the failures in economic development of the less advanced countries suggest that these larger systems are also in disarray and failing to serve mankind. On the global scale, growing population, pollution of the environment, and the conflict between a rising standard of living and limited natural resources again demonstrate that man has not yet come to terms with the systems within which he lives.

Are these difficulties unavoidable? Are we managing the best we can? Are we victims of circumstances beyond our control? Or, are these difficulties of our own making because we have created systems that we do not understand and that are unsuited to our purposes? The emerging answer suggests that man does not understand the nature of the complex systems within which he lives. Because he misjudges their character, man alters his social systems in directions that make these systems even less effective in serving his well-being.

Unsuccessful policies. Growing complexity in our social systems increases the probability that new laws, policies, and programs will produce results quite different from those anticipated. We build highways to reduce travel time but travel time is seldom reduced; the result is an increase in traffic. We put pressure on automobile companies to develop safer automobiles but the consequence is a raising of the speed limits, faster travel, and little change in the rate of fatalities. Rising land prices are used as an argument for allowing higher buildings in our cities, but zoning which permits higher buildings raises land prices still further and the result is not economy but congestion. Welfare systems are established to relieve the plight of the poor, but the consequence of the welfare program is often to actually generate welfare cases and to prevent escape from welfare status so that the result is to increase the welfare load rather than to reduce hardship. In a number of situations the increase in expenditure for police departments and crime control has aligned the public and the anti-social element against the police and thereby increased the crime rate by reducing cooperation between the community and the police.

New concepts. But a new science and methodology for dealing with complex systems is emerging. The concepts relate the structure of a system to its behavior. The theory and methods apply even if the system reaches across diverse intellectual fields. A system may combine aspects of engineering, medicine, management, economics, psychology, education, and sociology. In short, it is possible to relate all interacting functions that cause conditions to change with time.

Concepts forming a common foundation for dynamic behavior underlie all fields and can unify diverse disciplines. Divisions between intellectual fields are artificial and result from the happenstance evolution of research and education. Existing intellectual subdivisions accentuate the differences between fields rather than stressing their unity.

Past failure to understand our dynamic environment results partly from this compartmentalization and fragmentation of knowledge. Now, social, psychological, physical, biological, and humanitarian aspects can be unified into a single compatible system. The interactions emerge as more important than the separate fields of knowledge taken individually. Out of the interactions between diverse disciplines come the most conspicuous and most troublesome modes of behavior in our social-technical-psychological-humanitarian systems.

SYSTEM STRUCTURE

The word system as used here means feedback system or closed-loop system. The closed-loop configuration is the structure within which all decisions take place and all growth and control processes occur. To illustrate, Fig. 1 shows the simplest possible system structure. It is a single loop containing one level variable (the rectangle) and one rate variable (the valve symbol). The rate (valve) determines the flow (rate of change). The flow changes the system level (the state of the system). The level (or state) is the source of information that causes the rate to change. The rate (or policy) statement expresses how the information is used to determine the flow.

Figure 1 illustrates several general principles about system structure (see the author's book *Principles of Systems*). Only two kinds of variables (the levels and rates) are needed to describe any system, no matter how complex. Rates depend only on levels. Levels are caused to change only by rates. Information links always connect levels to rates. The information network connects dissimilar subsystems. All decisions are made in the feedback-loop structure. Beyond such concepts that help organize the structure of a system are principles that indicate how structure determines dynamic behavior.

Negative feedback. The circular structure in Fig. 1 and its endless elaborations describe the dynamic processes of man and nature. For example, assume that Fig. 1 represents a baby first learning to turn its head toward a light. The "level" is the position of his head. The "rate" is the rate of turning of his head. The information input is the perceived discrepancy between the angle of the head and the direction of the light. This is a negative-feedback loop, which is goal-seeking. The objective is to align the head and the light. We are immersed in feedback loops from our earliest development. Every decision, whether public or private, conscious or subconscious, is made in the feedback-loop context. The decision controls an action that alters the state of the surrounding system which presents new information to influence subsequent decisions.

But the same system structures that describe mental control and decision making operate in all natural processes. Consider Fig. 1 as representing a simple electric circuit in which a charged capacitor is discharging through a resistor. The level is the charge on the capacitor. The charge, combined with constants that describe the sizes of the capacitor and the resistor, determines the electric current (the rate of change, or flow, of charge). The current causes the charge (level) to change. The loop structure is one in which electric charge causes current which alters the charge. Such loops also represent the control process in biological systems.

Positive feedback. But feedback loops are more than control (that is, goal-seeking) processes. In their positive-feedback form they create growth. Furthermore, the nature of a loop, and the effect of other loops on it, can cause the loop to change from growth to goal seeking. This produces the growth-to-maturity and growth-to-stagnation transitions in natural and social change.

A rabbit population illustrates how a growth (positive) loop can become a goal-seeking (negative) loop. Figure 2 is the simple loop relabeled to represent rabbits in a fixed land area. The number of rabbits determines the net birth-minus-death rate. That rate changes the number of rabbits. But net birth-minus-death rate is not simply proportional to the number of rabbits. Instead it depends on crowding, as shown in Fig. 3. When the rabbit population is small, the population increase is proportional to the population. In other words, the more rabbits there are, the faster the rabbit population grows. But as the population density increases, the effects of crowding cause the birth rate to decrease and the death rate to increase so that the net rate of increase declines. With a sufficiently high population, shortage of food, increase in disease, and the psychological effects of crowding all combine to reduce the population growth rate and, under extreme circumstances, can result in the death rate exceeding the birth rate (as shown in Fig. 3 when there are more than 250 rabbits).

If the system of Fig. 2 starts with a small number of rabbits and is controlled by the relationship in Fig. 3, then the population increases to an equilibrium value as shown in Fig. 4. The rising section of Fig. 3 causes the upward exponential growth at the beginning of Fig. 4. The declining section of Fig. 3 reduces the growth rate and causes it to reach zero as the rabbit population reaches an equilibrium value. The rabbit population rises in the classic S-shaped growth curve.

Approach. Feedback loops produce growth. They also produce goal seeking in search of an equilibrium. They can produce oscillation as in a

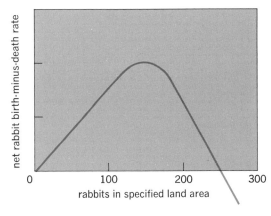

Fig. 3. Rate of change in rabbits depends on number of rabbits.

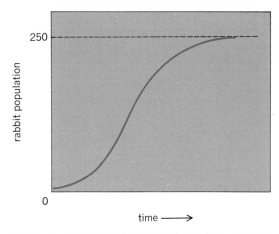

Fig. 4. Population growth curve generated by a positive (growth) loop that converts to a negative (equilibrium-seeking) loop.

simple swinging pendulum or in the rise and fall of business activities over an economic cycle. The behavior of feedback loops has been studied in engineering with the application of highly mathematical methods. But social systems are so complex that mathematical solutions are impossible. Progress in the field was hopeless until electronic computers became available. For the last three decades, computer simulation has been used to discover the characteristics of systems that are too complex for mathematical analysis. At first, computer simulation was developed for studying engineering systems. More recently, the same approaches have been extended to the much more complex structures of our social systems.

Stages of study. The process of studying a social system can be described in six stages. (1) The behavior modes of a system are observed in order to identify the symptoms of trouble. (2) The system is searched for the feedback structures that might produce the observed behavior. (3) The level and rate variables making up that structure are identified and explicitly described in the equations of a computer simulation model. (4) The computer model is then used to simulate, in the laboratory, the dynamic behavior implicit in the identified structure. (5) The structure is modified until components of the structure and the resulting behavior agree with the observed conditions in the actual system. (6) Modified policies can then be introduced into the simulation model in search of usable and acceptable policies that give improved behavior.

This design process brings the essential substance of a social system into the laboratory where the system can be studied. Laboratory representation of a social system can be far more effective than most people would expect. Anything that can be stated or described about a social system can be represented in such a laboratory model.

MODELS OF SOCIAL SYSTEMS

The use of models as a basis for policy design in social systems is not new. All human action and decision making are based on models. The mental image which one uses as a basis for decisions is a model. One does not have a real country or a real

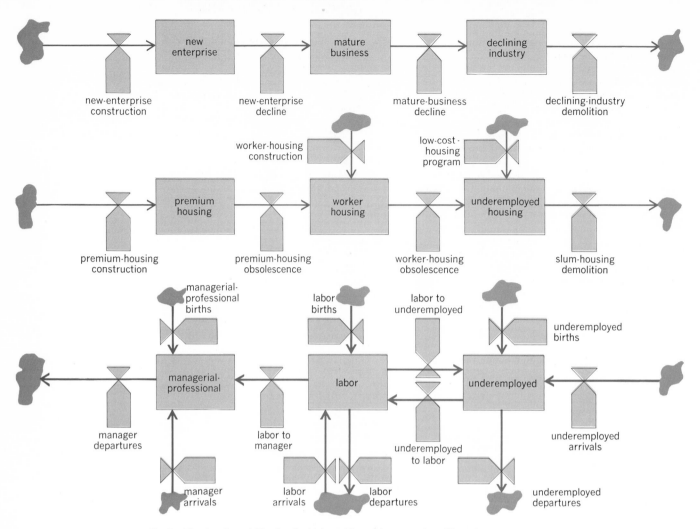

Fig. 5.   Nine levels and 22 rates that interact to produce an urban life cycle.

city in one's head, but only an abstraction of that real system. Each person has a different model and a different process and capability for estimating what the model implies for the future.

Mind and computer. One should compare the formal computer simulation modeling of a social system, not with perfection, but with the available alternatives. The alternatives are the mental models which are otherwise used by managers and political leaders. But these mental models are incomplete, unclear, shifting in structure, and hidden. Furthermore, they are naively simple compared to the computer models which are now possible and which have been demonstrated. But above all, the mental model is deficient in having no reliable way to establish the dynamic consequences of a given model structure. The human mind is good at perceiving the parts of a social system. The human mind is a good observer of the pressures, fears, objectives, information sources, and governing policies and actions at any one point in a system. But the human mind is misleading and ineffective in estimating what happens when a multiplicity of system components interact with each other.

Studies of industrial corporations have repeatedly shown that the very steps being taken to solve some major corporate problem can interact in such

a way as to cause that problem. Good intentions are not sufficient. The behavior of complex systems is counterintuitive. The steps adopted as corrective action often cause further degradation. This is true even when the individual parts of the system are well known and properly perceived.

The author's book *Urban Dynamics* illustrates how the general methods can be applied to a specific social system. This book examines the life cycle of a city as it grows, matures, and declines. Business, housing, and populations are interrelated with each other to produce the progressive changes in an urban area.

Urban model. Figure 5 shows nine system levels as rectangles; 22 flow rates are shown which cause the levels to increase and decrease. The information links from system levels to the system rates shown in Fig. 1 are not shown in Fig. 5 because of their complexity.

In Fig. 5 the nine levels are grouped into three subsystems. Across the top, the industrial sector contains commercial buildings in three categories distinguished primarily by age. Across the center are residential buildings in three categories, also distinguished by age and condition. Across the bottom are three economic categories of population. In this structure it is possible to see the reasons for urban decline. The age of a building tends to determine the character of its occupants. A new commercial building is occupied by a healthy, successful commercial organization that uses relatively more managers and skilled workers than those who are unskilled. As the building ages, it tends to house a progressively less successful enterprise with lower employment skills. In addition to the changing employment mix as the industrial building ages, there is a tendency for total employment per unit of floor space to decline. On the other hand, as residential buildings age there is a tendency for occupancy to increase as well as to shift to a lower economic category of population. Once the growth phase of the area is past and new construction is depressed because land area has become filled, the aging of buildings simultaneously reduces the opportunities for employment and increases the population. The average income and standard of living of the area decline.

Complex urban model. Figure 1 shows the information links from the system level as the input to the rate of flow. Figure 6 shows how the information links proliferate in a more complex system. Here the nine system levels reappear from the

previous figure but only one of the 22 flow rates is shown—the underemployed arrival rate. The figure traces how the various levels combine to create a composite "attractiveness" which determines the inflow rate to the area. If the area is more attractive than those from which people might come, a new inward population flow occurs. If the area is less attractive, an outward flow dominates. Five components of attractiveness are shown in Fig. 6. In the upper right corner, the underemployed-job multiplier relates the population to the available jobs and represents the income-earning attractiveness of the area. The underemployed-arrivals-mobility multiplier represents the attractiveness created by upward economic mobility. In other words, an area with high upward economic mobility is more attractive than one offering no hope of advancement. The underemployed-housing multiplier relates the underemployed population to the available housing. The area becomes more attractive as housing becomes more available. The underemployed-housing-program multiplier represents the attractiveness of a low-cost-housing program if such exists. And in the lower right corner, the public-ex-

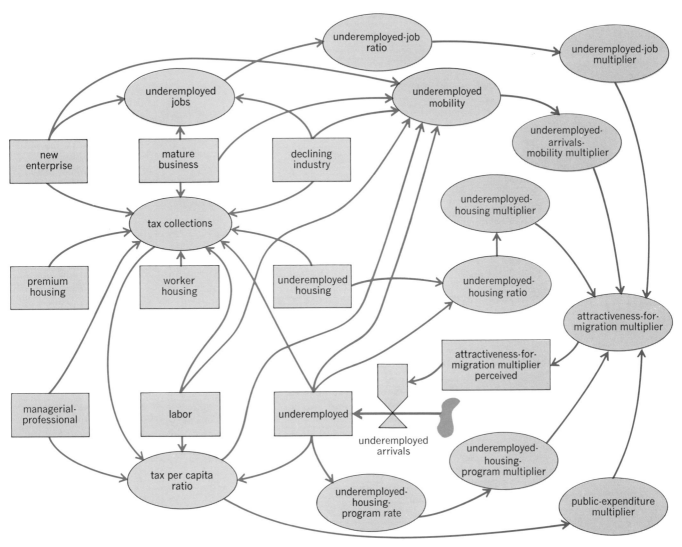

Fig. 6. Information links connecting the nine major system levels to the rate that describes underemployed arrival into the urban area. Inward population flow depends on "attractiveness" of area.

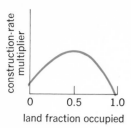

Fig. 7. Influence of land occupancy on building-construction rate.

penditure multiplier is the influence on attractiveness of the public expenditure per capita. As per capita expenditure rises, it means better public services, better schools, and higher welfare budgets.

The urban model contains behavioral relationships corresponding to Fig. 3 for the rabbit-population system. For example, Fig. 7 shows the effect of land occupancy on the construction of new housing and business buildings. When the area is empty, there are no services or workmen and construction is less likely. As the area begins to develop, it becomes progressively more attractive to new construction. But eventually, the land area starts to fill, the best sections have already been occupied, and the shortage of land depresses the construction rate. When all land is fully occupied, new construction is impossible. Demolition must first occur which opens available land before new construction can take place. Figure 8 shows the way in which the attractiveness of a city depends on the ratio of population to housing. The urban area becomes progressively less attractive as the population grows compared to the available housing.

The concept of attractiveness is fundamental to population flows. All of the characteristics of an area that make it attractive, the five in Fig. 6 and many more, combine to influence migration. An attractive area draws people. But almost every component of attractiveness is driven down by an increase in population. If there is an excess of housing, the area is attractive, but a rising population crowds the housing. If there is an excess of jobs, the area is attractive, but the incoming flow of people fills those jobs. In other words, migration continues until the attractiveness of the area falls and becomes equal to all other places from which people might come.

In a condition of population equilibrium, all areas must be equally attractive to any given population class, otherwise net migration would occur. If one component of attractiveness is increased in an area, other components must necessarily fall to establish a new equilibrium. Compensating changes in the components of attractiveness explain many past failures in our cities wherein we attempt to improve one aspect of the city only to discover that other aspects have become worse.

Theory and model. The system structure implied by Figs. 5 through 8 constitutes a theory that describes urban interactions. Each point in the system is described by an explicit statement in mathematical notation. Every relationship and assumption is clearly exposed. Each coefficient and each policy statement in the resulting model has real life meaning which can be discussed and compared with the observations and knowledge that may be available about the corresponding point in the real system.

Such a model consists of an unambiguous and explicit statement of relationships which might otherwise reside in a verbal description of the system. But verbal descriptions are not clear. Relationships are seldom stated explicitly. A model forces clarity of expression and allows others to perceive and evaluate the assumptions which are being made.

But a model is more than a theory. It also allows

the implications of the theory to be demonstrated. Figure 9 shows how this model generates the life cycle of a city. Growth occurs over the first 100 years. The land area becomes occupied and stagnation sets in during the next 50 years, with declining business activity and a shifting of population toward the lower income groups. Figure 9a at 100 years shows a labor population about double the underemployed population. The definitions of population categories and jobs make the 2:1 labor-underemployed ratio desirable. It gives good economic balance in the area and a high upward economic mobility for the underemployed. But after 150 years the two populations have become about equal. The underemployed group is excess compared to available jobs and the average standard of living is falling.

Figure 9b shows precipitous changes as growth in construction and population is arrested. The underemployed-job ratio rises rapidly, implying an increase in unemployment. The underemployed-housing ratio declines, indicating that housing for the lowest income group is becoming more readily available.

URBAN ECONOMIC BALANCE

Many people are surprised at the suggestion that depressed areas of our cities are areas of excess housing. But the housing is excess by almost any standard of reference. The area devoted to housing has become too large compared with the area devoted to job opportunities. This is because the industrial area has declined in employment at the same time that the residential area has increased in population density. But housing is excess even compared to the existing population in depressed areas of most American cities. Figure 10 shows conditions in New York City, where the high rate of abandonment of buildings indicates excess buildings for the economic conditions which exist. Simultaneously buildings are being abandoned while people live extremely crowded in others. But this crowding is an indication of the low standard of living and the failure of the economic balance in the area and is not an indication of housing shortage. Otherwise, if the economic condition of the population permitted, they could expand into the housing which is otherwise being allowed to deteriorate through lack of upkeep.

The stagnating urban area has become a social trap. Excess housing beckons people and causes inward migration until the rising population drives down the standard of living far enough to stop the population inflow. Anything which tends to raise the standard of living is defeated by a rise of population into the empty housing.

Reestablishment of balance. To reestablish a healthy economic balance and a continuous process of internal renewal, it appears necessary to reduce the inherent excess housing of depressed areas and to encourage the conversion of part of the land to industrial use. By doing so, a large enough wage and salary stream can be brought from the outside economy to make the area self-sustaining.

The way in which people interact with complex social systems has led to a steady evolution of laws and tax structures which make the city less viable. The book *Urban Dynamics* shows that most of the

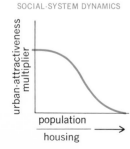

Fig. 8. Attractiveness declines as population exceeds available housing.

popular policies for reviving a city lie between neutral and detrimental. But policies can be designed which reestablish a healthy balance between housing and industry. The result is a cascading of mutual interactions which raise the economic activity of the area, increase upward economic mobility for the underemployed population, and shift the population internally from the underemployed to the labor class. This can be done without driving the existing low-income population out of the area.

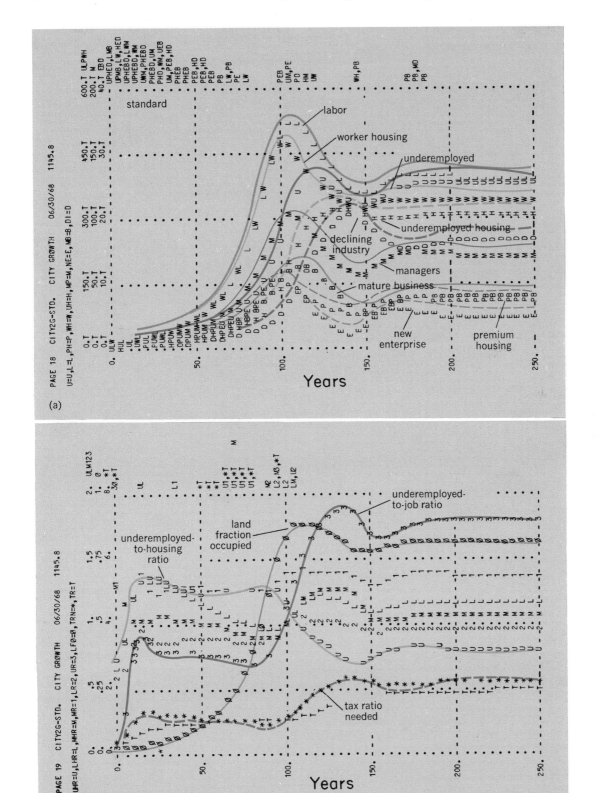

Fig. 9. Model for life cycle of a city. A period of growth is followed by stagnation. (a) Growth and stagnation in an urban area. (b) Social-pressures change at end of growth phase.

Fig. 10.   Excess housing beyond that which the economy of a depressed urban area can support.

Counterintuitive systems. The models of social systems in the laboratory, as well as actual systems in real life, show that complex systems are counterintuitive. They behave in ways opposite to what most people expect. Our experience and intuition have been developed almost entirely from contact with simple systems more like those in Figs. 1 and 2 than like complex systems of Figs. 5 and 6. Simple systems teach us that cause and effect are found close together in time and space. But this is not true of complex systems. In simple systems we think of action A causing result B without thinking of the information feedback loop, shown in Fig. 1, whereby result B influences future action A. We fail to realize that the behavior of the more complex systems does not arise from a cause-and-effect relationship but from the structure of the continuously interacting components.

In complex systems the short-term response to a policy change is apt to be in the opposite direction from the long-term response. A policy change which improves matters in the short run often degrades the system in the long run. This is especially treacherous. The short tenure of men in political office favors decisions which produce results quickly. These are often the very actions that

eventually drive the system to ever-worsening performance.

Short-run versus long-run reversal processes are all around us. If an agricultural country is to industrialize, it must accumulate railroads, factories, and steel mills. This capital accumulation can only be done by foregoing consumption and reducing the standard of living first in order that the standard of living may rise at a later time. If a company faces declining earnings because its products are obsolete, it must invest more heavily in product research and incur even deeper short-term losses if it is to recover in the more distant future to a profitable product stream. A person can accomplish more in the short run if he goes without sleep, but his effectiveness deteriorates later. This reversal between the short run and the long run occurs repeatedly.

In Fig. 9b the pressure from unemployment rose sharply as the system moved from growth into equilibrium. All models thus far examined which generate the life cycle of a social system show tremendous shifts in internal pressures as the system moves from the growth phase into equilibrium. Much of the worldwide stress at the present time is probably coming from the forces which are beginning to depress growth in geographical area, population, and standard of living. But the structure and rules under which a system operates determine the severity of the stresses that arise as equilibrium is entered. Various transitions between growth and equilibrium appear possible.

OBJECTIVE

One of the greatest forthcoming challenges to mankind is to enter a state of equilibrium by transitions that avoid catastrophe and still preserve individual values and purpose.

In social dynamics the objective is the design of improved social systems. Designing improved social systems is sometimes viewed as immoral based on the argument that one should not tamper with the natural human environment. But our social systems have not been given by nature. They have been designed by man. Constitutions of countries, laws, tax regulations, and religion combine to become the design of a social system. If we are not pleased with the way our social systems are operating, it implies that we prefer a different design.

[JAY W. FORRESTER]

John R. Zurbrick is a senior staff scientist in the Materials Applications Department of Avco Systems Division, Lowell, Mass., and a consultant on the characterization of fiber composite materials. He developed nondestructive evaluation techniques for directly predicting elastic moduli, failure strengths, and component volume fractions in fiber-reinforced plastics and metals. Currently he is developing new analytical methods and nondestructive testing techniques for predicting the adhesive and cohesive strengths of adhesively bonded joints.

Nondestructive Testing

I**F GIVEN AN OBJECT INTO WHICH** you cannot see, you may cut it apart to look inside. If requested to measure the mechanical properties of that object, you may bend it, distort it, or even break it. But when done, you no longer have the object you started with. Evaluation of the physical properties of an object without cutting, scratching, or otherwise causing it physical damage is called nondestructive testing. Such techniques began with the introduction of industrial radiography for the purpose of observing cracks, bubbles, slag, and other defects in heavy armored tank welds without damaging the welds. Whereas x-radiation and photographic film provided the first "eyes" in this technology, today all known forms of energy and every energy-producing or energy-detecting device imaginable serve to nondestructively probe engineering materials. When finished hardware components become so expensive that even a single component cannot be routinely sacrificed to assure quality and reliability, or so critical to system performance that each must be thoroughly inspected prior to use and during use, or each item must be rapidly measured for defects, nondestructive evaluation becomes technically necessary and economically wise.

The penetrating radiation, ultrasonic pulse echo, visual fluid penetrant, magnetic leakage field, and eddy current test methods and techniques are widely used today for observing defects and discontinuities in metal shapes. These evaluation capabilities have progressed from armored land vehicles, to ships and submarines, to nuclear devices and power plants, to rocket propellants and engines, to space vehicles, and to the special needs of commercial manufacturing and construction. The more recent challenges of new engineering materials (that is, fiber-reinforced plastics, high-strength ceramics, and coated refractory alloys) have been met with intensive applications research and development. The impact of research and development in nondestructive evaluation has already been felt in recent aerospace accomplishments. The future will see these hard-won capabilities applied widely in high-output, commercial production to assure uniform quality and minimize product liability.

hologram recording

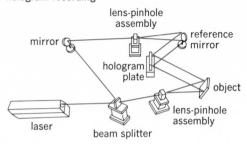

(a)

hologram viewing

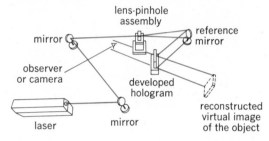

(b)

Fig. 1. Beam-path scheme for the laser hologram process. (a) Construction of a hologram diffraction pattern in a photographic emulsion. (b) Reconstruction of the three-dimensional image in space. (GC Optronics, Inc.)

The more recently developed techniques of nondestructive testing, the relationship of such testing to materials properties, particularly in design engineering, and the future direction of nondestructive testing in general are considered in this article.

TECHNIQUES

Recently developed techniques of nondestructive testing consist of holography, the light-scattering properties of liquid crystals, infrared scan-

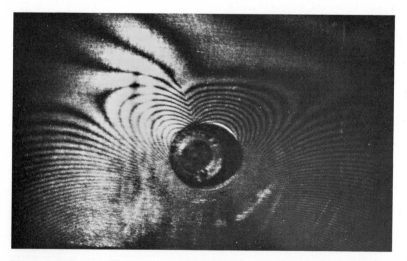

Fig. 2. Interference fringes which occur when the object is viewed, with the holographic image superimposed, under different levels of strain. (University of Michigan)

ning, microwave scanning, x-ray scanning, neutron radiography, and acoustic emission.

Holography. Holographic techniques for nondestructive evaluation have been advanced tremendously in the past few years and significant full-scale applications to product inspections have been accomplished. The laser stimulated and led this development activity, resulting in laser holographic interferometer instrument systems and test techniques.

Interferometry, in the form of optical interference microscopes and optical flats, has for years been common in nondestructive inspection. Holography, the creation of a phase and amplitude diffraction pattern in the plane of a photographic plate, had been studied for 10 years prior to the laser. Methods used near-coherent, monochromatic light to reconstruct x-ray diffraction holograms. The coherence length was so short that areas only a few angstroms square and one atom layer deep could be observed at any one time, so that light holography mainly provided a laboratory means for reconstructing and studying crystal atom lattice patterns.

The gas lasers, particularly helium-neon, provided power output and coherence lengths suitable for holograms of large-object curved surfaces. By combining three physical phenomena (the coherent light energy of the laser, the photographic diffraction hologram, and special double-exposure techniques to produce the defect information–bearing interference fringes) the laser holographic interferometer technique was created.

The optical layout for hologram construction is shown in Fig. 1a. The reference wavefront and the object-reflected wavefront combine at the photographic plate to expose it according to a three-dimensional diffraction pattern. Once the hologram plate is developed, it may be viewed as in Fig. 1b such that the object is reconstructed as a three-dimensional image, exact in size and position relative to the hologram. If the hologram is prepared with the object at rest and then if the holographic image and the real surface are viewed together while the object is mildly vibrated, interference fringes will reveal local areas of unusual deformation (Fig. 2). Knowing the construction and elastic properties of the object being inspected and the wavelength of the laser light, defects can be readily pinpointed. The holographic tire analyzer is the first truly commercial application (Fig. 3). Three viewing techniques have been developed: real time, where the stressed, illuminated object is viewed through its previously developed hologram; time lapse, where two holographic exposures are made on the same photographic plate with the object in two different states of stress, and later reconstructed for viewing; and time average, where one hologram exposure includes many cycles of sinusoidal vibration of the object, and is later reconstructed for viewing. Test object stressing may be accomplished by many means insofar as it remains a nondestructive evaluation. Acoustic vibration, vacuum, pressure, or local incremental force loading, and thermal stressing have been found easily and satisfactorily applied.

A more recent practical application of laser holographic interferometry has been to aircraft honeycomb sandwich constructions (Fig. 4), where face sheet – core adhesive bond condition and cell wall separations are critical defects. The ability to observe strain, particularly long-term plastic deformation at microinch levels in large objects from a considerable distance, is ready for application. Holographic detection of microcracks in metals has been demonstrated at the University of Michigan and Bendix Research Laboratories. Most significant results of these studies on stress corrosion cracking and fatigue cracking were the observations of characteristic interference fringes just prior to the formation of microscopically visible cracks (Fig. 5).

Holographic techniques, rapidly developed as a part of laser technology advances, established the general requirements for generating holograms and holographic interferograms using any form of propagating energy: spatial and temporal coherency, a suitable three-dimensional hologram plane, and a means for real object or hologram comparisons. Researchers in many fields have investigated holographic techniques based on microwaves, ultrasonics, and audible sound. Ultrasonic pulse echo and transmission techniques produce two-dimensional displays of defects in objects, which, with considerable labor, may be combined to give an approximate three-dimensional picture of the defect. By using continuous coherent ultrasonic waves to internally "illuminate" the object submerged in a tank of water, a hologram can be created at the water's free surface. By viewing, with a laser, the ripple pattern on the water or a photograph of it, an exact three-dimensional picture of the internal defect may be observed. Interferometric techniques using a nondefected standard have been developed.

A unique microwave holographic technique has been developed at Bendix Research Laboratories. A 35-GHz microwave beam illuminates the object, and transmitted or reflected waves join the reference beam at the holographic plane. Vapor-deposited Nichrome on Mylar converts the diffraction pattern into corresponding warm and cool zones. A thin layer of liquid crystals on the other side of the Mylar converts the thermal pattern into a color pattern which may be photographed for laser reconstruction at a later time. When a laser is used to reconstruct x-ray diffraction patterns (a hologram), fantastic magnification is obtained (26×10^7), while the laser reconstruction of microwave or ultrasonic holograms results in a great reduction in observed size (10^{-3}).

Liquid crystals. A number of liquid chemical derivatives of cholesterol exhibit, at certain temperatures, the selective light-scattering properties normally associated with crystals in solids. From these special molecular alignment capabilities has come the term liquid crystals. Many families of chemicals which exhibit this property continue to be discovered, and applications found for them in nondestructive evaluation. When used in thin films (for example, 1 mil thick), the liquid crystals serve to give visual indications of temperature gradients over an area of surface. Each specific liquid-crystal material has a unique range of temperatures over which it reflects the various colors of the rainbow. The cooler, or low-temperature, end of the range reflects red. As temperature increases, the colors pass through the visible spectrum, ending

with blue and violet at the warmer end of the range. Beyond these limits the liquid crystals remain colorless. The range, by design, may be as narrow as 1°C or as broad as 50°C to allow selection of desired temperature sensitivity.

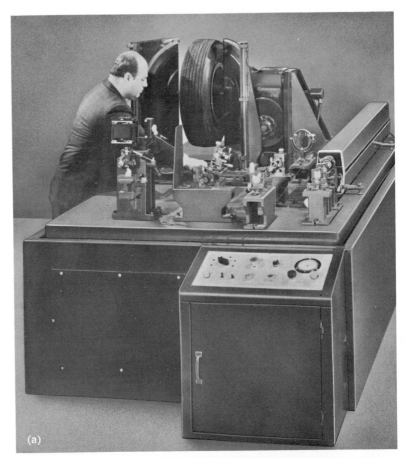

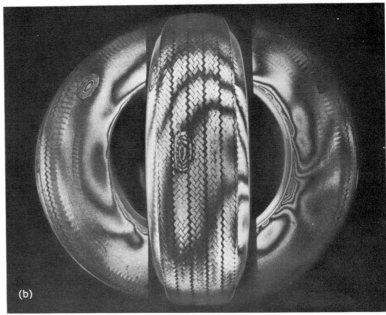

Fig. 3.   Testing of automobile tires. (a) Holographic tire analyzer reveals (b) hidden tire defects, such as tread- and shoulder-area ply separations, when the tire is inflated or subjected to various temperatures, captured here by double-exposure interferometry. Real-time viewing of a defect may be likened to observing the telltale moving ocean-wave patterns in the vicinity of a sunken rock. (*GC Optronics, Inc.*)

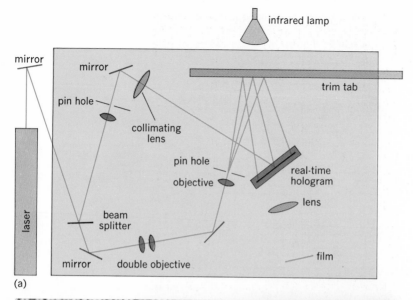

(a)

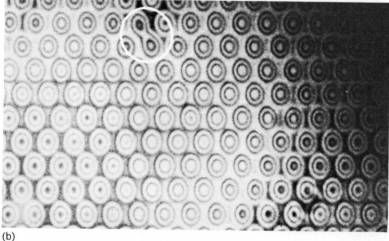

(b)

Fig. 4. Application of laser holographic interferometry for detection of unbonds in honeycomb sandwich panels. (a) One scheme for real-time viewing where thermal stressing provides the dimensional changes (*University of Michigan*). (b) A figure-eight fringe anomaly around two cells reveals a single cell-wall disbond in a titanium honeycomb structure following internal-pressure stressing (*GC Optronics, Inc.*).

Transient flow of heat by conduction in an object will follow the path of least resistance. By this means internal defects and discontinuities may be revealed at a surface as temperature gradients. For example, if one surface of a honeycomb sandwich panel is momentarily heated, conduction into the core through the adhesive bondline will rapidly cool the face sheet. Wherever a defective bond occurs, conduction will be blocked and a significant thermal gradient will exist around that area. Lateral heat flow will dissipate the gradient from some maximum, requiring a properly timed temperature measurement or observation. Liquid crystals provide one means to vividly indicate the defect-indicative gradients as abrupt changes in color. Beyond proper use of the liquid-crystal films themselves, test sensitivity depends on the nearness of the defect to the observation surface, thermal conductivity and heat capacity of the material, and conditions of heat input or extraction. Electronic

microcircuit devices, which generate heat in the resistive elements and at improper connections, may be nondestructively inspected with liquid crystals directly while the electronics are undergoing operational checks. The color changes of most liquid-crystal materials are reversible so that the heating or cooling cycle may be repeated as often as necessary. Determination that a certain temperature has been reached or exceeded is accomplished with "memory" type liquid-crystal systems, which exhibit a color until the conversion temperature is reached, at which point they turn black. Simple brushing restores the coating for reuse.

The cholesteric liquid crystals may be directly applied to a test surface or may be applied to one side of a very thin (0.00025 in.) plastic film, the other surface being brought in contact with the test surface. Encapsulated liquid crystals ease some handling and atmospheric deterioration problems associated with the free liquids. Expanded use of liquid crystals for visual observation of thermal gradients will be closely tied to advances in nondestructive thermal testing, where means to optimize defect indications and quantize their interpretation are currently being explored.

Infrared scanning. Considerable enthusiasm has been generated over the past 5 years for infrared thermal nondestructive evaluation. Cause for excitement has been the development and commercialization of sophisticated infrared radiometers, microscopes, and cameras (Fig. 6) based on high-sensitivity, optically immersed thermistor detectors. The incident radiant energy is absorbed in a thin blackened surface on the thermistor. The absorbed energy slightly raises the temperature of the thermistor, which has a high negative change of resistance with temperature (4% per degree Celsius). Means for concentrating the energy at the detector, chopping the incoming energy to establish an alternating-current detector output signal, and providing a carefully controlled internal blackbody for accurate data reference have been carefully balanced in commercial instruments. The accompanying advances in detected infrared radiation accuracy, reproducibility, and sensitivity have made such instruments suitable for the purposes of nondestructive testing.

Just as in the case of liquid crystals, the purpose here is to detect temperature gradients, on the surface of a structure, caused by defects or discontinuities beneath that surface. Since the various temperatures in the gradient radiate different combinations of infrared wavelengths, assuming constant emissivity, the surface temperatures may be detected and quantitatively recorded at some distance from the surface. Again it must be emphasized that success in finding and identifying defects in a structure depends primarily on heat conduction within the materials of the structure, and far less on the means to detect the resulting surface temperature changes. Because of this, current research and development investigations seek to relate defect size and location, as well as heat conduction controlling properties of the materials, to surface temperature profiles and defect depth.

Scanning systems which combine an infrared

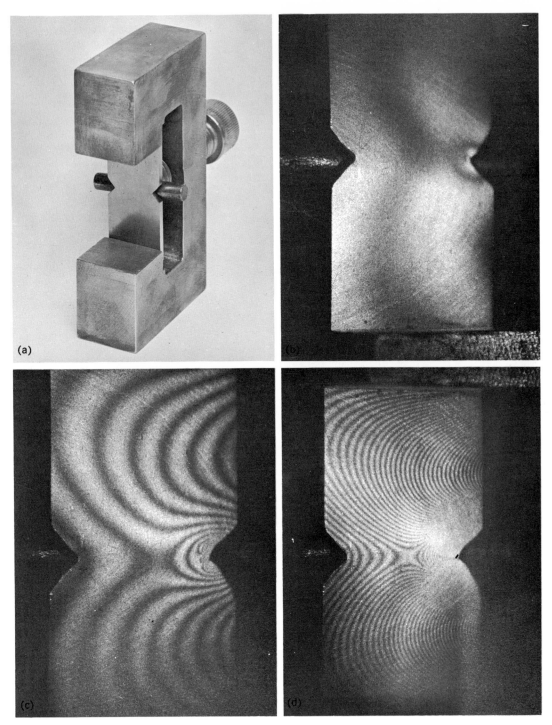

Fig. 5. Laser holographic interferometry used to reveal the progression of stress corrosion cracking in Ti-6AL-4V titanium alloy in a methanol environment. (a) Specimen-stressing fixture. (b) Fringes reveal crack initiation site prior to formation of microscopically visible crack. (c) Beginning of visible crack. (d) Visible crack extended. (*University of Michigan*)

heat source with an infrared radiometer to cover the curved surfaces of large structures (Fig. 7) have been the foremost outcome of development necessitated by practical needs. Detection and identification of unbonds between face sheet and core of honeycomb sandwich structures have been successfully demonstrated and used.

Thermally active systems, such as electronic circuit boards and microcircuits, require no exter-nal source of heat to produce the telltale tempera-ture gradients. Radiometer mechanical scanning detects overheated or cold resistors and improper high-resistance connections.

The Swedish company AGA Aktiebolag has developed and markets the most advanced ther-mographic imaging system. Its infrared camera supplies test object temperature and scan-syn-chronizing data to a separate cathode-ray-tube

(a)

(b)

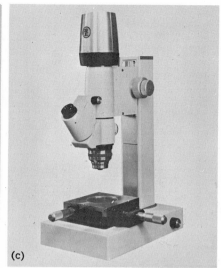
(c)

Fig. 6. Examples of infrared-radiation temperature-measuring equipment. (*a*) Precision radiometer for single point readings or mechanical scanning of temperature profiles. (*b*) Infrared scanning camera for viewing area thermographs on photographic film. (*c*) Infrared microscope for viewing electronic microcircuits. (*Barnes Engineering* Co.)

monitor unit, producing a real-time 100-line-scan thermogram with 100 picture elements per line, at 16 frames per second. Temperature resolution is 0.2°C. Accurate temperature evaluation is accomplished through the superimposed isotherm gray scale. Camera focusing is performed remotely at the display unit.

Microwave scanning. The absorption, reflection, scattering, and transmission characteristics of microwave energy offer particular promise for nondestructively detecting defects and measuring properties in nonmetallic materials. Microwave

Fig. 7 Automated mechanical scanner, infrared lamp heat source, and infrared radiometer detector are used to locate unbonds in a helicopter rotor blade section. (*Automation Industries, Inc.*)

instruments based on x-band (9.8 GHz) frequencies have been developed specifically for nondestructive evaluation and made commercially available in the late 1960s (Fig. 8). In typical measurements the object to be inspected is large and must be placed in a microwave beam which is propagating in "free space." Separate transmitter and receiver provide transmission capability, while a single unit combining both functions is suited for a reflection technique. Experience has shown that the microwave-producing klystron and the standing wave detector should be located in the sensing head, close to the measurement point, to allow short transmission lines and thereby minimize inherent instrument variables.

Every microwave signal is completely characterized by phase, amplitude, and frequency. Phase and amplitude values contributed by the instrument must be separated from the total response in order to observe and interpret the test object contributions. This is not so easily accomplished. Ideally, it is preferred that only one value, phase or amplitude, carry the required defect information without need to account for the other. Amplitude-sensitive, phase-diminished microwave-absorption instruments have been successfully applied to rapid-production monitoring of moisture content in bulk granular materials such as thermoplastic pellets and animal feed and in liquids and slurries. Phase-sensitive, amplitude-diminished instruments respond primarily to energy storage within a material and provide direct information on the plastic resin degree of cure, the changes in mixtures and composition, and the physical dimensions and defects.

The art of applying microwaves to specific nondestructive evaluation situations remains in its infancy, requiring that a skilled technologist develop special techniques by trial and error and empirical response signal versus material variable correlations. Commercial microwave instruments which detect output and plot both phase

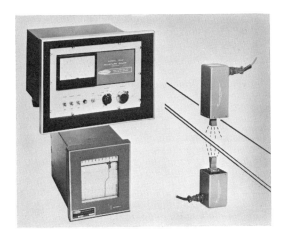

Fig. 8. A production-line through transmission microwave gage for detecting and displaying moisture content in strip-fed or bulk granular materials. (*Microwave Instruments Co.*)

(a)

and amplitude simultaneously meet these general-purpose needs very well.

X-ray scanning. Film radiography constitutes the major activity and market in nondestructive evaluation. Techniques for exposing the film to produce high-quality, maximum-sensitivity images of internal discontinuities have been perfected over the years for thick and thin objects having extremely high to very low absorption characteristics. But one inherent drawback has remained: the lengthy time required to expose, develop, and dry the radiographic film, even with once-through automatic film processors. Establishing correct exposure conditions (tube kilovolts, beam current, and operating time) can involve a number of cycles and most of an hour. Real-time exposure adjustment, observation, and accept/reject decision capabilities are required in high-output production for on-line 100% inspection. X-ray systems which do not rely on film exposure and processing have therefore received intensive development in recent years. Packaged units and custom-designed systems are now commercially available (Fig. 9a).

The active imaging device is typically a television camera, based on an x-ray-sensitive internal vidicon tube or in conjunction with an external fluoroscopic image-intensifier tube (Fig. 9b and c).

Fig. 9. X-ray inspection system. (a) Fully automatic system for small components. An electrostatic image intensifier provides data for electrical analog comparison, which in turn rejects or accepts the part by way of the chutes shown. Filmless radiography conditions are controlled and resulting images observed from the remote operator's console. Enlargement, focus, and contrast may be readily optimized for high-production component inspection. Mechanical manipulators, remotely controlled, provide six independent motions to allow convenient and rapid inspection of critical points in complex assemblies. (b) X-ray-sensitive vidicon tube offers extremely high resolution over a 1/2 × 3/8 in. field of view by using a 30-MHz bandwidth and 1050-line scan. (c) Fluoroscopic image intensifier–TV camera chain offers 3000–5000 times brightness gain, direct enlargement techniques, and zoom effects over a large field of view. (*Balteau Electric Corp.*)

(b)

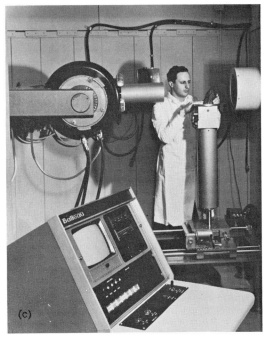

(c)

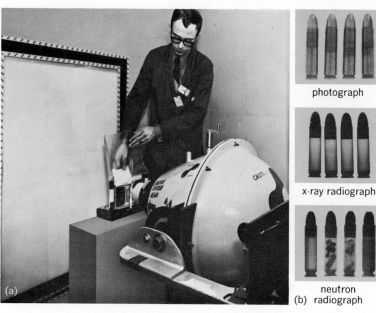

photograph

x-ray radiograph

neutron
(b) radiograph

Fig. 10. Neutron radiography. (a) Recently developed neutron-radiographic camera, based on californium-252, offers the significant advantage of portability (100-lb unit), so that the exposure may be conducted in a manufacturer's plant or at a field service facility, thus avoiding the necessity of shipping the item being neutron-radiographed to an atomic reactor facility. (b) Comparison of photograph, x-ray radiograph, and neutron radiograph of bullets. (*Battelle Northwest*)

The vidicon system is best suited for very small objects, $1/2 \times 3/8$ in. or smaller, such as electronic microcircuits, where magnifications up to 30× with a 1050-line-scan video output allow wires 0.0005 in. in diameter to be detected and resolved. Larger objects require the photoemissive, electrostatic image intensifier which supplies a 3000–5000 times gain in brightness from the large fluoroscope phosphor screen to the small anode phosphor screen. Optical and electrostatic adjustments provide up to 6× enlargement with zoom effect.

Beyond outstanding operator convenience at the control console, these real-time radiographic studies can be performed far removed from the source of radiation, for maximum operator safety.

The fully automatic x-ray inspection systems have been designed and placed in operation as a natural consequence of filmless imaging. In these, a component can be automatically loaded into the x-ray field, analyzed for defects according to a pre-programmed sequence and set of accept/reject criteria, and sent accordingly down paths to the next process step or to the scrap barrel. Inspection rates typically go as high as 1800 items per hour.

Neutron radiography. Polymeric and other organic compounds and materials and boron filaments have found their way into critical applications which require inspection for presence, alignment, and defects. Many applications place these materials within metal cases or with other metallic components. Since x-ray absorption is a direct function of atomic number, the hydrogen-, carbon-, oxygen-, and nitrogen-bearing materials absorb so few x-rays compared with the common engineering metals that they are unobservable radiographically. Nature has provided the means, however, to make the required observations through thermal neutron penetrating radiation.

Hydrogen and boron absorb or "capture" thermal neutrons far more readily than most metals, which are relatively transparent, even in thick sections.

Until recently the only reasonable source of thermal neutrons has been the atomic reactor. Neutron radiography required that the test item be shipped to the reactor, placed in the target area, and the reactor cycled. Recent improvements in reactor facilities at Aerojet-General and General Electric have been aimed at efficient handling and reactor operation for low-cost-production high-resolution neutron radiography on a contract basis.

A most recent and significant advance has been the invention and development at Battelle Northwest of a portable neutron radiographic camera, based on the isotope californium-252 as the source of neutrons (Fig. 10). The 268-μg source used yields 6.2×10^8 neutrons per second. A cadmium or gadolinium screen converts the neutron radiation pattern to gamma radiation, which exposes the radiographic film. Future development toward in-plant neutron radiographic facilities is anticipated as californium-252 becomes available in relatively large quantities from AEC's Savannah River facility, perhaps as early as 1972.

Acoustic emission. When a piece of wood is bent, audible snapping indicates that the wood is about to break. Such snapping consists of acoustic emissions in the audio frequency range from fairly large-scale, local failures. Local "failures" associated with dislocations in motion, fatigue cracks in progressive extension, and polymeric chains being broken emit snapping noises too, but in the ultrasonic frequency range. A sensitive ultrasonic transducer with electronic subtraction of a fixed carrier frequency, or slowing of a tape recording to translate the telltale emitted frequencies into the audio frequency range can be used to detect ultrasonic snapping. This is the general approach of acoustic emission methods and techniques in nondestructive evaluation.

Acoustic emissions result when loads are applied to engineering materials, some at low levels of stress and others at stresses near the failure point. Acoustic emissions which occur during fatigue studies, long-term creep studies, stress rupture studies, and quasi-static tests-to-failure are currently being investigated. High-strength/high-temperature metals, high-modulus/high-strength fiber composites, and ceramics are undergoing basic studies where particular emphasis is on learning the factors controlling brittle fracture mechanics and the causes of catastrophic failure. Development projects also include instruments for detecting fatigue cracks extending from rivet holes in aluminum aircraft structures and a means for in-service monitoring of acoustic emissions from ball bearings to detect incipient failure.

Developing or purchasing suitable ultrasonic emission detecting transducers and electronic signal processing and translating equipment is the first step in such studies. The resulting data, in the form of characteristic "noise," must be reduced to manageable quantities, and this is where the main research effort has been concentrated. The problem is one of pattern recognition. No pattern-recognition instrument/computer system has yet been devised which equals the human ear; hence the value of frequency translation. One straightforward method is frequency domain analysis, where a single characteristic frequency is sought from among all the frequencies occurring so that its amplitude may be quantitatively measured and related to an important material or structural change. Acoustic emission method and techniques have added a new dimension to nondestructive evaluation, and significant growth is anticipated in the future.

MATERIALS PROPERTIES

Much of historical nondestructive testing and current advances in nondestructive evaluation have been founded on the concept of the "defect," for example, a crack, an inclusion, a void, or a porosity. A defect connotes that something has deviated from the expected or the ideal and that it is "bad." Experience has taught us that not all defects measurable by nondestructive methods and techniques are detrimental to end-use performance. In some cases certain defects even improve performance. Out of this has come the term discontinuities. Nondestructive testing will reliably produce data concerning the size, shape, and location of discontinuities. The difficulty lies in drawing an empirical, indirect relationship between the degree of occurrence of discontinuities and their effect on some performance criterion. Responsibility for establishing the correlations is often placed in the hands of the practitioner.

Drawing such a correlation requires an accurate definition of the performance problem and its contributing defect factors before any reasonable data taking or nondestructive testing may begin. Only through a thorough understanding of both the problem and the selected nondestructive test technique will the correlation be evident. Most frequently these are indirect plots of instrument response versus extent of defect, with some indication of acceptance level. Rarely do they relate instrument response and end-use performance, even indirectly.

Design engineering. When applying complex design equations to calculate end-use performance, a design engineer uses materials property data, namely, elastic modulus, ultimate strength, fatigue life, and Poisson's ratio. To design properly and to select the most suitable material in the correct dimension, the design engineer requires materials property data exactly representative of the material to be produced in the shop. Destructive mechanical tests supply such data on a sampling basis and issue average values. In many of the advanced materials, and even in some more traditional materials, batch-to-batch, part-to-part, and within-part variability is so broad that average values are not representative and, in fact, may be misleading. Minimum values are wasteful and lead to costly overdesign. Destructive testing on anything approaching 100% of the product is impossible. But what about the nondestructive prediction of materials properties?

The concept of direct nondestructive relationships to materials properties is appealing because each item can be evaluated, actual materials properties determined, and acceptance or rejection made in terms of design requirements.

Research. Research over the past 5 years, by the author and others, has proved that direct, accurate predictions of materials properties can be readily accomplished in a surprisingly straightforward manner. First is the understanding of the materials-energy interaction involved: the physics of the problem. By characterizing energy in terms of frequency, phase, amplitude, plane of polarization, mode of oscillation, velocity of propagation, and so forth as it enters the material and again when it leaves the material, any changes which occur in the character of the energy are directly attributable to the properties of the material, in the region interrogated by the energy as it passes through.

If a change in energy character is correlated with a destructively determined materials property on a point-by-point basis over a range of inherent variability, the ability to predict that property nondestructively is accomplished.

The author developed a correlation relating ultrasonic compressional wave velocity V_L, bulk density ρ, and Young's modulus E to nondestructively predict in-plane modulus at any polar angle θ in fiber-reinforced composite materials. This relationship, shown below, holds for low-, medium-,

$$E_\theta = \frac{(V_L)_\theta^2 \rho}{0.0124}$$

and high-modulus fibers (glass, graphite, boron) and for any plastic resin or metal matrix. It was also found that 20 different properties and compositions varied randomly among these materials, even under supposedly tight process control. In like manner, 20 data (18 E_θs and 2 gamma radiation absorption values) completely characterized any given composite material. In addition, the "invariant" properties (E_{11}, $\overline{E}$, E_{22}, $\overline{G}$, σ_{12}, and G_{12}) could be calculated accurately. Component volume fractions of fiber, matrix, and porosity came from the gamma radiometric data (another predictive nondestructive relationship). Applications of these equations to gas turbine engine blades and primary aircraft structures are currently in development. This work was sponsored by the Air Force Materials Laboratory, Air Force Systems Command, United States Air Force.

In more recent work the author has attacked the problem of simultaneously predicting the adhesive strength and the cohesive strength of adhesively bonded joints, using a combination of four different nondestructive tests. A basic predictive equation for accomplishing this formidable goal has been developed. This work was sponsored by the Advanced Research Projects Agency, U.S. Department of Defense, through the Naval Air Engineering Center.

FUTURE DIRECTION

Two very recent analyses of technical and business prospects for nondestructive evaluation over the next 10–20 years have been published. The first is *Nondestructive Testing*, by the Long Range Planning Service, Stanford Research Institute (SRI), Report No. 399, February, 1970. The second is *Nondestructive Evaluation*, by the Ad Hoc Committee on Nondestructive Evaluation, National Materials Advisory Board (NMAB), National Research Council, Publication NMAB-252, January, 1969. The NMAB report, at the request of the U.S. Department of Defense, explores critical materials problems existing today and anticipated for the future, making recommendations for advances in specific nondestructive testing capabilities. The SRI report studies the business influences and technical applications expected by 1980.

Growth rate. The growth rate for nondestructive testing equipment and supplies in the United States market is expected to exceed an average 10% annually. The 1969 market of $85,000,000 should reach $250,000,000 in 1980. Each nondestructive testing technique will experience growth since techniques tend to complement rather than compete with one another. The newer techniques—infrared, microwave, ultrasonic, nonfilm x-ray, and eddy current—carry the potential for better than 15% annual growth, while the older established methods of magnetic particle, film x-ray, and dye penetrants will experience less than the 10% annual market increase. Even the most recently developed techniques—neutron radiography, liquid crystal thermography, eddy sonics, acoustic emission, and laser and acoustic holographic interferometry—will have established a strong industrial foothold by 1980.

Data processing. Qualitative rather than quantitative interpretation of output signals or images is a dominant feature in most nondestructive testing methods and applications today. As a result of current efforts to quantify and manipulate nondestructive testing information, data processing for analysis and storage of data will become an important feature of many future systems. The NMAB committee on nondestructive evaluation has pinpointed the urgent practicality of employing small digital computers for information processing. Applications include on-line processing and interfacing with analog systems and large general-purpose computers. The recognized need for data processing/correlation techniques led NMAB to recommend that highest development priority be granted to this aspect of nondestructive testing. The committee felt that many of the new computer processing techniques emanating rapidly from military and space programs were applicable directly to nondestructive evaluation. Nondestructive testing information processing applications now in use and emerging rapidly include digital readouts of ultrasonic flaw signals similar to systems currently available for thickness gaging, and digital computer plotting of radiographic film density and other facsimile recordings. Similarly, computer comparisons of frequency/amplitude "signatures" will be extended beyond current applications, where now amplitude/location thermal patterns in electronic components are compared, as pattern recognition techniques advance.

Development of high-speed inspection capabilities remains one of the notable trends in nondestructive evaluation. Equipment design is aiming toward complex, highly automated inspection systems based on efficient materials-handling facilities (Fig. 11). Handling requirements entail either a continuous transfer of test parts through a nondestructive testing instrumentation station, or, in the case of large unwieldy test objects (such as airframes and nuclear reactor vessels), the motion of the probe or equipment over the test surface. Station-transfer systems use nondestructive testing response via go-no-go logic circuits to operate signal lights, open chutes, or to activate auxiliary marking equipment.

Automation and multiple techniques. Automation of nondestructive testing equipment and the development of real-time viewing systems will contribute to the close integration of testing and fabrication functions. Nondestructive evaluation can be expected to evolve from its supplementary, supportive role in many manufacturing systems today toward an in-process quality assurance function, which will serve as a source of feedback control over production processes. Users, especially in the early stages of manufacturing, will gain greater economies by closely coordinating testing stations with flaw-producing points in the manufacturing process.

The SRI report made a strong point that the complementary nature of nondestructive testing techniques remains inadequately recognized, as does the fact that the applicability of any technique is impossible to determine without a complete definition of the testing problem. This is because such systems usually reveal only the specific kinds of defects they were designed to reveal and are "blind" to others. Multiple nondestructive testing methods in the future will serve in many testing situations where a single technique is now in use. The intrinsic, complementary nature of these methods will permit more thorough evaluation of materials by subjecting the component to two or more techniques, either simultaneously or successively.

Necessity of testing. The concern of manufacturers with reliability in large-scale engineering systems, such as those found in the aerospace, communications, nuclear power, energy transport, commercial transport, and chemical industries, has greatly stimulated the use of nondestructive testing. The proliferation of complex machinery

Fig. 11. This automated ultrasonic shear-wave inspection system for helicopter rotor-blade spar tubes is 75 ft long. The inspection station is centrally located and combines system controls, inspection equipment, data display, and discontinuity analysis and alarm circuitry. (*The Boeing Co., Vertol Division*)

representing enormous invested capital, manpower, materials, and the potential for causing catastrophic damage to human life and property has focused the public eye on product liability. Nondestructive evaluation of each component capable of crippling an entire system or causing bodily injury is becoming essential. The current trend toward stricter legal interpretation of producer liability is encouraging the use of nondestructive evaluation in industries previously unfamiliar with its special testing capabilities. Nondestructive evaluation will become, essentially, a general-purpose industrial tool, routinely found in every phase of the design-production-service cycle.

The SRI report looked to other important benefits of nondestructive evaluation in stating, "Research has indicated the growing value of NDE methods in materials science. Traditional NDT has been primarily concerned with the detection of relatively gross defects, essentially ignoring submicroscopic anomalies which may significantly affect product serviceability. Efforts are now being made to isolate quantifiable variables which will provide a direct measure of physical and mechanical properties. Correlation of nondestructive measurements with the mechanical properties of brittle materials, such as graphites, ceramics, and high modulus composites, has been highly successful."

The NMAB report, in summarizing the committee's recommendations, pointed out the need for more effective utilization and development of nondestructive testing through suitable procurement specifications, for increasing the supply of trained personnel, for further implementing the Information Analysis Center, for establishing centers of excellence to solve national problems through research and development, and for initiating a long-range interdisciplinary research program to utilize all possible energy-material interactions for nondestructive evaluation. The last recommendation has already been partially implemented through the Advanced Research Projects Agency.

[JOHN R. ZURBRICK]

Honor B. Fell is a graduate of Edinburgh University and was director of the Strangeways Research Laboratory, Cambridge, from 1929 to 1970. She is now doing full-time research in the Pathology Department, Cambridge University. Her research has concerned the developmental mechanics of the embryonic skeleton, and the direct action of vitamin A and various hormones on differentiated tissues that have been grown in organ culture.

Organ Culture

ORGAN CULTURE MAY BE DEFINED as the cultivation outside the body of tissues in a differentiated functional state similar to that of the organs from which they were derived.

For many years, organ culture was used almost exclusively for the study of morphological problems. This was due partly to the inadequacy of the microchemical techniques then available and partly to the use of complex semisolid culture media which were unsuitable for biochemical examination. With the spectacular development of microchemistry during the past 10–15 years, coupled with the improvement and simplification of tissue culture media, the organ culture technique can now be utilized in many aspects of biochemical and physiological research, while remaining a powerful tool for the experimental morphologist.

At the present time the organ culture method is being used to investigate two main groups of problems.

The first group of problems concerns the morphogenetic interactions between cells of different types in the formation of an organ; this is one of the most difficult and important fields of developmental biology and it is being explored mainly by experiments with organ culture.

The second group of problems is connected with the response of organs to changes in their environment; in the closed system provided by a culture, the investigator can examine in detail not only the effect of the environment on the tissue, but conversely that of the tissue on the environment, and it is thus that the researcher can obtain a more nearly complete picture of a physiological or pathological process than is possible in the intact animal.

In this article the history, methods, and chief applications of the organ culture technique are indicated, and examples are given of how it has been utilized in specific investigations. The advantages and disadvantages of the technique are discussed, especially in relation to future research.

HISTORY

Very early in the history of tissue culture, it was discovered that pieces of tissue cultivated in the laboratory showed two different types of growth which sometimes proceeded simultaneously in the same culture. In 1914 D. Thomson noticed that when he explanted feathers or toes from an embryonic chick in a nutritive clot in an ordinary hanging-drop culture, the explants enlarged as a whole; this he termed controlled growth and contrasted it with uncontrolled growth in which apparently undifferentiated cells crawled out of the edges of the tissue fragment to form a broad unorganized halo of proliferating cells around the original tissue. Others, notably A. Maximow, N. G. Chlopin, and A. Fischer, showed that controlled or, as Maximow preferred to call it, organotypic growth might be accompanied by progressive differentiation of the tissue. Surprisingly, however, for a long time organotypic growth aroused little interest, and attention became focused almost exclusively on uncontrolled or cytotypic (Maximow) growth.

Organotypic cultures, now known as organ cultures, were studied intensively at the Strangeways Laboratory, however, and it was here that the technique was mainly developed. In the first organ culture experiments to be made at the Laboratory, by T. S. P. Strangeways and H. B. Fell during 1925–1926, isolated limb buds and eye rudiments from 3-day embryonic chicks were grown on the surface of a semisolid medium at the bottom of a small centrifuge tube. The undifferentiated limb buds formed cartilage and a keratinizing epidermis, and in the eye rudiments a retina of almost adult complexity, pigment cells, ciliary epithelium, and lens fibers differentiated under laboratory conditions. Subsequently the centrifuge tubes were replaced by watch glasses enclosed in a moist chamber.

Some years after the early experiments by Strangeways and Fell, organ culture was taken up first by P. Gaillard in Leiden and then by E. Wolff in Strasbourg; both these workers fully appreciated the potentialities of the method and they and their pupils have applied it with great success to a wide range of problems. Nowadays organ culture is established in laboratories throughout the world, and its usefulness is fully recognized.

Cell cultures and organ cultures differ greatly both in their biological characteristics and in their applications. Cell culture is based on the cytotypic growth observed by the early investigators, and owes much to the pioneer work of A. Carrel and M. T. Burrows. In modern cell culture, the normal architecture of the tissue is deliberately destroyed by the enzymatic digestion of the intercellular material, and the isolated cells released in this way are seeded in tubes, flasks, or petri dishes in a suitable nutritive medium and incubated at body temperature. Every few days the used medium is replaced by fresh. The cells multiply and eventually reach saturation density, when growth stops; at this stage the cells are again freed by enzymatic treatment and reseeded in newly prepared culture vessels. Different lines of cells may be propagated for years in this way. Cell cultures are ideal for the study of cell growth or for the cultivation and investigation of viruses which are intracellular parasites, but the cells are very different from their

prototypes in the body and, with some exceptions, they cease to exercise their characteristic physiological functions.

On the other hand, as stated above, the primary object of organ culture is to maintain tissues in a differentiated, functional state resembling as closely as possible that of the corresponding tissues in the body. Explants derived from embryonic organs enlarge in culture, though much less than in the body, but those of adult organs do not increase in size and all that is desired of the latter is that, as in the body, cell multiplication shall just compensate for cell death. Organ cultures are used to investigate the functional activities of intact tissues, rather than the physiology of cell division, for which cell cultures are much more suitable.

In cell cultures, the vigorously proliferating cells adapt themselves fairly readily to the abnormal conditions of laboratory culture life, either by direct transformation or by the selection of favorable mutants; possibly both processes operate. It is this adaptability that enables the unorganized cells to survive and grow for months or even years. Differentiated tissues are much more sensitive to environmental influences, and consequently cannot be maintained in a normal state for more than a few weeks at most and sometimes not for more than a few days. This limitation, however, is of little practical importance because, as will be seen later, the explants respond so rapidly to experimental agents that longterm experiments are rarely needed.

METHODS

There are many different forms of organ culture, but most of them have certain features in common. The explants are grown in some kind of small dish and usually placed at or near the surface of the nutritive medium. Like all tissue cultures, they must be prepared under strictly aseptic conditions and incubated at body temperature. So far as possible, unorganized outgrowth of cells from the margin of the explant is discouraged because this disrupts the histological structure of the tissue. Since tissues in organ culture have no blood supply, they have to receive nutrients and oxygen and discharge waste products entirely through their outer surface. This means that only small explants can be grown because if their diameter exceeds 2–3 mm, their interior usually becomes necrotic.

A very wide variety of differentiated tissues, mainly from embryos and young animals, can now be grown in organ culture. Developing a suitable method for a given tissue is empirical, since it is difficult to predict what medium, gas mixture, or culture vessel will prove favorable. In all types of organ culture experiment, it is essential to monitor the condition of the tissue by careful histological examination of some of the explants.

Nutritive media. Different types of nutritive media are used according to the purpose for which the cultures are required. Originally the explants were grown on a clot composed of blood plasma mixed with a saline extract of embryonic tissues (embryo extract). This gave good results, but its semisolid consistency and enormous chemical complexity made it unsuitable for most biochemical studies; nowadays it is rarely employed, and most investigators use a chemically defined fluid

ORGAN CULTURE

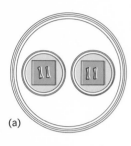

(a)

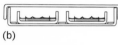

(b)

Fig. 1. Diagram of culture vessel now in use at the Strangeways Research Laboratory. (a) Surface view. (b) Sectional view. (From H. B. Fell, The effect of environment on skeletal tissue in culture, Embryologia, 10:181–205, 1969)

medium with or without the addition of some kind of serum. In Wolff's laboratory, however, tissues are cultivated on a semisolid medium consisting of an agar base to which various nutrients have been added; this has proved very favorable for the experimental study of many developmental problems.

The ratio between the volume of the explant and that of medium in the culture vessel may be important where fluid media are concerned. If this ratio is too high, the medium rapidly deteriorates and the explants degenerate, but if it is too low, important materials may be extracted from the tissue and so diluted by the medium that the cells fail to grow or differentiate. Determination of the optimum size of explant and volume of medium is usually a matter of trial and error.

Culture vessels. Many different types of organ culture vessels are employed in different laboratories and have been designed to suit a wide variety of experiments. The culture system used in the Strangeways Laboratory is a modification of the original watch glass method of Strangeways and Fell, and has been developed mainly by T. Fainstat. A pair of flat-bottomed culture vessels are enclosed in a large petri dish, the bottom of which is covered with several layers of filter paper (Fig. 1); two circular holes have been cut in the filter paper to accommodate the two culture dishes and to permit examination of the explants by transmitted light. Each culture vessel contains a shallow table made of stainless steel mesh (a feature originally introduced by O. A. Trowell) on which the explants are supported. The height of the grid is so adjusted that 1.5 ml of medium introduced into the vessel just wets the surface of the mesh. To maintain a moist atmosphere the filter paper is saturated with a sterile isotonic solution of sodium chloride.

Oxygen requirement. Different tissues vary greatly in their requirement for oxygen. Some, like cartilage, do best at the normal atmospheric level, while others, like the mammalian ovary or thyroid, thrive better in an oxygen concentration of 70–80%. Usually the medium is buffered with bicarbonate and, to preserve a neutral pH, the concentration of CO_2 in the atmosphere must be raised to an appropriate level. Some workers supply the entire incubator with a continuous flow of whatever gas mixture is needed. A simpler, more accurate, and more economical method, due to Fainstat, is to stack the unsealed petri dishes containing the culture vessels in a modified Fildes-MacIntosh jar (Fig. 2); these jars, which have a strong mechanical seal, are normally used by bacteriologists for the cultivation of anaerobic bacteria. The lid of each jar has an inlet tube and an outlet tube which are both fitted with taps, the jar is flooded with the desired gas mixture, the taps are closed, and the jar is incubated. This container preserves virtually intact whatever gas mixture is introduced, since the explants' consumption of oxygen is negligible. Every 2–3 days the cultures are removed from the Fildes-MacIntosh jar, the used medium is withdrawn, often for biochemical examination, and fresh medium is substituted.

MORPHOGENESIS

Organ culture has provided a new approach to the complex problems of tissue interaction in organ formation, thanks to the ease with which tis-

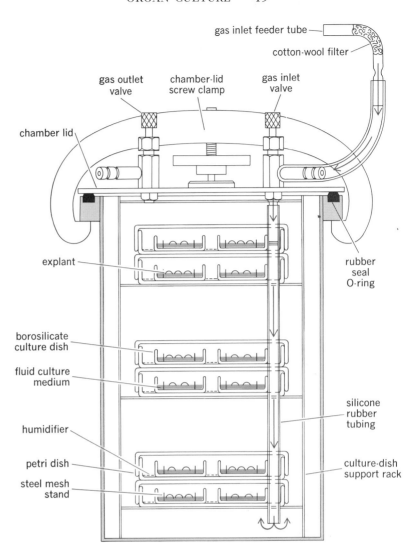

Fig. 2. Diagram of culture dishes stacked in an airtight Fildes-MacIntosh jar. (*From T. Fainstat, Organ culture of postnatal rat ovaries in chemically defined medium, Fert. Steril., 19:317–338, 1968*)

sues can be manipulated in this laboratory system. Much work has been done in this field, and the following investigations will be briefly described to show what can be achieved by experiments of this type.

Migration of germ cells. The first sign of embryonic development in a fertilized chicken's egg is the appearance of an oval disk of cells, a few millimeters in diameter, on the surface of the yolk. This disk is called the blastoderm, and the embryo forms along its midline. More than 30 years ago, C. H. Waddington showed that when the entire blastoderm is removed from an egg after 18–24 hr of incubation and grown as an organ culture, it continues to develop almost normally for a further 36 hr.

In Wolff's laboratory, D. Simon, using nutrient agar as culture medium, found that if two blastoderms were explanted side by side, they fused along the line of contact and developed a common blood circulation (Fig. 3). She made use of this fact to study the remarkable phenomenon of the migration of the germ cells. In normal development in the body the primordial germ cells at first occupy a crescentic area (the germinal crescent) (Fig. 4) in front of the future head of the embryo. When the

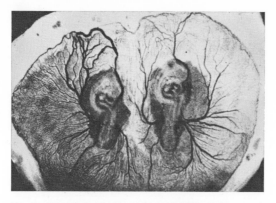

Fig. 3. Association between two chick blastoderms cultivated in the laboratory. The embryo on the left was deprived of its germinal crescent (see Fig. 4) before explantation. The blastoderms have fused and formed a common circulatory system. (*From D. Simon, Contribution a l'étude de la circulation et du transport des gonocytes primaires dans les blastodermes d'oiseau cultivés in vitro, Arch. Anat. Mic. Morph. Exp., 49:93–176, 1960*)

circulatory system develops, these cells crawl into the blood vessels and are carried passively in the bloodstream until they arrive at the site of the future gonads; this is a small region of the epithelium lining the body cavity (coelom) known as the genital epithelium. When the germ cells reach this site they emerge from the blood vessels and enter the genital epithelium, where they multiply actively and eventually give rise to sperm or ova. When Simon cut off the germinal crescent before explanting the blastoderm, no germ cells appeared in the germinal epithelium, as was to be expected, but when she explanted the mutilated blastoderm alongside a normal specimen, as usual the two fused along the line of contact (Fig. 3), formed a common circulatory system, and the genital epithelium of the defective embryo became populated by germ cells from its intact neighbor. When a defective chicken and a complete duck blastoderm were cultivated together in this way, the chicken genital epithelium became populated by duck germ cells and vice versa.

Fig. 4. Photographs of (a) normal chick blastoderm at the six-somite stage before a blood circulatory system has developed, and (b) same blastoderm as a except that the area of the germinal crescent has been marked in black. (*From D. Simon, Contribution a l'étude de la circulation et du transport des gonocytes primaires dans les blastodermes d'oiseau cultivés in vitro, Arch. Anat. Mic. Morph. Exp., 49:93–176, 1960*)

In other experiments Simon investigated the specificity of the genital epithelium as a goal for the migrating germ cells. To do this, she cut off the posterior part of the blastoderm, thus removing the future genital epithelium, and cultivated only the anterior part. The explanted portion formed a vascular system, and the germ cells entered the vessels and were carried round by the blood. Their proper destination, the genital epithelium, of course, was missing, but they refused to disembark elsewhere; eventually they died, and were discovered as a degenerate clump inside a blood vessel.

R. Dubois from the same laboratory, using the organ culture method combined with autoradiography, strikingly demonstrated the extraordinary attraction that the early genital epithelium exerts on the germ cells. He labeled the comparatively well differentiated gonad from a 6-day embryo with H^3-thymidine, then grafted a small piece of the labeled organ into the body cavity of an early (26–32 somite) embryo which had not been labeled, and cultivated the complex on nutritive agar in the usual way. There was a mass exodus of radioactive gonocytes from the graft into the unlabeled genital epithelium of the host, which thus became populated by H^3-labeled gonocytes from the implanted gonad. It was interesting that only the germ cells responded to the genital epithelium in this way.

Development of mouse salivary gland. In recent years, organ culture has played a key role in the study of the morphogenic relationship between epithelia and connective tissue, a subject that has become increasingly important in experimental biology. This work was inaugurated by C. Grobstein's classical experiments on the development of the salivary gland of the mouse. In the early embryo the gland arises from the mouth epithelium as a small simple bud which then grows downward into a relatively large oval condensation of mesenchyme representing the rudiment of the gland's stroma and capsule. As this epithelial downgrowth invades the mesenchymal mass, it branches to form the familiar glandular tree. Grobstein found that he could separate the epithelium from the mesenchyme at a very early stage by immersing the rudiment in a solution of the proteolytic enzyme, trypsin; this digested the intercellular material that cements the two tissues together. He then grew the epithelium in culture in various ways. Cultivated alone (Fig. 5a), it formed merely an unorganized sheet of cells; when associated with lung mesenchyme (Fig. 5b), it produced a round nodule in the middle of the connective tissue, but when recombined with its own mesenchyme, it developed into a typical branching, glandular tree (Fig. 5c).

This effect of salivary gland mesenchyme on the glandular epithelium is also produced when a millipore filter is interposed between the two tissues. Grobstein and collaborators have utilized this fact in an attempt to investigate the biochemical basis of the interaction between the tissues. Either the epithelium or the mesenchyme was labeled with a radioactive substance, such as proline or glycine, and the two tissues were then explanted on either side of the filter membrane; after a suitable culture period, they were fixed and sectioned and the distribution of the isotope was examined by autora-

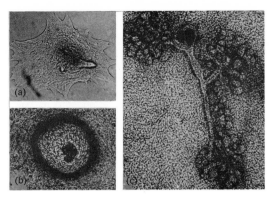

Fig. 5. The influence of mesenchyme on the development of the submandibular salivary gland of the mouse. (*a*) Isolated epithelial rudiment grown for 2 days without mesenchyme; the cells have spread into a thin undifferentiated sheet. (*b*) Similar epithelial rudiment combined with mesenchyme from a lung rudiment; after 4 days in culture a simple cyst has been formed. (*c*) Similar rudiment combined with the capsular mesenchyme of the salivary gland, 5 days in culture; a typical glandular tree has developed. (*From C. Grobstein, Epithelio-mesenchymal specificity in the morphogenesis of mouse submandibular rudiments in vitro, J. Exp. Zool., 124:383–414, 1953*)

diography. F. Kallman and Grobstein found that salivary mesenchyme transmitted material with labeled proline or glycine across the filter membrane and that this material then accumulated at the interface between the filter and the epithelium but not at the mesenchymal surface.

BIOLOGICAL ACTION OF CHEMICAL AGENTS

One important advantage of organ culture is that it enables the direct action of an environmental factor on a given tissue to be investigated in the absence of the complex systemic reactions that may hopelessly obscure the tissue in the body. It is often possible to obtain a complete picture of a synthetic or metabolic process in a culture because not only the tissue but, what is equally important, its humoral environment is available for detailed study. This is of crucial importance, since experience has shown that an error of several hundred percent can be made if the tissue only is examined and the used culture medium neglected.

These considerations have led many investigators to use organ cultures in their attempts to analyze the effects of such agents as hormones, vitamins, and drugs. Fortunately tissues in organ culture usually respond to these compounds in essentially the same way as they do in the body and their failure to do so often means that the effective agent is a metabolite of the original compound which in the body is produced by some organ other than that being investigated in the laboratory. It is also possible to study the effect of different organs on biologically active substances and thus discover which is the active derivative, but this potential application of the organ culture method has not yet been properly exploited.

In this section examples will be given of how organ culture may be used to study the effects on tissues of hormones, vitamins, and drugs, and conversely of the effect of a tissue on a biologically active compound.

Hormones. A hormone whose action has been investigated very extensively and fruitfully is parathyroid hormone (PTH). An effect of PTH was first obtained in organ culture by Gaillard, and most of the subsequent research on its biochemical and biophysical effects has been done in his laboratory in Leiden. In his original experiments, Gaillard explanted small pieces of parathyroid gland from man, mouse, and chicken in contact with parietal bone from late fetal mice and found that, after 3–8 days in culture, bone was being actively resorbed with the formation of many multinucleate osteoclasts. Later he obtained a similar result by adding parathyroid extract and finally purified PTH to the culture medium (Fig. 6). Limb bones from fetal mice also were drastically affected. These results showed clearly that the hormone had a direct action on skeletal tissue.

Gaillard and colleagues then proceeded to investigate the biochemical basis of these morphological changes. Histochemical studies by M. P. M. Hermann-Erlee showed that PTH severely inhibited a number of intracellular enzymes involved in energy (glucose) metabolism, such as isocitric dehydrogenase, glucose-6-phosphate dehydrogenase, lactic dehydrogenase, and NAD- and NADP-diaphorase. The production of citrate and lactate was increased, and protein synthesis also was affected, the uptake of glycine being diminished and the release of that already incorporated being enhanced. This last observation is probably correlated with G. Vaes's finding that PTH, like hypervitaminosis A, activates the lysosomal system of

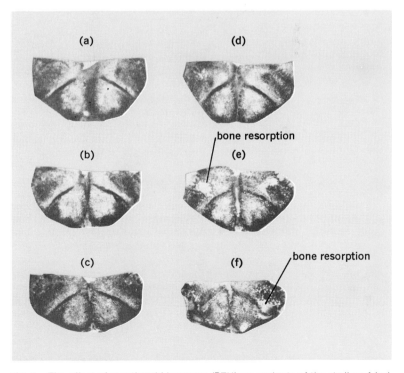

Fig. 6. The effect of parathyroid hormone (PTH) on explants of the skull roof (calvaria) of fetal mice. (*a*) Calvaria before explantation. (*b*) Control grown for 4 days without PTH. (*c*) Control grown for 10 days. (*d*) Calvaria cultivated for 2 days with PTH. (*e*) Calvaria cultivated for 4 days with PTH. (*f*) Calvaria cultivated for 10 days with PTH. Note bone resorption in e and f. Cultures prepared by P. Gaillard's method. (*From G. Vaes, in A. de Visscher (ed.), La résorption osseuse et l'hormone parathyroidienne, Libraire Maloine, 1967*)

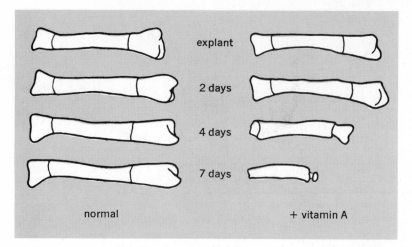

Fig. 7. Camera lucida drawings of a pair of living radii from a late fetal mouse, showing the effect of vitamin A added to the culture medium. During 7 days of cultivation the radius in normal medium enlarged somewhat and the bone and cartilage remained intact; the radius exposed to hypervitaminosis A also enlarged during the first 2 days, then suddenly the cartilage began to shrink, the bone to be resorbed, and by the 7th day both bone and cartilage matrix had almost disappeared. (*From H. B. Fell and E. Mellanby, The effect of hypervitaminosis A on embryonic limb-bones cultivated in vitro, J. Physiol., 116:320–349, 1952*)

bone cells and increases the synthesis and secretion of a number of lysosomal enzymes. Quite recently the Leiden group have shown that PTH inhibits the reaggregation of cartilage cells dispersed by tryptic digestion of the intercellular material. It is suggested that this effect, which occurs within 24 hr, may indicate that the active principle of the hormone becomes bound primarily at the surface of the target cells.

The effect of hormones on the structure and functional activity of the mammary gland in organ culture has been studied by several workers. In the absence of hormones, the postfetal gland degenerates rapidly. Insulin is essential for its survival, cortisol improves its condition still further, but secretory activity requires the presence of several hormones. For example, E. M. Rivera and H. A. Bern showed that the addition of mammotropin and somatotropin to medium containing insulin and cortisol maintained the secretory activity of late prelactating tissue (mouse) and stimulated secretion in early prelactating glands and those from nonpregnant animals. The optimum concentrations of mammotropin and somatotropin for inducing secretion in the explants were found to be four times as high for early prelactating as for late prelactating tissue. Mammary glands in organ culture have been extensively used for the biochemical analysis of the various synthetic changes induced by hormonal action.

It is interesting that the embryonic rudiments of mammary glands can differentiate in culture in a protein-free, chemically defined medium without any hormones, as shown by E. Y. Lasfargues and M. R. Murray, but after birth they become increasingly dependent on hormonal influence.

Vitamins. The effect of vitamin A has been more widely investigated in organ culture than that of any other vitamin. Much of the work concerns the action of hypervitaminosis A on the skeleton. If an animal is fed on a diet containing an excess of vitamin A, its skeleton is affected. Bone is resorbed; in

some species at least, intercellular material is lost from the cartilage; and the changes may be so severe that the bones spontaneously fracture. Originally it was not known whether these effects were due to a direct action of the vitamin on the cells of the skeleton or whether they were mediated through some other organ, possibly the parathyroid gland.

In the hope of settling this question, Fell and E. Mellanby cultivated limb bones from late fetal mice (Fig. 7) and embryonic chicks (Fig. 8) in medium to which they had added an excess of vitamin A; changes produced in this culture system from which all systemic reactions were automatically excluded could only be due to a direct action of the vitamin on the explants. The result of these experiments was unequivocal. In the presence of the added vitamin, bone (Fig. 7) and cartilage (Fig. 8) were rapidly resorbed and after 10 days of cultivation the intercellular material had almost disappeared, leaving an unorganized mass of viable ameboid cells. That this reaction required the presence of living, physiologically active cells was shown by the fact that high, toxic doses of vitamin A caused less breakdown of intercellular material than lower doses.

At that time (1952) nothing was known about the biochemical mechanism of these effects. One clue was noted, however. The explants were grown on a plasma–embryo extract clot, and those exposed to excess of vitamin A always liquefied this semisolid medium much more than their paired controls without the vitamin; this suggested that in some way vitamin A had increased the proteolytic activity of the cells. About 10 years later, the technical improvements in both microchemistry and organ culture to which reference has already been made enabled this clue to be followed up; a long series of experiments showed that this original idea was

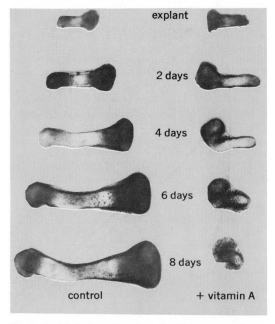

Fig. 8. Serial photographs of a pair of living cartilaginous femora from a 6-day embryonic chick, grown in culture for 8 days. Left: control explant. Right: explant grown in medium containing 10 iu vitamin A/ml; note almost complete dissolution of the intercellular material.

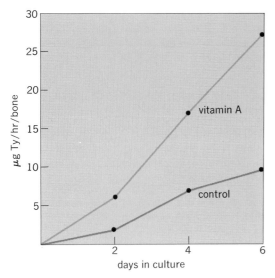

Fig. 9. Curves showing the effect of vitamin A (10 iu/ml of medium) on the release of lysosomal acid protease (cathepsin D) from limb-bone rudiments explanted from a 7½-day embryonic chick. During 6 days of cultivation, the vitamin A–treated explants released much more acid protease into the medium than their paired controls did. Ty represents release of acid-soluble tyrosine from hemoglobin substrate. (*From H. B. Fell, The effect of environment on skeletal tissue in culture, Embryologia, 10:181–205, 1969*)

well founded, and that treatment with vitamin A greatly increased both the synthesis and extracellular release of a lysosomal acid protease (Fig. 9), now known to be cathepsin D (J. T. Dingle, J. A. Lucy, and Fell) from both cartilage and bone (J. J. Reynolds) in organ culture. Recent results, from Dingle, A. J. Barrett, and P. D. Weston, indicate that this enzyme is mainly responsible for the breakdown of cartilage, but a collagenase appears to play an important part in the resorption of bone treated with vitamin A, as shown by J. M. Evanson and Reynolds; it is not known whether this collagenase is lysosomal.

The effect of vitamin A on epithelia has also been investigated in organ cultures. When animals are given a vitamin A–deficient diet, certain secretory epithelia, including those of the respiratory tract and various glands, lose their secretory activity, acquire a completely different histological structure (closely resembling that of the epidermis), and keratinize profusely. These changes are reversed if the vitamin is restored to the food; the epidermoid tissue is sloughed and a new secretory epithelium regenerated by the basal layer of germinative cells.

The reverse of this situation can be produced in some keratinized epithelia, such as the epidermis of the embryonic avian skin (Fell and Mellanby) or the epithelium of the rat's esophagus (I. Lasnitzki), by cultivating the tissue in organ culture in medium containing excess of vitamin A. The squamous, keratinizing cells are shed and replaced by a mucus-secreting epithelium (Fig. 10); if the A-hypervitaminotic medium is then replaced by control medium, mucus secretion is at first increased (Fig. 11a), mucous cells are lost, and the normal keratinizing epithelium is regenerated (Fig. 11b).

Another interesting example of the effect of vita-

min A was seen in Lasnitzki's experiments on the combined effects of a carcinogen (methylcholanthrene) and vitamin A on the prostrate gland in organ culture. When exposed to the carcinogen alone, the columnar secretory cells multiply, lose their orderly arrangement, and, though the epithelium does not keratinize, it develops an epidermoid structure which is maintained when the explant is transferred to control medium without methylcholanthrene. This effect of the carcinogen could be completely abolished, however, if the carcinogen-treated explant were transferred to medium containing vitamin A. Comparable results have since been obtained in the lungs of the intact animal by U. Saffiotti.

The effect of vitamin C on skeletal rudiments in organ culture has also been studied, by J. J. Jeffrey and G. R. Martin and by Reynolds. In the absence of this vitamin, embryonic chick cartilage is deficient in collagen and also becomes grossly hydrated, effects that can be reversed by restoring the vitamin.

Drugs. Some types of drug action can be examined in organ culture. A recent investigation of this type is C. K. Wildenthal's experiments on isolated hearts of fetal mice near term. Wildenthal developed a method for maintaining these hearts in organ culture in the contractile state for 3–4 weeks and found that they responded normally to a number of cardioactive drugs, such as acetylcholine and ouabain. These explants had the advantage over the simple perfusion systems commonly

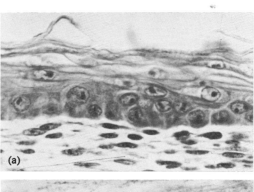

Fig. 10. The effect of vitamin A on skin from a 9-day embryonic chick. (*a*) Skin grown for 10 days in control medium; a squamous, keratinizing epidermis has differentiated in culture. (*b*) Similar explant grown for 10 days in medium containing 10 iu added vitamin A per milliliter; the epidermis has differentiated into an actively secreting mucous epithelium. (*From H. B. Fell and L. M. Rinaldini, in E. N. Willmer (ed.), Cells and Tissues in Culture, vol. 1, Academic Press, 1955*)

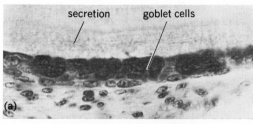

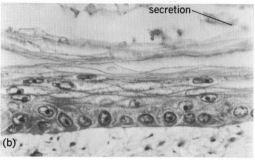

Fig. 11. Recovery of embryonic chicken skin in culture from the effects of vitamin A. (a) Skin from the metatarsal region of a 13-day embryo grown for 12 days in medium containing 10 iu added vitamin A per milliliter and then for 2 days in control medium; note goblet cells and profuse secretion. (b) Similar explant grown for 12 days in medium containing 10 iu added vitamin A per milliliter and then for 16 days in control medium; the basal cells have regenerated a squamous, keratinizing epithelium and the mucous cells have been sloughed; some secretion remains adherent. (From H. B. Fell and L. M. Rinaldini, in E. N. Willmer (ed.), Cells and Tissues in Culture, vol. 1, Academic Press, 1955)

used in cardiac physiology in that they enabled the longterm effects of chemical agents on the heart to be studied outside the body.

Effect of tissues on chemical agents. As already mentioned, organ culture can be used to study the effect of a given tissue on a biologically active compound, and thus to determine whether the substance is broken down by the cells and, if so, which is the effective component. This is an application of the method that deserves more attention than it has hitherto received. An excellent example of work of this type is the recent organ culture experiments of Lasnitzki, E. E. Baulieu, and P. Robel on the effect of the rat prostate gland on testosterone.

These investigators introduced tritium-labeled testosterone into the culture medium of explants of ventral prostate gland and, after incubation, analyzed radiochemically both the tissue and the used medium. The explants actively metabolized the hormone and a number of radioactive derivatives were identified in both the tissue and the medium. The main metabolite was dihydrotestosterone, which was preferentially bound to the nuclear fraction. This, in turn, was converted to 3α- and 3β-diol. In addition, some unidentified nonpolar compounds were found.

In other experiments, the different components were added singly to prostatic cultures, and their effects on the height, secretory activity, and mitotic rate of the glandular epithelium were examined. Dihydrotestosterone was by far the most active compound. It fully maintained epithelial differentiation at low concentrations but also promoted cell division and, at higher doses, induced

extensive epithelial hyperplasia. On the other hand, 3β-diol preserved and increased epithelial height but, even in high doses, did not induce hyperplasia.

CONCLUSIONS

To make the best use of organ culture it is important to realize its limitations as well as its advantages. As already explained, it is suitable only for fairly short experiments, because few differentiated tissues survive in a sufficiently healthy state for more than a few days or weeks. Since the explants have no blood supply, only small pieces of tissue can be grown, and hence their biochemical study demands accurate microchemical methods that often require the use of expensive isotopes.

It should be remembered that the chief advantage of organ culture, namely, the great simplification of experimental conditions that it provides, is also one of its most serious limitations. When a tissue responds to a given agent in organ culture in essentially the same way as it responds in the body, then in the writer's opinion it is justifiable to use organ culture for the detailed investigation of the response and to assume that the results thus obtained apply, at least qualitatively, to the same phenomenon in the living body. On the other hand, it would be rash to proceed in the reverse direction: to test an agent on a tissue in organ culture and then conclude that the effect produced under these abnormally simple conditions would necessarily apply to the same tissue in the body. In the body, the action of the agent on the target cells might be profoundly modified by complex systemic reactions. Thus the intact organism might have some mechanism for inactivating the agent, in which case, though active in culture, the agent might be ineffective in the body; alternatively some structure other than the organ studied in culture might be affected in the body, perhaps disastrously.

It has often been suggested that organ cultures might be used for experiment instead of animals, but this would be impracticable except for very restricted purposes. For example, a drug tested on organ cultures might appear to have very desirable properties from a therapeutic standpoint, but when given to a patient it might either be inactivated in some way and so rendered useless or produce dangerous side effects that would not be manifested in the simple laboratory system. Hence tests on animals are essential.

The writer has long cherished the hope that future work may bridge the existing gap between cell culture and organ culture and produce a technique that will enable a much larger volume of tissue to be grown in a differentiated functional state for much longer periods than is possible at present. There are signs that this approach may not be altogether impracticable. A. Moscona has shown that embryonic tissue can be dissociated into its component cells and that, when reaggregated, the cells differentiate into a tissue similar to that of the parent organ. An ideal experimental system would be one in which the dissociated cells were seeded on the floor of a petri dish as in a cell culture, but instead of being maintained in a state of active, unorganized proliferation, were cultured under conditions that would enable them to form a large, thin sheet of differentiated tissue that would survive for a long time in a functional state.

Some progress in this direction has already been made with fibroblasts, muscle cells, and cartilage cells, all of which can be kept in culture as a sheet of differentiated tissue. A striking example of such work is provided by recent (unpublished) experiments by Sylvia Fitton Jackson. Using a method similar to that indicated above, she has been able to produce hard plates of cartilage more than a centimeter in diameter in a chemically defined medium and to keep the tissue in a healthy state for at least 7 weeks. There seems to be no obvious reason why more complex tissues also should not be grown on a relatively large scale in this way, if the cells were supplied with a suitable basic medium appropriately supplemented with vitamins and hormones in physiological concentrations, and given the right gaseous atmosphere.

Recently G. G. Rose and colleagues, using a complex perfusion system, maintained the fetal thyroid and ovary (rat) in culture for many weeks as thin sheets of healthy, histologically differentiated tissue. The amount of tissue produced was not very large as compared with that obtainable in ordinary cell cultures, but the experiments are important and encouraging, since they indicate that the hybridization of cell culture and organ culture is not inherently impossible. The successful development of such a technique would greatly expand the potentialities of organ culture as a research tool.

To sum up, the essential value of the organ culture technique in biological and medical research is in the analysis of phenomena observed in the body which cannot be fully resolved under the complex conditions obtaining in the body; experiments on organ cultures should always be closely correlated with experiments on animals, for which they cannot be a substitute.   [HONOR B. FELL]

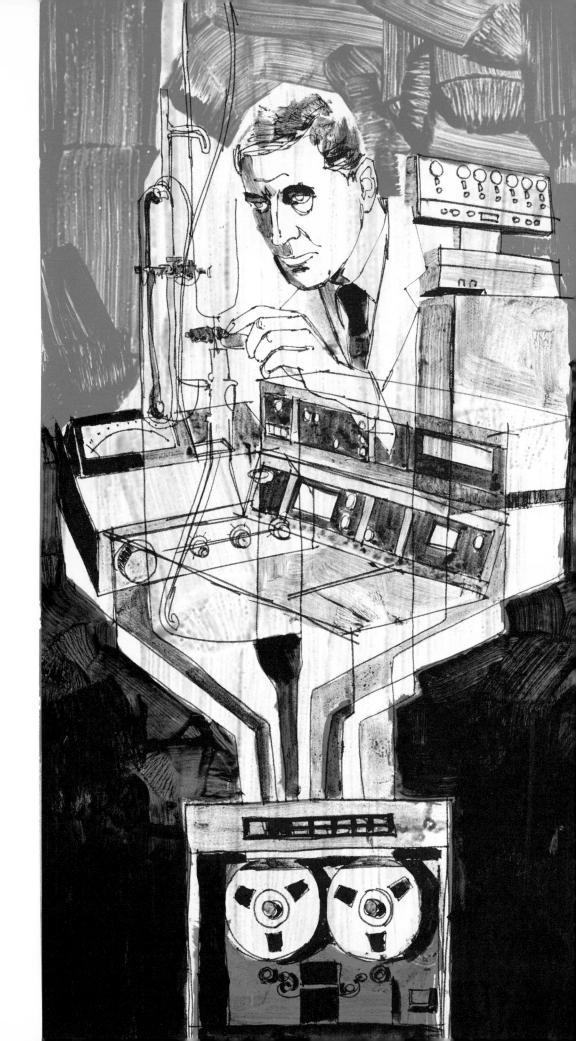

Max Tochner is coordinator for analytical development at the General Electric Co., Silicone Products Department, Waterford, N.Y. He is a graduate of the State University of New York at Albany and joined General Electric in 1963, where he has worked on the physical chemistry aspects of fuel cells and on the application of timesharing computers to analytical chemistry problems.

Computer-Assisted Analytical Chemistry

ANALYTICAL CHEMISTRY, A SCIENCE devoted to measuring the composition of material substances, encompasses a wide degree of complexity, ranging from a simple volumetric titration in which an end point is determined by the color change of an indicator to sophisticated electronic instruments such as an x-ray spectrometer which automatically (under computer control) performs the analysis and provides a printout of the analytical results.

The fundamental problem which must be resolved by the analytical scheme is the conversion of measured parameters to meaningful physical quantities.

The past half-decade has been a period of much growth in which the electronic computer has become a ubiquitous part of the analytical system.

Recent meetings of the professional chemical societies have devoted large portions of their technical program to various aspects of the continuing developments in computer-assisted analytical chemistry. The exhibition areas used by vendors for the display of new instrumentation continue to be dominated by the myriad analytical devices which are coupled to digital computers to play an integral role in the total analytical system. For all practical intents and purposes, the computer developments discussed in this article are limited to digital computers. Analog computers have not really received extensive utilization in the field.

ASPECTS OF COMPUTER ASSISTANCE

Digital computer technology has been applied to several distinct problem areas of analytical chemistry. In a broad sense, some of the major segments can be categorized as mathematical analysis, control of instrumentation, and information retrieval.

Before launching a discussion of the categories, a brief digression to the jargon of computer technology should prove helpful. Computer hardware, or simply hardware, is defined as the computer itself; that is, the magnetic, mechanical, electrical, and electronic devices from which the computer is constructed.

Types of computer systems used for analytical chemistry

Mode	Type	Instrument control	Turnaround time	Relative cost
Off-line	Conventional batch processing	No	Hours to overnight	Small
Off-line	Time-sharing utility	No	Minutes	Small
On-line	Dedicated minicomputer	Yes	Immediate	Moderate
On-line	Laboratory midicomputer	Yes	multiple access	Large

Software is defined as the total set of instructions which cause the computer hardware to operate on a problem. Software is the general term used to describe various levels of the language of computer instructions; it includes compilers and assemblers, as well as application programs in high-level languages such as FORTRAN.

An interface is the boundary between portions of the computer system. For instance, the interface between a human and the computer can be a mechanical card punch or a teletype terminal, while the interface between an analytical instrument and a computer is often an analog-to-digital converter (an electronic device which measures a voltage at the instrument and transmits its value as a digital signal to the computer).

Mathematical analysis. The early applications of computers to typical analyses facilitated a significant advance in the mathematical treatment of the data-reduction step. The data-acquisition stage was achieved by the operator in the conventional manner, manually logging the data from the instrument to a notebook. The advent of high-speed electronic computers which perform single arithmetic operations in millionths of a second permits more complicated calculations than were feasible with a desk calculator. To convert the data to a form intelligible by the computer, punched cards are prepared from the raw data which had been logged in a notebook. Removing the time burden permits more complicated calculations than were previously feasible. Theoretical models that had been previously limited to simple linear functions were expanded to account for multiple parameters adjusted to fit experimental observations. Of particular significance is the ability to apply mathematically sophisticated curve smoothing and curve fitting techniques to reduce the instrumental data to physical models.

A significant extension of the manually prepared data-acquisition system was achieved by the introduction of interfacial devices. They mechanically and electronically log the data generated by the instrumentation and transform it to a computer-compatible format by means of punched cards or paper tape. Although the obvious advantage ap-

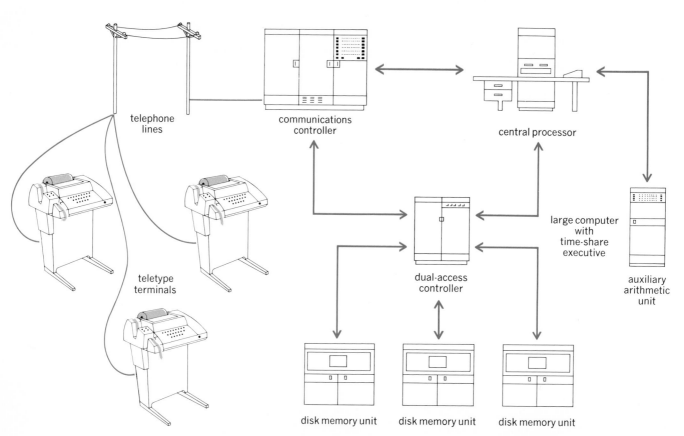

Fig. 1. Commercial time-sharing utility configuration.

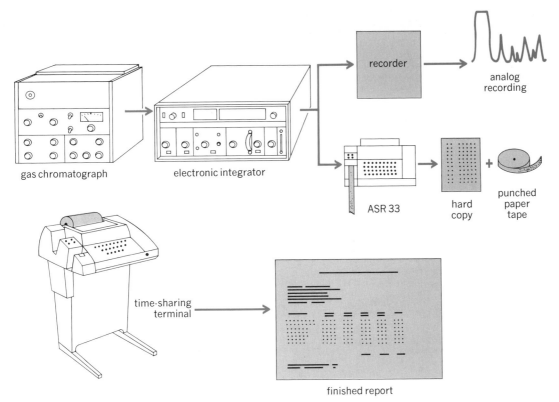

Fig. 2. Off-line data-acquisition system. ASR 33 is the teletype model number. This particular application is for gas chromatography. (*From Vidar Corp. Autolab Div. Bull., 100:5, 1970*)

pears to be in the increased speed and reduced labor cost of the data logging, a more significant factor is the increased accuracy by the elimination of human error in data transcription.

Control of instrumentation. The next step along the path of evolutionary development was the linking of the digital computer directly to the analytical instrument to aid in the control of the experiment, in addition to the first-mentioned task (that is, the mathematical treatment of the output data). It represented a substantial increase in the complexity of the system, particularly in the realm of the communication link between the computer and the experiment. It permitted real-time interaction with the experiment, in which the digital computer acquired data, performed calculations immediately, and transmitted back to the experiment appropriate instructions to modify the conditions of the continuing experiment. Some manufacturers incorporate small digital computers as a peripheral component of the analytical system and market the entire system, including the computer software. Another approach involves the use of a larger computer to simultaneously service several instruments of the analytical laboratory. Some specific examples of both types are discussed later in this article.

Information retrieval. The great speed with which large-scale digital computers can scan a massive memory core has fostered the development of information-retrieval systems for chemical information.

The large data base systems designed to search and retrieve information from the vast bank of chemical literature to facilitate the preparation of current-awareness journals, such as *Chemical Titles*, by means of a key word in context computer program, are significant but beyond the scope of this discussion. Some systems directly concerned with interactive searching of large files of analytical data, particularly those of infrared spectroscopy, will be discussed in a later section.

TYPES OF COMPUTER SYSTEMS

Many different approaches to computer-assisted analytical chemistry have been conceived and implemented by the innovators in the field. In general, they can be categorized by the size of the computer system used for the application (see table). The major constraint in this hierarchy was the financial resources available for the computer system. In the simplest and least expensive case, the computer was used for mathematical analysis in an off-line mode, using either conventional batch processing at a computer center or a commercial time-sharing utility over telephone lines. At some increase in expenditure, the scientist approached the problem by using a small digital computer dedicated to a particular analytical instrument or experiment. The most elaborate and expensive system has consisted of linking a medium-sized process computer to multiple analytical instruments for simultaneous instrument control, data acquisition as well as data reduction. Each of the approaches shall be examined in greater detail.

Off-line systems. The off-line approach is one in which the experimental data from an analytical device is entered into a computer by manual intervention. The computer accepts the data in one of several possible forms—punched cards, magnetic

Fig. 3. Varian 620/i general-purpose digital minicomputer in a chemistry laboratory. To the right of the minicomputer (shown being programmed by one of the students) is a Varian HA-100 high-resolution nuclear magnetic resonance spectrometer. The SpectroSystem 100 control board is immediately to the left of the minicomputer. (*Varian, Inc.*)

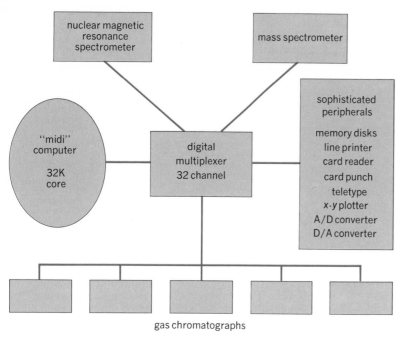

Fig. 4. Analytical laboratory automation.

cards are prepared and placed in the queue. Depending on the scientist's priority level, the output is available for pickup at the computer center in a time period that ranges from a few hours to the next day.

Batch processing exhibits significant importance when applied to the lengthy complex mathematical calculations required for many theoretical problems. These include regression fits of mass spectrometer data, inversion of large matrices, solution of fast Fourier transforms, and so forth.

Although batch processing permits a more sophisticated mathematical analysis, the significant time delays in obtaining the output inhibit rapid interaction of the scientist with the experiment.

Recognition of the need for a more user-oriented computer system led to the development of the time-sharing computer utility concept. John Kemeny, president of Dartmouth College, directed the efforts of faculty, undergraduates, and General Electric engineers to assemble a practical time-sharing system that simultaneously serviced some 30 users at remote locations. The rapid response was such that each user was unaware that the computer was not dedicated exclusively to his terminal. Since the first system was operational in 1965, many substantial improvements have been implemented, enabling more simultaneous users (over 100) to operate much larger programs (Fig. 1). The scientist in his own work area has a terminal device which permits access to the computer over commercial telephone lines. The local time-sharing center uses a small computer as a multiplexer and communication concentrator to connect the users terminal device to the main computer. The central computer has massive memory capabilities as peripherals, such as multiple disks and drums. The central computer operates under a complex executive program which allocates slices of time to each user, as well as keeping track of the users' files and providing the accounting system. The major advantage of the concept is that a powerful high-level computer is available to the scientist without the enormous price for purchase and maintenance of the system. When allocated over many users, the actual cost for each application is quite nominal.

Figure 2 shows a data-acquisition system for the off-line approach. The particular application illustrated is for gas chromatography. The chromatographic instruments were interfaced to an electronic integrator which generates its output with a visual-display, digital printer, and punched paper tape. In typical routine use, the chemist would remove the punched paper tape from the integrator and enter the data tape in the time-sharing computer at a teletype terminal. After having specified the particular stored files pertinent to the analysis, the computer would calculate the composition of the analytical sample and type out the results on the terminal in a matter of minutes. The off-line approach to data acquisition, coupled with a reduction by a time-sharing utility, offers an introduction to laboratory automation at a minimal capital investment.

On-line systems. An on-line computer is one that is interfaced directly to the analytical instrument or experiment. The data is logged, analyzed, and reduced in a real-time sequence while the

tape, punched paper tape, or even entry from a teletypewriter. The conventional batch mode of operation has its efficiency optimized for the computer center. Each job awaits its place in the queue, for the sole possession of the central computer, until the computation is completed. From the viewpoint of the scientist who uses the system, it is far from optimum. The raw data must be transported to the computer center, where punched

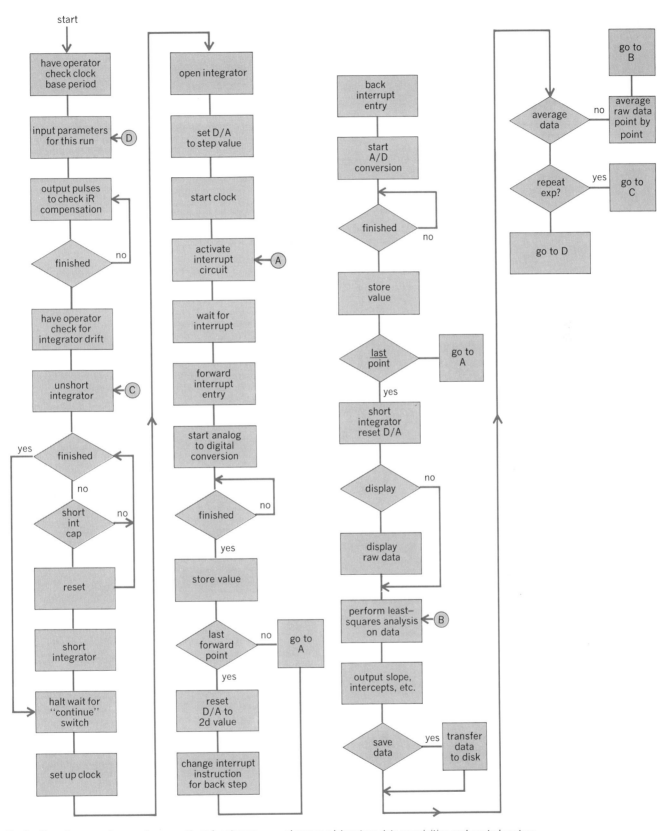

Fig. 5. Flow diagram of computer operations for chrono-coulometry. (*From G. Lauer and R. A. Osteryoung, A gener-* *al purpose laboratory data acquisition and control system,* *Anal. Chem., 40:41A, August, 1968*)

experiment is taking place. Often results can be determined within a time frame that permits investigation of unique phenomena which occur during the experiment. The turnaround time between the end of the experiment and the presentation of re-

sults is greatly reduced. The linking of an analytical instrument to an on-line computer has been accomplished in several different schemes which can be categorized by the size of the computer. In general, they fall into two categories: the minicompu-

```
GE.SPD.R&D..                        TELETYPE IDENTITY


GE TIME-SHARING SERVICE             COMPUTER IDENTITY


ON AT  21:41   28 THU 08/14/69  TTY 31    HEADING

USER NUMBER--U19007
PROJECT ID---000045 TOCHNER         CONVERSATIONAL SEQUENCE
SYSTEM--BASIC
NEW OR OLD--OLD
OLD FILE NAME--GCSAM1               RETRIEVAL OF GCSAM1


READY.

SCRATCH                             GENERATION OF EMPTY FILL
READY.

TAPE                                INITIATION OF TAPE MODE
READY.

DSM                                 INITIATION OF DATA
READY.                              STORAGE MODE

 0002,0000145
 0045,0000010
 0051,0008289
 0149,0000103
 0167,0032552
 0262,0006099
 0344,0040865
 0427,0003295
 0537,0037609
 0614,0000591
 0718,0031505                       ENTERING DATA TAPE
 0798,0000070
 0881,0023654
 0028,0015186
 0093,0000167
 0161,0008355
 0282,0003663
 0397,0001284
 0504,0000423
 0605,0000159
 0672,0000599
 0791,0000556


STOP.                               TERMINATION OF DATA
READY.                              STORAGE MODE

SAVD                                TRANSFERRING DATA
                                    TO DISK MEMORY

READY.
```

Fig. 6. Conversational sequence for entering a data file. (*From M. Tochner, J. A. Magnuson, and L. Z. Soderman, Time-sharing: A powerful approach for gas chromatographic data reduction, J. Chromatogr. Sci., 7:743, 1969*)

ter dedicated to a single application, and the larger midicomputer with associated peripherals shared among several simultaneous analytical applications.

Dedicated minicomputer. Advances in electronics, in particular, integrated circuit technology, have brought the cost of a small digital computer down to such an extent that it can be dedicated to a single analytical instrument. Such computers have been applied in several ways.

One type of dedicated minicomputer is a built-in computer which is an integral portion of the analytical instrument, perhaps not even identifiable as a computer and generally unavailable for tasks external to the instrument. This type of instrument often has analytical capabilities that would not be technically feasible without a computer. Such is the case with the recently introduced fast Fourier transform infrared spectrometers.

Another type is a dedicated computer which is an auxiliary peripheral to the analytical instrumentation. In many cases, systems are sold by the manufacturers of the analytical instruments on a turn-key basis; that is, the manufacturer takes the responsibility for interconnecting the computer to the instrument and providing the necessary software to make the combination work. Such systems are available from several vendors for gas chromatography, nuclear magnetic resonance (NMR) spectrometry, mass spectrometry, and x-ray spectroscopy. Figure 3 shows an NMR spectrometer automated with a dedicated Varian 620/i computer.

A third type is a general-purpose minicomputer that can be dedicated to any one of a number of analytical instruments or other experiments at the discretion of the scientist performing the research. In this case, the scientist develops the necessary software himself or has programming assistance to provide the computer sequencing for control, acquisition, and analysis of data. Jack Frazer, of the Lawrence Radiation Laboratory, has been a pioneer of this approach. He believes that it allows the scientist complete freedom of action within the computer's limitations. He advocates that the scientist conducting the experimentation is in the best position to design the software. It is much more reasonable for the motivated scientist to acquire the programming skills required to design software, even in machine language, than it is for a professional programmer, trained in business applications, to understand the control of a laboratory experiment.

Midicomputer for laboratory automation. Charles Sederholm of IBM Corp., Palo Alto, Calif., has pioneered the effort for the laboratory automation concept. Although it is substantially more expensive ($200,000–300,000), he believes that a properly designed shared-time system offers substantially more capability for each instrument than a dedicated minicomputer. A major difference is the availability of the more sophisticated peripheral devices of the larger system, which cannot be cost-justified for a single application.

Figure 4 shows a schematic of the configuration which was demonstrated at a recent chemical meeting. An IBM-1800 process computer with 32K words of core memory was linked by a 32-channel digital multiplexer to a Varian A60 NMR spectrometer, a Varian M66 mass spectrometer, and several Varian Aerograph gas chromatographs. The sophisticated peripherals included an auxiliary disk memory capacity of 512K 16-bit words, an analog-to-digital converter, a digital-to-analog converter, a line printer, an *x-y* plotter, a card punch, a card reader, and a teletype terminal. The system provides for the simultaneous and independent operation of the multiple instruments. Data are transmitted in a demand/response mode through the multiplexer when each instrument is ready to transmit data. The shared-time system is able to function as such principally because the instruments have a relatively low duty cycle, in spite of the fact that they require large system capacity in terms of input/output, peripheral memory, and compute power. Accordingly, an elaborate software monitor was designed to time-share the instruments. The monitor provides the necessary isolation for the various applications. Variable core is dynamically allocated to applications programs in blocks of 512 words, as required for each program. Real-time programs receive 5-msec time slices every 50 msec. Applications programs are run in 100-msec blocks at lower level priority in background when time is available. All control and data-acquisition functions can be controlled at the individual devices, through the multiplexer.

Advocates of the approach claim that although the initial investment for automating the entire laboratory is significant, it is economical in the long run.

ANALYTICAL CHEMISTRY APPLICATIONS

Although a detailed and comprehensive discussion of all the specific applications of computer-assisted analytical chemistry is beyond the scope of this article, several applications which typify the current state of the art will be presented. The following areas of application have received considerable attention and are discussed in depth: electrochemistry, gas chromatography, infrared spectroscopy, and mass spectrometry.

Electrochemistry. G. Lauer and R. A. Osteryoung utilized a general-purpose dedicated minicomputer, at the North American Rockwell Corp., Thousand Oaks, Calif., for on-line chronocoulometric measurements. It is an analytical technique for the measurement of the adsorption of electrochemically active species at an electrode surface. Experimentally the electrochemical system is perturbed by application of a potential on the electrode for a period of time, followed by a return to the original electrode potential. The response function measured as a function of time is the integral of the electrochemical current. The minicomputer, a PDP-8I manufactured by the Digital Equipment Corp., controls a group of analog operational amplifiers, which in turn control the cell potential. Figure 5 shows a detailed flow chart of the computer-controlled experimental system. The same computer is programmed for immediate analysis of the data, using regression analysis by least-squares fitting techniques, to test the experimental model. The computational time required to complete an analysis is of the order of 30 sec, enabling the experimenter to quickly interpret the results and immediately input new parameters, by means of the teletype terminal, for a subsequent experiment. The system is also used for preparing appropriate plots of raw data, fitted data, and the theoretical model for visual comparison.

Gas chromatography. Gas chromatography is an analytical technique which has received considerable attention in the realm of computer assistance. Virtually all of the types of computer systems discussed in the previous section have been applied to gas chromatography. For purposes of illustration, two will be discussed: off-line data acquisition using a time-sharing computer utility for data reduction, and a minicomputer dedicated to servicing multiple gas chromatographs.

The off-line approach (Fig. 2) yields a punched paper tape which must be input to a time-sharing utility.

The conversational sequence which is employed is shown in Fig. 6. The sequence is initiated by a local telephone call to the computer. The computer answers with a high-pitched beep signal, and the teletype identity is transmitted to the computer. The computer and operator alternately contribute to the conversational sequence. The

```
AUTØGC     21:47    28 THU 08/14/69

EQUILIBRATED SILICØNE FLUID
MIXTURE ØF LINEAR AND CYCLIC PØLYDIMETHYLSILØXANES
G.C. GRADE SE-30 CØLUMN
```

COMPONENT	RESPONSE FACTOR	RETENTION TIME	INTEGRATOR AREA	FACTORED AREA	WEIGHT PERCENT	M-UNIT PERCENT
AIR	.000	2	145	0	.00	.00
UNKNØWN	1.000	45	10	10	.00	.00
M-M	.735	51	8289	6092	3.01	3.01
D3	.880	149	103	91	.04	.00
M-D-M	.810	167	32552	26367	13.04	8.95
D4	.950	262	6099	5794	2.87	.00
M-D2-M	.874	344	40865	35716	17.66	9.23
D5	1.000	427	3295	3295	1.63	.00
M-D3-M	.930	537	37609	34976	17.30	7.30
D6	1.100	614	591	650	.32	.00
M-D4-M	.981	718	31505	30906	15.28	5.41
D7	1.100	798	70	77	.04	.00
M-D5-M	1.030	881	23654	24364	12.05	3.67
M-D6-M	1.070	1028	15186	16249	8.04	2.15
D9	1.200	1093	167	200	.10	.00
M-D7-M	1.110	1161	8355	9274	4.59	1.09
D10	1.200	1193	150	180	.09	.00
M-D8-M	1.150	1282	3663	4212	2.08	.45
M-D9-M	1.190	1397	1284	1528	.76	.15
M-D10-M	1.220	1504	423	516	.26	.05
M-D11-M	1.260	1605	159	200	.10	.02
M-D12-M	1.290	1672	599	773	.38	.06
M-D13-M	1.330	1791	556	739	.37	.05
TØTALS			215329	202211.	100.01	41.59

Fig. 7. Output from program AUTOGC. (*From M. Tochner, J. A. Magnuson, and L. Z. Soderman, Time-sharing: A powerful approach for gas chromatographic data reduction, J. Chromatogr. Sci., 7:744, 1969*)

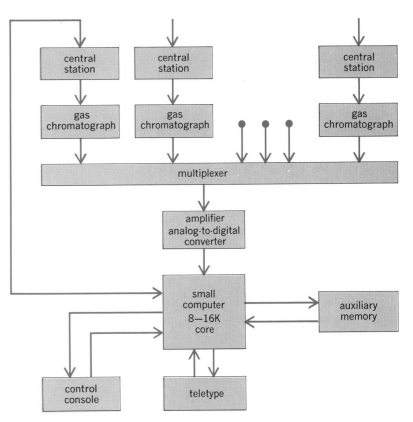

Fig. 8. Multichannel dedicated system. (*From J. M. Gill, Automation of gas chromatography, J. Chromatogr. Sci., 7:733, 1969*)

sequence continues in order to supply project identification, select the proper language, and retrieve file GCSAM1 from memory. Each time a command is completed, the computer responds that it is READY to continue. An empty file is generated by the SCRATCH command, followed by initiation of

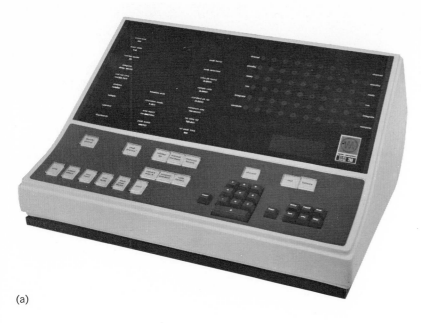

(a)

CONTROL CONSOLE

PARAMETER DISPLAY — **CHANNEL MONITOR**

	ILLEGAL	
TIME		
EVENT		PK FACT
CYCLES	SHOULDER	PK RET
DELAY	SEP RATIO	UNK FACT
RUN TIME	MIN AREA	WT RATIO
ANAL NO	SCALER	STD RET
SAMPLE	WIDTH D	REF SIZE
OPER	PK HEIGHT	REF TOL
DATE	PK WIDTH	REF TIME

CHAN ① ② ③ ④ ⑤ ⑥ ⑦ ⑧ ⑨ ⑩

STBY ● + + ● + + + + + +
RDY + + + + + ● + ● + +
RUN + ● + + + + ● + + ●
INT + + ● + + + + + + +
COMP + + + + ● + + + ● +

3 + 2 8 3 5

KEYBOARD DISPLAY

IDENT	PROC	CORR %	INT STD	EXT STD
		REL RET	RESP FACT	CALIB FACT
PROG	OUT-PUT	TAPE	END CYC	CLR	PRT

CHAN | TEST | BAL

7 8 9
4 5 6
CLR 1 2 3
0 LOAD SET VFY SKIP
STBY RDY END
VFY SKIP

(b) **CONTROLS/KEYBOARD**

Fig. 9. Control console for dedicated minicomputer servicing simultaneously operating chromatographs. (a) Photograph (*from Interfacing chromatographs to the computer, Varian Aerograph Tech. Bull., 138:19, 1969*). (b) Diagram (*from F. W. Baumann, A. C. Brown, and M. B. Mitchell, Chromatographer and the computer, J. Chromatogr. Sci., 8:21, 1970*)

TAPE transmission for data-storage mode (DSM). The user enters the data tape in the form Retention Time (sec), Area for each species. Following the transmission, the data are saved in memory and then the application program is brought into working core, where it reads the designated standard file (GCSTD1) and the data file (GCSAM1), calculates the composition, and provides the printed output (Fig. 7).

The off-line approach is best suited for small laboratories with a few simultaneously operating chromatographs.

The dedicated minicomputer is more appropriate for servicing many (5–40) simultaneously operating chromatographs. F. Baumann, of Varian Aerograph, Walnut Creek, Calif., has led a team of workers in the development of this type of system. It has been shown to be of significant importance in a research-and-development environment, as well as a system for the quality control laboratory. Figure 8 shows the overall organization of the system. Each gas chromatograph is scanned sequentially by the multiplexer. The signal is amplified and converted to digital form suitable for processing by computer. Control stations mounted at each chromatograph provide the link connecting the chromatographs to the computer. They provide the operator with the status information (RUN, READY, STANDBY, COMPUTE) pertinent to the particular chromatograph. A push button places the channel from a ready state to the run mode, thus beginning the acquisition of peak data and measuring of retention time.

The control console (Fig. 9) provides the interface for the chromatographer with the computer system in terms that are familiar to him. The upper left section labeled parameter display provides a means for entering data from the lower keyboard for run identification and data computation. The channel monitor indicates the mode of each channel. The functional modes include: standby (STBY) for input of parameter; ready (RDY) for sample injection; RUN, analysis in progress; integrate (INT) peaks; and computation (COMP) and printout of results. The parameter display legends are illuminated automatically and sequentially to facilitate the simplicity of operations. Data-processing operations for peak detection, such as fused-peak resolution and baseline-drift correction, as well as the mode of computation, such as internal standard, external standard, and response factors, are also entered from the control console.

The concept designed into this system is to make it easy for the chromatographer to work with the system, without becoming an electronics engineer or computer expert.

The system described above is not the only one of its type, but is typical of the concept. Competition will evolve substantial improvements in the future.

Infrared spectroscopy. Infrared spectroscopy is another branch of analytical chemistry that has received considerable attention in the realm of computerization. R. Norman Jones, of the National Research Council of Canada, has effectively contributed to the advancement of infrared spectroscopy by development of extensive computer software. His approach was the encoding of spectral data in digital form, followed by the computer application of the principles of numerical analysis for digital smoothing, and fitting infrared band envelopes with analytical functions by using nonlinear least-squares approximations. He found that mathematical relationships could be determined which would fit the absorption bands with a close fit approximating the uncertainty of the measurements. Although a substantial quantity of computer power is required to perform the mathematical analysis, the coefficients of the analytical expressions provide a more efficient means for the storage of high-resolution spectral data, for subsequent computer search, than recording the intensity and frequency data over the entire spectral range.

Infrared spectrophotometers have also been incorporated into on-line computer systems of various sizes. One case worth noting was the linking of a Perkin Elmer 621 grating instrument to an IBM-1800 midicomputer by means of a digital-encoder interface. A large system of this type, although expensive, offers some exciting possibilities to the researcher in the area of closed-loop control.

Another advance in infrared spectroscopy utilizes a dedicated minicomputer as an integral component along with a Fourier transform interferometer optical system. The DIGILAB FTS-14 (Fig. 10) is the first instrument of its type designed to exploit the maximum capabilities of both the computer and the optical system. The computer not only records and processes the data, but also controls and calibrates the spectrometer. The usual complex mechanical and electrical instrument controls are not required because the computer provides digital control capabilities. The system provides capability for a very rapid scan of the spectrum. The data are stored in digital form, and can be processed by powerful mathematical techniques. High-quality spectra can be obtained in less than 1 sec, permitting the analysis of gas chromatographic species "on-the-fly." The free-standing digital computer can be used for other data processing and mathematical analysis, as well as be inter-

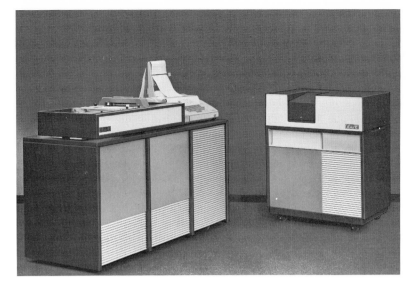

Fig. 10.   FTS-14 infrared spectrometer. (*From DIGILAB News, p. 16, February, 1970*)

faced to different analytical heads, such as pulsed NMR.

Infrared spectroscopy information retrieval. Duncan Erley, of Dow Chemical Co., Midland, Mich., has developed a fast-search system for

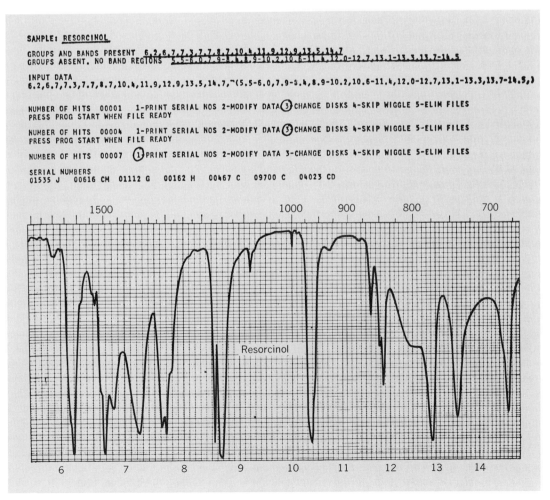

Fig. 11.   Infrared spectrum of resorcinol and typical input/output. (*From D. Erley, Anal. Chem., 40:897, 1968*)

scanning the American Society for Testing and Materials (ASTM) infrared data file. An IBM-1130 computer was programmed to search the spectral information from over 90,000 infrared data cards of the ASTM file. The data are entered in a conversational mode from a teletype terminal. The user enters the code for groups known to be present and the wavelength (or wave number) of bands in the infrared spectrum, followed by entry of groups known to be absent and no band regions. After the search the number of hits are printed and several options are provided: print of the ASTM serial numbers, modification of the data and a repeat search, search of another disk, modification of the acceptance zone, or elimination of a subfile. Figure 11 shows an example of the typical input/output interaction for the search of a spectrum of resorcinol. The search algorithm takes about 90 sec to complete a scan of the 90,000 file entries on the dedicated IBM-1130. Recently an analogous search system has been introduced for remote operation on a time-sharing utility.

Mass spectrometry. Computer capabilities have been particularly attractive for several aspects of the analysis of mass spectral data. In the field of high-resolution mass spectrometry, the quantity of data precludes manual treatment and readily lends itself to computer processing. Digital computers have been used for the identification of unknown compounds by searching a library file of low-resolution mass spectra to compare to the unknown spectrum. Off-line computer approaches have been utilized for analysis of data which was recorded on photographic film or even directly digitized and recorded on magnetic recording tape. Recent innovations by A. L. Burlingame, of the University of California, Berkeley, have provided on-line real-time high-resolution mass spectrometry. The data are collected, processed, displayed, and stored while the scan is in progress. The response time of current digital computers is of sufficient speed to be compatible with the fast data rates generated by a high-resolution mass spectrometer.

The research efforts of K. Biemann, of MIT, and associates have led to the computer assist to mass spectrometry called element mapping. It is an automated method in which all of the peaks of a high-resolution mass spectrum are displayed in a con-

cise manner, retaining much valuable information at a great saving of interpretation time.

The computer software calculates all possible empirical formulas corresponding to each computed mass and provides an appropriate printed output. The empirical formulas are calculated by the computer, which subtracts the sum of the masses being considered from the observed mass and tests the remainder for correspondence to an integral number of hydrogen atoms. The calculations continue for each mass until all possible combinations of atoms have been found to be within a prescribed tolerance of the observed mass. Tools of this type greatly aid the skilled spectroscopist, particular with interactive systems in which his experience and intuition can be coupled to the speed and thoroughness of the computer.

FUTURE

The particular areas of application which have been discussed are by no means completely comprehensive. The applications discussed were selected because they are typical of the current state of the art. Space limitations prohibited inclusion of many other application areas, such as kinetic studies, NMR spectrometry, and molecular simulation.

For the future, certain advances in the field can be anticipated with the continued advancement of electronics technology, particularly in integrated circuits. It is not unreasonable to forecast continued decreases in the price of small digital computers. The price range is likely to drop sufficiently for a dramatic increase in new instrument offerings which contain dedicated computers as integral components. More emphasis will be placed on manufacture of analytical instruments that are designed from scratch to take full advantage of computer capabilities. One can also foresee developments of communications systems that will permit linking of the analytical instrument's computer with a much larger computer in a remote location over telephone lines. Larger computers with sophisticated peripherals will thus be a mere "touch tone sequence" away from the analyst in the laboratory. The analytical chemist of the next decade can certainly look forward to the digital computer as an eager partner in the laboratory.

[MAX TOCHNER]

Photographic Highlights

These photographs have been chosen for their scientific value and current relevance. Many result from advances in photographic and optical techniques as man extends his sensory awareness with the aid of the machine, and others are records of important natural phenomena and recent scientific discoveries.

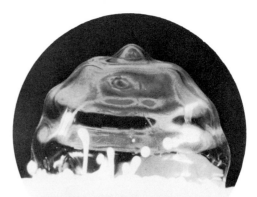

Visible effects of sound on matter. Right, a soap bubble vibrating at 110 Hz and strong amplitude. Below, quartz sand vibrating on a steel plate at 16,800 Hz, forming a high-frequency Chladni figure. *(Photos by J. C. Stuten, in H. Jenny, Cymatics, Basilius Presse, 1967)*

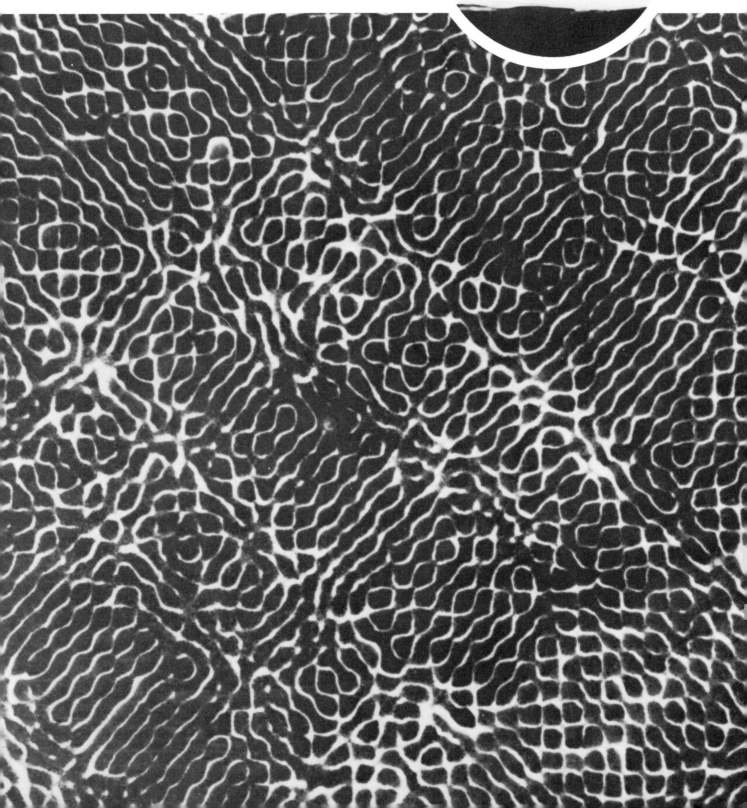

Devastation in the suction path of a major tornado which passed through Lubbock, Tex., on May 11, 1970. *(T. T. Fujita, Weatherwise, 23(4):160–173, 1970)*

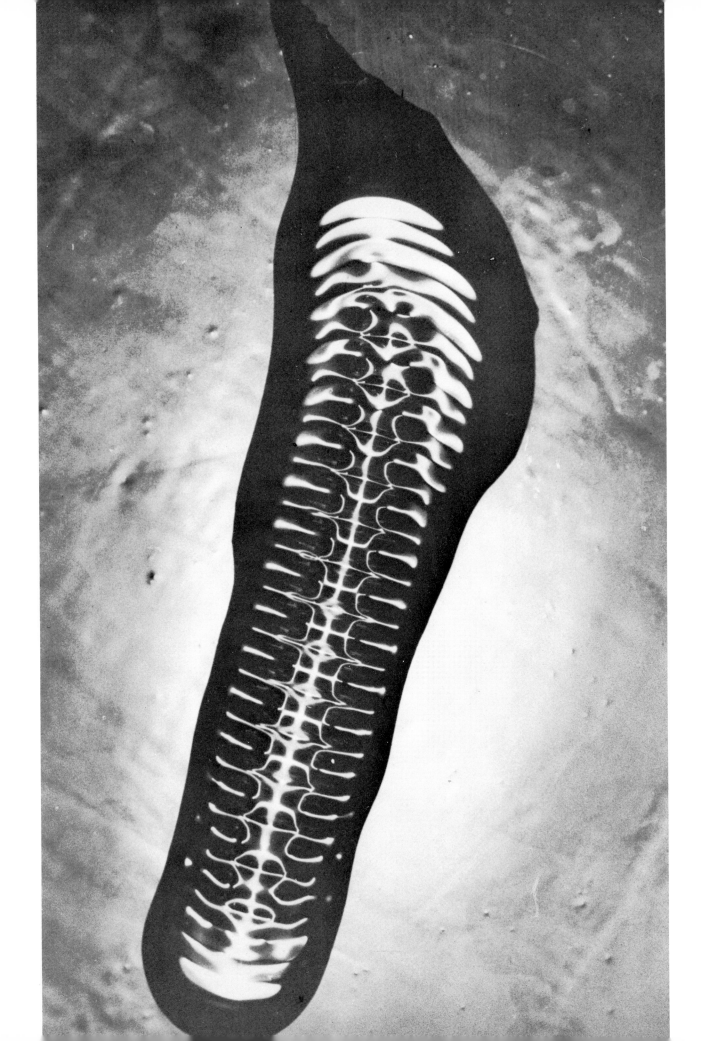

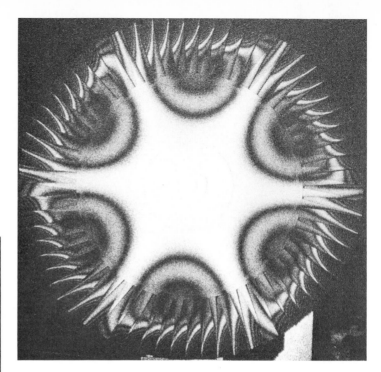

Above. Holographic visualization of
vibration in engineering components, in
this case a typical turbine disk.
The fringes indicate the mode
of vibration, the brightest parts of
the image representing nodes.

Opposite page. Air flowing past
a cylinder rolls up into a regular
succession of vortices. A cross
section of these vortices is made visible
by injecting a sheet of smoke in
the center of a wind tunnel. *(Photograph
taken by Gary Koopmann at the Naval
Research Laboratory, Washington, D.C.)*

Below. Holographic inspection of
the performance of a diaphragm
typical of those used in pressure
transducers. A defect is revealed
by distortion of the fringes.

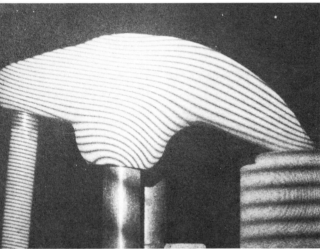

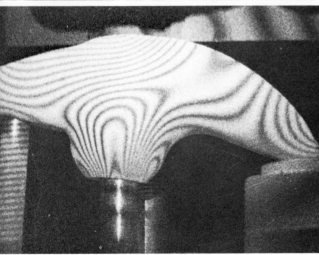

Investigation of the design of
engine rockers using holographic
interferometry. In upper photograph, an
even distribution of fringes
indicates good design. Lower photograph
shows a rocker prone to early
fracture where the fringes are bunched.

*Photographs from D. Denby and
J. N. Butters, Holography as an
engineering tool, New Scientist,
pp. 394–396, Feb. 26, 1970.*

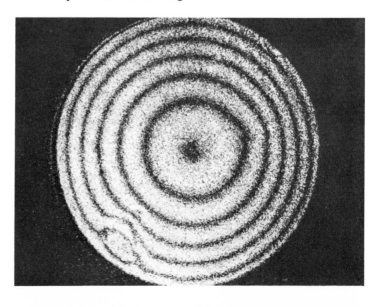

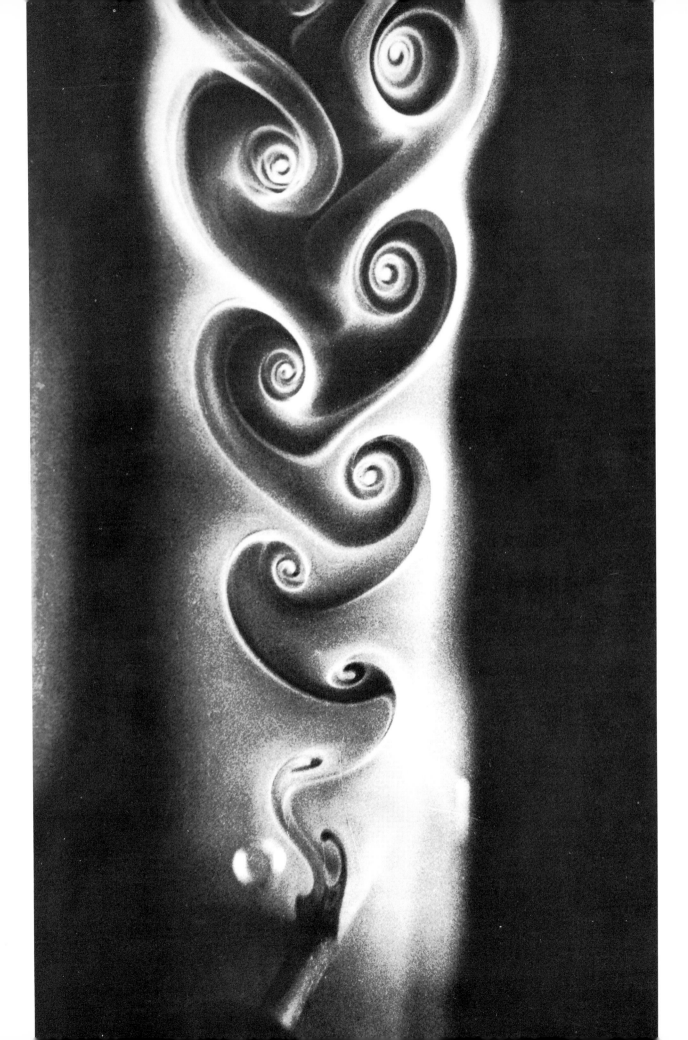

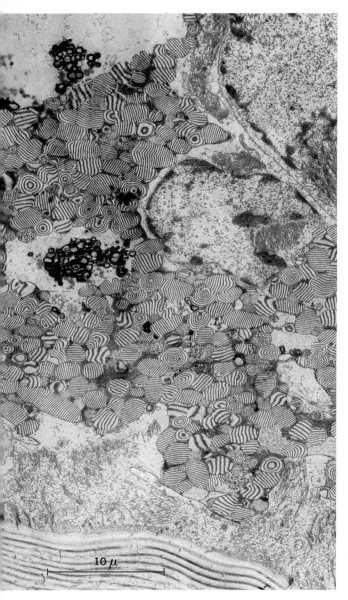

10 μ

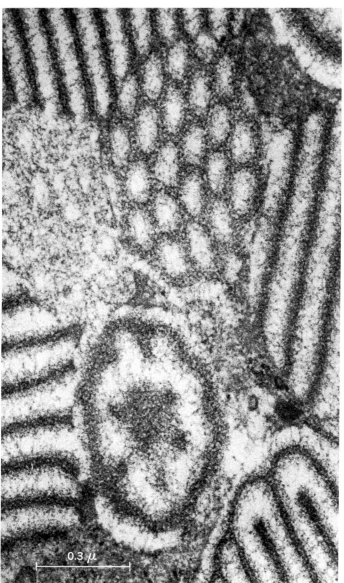

0.3 μ

Above and opposite page. Electron micrographs
of periodically layered intracellular
inclusions in crayfish statocyst epidermis. Although
the function of these "op-art-bodies"
is unknown, they are suspected of being involved
in moulting processes. (R. A. Steinbrecht,
Max-Planck-Institut für Verhaltensphysiologie)

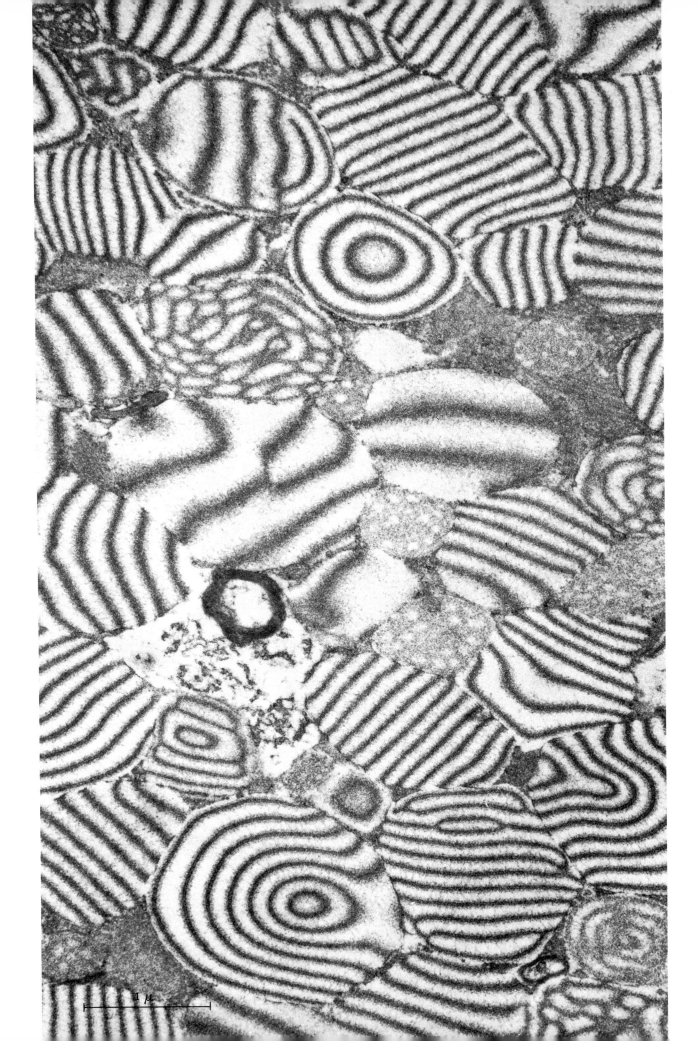

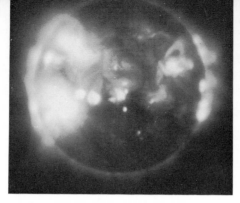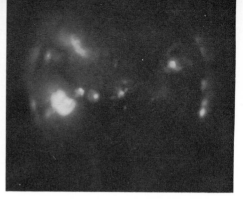

X-ray photographs of the solar corona
taken from an Aerobee rocket
payload during the Mar. 7, 1970, eclipse.
Photograph at left is an exposure
in the bands 3–30, 44–55 A; photograph at
right, in the bands 3–13, 19–24 A.

Opposite page. Superposition of
the x-ray exposure shown at upper left
on a white-light coronal photograph
taken by G. Newkirk and L. Lacey.

A region of diffuse x-ray emission on the
solar disk as observed (from left
to right) in soft x-rays, photospheric
magnetic fields, Hα, and CaK.
(L. P. VanSpeybroeck, A. S. Krieger,
and G. S. Vaiana, X-ray photographs
of the Sun on March 7, 1970,
Nature, 227(5260):818–822, 1970)

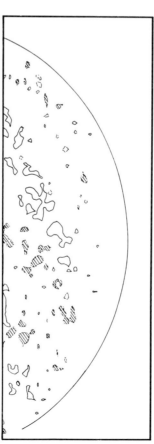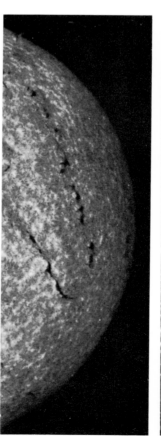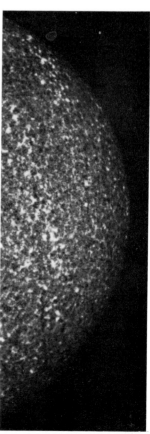

Opposite page. *Glaucilla*, Australian floating marine nudibranch mollusks, store the stinging cells of the Portuguese man-of-war jellyfish and can inflict painful stings on bathers. *(Photo by F. G. Myers, in T. E. Thompson and I. Bennett, Physalia nematocysts: Utilized by mollusks for defense, Science, 166(3912):1532–1533, 1969)*

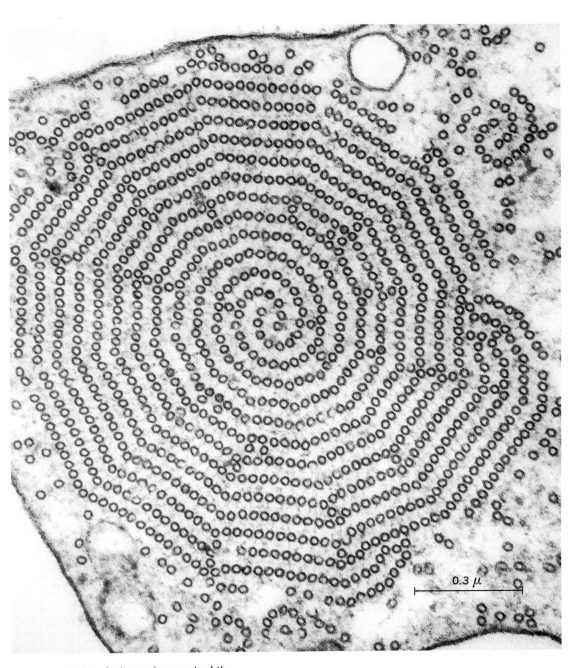

0.3 μ

Cross-section electron micrograph of the microtubule system in a long axopodial extension of the protozoan cell *Echinosphaerium nucleofilum.*
(L. E. Roth, D. J. Pihlaja, and Y. Shigenaka, Journal of Ultrastructure Research, 30:7–37, 1970)

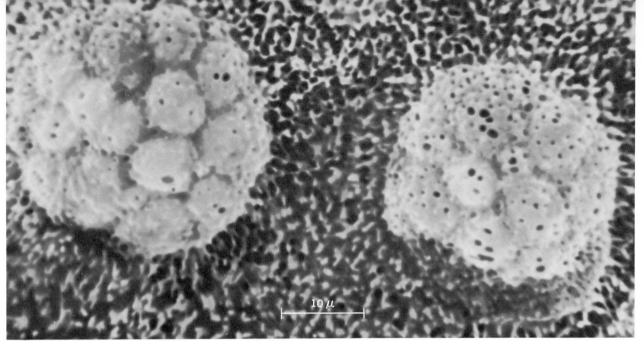

Above. Scanning electron micrograph of surface of an egg of the stone fly *(Pteronarcys dorsata)* showing plastron type of gills which allow the insect egg to breathe oxygen even when immersed in water.

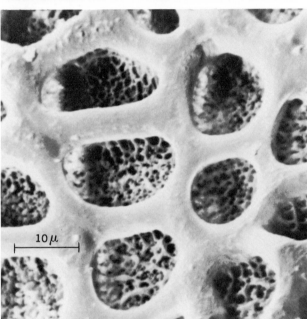

Left. Plastron structures on front end of egg of the long-horned grasshopper *(Plagiostira gilletti).*

Below. Plastron "crater" on side of egg of the Australian bush fly *(Musca vetustissima).*

Opposite page. Ovarian follicles and incomplete egg of the fly *Fannia armata.* Upper photograph shows surface cells of the ovarian follicles, which secrete the eggshell. Lower photograph shows incomplete egg removed from the ovary.

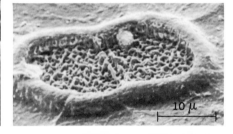

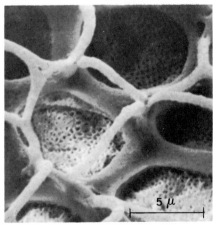

Above. Side of egg of the fly *Fannia armata.*

Right. Side of egg of one of the scorpion flies *(Panorpa anomala). (H. E. Hinton, Insect eggshells, Scientific American, pp. 84–91, August, 1970)*

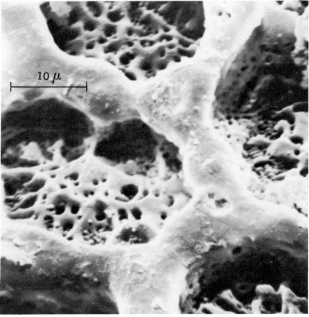

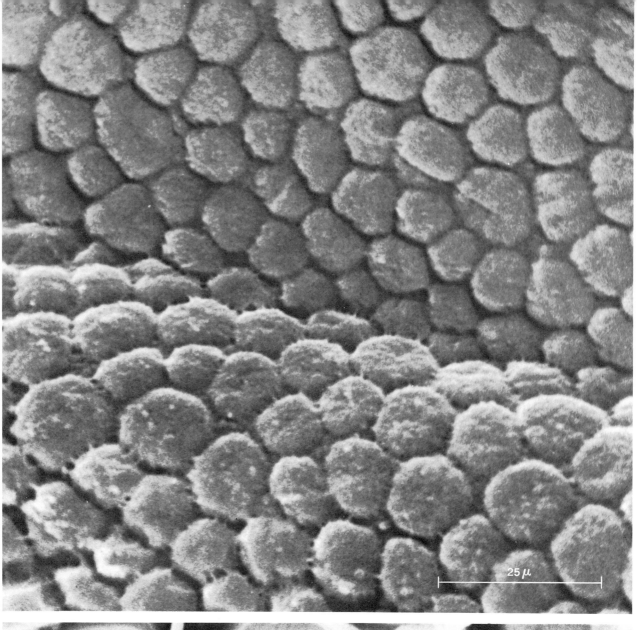

25 μ

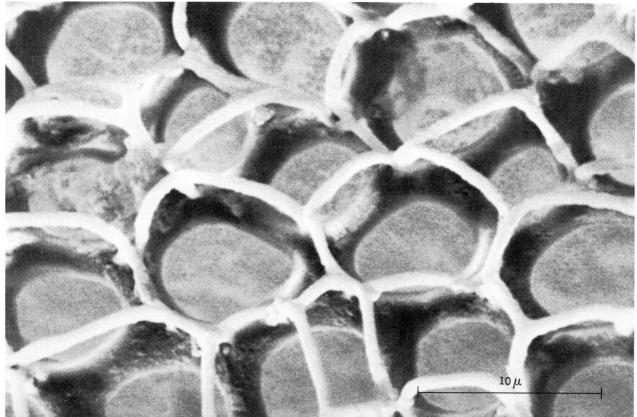

10 μ

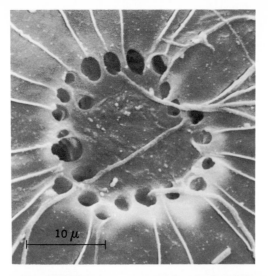

10 μ

Small tunnels, or micropyles,
in the eggshell of the
puss moth *(Cerura vinula)*
provide a way for the
fertilizing sperm to reach
the interior of the egg.
*(H. E. Hinton, Insect
eggshells, Scientific American,
pp. 84–91, August, 1970)*

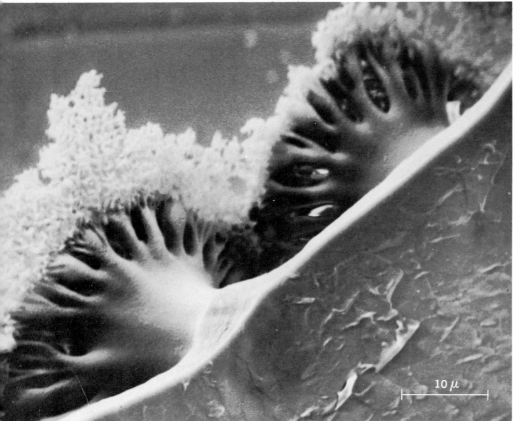

10 μ

Opposite page. Nannoplankton
collected from the
southwestern Pacific. Upper
photograph shows
Umbellosphaera cf. *tenuis;*
lower photograph,
Umbillicosphaera mirabilis.
*(S. Honjo and H. Okada,
Nature, 224(5222):845, 1969)*

Oblique view from interior
of the torn plastron of
the stone fly *(Pteronarcys
dorsata).* The finely
ramified tops of the
treelike structures that
comprise the plastron
intermesh to form the egg's
water-air interface.
*(H. E. Hinton, Insect
eggshells, Scientific American,
pp. 84–91, August, 1970)*

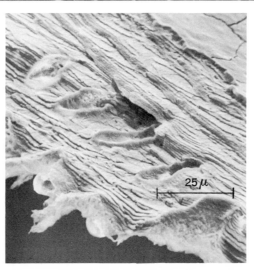

25 μ

The eggshell of the puss
moth *(Cerura vinula)* is
made up of multiple layers.
*(H. E. Hinton, Insect
eggshells, Scientific American,
pp. 84–91, August, 1970)*

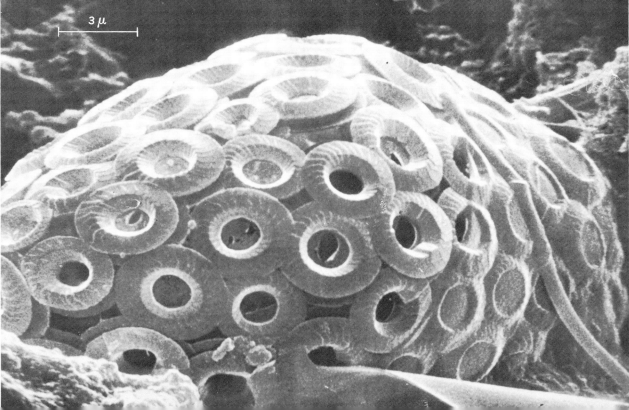

Scanning electron micrographs of chains of thorium atoms separated by an
organic molecule. These chains were placed on a thin carbon film a tenth of
a millionth of an inch thick. In each chain, it is the smallest white
dots which represent single thorium atoms. The larger white dots are probably
aggregates of a few thorium atoms very close together. (A. V. Crewe, J. Wall,
and J. Langmore, University of Chicago; *Visibility of single atoms,*
Science, 168(3937):1338–1340, 1970)

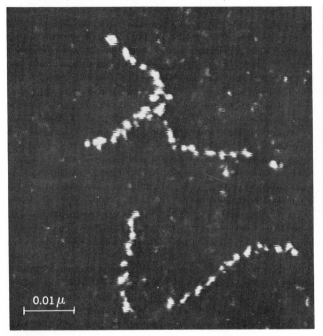 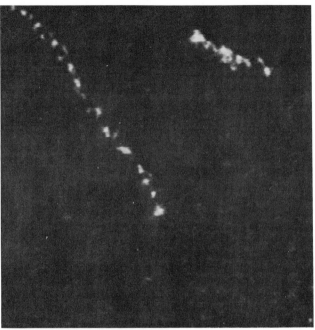

Unique scanning electron microscope built
by University of Chicago scientists.
Single atoms of thorium and uranium have
been revealed on the display panel
on the left. (A. V. Crewe, J. Wall, and
J. Langmore, University of Chicago)

Single atoms were viewed on the oscilloscope
of the display panel. (A. V. Crewe,
J. Wall, and J. Langmore, University of Chicago)

A-Z

Absolute zero

According to the third law of thermodynamics, the absolute zero (0°K, −273.15°C, −459.67°F) can never be reached. However, with improvements in cryogenic technology it is possible to come closer and closer to this point. Recently, three new procedures became available for attaining temperatures in the millidegree and submillidegree region. These are the He^3/He^4 dilution process, the Pomeranchuk method, and nuclear cooling. It should be emphasized that, near the absolute zero, temperature ratios, instead of temperature differences, are important. A logarithmic scale, rather than a linear one, is thus more appropriate in this domain; on a log T scale the temperature interval from 1°K to 0.1°K is just as long as the interval from 0.001°K to 0.0001°K.

He^3/He^4 dilution refrigerator. This refrigerator, first proposed by H. London, G. R. Clarke, and E. Mendoza in 1962, has made many new types of experiment possible at ultralow temperatures, that is, in the region between 0.01 and 0.3°K. The dilution process is characterized by two important advances over adiabatic demagnetization of a paramagnetic salt, which until recently was the only practical method for reaching temperatures below 0.3°K. First, a dilution refrigerator operates continuously for long periods of time and second, quite large cooling powers are obtainable even with relatively modest pumping arrangements.

Furthermore, a properly designed dilution refrigerator can be operated by any competent technician after short training. Dilution cryostats, capable of reaching 10 m°K (0.01°K) are commercially available in the United States from the S.H.E. Manufacturing Co.

The working substances in a dilution refrigerator are He^3 and He^4. At ultralow temperatures a liquid mixture of these two isotopes separates into two phases, one of which is practically pure He^3, the other containing about 6% He^3 in a He^4 background. Because of its lower density the He^3-rich, or concentrated, phase floats on top of the He^4-rich, or dilute, phase. When He^3 atoms cross the phase boundary from the concentrated phase to the dilute phase, cooling is produced, according to the same principle as that of an ordinary evaporation-type refrigerator when molecules cross the boundary from the liquid to the vapor phase.

The principal parts of a continuously operating dilution refrigerator are shown in Fig. 1. The phase separation occurs in the mixing chamber, and cooling is produced there by causing He^3 atoms from the upper phase to move across the boundary to the lower phase. The continuous dissolving of He^3 from the concentrated to the dilute phase is obtained by circulating He^3 in the system by means of a pump at room temperature. Incoming gas is first precooled in a condenser attached to a pumped He^4 pot at 1.0–1.2°K. The pressure of He^3 is kept sufficiently high for condensation to occur

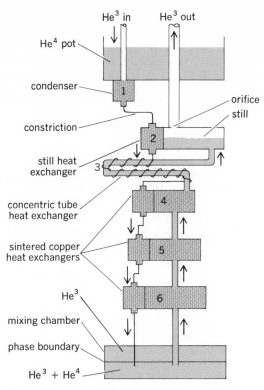

Fig. 1. Schematic drawing of the low-temperature parts of a He³/He⁴ dilution refrigerator.

by means of a flow-limiting constriction. The liquid then enters the still heat exchanger at 0.6–0.8°K, a concentric tube heat exchanger, several sintered copper exchangers, and finally the mixing chamber. After crossing the phase boundary, He³ atoms, driven by an osmotic pressure gradient, proceed in reverse order through the heat exchangers to the still, where vapor is removed by pumping. Well over 90% of the outcoming gas is He³ because He⁴ atoms are, even at the still temperature, relatively inert.

The main reason for the success of the dilution refrigerator is the 6% solubility of He³ into He⁴ even at $T = 0°$K. When the temperature of the mix-

ing chamber is lowered during the operation of the refrigerator, one does not "run out" of He³ in the dilute phase as one runs out of molecules in the vapor phase when using an ordinary evaporation-type refrigerator. The circulation rate of He³ is thus independent of the mixing chamber temperature. Enthalpy considerations show, however, that the cooling power of a given refrigerator is proportional to T^2. The lowest temperatures so far reached in a continuously operating dilution refrigerator are about 6 m°K achieved by B. S. Neganov and coworkers, Joint Institute for Nuclear Research, Dubna, Soviet Union, and J. C. Wheatley and coworkers, University of California, San Diego.

Pomeranchuk method. First proposed in 1950, the Pomeranchuk method is based on the unusual thermodynamic properties of a solid-liquid mixture of He³ at low temperatures. In simple terms this type of refrigerator can be explained by using the phase diagram shown in Fig. 2. The melting curve of He³, separating the solid and liquid phases, displays a peculiar feature: It has a rather deep minimum, at $T = 0.318°$K and $p = 28.9$ atm, below which dp/dT of the melting curve is negative. Consequently, if a mixture of solid and liquid He³, initially in equilibrium at point A on the melting curve, is compressed, liquid will gradually be converted into solid and the system will cool along the melting curve until, at B, only solid is left, whereby the cooling capacity has been exhausted. Above 0.318°K, compression of He³ results in warming, which is the normal behavior observed in other substances.

The main difficulty in the practical realization of Pomeranchuk cooling is connected with the question of how to compress He³ adiabatically, without frictional heating. The minimum in the melting curve makes it difficult to increase the pressure. A tube filled with He³ will become blocked with solid at 0.318°K, as soon as the pressure has been increased to 28.9 atm. A cell containing He³ at a temperature below 0.318°K will thus be cut off from the outside He³ supply at this pressure. Further compression, therefore, cannot be obtained by admitting more He³ into the cell but must be achieved by other means, for example, by "squeezing" the cell in some way. Another experimental difficulty is that, for useful results, the starting temperature (point A) must be well below the minimum of the melting curve, preferably not higher than 30 m°K. Efficient precooling is thus very important.

There is not sufficient space to discuss the Pomeranchuk method in detail here. The first successful experiments were performed just a few years ago, and Wheatley and coworkers now can reach 2 m°K by this procedure. The starting temperature was 24 m°K, obtained by a dilution refrigerator.

It is too early to predict the impact that the Pomeranchuk method will have on ultralow-temperature refrigeration. It has a good chance of becoming important between 2 and 10 m°K; its high cooling capacity, approximately proportional to T and surpassing that of a dilution refrigerator by an order of magnitude, is of particular practical importance.

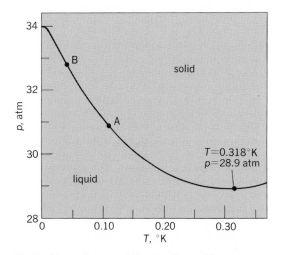

Fig. 2. Phase diagram of He³ near the melting curve.

Nuclear cooling. This is adiabatic demagnetization utilizing the nuclear magnetic properties of a suitable metallic specimen. It proceeds, in principle, the same way as does conventional adiabatic demagnetization of a paramagnetic salt. However, because the nuclear magnetic moments are about 2000 times smaller than the electronic moments in paramagnets, the initial conditions are quite different. Electronic demagnetization can be started from $T = 1°K$ and magnetic field strength $H = 20$ kilogauss (kG); nuclear demagnetization requires that $T = 10$ to 20 m°K and $H = 40$ to 60 kG. The necessary initial conditions for nuclear cooling experiments are thus difficult to achieve. However, there is an important compensation. The ultimate temperature that can be reached after demagnetization depends on the mutual interactions between the elementary magnets; it is impossible to cool to a temperature lower than that at which spontaneous ordering of the elementary magnets occurs. In the electronic case the limiting temperature is about 3 m°K, whereas in the nuclear case it is below one-millionth of a degree kelvin!

Nuclear cooling was first proposed by N. Kurti, F. E. Simon, and C. J. Gorter in 1934; the first partially successful experiments were performed in 1956. With the advent of the dilution refrigerator and high-field superconducting magnets, the initial conditions required for nuclear cooling have become much more readily accessible than before. There are, however, many other practical difficulties that still must be overcome. These are mainly connected with heat transfer, thermal isolation, and thermometry.

At low temperatures the nuclear elementary magnets, that is, the nuclear spins, are, even in a metal, rather weakly coupled to the conduction electrons and the lattice. It is thus possible to reach very low temperatures in the nuclear system without cooling the rest of the specimen; recently a spin temperature of 0.2 $\mu°K$ was achieved by D. A. Spohr, U.S. Naval Research Laboratory, Washington, D.C. More important for actual experiments is the true temperature of the whole specimen. The lowest equilibrium temperature reached so far is 0.54 m°K achieved by O. V. Lounasmaa and coworkers, Technical University of Helsinki, Otaniemi, Finland; this temperature was obtained by employing a cryostat in which a dilution refrigerator and a copper nuclear stage operate in series.

Of the three new methods of cooling described, only the dilution refrigerator is fully operational at present. It has largely replaced the conventional adiabatic demagnetization cryostat and even the simple He³ refrigerator. Many experiments that formerly would not have been feasible below 0.3°K have already been carried out with dilution machines, and much more work will be undertaken in the immediate future.

In the past the quest for progressively lower temperatures has been amply rewarded by new and fundamentally important discoveries; it seems equally clear that much of importance lies ahead in the yet unexplored territory. Pomeranchuk cooling and nuclear demagnetization, even though they are not continuous modes of refrigeration, are the likely tools that will open this new territory.

For background information *see* ABSOLUTE ZERO; CRYOGENICS; LOW-TEMPERATURE PHYSICS in the McGraw-Hill Encyclopedia of Science and Technology.   [O. V. LOUNASMAA]

Bibliography: R. A. Hein, D. U. Gruber, and E. H. Takken (eds.), *Proceedings of the 1970 Ultralow Temperature Symposium*, U.S. Naval Research Laboratory Report no. 7133, p. 113, 1970; R. T. Johnson et al., *J. Low Temp. Phys.*, 1:451, 1969; O. V. Lounasmaa, *Sci. Amer.*, 221:26, December, 1969; O. G. Symko, *J. Low Temp. Phys.*, 1:451, 1969; J. C. Wheatley, in C. J. Gorter (ed.), *Progr. Low Temp. Phys.*, 6:77, 1970.

Actinomycetales

Most microbiologists are unfamiliar with the organisms which belong to the family Dermatophilaceae, of the order Actinomycetales. Until recently, the Dermatophilaceae consisted of a single genus, *Dermatophilus*. This bacterium possesses a unique, complex morphogenetic growth cycle. Its confusing polymorphism may be the major cause for the unstable taxonomic status of this group, which is reflected by the numerous synonyms compiled for *Dermatophilus*. These include *Streptothrix*, *Rhizobium*, *Polysepta*, *Tetragenus*, *Actinomyces*, and *Nocardia*. The situation has been further complicated by G. M. Luedemann's recent description of a second member of the Dermatophilaceae, the genus *Geodermatophilus*.

Geodermatophilus exhibits a morphogenetic growth cycle which is superficially similar to that of *Dermatophilus*. An important aid in future taxonomic studies would be the development of methods for controlling morphogenesis of these microorganisms under defined, reproducible conditions. In this respect, *Dermatophilus* and *Geodermatophilus* represent excellent model systems for studies on microbial morphogenesis since most strains can be cultured readily on simple defined media. Such studies are in progress, but the biochemical bases for differentiation in these organisms are still not understood.

Dermatophilus. The genus *Dermatophilus* is represented by a single species, *D. congolensis*. It is an animal pathogen and is the etiological agent of mycotic dermatitis, or cutaneous streptotrichosis. The infection involves invasion by the organism of the cornifying layer of the skin, with subsequent development of scabs or crusts. A wide variety of animals have been found to be susceptible to the disease, particularly sheep, cattle, horses, and goats. Two human infections have been reported. *Dermatophilus* has not been isolated from sources other than diseased animals. D. S. Roberts concluded that the skin of animals may be the specific habitat for this organism and that survival of the species may be dependent on transmission of the disease from one animal to another. However, it is possible that *Dermatophilus* inhabits other ecological niches in nature, such as soil, and failure to detect it may be due to lack of suitable selective enrichment techniques. A susceptible animal may thus be considered a natural selective medium for the organism.

Morphogenesis of *Dermatophilus* has been studied with the light microscope by Roberts. In the so-called zoospore stage, the organism occurs

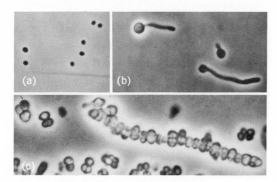

Fig. 1. Morphogenesis of *Dermatophilus*. (*a*) Active spherical cell. (*b*) Nonmotile cell germinating to form hypha. (*c*) Mature mycelium.

as a spherical cell which is actively motile by means of a tuft of flagella (Fig. 1*a*). As differentiation progresses, the cell becomes nonmotile and germinates to form a hypha (Fig. 1*b*). The hypha then elongates and branches. This stage is accompanied by hyphal segmentation by means of transverse fission. Further segmentation later occurs along the longitudinal axis of the hypha. Fission along the transverse and longitudinal axes continues in two or more planes. The resulting mature mycelium is composed of coccoid subunits apparently held together by a thick capsule (Fig. 1*c*). The growth cycle is completed when the coccoid cells assume motility and escape from the surrounding gelatinous capsular matrix. These observations were confirmed by the ultrastructural studies of M. A. Gordon and M. R. Edwards.

Distribution of the organism in infected tissue has been studied by Roberts. The mycelial stage was found in the depths of the lesion, invading the cornifying cells in the stratum lucidum. The motile zoospore stage was distributed exclusively on the surface of the scab. It was proposed that morphogenesis is controlled by oxygen tension, temperature, and concentration of nutrients. More defi-

nitive work is required to establish the biochemical basis for differentiation in *Dermatophilus*.

Geodermatophilus. Organisms described as the new genus *Geodermatophilus* were isolated from American desert soils by Luedemann. All of his isolates produced an unidentified black pigment and were classified under a single species, *G. obscurus*. Cultures of *Geodermatophilus* have been also obtained from high-altitude Mount Everest soil samples and from Baltic Sea water, and these organisms either produced a red pigment or were nonpigmented. R. Ahrens and G. Moll have tentatively identified the red pigment as a carotenoid. All strains thus far studied are aerobic and do not appear to have unique physiological characteristics. Further work is required to determine whether the differences in pigmentation are sufficient to warrant the establishment of new species. To date, all attempts to demonstrate animal pathogenicity have been unsuccessful. Despite only three reports of isolation of *Geodermatophilus*, this bacterium is thought to be widely distributed in nature. Its apparent infrequent occurrence can again be attributed to unavailability of selective enrichment methods.

In the C-form stage of growth, *Geodermatophilus* occurs as a nonmotile, irregularly shaped clump of coccoid cells (Fig. 2*a*) on which buds form (Fig. 2*b*). Mature buds detach as motile rods that have polar tufts of 1–4 flagella (Fig. 2*c*). In this stage, known as the R-form, cells multiply exclusively by budding. During later stages of differentiation, rods become enlarged and budding reproduction ceases. Segmentation of the enlarged rods then occurs. This process follows a definite pattern. Initial segmentation occurs with several septa being formed by fission along the transverse plane. Further segmentation then takes place along the longitudinal plane. The resulting compartmentalized rod is made up of coccoid subunits (Fig. 2*d*). To complete the cycle, irregularly shaped aggregates of coccoid cells are formed when the pattern of fission becomes random in all planes.

E. E. Ishiguro and R. S. Wolfe recently described a system for controlling differentiation of *Geodermatophilus*. In this method, morphogenetic events take place synchronously. Morphogenesis appears to be controlled by a single, unidentified nutritional factor. This factor is required for differentiation of *Geodermatophilus* to the C-form. The only source found thus far for this compound has been animal tissue, and purification and characterization of the factor are in progress. Using this system, these workers have studied the entire growth cycle by electron microscopy. The major ultrastructural difference between the C-form and the R-form was found to be in the structure of the cell envelope. R-form cell walls consist of two layers and are approximately 20 nm in thickness. In addition to these layers, the C-form cell wall has a thick outer layer which is about 30 nm in thickness. This layer apparently serves as a cementing substance which holds the individual coccoid cells together. The unidentified morphogenetic factor may play an important role in the biosynthesis of this cementing layer since the structure is present only in cells grown in media containing this factor.

For background information *see* ACTINOMYCE-TALES in the McGraw-Hill Encyclopedia of Sci-

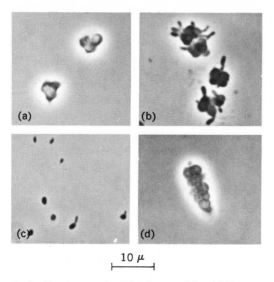

10 μ

Fig. 2. Morphogenesis of *Geodermatophilus*. (*a*) Nonmotile, irregularly shaped clump of coccoid cells. (*b*) Buds form on coccoid cell. (*c*) Buds detach as motile rods. (*d*) Coccoid subunits form in enlarged rods as a result of segmentation and compartmentalization.

ence and Technology.   [EDWARD E. ISHIGURO]

Bibliography: R. Ahrens and G. Moll, *Arch. Mikrobiol.*, 70:243, 1970; M. A. Gordon and M. R. Edwards, *J. Bacteriol.*, 86:1101, 1963; E. E. Ishiguro and R. S. Wolfe, *J. Bacteriol.*, 104:566, 1970; G. M. Luedemann, *J. Bacteriol.*, 96:1848, 1968; D. S. Roberts, *Aust. J. Exp. Biol. Med. Sci.*, 39:463, 1961.

Adenoviruses

Some recent aspects of research on adenoviruses have been concerned with the serological structure of the virus. Adenoviruses have been isolated from various mammals and from birds. A total of about 80 serotypes have been identified. Some of these are of medical importance, since they cause respiratory illnesses, eye infections, and so on. Furthermore, it has been found that certain serotypes are capable of transforming rodent cells in the animal and in tissue culture. These cells do not allow a complete multiplication of the virus type concerned.

Adenoviruses represent an excellent model system for studying the occurrence of different structural components (building stones) in virus particles. They are medium-sized, with a diameter of 80 nanometers (nm), and in their center contain deoxyribonucleic acid (DNA) with a molecular weight of 20,000,000–25,000,000 daltons. This amount of DNA maximally could carry information for the production of 30–70 medium-sized proteins. In the virus particle, DNA occurs associated with two or three arginine-rich (basic) proteins. This nucleoprotein complex forms the core of the virus particle and is surrounded by a shell (capsid) which has the strict symmetrical shape of an icosahedron: 252 morphological units (capsomers) participate in the formation of the capsid (Figs. 1 and 2a). These capsid components have been structurally and biologically well characterized during recent years. Surplus components produced during virus multiplication in cells represent a good source of material for studies of this kind.

In the capsid the 240 nonvertex capsomers occur surrounded by six neighboring capsomers and therefore have been named hexons. For corresponding reasons, the combined structure of a vertex capsomer plus the projection carried by this component is called a penton. The individual parts

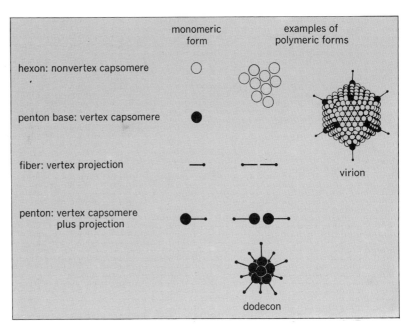

Fig. 1. Schematic description of monomeric and some experimentally identified polymeric forms of adenovirus capsid components. (*From E. Norrby, The structure and functional diversity of adenovirus capsid components, J. Gen. Virol., 5:221, 1969*)

of the penton are referred to as the penton base and fiber.

Hexons. Hexons have a tubelike shape, with a diameter of 8–9 nm. They can occur isolated and occasionally in groups of nine hexons arranged in a specific way (Fig. 1). These groups possibly represent functional aggregates, since assembly of 20 groups of nine hexons, each group in one of the triangular facets of an icosahedron, can give a particle lacking only vertex capsomers and the five surrounding hexons. The hexon is composed of six polypeptides, each with a molecular weight of 60,000–70,000.

A wide spectrum of immunological specificities can be identified in hexons, ranging from antigen components shared by all mammalian adenoviruses to those specific for each type. Only the latter kind of antigen component is available at the surface of virus particles, as demonstrated by the attachment of type-specific antibodies against hexons (Fig. 2b).

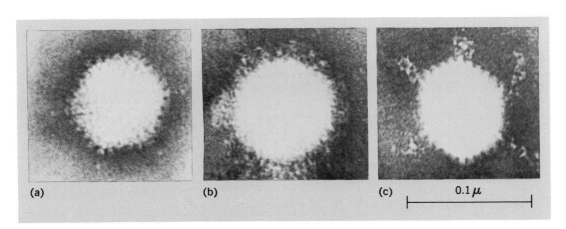

(a)   (b)   (c)   0.1 μ

Fig. 2. Ultrastructure of adenovirus particles. (a) Untreated. (b) Combined with type-specific antibodies against hexons. (c) Combined with type-specific antibodies against fibers.

Vertex capsomers. Vertex capsomers have a more wedgelike shape than do hexons. The number of polypeptide chains from which they are formed is not known at present. Their amino acid composition is different from that of hexon polypeptides, indicating at least a partial dissimilarity in amino acid sequence. Vertex capsomers also carry a range of different antigenic specificities, all of them distinct from those occurring in hexons. In preparations of most types of adenoviruses, excess vertex capsomers occur in association with fibers in penton structures. Isolated pentons, but not fibers, of some serotypes carry a capacity to detach cells from glass surfaces, a phenomenon not dependent upon virus multiplication. Treatment with trypsin destroys the cell detachment activity. The destruction is due to a relatively higher degree of sensitivity of vertex capsomers, as compared to other structural components, to treatment with proteolytic enzymes.

Fibers. Fibers have an elongated form. Their length can vary from one type of adenovirus to another, and in parallel with this variation there is a variation in the number of antigenic specificities which can be demonstrated. Minimum and maximum lengths are 10 and 38 nm. Fibers are built up of polypeptides different from those of hexons and vertex capsomers. The tip of the fiber is type-specific, as demonstrated by the selective attachment of certain antibodies to this antigenic component (Fig. 2c). Different antigenic specificities occur in other parts of the fiber, and in some cases these have been found to be shared between adenoviruses of different species origin.

Hemagglutinins. The tip of the fiber can interact with receptors on cells susceptible to infection by the virus or on certain kinds of red blood cells. The former event represents the first step in the process of virus multiplication in cells. The combination of fiber structures with receptors of red blood cells can lead to an aggregation of these cells (hemagglutination). Isolated fiber or penton components are monovalent with respect to their capacity to interact with receptors. Therefore they represent incomplete hemagglutinins. These incomplete hemagglutinins can be demonstrated by combining their nontip parts with antibodies, which represent polyvalent structures. The aggregates formed can agglutinate red blood cells. However, spontaneously occurring oligomers of fiber and penton structures have been identified (Fig. 1). Dimers and an interesting aggregate of 12 pentons (a dodecon) were found. Because of their polyvalent nature, these oligomers represent complete hemagglutinins.

Other components. The fact that a minimum number of nine polypeptides of different molecular weights have been demonstrated to occur in adenovirus particles indicates that structural components, in addition to those discussed above, remain to be characterized.

Structural-functional relationships. Knowledge of the structural-functional relationship between different components in virus particles is of value in a number of connections. Such knowledge is needed as a background for analysis of the early steps of interactions between virus (disassembly) and cells during the process of infection, and also for analysis of the sequence of events during as-

sembly of components into virus particles. As far as infection is concerned, it appears that virus particles lose their pentons and the five neighboring hexons during their passage into the cytoplasm of cells. Remaining parts of the capsid (180 hexons) are lost during passage of the nucleoprotein component of the virus into cell nuclei. In the cell nucleus, the virus nucleic acid can take over the cell's metabolic machinery and cause the production of several thousands of new, complete virus particles per cell.

Comparative immunological relationships. A comparison of immunological relationships between hexons, vertex capsomers, and fibers of types of different species origin allows some speculation on the evolutionary differentiation of capsid components. The derivation of components of increasing specificity seems to occur in the order of hexons to penton bases to fibers. An additional "axis" of evolutionary differentiation presumably also passes through hexons, as evidenced by the fact that these components expose only their type-specific part or parts at the surface of virus particles.

Vaccines. A thorough knowledge of the immunological nature of different structural components offers opportunities for the introduction of virus vaccines of a new kind. The possible utilization of a purified and isolated virus component (biochemically defined) for immunization seems to be within reach. Therefore the goal of applying microbiological products for immunization under conditions of maximum medical safety seems to be attainable.

For background information *see* ADENOVIRUS; DEOXYRIBONUCLEIC ACID (DNA); VIRUS in the McGraw-Hill Encyclopedia of Science and Technology. [ERLING NORRBY]

Bibliography: K. Lonberg-Holm and L. Philipson, *J. Virol.*, 4:323, 1969; E. Norrby, Adenoviruses, in K. Maramorosch and E. Kurstak (eds.), *Comparative Virology*, in press; E. Norrby, The structure and functional diversity of adenovirus capsid components, *J. Gen. Virol.*, 5:221, 1969; E. Norrby, H. Marusyk, and M. L. Hammarskjöld, *Virology*, 38:477, 1969.

Air pollution

Air pollution is a source-transport-effect phenomenon; that is, pollutants originate at a source and are transported through the atmosphere to a point at which they cause an unwanted effect. In the past, there have been many attempts to define air pollutant and air pollution. Air pollutant is often defined as any constituent of the atmosphere that causes an unwanted effect. This is a rather general definition and includes gases, particulate matter, and other factors, such as noise and odors mediated by the atmosphere. It is erroneous to define an air pollutant as any unnatural constituent of the atmosphere, for, much of the time, pollutants such as ozone (O_3) occur as a natural constituent of the atmosphere. In fact, at the present time in the United States, O_3 is considered to be the most important plant-pathogenic air pollutant. There are also natural sources of other pollutants, such as sulfur dioxide (SO_2). Consequently, an air pollutant cannot be accurately defined as any unnatural constituent of the atmosphere. Since the unwanted effects caused by air pollution represent the basis for the interest in air pollution, it seems logical to define an air pollutant as any constituent of the atmosphere that causes such effects.

Air pollutants affect materials, plants, and animals, including man. The deleterious effects of air pollution on vegetation have been recognized for more than 100 years. Since vegetation, through the process of photosynthesis, represents the basis for life on this planet, it is essential that effects of various air pollutants on vegetation be recognized and understood. To obtain a clear understanding of the effects of air pollution on vegetation, one must know (1) the types of compounds that can damage plants; (2) how and where these compounds originate; and (3) their effects, both on the whole plant and the cells of plants. One must also have some idea of what problems to expect in the future.

Plant-pathogenic air pollutants. Air pollutants that are pathogenic to plants arise from a variety of natural and man-related phenomena. They can be divided into primary and secondary types. Primary pollutants originate at the source in a form toxic to plants; SO_2 and hydrogen fluoride (HF) are examples of this type. Secondary pollutants develop as a result of reactions among pollutants. The photochemical pollutants peroxyacetyl nitrate (PAN) and O_3 are examples of this type. O_3, SO_2, HF, PAN, oxides of nitrogen, particulates, aldehydes, chlorine, hydrogen chloride, ethylene, hydrogen sulfide, and silicon tetrafluoride are the most common plant-pathogenic air pollutants in the United States today. Of these, the first four are by far the most important and cause the most damage. In addition to the above, there are many pollutants of rather minor importance.

Primary pollutant source. SO_2, a pollutant known to be toxic to plants for well over 100 years, is emitted principally from the combustion of coal; the production, refining, and utilization of petroleum and natural gas; manufacturing and industrial utilization of sulfuric acid and sulfur; and the smelting and refining of ores, especially copper, lead, zinc, and nickel.

Sulfur dioxide. The combustion of coal represents the major source of SO_2. The amount of SO_2 emitted depends upon, among other things, the sulfur content of the coal. The sulfur content of coal in the United States ranges from less than 1% to as much as 6%, with 2% as an average. Coal-burning power plants are the most important single source of SO_2, the numerous instances of vegetation damage have been associated with this source.

Fluorine compounds. Fluorine-containing compounds, such as HF and silicon tetrafluoride (SiF_4), have been recognized as being pathogenic to plants for over half a century. They originate principally from aluminum reduction processes, manufacture of phosphate fertilizer, steel manufacturing plants, brick plants, pottery and ferroenamel works, and refineries.

In general, the fluorides originate from the molten cryolite bath in the manufacture of aluminum and from impurities in the raw materials used in the other industries. Fluorides are toxic at much lower concentrations than are most other plant-pathogenic air pollutants. Therefore, although the annual tonnages produced do not compare with those of pollutants such as SO_2, the fluorides still represent a major problem as phytotoxicants.

Ozone. Usually a secondary-type pollutant, O_3 has been recognized as a phytotoxicant for at least 100 years. Primary sources of O_3 include the upper atmosphere and electrical storms.

Some meteorologists and pathologists believe that tropospheric O_3 may be brought to the Earth's surface during atmospheric disturbances such as violent storms. It has also been established that O_3 may form during thunderstorms as a result of the splitting of molecular oxygen by electrical discharge causing the formation of atomic oxygen, which subsequently reacts with molecular oxygen to form O_3. The ground-level concentration of O_3 increases sufficiently during many of these thunderstorms to cause injury to vegetation. Weather fleck of tobacco, a disease which has been associated with atmospheric disturbances for many years, is now known to be caused by O_3.

Nitrogen oxides. Recently, oxides of nitrogen have been cited as a potential problem as phytotoxic primary pollutants, especially in view of the amounts of these compounds emitted annually. Oxides of nitrogen originate from a variety of sources which include gasoline combustion in motor vehicles, refining petroleum, combustion of natural gas, fuel oil and coal, and incineration of organic wastes.

Chlorine compounds. Hydrogen chloride (HCl) was an important plant-pathogenic air pollutant in Europe approximately 100 years ago. However, technological changes replacing the Le Blanc soda process resulted in the disappearance of the HCl problem. In recent years there has been a reappearance of injury and damage attributed to HCl. Chlorine (Cl_2), which is also highly toxic to plants, has been implicated as a cause of damage to vegetation. These two chemicals are increasing in importance as phytotoxicants. Current major sources of HCl and Cl_2 include refineries, glassmaking, incineration and scrap burning, and accidental spillage. Polyvinyl chloride is used in large quantities in manufacturing packaging materials and wire insulation. Combustion of polyvinyl chloride results in the emission of HCl and a myriad of other compounds. The increased use of this material is one of the main reasons why HCl will become more important as an air pollutant. However, HCl is soluble in water, and techniques are available for efficiently scrubbing it from flue gases.

Secondary pollutant sources. Some of the secondary pollutant sources are discussed below.

Ozone. Probably the most important source of O_3 today is photochemical reactions in polluted atmospheres. Oxides of nitrogen, which are emitted into the atmosphere by automobiles and by a variety of industries and utilities, may react in the presence of light (sunlight) with oxygen to form O_3. Likewise, the irradiation of mixtures of oxides of nitrogen and various hydrocarbons often results in O_3 formation. Hydrocarbons and oxides of nitrogen emitted by motor vehicles, industrial processes, and the generation of electrical power can react to form the secondary pollutant O_3, which is a major phytotoxic ingredient of urban smog.

PAN. PAN has been recognized as a common phytotoxic constituent of the smog over cities for the past 10–15 years. It is a member of an homologous series of compounds which originate principally from the reaction of olefin-type hydrocarbons and oxides of nitrogen in the presence of light. PAN, its relatives, and a number of other compounds, such as aldehydes, can form during this reaction. Their formation depends on the type of olefin involved, on whether NO or NO_2 is present, and on the duration of the time of irradiation. Again, the automobile and other forms of transportation represent the major sources of hydrocarbons and oxides of nitrogen that function as reactants to form PAN.

Terpenes that evolve from coniferous vegetation can react photochemically with oxides of nitrogen to form O_3 and PAN. Consequently, natural sources of hydrocarbons in the vicinity of man-related sources of oxides of nitrogen, such as power plants, could result in the formation of photochemical air pollutants and thus complicate control efforts.

Particulates. Particulates have been recognized as plant-pathogenic air pollutants for many years. However, there has been relatively little research on the problem. The major sources of atmospheric particulates include combustion of coal, gasoline, and fuel oil; cement production; lime kiln operation; incineration; and agricultural burning and agricultural-related activities. Bag houses and various types of precipitators have been developed for controlling particulate emissions from a variety of processes; this can be accomplished with efficiency rates up to 99%.

Effects of air pollution on plants. In general, the symptoms of air-pollution injury to plants can be classified as necrotic, chlorotic, or atrophic. Specific effects include changes in cell-wall permeability, plasmolysis, changes in tissue pH, interference with cell-wall synthesis, acceleration of respiration, inhibition of enzymes, and a reduction in the rate of photosynthesis. Effects are often classified as either acute or chronic, with acute implying some degree of necrosis of tissue. In contrast, chronic refers to all nonnecrotic, usually more subtle, effects. From a diagnostic viewpoint, the effects that specific pollutants have on a given plant can often be unraveled in the field. However, with two or more pollutants acting in consort, the differences are often masked and the resultant symptom picture is confused. Consequently, in most instances it is impossible on the basis of field observations alone to make a positive identification and to implicate a specific pollutant.

SO_2 causes an interveinal chlorosis and necrosis of the leaves of broad-leaved plants and usually a tip necrosis of the leaves of evergreens. There is often a reddish banding of the leaves of some evergreens such as larch. On broad-leaved plants such as pinto bean and alfalfa, the interveinal necrosis is usually ivory-colored; on broad-leaved plants such as red maple, the interveinal necrosis is reddish-brown. Tip necrosis of evergreens is characterized by a dead needle tip with a chlorotic band between the necrotic area and the healthy tissues. On many evergreens there is often a premature defoliation of older needles which ultimately results in a reduction in growth rate.

In contrast to SO_2, HF causes a marginal chlorosis and necrosis of the leaves of broad-leaved

Fig. 1. Chlorotic stipple of white ash caused by ozone.

plants and tip chlorosis and necrosis of ever-greens. Fluorides in general are mobile within the plant. Instead of causing injury at the point of entry, for example in the interveinal areas, they often move to the margin of the leaf, where they accumulate and eventually cause damage to sensitive plants. Because of their mobility and ability to accumulate in certain plant tissues and because of their extreme toxicity (atmospheric concentrations of one part per billion may result in injury to some plants), fluorides represent a special type of problem.

The ability of fluorides to accumulate in vegetation and their toxicity to animals have resulted in an additional problem. Cattle that graze on vegetation in the vicinity of fluoride sources often develop a disease known as fluorosis. This is caused by the fluoride ingested during foraging. The fluorides interfere with blood and bone metabolism, causing a blackening of the teeth and an overall debilitation of the animal. There have been some serious outbreaks of fluorosis in cattle herds in the western United States.

O_3 damage to plants is often characterized by a metallic flecking or necrotic stippling of the upper leaf surface of broad-leaved plants (Fig. 1). The upper-leaf-surface effect is due primarily to the fact that the palisade cells are much more sensitive to O_3 than are the spongy mesophyll cells of the leaf. On some conifers, O_3 causes a tip necrosis of current needles (Figs. 2 and 3), whereas on others it causes a chlorotic mottle of either the current or older needles. The greater sensitivity of the older needles to O_3 corresponds to the situation in broad-leaved plants, in which the older, more senescent leaves are often most sensitive to O_3.

As was mentioned earlier, O_3 may originate during certain storms and, as a consequence, relatively high concentrations of O_3 may develop in remote areas. This "naturally" occurring O_3 has been implicated as at least a part of the causal complex of the needle blight syndrome of eastern white pine and of weather fleck of tobacco.

O_3 that develops photochemically in the atmosphere over cities is not necessarily confined to the city. In recent years, the chlorotic decline of ponderosa pine in the San Bernadino Mountains east of Los Angeles, Calif., has been attributed to O_3 that originates in the smog cloud over Los Angeles. The chlorotic mottle observed on the older needles in the field has also been duplicated in fumigation chambers with O_3. Hence, photochemical air pollutants such as O_3, which develop in the atmosphere over urban centers, may be carried to adjacent forest areas and cause damage as far away as 50–70 mi. This prospect can be quite alarming when one considers the millions of acres of forest land within a radius of 60–70 mi of the megalopolis that extends from Boston, Mass., to Richmond, Va.

In contrast to O_3, PAN causes symptoms to develop for the most part on the lower surface of the leaf. The symptom that occurs most often is a glazing or silvering of the underleaf surface which may develop into a necrotic symptom. This glazing or silvering is due primarily to a destruction of spongy mesophyll cells and the subsequent movement of air into the area of the leaf between the lower

Fig. 2. Tip necrosis of Austrian pine caused by 24-pphm ozone for 4 hr.

epidermis and the palisade cells.

As far as relative toxicity is concerned, SO_2 is the least toxic and HF the most toxic of the four most important pollutants. O_3 and PAN are intermediate in toxicity, with PAN slightly more toxic than O_3. Of course, many factors are important in determining the severity of plant response, including degree of susceptibility of a given plant, the nature of environmental conditions during fumigation, the concentration of the pollutant during fumigation, and the length of the fumigation.

In addition to the above effects, air pollutants may also cause synergistic, potentiative, and predisposing effects. For example, it has been shown that SO_2 and O_3 have a true synergistic effect on tabacco plants. Tobacco plants exposed to very low concentrations of SO_2 and O_3 simultaneously were more severely damaged and the effect was greater than the additive effects on plants exposed to the pollutants individually. The exposure of long-lived plants such as trees to low and continuous or intermittent concentrations of various air pollutants may result in a gradual weakening and predisposition of such plants to attack by relatively innocuous pathogens. It is relatively safe to conclude that air pollutants may have a variety of effects on vegetation, the types and complexities of which need be limited only by the imagination.

The development of new sources of currently important plant-pathogenic air pollutants and the appearance of totally new plant-pathogenic air pollutants in the future have not been mentioned.

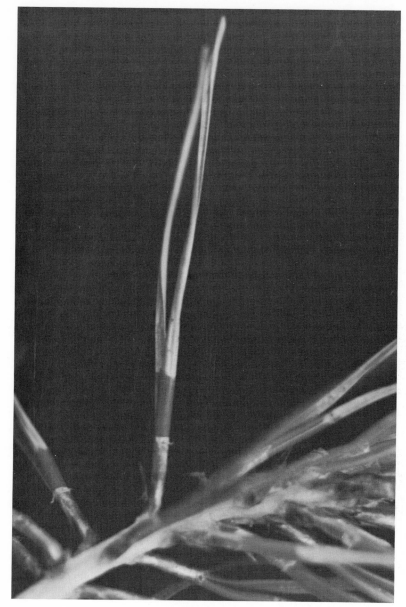

Fig. 3.   Tip necrosis of Virginia pine caused by ozone.

Fluorides are currently being used in rocket fuels, and they represent a basic ingredient of some of the most efficient fuels. The National Aeronautics and Space Administration (NASA) has developed a policy of not using fluoride-based fuels in the first stage of the missile systems. When these fuels are burned in the second- and third-stage systems, they are exhausted into the atmosphere at very high altitudes and theoretically do not constitute a contamination hazard at the Earth's surface. NASA also does not contemplate any major increase in the number of launchings in the next 30 to 40 years. Unless there is accidental spillage of these fluoride-based fuels at the missile sites, it is unlikely that there will be any significant contribution by space exploration activities to the fluoride air-pollution problem. As technology advances and more exotic combinations of chemical elements are utilized in everyday life, the ultimate disposal or incineration of these materials will undoubtedly result in new and important plant-pathogenic air pollutants.

For background information *see* ATMOSPHERIC POLLUTION in the McGraw-Hill Encyclopedia of Science and Technology.   [FRANCIS A. WOOD]

Bibliography: E. F. Darley, Symposium on trends in air pollution damage to plants, *Phytopathology*, 58:1075–1113, 1968; I. J. Hindawi, *Air Pollution Injury to Vegetation*, National Air Pollution Control Administration Publ. no. AP-71, U.S. Department of Health, Education, and Welfare, 1970; J. S. Jacobson and A. C. Hill, *Recognition of Air Pollution Injury to Vegetation*, Inform. Rep. no. 1, TR-7 Agricultural Committee, Air Pollution Control Association, 1970; A. C. Stern (ed.), *Air Pollution*, vol. 1, 1968.

Air samplers, rocket-borne

The collection of large, undifferentiated high-altitude air samples has become a practical possibility with the recent development of a rocket-borne cryogenic air sampler at the National Center for Atmospheric Research in Boulder, Colo.

The earliest application of rockets to high-altitude air sampling involved the collection of small air samples for investigating the diffusive separation of gases. In a series of rocket flights that began on July 29, 1947, evacuated steel bottles of 8.3 liters volume were carried in various unused spaces of V-2 rockets and the sample bottles were recovered from the wreckage after flight. Subsequently, Aerobee rockets were used, and a number of changes were made to eliminate sample contamination, to extend the altitude of sampling up to ~100 km, and to improve recovery by use of parachutes. Analysis of helium, neon, argon, and nitrogen in these samples gave results which indicated a significant degree of gravitational separation of gases above 59 km altitude. Subsequently, it became clear that the observed separations could be explained by mass discrimination at the sampler inlet.

An improved method of small air sample collection was developed in the Soviet Union. Evacuated glass cylinders of large inlet diameter were ejected from the rocket near apogee, were opened and closed automatically at speeds <50 m/sec, and were recovered by parachute. Results from rocket flights between 1951 and 1956 suggested that there was some diminution of argon relative to nitrogen at heights of 85–95 km. Similar techniques were used in two rocket flights at White Sands Missile Range in August, 1956. In this case the results gave inconclusive evidence for the presence of diffusively separated air between 60 and 90 km altitude.

Because of experimental difficulties and large errors in the analysis of the small air samples obtained in these early rocket sampling experiments, the results were qualitative and limited to a few major constituents of air.

Cryogenic sampler operation. The new cryogenic sampler is unique in that it employs a simple normal shock diffuser inlet and a low-temperature heat exchanger for quantitative recovery of undifferentiated air samples collected at supersonic velocities during rocket ascent. The cryogenic sampler was first used on Sept. 4, 1968, on an Aerobee-150 rocket at White Sands; pressurized liquid hydrogen was used as the refrigerant. In principle, liquid helium or liquid neon also can be used.

A schematic diagram of the cryogenic air sampler is given in the illustration. The central heat exhanger consists of 32 double spiral coils which provide 10 m² of condenser surface held at low temperatures by pressurized liquid hydrogen circulated inside the coils. Each double spiral coil is formed from continuous, seamless, thin-walled stainless tubing welded only to the outside wall of the heat exchanger. The coils are held in position by spacers (not shown), also welded to the outer cylinder. The absence of internal welds across the thin-walled heat exchanger minimizes the possibility of developing inaccessible leaks, making the heat exchanger a practical and reusable system. Prior to launch the heat-exchanger coils and refrigerant reservoir are filled with liquid hydrogen. In addition, the helium cylinder is filled with high-purity helium to the desired operating pressure. Immediately prior to launch, the liquid hydrogen fill and vent probes are withdrawn remotely.

Immediately before the rocket reaches its sampling altitude, a squib valve is actuated either by radio command signal or by preset timer, initiating helium pressurization. Thereupon hydrogen refrigerant is pressurized to a point just above its critical pressure, the pressure at which the relief valve in the refrigerant vent system opens. Thereafter, refrigerant flows continuously from the reservoir through the supply manifold, through control orifices into each of the coils in the forward portion of the heat exchanger, on through the coils and vent manifold, and finally out through the in-flight refrigerant vents. The orifices controlling the flow into each coil are designed and calibrated to supply refrigerant at a rate in excess of that required to condense air entering the sample chamber. The flow of refrigerant continues throughout the period of air sampling and thereafter until the supply is exhausted.

Air sampling is initiated at the selected threshold sampling altitude by deploying the split nose cone and inlet cover by means of a spring ejection mechanism, actuated by a hermetically sealed pin puller. The sampled air passes through the divergent inlet diffuser into the sample chamber at supersonic speeds. Just forward of the first heat-exchanger coil the incoming air undergoes a normal shock-wave transition, where it is heated and pressurized for a period of milliseconds. Downstream of the shock wave the air flows subsonically over the heat-exchanger coils, where it is cooled and condensed. The noncondensable air constituents, hydrogen, helium, and neon, which make up only 24 parts per million by volume (ppmv) of air, accumulate in the sample chamber. As long as supersonic flow is maintained in the sampler inlet, the noncondensables are confined to the region of subsonic flow, downstream of the normal shock, and thus are quantitatively retained with the collected air sample.

At the end of the sampling interval, the intake valve is closed. The valve is rotated 90°, and the intake valve cap is forced down against a Viton O-ring. Latching lugs slide into position, securing the air sample inside the sample chamber. Immediately after parachute recovery of the sampler, residual hydrogen is flushed out of the refrigerant system with pure helium. The condensed air is allowed to reevaporate in the sampler before it is withdrawn via the sample-removal line.

Results. The first rocket flight of the cryogenic air sampler, carried out at White Sands on Sept. 4, 1968, was successful in collecting a column of air 150 cm² in cross-sectional area between 43.6 and 62.3 km altitude, a total of 9.3 moles of air. Measured concentrations of neon, argon, and krypton in the sample were identical to the concentrations of these noble gases in surface air, confirming that, up to the altitude of sampling, the atmosphere is very well mixed and demonstrating that the sam-

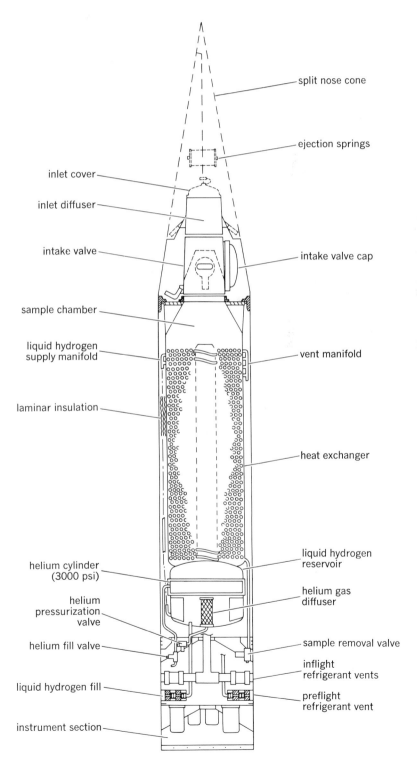

Rocket-borne cryogenic air sampler.

pler provides representative air samples, including the quantitative recovery of the noncondensable gases. Principal results of trace gas measurements were as follows: carbon dioxide, 322.8 ± 2.5 ppmv; methane, ≤ 0.05 ppmv; molecular hydrogen, ≤ 0.45 ppmv; and water vapor, ≤ 10 ppmv. The results also showed that tritium and carbon-14 concentrations were greatly in excess of natural levels because of thermonuclear weapons tests in the atmosphere.

Applications. The design of the cryogenic sampler incorporates many features which make it a versatile system, with a wide range of research applications in the upper atmosphere. Its special advantages stem from its ability to collect and retain large, undifferentiated air samples from a wide range of altitudes. The sampler was designed to retrieve samples of size sufficient to allow isotopic analysis of carbon dioxide, water vapor, and molecular hydrogen in high-altitude air. The samples also are ample for a variety of nuclear debris studies. In addition, the large air samples may be liquefied and fractionally distilled to enrich the trace gases for improved detection sensitivity or to separate them for isotopic analysis.

Measurements of large air samples collected by the cryogenic air sampler will yield reliable information on the gaseous composition of the upper atmosphere, with by-product benefits in improved knowledge of other high-atmosphere properties and processes. Accurate measurement of the temperature and pressure in the upper atmosphere depends upon knowledge of the gaseous composition. Determination of the altitudes at which diffusive separation of various trace gases takes place also is important in relation to such questions as the rate of dissipation of atmospheric hydrogen and helium into interplanetary space. Direct measurements of the distribution of carbon dioxide, water vapor, methane, nitrous oxide, and other trace gases also provides information needed for calculations of the radiation balance in the upper atmosphere.

In the collection of particulate debris from nuclear tests, the cryogenic sampler has advantages over other rocket sampling systems. Most radioactive aerosols which persist at altitudes above 25 km have radii less than 0.01μ. For collection of aerosols ranging from particles of 0.01μ radius down to individual molecules, impaction collectors are inapplicable, and even the best filter techniques are of uncertain efficiency. However, the cryogenic sampler is 100% efficient for the collection of all particles in the intercepted air column, regardless of size. The sampled volume is accurately determined by direct measurement of the recovered air sample. The large inlet area provides for the collection of samples of ample size for the measurement of significant levels of nuclear bomb debris as well as for nuclear detection applications.

Other high-altitude applications of the rocket-borne cryogenic sampler include the determination of residence times of gaseous and aerosol trace constituents, the accumulation of pollutants, and measurement of micrometeorites and other atmospheric particles, the redistribution of atmospheric tracers, and perhaps also the determination of the concentrations of atomic oxygen, ions, and free radicals.

For background information *see* ATMOSPHERIC CHEMISTRY in the McGraw-Hill Encyclopedia of Science and Technology.   [EDWARD A. MARTELL]

Bibliography: R. H. Bieri et al., *J. Geophys. Res.*, no. 75, 1970; L. M. Jones, in R. L. F. Boyd et al. (eds.), *Rocket Exploration of the Upper Atmosphere*, 1954; E. A. Martell, *J. Appl. Meteorol.*, 9:170–177, 1970; G. R. Martin, in R. L. F. Boyd et al. (eds.), *Rocket Exploration of the Upper Atmosphere*, 1954; B. A. Mirtov, in *Russian Literature of Satellites*, pt. 2: *Telemetry*, Progr. Phys. Sci., 1958; F. A. Paneth, *J. Chem. Soc.*, pp. 3651–3662, 1953; T. G. Scholz et al., *J. Geophys. Res.*, 75:3049–3054, 1970; E. A. Wenzel et al., *Ann. Int. Geophys. Year*, 1:407–421, 1960.

Algae

Blue-green algae have interested man not only because of their tolerance to extreme environmental conditions (including a simulated atmosphere of the planet Mars) but also because they can be both beneficial and troublesome. They are beneficial in the sense that many genera have the ability to fix nitrogen and are of widespread occurrence, but they can also be troublesome in the form of heavy water blooms.

Can their nitrogen fixation be improved and can their sometimes undesired development be controlled? These questions might be answered by studying their viruses. An increasing number of blue-green algae have been reported to be sensitive to virus infections. These viruses are called cyanophages because of their great resemblance in morphology and behavior to bacteriophages. Some of the cyanophages, including the one that attacks *Anabaena variabilis*, are similar to coliphages of the T group, which are among the most complicated viruses known. Several cyanophages have already been used in genetic experiments and in the control of algal growth.

Blue-green algae are considered to link a postulated prokaryotic group with the higher plants. Originally, they were classified as algae because of

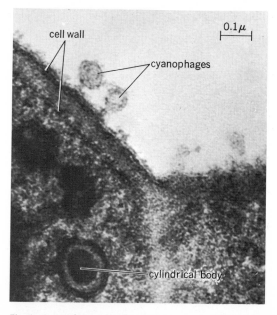

Fig. 1. Longitudinal section of *Anabaena variabilis* showing cyanophages attached by their tails to the outer layer of the cell wall.

their plantlike photosynthesis, but they are better grouped with the bacteria in the light of their typically prokaryotic nature. The prokaryotic cell is characterized by the lack of a nuclear membrane and membrane-bound organelles and by the presence of unique cell-wall constituents.

Properties of viruses. Reported properties of viruses and their host-ranges are given in the table. Three main types of cyanophages have been recognized: polyhedrons with tails, polyhedrons without tails, and infectious raphidosome-like rods. The last type, reported by R. N. Singh and coworkers, is suspected to be an RNA virus, and the others are probably DNA viruses. Most cyanophages are restricted to one algal genus or to closely related genera. An apochlorotic member of the Oscillatoriales, *Spirochaeta rosea* nom. prov., was reported by A. Lewis in 1961 to be lysed by a virus isolated from marine mud. Morphologically, the virus resembles T2 phages.

Viral aggregates have been suspected to be present in *Tolypothrix tenuis*.

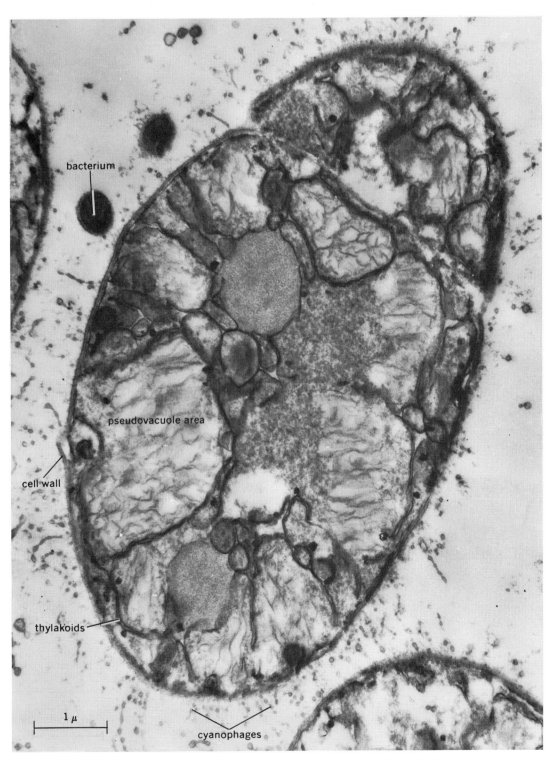

Fig. 2.   Longitudinal section of *Aphanizomenon flos-aquae* showing many cyanophages attacking the cell wall.

Host-ranges and properties of viruses which infect blue-green algae

Algal family	Infected genus	Virus	Properties
Nostocaceae	*Anabaena*	An–1	Polyhedron (40–50 nm) with tail
	Anabaena	A–1	Headless rod; sheath (20 nm), tail (130–200 nm); restricted to *A. cylindrica*
	Anabaenopsis *Raphidiopsis*	AR–1, AR–2	Lysogenic strains occur
	Aphanizomenon	Ap–1	Polyhedron (50–60 nm) with tail
Oscillatoriaceae	*Cylindrospermum* *Lyngbya* *Plectonema* *Phormidium*	C–1 LPP–1, LPP–2, etc.	Polyhedron (56 nm) with tail; unstable in dist. H_2O; lysogenic strains occur
Chroococcaceae	*Microcystis* *Synechococcus*	SM–1	Polyhedron (88 nm) without tail; stable in dist. H_2O

Alga-virus relationship. The most thorough investigations of the morphology and replication cycles of cyanophages have been by K. M. Smith and colleagues and by R. S. Safferman and R. M. Morris on cyanophages of the LPP group (see table). The following description of the alga-virus relationship is based on investigations carried out by three groups (Smith and coworkers; Safferman and Morris; and U. Granhall) with cyanophages like An-1 (which attacks *Anabaena variabilis*). Like bacter-

iophages with tail structures, these cyanophages attach themselves to the outer layer of the cell wall by their tails. Special receptor sites within the cell wall are recognized by the tail structures by means of groups of opposite polarities. Attachment then takes place by the formation of ionic bonds. The virus acquires a characteristic position, perpendicular to the cell wall with the head sticking out (Fig. 1). Infection occurs by penetration of the viral nucleic acid into the cell. This is mediated by a contractile sheath of the tail, the whole virus functioning much like a microsyringe, injecting its nucleic acid into the cell. After injection, the viral nucleic acid seems to move into the nucleoplasm, where it multiplies. Assembly of the viral nucleic acid, presumably deoxyribonucleic acid (DNA), occurs in association with the thylakoids (photosynthetic lamellae). Viral coat proteins are then condensed around the viral DNA, and "virus heads" are formed. Tail structures are added late in the maturation process. The vegetative cyanophage of the virus An-1 shares with some complicated bacteriophages the remarkable feature of breaking down host DNA. It is not unlikely that precursors for virus synthesis are derived by this process. The release of fully matured viruses occurs by lysis, which is probably caused by virus-induced cellular lysozymes. Easily detected metabolic changes during the replication cycle are decreased photosynthesis and nitrogen fixation and increased respiration.

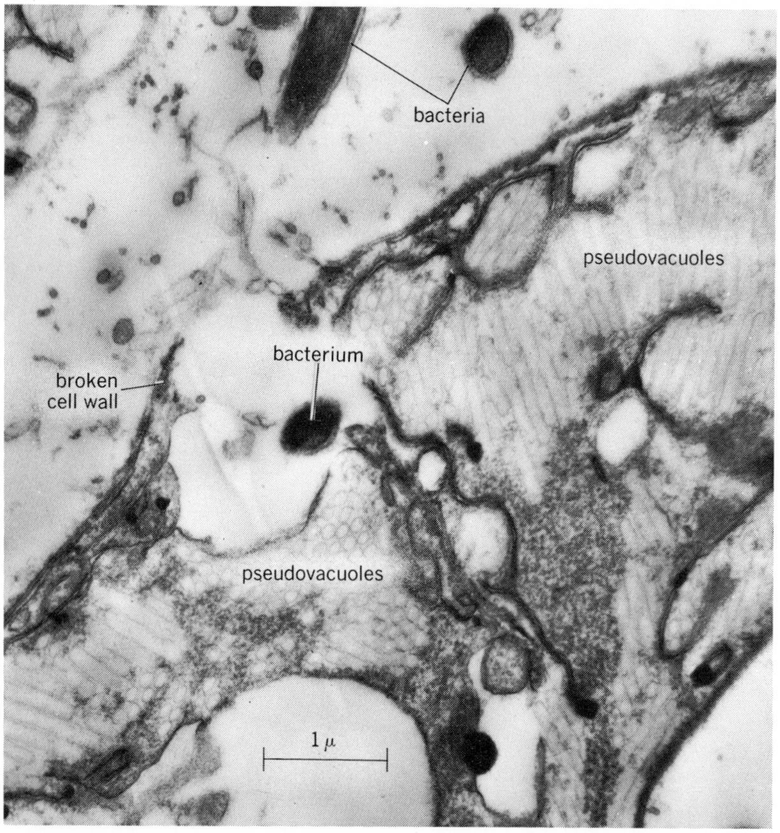

Nitrogen-fixing blue-green algae have heterocysts and spores (akinetes). These cells are not infected by viruses. Judging by Granhall's electron-microscope studies, the viruses probably do not adsorb to these cells. This might be because of the absence of receptor sites in their cell walls, which are chemically different in composition from those of the vegetative cells. The same phenomenon has been noticed with a cyanophage infecting *Aphanizomenon flos-aquae*. This virus, Ap-1, was recently isolated in Granhall's laboratory from the final stage of a water bloom. It was present in such high concentrations in the lake samples that lysis from without probably accounted for part of the total algal degradation (Fig. 2). Intracellularly, it causes

Fig. 3. Longitudinal section of *Aphanizomenon flos-aquae* showing bacteria entering algal cell previously lysed by cyanophages.

breakdown of host DNA, vacuolization, and distortion of thylakoids. The morphology of this virus was similar to that of the An-1 type.

Possible applications of cyanophages. Cyanophages might be used to clarify the genetic regulation of nitrogen fixation. Singh and associates have supposed that there is one regulator gene only for two linked operator genes of nitrogenase and heterocyst formation, respectively. If such functions could be transduced, it would be of great importance, since blue-green algae are widespread as nitrogen fixers, but the mechanism of fixation is still obscure from genetic and enzymologic viewpoints. Genetic experiments have been performed by Singh and associates with the LPP and AR viruses to transduce resistance to antibiotics, to induce lysogenic strains, and in studies of photoreactivation. Plaque mutants, for example, rapid lysis (*r*), have been isolated from several cyanophages, including Granhall's An-1 virus. Most bacterial strains freshly isolated are lysogenic, and this may well be the case also with blue-green algae. Transduction by cyanophages is likely to be the main route of genetic recombination in nature, since both sexduction and conjugation are rare phenomena among blue-green algae.

Apart from the possible use of cyanophages for genetic studies, they could also have application in the control of algal growth. This has been shown by Safferman and Morris in waste-stabilization ponds. The often sudden and inexplicable, rapid decay of water blooms may be caused by viruses, since cyanophages have been easily detected and isolated at this stage. Autolysis and bacterial degradiation seem unlikely. The observed rise in bacterial counts associated with decaying algae seems to occur only after their lysis. In *A. flos-aquae*, bacteria could be seen in the electron microscope entering cells previously lysed by cyanophages (Fig. 3).

If cyanophages active against bloom-forming algae could be used to control algal growth on a larger scale, they certainly would have great practical value.

For background information *see* ALGAE; BACTERIOPHAGE; NITROGEN FIXATION in the McGraw-Hill Encyclopedia of Science and Technology.

[ULF GRANHALL]

Bibliography: U. Granhall and A. V. Hofsten, *Phys. Plant*, vol. 22, 1969; R. S. Safferman and R. M. Morris, *Appl. Microbiol.*, vol. 15, 1967; R. S. Safferman and R. M. Morris, *Virology*, vol. 37, 1969; R. N. Singh et al., *Proc. 1st Int. Symp. Taxon. Biol. Blue-Green Algae*, University of Madras, 1970; K. M. Smith et al., *Virology*, vols. 28 and 30, 1966, vol. 31, 1967.

Animal evolution

Modern studies of animal evolution are breaking down the disciplinary wall that once separated the zoologist from the paleontologist. To discover why animals change over long periods of time, paleontologists are increasingly turning to the study of living animals. It is now not unusual for a paleontologist, having hammered and dug through a section of marine rocks, to step from jeep to boat, from field boots to flippers, and from backpack to scuba tank to dive into the undersea environment for answers to long-persisting questions.

T. R. Waller has published an account of the evolution of the *Argopecten gibbus*, a group of closely related marine bivalved mollusks that includes the common bay and calico scallops living along the Atlantic and Gulf of Mexico coasts of North America and other species presently living in the Caribbean and along the Pacific coasts of both North and South America. Exemplifying a method for obtaining a potentially vast amount of information about the biological history of the oceanic environment, this study begins with a review of the zoology of the living species of the stock, including their functional morphology, habi-

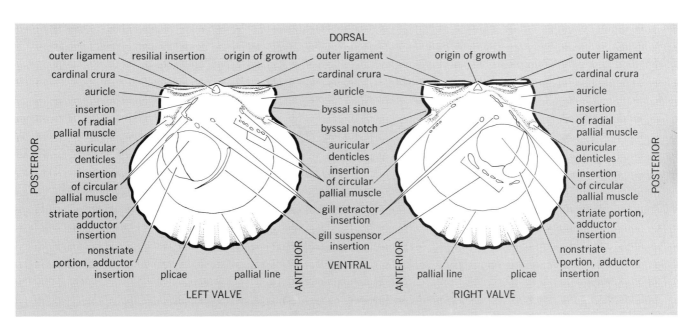

Fig. 1. Internal shell morphology of *Argopecten gibbus*. (From T. R. Waller, The Evolution of the Argopecten gibbus Stock, with Emphasis on the Tertiary and Quaternary Species of Eastern North America, Paleontol. Soc. Mem. no. 3, J. Paleontol., vol. 43, no. 5, suppl., 1969)

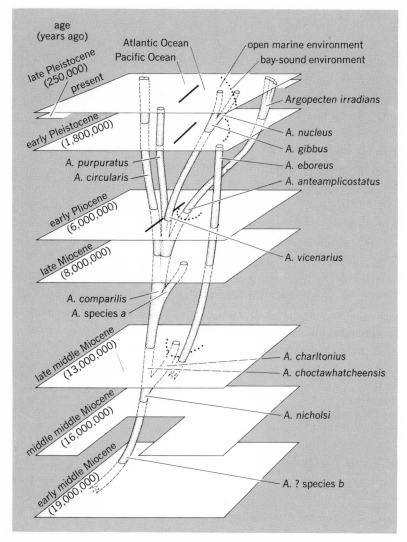

Fig. 2. The Cenozoic evolution of the *Argopecten gibbus* stock. Chronological planes are spaced according to absolute time. The heavy solid line on each of the three top planes represents the geographic barrier separating the Atlantic, Gulf of Mexico, and Caribbean from the Pacific after the Miocene; the heavy dotted lines represent the ecological barrier separating bay-sound environments (upper right) from open-marine environments (lower left) on the Atlantic side of the geographic barrier. (*From T. R. Waller, The Evolution of the Argopecten gibbus Stock, with Emphasis on the Tertiary and Quaternary Species of Eastern North America, Paleontol. Soc. Mem. no. 3, J. Paleontol., vol. 43, no. 5, suppl., 1969*)

Using an electronic digital computer, data were subjected to univariate and bivariate analyses, and samples were compared using machine-plotted bivariate scatter diagrams, lines of best fit, and other graphical techniques. Data from right and left valves were treated separately, except that they were recombined in the study of characters that differ between valves, thereby furnishing new information on intervalve features.

Phylogeny. The postulated phylogeny of the *A. gibbus* stock began in the early middle Miocene of Florida, approximately 18,000,000 years ago (Fig. 2). A poorly known fossil species (*A.? sp. b*) of that age, which is apparently very near the origin of the stock, gave rise through other species shown in Fig. 2 to *A. comparilis* in the late Miocene, about 12,000,000 years ago. *A. comparilis* was broadly adapted and widely distributed, living in bays, sounds, and open marine waters in the western Atlantic, Gulf of Mexico, and Caribbean and extending through seaway passages to the Pacific. By the end of the Miocene on the eastern side of the Americas, about 7,000,000 years ago, the variable and widely distributed *A. comparilis* began to split, giving rise to a primitive bay scallop that, like its living descendant, was restricted to the semienclosed waters of bays and sounds, and to a primitive calico scallop that was restricted to open marine waters. The early bay scallop was apparently unable to reach the Pacific, but the primitive calico scallop seems to have given rise to both of the living Pacific coast species. The living *A. circularis* is morphologically primitive in that it resembles the Miocene *A. comparilis* more than it does any of the later species on the eastern side of the Americas, and it is ecologically primitive in that it is broadly adapted and able to live in both bays and sounds and in open marine waters. *A. eboreus*, a common scallop broadly distributed in temperate waters on the eastern sides of the Americas in the Miocene and Pliocene, represents a highly variable, yet morphologically persistent, lineage that neither split nor gave rise phyletically to other species and that became extinct during the early Pleistocene.

The living species of the *A. gibbus* stock are separated ecologically by both hydrographic and climatic factors. *A. irradians*, the common Atlantic bay scallop, lives only in semienclosed bays, sounds, or estuaries throughout its geographic range, which extends from Massachusetts to around the Florida peninsula and into the Gulf of Mexico as far west as the coast of northern Mexico. Mature bay scallops do not survive under open marine conditions. In contrast, the calico scallop is restricted to the open sea throughout its range, which for mature populations extends from just north of Cape Hatteras, N.C., to the Atlantic side of the Antilles and throughout the Gulf of Mexico. It lives at depths of 5–200 fathoms in warm waters adjacent to the Gulf Stream. Apparently, mature calico scallops are unable to survive in bays and sounds along the present coast of North America. A third species, *A. nucleus*, closely related to the calico scallop, from which it evolved within comparatively recent time, occupies a habitat like that of the Atlantic bay scallop. It lives only in bays and sounds in warm-water areas of southern Florida, the Bahamas, and the Antilles, although it exhibits

tat preferences, behavior (ethology), distribution, and geographic variation. The study forms this information into a present-day model against which paleontological information can be compared; morphological and distributional data from fossil samples from progressively older strata dating back to about 18,000,000 years are analyzed and compared to the zoological model.

Muscle and shell study. In the scallop, numerous muscles leave their marks on the shell (Fig. 1); an understanding of the functions of the muscles that leave these marks is an obvious prerequisite to understanding changing spatial relationships of muscle insertions in fossil shells. In the study of fossil scallops, differences between samples were evaluated by means of morphometric data consisting of 70 measurements and form ratios of the outline, ligamenture, and musculature of each valve.

a trend of living at increasing depths southward in the Caribbean. Along the Caribbean coast of South America, *A. nucleus* lives in open marine waters as does the calico scallop of less tropical regions.

On the Pacific coast the two living members of the stock are the California scallop (*A. circularis*), which, before pollution of inshore waters, extended as far north as Monterey Bay but is now found mainly along the coasts of southernmost California and Mexico as far south as Peru. It lives primarily in bays in the northern part of its range and in both bays and the open sea to depths of 75 fathoms in the southern part. *A. purpuratus*, the fifth living species of the stock, favors the open, cool waters of the Humboldt Current and occurs most abundantly along the coast of Peru.

Evolutionary changes. The picture that emerges is that the evolution of the *A. gibbus* stock has involved both the origin and transformation of species in response to hydrographic and geographic changes. Initially broadly adapted species have become more specialized in that they have become adapted to more restricted environments. Among phylogenetic trends are examples of convergence, which seems to be the result of adaptation to the same environment by species that were previously ecologically separated. There are also extinctions of lineages, one of which terminated a morphologically persistent species that was less variable than species that continued to evolve. Rates of evolutionary change (both the splitting and the gradual change of lineages) have been more rapid on the Atlantic side of the Americas than on the Pacific side since the time when scallop populations in these regions became separated by rising land barriers in the Pliocene, about 4,000,000 years ago. Presumably the causes of regional differences in evolution are the more intensive tectonism and coastline change in the Pacific than in the Atlantic – Gulf of Mexico – Caribbean region. This has resulted in selection for evolutionarily conservative, broadly adapted, and more adaptable species on the Pacific side, and for more specialized, narrowly adapted species, which evolved under more stable conditions, on the Atlantic side.

For background information *see* ANIMAL EVOLUTION; EVOLUTION, ORGANIC; PALEONTOLOGY in the McGraw-Hill Encyclopedia of Science and Technology.   [THOMAS R. WALLER]

Bibliography: T. R. Waller, *The Evolution of the Argopecten gibbus Stock, with Emphasis on the Tertiary and Quaternary Species of Eastern North America*, Paleontol. Soc. Mem. no. 3, *J. Paleontol.*, vol. 43, no. 5, suppl., 1969.

Antibiotic

Rifampicin is a recently developed semisynthetic antibacterial and antiviral drug. It is composed of the rifamycin moiety, a natural fermentation product of a mold, and a hydrazone side chain. Low concentrations of rifampicin are effective in inhibiting the growth of bacterial cells. The antibacterial activity of rifampicin resides in the macrocyclic ring and is enhanced by the hydrazone side chain. Rifampicin molecules bind irreversibly to the DNA-dependent RNA polymerase molecules and inhibit RNA synthesis. Because of this property, rifampicin also inhibits the development of trachoma agent, a microbial obligate parasite, the causative agent of a human eye disease.

Rifampicin also interferes with the replication of viruses. The rifampicin-mediated inhibition of the activity of the DNA-dependent RNA polymerase molecules in bacterial cells results in the prevention of the development of certain bacteriophages. At a high concentration (100 μg/ml), rifampicin also inhibits the replication of poxviruses, such as Shope fibroma virus, in infected mammalian cells. The antipoxvirus activity of rifampicin is due to the hydrazone side chain, which interferes with the formation of virus particles. In addition, rifampicin affects the synthesis of poxvirus and mammalian cell DNA.

Structure. The mold *Streptomyces mediterranei*, isolated in 1957 from the soil in the pine forests in southern France, produces under normal fermentation conditions a series of five different rifamycins, among which rifamycin B is the least toxic. Addition of diethylbarbituric acid to the medium of the fermenting mold results in the production of only rifamycin B. Rifamycin B is easily oxidized to rifamycin O, which undergoes hydrolysis (with the loss of glycolic acid), giving rise to rifamycin S, which has an antimicrobial property and a low toxicity. A reduction product, rifamycin SV, with a formyl group attached to carbon 3 of the chromophoric ring (3-formylrifamycin SV), enabled the synthesis of 42 derivatives with different side chains. The derivative number 31, which has a hydrazone side chain, was named rifampicin (Fig. 1). This compound was selected as an antibacterial drug on the basis of its solubility in water, low toxicity, and antibacterial activity when given orally to bacteria-infected mice. In addition to rifampicin, other hydrazone derivatives of rifamycin have antitrachoma and antipoxvirus activities.

Antibacterial activity. As demonstrated in Fig. 1, the rifampicin molecule is made of three major components: (1) the chromophoric (naphthahydroquinone) ring, which is responsible for the red color of the antibiotic; (2) the macrocyclic ring, which is responsible for the binding of the antibiotic to the bacterial-type DNA-dependent RNA polymerase; and (3) the hydrazone (methylpiperazine) side chain attached to carbon-3 in the chromophoric ring, which enhances the antipolymerase activity of the antibiotic and is responsible for the antipoxvirus activity of rifampicin.

Bacteriophage inhibition. Rifampicin binds irreversibly to the DNA-dependent RNA polymerase in bacterial cells. The antibiotic binds to a specific region in the bacterial RNA polymerase core, which is composed of two α, one β, and one β' subunits (designated $\alpha_2\beta\beta'$). Mutants of bacteria were obtained which were resistant to rifampicin because of a change in the receptor site on the core enzyme. Treatment of the bacterial host (*Bacillus subtilis*) with rifampicin prior to infection with a bacteriophage (SPO1) results in the inhibition of the bacteriophage development, thus indicating that the DNA-dependent RNA polymerase of the bacterial host is essential for the early development of the bacteriophage. It was subsequently demonstrated that the bacteriophage utilizes the host core enzyme for the synthesis of the early viral messenger RNA molecules. Later, the phage is responsible for the synthesis of new protein molecules (a sigma factor) which can attach to

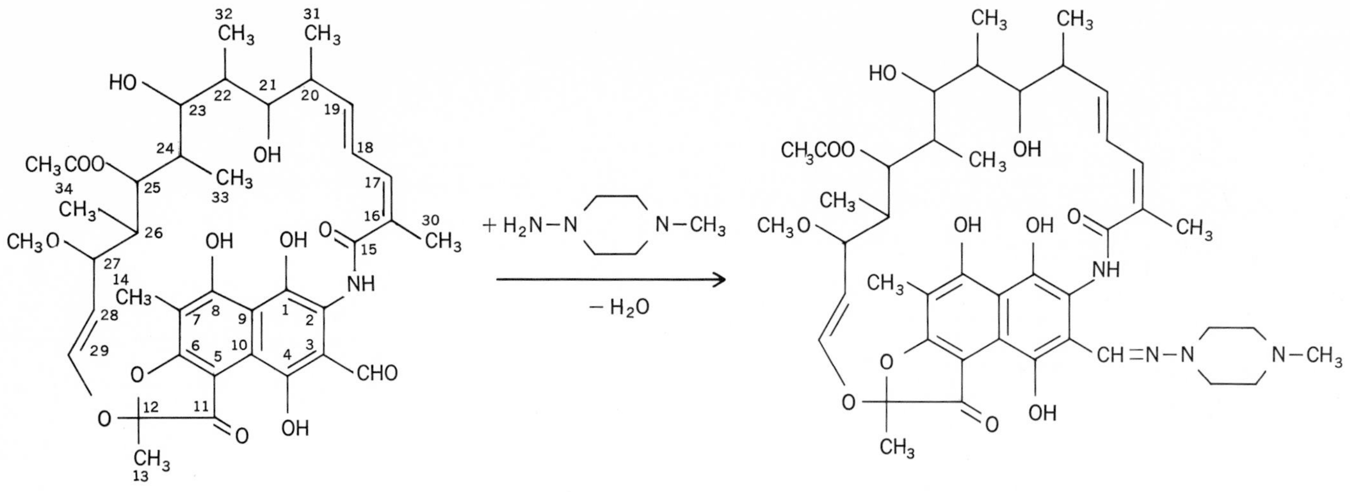

3-formyl rifamycin SV

Rifampicin

Fig. 1. Synthesis of rifampicin. Rifampicin molecule is made of three major components. (*After P. Sensi et al.,* *Antimicrobial Agents and Chemotherapy, 1966, American Society for Microbiology, 1967*)

the core enzyme of the host cell and determine its binding to new sites on the phage DNA. All the steps in the replication of the bacteriophage in which the host cell polymerase participates are sensitive to rifampicin inhibition. *See* BACTERIO-PHAGE.

Inhibition studies. Inhibition studies with rifampicin demonstrated that the antibiotic interferes with the growth of the trachoma agent, RNA synthesis in mitochondria and chloroplasts, replication of poxviruses, and DNA synthesis in cultured mammalian cells.

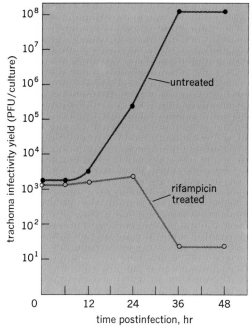

Fig. 2. The effect of rifampicin, added immediately after infection, on the growth cycle of the T'ang strain of trachoma agent in FL cell; PFU, plaque-forming unit, is an infective unit responsible for the formation of one infection center. (*From Y. Becker and Z. Zakay-Rones, Rifampicin: A new antitrachomic drug, Nature, 222:851, 1969*)

Antitrachoma activity. Rifampicin is capable of penetrating mammalial cells and can interfere with the development of a microbial parasite within the host cell cytoplasm. This is shown in Fig. 2 by the effect of rifampicin on the growth cycle of trachoma agent. The infectious entities of trachoma agent are elementary bodies, 0.3 μ in diameter, which contain DNA, RNA, proteins, and a rigid outer wall. These particles enter the mammalian host cells in a manner similar to that of viruses but, contrary to the viruses, they retain their cellular structure and have the ability to synthesize their own proteins by their specific ribosomes. The progeny of infectious elementary bodies is synthesized within the developing inclusion bodies, which are large, membrane-bound structures. Addition of rifampicin to infected cells irreversibly prevented the development of the trachoma elementary bodies in cell cultures, embryonated eggs, and laboratory test animals. Monkeys which developed trachoma disease in the eyes because of experimental infection with a trachoma agent were cured by topical application of rifampicin into the eyes. The presence of DNA-dependent RNA polymerase molecules, both in the infectious elementary bodies and in the developing trachoma inclusion bodies, makes trachoma agent highly sensitive to the antibiotic. The binding of rifampicin to the RNA polymerase molecules of the developing trachoma agent is irreversible, and further development of the agent is prevented even after the removal of the antibiotic.

RNA synthesis. Mitochondria and chloroplasts are organelles which are present in the cytoplasm of nucleated cells. They were found to contain DNA molecules and enzymes capable of RNA synthesis. Addition of low concentrations of rifamycin or rifampicin completely blocked RNA synthesis in both organelles. These experiments suggested that the DNA-dependent RNA polymerase molecules in the two organelles are affected by the antibiotic.

Poxviruses. Rifampicin inhibits the replication of poxviruses in mammalian cells, but members of other virus groups are not affected. The poxvirus

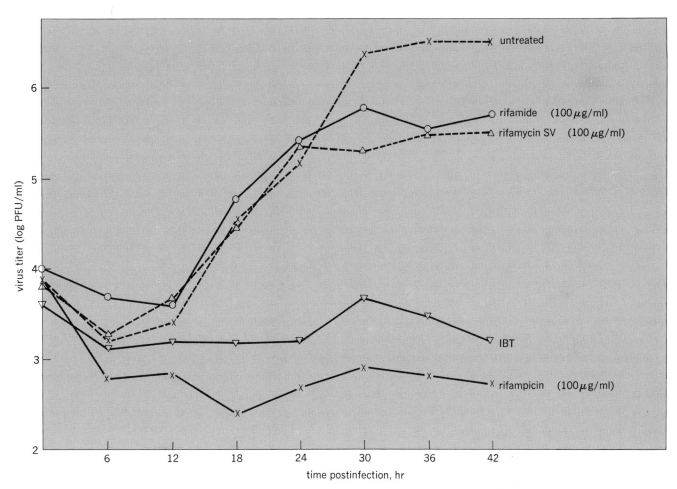

Fig. 3. The effect of rifampicin, rifamycin SV, rifamide, and isatin-β-thiosemicarbazone on the replication of Shope fibroma virus. (*From Z. Zakay-Rones and Y. Becker,* *Rifampicin-antipoxvirus activity due to the hydrazone sidechain, Nature, 226:1162, 1970*)

particle has a DNA molecule with a molecular weight of 150×10^6 daltons which contains genetic information for a large number of proteins. Although the length of the DNA molecule is 75 μ, it is tightly packed within a brick-shaped virion, 250 mμ in length. In order to replicate in the cytoplasm of the infected mammalian host cell, the poxvirions contain DNA-dependent RNA polymerase molecules which are attached to the viral DNA genomes. After entry to the host cell and the removal of the lipid envelope which coats the protein-wrapped viral DNA, the viral RNA polymerase molecules are capable of synthesizing the virus-specific initial messenger RNA molecules. These molecules attach to ribosomes and are responsible for the synthesis of the viral proteins necessary for the synthesis of viral enzymes and the replication of the viral DNA genomes and some virus structural proteins. Later, additional messenger RNA molecules are synthesized by the viral enzymes which enable the synthesis of more structural proteins necessary for the coating of the synthesized DNA and the formation of the poxvirions.

Addition of rifampicin at high concentrations (80–100 μg/ml) to poxvirus-infected mammalian cells immediately after infection resulted in the complete inhibition of poxvirus replication (Fig. 3). Studies on the antipoxvirus drug action demonstrated that (1) although poxvirions carry DNA-dependent RNA polymerase molecules, the viral enzyme was highly resistant to the antibiotic, and resembled in this respect the nuclear DNA-dependent RNA polymerase molecules of the mammalian host cells; (2) with the aid of different derivatives of rifamycin, which contain modified hydrazone side chains attached to carbon 3 of the chromophoric ring, it was found that the hydrazone side chain, not the rifamycin molecule, was essential for the inhibition of poxvirus replication; (3) rifampicin did not affect the synthesis of virus-specific messenger RNA or its binding to cellular ribosomes or the synthesis of early viral proteins; (4) rifampicin affected the rate of viral DNA synthesis in the cytoplasm of the host cells and reduced it to about two-thirds of the regular level; (5) although the early viral processes were not affected by rifampicin, the translation of late viral messenger RNA was affected in the presence of rifampicin (100 μg/ml); (6) the synthesis of the virus-specific DNA-dependent RNA polymerase late in the virus growth cycle was eliminated in the presence of the antibiotic; and (7) rifampicin seems also to affect the formation of the viral envelopes and therefore prevents the assembly of the poxvirions.

The inhibition of poxvirus replication by rifampicin takes place only in the presence of large amounts of the antibiotic. Removal of rifampicin from the treated infected cells results in the synthesis of infectious virus progeny, indicating that the antipoxvirus inhibitory effect of rifampicin is reversible. It was also noted that mutants of vaccinia virus (a poxvirus used for vaccination against smallpox) were obtained and that their replication was not affected by rifampicin.

Cultured mammalian cells. Rifampicin interferes with DNA synthesis in synchronized mammalian cells in cell cultures. The treated cells synthesize DNA but at a lower rate than that of the untreated controls.

The ability of rifampicin to interfere with various functions of mammalian cells permits it to interfere with the transformation of mammalian cells by the RNA virus Rous sarcoma without affecting the replication of the virus in the antibiotic-treated host cells.

Conclusions. The various studies demonstrated that, because of its complex structure, rifampicin has several different activities: (1) Because of the macrocyclic ring, rifampicin binds to the bacterial DNA-dependent RNA polymerase and inhibits the growth of microorganisms and the replication of bacteriophages in them. (2) Because of the hydrazone side chain, the antibiotic is also an antipoxvirus drug. (3) Rifampicin interferes with processes in the mammalian cells, affects DNA synthesis, and prevents cell growth.

For background information *see* ANTIBIOTIC; DEOXYRIBONUCLEIC ACID (DNA); MITOCHONDRIA; VIRUS in the McGraw-Hill Encyclopedia of Science and Technology. [YECHIEL BECKER]

Bibliography: Y. Becker and Z. Zakay-Rones, *Nature*, 222:851, 1969; Z. Ben-Ishai et al., *Nature*, 224:29, 1969; R. D. Brown, D. Bastia, and R. Haselkorn, *Proceedings of the 1st International Lepetit Colloquium*, 1970; H. Diggelmann and C. Weissmann, *Nature*, 224:1277, 1969; E. P. Geiduschek and J. Sklar, *Nature*, 221:833, 1969; B. McAuslan, *Biochem. Biophys. Res. Commun.*, 37:289, 1969; B. Moss et al., *Nature*, 224:1280, 1969; P. Sensi et al., *Antimicrobial Agents and Chemotherapy, 1966*, American Society for Microbiology, 1967; Z. G. Shmerling, *Biochem. Biophys. Res. Commun.*, 37: 965, 1969; J. H. Subak-Sharpe, M. C. Timbury, and J. F. Williams, *Nature*, 222:341, 1969; W. Wehrli and M. Satehelin, *Biochim. Biophys. Acta*, 182:24, 1969; Z. Zakay-Rones and Y. Becker, *Nature*, 226:1162, 1970.

Antibody

Recent studies have demonstrated that it is possible to initiate antibody formation prematurely in newborn animals and thus permit them to respond to certain foreign materials (antigens) well in advance of the age at which such responses normally occur. This finding has aided in the understanding of the steps that occur normally in mature animals between the time that such animals, that is, higher vertebrates, are exposed to an immunizing antigen and the time that antibodies capable of reacting specifically with certain portions of the antigenic molecule are formed in detectable amounts.

Steps in antibody formation. It is now established that the events that lead to antibody formation in mature animals are complex and involve cooperative interactions among several cell types, all of which are derived from the bone marrow but some of which are further modified by the thymus. This participation of at least two major groups of cells, bone-marrow-derived and thymus-modified, is well documented by experimentation in which animals either have been depleted surgically of their thymus cells or have been deprived of functional bone-marrow cells by irradiation. Subsequent repopulation of such animals with bone-marrow or thymus cells, or both, has clearly indicated the critical and cooperative role of both these cell types in antibody formation. The specific morphological and functional features of the bone-marrow-derived and thymus-modified cells that participate in immune responses are, however, not yet fully identified, and a different line of experimentation has suggested that actually three different cell types may be involved in antibody formation.

Available evidence indicates that the initial meaningful encounter between a foreign substance, that is, an immunogenic antigen, and cells of the host involves primary antigen-handling (PAH) cells that are present in so-called macrophage populations. These cells concentrate, and possibly even modify, antigens and then present the antigen to the precursors of antibody-forming cells which are, like the antibody-forming cells to which they give rise, members of the lymphocyte population. The figure shows two possible alternative events that, on the basis of current thinking, may follow the activation of lymphocytic stem cells: (1) The activated precursor cells may start to multiply, giving rise to long-lived "memory cells," which, in the presence of an appropriate amount of antigen, differentiate into antibody-forming cells; or (2) the precursor cells (thymus-modified cells) may start to multiply and, in the presence of antigen, interact with, and specifically activate, still other cells (bone-marrow-derived) that are the actual producers of antibodies (that is, of specific proteins or, more precisely, specific globulins). Which of the many preexisting lymphocytic stem cells (cells at encircled numeral 2 in the figure) will be activated to yield, by division, micropopulations (clones) of potentially antibody-forming cells seems to depend on the antigen involved.

The stem cells (often referred to as antigen-reactive or antigen-sensitive cells) appear to be predisposed for the formation of one, or a few, types of antibodies; that is, antigen A activates stem cell B, and so on. The activation itself appears to take place by an interaction between an antigen and an antibodylike receptor site on the cell membrane. To account for the occurrence of specific nonresponsiveness to a given antigen, it has been proposed that the antigen must be processed by the PAH cells so that a complex of antigen and nonspecific activator will result. The antigenic portion of this complex may then serve to guide the activator to the appropriate lymphocytic stem cell. Therefore, when the antigen bypasses the PAH cells, as it apparently does under certain conditions, no activator-containing complex is formed and thus no activation of lymphocytes occurs. Instead, activator-devoid antigen reacts with the receptor on lymphocytes, and although it activates nothing, this event blocks the subsequent availability of the receptor site. This scheme, as well as several other possible schemes, can explain why it is possible, by blocking the participation of PAH cells at the time of immunization, to induce a

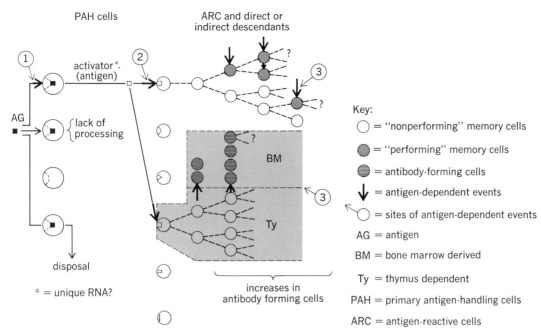

Key:

◯ = "nonperforming" memory cells

◉ = "performing" memory cells

⬭ = antibody-forming cells

↓ = antigen-dependent events

↰◯ = sites of antigen-dependent events

AG = antigen

BM = bone marrow derived

Ty = thymus dependent

PAH = primary antigen-handling cells

ARC = antigen-reactive cells

An outline of probable steps in antibody formation. The shaded and nonshaded portions on the right side represent alternative possibilities. The various marginal indentations in the circles represent membrane recognition sites for antigen, which have been demonstrated to exist on lymphocytes; the existence of such sites on macrophages is still hypothetical.

specific and fairly lasting immunological nonresponsiveness in adult animals, that is, a nonresponsiveness to the antigen presented at the time of impaired functioning of PAH cells.

Tolerance and premature initiation. A similar, naturally occurring process may be responsible for the fact that, as a rule, animals and man do not form antibodies to their own antigens, that is, to substances, principally cell components, that would cause antibody formation in unrelated members of the same species. The prediction has been made that the normal process of development of nonresponsiveness (tolerance) to self-antigens might be associated with an early maturation of lymphocytic cells and a late maturation of PAH cells. Such a situation could assure that, early in life, self-antigens would induce nonresponsiveness because they would react directly, and devoid of activator, with available lymphocytic stem cells and eliminate them, by receptor clogging, from future participation in immune responses.

It is known that the capacity to form antibodies to different types of foreign materials develops for different antigens at different times during the development of an individual, during a period extending from the last trimester of fetal life to shortly after birth. Newborn mice will not make antibodies to sheep red blood cells until they are 10–12 days old. This situation has provided an opportunity to test the idea that the normal process of development of immune responses may involve a late maturation of the function of appropriate PAH cells because it permitted the testing of whether such normally nonresponsive animals would become responsive if supplied with mature PAH cells from adult animals. A positive result would indicate that all that was lacking was appropriate PAH cells, the other cells required for antibody formation being present and functional.

Indeed W. Braun and L. Lasky reported in 1967 that the transfer of mature macrophages from adult animals of the same mouse strain into 3- or 4-day-old mice resulted in a premature initiation of antibody formation to sheep red blood cells, and this finding was confirmed and extended by B. F. Argyris. There were comparable results from similar tests with several other antigens to which newborn mice normally fail to respond. Also, R. Winchurch and Braun found that agents known to stimulate macrophage functions, such as bacterial lipopolysaccharides and synthetic polynucleotides, will induce premature antibody responses in newborn mice when they are given at the time of antigen administration. The polynucleotides capable of supporting this effect are known stimulators of immune responses in adult animals and include a double-stranded complex of polyadenylic and polyuridylic acid (poly A:U) and a similar complex of polyinosinic and polycytidylic acid (poly I:C). The results achieved with these stimulators of macrophages in inducing premature responses in newborn animals have been interpreted as reflecting a "forced maturation" of PAH cells by agents that enhance the functions of such cells.

Consequences. It is well established that exposure of newborn animals to an antigen can cause a lasting specific nonresponsiveness to that antigen. In contrast, the premature initiation of antibody formation to an antigen, by administering it to newborns in conjunction with a stimulator of macrophage functions, seems to lead to an unusually high response to that antigen in adult animals. Thus events shortly after birth may have a striking influence on the immunological reactivity of the animal later in life. *See* IMMUNOLOGY.

The phenomenon of premature initiation of antibody formation may have a number of practical implications. First of all, it suggests that natural exposure of newborns to gram-negative bacteria (such as *Escherichia coli* and *Salmonella*), that is, to agents that contain the stimulatory lipopolysaccharides, may make the newborn immunologically reactive to stimuli to which he is usually unresponsive, for example, to foreign materials in transfused blood. More important, exposure of the newborn to the stimulation that can be provided by gram-negative bacteria may render the individual subsequently hyperreactive to antigens that were simultaneously present in the environment of the newborn. *See* BLOOD.

Another facet of such considerations could be a beneficial one: It is suspected that the poor immunological responsiveness of the adult to tumor viruses may be due to an immunological tolerance that results as a consequence of the presence of such viruses in the newborn. The same situation may apply to tumor cell antigens. Since the exposure of newborns to an antigen in the presence of stimulating polynucleotides or lipopolysaccharides results in what appears to be the opposite of tolerance, the premeditated exposure of newborns to tumor viruses in the presence of stimulators of PAH cells may render the adult immunologically more responsive to such viruses and may thus lead to a suppression of tumor formation. This possibility is now being explored in experiments with C_3H mice that spontaneously develop virus-dependent mammary tumors when they are more than 12 months old.

Thus studies on the premature initiation of antibody formation in newborn animals have confirmed the critical role of members of the macrophage population in the initiation of antibody formation and have indicated the importance of early environmental conditions in life for the subsequent immunological reactivity of the individual.

For background information *see* ANTIBODY; ANTIGEN; ANTIGEN-ANTIBODY REACTION in the McGraw-Hill Encyclopedia of Science and Technology.   [W. BRAUN]

Bibliography: B. F. Argyris, *J. Exp. Med.*, 128: 459, 1968; M. Landy and W. Braun (eds.), *Immunological Tolerance*, 1969; J. Sterzl and A. M. Silverstein, *Advan. Immunol.*, 6:337, 1967; R. Winchurch and W. Braun, *Nature*, 223:337, 1967.

Arctic biology

Life of the arctic region, with its small land areas surrounding the frozen arctic seas, has been exploited from the south for its yield of oil and furs of animals. Until the recent prospect of arctic petroleum and minerals, only traders and explorers from temperate lands came as transients among the scant population of indigenous arctic people. Terence Armstrong described how the recent development of arctic natural resources has multiplied invasion of the Asian Arctic by strangers from the south. Many new residents are now moving northward to exploit American arctic petroleum and minerals. This new and still unsettled movement of people into the Arctic encounters life that is strange to them in the long cold and darkness of arctic winters. Interesting biological, social, and economic conditions, as well as problems of adjustment to the Arctic, face the newcomers.

Life on land. Only about 20 species of mammals of the more than 3000 species in the world live on the treeless arctic tundra, where production of plants is so sparse and specialized that few kinds of animals can find a living. The only large arctic herbivores are musk-ox and caribou or reindeer, but many small herbivorous mice and lemmings live in obscure ways under winter snow. These small herbivores sustain carnivorous bears, wolves, foxes, weasels, and a surprising variety of small shrews.

The sparse vegetation on arctic lands must accomplish the annual production on which all arctic life depends during a short cool summer in which periodic freezing requires that all plants must retain resistance to frost and adjust reproduction opportunistically to brief spells of sufficient warmth. Spiders, insects, and many cold-blooded animals meet requirements of the Arctic by devices that could not be projected by imagination based on southern experience. Their success demonstrates the surprising adaptability of some species. Their few kinds show the rigorous selection by the arctic seasonal regimes.

No vertebrate animal seems to endure freezing, as do so many plants and invertebrates. Arctic flowers protruding through snow in early spring freeze to brittle hardness in repeated episodes of cold with only brief interruptions of the progress of forming seeds. John Baust finds that a small boreal beetle that winters in stumps above snow contains over 20% of glycerol in its fluids. The attractive view that the substance serves as an antifreeze is spoiled by the observed freezing of the beetle in arctic temperatures. In warm summer the beetle contains no glycerol, and is killed at the temperature at which water freezes. It is possible that glycerol protects against the destructive consequences of freezing as glycerol and some other cryoprotective agents allow long frozen storage of bull's sperm for artificial insemination and tissues like cornea for surgical repair of damaged eyes. Natural protection from injury by freezing occurs by diverse ways in the Arctic. Results of research into natural adaptation to arctic temperatures may lead to cryoprotective ways for preservation of cells and tissues.

Life on seas. Peter Freuchen and Finn Salomonsen narrated how the arctic seas support the wandering polar bear and the small arctic fox that often accompanies the bear far out on winter ice-covered seas. These ice-going mammals nevertheless return to shore to breed. The main food of the bear is the small arctic ringed seal, but it also takes eagerly to carrion remains of sea mammals. On sea ice the arctic foxes take remnants of the prey of bears or any flesh. On land in summer their prey is mice and birds.

The large bowhead whale drew whalers into the margins of Atlantic and Pacific ice with such success that the Atlantic bowhead is very nearly extinct. Fur seals and sea lions do not enter arctic life, but large numbers of hair seals breed at the margins of arctic ice. In fact, about 90% of the world's hair seals bear their pups on antarctic or arctic ice, attesting the productivity of polar seas and the capability of warm-blooded seals for living in icy waters.

Preservation of warmth. The preservation of warmth in arctic winter is a matter of real concern. Early explorers recorded that some arctic birds and mammals were as warm as those of warmer regions. L. Irving and J. Krog found that arctic and tropical birds and mammals are similarly warm in their interiors. Exposed experimentally to a temperature of −40° an arctic fox is comfortable in its thick fur. It generates no more heat in cold and so conserves heat by its insulation. This model of adaptation to arctic cold through insulation indicates an economy of heat that sustains a large arctic mammal at an expenditure of metabolism no greater than that required for mammals in a milder climate. The large arctic mammal is thus not handicapped by cold.

Arctic people. Arctic man, with meager natural insulation, wore fur clothing skillfully made from arctic animals, built shelters, and utilized fire. From Greenland across North America to the edge of Siberia, Eskimos lived with a similar language and a culture distinct from those of northern American Indians. Across arctic Eurasia, people of different cultures and languages reflected the fact that they were derived from peoples of steppes and forests who had migrated northward along great rivers and coastal routes. Indigenous Eurasian people are now far exceeded in numbers by fresh migrants who perform technical tasks for exploitation of minerals. Migration of strangers into arctic America seems to have just begun.

Migration of life into Arctic. Throughout the last Pleistocene glaciation, arctic plants and animals were confined to limited refuge areas free from ice. David Hopkins showed that 10,000 years ago melting of the ice sheets allowed spread of plants and animals from arctic refuges and invasion from the south by plants and animals that had retained versatility to adapt for arctic life. The distribution and many forms of life are modern characteristics of the recently changing Arctic.

Spectacular in the arctic spring are migrations of birds coming to nest for a few summer months on arctic lands and coasts after wintering on temperate lands and seas and even over South America and Africa. These great migrations visibly demonstrate the inclinations of birds to occupy opportunistically suitable parts of the world and their capability for organized flights and long navigation. In the seas fishes, whales, and seals annually migrate northward to breed and harvest the production of the arctic year. On land a residue of American caribou still migrate in summer into the Arctic. As arctic migrants return south for winter, they export from the Arctic the annual increment in their populations, an exploitation of arctic production.

It is an interesting speculation that these visible annual migrations in some ways recapitulate the postglacial resettlement of the Arctic. Plant, animal, and human life seem to have been ever pressing their capability for life in the Arctic and developing adaptability for the changing conditions of the arctic seasons.

For background information *see* BIOCLIMATOLOGY; BIOGEOGRAPHY; VEGETATION ZONES, WORLD in the McGraw-Hill Encyclopedia of Science and Technology.

[LAURENCE IRVING]

Bibliography: T. E. Armstrong, *Russian Settlement in the North*, 1965; J. Baust, Seasonal variations in the glycerol content and its influence on cold hardiness in the Alaskan carabid beetle, *Pterostichus brevicornis, J. Insect Physiol.*, in press; P. Freuchen and F. Salomonsen, *The Arctic Year*, 1958; D. M. Hopkins (ed.), *The Bering Land Bridge*, 1967; L. Irving, Adaptations to cold, *Sci. Amer.*, 214:94-101, 1966.

Astronomical spectroscopy

One recent development in astronomical spectroscopy has been the use of the Celescope Experiment of the Smithsonian Astrophysical Observatory to conduct a survey of ultraviolet radiations from space. This survey has been devoted primarily to observations of the ultraviolet brightnesses of about 25,000 stars. The Smithsonian observations, however, include several thousand measurements of the brightness of Lyman-alpha radiation from the Earth's outer atmosphere (geocorona), as well as a few special observations of the Moon and of the comet Tago-Sato-Kosaka. Celescope was designed to measure faint objects such as stars; it cannot be used to look directly at either the Sun or the Earth, both of which are extremely bright.

The word experiment, when used in the term Celescope Experiment, has the very specialized meaning of "scientific instrument mounted in an orbiting spacecraft." Celescope is really more of an astronomical observatory than an experiment in the ordinary sense of the word. It is mounted in NASA's second Orbiting Astronomical Observatory. The Celescope is an astronomical instrument consisting of four telescopes, each equipped with a television camera. These cameras are sensitive to four ultraviolet spectral regions, allowing pictures of selected regions of the sky to be taken and thus the brightnesses of the stars and other objects at wavelengths too short to penetrate the Earth's atmosphere to be determined. The size of the field of view is 2° square. The four regions of spectral sensitivity are 2100−3200, 1550−3200, 1350−2150, and 1050−2150 A.

The Orbiting Astronomical Observatory, containing both the Celescope Experiment and a set of seven ultraviolet telescopes from the University of Wisconsin, was launched on Dec. 7, 1968, from Cape Kennedy. During the 16 months that the Smithsonian astronomers operated this instrument, it produced more than 8000 ultraviolet television pictures covering about 10% of the sky. It was finally turned off in April, 1970, when the sensitivities of the television cameras dropped to the point at which they were recording only four or five stars in each picture. The Orbiting Astronomical Observatory continues to operate, now devoting full time to the experiment of the University of Wisconsin, which measures spectra and brightnesses of individual stars. *See* OBSERVATORY, ASTRONOMICAL.

Television pictures. For a typical picture, three of the spectral bands show images of several stars. An average picture contains 10 stars, but some have as many as 100, and a few contain no stars at all. Most of those stars observed are hot stars, with effective temperatures between 10,000 and 50,000°K. For these stars, most of the energy is emitted in the ultraviolet part of the spectrum. Observations at wavelengths shorter than 3000 A are necessary to give a clear picture of the physical conditions in the atmospheres of these stars.

Figures 1 and 2 are typical examples of Celescope television pictures. Figure 1 contains pictures of a star field in the southern constellation Vela, taken on Jan. 5, 1969. This star field, near the

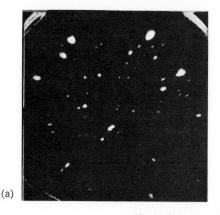 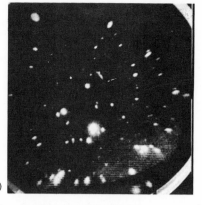 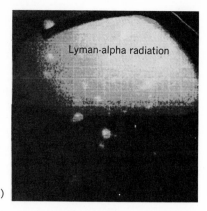

(a)   (b)   (c)

Fig. 1. Celescope television pictures of a star field in the southern constellation Vela on Jan. 5, 1969. (a) Picture taken with camera 1 in wavelength bands 2200–3000 and 1600–3000 A. (b) Picture taken with camera 3 in wavelengths as in a. (c) Picture taken with camera 4 in shortwave bands 1350–2000 and 1100–2000 A.

plane of the Milky Way, contains large numbers of hot stars of spectral types O, B, and A. Figure 1a and b, taken with cameras 1 and 3, respectively, show this star field in two long-wavelength bands (2200–3000 and 1600–3000 A). These pictures contain information on the ultraviolet brightnesses of stars as faint as the tenth magnitude. Figure 1c, taken with camera 4, shows this same star field in two short-wavelength bands (1350–2000 and 1100–2000 A). The bright glow in the upper half of Fig. 1c is Lyman-alpha radiation from hydrogen in the geocorona, discussed more fully below. Only the hottest stars (of spectral types O and B) radiate strongly enough at these short wavelengths to be detected by Celescope.

Figure 2 contains pictures of the Pleiades star cluster, taken on Jan. 18, 1969, with the long-wavelength cameras, 1 and 3. The brighter stars are of spectral type B; the fainter stars, spectral type A.

The television signals from which these pictures were created have also been recorded on magnetic tape for processing by a digital computer. This processing is a complicated procedure in which each star must be located and identified and its ultraviolet brightness computed. This data reduction is expected to continue for another year, after which the Smithsonian astronomers will be able to evaluate more fully the implications in regard to astrophysical theory. However, the data thus far analyzed give strong confirmation of the most re-

cent theories of stellar atmospheres and of the interstellar medium. With very few exceptions, the objects that the Smithsonian astronomers have examined from the Celescope pictures are normal stars, emitting the amount of ultraviolet light expected on the basis of the latest theories of stellar atmospheres and of interstellar absorption.

Lyman-alpha radiation. The Lyman-alpha observations shed additional light on the distribution of hydrogen in the geocorona, or Earth's outer atmosphere. Lyman alpha is the resonance line of the hydrogen atom, at 1216 A. The amount of such radiation received by the Celescope Experiment depends almost entirely upon the position of the satellite within the Earth's shadow. Lyman alpha, a very bright line in the ultraviolet solar spectrum, is scattered by hydrogen in the geocorona so effectively that even when the satellite is in the center of the Earth's shadow, this radiation is strong enough to obliterate the images of most of the stars falling in that half of the camera 2 and camera 4 pictures that is sensitive to that wavelength. When the satellite is outside the Earth's shadow and receiving direct sunlight, Lyman alpha is too strong to allow safe operation of the cameras. Just inside the Earth's shadow, it is about three times as intense as it is at the center of the Earth's shadow. The altitude of the satellite is 800 km; this radiation comes from hydrogen above that altitude. The amount of Lyman-alpha radiation observed is consistent with current theories of the structure and composition of the Earth's outer atmosphere.

The Celescope project has been supported in part by contract NAS 5-1535 from the National Aeronautics and Space Administration.

For background information *see* ASTRONOMICAL PHOTOGRAPHY; ASTRONOMICAL SPECTROSCOPY; SUN in the McGraw-Hill Encyclopedia of Science and Technology.   [ROBERT J. DAVIS]

Bibliography: R. J. Davis et al., *Smithson. Astrophys. Observ. Spec. Rep. No. 310,* in press; R. R. Meier, *J. Geophys. Res.,* vol. 74, 1969.

Atmosphere

President Nixon's signing of the National Environmental Policy Act in January, 1970, created a firm footing for the fight against pollution. He stated that "the 1970s absolutely must be the years when America pays its debt to the past by reclaiming the

Fig. 2. Celescope television pictures of Pleiades star cluster taken on Jan. 18, 1969, with (a) camera 1 and (b) camera 2.

purity of its air, its waters, and our living environment. It is literally now or never." Man has finally realized the necessity of curbing the pollution of the atmosphere and water for self-preservation.

A fundamental understanding of atmospheric behavior is important in the fight against air pollution. Remote sensing of atmospheric phenomena is a relatively new technique used in investigating tropospheric pollution and probing the stratosphere and mesosphere. Lidar (light detecting and ranging) is an instrument that can make such measurements.

Lidar soundings of atmospheric phenomena were the central theme of two interesting national conferences held at the National Center for Atmospheric Research, Boulder, Colo., in April, 1968, and at the Brookhaven National Laboratory in April, 1969. The capability and versatility of lidar research were well established at these meetings, as was its potential use in future scientific endeavors.

Essentially, lidar is an instrument analogous to radar in that it is composed of a transmitter which emits energy to space and a receiver which detects that portion backscattered by obstacles in its path. Recent applications in the lower atmosphere in-

clude cloud height measurements, haze layer and visibility determinations, dimensions of plumes and clouds of particulates, water vapor profiles, accurate distance measurements, clear-air turbulence, and other aeronautical problems. Stratospheric probing provides new information on aerosol content, atmospheric density and its variations, the twilight phenomenon, and composition of stratospheric matter.

Cloud heights. A relatively simple application of lidar is the determination of cloud heights. A laser cloud ceilometer is commercially available for automatic field operation. Figure 1 shows the cloud structure as recorded by a cloud ceilometer manufactured by the Swedish firm ASEA. Although not widely used, it is perhaps the most accurate method of determining cloud heights remotely from the ground both by day and night.

Haze and visibility. The Stanford Research Institute, Menlo Park, Calif., continues to be one of the leading lidar investigation groups in the world. It recently participated in the Barbados Oceanographic and Meteorological Experiment (BOMEX). The lidar was mounted in an aircraft which was flown over the experimental area near the island of Barbados during the third period of the investiga-

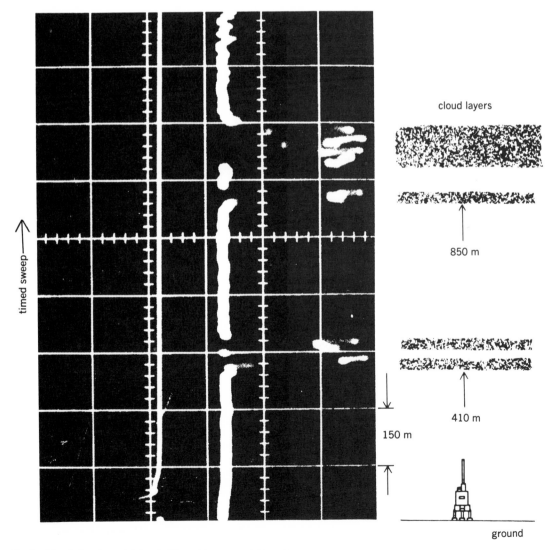

Fig. 1. Cloud structure obtained with oscilloscope and camera.

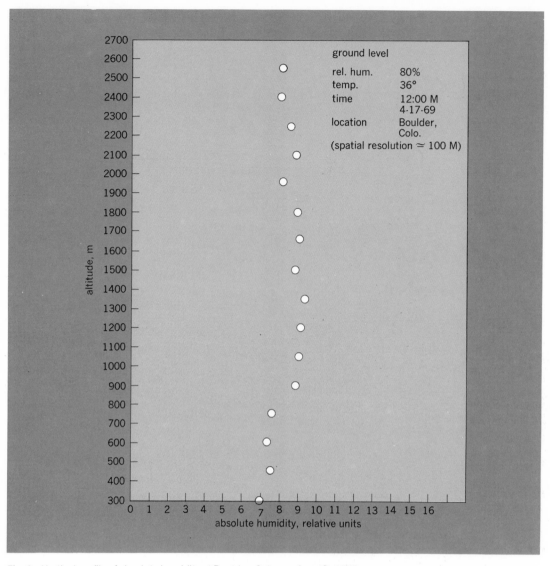

Fig. 2.   Vertical profile of absolute humidity at Boulder, Colo., on Apr. 17, 1969.

tion. The lidar used a neodymium laser (1.6 μ wavelength) having a firing rate of one pulse every 3.5 sec. A total of 5192 lidar soundings was collected during the flight missions, and analysis of the data clearly revealed the existence of well-defined haze layers associated with a subcloud layer and the trade-wind temperature inversion. The data were of sufficient spatial extent to allow presentation in terms of vertical cross sections of optical density or visibility.

Stack plumes. Models describing the diffusing power of the atmosphere were developed before the advent of the tall stacks used by most large industrial plants today. Such tall stacks are increasing in number in order to lower ground-level pollution concentrations. The accuracy of theoretical predictions based on the models has not been adequately verified, and modifications are in order to facilitate advances in the theory of tall stack diffusion.

P. Hamilton of the Central Heating Research Laboratories, Leatherhead, Surrey, England, has been observing tall industrial stack plumes using a ground-based pulsed ruby laser for several years.

These large plumes can be detected several kilometers from their points of origin as well as several kilometers from the lidar site. A unique recording technique provides a brightness-modulated display on an oscilloscope for time averaging of plume parameters. The measurements have provided new data to test the adequacy of recent models concerned with predicting ground-level concentrations from emissions by tall stacks.

The National Air Pollution Control Administration has been conducting a Large Power Plant Effluent Study (LAPPES) at the coal-burning Keystone Generating Station near Shelocta in western Pennsylvania. Stanford Research Institute researchers have utilized a truck-mounted lidar system during some of these experiments to examine the plume's behavior under different atmospheric turbulent conditions. The data they collect are presented to show the spatial and temporal variations in plume geometry and relative particulate concentration distributions. Modification of their lidar will permit immediate inspection by means of a video recording system display in future experiments at Keystone.

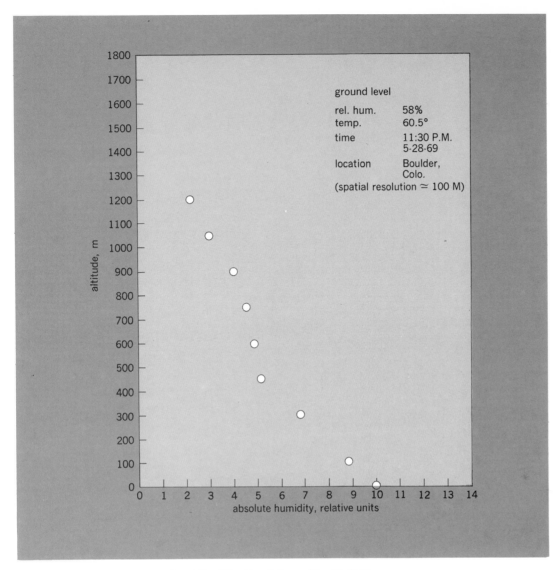

ground level

rel. hum.	58%
temp.	60.5°
time	11:30 P.M. 5-28-69
location	Boulder, Colo.

(spatial resolution ≃ 100 M)

Fig. 3. Vertical profile of absolute humidity at Boulder, Colo., on May 28, 1969.

Water vapor profiles. Remote measurements of atmospheric water vapor profiles utilize the fact that the backscattered radiation is frequency-shifted from the incident radiation by an amount corresponding to energy-level differences in the scattering molecule. This is called Raman scattering and can be used to identify one molecule from another by the amount of frequency shift. J. Cooney, of the National Center for Atmospheric Research, has been making such measurements from ground level to altitudes of 2.5 km. His receiver system is different from others in that it is set up to monitor four returning wavelengths. Two are used as calibration checks, and the others are used to determine water vapor profiles. Figures 2 and 3 show two typical profiles. Researchers R. M. Schotland and J. T. Bradley, of New York University, use a spectrometer in combination with a lidar to determine water vapor profiles in a manner somewhat similar to that used by Cooney.

These profile measurements can be made in a matter of microseconds, as compared to 90–120 min using conventional radiosondes. Another ad-

vantage is that the profiles can be made at whatever elevation angle is desired. A 90° vertical profile is rarely acquired with a balloon-borne unit.

Detecting gaseous pollutants. As mentioned above, the Raman process can be used to identify the scattering species. This fact is important in developing lidar systems to detect gaseous pollutants. S. Zaromb, of Zaromb Research Foundation, has developed a system which is capable of detecting SO_2, O_3, and NO_2 in commonly encountered plume concentrations from a distance of 3 km. D. Leonard, of Avco-Everett Research Laboratories, heads a team that uses a pulsed nitrogen laser to obtain Raman cross sections to identify some of the nitrogen and sulfur oxides found in polluted atmospheres.

It is anticipated that present work on tunable lasers will provide future researchers with the ability to transmit the variety of wavelengths necessary for pollution detection and eventual control.

Distance measurements. Distance measurements are a built-in feature of lidar systems. The transmitted pulse length determines the distance resolution of the particular system, which may be shortened by altering system components.

S. K. Poultney, of the University of Maryland, and J. Brault, of the Kitt Peak National Observatory, proved the feasibility of making accurate distance measurements from some known site on Earth to another site on the Moon. Their experiment culminated in a picture taken by *Surveyor 7* on the Moon of laser light from the Earth. The *Apollo 11* crew left retroreflectors at their landing site on the Moon for a lunar ranging experiment which is being undertaken at the present time by scientists from the University of Maryland, the NASA Goddard Space Flight Center, and the McDonald Observatory at the University of Texas. Successful measurements have already been accomplished, and continued monitoring of point-to-point distances will provide new information on the dynamics of the Earth-Moon relationship. *See* MOON.

Upper atmospheric studies. Below 20 km a rather comprehensive knowledge exists of the variations in atmospheric composition and motion. Above 20 km little is known of day-to-day atmospheric fluctuations.

Several lidar systems are in operation for the study of the various properties of these upper atmospheric fluctuations. The lidar at the University of the West Indies in Jamaica, under the direction of R. W. Wright and G. S. Kent, has provided some information on stratospheric aerosol layers and their variation with time. Density and temperature data have been determined by C. Sandford, at Winkfield, England, and he has been instrumental in using resonant scattering to measure atmospheric sodium between 80 and 110 km.

F. Volz's determination of twilight sky brightness shows that appreciable dust layers must exist above the 30-km height; however, these layers have not been adequately investigated to date. *See* AIR SAMPLERS, ROCKET-BORNE.

It is generally accepted that noctilucent clouds exist around the 80-km height. This is an area of interest and conflict to lidar researchers because of the absence of reliable data concerning the composition of these clouds.

The highly developed and powerful lidar system of today should provide statistically significant information in the near future on the upper atmosphere.

For background information *see* ATMOSPHERE; ATMOSPHERIC POLLUTION; MOON; NOCTILUCENT CLOUDS; RADAR in the McGraw-Hill Encyclopedia of Science and Technology. [ROBERT M. BROWN]

Bibliography: C. O. Alley, H. H. Plotkin, and B. Warner, *Science,* vol. 167, 1970; J. T. Bradley and R. M. Schotland, *Optical Sounding V,* Tech. Rep. no. ECOM-0267-F, U.S. Army Elect. Command, Ft. Monmouth, N.J., 1969; J. Cooney, *J. Appl. Meteorol.,* vol. 9, 1970; P. M. Hamilton, *Phil. Trans. Roy. Soc. Lond. Ser. A,* vol. 265, 1969; W. B. Johnson and E. E. Uthe, *Lidar Observations of the Lower Troposphere During BOMEX,* SRI Rep., 1969; W. B. Johnson and E. E. Uthe, *Lidar Study of Stack Plumes,* SRI Rep., 1969; G. S. Kent and R. W. Wright, *A Review of Laser Radar Measurements of Atmospheric Properties,* Rep. no. 22, University of the West Indies, 1969; S. K. Poultney and J. Brault, *Surveyor 7 Illumination Test,* Tech. Rep. no. 884, University of Maryland, 1968; S. Zaromb, *Proc. Elec.-Opt. Sys. Des. Conf.,* 1970.

Atmospheric particles

Recent studies have elucidated the composition of atmospheric particles formed by volcanoes. Volcanoes provide an important natural source of fine particles carried by the Earth's atmosphere. The suggestion has often been made that volcanic dust in the atmosphere may decrease the temperature at the Earth's surface to an important extent. This may have been especially true during Tertiary times, when volcanic activity was very much greater than it is today. Furthermore, the particles may serve as nuclei on which cloud droplets form, and they may constitute an important portion of the particles suspended in the stratosphere.

One of the most famous eruptions was that of Krakatoa in the East Indies in 1883. It produced eruption clouds 18 mi high and turned day into night in Batavia, 100 mi away. A more recent, very violent eruption was from the volcano Gunung Agung in Bali in 1963. The Gunung Agung eruption, like that of Krakatoa, injected eruption clouds into the stratosphere as well as into the troposphere and produced spectacular sunsets throughout the world. *See* VOLCANO.

M. P. Meinel and A. B. Meinel studied the late twilight glows produced by the airborne ash from the Gunung Agung eruption. They pointed out that the appearance and intensity of the sunset glow changed from day to day, and they suggested that a study of the appearance of the glow and measurements of the height over an extended period of time and from many places might be useful in the study of atmospheric circulation. Other investigators, by making turbidity measurements of the Earth's atmosphere, obtained much useful information concerning the circulation of the atmosphere. Much less violent eruptions may also be important contributors to the particle content of

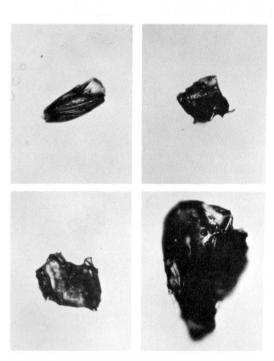

Fig. 1. Particles collected from the cloud produced by the eruption near Hekla volcano in Iceland in 1970. These particles were a few hundred microns in size.

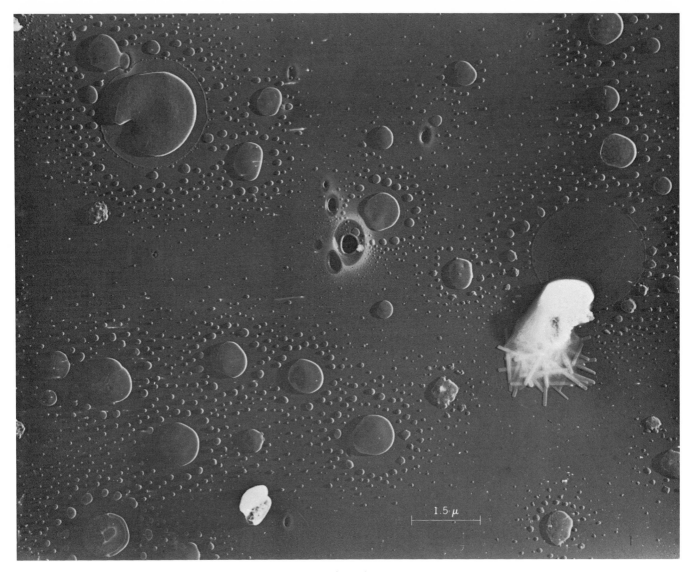

Fig. 2. Electron micrograph of particles from fume from Kilauea lava fountains.

the atmosphere.

Until the last few years, little has been known concerning the nature of particles formed by volcanoes. Some of them are merely finely divided lava. Others, especially those from highly explosive eruptions, are produced by attrition of the crater walls. Other particles are droplets of sulfuric acid and dissolved crystalline material consisting largely of inorganic sulfates and halides that have probably sublimed from the magma.

Iceland volcanoes. One of the first attempts to obtain information concerning the chemical nature of the particles in eruption clouds was undertaken by L. A. McClaine and coworkers in 1965. They collected samples on filters from the eruption cloud of the volcano Syrtlinger, off the southern coast of Iceland. A variety of particle compositions was observed. Both nickel and chromium were strongly evident, although these elements occur only in trace quantities in magma. They concluded that the particles were formed by a process of selective vaporization from the magma and subsequent condensation from the vapor phase. Particles have also been collected for studies at the National Center for Atmospheric Research from

an eruption near Hekla volcano in Iceland in 1970. These samples are still being analyzed, but the larger particles, at least, seem to consist of fragments of lava (Fig. 1).

Hawaii volcanoes. One of the most thoroughly studied volcanoes in the world is Kilauea on Hawaii Island. It is accessible, and the eruptions are relatively mild. During the last few years, R. D. Cadle and associates have made a number of collections of the particles in the fume from this volcano, both during and between eruptions. These eruptions may take the form of lava fountaining or merely outwelling, and although lava particles are emitted into the air, they are large enough so that they generally settle out very rapidly. The particles that remain suspended are mainly droplets of submicron size and have the composition indicated above.

When the particles are collected on microscope slides, crystals of various salts, such as calcium and ammonium sulfate, may separate from the droplets. An electron micrograph of particles collected from the fume from the 1967–1968 eruption is shown in Fig. 2. The droplets have partially evaporated, giving rise to configurations consisting

Weights of various constituents of Arenal eruption fume collected on polystyrene fiber filters

Sample	Si, μg	NH_4^+, μg	SO_4^{--}, μg	Na^+, μg	Cl^-, μg	Ca^{++}, μg	Mg^{++}, μg	K^+, μg	Approximate volume of air sampled, m^{3*}
431	26,500	41	1630	184	387	590	220	60	2×10^2
451	23,000	10	1160	163	401	470	200	57	1.0×10^2
459	33,400	26	2300	189	480	810	330	89	3×10^2

*Values corrected to sea level.

of nuclei surrounded by satellite droplets, a pattern which is characteristic of sulfuric acid. Inorganic substances that have crystallized from the droplets can also be seen.

Thermodynamic considerations of the equilibria existing in magmas indicate that the gaseous sulfur compounds must be largely in the form of hydrogen sulfide and sulfur dioxide. Studies at Kilauea by Cadle and others from the National Center for Atmospheric Research have recently demonstrated that the proportion of sulfuric acid in the eruption clouds is much greater than that in the fume from primary fumaroles. This suggests that considerable oxidation occurs when the hot eruptive gases mix with oxygen in the air. If this process also occurs, as seems likely, in more explosive eruptions, such eruptions may inject tremendous amounts of sulfate directly into the stratosphere.

Philippines and Costa Rica volcanoes. Recently, investigations have also been made of particles collected from the eruption clouds produced by the highly explosive volcanoes Mayon in the Philippines and Arenal in Costa Rica. These volcanoes are very similar with regard to type of eruption and of lava emitted. The finely divided lava particles, the so-called ash, are largely irregular in shape and angular, but a small percentage of spheres is also present. Nonetheless, much of the fume consists of droplets containing sulfuric acid and dissolved inorganic salts, although the percentage of sulfate seems to be lower than in the case of Kilauea. The table is an analysis based on aqueous extracts of filters containing fume from Arenal.

Summary. Much remains to be learned concerning the nature of particles in the clouds produced by volcanic eruptions, although scientists now know that relatively simple chemical compounds such as sulfuric acid, sulfates, and halides are much more important constituents than had previously been suspected. In particular, more knowledge is needed concerning the nature of particles produced by various types of volcanoes and the variation in composition with the stage of eruptive activity.

For background information *see* ATMOSPHERE; ATOMIZATION; LAVA; MAGMA; VOLCANO in the McGraw-Hill Encyclopedia of Science and Technology.   [RICHARD D. CADLE]

Bibliography: R. D. Cadle, *EOS Trans. Amer. Geophys. Union*, 51:440, 1970; R. D. Cadle and E. R. Frank, *J. Geophys. Res.*, 73:4780–4783, 1968; R. D. Cadle, A. L. Lazrus, and J. P. Shedlovsky, *J. Geophys. Res.*, 74:3372–3378, 1969; L. A. McClaine, R. V. Allen, and R. K. McConnell, Jr., *J. Geophys. Res.*, 73:5235–5246, 1968.

Atomic beams

Contemporary work on atomic beams has been concerned mainly with the study of fundamental problems in atomic and subatomic physics. Seen from a historical perspective, this work represents a continuation of the trend started by the pioneering experiments of Otto Stern and collaborators on space quantization, atomic angular momenta, and the magnetic moments of atoms. Stern's measurement technique was based on the deflection of the atomic magnetic moment in an inhomogeneous magnetic field. The modern technique of atomic-beam measurement is based on the magnetic-resonance method introduced by Isadore I. Rabi. This method is still the basic tool for studying atoms, as is discussed below. Moreover, this method was an early experimental realization of the idea of quantum-mechanical resonance, a concept that plays a significant role in almost every field of contemporary physics.

Electric dipole moments. Rabi's original idea is still used in atomic-beam experiments in a basically unmodified form. Advances in the resonance method have been concerned mainly with improvement in the precision of measurement. A striking demonstration of the precision currently obtainable is illustrated by recent experiments in the search for a permanent electric dipole moment (EDM) of particles. The existence of an EDM would definitely indicate a failure of both parity (P) and time reversal (T) as symmetry properties of the atomic interaction. The breakdown of both P and T has already been established in the weak interactions, and the presence (or absence) of an EDM on atomic and subatomic particles will help establish the nature of this breakdown. The most precise upper limits on the EDM of fundamental particles have, to date, all been established by continuing atomic-beam experiments. The best upper limits available are: electron, 3×10^{-24} cm; neutron, 4×10^{-23} cm; and proton, 1.6×10^{-20} cm.

The basic technique in each case is to induce magnetic resonance between the Zeeman sublevels of an appropriate particle, and then search for a shift in the transition frequency in the presence of an applied electric field that is linear in the magnitude of the applied field. The presence of such a shift immediately establishes a P- and T-violating term in the Hamiltonian. For the electron, a shift is looked for in the Zeeman resonance of the ground state of the cesium atom. Shifts less than a millionth of the line width can be observed by a remarkable alternating-current technique. The failure to observe shifts of this order immediately establishes an upper limit to the dipole moment of the cesium atom. Theoretical analysis has shown that, if the electron possesses an EDM (d_e), the cesium atom will have an EDM (d_{Cs}) such that $d_{Cs} \approx 100\, d_e$. Hence this experiment is very sensitive to d_e. The proton EDM is established by resonance of the thallium fluoride molecule. The internal molecular electric field is used to advantage in this experiment. See ELECTRON PARAMAGNETIC RESONANCE (EPR).

Proton and electron charges. A fundamental assumption of accepted physical theory is that the proton and electron charges are exactly equal and opposite. However, from time to time there has been speculation about the possible consequences of a small charge difference. For example, the laws of conservation of baryons and conservation of charge are regarded as independent conservation laws. Should there be a small proton-electron charge difference (δ_{p-e}), the baryon conservation law could derive from charge conservation. Moreover, an earlier speculation implied that, if δ_{p-e} were $\geqq 2 \times 10^{-18}e$, where e is the charge on the electron, then the observed rate of expansion of the universe could be ascribed to the Coulomb forces between like charges.

A continuing atomic-beam experiment to search for a small charge difference is being carried out at Yale University. If a small proton-electron charge difference were to exist, atoms would possess a small electric charge and would deflect in an external electric field. The Yale experiment attempts to measure the deflection of a well-collimated beam of atoms when an electric field is applied. Again, precision techniques are employed to measure as precisely as possible an upper limit to the deflection. To date, the best results indicate that $|\delta_{p-e}| \leq 1.3 \times 10^{-20}e$.

Spectroscopic studies. Classical atomic-beam studies continue to play an important role in atomic and nuclear spectroscopy. Through the study of hyperfine structure, the nuclear spin, magnetic moment, and nuclear quadrupole moment can be measured for the nuclear ground state. Quantities pertaining to the electrons that can be determined are the Landé g factor, the electronic angular momentum, and the magnetic field at the nucleus produced by the electrons. The measurement of the foregoing quantities in a large number of isotopes, both radioactive and stable, has been made possible by important advances in instrumentation and methodology. The atomic-beam method has been successfully applied to measurements of several hundred isotopes of more than 70 elements. Moreover, techniques of increasing sensitivity have made possible the extension of the method to many electronic states in the same atoms.

A recent variation of the method has been the combination of optical techniques with the atomic-beam method to make possible measurements of isotope shifts and optical Stark effect in radioactive isotopes. In this method, an optical resonance line with two or more hyperfine components irradiates the atoms in the transition region of an atomic-beam apparatus. If the atoms in the atomic beam are identical to those in the lamp, then resonance absorption takes place and the beam atoms can be excited to the upper atomic state. In the subsequent decay, some of the atoms undergo spin flip and contribute to the signal. By applying an electric field, an absorption line can be shifted through an amount equal to the hyperfine structure, and the Stark effect can be measured. Isotope shifts are measured by placing a different isotopic species in the beam. In general, at zero electric field, there will not be coincidence between the absorption lines of beam atoms and the emission line of the lamp. However, by applying a suitable electric field and inducing the Stark effect, coincidences can be induced. By measuring the electric fields at which they occur and using the Stark effect previously determined, the isotope shift can be found. The method has been successfully employed in studying isotope shifts in a large number of cesium isotopes.

For background information see ATOMIC BEAMS; MOLECULAR BEAMS; STARK EFFECT; SYMMETRY LAWS (PHYSICS) in the McGraw-Hill Encyclopedia of Science and Technology.   [RICHARD MARRUS]

Bibliography: J. K. Baird et al., *Phys. Rev.*, 179: 1285, 1969; L. J. Fraser et al., *Bull. Amer. Phys. Soc.*, 13:636, 1968; G. E. Harrison et al., *Phys. Rev. Lett.*, 22:1263, 1969; R. Marrus et al., *Phys. Rev.*, 177:122, 1969.

Atomic constants

The numerical values of the fundamental physical constants, or so-called atomic constants, such as α, e, m_e, h, c, and N, are periodically revised as new measurements based on new measuring techniques become available. Here α represents the fine-structure constant, e the electron charge, m_e the electron rest mass, h Planck's constant, c the velocity of light, and N Avogadro's number. It is usually necessary to revise the values of most of the constants even when a new value for only one particular constant is obtained because few of the constants can be measured to high accuracy directly; the majority, and usually those of most interest, must be obtained by "combining" those directly measured. This combining is done by a mathematical method known as least squares, and the resulting values of the constants are known as least-squares adjusted values. The least-squares technique provides a self-consistent procedure for calculating "best" compromise values of the constants from all the available measurements.

R. T. Birge first pioneered least-squares studies of the constants in the late 1920s and continued them into the 1940s. Similar studies have been continued to the present by several workers. The most recent critical analysis and least-squares adjustment of the constants were carried out in 1969 by B. N. Taylor, W. H. Parker, and D. N. Langenberg. This work was based to a large degree on their measurement (with A. Denenstein) of $2e/h$ using the alternating-current (ac) Josephson effect in superconductors.

Josephson effect. In 1962 B. D. Josephson predicted that, if two weakly coupled superconductors are maintained at a potential difference V, an ac supercurrent of frequency $\nu = 2eV/h$ will flow between them. (The two superconductors may be weakly coupled by the tunneling of electron pairs through a thin insulating barrier separating the two superconductors. Such structures are called Josephson junctions.) This equation, known as the Josephson frequency-voltage relation, follows directly from the macroscopic phase-coherent nature of the superconducting state and is exact, independent of experimental conditions.

To determine $2e/h$, junctions are irradiated with microwave radiation of frequency ν, and the voltages at which constant voltage current steps appear in the current-voltage curves of the junctions are measured. The step voltages V_n are related to ν by the generalized frequency-voltage equation $2eV_n = nh\nu$, where n is the step number. Thus knowing n, ν, and V_n permits $2e/h$ to be determined. The result of the measurement was $2e/h = 483.5976$ MHz/μV_{NBS}, where V_{NBS} means the volt as maintained by the National Bureau of Standards (NBS). The error in the experiment was ± 2.4 parts per million (ppm).

The Josephson effect measurement of $2e/h$ has had its greatest impact in the field of quantum electrodynamics (QED). To compare theoretical predictions of QED with experiment requires an accurate value of α, the fine-structure constant.

Before the $2e/h$ measurement, the most accurate values of α were obtained from experiment with the aid of theoretical equations containing significant contributions from QED. It was thus difficult to compare QED theory and experiment unambiguously since the theory had to be evaluated using values of α derived from the experiments themselves. Such comparisons were therefore limited to the testing of internal consistency. Now, however, by combining the value of $2e/h$ with the measured values of certain other constants, a highly accurate indirect value of α can be obtained without any essential use of QED theory. Consequently, direct and unambiguous comparisons can be made between QED theory and experiment.

In practice, there are several methods of obtaining α from $2e/h$. One method involves the gyromagnetic ratio of the proton γ_p. Another involves the Faraday constant F, the magnetic moment of the proton in Bohr magnetons μ_p/μ_n, and the conversion factor relating the ampere as maintained by the NBS to the absolute ampere. In addition, other quantities are required, such as the velocity of light and the Rydberg constant for infinite mass. These are sufficiently well known (with uncertainties of only a few tenths of a part per million) to be considered exactly known. To obtain the best possible value of α from all these data, it is combined by the method of least squares. The result is $1/\alpha = 137.03608 \pm 1.9$ ppm. This QED-independent value of α, derived mainly from the Josephson effect measurement of $2e/h$, can be used to critically compare QED theory and experiment.

Hyperfine splitting of hydrogen. Among the quantities for which such a comparison is of interest is the hyperfine splitting (hfs) in hydrogen. It can be measured experimentally to the phenomenal accuracy of 1 part in 10^{12} (one in a thousand billion) using the hydrogen maser. The theoretical QED equation for the hfs involves only well-known constants and α, but is limited to an accuracy of a few parts per million because of the difficulty in calculating some of the terms in the equation from theory.

One such term is called the proton polarizability correction δ_N. This correction arises from the fact that the proton in the hydrogen atom cannot be regarded simply as a charged spinning baseball —the proton has an internal structure of its own

and can be excited to different energy states. This can affect the magnitude of the hfs. However, all calculations of δ_N show it to be rather small, 1 or 2 ppm at most.

The small size of the theoretical value for δ_N is in marked contrast to that implied by the previously accepted value of α as obtained from a measurement of the fine-structure splitting in deuterium and a theoretical equation for the splitting: $1/\alpha = 137.0388 \pm 4.5$ ppm. If this value of α is used to calculate a theoretical value for the hydrogen hfs, if the hydrogen hfs is then compared with the experimental hydrogen maser value of the hfs, and if their difference is assumed to arise solely from the existence of a polarizability correction, then it is found that $\delta_N = 43 \pm 9$ ppm. This means that the probability for δ_N to be as small as predicted by theory is only 1 in 20,000. The discrepancy and resulting challenge to QED are clear. On the other hand, if the value of α obtained from the Josephson effect measurement of $2e/h$ is used, it is found that $\delta_N = 2.5 \pm 4.0$ ppm. This is quite consistent with theoretical predictions. Thus the Josephson effect value of α removes the discrepancy and resulting challenge to QED. (Recent measurements of the fine-structure splitting in hydrogen strongly support the Josephson effect value. But the source of error in the deuterium fine-structure measurements has yet to be established.)

This case is an excellent example of how fundamental-constants experiments carried out in one field of physics can have important implications for other fields—a low-temperature solid-state physics experiment has given information about the excited states of the proton. This is a subject usually associated with the field of fundamental-particle or high-energy physics. One could not imagine two areas of a natural science further removed from one another! It is therefore a good example of the overall unity of physics as well as the important role that can be played by high-accuracy measurements of the fundamental constants.

Other measurements. Other QED quantities for which it is of interest to compare theory and experiment using the new non-QED value of α are the Lamb shift and the anomalous moment of the electron. For the Lamb shift in hydrogen and deuterium, it was initially found that QED theory and experiment were in very poor agreement; the theoretical and experimental values differed by 3–5 times the combined error. However, the cause for this discrepancy was recently discovered, by T. Applequist and S. J. Brodsky, to be a miscalculated term in the theoretical equation for the Lamb shift. When this term is corrected, theory and experiment agree well.

For the anomalous moment of the electron, the QED theoretical value exceeds the experimental value by 77 ± 26 ppm. The probability for this to occur by chance is about 1 in 400. Although the discrepancy is rather disturbing, there is but one experimental value for the anomalous moment, and it was derived about 10 years ago. Thus several new measurements of the moment currently under way will have to be completed before it can be said that QED is faced with a serious challenge. Indeed, the preliminary result of such a measurement by J. C. Wesley and A. Rich is in good agree-

Comparison of output values for some of the more important constants*

Quantity	Units	Value 1969 adjustment	Error, ppm	Value 1963 adjustment	Error, ppm	Change, ppm
α^{-1}		137.03602(21)†	1.5	137.0388(6)	4.4	−20
e	10^{-19} coulomb	1.6021917(70)	4.4	1.60210(2)	12	+57
h	10^{-34} joule-sec	6.626196(50)	7.6	6.62559(16)	24	+91
m_e	10^{-31} kg	9.109558(54)	6.0	9.10908(13)	14	+52
N	10^{26} kmole^{-1}	6.022169(40)	6.6	6.02252(9)	15	−58

*From 1969 adjustment of B. N. Taylor et al. and 1963 adjustment of E. R. Cohen and J. W. M. DuMond.
†Numbers in parentheses are the uncertainties in the last digits of the main number.

ment with the theoretical value of the moment as calculated with the Josephson effect value of α, thereby apparently removing the discrepancy.

Recommended set of constants. To obtain a final best or recommended set of fundamental constants, Taylor and colleagues pooled the most reliable QED data with the data used to obtain the QED-independent value of α and carried out a second least-squares adjustment. This was not done without some difficulty because most of the QED data, which consisted of several values of α derived from appropriate theoretical QED equations and measurements, were of large uncertainty, inconsistent among themselves, or of questionable reliability both experimentally and theoretically. However, after a careful analysis of the situation, it was decided to include the value of α derived from the hydrogen hfs. The proton polarizability correction δ_N was taken to be 0 ± 5 ppm and yielded $1/\alpha = 137.0359 \pm 2.6$ ppm.

The resulting values of this final adjustment for some of the more important constants are given in the table, along with the previously accepted best values resulting from the 1963 adjustment of E. R. Cohen and J. W. M. DuMond, which was based on the deuterium value of α. Clearly, the values of the constants have changed by several times their assigned errors. Such a situation is quite common in the fundamental-constants field because of the intimate relationship which exists among the constants: A large shift in one will generally cause significant changes in the others. Thus no set of constants should be considered irrefutable.

For background information *see* ATOMIC CONSTANTS; JOSEPHSON EFFECT; LEAST SQUARES, METHOD OF; QUANTUM ELECTRODYNAMICS in the McGraw-Hill Encyclopedia of Science and Technology. [BARRY N. TAYLOR]

Bibliography: T. Applequist and S. J. Brodsky, *Phys. Rev. Lett.*, 24:569, Mar. 16, 1970; T. F. Finnegan, A. Denenstein, and D. N. Langenberg, *Phys. Rev. Lett.*, 24:738, Mar. 30, 1970; W. H. Parker et al., *Phys. Rev.*, 177:639, Jan. 10, 1969; B. N. Taylor, W. H. Parker, and D. N. Langenberg, *Rev. Mod. Phys.*, 41:375, July, 1969.

Bacteria, taxonomy of

Bacterial taxonomy concerns the characterization, differentiation, description, classification, and labeling of the different kinds of bacteria for transmission and compilation of information from research, medical, biochemical, and other laboratories. It is only on the basis of thorough characterization that accurate differentiation and identification or recognition of each type or species are possible. Bacteria are classified by species and genera, as are higher plants and animals. However, since bacteria reproduce primarily by an asexual process, the classic characteristic that defines the boundaries of a plant or animal species (sexual production of fertile offspring) cannot be used to define bacterial species. Nevertheless, it is necessary to classify the several thousand different kinds of bacteria in order to study them in a rational manner.

Historically, bacteriologists have assigned bacteria to different species solely on the basis of phenotypic characteristics. The distinction between different species usually has been subjective and often a matter of controversy. Application of new techniques which detect previously unmeasured properties of bacteria often demonstrates that some bacteria that had been thought to be similar are only superficially so, whereas other groups that had previously been thought to be dissimilar are basically the same. Recognition that two different organisms have been confused as one is important when their interaction with their environment differs. Of equal importance is the converse discovery that bacteria thought to represent several different species actually are variants of a single species, and that information about one also pertains to the others.

Type strains. Many types of bacteria have not been well characterized, and information regarding their significance and occurrence is fragmentary or lacking. As a result, it is often impossible to determine whether a newly isolated strain is a member of a species that has been incompletely or inaccurately described in the past. Therefore any previously known information cannot be related to this isolate. For this reason, strains upon which original descriptions were based are very important. They are designated as type strains and preserved to represent the group so that they can be reexamined whenever more complete information concerning their properties is needed. If the type strain has been lost, an attempt is made to find a new strain that conforms to the original description. Such a strain may be designated as a neotype, which then represents the species. Type strains or proposed neotype strains are deposited and maintained in national collections to serve as reference material. Although this procedure is invaluable, it is not without pitfalls, because such cultures may become contaminated by other bacteria or mislabeled at any step in handling. This problem is well expressed in a statement by I. C. Hall: "One of the first duties of any author who refers to a living organism by its generic and specific names is to make sure by all means at his command of the identity of the form in question. . . . There is no practice so conducive to confusion or so detrimental to progress in the science of bacteriology as the willingness of workers to accept uncritically the names upon so-called 'authentic stock cultures.' "

Since the type strain is the first, or one of the first, strains of a species to be isolated and described, it does not necessarily represent the average organism of the species. Many of the strains that are later isolated from nature may differ in a few of their characteristics but may still be considered to be similar enough to be members of the same species. Some workers assign organisms with divergent characteristics to the same species, whereas others have designated new species on the basis of comparatively minor differences. The first approach causes confusion, since not all members of the species have the same reactions. The second approach leads to considerable difficulty, since it appears that no two strains of bacteria are truly identical if enough of their characteristics are considered.

Genus characterization. Groups of species that share certain characteristics are placed together in a genus. Previously, some genera were differentiated on the basis of shape or arrangement of the cells. Although this may have been the best basis available at that time, cellular morphology often depends upon the cultural conditions and therefore is not always repeatable. Nor is there always a clear-cut division between morphological groups in terms of cellular shape or arrangement. For example, among the anaerobic, gram-negative, non-spore-forming rods, the fusobacteria were described as having pointed ends and thus were differentiated from bacteroides (described as having rounded ends) and sphaerophorus (described as pleomorphic or variable in form). In reality, many strains have cells of intermediate shape and cannot be accurately assigned to any of these groups. Improved culture methods for anaerobic bacteria and more thorough examination of their metabolic, chemical, and genetic properties have modified ideas concerning the relationships of these organisms to each other and their significance to man and the environment.

Routine application of gas chromatographic analysis of fermentation products has demonstrated that organisms first described as *Bacteroides* species produce metabolic products other than butyric acid. The organisms that were first considered *Fusobacterium* uniformly produce butyric acid as a major product of metabolism. Most species in the genus *Sphaerophorus* also produced butyric acid, but the energy metabolism of a few was like that of the *Bacteroides*. Assignment of these species to two genera on the basis of metabolic products provides a reproducible method for differentiation and identification. Similarity in metabolic pathways is often correlated with genetic similarity, and with similarity of cellular composition as well. Thus metabolic products more accurately reflect relationships among these organisms than do the cellular shapes. Metabolic activity may also be of greatest importance in the interactions between the bacteria and the host.

Metabolic activity. Definition of genera on the basis of products is not a new concept. The genus *Lactobacillus* has long been recognized as containing the lactic acid–producing, gram-positive, non-spore-forming rods, and the genus *Propionibacterium* as containing the propionic acid–producing, gram-positive, non-spore-forming rods. With the advent of simple gas-chromatographic techniques, it is practical for nearly any bacteriology laboratory to measure metabolic products rapidly on a routine basis for aid in identification. In the eighth edition of *Bergey's Manual of Determinative Bacteriology,* several genera of anaerobes are to be defined in part according to the products of energy metabolism. These modified definitions are based on the results of analyses from cultures of large numbers of strains in many species of anaerobes. These data demonstrate the reliability of this characteristic.

In describing a genus, some microbiologists prefer to list many characteristics that are found among the organisms in that genus. When using such a description, other microbiologists are expected to decide whether or not strains representing new species are sufficiently like the genus description to be placed in the genus. Included among the characteristics describing the genus *Lactobacillus* was the statement that its members were nonmotile. Later a motile strain was found that otherwise conformed to the description of the genus. The purposes of taxonomy would not be furthered by creating a new genus for this single species; instead, the description of the genus will be modified to include motile species.

Definition of genus. Rather than list a large number of characteristics to describe a genus, there is some tendency now to provide a definition. The authors of the section on *Clostridium* for the forthcoming eighth edition of *Bergey's Manual of Determinative Bacteriology* propose that any anaerobic, spore-forming rod be considered a member of the genus *Clostridium.* A separate genus has been proposed by other authors for those members of this group that reduce sulfates to hydrogen sulfide. The sulfate-reducing organisms differ from most, but not all, clostridia in their reaction to Gram's stain and in the percentage of guanine and cytosine (G + C) in their deoxyribonucleic acid (DNA). Only experience will tell whether or not the separation of sulfate-reducing strains into a distinct genus will be useful, since many species in the genus *Clostridium* have not been tested for this property. It may well be that the G + C percent will be a better basis for division of these organisms into two or more genera.

DNA homology. One of the most important advances in microbial taxonomy is the development of tests to measure the similarity of the arrangement of the base pairs in microbial DNA by homology competition experiments. In this procedure, the DNA of organism A is split into single strands, which are then immobilized on a membrane. DNA from organism B is split into single strands, which are then broken into small fragments. The immobilized strands of A are submerged in a solution containing the fragments, which reunite with the immobilized strands if the sequence of base pairs is the same. If, in recombining with the immobilized single strands, fragments of DNA from organism B can compete equally with radioactively labeled fragments from organism A, then it is certain that the arrangement of the base pairs in the

DNA of the two organisms is essentially the same; that is, they are genetically alike. The degree of homology, or genetic similarity, is measured by the degree of competition.

In the past, some bacteria have been differentiated on the basis of phenotypic tests such as the ability or inability to produce indol. Now, by measuring the homology of the DNA of these two organisms, it is possible to determine whether or not this single characteristic is associated with additional (perhaps unmeasured) differences, or whether it represents only a minor variation between strains that are essentially genetically the same. Bacteriologists have never been sure how many of the unmeasured or unmeasurable properties of bacteria were correlated with the characteristics that have been used for identification and taxonomy. The similarity or dissimilarity of the DNA of two species reveals in a single step the overall relationship of two organisms.

Analysis of DNA homology by competition experiments is time-consuming, and only a few of the problem groups have been analyzed. Recently homology experiments demonstrated two distinct groups among 24 strains that had been thought to be members of a single species. The strains were similar for an extensive list of phenotypic characteristics. Strains within each group were essentially identical, but there was little or no homology between the groups. As a result of the homology experiments, the strains were reexamined, and differences were found to exist in the nutritional requirements of the two groups.

An opposite result came from the analysis of a taxonomically confusing group of anaerobes bearing more than 30 different names. This entire group was found to comprise only nine genetically distinct species. Once the genetic species were established by homology experiments, it was possible to identify those phenotypic characteristics which correlated with the genetically distinct species and which could be used to identify them. The characteristics which vary among strains within each genetic species are now known to represent minor variations rather than major differences in genetic composition. Although most minor genetic differences are inconsequential, a minor difference (for example, absence of toxicity because of a single enzyme mutation) might greatly affect the significance of an organism in a given environment. Such a single genetic difference should not be used as the basis for establishing a different species.

The developments in DNA homology studies are extremely important. It will now be possible to determine, for example, whether what has been considered to be one organism is really two that may react differently in infections, or whether bacteriologists have been differentiating species that are genetically so similar that one can expect them to have equal significance or produce the same effects.

Nomenclature. Different laboratories may place different emphasis on the characteristics that they use to identify the many or few kinds of bacteria with which they work. Each laboratory usually has a specialized ability, and therefore may define species using different sets of characteristics. Two authors may use the same name for a bacterial species, but sometimes they are not discussing the same organism or accepting the same range of bacteria in the species. In other cases, different laboratories have used different names for the same bacterial species, and each is sometimes unaware that the other laboratory has additional information concerning the same organisms.

In an attempt to attain uniformity of nomenclature so that communication can be improved, there is an international commission on bacterial taxonomy and nomenclature which publishes a code of taxonomy and nomenclature. The code has been modified several times, as may be required when new methods and information are introduced. A judicial committee of the international commission rules on controversial problems in a further attempt to establish uniformity of nomenclature. For example, as a general rule, the earliest name applied to an organism is the only valid name. Sometimes such names have gone unnoticed and the organism is widely recognized under another, later, name. In several such cases the commonly recognized name has been officially accepted by the committee to avoid the unnecessary confusion of reestablishing the obscure but earlier name.

Anaerobes. It is clear that much confusion concerning bacterial species stems from insufficient characterization. The new information is serving to clear away much confusion, especially about the non-spore-forming anaerobes, which increasingly are found to be of importance in soft tissue and organ infections in man and animals. The new analyses have supported widely held opinions concerning the well-studied and easily recognized groups of bacteria.

Differentiation among some genera of anerobic bacteria

Rods
  Form spores . . . . *Clostridium*
  Do not form spores
    Gram-positive
      Propionic and acetic acids as the major volatile acid products . . . . *Propionibacterium* and *Arachnia*
      Acetic and lactic acids (1 + to 1) . . . . . *Bifidobacterium*
      Lactic acid sole major product . . . . *Lactobacillus*
      Moderate acetic, ± formic. major succinic with lactic (1 acetic to 2+ lactic) acids . . . . *Actinomyces*
      Other: butyric + others, acetic and formic, or no major acids . . . . *Eubacterium*
    Gram-negative
      Peritrichous flagella or nonmotile
        Produce butyric acid (without much isobutyric and isovaleric acids) . . . . *Fusobacterium*
        Produce only lactic acid . . . . *Leptotrichia*
        Not as above . . . . *Bacteroides*
      Polar flagella, fermentative
        Produce butyric acid . . . . *Butyrivibrio*
        Produce succinic acid
          Spiral-shaped cells . . . . *Succinivibrio*
          Ovoid cells . . . . *Succinimonas*
      Tufts of flagella on concave side of crescent-shaped cells . . . . *Selenomonas*
      Spiral-shaped cells with axial filaments . . . *Treponema*
Cocci
  Gram-positive
    Chains (usually strongly fermentative) . *Peptostreptococcus*
    Singly or in clumps . . . . *Peptococcus*
  Gram-negative
    Produce propionic and acetic acids . . . . *Veillonella*
    Produce butyric and acetic acids . . . . . *Acidaminococcus*

SOURCE: E. P. Cato et al., *Outline of Clinical Methods in Anaerobic Bacteriology*, 2d rev., Anaerobe Laboratory, Virginia Polytechnic Institute, 1970.

Although many groups of anaerobic bacteria have not yet been characterized and cataloged, the information that is now available has been incorporated in manuscripts by many authors to be published in the next (eighth) edition of *Bergey's Manual.*

The key developed by E. P. Cato and colleagues for differentiation of genera of anaerobes is presented in the table.

For background information see BACTERIA, TAXONOMY OF in the McGraw-Hill Encyclopedia of Science and Technology.

[W. E. C. MOORE; L. V. HOLDEMAN]

Bibliography: J. E. Blair, E. H. Lennette, and J. P. Truant (eds.), *Manual of Clinical Microbiology*, American Society for Microbiology, 1970; R. S. Breed, E. D. G. Murray, and N. R. Smith (eds.), *Bergey's Manual of Determinative Bacteriology*, 7th ed., 1957; E. P. Cato et al., *Outline of Clinical Methods in Anaerobic Bacteriology*, 2d rev., Anaerobe Laboratory, Virginia Polytechnic Institute, 1970; L. DS. Smith and L. V. Holdeman, *The Pathogenic Anaerobic Bacteria*, 1968.

Bacterial genetics

Even the best antibiotics have drawbacks—problems of toxicity, sensitivity, and the ever-increasing occurrence of resistance among the target organisms that the antibiotic is intended to kill. In recent years the emphasis in the search by major drug companies for new and better antibiotics has shifted. The massive screening programs that annually evaluated many thousands of organisms for antibiotic-producing potential are being replaced by programs aimed at chemically modifying known antibiotics of proved therapeutic value. The latter approach also has limitations, since the structures of many antibiotics are too large and complex for modification by standard chemical methods. W. Thomas Shier, Kenneth L. Rinehart,

Jr., and David Gottlieb, at the University of Illinois, have recently developed a novel approach to the modification of even very complex antibiotics. In their method, modified antibiotics are produced by altering the genetic structure of the microorganisms which produce them. In its level of sophistication the method represents a new phase in the development of applied bacterial genetics.

In their first application of the method the Illinois group employed the organism *Streptomyces fradiae*, which synthesizes neomycin, one of the most potent antibiotics yet discovered. Like other antibiotics, neomycin possesses certain drawbacks. After prolonged internal usage, it causes extensive ear and kidney damage, and target organisms develop resistance to it more rapidly than they do to many other antibiotics.

Metabolism of S. fradiae. Neomycin is an oligosaccharide with four sugar units linked together in a linear fashion (Fig. 1). One of the subunits, the deoxystreptamine moiety, differs from the others in that the ring is composed only of carbon atoms. This chemical difference is reflected in biochemical differences. *S. fradiae* readily takes up radioactively labeled deoxystreptamine from its growth medium and specifically incorporates it into the deoxystreptamine ring of the neomycin being produced. In contrast, none of the other rings is taken up effectively, nor is any of these rings specifically incorporated into the corresponding ring of the neomycin being produced. The Illinois group sought to capitalize on this difference and use it to prepare modified antibiotics.

They reasoned that the markedly different ability of the organism to convert deoxystreptamine into neomycin compared to the conversion of the other subunits into neomycin suggests that the deoxystreptamine subunit is synthesized by a different set of enzymes from those that are involved in the synthesis of the other subunits. Hence it should be possible to isolate a mutant of *S. fradiae* in which one of the enzymes involved only in the synthesis of the deoxystreptamine ring is defective. Such a mutant should be capable of synthesizing neomycin only in the presence of an outside source of deoxystreptamine; that is, it should take the deoxystreptamine provided and incorporate it into the deoxystreptamine moiety of the neomycin molecule. The next step in the procedure would be to prepare a series of analogs of deoxystreptamine by chemical synthesis, a very much simpler task than the chemical synthesis of analogs of neomycin itself, and to present these analogs one at a time to the mutant in the hope that the enzyme systems in the mutant will not be able to distinguish some of them from deoxystreptamine. In such cases the mutant should take up the deoxystreptamine analogs and convert them into analogs of neomycin.

Isolation of mutant. The isolation of the desired mutant was a difficult task, since *S. fradiae*, like other streptomycetes, is not ideally suited to genetic experiments. The long, filamentous structures that grow out from germinating spores of streptomycetes impose limitations not found in the practice of bacterial genetics in more typical bacteria, such as *Escherichia coli*. By exploiting the unique characteristics of the chemical mutagen *N*-methyl-*N'*-nitro-*N*-nitrosoguanidine (NG), the Illinois

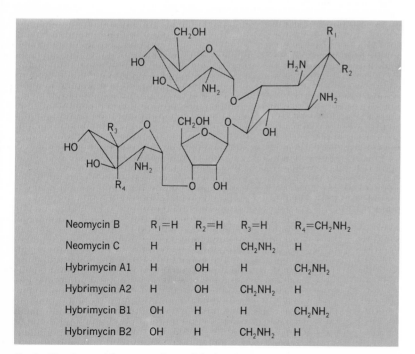

	R_1	R_2	R_3	R_4
Neomycin B	R_1=H	R_2=H	R_3=H	R_4=CH$_2$NH$_2$
Neomycin C	H	H	CH$_2$NH$_2$	H
Hybrimycin A1	H	OH	H	CH$_2$NH$_2$
Hybrimycin A2	H	OH	CH$_2$NH$_2$	H
Hybrimycin B1	OH	H	H	CH$_2$NH$_2$
Hybrimycin B2	OH	H	CH$_2$NH$_2$	H

Fig. 1. Structures of the neomycins and the hybrimycins.

group was able to define conditions under which the streptomycetes could be handled as effectively as are the more typical bacteria. They treated a suspension of individual spores of *S. fradiae* under normal growing conditions with a concentration of NG sufficient to kill 99.9% of the germinating spores over a 12-hr period (Fig. 2). They withdrew a small quantity of the resulting suspension, subjected it to serial dilution, and transferred portions of the appropriate dilution to a large number of petri dishes containing a type of solid agar medium on which *S. fradiae* grows as small colonies with a dense aerial sporulation. Using the standard replica plating technique, they transferred the spores which formed on these plates to two plates of nutrient agar medium that differed only in that one contained deoxystreptamine and the other did not. The replica plating technique, introduced in 1952 by Joshua Lederberg, employs a piece of sterile velvet, stretched over an appropriately sized piece of wood, to pick up spores when pressed against the master plate and then deposit them in identical patterns when pressed against other plates. When the resulting replica plates were incubated, the competent colonies produced antibiotic, which diffused into the surrounding medium. The *S. fradiae* colonies were then covered with a second layer of nutrient agar medium containing *Bacillus subtilis*, an organism that is killed effectively by neomycin. Antibiotic produced by the competent colonies diffused from the lower agar layer into the upper agar layer, killing the *B. subtilis*. The remaining *B. subtilis* was free to grow, rendering the medium opaque except for zones of inhibition of growth around colonies that had produced antibiotic. Approximately 4000 colonies were screened before a mutant colony was found that was capable of producing antibiotic on the medium containing deoxystreptamine but not on the medium lacking it.

Production of neomycin analogs. Once isolated, the mutant produced the structures that had been predicted. In the presence of adequate deoxystreptamine it produced neomycin almost as well as does wild-type *S. fradiae*. Of a total of 34 analogs of deoxystreptamine that have been prepared and tested at the University of Illinois, only two analogs were converted by the mutant into active antibiotics. Since wild-type *S. fradiae* produces two neomycins which differ only in the stereochemistry about one carbon atom (Fig. 1), it was expected, and ultimately observed, that two new antibiotics were produced for each deoxystreptamine analog taken up by the mutant. It is interesting that these two deoxystreptamine analogs, streptamine and 2-epistreptamine, are in fact parts of two other antibiotics, streptomycin and spectinomycin, respectively. Hence the new antibiotics were given the name hybrimycins, since they can be considered hybrid antibiotics, with the streptamine or 2-epistreptamine ring coming from streptomycin or spectinomycin, respectively, and the remaining rings coming from the neomycins.

More than 1000 antibiotics are already known, and only a very small percentage of them are useful in medical treatment. The hybrimycins face a large battery of tests to establish whether or not they will have any uses in medicine. To date it has been shown in laboratory tests that they are active

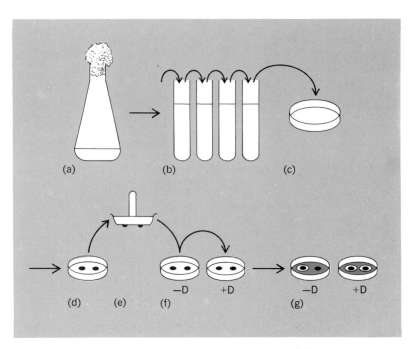

Fig. 2.   Isolation of the deoxystreptamine negative mutant of *Streptomyces fradiae*. (*a*) Spores treated with mutagen. (*b*) Aliquot subjected to serial dilution through tubes of sterile water. (*c*) Aliquots of appropriate dilution added to plates of a sporulating agar medium. (*d*) Plates incubated until colonies with spores develop. (*e*) Spores transferred by replica plating technique to two plates of nutrient agar medium, one containing deoxystreptamine (+D) and the other lacking it (−D) in *f*. (*f*) Plates incubated until competent colonies produce antibiotic, and then overlayed with a second layer of agar medium containing the test organism, *Bacillus subtilis*. (*g*) After further incubation, antibiotic-producing colonies are surrounded by a clear zone of inhibition of *B. subtilis* growth. Desired mutant (colony on right) shows a zone of inhibition on +D medium but not on −D medium.

against species of *Staphylococcus*, *Salmonella*, *Streptococcus*, and numerous other pathogens, although not as active as neomycin. Larger doses of hybrimycins would be needed in medical use, but if these hybrimycins, or hybrimycins yet to be discovered, lack the adverse side effects of existing drugs, higher dosage would not be a problem.

Use of method with other antibiotics. The new method for modifying antibiotics has already been extended to paromomycin, an antibiotic closely resembling neomycin. Presumably, it could be applied successfully to any antibiotic that is biosynthesized from subunits that can be conveniently prepared and provided to the appropriate mutant. A large number of antibiotics fit this description, including the medicinally important antibiotics kanamycin, novobiocin, and gentamicin, as well as a long list of antitumor antibiotics. In the last group the need for new drugs is great and the requirements for a satisfactory drug are less stringent. What has been developed is not just a series of new antibiotics but an application of bacterial genetics to give a new method of broad potential applicability to the preparation of new and, it is hoped, superior antibiotics.

For background information *see* ANTIBIOTIC; BACTERIAL GENETICS; NEOMYCIN in the McGraw-Hill Encyclopedia of Science and Technology.

[W. THOMAS SHIER]

Bibliography: K. L. Rinehart, Jr, *The Neomycins and Related Antibiotics*, 1964; W. T. Shier et al., *Proc. Nat. Acad. Sci. U.S.*, 63:198, 1969.

Band theory of solids

A new technique, referred to as the piezo soft x-ray effect, has been developed to determine the energy-band structure of solids. Essentially this technique requires the observation of the modulation of the soft x-ray emission band of a solid due to an alternating elastic strain. Structure, unresolved in the normal emission band, is observed in the modulation curve, which is directly related to the electronic structure of the solid.

Soft x-ray technique. Soft x-ray emission appears to be a simple and easily interpretable technique to probe the energy band structure of solids below the Fermi energy. The states involved in the emission process, one being an excited core level and the other a transition from the conduction or valence band, are fairly well defined. On closer investigation, the interpretation of the emission spectra is not as simple as might be inferred from the one-electron description of a solid. The shape of the emission spectrum is not a good representation of the density of states because of effects such as variation of the transition probability within the band, lifetime broadening distortions mainly due to Auger transitions, and many body effects. Peaks in the density of states in the electronic structure which correspond to the high symmetry points in reciprocal space (Van Hove singularities) are usu-

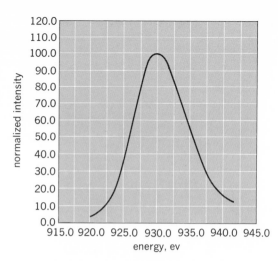

Fig. 2. Copper L$_{III}$ emission band.

ally unresolved. Also, except for a few simple metals such as aluminum and magnesium, the beginning and terminus of the band are masked by the low- and high-energy tails.

Within the past 5 years, derivative techniques, which are the result of modulating a measured property by an external parameter, have come into prominence because of the enhanced structure which can be obtained. The piezo soft x-ray effect is in the category of modulation of the emission spectrum due to an alternating mechanical strain. The intent is to reveal structure which is unresolved in the normal emission spectrum but which can be correlated with Van Hove singularities or critical points and the Fermi energy.

Experimental apparatus. Three experimental requirements must be satisfied in order to observe the piezo soft x-ray effect. First, the order of magnitude of the effect (between 1 part in 1000 to 1 part in 10,000) requires that the emission spectrum be excited to a high counting rate in order to perform the experiment within a reasonable time. This means that direct excitation by electrons is preferable. Second, since a single scan of an emission band requires several days, the rate of contamination buildup on the specimen surface must be reduced to a minimum. Third, care must be taken to minimize any distortion strains transmitted to the spectrometer by the specimen-stressing device.

The spectrometer (Fig. 1) is of the focusing type with the analyzing crystal bent to the approximate focusing radius. The sample remains stationary and the analyzing crystal mount moves in a linear motion with the correct angular motions transmitted to crystal and detector to maintain the focusing conditions. The x-ray emission from the specimen is excited directly by electrons with an incidence angle of about 75 to 80°. The detector is a flow-proportional counter with a stretched polypropylene window. The detector and analyzing crystal are located in a low-vacuum chamber, about 10 μ pressure. The sample chamber box (not shown) is isolated from the low-vacuum chamber and pumped to a vacuum of 10^{-9} mm by ion and titanium sublimation pumps. The x-rays from the sample pass through a stretched polypropylene window which separates the two vacuum systems. A liquid-nitro-

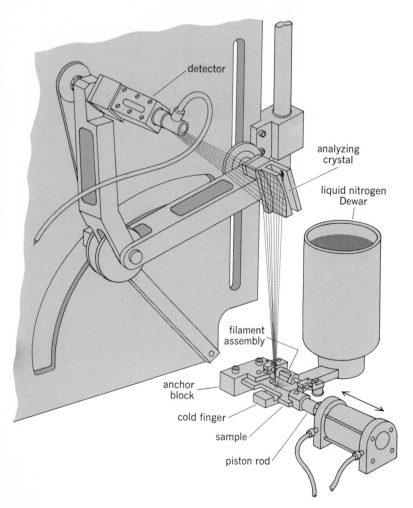

Fig. 1. Schematic of piezo soft x-ray apparatus.

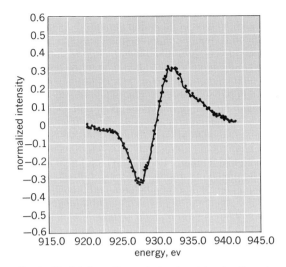

Fig. 3. Modulation of L$_{\text{III}}$ emission band due to alternating elastic strain.

gen cold finger in the proximity of the sample is necessary to reduce the contamination rate to a vanishing level. The sample is stressed by the force from a hydraulic cylinder which is actuated by a high-speed solenoid valve. The loading system can apply a maximum stress of 40,000 psi; the specimen cross section is 1/8 × 1/4 in. The time required to load or unload the sample is about 15 msec. The loading of the sample is done in a coaxial manner so that negligible distortion strains are transmitted to the spectrometer, which might introduce spurious emission intensity changes.

Experimental results. The first experiments were performed on polycrystalline copper. The L$_{\text{III}}$ emission band, which has a maximum intensity at 930.1 ev and corresponds to transitions between the 3d-4s levels to the 2p$_{3/2}$ level, was step-scanned in wavelength increments of 0.002 A with a potassium acid phthalate (KAP) (2d = 26.8 A) analyzing crystal.

At each position the sample was alternately loaded, to a stress of 12,000 psi in tension, and unloaded 100 times at 10-sec intervals. The difference in intensity under the strained and unstrained states was determined by subtracting the averaged corresponding readings. The magnitude of the strain effect required between 10^7 and 10^8 counts to be accumulated at each step position to assure statistical significance.

Figure 2 shows the observed emission spectrum normalized to a peak value of 100. It is completely structureless without any indication of the location of the Fermi energy, which should be located somewhere in the region of 932–934 ev, as determined by the x-ray absorption measurements of Y. Cauchois. The lack of any indication of the Fermi energy results probably from masking by satellite emissions or many body effects. There should be only natural lifetime and instrumental broadening at this energy, since Auger processes would not contribute to the lifetime broadening. Figure 3 shows the change of the emission intensity which results when the copper specimen is stressed in tension. The modulation curve appears to be fairly representative of the derivative of the emission peak. Apparently there is also additional

structure present. One prominent feature, besides the small wiggles and peaks, is the abrupt change in slope that occurs at about 934 ev.

If the modulation curve were the result of a rigid shift of the emission band, the curve would be exactly the derivative of the emission band. To check this possibility, the product of the derivative of the emission band and a suitable constant, which would correspond to the energy shift of the band, is subtracted from the modulation curve in an attempt to produce a null curve (Fig. 4). Curve 1 is the original curve, and curves 2 and 3 are the resultant subtraction curves, assuming shifts of 0.010 and 0.015 ev, respectively. It is apparent that no appropriate uniform or smoothly varying band shift can account for the total modulation curve. The energy shift which reduces most of the effect to zero is 0.010 and 0.015 ev at the bottom and top of the band, respectively.

The portion of the modulation curve which cannot be nulled by the derivative subtraction technique must be associated with regions in the band which are extrasensitive to strain as regards altering the x-ray emission intensity. This could be at the Fermi energy and Van Hove singularities or critical points. There is some arbitrariness between the energy scale as measured by x-ray emission and the scales used in the theoretical band-structure calculations. G. A. Burdick correlated his energy scale to the x-ray emission scale by matching the peak in the density of states with the peak in the emission band. Burdick's calculated values for the Fermi energy E_F and Van Hove singularities are indicated in Fig. 4. The abrupt slope change at 933.8 matches the Fermi energy, and much of the fine structure matches the Van Hove singularities.

Conclusions. A new experimental technique has been developed to probe the energy-band structure of solids. Although the interpretation of the results is still in a primitive state, it is evident that this method can be a useful means for studying the electronic structure of solids. One of the notable

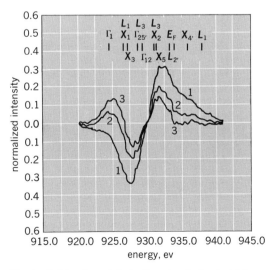

Fig. 4. Subtraction of derivative of emission band from modulation curve. Curve 1 is original curve. Curves 2 and 3 are subtractions, assuming shift of 0.010 and 0.015 ev, respectively.

features of these experiments is that a structure-less emission spectrum has been transformed into a modulation curve which contains considerable structure.

For background information *see* BAND THEORY OF SOLIDS; CRYSTAL STRUCTURE; PIEZOELECTRIC CRYSTAL; X-RAY CRYSTALLOGRAPHY in the McGraw-Hill Encyclopedia of Science and Technology.   [RONALD H. WILLENS]

Bibliography: R. H. Willens et al., *Phys. Rev. Lett.*, 23:413, 1969.

Barnard's star

Existing optical equipment cannot directly reveal a planetlike object such as Jupiter outside the solar system even if attached to the nearest star. However, such an object, or objects, can be revealed through a study of a perturbation in the star's proper motion. This is done by means of the techniques and methods of long-focus photographic astrometry, for which the traditional long-focus refractors are eminently suitable. Evidence for extrasolar planets has thus been found for Barnard's star, the nearest star in the northern celestial hemisphere.

In June, 1916, the astronomical community was surprised by the discovery of a large motion along the sky for a faint star. The discoverer, Edwin Emerson Barnard (1857–1923), was a distinguished amateur astronomer who, at that time, was a member of the staff of the Yerkes Observatory. That the stars are not fixed on the celestial sphere but have motions relative to each other was first noted in 1718 by Edmund Halley. The amount of proper motion proves to be an excellent criterion for distance; in general the larger its proper motion the nearer the star.

Barnard's star, of visual magnitude 9.5, has a proper motion of 10.31 seconds of arc annually, still the largest known for any star, corresponding to an apparent angular motion of the Moon's diameter in about 170 years; it has often been referred

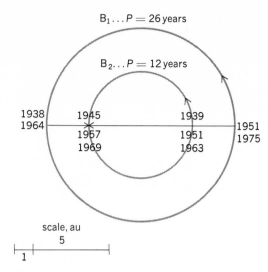

Fig. 2. The circular, corevolving, coplanar orbits of the two companions around Barnard's star. The perturbations caused by the companions, B₁ and B₂, reinforce each other most conspicuously about 1951.

to as Barnard's "runaway star." Its parallax is $0\overset{''}{.}552 \pm 0\overset{''}{.}001$, corresponding to a distance of 5.90 light-years, with an uncertainty of about 1 light-week. The tangential velocity is 89 km/sec. Observations of the Doppler shift in the spectrum of Barnard's star have yielded a radial velocity of −108 km/sec, that is, Barnard's star approaches the solar system at this rate. The total space velocity is 140 km/sec. Barnard's star is the second nearest star (the nearest is the triple system Alpha Centauri in the southern celestial hemisphere, at a distance of 4.3 light-years); it is the nearest known star in the northern celestial hemisphere. The intrinsic luminosity of Barnard's star is only 1/2300 of that of the Sun (in the visual range of light); it is a so-called red dwarf star. The distance to Barnard's star will gradually diminish to a minimum of 3.75 light-years in about A.D. 11,800 (Fig. 1). At that time its brightness will have increased 2.5 times, and Barnard's star will appear as a star of visual magnitude 8.5.

The proximity of Barnard's star makes it the ideal object for a precision study of parallax, proper motion, secular perspective acceleration, and other possible effects, in particular, any possible deviation from uniform rectilinear motion caused by an unseen companion, or companions, revolving around it.

Long-focus photographic technique. Barnard's star was given highest priority on the systematic program begun in 1937 at the Sproul Observatory for the specific purpose of establishing possible perturbations. The required high precision is obtained by a combination of the following factors. (1) Long focal length, that is, large-scale portrayal of a small field of the sky containing the star one is interested in and a small number of surrounding background, or reference, stars. For the Sproul 24-in. visual refractor of 10.93 m focal length, the scale in the focal plane is 1 mm = $18\overset{''}{.}87$, or $1'' = 53\ \mu$. (2) The inherent positional quality and stability of photographic plates; for visual telescopes, a yellow filter is used to yield sharp images. (3) A precision-measuring engine, for the greater part

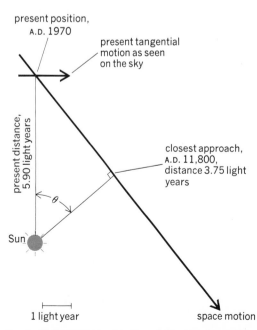

Fig. 1. Path of Barnard's star relative to solar system.

still making use of precision long screws and personal bisection of star images.

Astrometric study. The position of Barnard's star is measured against a reference background of three stars. By combining observations from several nights, an accuracy of ±0″.01 is reached. More than 30 years of intensive photographic coverage, with well over 10,000 separate exposures, has yielded precise values for proper motion, parallax, and secular acceleration. After allowing for these effects, there are still systematic deviations, mostly in right ascension, with a cycle of approximately one-fourth of a century. There appears to be no better way to explain these deviations than by interpreting them as a perturbation caused by an unseen companion object (or objects) of planetlike nature.

Although this perturbation was first recognized as early as 1956, it was not till 1963 that it was announced and a first interpretation given. On the assumption that the mass of Barnard's star is 15% of the Sun's mass, the analysis yielded an unseen companion with a mass of 0.15% the Sun's mass, or 1.6 times the mass of Jupiter, revolving around Barnard's star with a period of 24 years at a mean distance of 4.5 astronomical units (au). The orbit proved to have an appreciable eccentricity (0.6). Early in 1968, with 5 additional years of observations, a new analysis was made which essentially confirmed the earlier results, except that the eccentricity appeared to be 0.75, or possibly even higher. This new information, together with increasing evidence of the presence of a secondary perturbation, led to an alternate dynamical analysis involving two perturbations.

Two-orbit interpretation. Simple trial and error showed that two circular, corevolving, nearly coplanar orbits with periods P of 26 and 12 years and radii of 4.7 and 2.8 au, respectively, represent the observations very well, the masses of the two perturbing companions B_1 and B_2 being somewhat less than that of Jupiter (Fig. 2). This alternate dynamical analysis represents the observational data slightly better than one-planet orbit with high eccentricity does.

The present technique does not permit any interpretation which would involve more than two planets. It would be impossible to discover an Earthlike planet; the Earth has a mass of 1/319 times that of Jupiter, which would result in a completely negligible perturbation.

The total amplitude of the perturbation of Barnard's star is 0″.04; the accuracy of the yearly positions, based on hundreds of exposures, is about ±0″.004. Figure 3 is a graph of the actual deviations (magnified 2000× relative to the proper motion) that lead to the perturbation hypothesis. The barycentric path in the figure is the path as seen from the center of mass of the solar system. RA and Decl represent the traditional equatorial coordinates in which astronomical positions are defined, determined, and measured. Thus the perturbations appear to be well above the threshold value of errors, or "noise." Whether the one- or two-planet hypothesis is to be preferred on the basis of observations may be decided after several more years, when present predictions can be tested by additional observations. Further improvement of the results may be expected from

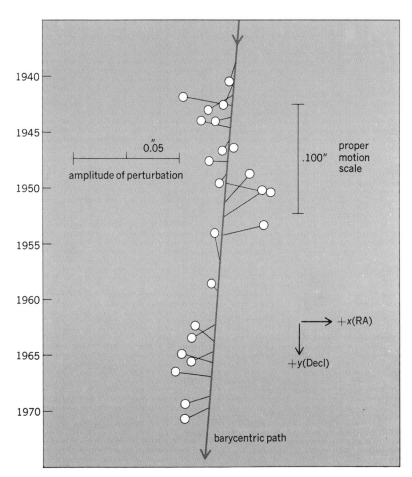

Fig. 3. The observed positions (that is, what is left after secular perspective acceleration and parallax are removed) and the corresponding positions in the path of the center of mass of Barnard's star and its companions, B_1 and B_2.

more impersonal measuring techniques. Meanwhile, it is interesting to reflect that there may be a planetary system elsewhere in which the two principal planets show analogy with the situation in this solar system, in terms of space, with the orbits of Jupiter and the average asteroid, and in terms of time, with the orbits of Saturn and Jupiter.

For background information *see* CENTER OF MASS; ORBITAL MOTION; PERTURBATION (ASTRONOMY); PLANET in the McGraw-Hill Encyclopedia of Science and Technology.

[PETER VAN DE KAMP]

Bibliography: P. van de Kamp, Alternate dynamical analysis of Barnard's star, *Astron. J.*, 74: 757, 1969; P. van de Kamp, Parallax, proper motion, acceleration, and orbital motion of Barnard's star, *Astron. J.*, 74:238, 1969; P. van de Kamp, The search for perturbations in stellar proper motions, *Vistas Astron.*, 8:215, 1966; P. van de Kamp, *Unresolved Astrometric Binaries*, Astron. Soc. Pac. Leafl. no. 470, 1968.

Bioluminescence

Light production is a common characteristic of deep-water mesopelagic and bathypelagic fishes, but many shallow-water species are luminous also. Among the most interesting of the shallow-water fishes are the apogonids, or cardinal fishes, which belong to the family Apogonidae. Recently, six more luminous species of the family Apogonidae

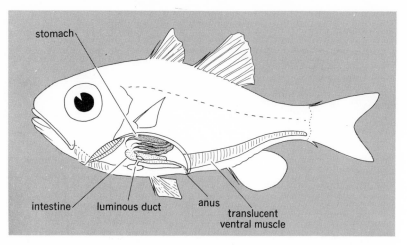

Fig. 1. Schematic diagram of the luminescent organ system of *Siphamia versicolor* from the left lateral aspect.

were discovered, five in the Philippines and one in New Guinea.

The apogonids are small- to moderate-size fishes that are widely distributed, especially in the Pacific and Indian oceans. Approximately 100 different species are known, and of these about 40 species are found in the warm waters of the Philippine Islands. Large numbers also occur in the waters of the Indonesian archipelago and neighboring islands of the western Pacific. Smaller numbers are further distributed in the coastal waters of Japan, the Hawaiian Islands, Australia, and the Comoro Islands. However, not all are shallow-water species. Some pelagic forms are known, and a few have been caught in waters as deep as 300 m. Although the occurrence of luminosity in the family has not been completely investigated, at least 12 species are known to be luminous. In all these species, the light originates in from 1–3 luminous organs located near the ventral surface of the body. In five species, the luminescence is due to symbiotic luminous bacteria infecting the photogenic organs. In the remaining seven species, the luminescence is due to the oxidation of a complex organic compound.

Symbiotic luminous bacteria. Species possessing symbiotic luminous bacteria appear to be confined to the genus *Siphamia*. The five luminous species are *S. versicolor*, *S. majimai*, *S. elongata*, *S. cuneiceps*, and *S. roseigaster*.

The standard length of *S. versicolor* is about 3 cm. It lives symbiotically among the long spines of the sea urchin *Diadema setosum* in the littoral waters of Okinawa Island. The luminescent organ system of *S. versicolor* (Fig. 1) consists of three structural elements: a luminous organ or duct, a small tube connecting the organ with the intestine, and a thick layer of translucent ventral muscles. The luminous organ is oval-shaped and lies against the translucent ventral muscles, just below the anteroventral surface of the liver and at the level of the base of the ventral fins. In a specimen 2.6 cm long, the luminous organ is 2.0 mm long, 1.5 mm wide, and 0.3 mm thick. The dorsal surface of the organ faces the abdominal cavity and is overlaid by an opaque stratum of fibrous connective tissue that acts as a reflector. The luminous organ itself is

lined with many ducts that are interlaced by a network of blood capillaries. Large numbers of luminous bacteria are found within the ducts. A slender tube, which passes through the reflector, connects the posterior end of the luminous organ with the intestine. Ventrally, the luminous organ is underlaid by the translucent keel muscle and a pair of longitudinal muscle bundles that extend posteriorly along the ventral side of the body. These muscles are translucent and serve to transmit and diffuse the luminous light. The skin also contains many chromatophores, especially around the area of the translucent muscles, which appear to regulate the intensity of light.

When *S. versicolor* is placed in a viewing tank in a dark room, no luminescence is visible laterally. However, when the fish is observed ventrally, a weak, diffuse bluish-white luminescence can be seen. When the fish is removed from the tank, the lower section of the thoracic region lights up strongly. Since bacteria are continuously luminous, it must be presumed that the light intensity is controlled by the expansion and contraction of skin chromatophores.

Much evidence exists that the *Siphamia* luminescence is due to bacteria. Bacterial luminescence itself is due to the oxidation of reduced flavin mononucleotide ($FMNH_2$) by molecular oxygen, catalyzed by bacterial luciferase in the presence of a long-chain aldehyde (9–14 carbon atoms long). Thus the rich blood supply to the luminous organ provides not only nutrients essential for growth and maintenance of the bacteria but also oxygen, which is necessary for luminescence.

In *S. versicolor* and most apogonid fishes, the adult male keeps the eggs and newly hatched larvae in its mouth. Therefore, during the breeding season, great numbers of eggs and larvae can be collected. When large masses of such eggs or larvae are examined in the dark, they are found to be nonluminous. The luminous bacteria in the organ appear to be acquired after birth and not transmitted through the egg. The slender tube connecting the luminous organ with the intestine appears to serve two purposes: for infecting the luminous organ and for getting rid of dead bacteria.

The luminescent organ systems of *S. cuneiceps*, *S. roseigaster*, and *S. majimai* are similar to that of *S. versicolor*, except that the first two have thicker layers of translucent muscles. The system of *S. elongata* has not been carefully studied but probably resembles that of *S. versicolor*.

Fish's own light-emitting components. Luminescence in this group is probably due to the oxidation of a compound containing an indole nucleus. The seven luminous species known are *Apogon ellioti*, *A. striata*, *A. poecilopterus*, *Archamia lineolata*, *A. fucata*, *A. zosterophora*, and *Rhabdamia cypselura*. Except for *A. ellioti*, the luminescence in these species was discovered only in 1969. Adult *A. ellioti* and *A. poecilopterus* range in size from approximately 5 to 9 cm and 7 to 10 cm, respectively, whereas the standard length of the others ranges from approximately 4 to 7 cm. The luminescent organ systems are of three types.

A. ellioti is the most studied member. It is found in the coastal waters of southern Japan, the Philippine Islands, New Guinea, eastern Australia, and Madras, India. It was the first apogonid reported to be luminous. The luminescent organ system (Fig. 2) consists of one anterior and two posterior organs, all ventrally located, and their accessory structures. The anterior organ or the thoracic luminous duct is a round lemon-yellow body. It lies in the ventral, translucent keel muscle and is connected by a small duct to the second bend of the intestine. Lying above the keel muscle is a white, opaque membrane forming the reflector. The light is diffused to the outside through the translucent muscle. The posterior organs or anal luminous ducts are bean-shaped and are embedded in the muscle, one on either side of the rectum. Each is connected to the rectum by a small duct. In a specimen 7.5 cm long, the anterior organ is 2–3 mm in diameter, and each of the posterior organs is about 2.8 mm long, 1.5 mm wide, and 0.5 mm thick. The luminescence is readily visible externally.

The luminescent organ system of *Apogon striata* is the same as that in *A. ellioti*, except that the anal organs are missing. The *A. poecilopterus* system also appears to be similar to that of *A. ellioti* but lacks anal organs.

The second type of luminescent organ system is found in *Archamia lineolata*, *A. fucata*, and *A. zosterophora*. In these, the luminous body is represented by a section of the gut, namely, the juncture of the pyloric ceca and the intestine. No pouchlike or anal organ is present. The luminescence is visible from the outside through the translucent ventral muscle.

The third type of luminescent organ system is in *R. cypselura* (Fig. 3). It possesses different features: The distal ends of a pair of pyloric ceca are transformed into luminous organs or ducts. A pair of transparent lenslike organs, encircled with black pigments, are situated in the ventrolateral wall of the body cavity. The luminous organs are attached to the lenslike organs, and light is transmitted to the outside through them.

Since the above three distinct types of luminescent organ systems differ markedly from that in *Siphamia*, at least four types of luminescent systems are present in apogonids.

The color of the luminescence is blue and appears to be identical in all seven species. The bioluminescence emission maximum of *A. ellioti* is 460 nm. The light is produced by a luciferase-luciferin (enzyme-substrate) type of reaction, in which luciferin is oxidized by molecular oxygen, catalyzed by the enzyme luciferase. Each organ contains its own supply of luciferin and luciferase. The chemical structure of luciferin is unknown, but it appears to be similar, if not identical, to the luciferin of the small (2–3 mm long) luminous marine ostracod crustacean *Cypridina hilgendorfii*. The structure of *Cypridina* luciferin (molecular weight 405) consists of an indole group joined to isoleucine and arginine moieties. Only limited information is available concerning the luciferases. Because luciferin concentration is in great excess in the organ, luminescence is continuous and the intensity is fairly constant. However, in a living fish, the light intensity varies depending on the expansion and contraction of the skin chromatophores. When a dark, cold-water extract (luciferase) and a hot-water extract (luciferin) of an organ are mixed, light is produced. The luminescence intensity then decreases exponentially with time.

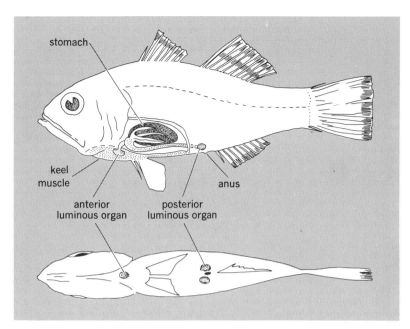

Fig. 2. Schematic diagram of the luminescent organ system of *Apogon ellioti* from the left lateral and ventral abdominal aspects.

When luciferin and luciferase from different organs of the same fish or of two different species are mixed, light also results. Thus reciprocal cross-reactions exist between the luciferins and luciferases of all pairs of these seven apogonid species.

Biochemical relationship. The essential components in bacterial luminescence are the same for all species of bacteria. However, in the seven apogonid species that do not contain luminous bacteria, the luciferins and luciferases show a striking luminescent cross-reaction with those of *C. hilgendorfii*. There is also a further luminescence cross-reaction between the luciferin and luciferase in the foregoing species and the luciferin and luciferase of a small pempherid fish, *Parapriacanthus*

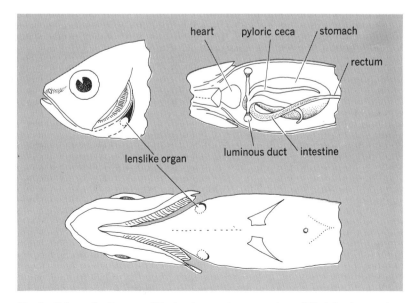

Fig. 3. Schematic diagram of the luminescent organ system of *Rhabdamia cypselura* from the left lateral and ventral abdominal aspects.

ransonneti. The chemical and physical properties of the luciferins of *A. ellioti* and *C. hilgendorfii* are nearly identical, and the emission spectra of the bioluminescence cross-reactions are the same. However, the two luciferases appear to be chemically different. Since *C. hilgendorfii* and other luminous members of the family Cypridinidae are distributed in the same waters as the luminous apogonids, it is possible that ingestion of *Cypridina* provides the luciferin and luciferase for the apogonids. Since luciferase is a protein, some modification may occur during digestion. The luciferin and modified luciferase may then enter the luminous organ from the gut through the connecting duct. If such a mechanism exists, the luminescent organ systems in *Apogon*, *Archamia*, and *Rhabdamia* have had a radically different evolutionary development from that in *Siphamia*.

Function. Fishes represent the highest forms of life known to be luminous. Among them, only the marine forms are luminous. Thus it is of some importance to know whether luminescence has any useful purpose. Many theories have been proposed. These include the use of luminescence for frightening a predator, a warning signal, attracting prey, species identification, schooling, a mating signal, protective countershading against predators, distracting predators, and illumination. None of these theories has been proved experimentally.

For background information *see* BIOLUMINESCENCE in the McGraw-Hill Encyclopedia of Science and Technology.

[FREDERICK I. TSUJI; YATA HANEDA]

Bibliography: Y. Haneda, *Sci. Rep. Yokosuka City Mus.*, 11:1–12, 1965; Y. Haneda, F. I. Tsuji, and N. Sugiyama, *Science*, 165:188–190, 1969; E. H.-C. Sie et al., *Arch. Biochem. Biophys.*, 93:286–291, 1961; F. I. Tsuji and Y. Haneda, in F. H. Johnson and Y. Haneda (eds.), *Bioluminescence in Progress*, 1966.

Biopotentials and electrophysiology

The messages transmitted through the nerve fibers are composed of small, brief electric potential changes called impulses or action potentials. Particularly notable among a number of studies on impulse conduction are (1) the discovery of an overshoot of the action potential by two independent groups about 30 years ago, A. L. Hodgkin and A. F. Huxley in England and H. J. Curtis and K. S. Cole in the United States; and (2) the development of the ionic theory by Hodgkin and Huxley in the early 1950s. Recent progress in the study of nerve conduction, which is directed toward the molecular mechanisms, depends greatly on these two developments.

Membrane potentials. The squid possesses exceptionally large nerve fibers called giant nerve fibers which have a diameter of about 500 microns (μ). There is an electric potential difference of 50–100 millivolts across the nerve membrane in the direction of the inside negative with respect to the outside. This is the resting potential. When the nerve fiber is stimulated by a brief electric shock, an action potential of about 1 msec duration and 100–120 millivolts amplitude is recorded. During the peak of the action potential, the membrane potential is reversed by an amount of 20–50 milli-

volts. An outward electric current is effective in stimulating the nerve membrane, and an action potential is initiated when the membrane potential is decreased or when the membrane is depolarized beyond a certain threshold level. *See* EYE (INVERTEBRATE).

How impulses conduct. Nerve conduction is explicable by local circuit current theory. When an impulse is propagating along the nerve fiber, a potential difference appears between the active site and the adjacent sites in such a way as to cause a local circuit current to flow outwardly across the adjacent sites. Thus an action potential is produced from the site ahead of the active site. The site behind the active site can no longer produce the action potential because it is in the refractory state. In the myelinated nerve fiber, the local circuit current can flow only across the node of Ranvier, where the myelin sheath is lacking; the action potential is transmitted from one node to the next. This is called saltatory conduction, and it speeds up the velocity of conduction. The conduction velocity is higher in larger nerve fibers, the values ranging from about 100 m/sec for large myelinated nerve fibers to less than 1 m/sec for small nonmyelinated nerve fibers.

How excitation occurs. The mechanism of production of the resting and action potentials is satisfactorily explained by the ionic theory. As shown in Fig. 1, the concentration of potassium ions is higher in the axoplasm than in the blood serum, whereas the reverse situation holds for sodium ions. The nerve membrane at rest is permeable to potassium but almost impermeable to sodium. Therefore the membrane potential approaches the equilibrium potential for potassium (inside negative with respect to outside). When the nerve membrane is depolarized, the permeability of the membrane to sodium rapidly and greatly increases, so that the membrane potential approaches the equilibrium potential for sodium (inside positive with respect to outside), forming the rising phase of the action potential. The increased sodium permeability starts decreasing soon, and the potassium permeability starts increasing beyond the resting level, so that the membrane potential is brought down to the resting level. The local circuit current across the nerve membrane stimulates the resting region ahead of the active region, thereby enabling the action potential to be propagated along the nerve fiber.

Metabolic energy plays only an indirect role in nerve excitation and conduction. The internal sodium is pumped out, and the internal potassium is retained by the metabolically supported mechanism called the sodium pump. The internal sodium concentration is slightly increased immediately after excitation, and thus the sodium pump is stimulated to restore the concentration gradient.

Elucidating molecular mechanisms. The physicochemical or molecular mechanisms whereby the nerve membrane undergoes increases in permeability to sodium and potassium upon stimulation have not been determined. Among a number of attempts to elucidate the molecular mechanisms, three new approaches that were widely used during the 1960s are worthy of note.

One approach is intracellular perfusion of squid giant axons, which was originally developed in

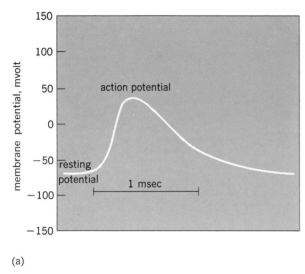

(a)

(b)   ⟵ impulse propagation

Fig. 1. Diagram illustrating the action potential mechanism. (a) Time course of an action potential. (b) Mechanism of the propagation of the action potential along a nerve fiber. Sodium ions enter and potassium ions leave the nerve fiber during the action potential.

1961 by I. Tasaki's group in the United States and by Hodgkin's group in England. These techniques permit freely changing intracellular environments, as well as extracellular ones, and allow the study of the effects of a wide variety of experimental conditions on nerve excitability and its molecular mechanisms. In Tasaki's method, two glass capillaries are inserted into a squid giant axon from both ends, and the axoplasm is removed by suction. In Hodgkin's method, the axoplasm is squeezed out by means of a small roller, and the crushed axon is inflated with an artificial internal perfusate, which usually contains a potassium salt. Surprisingly, the internal anion that is most successful in maintaining the excitability of the perfused axon is fluoride, which is a metabolic inhibitor and is toxic when applied outside the axon. When the potassium concentration gradient across the nerve membrane is altered, the resting potential changes, as is to be expected from the behavior of the potassium electrode. The action potential changes with the change in sodium concentration gradient, as is to be expected from the behavior of the sodium electrode.

Several new observations can be made only with the internal perfusion method. For example, when the squid giant axon is internally perfused with a solution containing a monovalent cation, such as sodium or choline, and bathed in a calcium chloride solution, action potentials are produced upon stimulation. This was reported by Tasaki, L. Lerman, and A. Watanabe in 1969. Full explanation for this phenomenon has not been given, but it is possible that the inward calcium flow across the nerve membrane, which is a negligible amount in normal axons, produces the action potential. Another result is the production of the normal-sized action potential in the virtual absence of the resting potential when the squid giant axon is internally perfused with a solution having a very low ionic strength, for example, 6 millimoles per liter of potassium added with sucrose to maintain the normal osmolarity. This was reported by T. Narahashi in 1963. Under these conditions, the boundary poten-

tial at the inner edge of the nerve membrane was calculated to be very high (about 80 millivolts). This creates a potential gradient within the nerve membrane, thereby maintaining the excitability, although the apparent resting potential is very small. However, owing to the difficulty of measurement of the boundary potential experimentally, this hypothesis remains speculative.

In another approach, a number of studies of artificial bimolecular membranes made of lipids and proteins is being carried out. Monomolecular lipid film spread over an aqueous solution has long been studied in surface chemistry. In 1967 P. Mueller and D. O. Rudin formed a thin bimolecular lipid membrane between two aqueous phases. When a crude proteinaceous excitability-inducing material obtained from *Aerobacter cloacae* is added, the artificial membrane becomes capable of producing a variety of electrokinetic phenomena such as those observed with the nerve membrane. These phenomena include electric potential changes associated with conductance changes, thresholds, refractoriness, anode break excitation, and spontaneous firing. These studies opened a new approach toward the molecular aspects of nerve membrane excitation.

In a third approach, certain chemicals have been found to produce highly specific actions on nerve fiber excitation. Tetrodotoxin (Fig. 2), the active principle of puffer fish poison, is particularly notable among many such chemicals, and selec-

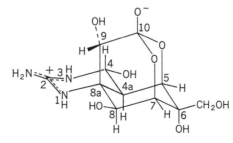

Fig. 2. Chemical structure of tetrodotoxin.

tively inhibits the sodium permeability increase without affecting any other parameters at a very low concentration ranging from 10^{-8} to 10^{-9} mole per liter. Since 1964, when this selective action was first reported by Narahashi, J. W. Moore, and W. R. Scott, tetrodotoxin has become a popular tool to study nerve excitation.

When a crayfish stretch receptor is treated with tetrodotoxin, the discharge of action potentials is completely inhibited but the receptor potential remains unchanged. Therefore the receptor potential can be studied in detail without being disturbed by action potentials. In a frog neuromuscular preparations exposed to tetrodotoxin, the release of the transmitter substance from the nerve terminal and the sensitivity of the end-plate potential to the transmitter are unaffected, whereas the action potentials in the nerve and muscle are completely blocked. This gives scientists an excellent opportunity to study the mechanism of neuromuscular transmission. By the measurement of the amount of tetrodotoxin bound to the membrane, the number of sodium channels in the nerve membrane was estimated to be a maximum of 13 per square micron of membrane. These studies of tetrodotoxin stimulated the use of various other chemicals as tools. They include tetraethylammonium, DDT, batrachotoxin, scorpion venoms, and saxitoxin.

Electrophysiology, interdisciplinary between physiology and pharmacology, is expected to continue to contribute to the study of the nervous function.

For background information *see* BIOPOTENTIALS AND ELECTROPHYSIOLOGY; NERVOUS SYSTEM (VERTEBRATE) in the McGraw-Hill Encyclopedia of Science and Technology.   [TOSHIO NARAHASHI]

Bibliography: H. J. Curtis and K. S. Cole, *J. Cell. Comp. Physiol.*, 15:147–157, 1940; A. L. Hodgkin, *The Conduction of the Nervous Impulse*, 1964; A. L. Hodgkin and A. F. Huxley, *Nature*, 144:710–711, 1939; P. Mueller and D. O. Rudin, *Nature*, 213:603–604, 1967; T. Narahashi, *J. Physiol.*, 169:91–115, 1963; T. Narahashi, J. W. Moore, and W. R. Scott, *J. Gen. Physiol.*, 47:965–974, 1964; I. Tasaki, L. Lerman, and A. Watanabe, *Amer. J. Physiol.*, 216:130–138, 1969.

Biotelemetry

Radio telemetry has become an invaluable technique for obtaining physiological information from active unrestrained animals and from man. It has been used for such diverse applications as monitoring astronauts in space, for observations of hospitalized patients, and for observations of experimental animals in the laboratory or their natural environment. Biotelemetry is being used to monitor human response to environmental stress (exercise, heat, cold, space flight), critical illness (myocardial infarction, postsurgery), birth (fetal heart sounds), during surgery and its associated anesthesia, and any situation in which the elimination of encumbering wires or the direct connection of the subject to a recorder is desirable. For animal research, biotelemetry is essential if meaningful physiological data are to be obtained under normal conditions of rest and exercise when the animal is awake, unanesthetized, and without restraint. Both single and multichannel system capability

now exist for transmission of signals from the body surface (backpack) or from surgically implanted units. Certain transmitters need send their signals only a few centimeters; others provide useful information from distances of 20 mi or more. Work with implantable units is now being directed at continued reduction in unit size, improved reliability of operation, improved sealing from body fluids, improved power sources, and the ability to measure a wider variety of parameters.

All presently planned in-flight experiments using subhuman primates or large vertebrates incorporate the use of biotelemetry systems. When a weak signal is received from an internal transmitter, a booster transmitter is carried by, or provided for, the subject to reradiate the signal for a great distance or to a central receiving point. Plans are in progress not only to make biologic observations from satellites but to use satellites to track migration patterns or biologic behavior of various animals from externally and internally implanted transducers.

Technology. The physiological parameters that can be measured by telemetry are (1) bioelectric potentials such as the electrocardiogram (ECG), electromyogram (EMG), and electroencephalogram (EEG), where the signal is obtained directly in electrical form, and (2) transduced variables such as pressure, flow, and temperature, where the signal is obtained indirectly from a transducer after appropriate excitation.

In a typical system the appropriate analog signal resulting from the generated biopotential or transducer is put into a form or code for transmission, transmitted, and then decoded at the other end in its original form, as schematically illustrated in Fig. 1. The choice of operating frequency, transmitting and receiving antennas, modulation scheme, and receiver must all be considered in optimizing the system. Monitoring a small caged animal presents a different situation from that of monitoring a human in an orbiting vehicle or an animal moving in its natural habitat. The use of subcarrier modulation, such as frequency modulation (FM), pulse-code modulation, or pulse-interval modulation, is critical and provides data transmission without significant loss of accuracy. The choice of receiver depends on the selected transmission frequency. The commercial FM band has proved most satisfactory for laboratory experiments in providing reliable operation at low cost. The more expensive telemetry receivers (tunable from 30 to 250 MHz) allow interference-free selection of many frequencies and are desirable for ap-

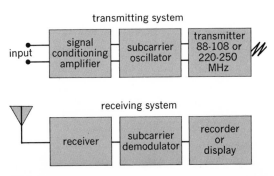

Fig. 1. Transmitting and receiving biotelemetry system.

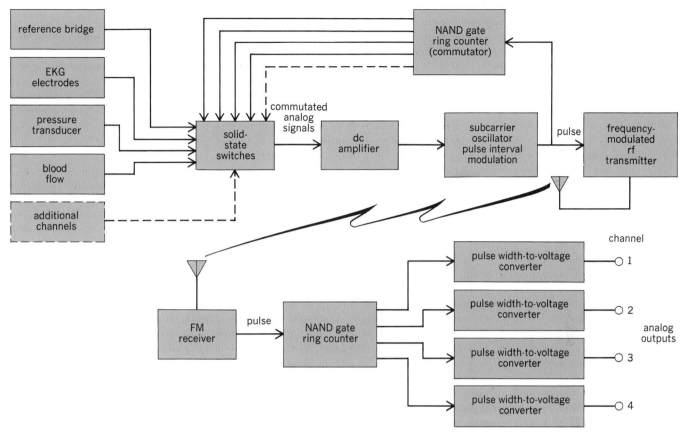

Fig. 2. Block diagram of a multichannel telemetry system.

plications needing maximal range and where critical measurements are being made.

Physiological data from astronauts in the Gemini and Apollo programs are presently obtained by surface electrodes or sensors, and have consisted of the ECG, body temperature (oral or rectal thermistors), respiration (impedance pneumograph), blood pressure (arm cuff and microphone), and EEG (surface electrodes). The recent biosatellite flights have extended such physiological measurements from transducers placed on the body surface to sensors chronically implanted into the central nervous system (EEG, temperature) and to catheters for measuring intravascular pressures and sampling blood. The variables measured by such advanced devices have been collected by appropriate on-board physiological monitoring systems and subsequently relayed to Earth for analysis and interpretation during flight.

Implantable telemetry units. Recent developments in transistors and integrated circuits have permitted the design of sophisticated circuitry which is small in size and low in power, with high reliability of operation. Units are now available which can be swallowed by animals or man to monitor intestinal mobility, pH, or deep-body temperature or any combination of the three. Single-channel transmitter units have also been surgically implanted in animals (ranging in size from a mouse or a bird to a chimpanzee) for measurement of ECG, EEG, body temperature, intraocular or intravascular pressures, cardiac dimensions, and blood flow. Proper sealing of these circuits from moisture and attack by body fluids is an important and

critical task. System failure due to breakage of wires connecting the transducer or battery pack to the transmitter and corrosion of internal connections by the presence of body fluids are major problems that have plagued developers of implanted systems. High-impedance circuit design, which markedly reduces power consumption from batteries, makes proper sealing even more important. At present the only successful long-term (at least 1–2 years) sealing techniques are accomplished by the use of a hermetically sealed glass, metal, or ceramic container. Plastics (epoxy or vinyl) and waxes have proved useful for shorter-term sealing or where critical high-impedance circuitry is not involved. An outer coating of silastic is usually used, not to seal against moisture but to minimize the reaction of body tissues to the implant.

Multichannel implantable unit. For investigations in which more information than ECG, body temperature, or simple location and tracking of an animal is needed, multichannel capability is required. Many multichannel systems have been reported for use as backpack devices but either are too large or consume excessive power for implant application. Although several single-channel units can be chronically implanted, this approach is impractical if a large number of channels are to be recorded, as during neurological, circadian rhythm, or cardiovascular studies. A chronically implantable multichannel unit (3–10 channels) capable of accepting inputs from a wide variety of sensors has been developed for these purposes. The operation of such a unit is schematically illus-

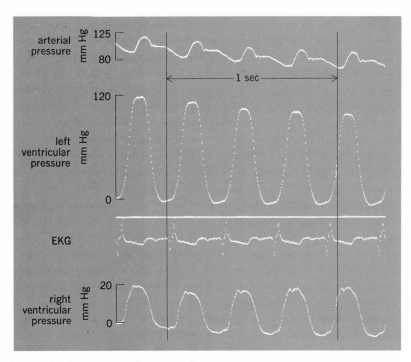

Fig. 3. Recordings from a five-channel implanted system.

trated in Fig. 2. Multichannel operation is provided by a time-sharing multiplex scheme for commutation between various sensor inputs. Data are transmitted on a single carrier frequency, and only one receiver is required for recording 3–10 channels of data. Sampling rate is approximately 0.7 msec per channel. The commutated analog voltages control the period between subcarrier pulses. One long duration pulse is added to each data cycle to provide synchronization. The received demodulated pulses are decommutated and converted from pulse period to an analog signal by subsequent integration and a sample and hold circuit. A typical recording is shown in Fig. 3. This oscilloscopic tracing was obtained in a dog 6 months after implantation. The interruptions in the recordings are due to the time-shared multiplex operation. To conserve battery power and extend operating life, a remotely actuated radio-frequency (rf) switch has been incorporated into these units. With intermittent samplings, the 500 hr of continuous operation with two pacemaker cells (1000 ma/hr) can be extended to periods in excess of 1–2 years.

Power source. The key component in a biotelemetry system is the power source. It represents the largest component and the one most likely to fail. Mercury cell and silver oxide batteries are the power sources of choice at present. Other future sources of power are atomic cells, conversion of biopotentials (implanted stainless steel or platinum—platinum black electrodes), mechanical conversion devices (a piezoelectric crystal in conjunction with a self-winding watch or accelerometer), and radiation of energy into the body from an external source (rf or optical). The proposed use of the telemetry system (in a large or small animal, caged or uncaged) dictates the choice of battery size or power supply to a large extent.

Future. The use of telemetry for biomedical purposes has grown rapidly during the past few years. The potential for the use of associated techniques during space flight, in the laboratory, at the bedside, or in clinical medical practice has not been even partially realized.

For background information *see* BIOTELEMETRY in the McGraw-Hill Encyclopedia of Science and Technology.   [HAROLD SANDLER]

Bibliography: T. B. Fryer, *Implantable Biotelemetry System Developments at Ames Research Center*, NASA-SP 5094, 1970; T. B. Fryer, H. Sandler, and B. Datnow, *Med. Res. Eng.*, 8:9–15, 1969; R. D. Lee and H. Sandler, *J. Appl. Physiol.*, 28: 110–112, 1970; R. S. Mackay, *Biomedical Telemetry*, 1968.

Blood

Numerous clotting factors in the fluid portion of the blood, called plasma, interact sequentially during blood coagulation. Recently, clarification of the means of activating these clotting enzymes and the order of their interactions with one another has shown that some of these enzymes form complexes with lipids in order to be effective in coagulation. Much information has been obtained through studies of blood from persons with inherited deficiencies of a single clotting factor, such as hemophiliacs who are deficient in antihemophilic factor activity, either because they cannot make this protein or because they make the wrong protein. Blood coagulation can affect inflammatory processes, for substances which can mediate inflammatory changes may be released once coagulation is initiated.

Clot formation. In the body, small fragments of cells, called platelets, cover minor injuries that disrupt the linings of blood vessels, since they readily stick to tissues exposed by disruption. Although this tiny plug has limited ability to stop bleeding, the clotting of plasma can reinforce it and aid in preventing hemorrhage.

When withdrawn from the body, blood normally forms a gel in a few minutes, even though it remains liquid within the blood vessels. This transformation to a fibrin clot occurs because polymers are formed of microscopic fibrils. For these fibrils to form, the enzyme thrombin attacks fibrinogen, which is the soluble precursor of fibrin normally present in the circulating blood. Thrombin is a protein-splitting enzyme which removes two pairs of chains of amino acids (fibrinopeptides) from fibrinogen, leaving behind larger fragments called fibrin monomers. These monomers are only transiently soluble and quickly polymerize with one another, making a meshwork of fibers which compose a blood clot. The fibrin is normally reinforced by a fibrin-stabilizing enzyme (FSF, factor XIII) which creates cross-links between fibrin monomers in a chemical reaction which increases the strength of the blood clot and increases its ability to repair a wound.

Intrinsic mechanism of plasma coagulation. One way of initiating the many steps to thrombin formation is by exposing the blood to glass or a similar surface. A specific plasma-clotting factor known as Hageman factor (factor XII) must be present for glass to promote clotting. This factor is markedly deficient in the blood of persons who have inherited Hageman trait; their blood clots very slowly in a glass test tube, in contrast to normal blood, which contains Hageman factor. Other substances which can convert Hageman factor to a clot-promoting substance include kaolin, diatomaceous earth, barium carbonate, cellulose sulfate, ellagic acid, heparin, chondroitin sulfate, and sebaceous secretions of the skin. When the inner lining of the blood vessels is disrupted, the underlying collagen which is then exposed to the flowing blood can also activate Hageman factor, and may be an important factor in starting clotting in the body following injury to the blood vessels. The negative charge of these substances is important in Hageman factor activation.

In addition to promoting clotting, activated Hageman factor can initiate the function of other enzymes in the blood plasma which ultimately increase vascular permeability, contract isolated smooth muscles, induce pain and dilatation of the blood vessels (thus increasing blood flow to the area), and even cause the white blood cells to stick to injured areas of blood vessels and subsequently traverse this barrier into surrounding tissues. Since these changes characterize inflammation, activated Hageman factor is clearly important in the genesis of inflammation. The substances which actually cause the inflammatory changes are, in part, small polypeptide molecules split away from certain plasma proteins by enzymes secondarily activated by activated Hageman factor. These polypeptides are referred to as kinins. Activated Hageman factor is possibly as important to the defense of the body through its initiation of inflammation as it is through its coagulation function.

During clotting a complex series of events follows the activation of Hageman factor, for it next interacts with another clotting factor called plasma thromboplastin antecedent (PTA or factor XI), converting it to activated PTA (Fig. 1). Activated PTA then interacts with a clotting factor called Christmas factor (factor IX), provided a suitable concentration of calcium is available. Activated Christmas factor then forms a complex with antihemophilic factor (factor VIII), phospholipid, and calcium, and the activated complex converts Stuart factor (factor X) from an inactive to an active form. Next, another complex is formed of phospholipid, activated Stuart factor, calcium, and proaccelerin (factor V) which then acts upon prothrombin (factor II), apparently splitting the prothrombin molecule into thrombin, which attacks fibrinogen, and an inactive portion.

Thrombin can regulate the rate of its own formation by affecting at least two steps in the coagulation mechanism either in a positive or a negative way, depending on the amount of thrombin present. A small amount of thrombin can change antihemophilic factor and proaccelerin so that each is more reactive in the particular step in which they function during coagulation. Thus thrombin can enhance formation of more thrombin. On the other hand, when a large amount of thrombin is present, it can destroy both antihemophilic factor and proaccelerin activity, thus inhibiting the formation of more thrombin. This negative feedback may inhibit the formation of excess thrombin likely to form large obstructive clots in the blood vessels. The precise mechanism of these thrombin effects remains obscure.

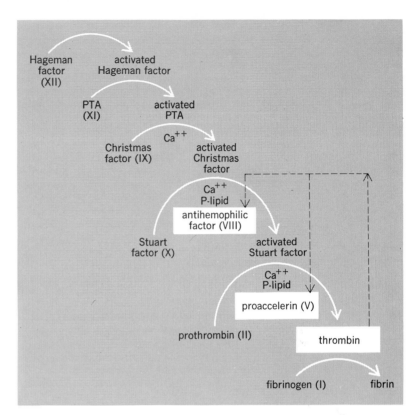

Fig. 1. The sequence of activation of plasma-clotting factors in the formation of thrombin after Hageman factor has been activated; the broken lines indicate the effect of thrombin on earlier-acting clotting factors.

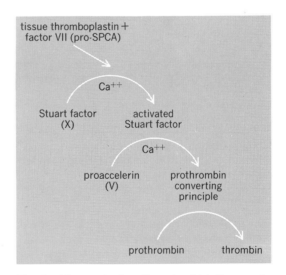

Fig. 2. The mechanism through which tissue substances (thromboplastins) cause clotting of plasma.

Extrinsic mechanism of plasma coagulation.
A shortcut exists through which body tissues can initiate the formation of thrombin in the blood without apparent activation of Hageman factor, Christmas factor, and antihemophilic factor (Fig. 2). The effective tissue fractions are complexes of lipids and proteins which interact with calcium and another plasma clotting factor, called factor VII, to convert it to an activated form. This activated complex converts Stuart factor into activated Stuart factor, which then combines with proaccel-

erin, phospholipid, and calcium to convert pro-thrombin to thrombin, in a manner probably identical to that of the late steps in clotting induced by activated Hageman factor.

The intricate steps of clotting described here have been defined through many experiments, necessarily done in the test tube instead of the living body. It is therefore difficult to judge the directness of the relationship of these observations to physiological events in a living person, since other substances in the living body may complicate clotting. Even so, the importance of these clotting steps to survival is illustrated by the bleeding tendencies associated with acquired or inherited deficiencies of individual clotting factors. With the exception of Hageman factor, those persons sufficiently deficient in one or more clotting factors are subject to repeated hemorrhages following minimal injury.

It is easy to imagine that the stepwise interacting enzymes might amplify the effect of a tiny amount of activated Hageman factor many thousands of times, since a molecule of activated Hageman factor may in turn activate many molecules of PTA, and so on. The formation of complexes may limit endless augmentation of a small amount of activity and thereby limit the rate of formation of thrombin. Additional regulators exist in the blood plasma, because multiple inhibitors can impair the action of most, if not all, of the active clotting substances. In addition, activated complexes can probably be removed from the blood by the reticuloendothelial system, which is a complex network of cells in many of the body's organs that can ingest noxious substances brought to it by the circulating blood.

For background information *see* BLOOD; HEMORRHAGE; INFLAMMATION in the McGraw-Hill Encyclopedia of Science and Technology.

[VIRGINIA H. DONALDSON]
Bibliography: V. H. Donaldson, *Ser. Haematol.*, 3:1, 1970; M. P. Esnouf and R. G. Macfarlane, *Advan. Enzymol.*, 30:255–315, 1968; L. Poller (ed.), *Recent Advances in Blood Coagulation*, 1969; O. D. Ratnoff, in *Advances in Immunology*, vol. 10, 1969.

Calcitonin

Calcitonin is a polypeptide hormone that controls bone resorption. Recently, human and salmon calcitonin, both of which are active in man in controlling excessive bone turnover, have been characterized and synthesized.

Discovery. In 1961 D. H. Copp and coworkers originally postulated the existence of a hormone that controls hypercalcemia, to which they gave the name calcitonin because it was concerned with "calcium tone." This hormone was first extracted from the thyroid gland of rats by P. L. Munson, P. F. Hirsch, and associates in 1963. Shortly afterward G. V. Foster, I. MacIntyre, and A. G. E. Pearse showed that in mammals the principal source of calcitonin is the parafollicular or calcitonin (C) cells of the thyroid. In lower vertebrates the C cells form a distinct organ, the ultimobranchial body, from which the hormone can be extracted; in mammals most of the C cells migrate to the thyroid during development but some cells also migrate to the parathyroids and the thymus.

Chemistry. Recent rapid progress has led to the isolation, elucidation of structure, and total synthesis of several calcitonins. The amino-acid sequences of porcine, bovine, human, and salmon calcitonin are shown in Fig. 1. They are single chains of 32 amino acids, with a proline amide at the C terminus and a 1–7 disulfide bridge at the N terminus. There are great similarities in the sequences of the different calcitonins along the first third of the chain from the N terminus, but large differences along the rest of the chains, except that all have a glycine residue at position 28.

Assay. The principal assay has been to measure the degree of hypocalcemia that a given preparation produces after injection into a young rat, compared with a standard preparation. Recently, sensitive radioimmune and tissue-culture assays have been described which should be valuable adjuncts to the standard method.

Action. In young animals calcitonin causes a rapid fall in the level of plasma calcium by inhibiting the resorption of bone. John J. Reynolds showed that calcitonin prevents the removal of bone mineral by multinucleate osteoclasts, but the precise biochemical mechanism of action is unknown. Calcitonin has a renal effect that is similar to that of parathyroid hormone; both human and porcine calcitonin cause a marked phosphaturia in parathyroidectomized rats. These two calcitonins also have a phosphaturic effect in man, but because the doses needed are usually larger than those necessary to lower plasma calcium, there is still controversy as to whether this renal action is physiological.

Although it has been conclusively demonstrated that calcitonin blocks bone resorption, the precise physiological role of the hormone remains uncertain. Figure 2 shows the possible functions of calcitonin and parathyroid hormone in regulating the level of plasma calcium. The two hormones oppose each other in controlling the release of calcium from bone, and since the level of plasma calcium operates a feedback system in controlling the release of the hormones, the plasma calcium usually remains constant around 10 mg %. The uncertainty as to the physiological role of calcitonin lies in the fact that although bone turnover is large enough to make a significant contribution to the pool of plasma calcium in young animals, bone remodeling decreases with age. Thus it seems likely that in mature animals the principal control of plasma calcium is by the actions of parathyroid hormone on the kidney to increase tubular reabsorption of calcium, and on the gut to increase absorption. However, calcitonin circulates in the plasma of normal animals, and since stresses of hypercalcemia are relatively uncommon, perhaps it continues to have a physiological role in mature animals in the control of bone-cell modulation. Calcitonin is known to block the differentiation of new bone-resorbing cells, and perhaps this action, which counteracts the effect of parathyroid hormone on bone, is an important function of calcitonin in all stages of life.

Recently, D. H. Robertson removed the ultimobranchial gland from frogs and observed that the level of plasma calcium rose gradually during 2

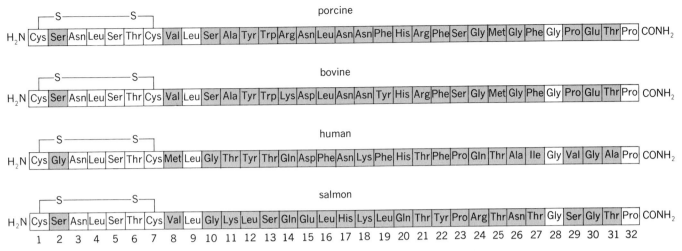

Fig. 1. Amino-acid sequences of calcitonins. Unshaded residues represent amino acids common to all four molecules.

weeks in which the paravertebral lime sacs became depleted of calcium carbonate. These results suggest that calcitonin inhibits the flux of calcium from the lime sacs and that the hormone may be concerned in the transport of calcium in cells other than those of bone.

In man. The effect of calcitonin injection on the level of plasma calcium in man is small because of low bone turnover in the normal adult but is much more dramatic in diseases of excessive bone resorption.

Therefore several groups have been able to show that calcitonin reduces the hypercalcemia of malignancies of bone, hyperparathyroidism, hypervitaminosis D, idiopathic hypercalcemia of infancy, and Paget's disease. O. L. M. Bijvoet and colleagues found that in Paget's disease porcine calcitonin produced an acute fall in the urinary excretion of hydroxyproline, indicating that bone resorption had been blocked. MacIntyre and colleagues showed that human calcitonin is effective in producing remission in patients with Paget's disease, and J. Potts and coworkers reported that salmon calcitonin is even more effective than human calcitonin in man, but it is not certain whether long-term therapy with nonhuman calcitonins will lead to the development of antibodies. Recent knowledge of the structures of calcitonins should allow more potent analogs of human calcitonin to be synthesized.

Medullary carcinoma of the thyroid is a cancer of the C cells. This tumor secretes calcitonin in large quantities and produces high circulating levels of the hormone, but such patients rarely have an abnormal level of plasma calcium. Probably any tendency toward hypocalcemia is counteracted by a compensatory hypersecretion of parathyroid hormone; this would explain why these patients do not always become severely osteopetrotic. L. Kraintz and K. Intscher have described a corresponding situation in the domestic fowl: Calcitonin was inactive in chickens unless they had been partially parathyroidectomized, thereby reducing the circulating level of parathyroid hormone. The compensatory effects of parathyroid hormone may also be a problem in long-term therapy of patients

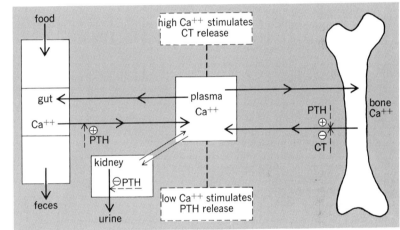

Key:
⊖→ = hormone action to decrease calcium flux
⊕→ = hormone action to increase calcium flux
→ = calcium movement
CT = calcitonin
PTH = parathyroid hormone

Fig. 2. Schematic representation of the actions of calcitonin and parathyroid hormone in controlling the level of plasma calcium.

with osteoporosis.

Much interest has been aroused in the possible use of calcitonin in the osteoporosis of old age and of steroid therapy; however, it is still too early to judge whether bone strength will be significantly improved.

For background information *see* BONE; HORMONE, PARATHYROID; PARATHYROID GLAND (VERTEBRATE) in the McGraw-Hill Encyclopedia of Science and Technology.

[JOHN J. REYNOLDS]

Bibliography: D. H. Copp, *Amer. Rev. Physiol.*, 32:61, 1970; P. F. Hirsch and P. L. Munson, *Physiol. Rev.*, 49:548, 1969; J. J. Reynolds, in J. T. Dingle and H. B. Fell (eds.), *Lysosomes in Biology and Pathology*, vol. 2, 1969; S. Taylor and G. V. Foster (eds.), *Calcitonin 1969*; *Proceedings of the 2d International Symposium*, 1970.

Caulobacteraceae

Stalked bacteria, which form the family Caulobacteraceae and which occur commonly in soil and water, have received attention in recent years with respect to their physiology, taxonomy, and interactions with other microorganisms. Bacteriophages specific for bacteria of the family Caulobacteraceae have been isolated, and among these, phages of two different nucleic acid types, double-stranded deoxyribonucleic acid (DNA) and single-stranded ribonucleic acid (RNA), were found. The host bacteria for these viruses, representing several species of the genera *Caulobacter* and *Asticcacaulis*, each produce a cellular stalk or prostheca, and are of special interest because of their dimorphic life cycle. Reproduction in the Caulobacteraceae, which occurs asexually by asymmetric binary fission, gives rise to morphologically distinct daughter cells. One of the daughter bacteria, the swarmer, is motile by means of a single polar flagellum. The other daughter cell is nonmotile and possesses the stalk of the parent. The swarmer does not undergo division until it has initiated stalk formation, and the dividing caulobacters always possess a recognizable stalk at the time fission occurs.

The first large-scale attempts to isolate *Caulobacter* bacteriophages were undertaken to provide phages for host-range studies in conjunction with a taxonomic study of the Caulobacteraceae. When both DNA and RNA phages were found among the isolates, the infective processes of these two types of viruses and their effects on the host bacterium's morphogenetic process also invited investigation, since the life cycle of the caulobacters introduced some complexities not encountered in many other phage-bacterial host systems.

Bacteriophage isolation. Many lytic bacteriophages specific for caulobacters have been isolated from fresh water and sewage. Enrichment techniques usually have been required for the phage isolations, since the *Caulobacter* bacteriophages are not plentiful enough to be detected by direct sampling methods. Species for which DNA phages have been isolated include *C. crescentus*, *C. bacteroides*, and *C. vibrioides*; DNA phages that lyse a strain of the excentrally stalked bacterium *A. excentricus* have also been obtained. The *Asticcacaulis* phages do not attack species of *Caulobacter*, and in the reciprocal situation, *Caulobacter* phages do not lyse *Asticcacaulis* strains. These observations tend to support the taxonomic separation, made on the basis of morphological characteristics and DNA base ratio analyses, of these stalked bacteria into different genera. A number of gram-negative bacteria of other genera (*Pseudomonas*, *Xanthomonas*, and *Erwinia*) have been examined for susceptibility to *Caulobacter* and *Asticcacaulis* phages, and phages attacking these other genera have been tested against the stalked bacteria; in all cases, negative results were obtained. The stalked bacteria appear to be unrelated to these other bacteria on the basis of phage susceptibility.

Bacteriophages containing RNA have been isolated from sewage. Three serologically distinct phage types attack respectively *C. fusiformis*, *C. bacteroides*, and *C. crescentus* and *C. vibrioides*; their host ranges are very limited. The swarmers attacked by the RNA phages possess filamentous structures, called pili or fimbriae, which emanate from the flagellated pole. They are approximately 4 nm in diameter (much thinner than bacterial flagella) and several micrometers in length. The pili serve as the adsorption site for the specific RNA phage attacking that species. The pili are usually lost during morphogenesis of the swarmer into a stalked bacterium, and the RNA phages adsorb only negligibly to stalked cells that have been mechanically separated from swarmers.

The DNA-containing *Caulobacter* phages adsorb by their tail to the cell wall of their gram-negative hosts, and they are not associated specifically with flagella or pili during their adsorption process.

Lysogeny. Bacteriophages that are apparently temperate have been induced from *C. crescentus* and *C. vibrioides* strains using Mitomycin C or ultraviolet irradiation as inducing agents. Suitable bacterial indicators on which these phages could undergo lytic replication included strains of *C. crescentus* and *C. henricii*. The strains from which the phages were induced were resistant to infection by the same phage, and antisera effective in neutralizing the lytic activity of the free phage particles did not eliminate the capacity of the inducible strains to produce the phages, suggesting that the bacteria were lysogenic. Several other surveys for lysogeny that used many *Caulobacter* strains have given negative results, indicating that lysogeny may seldom occur or may be difficult to detect in the stalked bacteria. Some defective phage-like particles for which no suitable productive host strain has been found have also been induced from caulobacters.

Morphology. The *Caulobacter* and *Asticcacaulis* phages have been examined with the electron microscope using the negative-contrast method. The *Caulobacter* RNA phages are polyhedral, 23 nm in diameter, and tailless. They resemble the male-specific RNA phages of *Escherichia coli*. One type of DNA phage has an elongated head, 170×50 nm, and a tail which, when stained at a neutral pH, appears flexible. The tail is 10 nm in diameter and 200–250 nm long. Other DNA phages have more conventional morphologies, the heads having a hexagonal outline, with a size range of 55–70 nm. The tails are rigid and noncontractile.

Phage production. Latent periods of the *Caulobacter* phage–host bacterium systems, determined in one-step growth experiments, indicated that phage release began 2–3 hr after infection. The generation times of the hosts were 2–5 hr, and the kinetics of phage production may reflect the slow growth rates. A double-stranded RNA replicative intermediate has been found in RNA phage–infected bacteria. Stalk formation can proceed in the RNA phage–infected swarmers. RNA phage release occurs slowly, either from localized sites of rupture in the cell wall, without generalized loss of the murein rigidity layer, or by passage through the cytoplasmic membrane and wall without obvious rupture. The DNA phages mature near the periphery of the bacterial cell and cause loss of the murein wall layer of the lysing bacterium. The average burst sizes (yield of phage particles per bacterium) are very small.

It is possible that bacteriophage studies may lead to the discovery of a genetic transfer system, transduction, in the stalked bacteria.

For background information *see* BACTERIO-PHAGE; CAULOBACTERACEAE in the McGraw-Hill Encyclopedia of Science and Technology.

[JEAN M. SCHMIDT]

Bibliography: L. J. Driggers and J. M. Schmidt, *J. Gen. Virol.*, 6:421–427, 1970; J. S. Poindexter, *Bacteriol. Rev.*, 28:231–295, 1964; J. M. Schmidt, *J. Gen. Microbiol.*, 45:347–353, 1966; J. M. Schmidt and R. Y. Stanier, *J. Gen. Microbiol.*, 39: 95–107, 1965.

Channeling in solids

The study of channeling, the steering of energetic ions by the atomic rows or the atomic planes of a solid, began in 1963. M. Y. Robinson and O. S. Oen made calculations of the slowing down of 5-kev copper ions in the face-centered cubic copper lattice. They found that copper ions shot into a single crystal along the $\langle 110 \rangle$ or the $\langle 100 \rangle$ direction penetrate deeper than if they are shot in a random direction. Experimental corroboration came in three experiments also performed in 1963. G. R. Piercy, F. Brown, J. A. Davies, and M. McCargo and H. O. Lutz and R. Sizmann made range measurements of radioactive ions shot into single crystals. They found deep penetration into aluminum and copper single crystals when the ions were incident in the $\langle 110 \rangle$ or the $\langle 100 \rangle$ direction. R. S. Nelson and M. W. Thompson measured the ion current transmitted through 3500-A-thick single crystals of gold as a function of orientation (Fig. 1). Well-collimated 75-kev protons were incident. The following sections summarize the current status of theoretical and experimental investigations, and the applications of these studies.

Theory. Two kinds of channeling exist: channeling along certain crystallographic directions and channeling between atomic planes.

J. Lindhard has used classical theory to calculate the average potential $\bar{V}(r)$ (that is, averaged along a direction parallel to the row) seen by a channeled ion moving near a row of atoms. Here r is the distance of the ion from the row represented by Eq. (1), where d is the separation of atoms along

$$\bar{V}(r) = \frac{Z_1 Z_2 e^2}{d} \log \left\{ \left(\frac{a\sqrt{3}}{r} \right)^2 + 1 \right\} \quad (1)$$

the row; a is the Thomas-Fermi screening radius, which is 0.46 A/$\sqrt{Z_1^{2/3} + Z_2^{2/3}}$; Z_1 and Z_2 are the atomic number of the channeled ion and the crystal atom, respectively; and e is the elementary charge. Lindhard found that the critical angle ψ_c between the path of the ion and the row direction for which channeling occurs is given by Eq. (2),

$$\psi_c = \sqrt{\frac{2Z_1 Z_2 e^2}{dE}} \quad (2)$$

where E is the kinetic energy of the ion. In planar channeling Lindhard found that the average potential at distance r from the plane is given by Eq. (3), and that the critical angle above which no channeling occurs is given by Eq. (4). In Eqs. (3)

$$\overline{V_p(r)} = 2\pi Z_1 Z_2 e^2 N_p \left\{ \sqrt{r^2 + 3a^2} - r \right\} \quad (3)$$

$$\psi_p = \sqrt{\frac{2\pi N_p Z_1 Z_2 e^2 a}{E}} \quad (4)$$

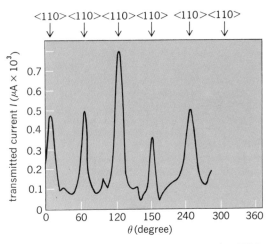

Fig. 1. Transmission of 75-kev protons through a 3000-A-thick gold single crystal as a function of orientation. The foil was rotated about its normal, which coincided with a $\langle 111 \rangle$ axis. The proton beam was incident at 35° from the normal. (*After R. S. Nelson and M. W. Thompson, Phil. Mag., 8:1677, 1963*)

and (4), N_p is the atomic density in the plane. The equations given above are for high energy, that is, where E has the value given in Eq. (5).

$$E > \frac{M_1 Z_1^{4/3} e^4}{2\hbar^2} \quad (5)$$

O. B. Firsov calculated the energy loss of a well-channeled ion. Typically it is about a third or a quarter of the energy loss of a similar ion moving in a random direction through a crystal. Two processes are responsible for the stopping of a slow heavy ion: elastic encounters with atoms as a whole (nuclear collisions); and inelastic collisions, in which energy is transferred to electrons as excitation or ionization energy (electronic collisions). At

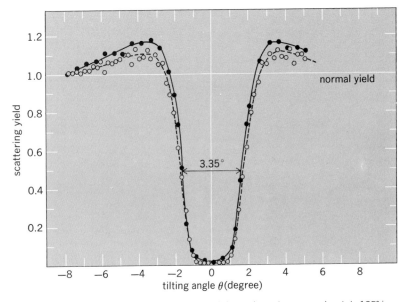

Fig. 2. The yield of 500-kev protons scattered through angles approximately 135° in a tungsten single crystal versus the angle θ between the incident proton beam and the $\langle 100 \rangle$ direction in tungsten. The broken curve is a theoretical prediction based on J. Lindhard's theory. (*After J. U. Anderson and E. Uggerhoj, Can. J. Phys., 46:517, 1968*)

sufficiently high energies the electronic losses predominate. Then Eq. (6) holds. Here η is the

$$\left(\frac{dE}{dx}\right)_e = -\frac{6.1 \times 10^{-8}\eta(Z_1+Z_2)^{\frac{5}{3}}\sqrt{\dfrac{E}{M_1}}}{\{1+0.31\,b\,(Z_1+Z_2)^{\frac{1}{3}}\}^5} \quad (6)$$

number of electrons encountered per unit path length and b is half the channel width in angstroms.

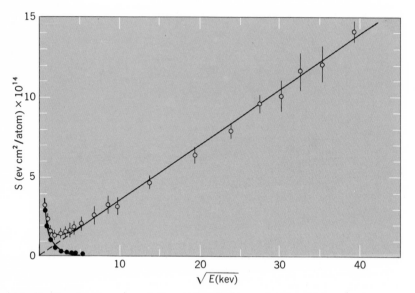

Fig. 3. The stopping cross section derived from maximum range data for xenon in the $\langle 100 \rangle$ direction of tungsten. Open circles represent experimental points; dashed line represents extrapolated electronic stopping; and solid circles represent derived nuclear stopping. If E, the incident kinetic energy, exceeds 25 kev, the stopping is mainly electronic. (After L. Eriksson et al., Appl. Phys. Lett., 10:323, 1967)

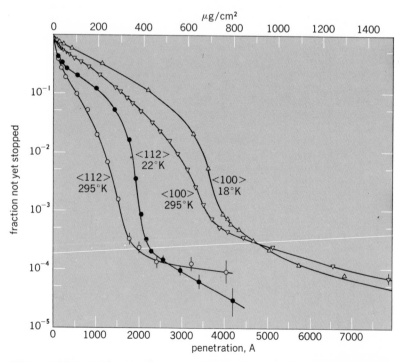

Fig. 4. Fraction of 40-kev xenon ions not yet stopped in the $\langle 100 \rangle$ and the $\langle 112 \rangle$ directions in tungsten as a function of depth. Note the influence of temperature. (After C. M. Howe and O. A. Channing, Can. J. Phys., 46:2467, 1967)

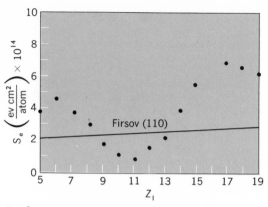

Fig. 5. The stopping cross section for various ions in the $\langle 110 \rangle$ direction of silicon measured by studying the energy loss suffered by ions transmitted through a thin foil. All ions have an incident velocity of 1.5×10^9 cm sec^{-1}. (After F. H. Eisen, Can. J. Phys., 46:561, 1968)

Experimentation. In general, the observations are in good agreement with the theory. Figure 2 shows the number of protons backscattered into a direction near the incident direction from a tungsten single crystal. Note that protons incident along the $\langle 100 \rangle$ direction are not backscattered; they are channeled. The dashed curve is theoretical ($2\psi_c = 3.35°$). Figure 3 shows the stopping cross sections derived from maximum range data for xenon along the $\langle 100 \rangle$ direction in tungsten. At low energy, nuclear collisions provide most of the stopping; at high energy, electronic effects predominate.

It is expected that channeling should improve at low temperatures, since at low temperature the guiding atomic rows should be more perfect. Figure 4 shows results for 40-kev xenon in tungsten. Some increase in range is found at low temperature. The extremely long tails at about the 10^{-4} fraction, which have not yet stopped, probably result from the diffusion of interstitial xenon.

Recently F. H. Eisen measured the electronic stopping cross sections $S_e = 1/N(dE/dX)_e$ for various ions incident in the $\langle 110 \rangle$ direction of a silicon crystal. The ion energies were adjusted so that the incident velocity was constant (1.5×10^9 cm/sec) since the electronic energy loss is proportional to the velocity (that is, to $\sqrt{E/M_1}$) in this energy range.

Note the oscillation in S_e versus Z_1 as shown in Fig. 5. I. M. Cheshire, G. Dearnaley, and J. M. Poate explained this oscillation as arising from the variation in the ion core size with Z_1.

Channeling of electrons and positrons was observed by E. Uggerhoj and J. U. Anderson. At energies of millions of electron volts, such particles are expected to obey the above classical theories, but at lower energies, quantum effects (that is, diffraction effects) should be expected.

Applications. Backscattering of channeled ions was used by J. W. Mayer and coworkers and J. A. Davies and coworkers to locate interstitial impurities. Accurate energy measurements of the backscattered ions can give information about the depth of the scattering event below the incident surface, as shown by J. E. Westmoreland and coworkers. Channeling is being used to insert known amounts of various impurities. Davies and coworkers found evidence that the track left behind the channeled ion consists of amorphous material in semiconducting crystals.

For background information *see* CRYSTAL DEFECTS; CRYSTAL STRUCTURE; CRYSTALLOGRAPHY; IONIC CRYSTALS in the McGraw-Hill Encyclopedia of Science and Technology.

[JAMES S. KOEHLER]

Bibliography: R. S. Nelson, *The Observation of Atomic Collisions in Crystalline Solids*, 1968.

Chelation

The use of chelates in analytical chemistry has been widespread for many years. The greatest number of applications have been as titrating agents, extractants, and masking agents using compounds such as ethylenediaminetetraacetic acid (EDTA). Recently, the syntheses and characterizations of several β-diketones, particularly derivatives of acetylacetone, many fluorinated, have allowed new applications of chelation methods to the trace analysis of metals using instrumental techniques.

Metal complexes of these compounds can often be formed quantitatively by direct reactions and are sufficiently stable and have high enough vapor pressures to be used, at least for purposes of analysis, in the vapor state. For these reasons, β-diketonates are excellent compounds for the trace analysis of metals by gas chromatography and conventional mass spectrometry. For some elements the mass spectrometric and gas chromatographic techniques are more sensitive than any other analytical method known, including spark-source mass spectrometry and activation analysis. This article is devoted to a discussion of the applications of β-diketones to the trace analysis of metals by these techniques.

The β-diketones of interest are derivatives of acetylacetone, as shown in formula I, where R_1

$$R_1 - \underset{\underset{O}{\|}}{C} - CH_2 - \underset{\underset{O}{\|}}{C} - R_2 \qquad (I)$$

and R_2 can be a variety of alkyl and fluorinated alkyl groups.

A list of β-diketonates which have been used for analytical purposes follows:

Pentane-2,4-dione
1,1,1-Trifluoropentane-2,4-dione
1,1,1,5,5,5-Hexafluoropentane-2,4-dione
2,2,6,6-Tetramethylheptane-3,5-dione
1,1,1-Trifluoro-5,5-dimethylhexane-2,4-dione
3-(2-Thenoyl)1,1,1-trifluoroacetone
1,1,1,2,2-Pentafluoro-6,6-dimethyl-heptane-3,5-dione
1,1,1,2,2,3,3-Heptafluoro-7,7-dimethyl-octane-4,6-dione
1-Phenylpentane-2,4-dione
1,1,1-Trifluoro-5-phenylpentane-2,4-dione
1,5-Diphenylpentane-2,4-dione
1,1,1,2,2,6,6,6-Octafluorohexane-3,5-dione
1,1,1,2,2,3,3,7,7,7-Decafluoroheptane-4,6-dione

These chelates readily react under a variety of conditions, losing one methylene hydrogen and forming complexes of the type shown in formula II.

$$M^{n+} \left(\begin{array}{c} R_1 \\ O-C \\ \vdots \quad CH \\ O-C \\ R_2 \end{array} \right)_n \qquad (II)$$

Specific reaction conditions vary with the particular ligand and with the form of the metal. Many quantitative reactions, however, can be carried out in capillary tubes with the metal, metal oxide, or alloy. Typical methods of preparations include direct reaction by heating the sample in the neat ligand (ligand used in a pure form), reaction with the ligand after an oxidation step and reaction with the ligand in the presence of a small amount of water. Often an extraction step is employed to obtain the pure β-diketonate.

Solvent extraction. Because the complexes formed from these ligands are neutral and are composed mainly of a carbon-fluorine structure, they readily extract from water solution with a variety of organic solvents. Although the extraction itself is not usually a complete analysis, it often acts as an important intermediate step for reducing interferences for trace analysis. Extractants such as benzene, chloroform, hexane, pentane, and 4-methyl-2-pentanone have been used with extraction efficiencies varying widely with solvent type, pH, and the particular chelate. For the various ligands studied, extraction efficiency seems to generally increase with the length of the fluorocarbon chain, and marked improvement in extraction efficiency for divalent metals has been accomplished by the addition of isobutylamine to the organic phase.

The extraction step may be highly efficient over a wide range of concentrations. For example, $Be(tfa)_3$ has been successfully extracted into an equal volume of benzene from 0.1 M $Cu(tfa)_3$ with 98.5% efficiency over the range from 10^{-6} to 10^{-13} g Be, where $H(tfa)=1,1,1$-trifluoropentane-2,4-dione. Furthermore, the successful extraction at the trace level allows many otherwise difficult analyses to be performed. For example, aluminum has been determined by the gas chromatographic method at the 0.1-ppm level after benzene extraction from uranyl nitrate solutions as $Al(tfa)_3$.

Gas chromatography. The gas chromatography of β-diketonates is greatly enhanced by the use of electron-capture detectors. The electron-capture detector works on the principle that free electrons, usually produced from a beta-ray source such as a tritium foil, have a low efficiency for recombination with positive charges and therefore can be readily collected and measured in an ion chamber. However, if a gas with a high affinity for electrons is present, the electrons will be captured to form negative ions which have much higher efficiencies for recombination with positive charges. Hence the electron-capture detector exhibits a decrease in current whenever a gas with a high electron affinity enters the ion chamber. This effect is practically optimized in the case of fluorinated β-

diketonates because of the very great electron affinity of compounds containing fluorine and the very low affinity of the typical carrier gases: nitrogen, argon, and helium.

Columns used with metal chelates are typically made of Teflon or borosilicate glass. (Metal columns show some undesirable properties, including distorted peak shapes and the possibility of formation of complexes with the column material itself.) Although a variety of liquid phases have been employed, silicones are generally credited with giving the best results. Column temperatures on the order of 100°C are typical, with retention times on the order of several minutes.

The most sensitive analysis known for beryllium (4×10^{-13} g) has been accomplished by the gas chromatography of Be(tfa)$_3$. There are numerous other applications, for example, the analysis of chromium in urine.

Mass spectrometry. Once the preparation of the metal chelates has been accomplished, mass spectrometric analysis can often be employed. In certain cases, a conventional mass spectrometer can be used with a high degree of accuracy because the problem of the generation of a wide range of kinetic energies in the source does not exist, as it does in a spark-source mass spectrometer. Hence the formation of the chelate replaces the use of the radio-frequency spark, at least in an operational sense, as a method of getting the metal into a vapor state.

In general, β-diketonates are excellent subjects for study in a mass spectrometer. All those studied for analytical purposes produce parent peaks and have fairly simple fragmentation patterns, usually including the *mono, bis,* and, in the case of trivalent metals, *tris* fragments along with various other fragments resulting from the loss of F$^-$, CF$_3{}^-$, or other groups. Fluorinated chelates have good ionization efficiencies and, in a double-focusing mass spectrometer in which an electron-multiplier detector is used, sensitivity limits in the picogram range and below have been attained. In addition, the high mass range of the main fragments is well above that of the typical strong organic background, and the differences in mass and isotopic distributions for various metals make identification relatively easy as long as there are not, simultaneously, widely different concentrations of neighboring elements in the source. Isotope peaks from C^{13} must be considered as possible interferences, since many of these complexes contain as many as 20 or more carbon atoms.

The main problem in mass spectrometry is the introduction of the sample into the instrument. A number of the metal chelates are sufficiently volatile to be introduced directly through a gas inlet system. Usually, introduction is made using an inert (at least as far as the chelates are concerned), very volatile solvent such as pentane. A measured volume of solution can then be introduced and the solvent pumped off before allowing the chelate to vaporize into the source. A heated gas inlet system considerably increases the range of chelates which may be handled in this manner. Although this is a convenient method of sample introduction, a sensitivity loss of a factor of 100 to 1000 is always encountered in using a ballasted system. For the less volatile chelates, a method of introduction that

gains in sensitivity, although it is less convenient and usually less accurate, is the use of a direct insertion probe. The sample is deposited on the tip of the probe, which is usually made of ceramic or quartz, and then inserted through a vacuum lock directly into the ion source. When the tip is inserted into the source, rapid heating occurs, causing vaporization and ionization. Chelates which require a few hundred degrees to vaporize can be analyzed in this manner. Because the sample is being simultaneously vaporized and pumped off in the source, no steady-state signal may be observed, and it is necessary to analyze the quantitative results by integrating the signal from a single peak.

A number of rare-earth complexes and many compounds of other transition metals have been studied by mass spectrometry of volatile chelates. For example, chromium can be generally analyzed in various alloys such as Cr(hfa)$_3$ after a preparation which uses a perchloric acid oxidation step. Here   H(hfa) = 1,1,1,5,5,5-hexafluoropentane-3,4-dione.

The notable activity in the use of β-diketonates for trace analysis during the few years since the development and characterization of most of these compounds implies that metal chelates should play an increasingly important role in trace analytical chemistry in future years.

For background information *see* CHELATION; EXTRACTION; MASS SPECTROMETRY; TRACE ANALYSIS in the McGraw-Hill Encyclopedia of Science and Technology.   [THOMAS L. ISENHOUR]

Bibliography: R. Belcher, et al., *Anal. Chem. Acta,* 43:451, 1968; J. L. Booker, T. L. Isenhour, and R. E. Sievers, *Anal. Chem.,* 41:1705, 1969; B. R. Kowalski, T. L. Isenhour, and R. E. Sievers, *Anal. Chem.,* 41:998, 1969; R. W. Moshier and R. E. Sievers, *Gas Chromatography of Metal Chelates,* 1965; W. D. Ross and R. E. Sievers, *Gas Chromatography 1966,* 1966; J. Savory, P. Mushak, and F. W. Sunderman, *Clin. Chem.,* 15:790, 1969; W. G. Scribner, *Applied Research Laboratories Publ.,* no. ARL 68-0193, Johns Hopkins University, November, 1968.

Chelonia

Turtles have long been considered a "primitive" or relict reptile group that has maintained a consistent morphologic structure since its origin. Recent phylogenetic and functional work on the skulls of fossil and modern turtles suggests that this approach is incorrect. Two major groups of turtles, the Cryptodira and the Pleurodira, probably represent a very early dichotomy which took place soon after the origin of the turtles. These two groups passed through similar stages during their phylogeny and repeatedly evolved parallel adaptations. The origin of chelonians is unknown. *Eunotosaurus,* previously thought to represent the ancestor of turtles, is now known to be a turtlelike reptile unrelated to true turtles.

Phylogeny. The basicranium of turtles has rarely been used in a comparative sense, and the morphology of this region is still poorly known in many families of turtles. Nonetheless, the basicranium is one of the most useful regions for phylogenetic work because of its extreme complexity and relative conservativism compared with other areas.

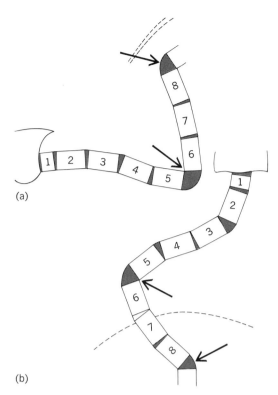

(a)

(b)

Fig. 1. Centers of motion in the testudinate neck. (a) Vertical neck retraction of the Cryptodira. (b) Horizontal neck retraction of the Pleurodira. Arrows indicate points of greatest flexure or centers of motion. Broken lines indicate edges of shells.

The main arterial and venous circulation of the head, the cranial nerves, the brain, and the neck and jaw musculature are all related to features in the basicranium. Changes in any of these systems can usually be seen in the basicranium. Although skulls of turtles are rare compared to shells, the fossil record contains a good series of skulls with basicrania from the Upper Triassic to the Recent.

Previous phylogenies, however, have been based primarily on shell criteria and vertebral characters. Two major living groups are usually distinguished: the Cryptodira, with vertical neck retraction, and the Pleurodira, with horizontal neck retraction (Fig. 1). A few shell characters also distinguish the living forms: pelvis sutured to carapace and plastron in pleurodires, pelvis free in cryptodires. These features rather clearly separate living turtles into two groups, but when fossils are examined it becomes difficult to distinguish pleurodires and cryptodires. The neck retractile mechanisms had not evolved until the end of the Mesozoic, and therefore the earlier phylogeny of the two groups is difficult to determine by the use of neck characters. This problem has usually been solved by the erection of the group Amphichelydia for Mesozoic turtles that could not be identified as cryptodires or pleurodires. The classification of turtles, therefore, has usually looked like this:

Order Testudinata
Suborder Amphichelydia
Cryptodira
Pleurodira

Phylogenies of turtles usually show the Pleurodira and Cryptodira evolving from the Amphiche-

lydia sometime in the Cretaceous, when specialized retractile neck vertebrae begin to appear in the fossil record.

Work on the skulls of fossil and recent turtles, however, has begun to determine the true relationships of cryptodires and pleurodires and unfold hitherto unknown chapters in the history of turtles. Work beginning with S. B. McDowell in 1961 and P. W. Albrecht in 1967 has resulted in a reassessment of the early phylogeny and evolution of the Chelonia by E. S. Gaffney in 1969. The pleurodires and cryptodires can be recognized well into the Mesozoic as distinct lineages, and the divergence of the two groups from a common ancestor probably took place before the Late Jurassic and after the Late Triassic.

The new phylogenetic information has enabled functional interpretations to be made concerning the biologic significance of the differences between cryptodires and pleurodires. These interpretations show that the two lineages have developed similar structures in parallel during their evolution (Fig. 2). A similar but nonhomologous jaw mechanism involving a trochlea to compensate for the expanded otic chamber (Fig. 3) and a method of bracing the palatoquadrate against the braincase are complex features evolved independently in both groups. Later in their evolution the neck retractile mechanisms evolved, but by this time the

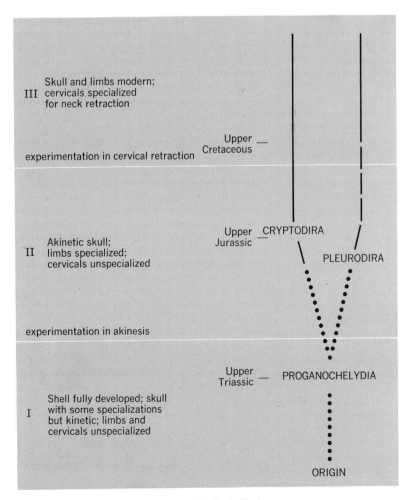

Fig. 2. Levels of organization in the order Testudinata.

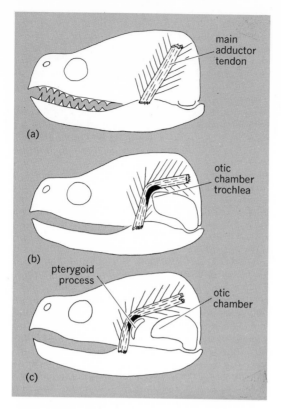

Fig. 3. Position of main jaw muscle tendon in (a) early reptiles, (b) cryptodires, and (c) pleurodires. The trochlea or pulley formed in turtles to get around the expanded otic chamber is independently derived in pleurodires and cryptodires.

two groups had been separate for many millions of years.

Early diversification. These ideas concerning the early diversification of turtles are directly pertinent to the origin and ancestry of the group. Examination of a description, by T. S. Parsons and E. E. Williams in 1961, of the oldest known turtles (Upper Triassic, Germany) strongly suggests that these Triassic forms were intermediate between cryptodires and pleurodires. The structure of the jaw mechanism and basicranium (as far as it is known) demonstrates this. A phylogenetic diagram (Fig. 2) summarizes this information into a series of organizational levels from the ancestral Triassic proganochelydians, the oldest known turtles, through an early period during which the pleurodires and cryptodires diverge and develop trochlear jaw mechanisms, and finally to a stage when the two groups attain the features by which they are recognized in the living fauna.

Origin. The origin of turtles is more in doubt than ever. Previous workers have emphasized the peculiar adaptations of a Permian reptile from South Africa, *Eunotosaurus*. This animal, represented by postcranial remains but very little skull material, was suggested as an ancestor for turtles on the basis of its expanded ribs, which form a rudimentary shell. However, Parsons and Williams in 1961 and, in particular, C. B. Cox in 1969 showed that *Eunotosaurus* is not related to turtles. Cox stated that *Eunotosaurus* is a captorhinomorph and that its Chelonia-like features are due to structural convergence with turtles. *Eunotosau-*

rus lacks separate ribs on its last dorsal vertebra, and Cox held that this feature prevents *Eunotosaurus* from being a chelonian ancestor. No reptile known to science definitely possesses specialized characters in common with turtles that would allow it to be a close relative or ancestor of that order. Presumably the stem reptile group Captorhinomorpha includes the eventual forebears of turtles since captorhinomorphs probably were ancestral to all reptiles, but the intermediate forms are unknown.

For background information *see* CHELONIA; EVOLUTION, ORGANIC; TURTLE in the McGraw-Hill Encyclopedia of Science and Technology.

[EUGENE S. GAFFNEY]

Bibliography: P. W. Albrecht, *Tulane Stud. Zool.*, 14:81–99, 1967; C. B. Cox, *Bull. Brit. Mus. (Nat. Hist.)*, 18(5):167–196, 1969; E. S. Gaffney, unpubl. Ph.D. thesis, Columbia Univ., 1969; S. B. McDowell, *Bull. Mus. Comp. Zool.*, 125(2):23–39; 1961; T. S. Parsons and E. E. Williams, *Bull. Mus. Comp. Zool.*, 125:43–107, 1961.

Chromatophore

Chromatophores are cells and organs that effect color changes of the animal which bears them by dispersion and aggregation of the pigment they contain. Although studies of chromatophores at the organismic level continue to produce significant new information about animal color changes, recent studies of chromatophores at the cellular level have been especially successful in providing new insights into the mechanisms of action of the hormones and transmitter substances that control the pigment migration.

Cephalopods. Chromatophores of cephalopods (such as squid and octopus) are organs consisting of a pigment-containing cell and several radial muscle cells. The detailed structure of chromatophores in a squid was elucidated recently by means of the transmission electron microscope. The pigment granules are in an elastic sacculus within the pigment cell. The sacculus is attached around the equator of the pigment cell to the cell membrane. In turn, the cell membrane is attached to the muscle cells. Contraction of the muscle cells expands the pigment-containing sacculus. The muscle cells are stimulated by excitatory nerves. Recent electrophysiological investigations revealed no evidence for inhibitory fibers, whose presence had been postulated earlier by E. Bozler. The transmitter substance of the motor neurons is unknown. Possibly it is acetylcholine, application of which results in tonic contractions of the muscle cells. E. Florey and M. E. Kriebel found a diminution of the responses of the muscle cells after nerve stimulation in the presence of high concentrations of the cholinergic blocking agents atropine and banthine; on the other hand, acetylcholine had no effect on the electrical properties of the muscle cells. Acetylcholine is capable of causing release of transmitter from these nerve terminals. If acetylcholine is not the natural transmitter, its tonic action is an indirect one only.

Crustaceans. Chromatophores of animals other than cephalopods are unicellular. Those of crustaceans are noninnervated; they are controlled by blood-borne substances that appear to be polypeptides of neurosecretory origin. K. R. Rao showed

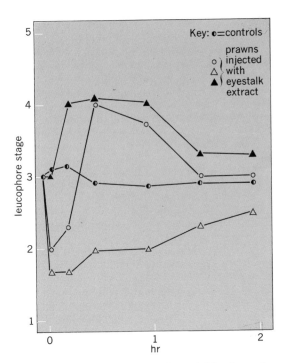

Fig. 1. Responses of the leucophores (white chromatophores) of the eyestalkless prawn *Palaemonetes vulgaris* to injection of eyestalk extracts. Each extract was prepared from a single eyestalk. Stage 1 represents maximal concentration of the pigment, and stage 5 maximal dispersion. (*From M. Fingerman, Dual control of the leucophores in the prawn, Palaemonetes vulgaris, by pigment-dispersing and pigment-concentrating substances, Biol. Bull., 138:26–34, 1970*)

phosphate) in proportion to the amount of RPDH present and the amount of calcium that entered. Cyclic AMP then triggers dispersion in proportion to the amount synthesized. RPCH is thought to exert its effect by stimulating a "pump" in the cell membrane that exchanges internal sodium for external potassium, resulting in a high internal K^+/Na^+ ratio which in some manner triggers the red pigment – concentrating mechanism.

Fishes. Fishes show considerable variation with respect to chromatophore control. Lampreys and some elasmobranchs have entirely hormonal control of their chromatophores, whereas the rest of the fishes typically show some combination of nervous and endocrine control. The melanophores (brown, brown-black, and black chromatophores), the most common and conspicuous type, have been the most carefully investigated. When melanophores are singly innervated, it is always by a melanin-concentrating nerve fiber. In some teleosts there is evidence for a dispersing fiber also. R. R. Novales and R. Fujii found that cyclic AMP induced melanin dispersion in the killifish (*Fundulus heteroclitus*) and postulated that the transmitters control the cyclic AMP level within these dineuronic melanophores, the dispersing transmitter inducing an increase and the concentrator a decrease.

Amphibians. Chromatophores of amphibians, as of crustaceans, give no evidence of innervation.

the necessity of performing assays of crustacean chromatophorotropic hormones at controlled temperatures in order to obtain quantitatively reproducible results. He found that the rates of the initial pigment-dispersing responses of both the black and the white chromatophores of the crab *Ocypode macrocera* to an eyestalk extract were directly related to temperature, but the standard integrated responses, which take into account both the amplitude and duration of the responses, were inversely related to temperature.

A seeming conflict in earlier published data relating to the endocrine control of the white chromatophores of the prawn *Palaemonetes vulgaris* was resolved. One pair of investigators had reported that eyestalk extracts caused white pigment dispersion only; two others who worked independently later reported that such extracts caused concentration alone. M. Fingerman showed that some eyestalk extracts from this prawn produce white pigment concentration alone, some produce dispersion alone, and some produce concentration followed by dispersion (Fig. 1). By means of gel filtration the white pigment disperser was separated from the concentrator. Possible mechanisms of action of the red pigment – dispersing hormone (RPDH) and red pigment – concentrating hormone (RPCH) of this prawn have also been reported by Fingerman (Fig. 2). RPDH is visualized as causing an increase in the rate of influx of calcium ions into the red chromatophores. The higher concentration of calcium ions inside then stimulates synthesis of cyclic AMP (adenosine 3′,5′-mono-

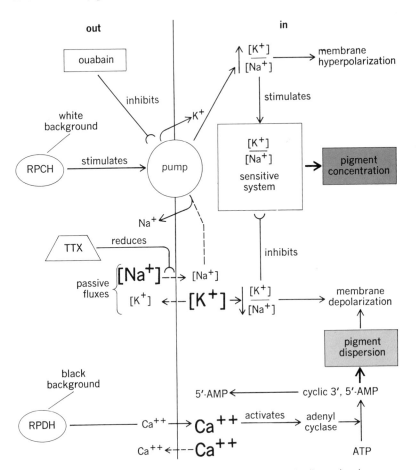

Fig. 2. Possible mechanisms of action of the red pigment–dispersing hormone (RPDH) and red pigment–concentrating hormone (RPCH) of the prawn *Palaemonetes vulgaris*. (*From M. Fingerman, Cellular aspects of the control of physiological color changes in crustaceans, Amer. Zool., 9:443–452, 1969*)

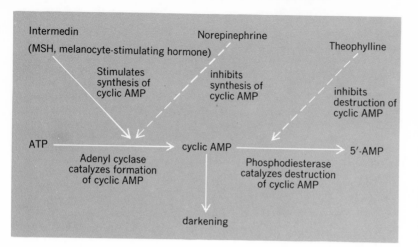

Fig. 3. The proposed mechanism of darkening of the skin of the leopard frog (*Rana pipiens*). (*From K. Abe et al., Adenosine 3',5'-monophosphate (cyclic AMP) as the mediator of the actions of melanocyte stimulating hormone (MSH) and norepinephrine on the frog skin, Endocrinology, 84:362–368, 1969*)

However, hormones are involved. Much of the recent research on amphibians has utilized the common leopard frog (*Rana pipiens*) and has emphasized the role of cyclic AMP in the darkening of amphibian skin. The chromatophorotropic action of cyclic AMP was in fact first demonstrated with this frog. Intermedin (Fig. 3) from the pituitary gland stimulates synthesis of cyclic AMP in these melanophores, and darkening of the skin results. Cyclic AMP has been referred to as the second messenger, that is, the intracellular mediator responsible for the ensuing responses, the first messenger being the nerve transmitter (as in *Fundulus*) or hormone that stimulates cyclic AMP synthesis. As in most amphibians, epinephrine and norepinephrine cause lightening of intermedin-darkened skin of this frog. Norepinephrine (and presumably epinephrine) appears to exert its lightening effect following interaction with an α-adrenergic receptor by inhibiting cyclic AMP synthesis. Theophylline, an inhibitor of phosphodiesterase, acts synergistically with intermedin.

The skin of the toads *Xenopus laevis* and *Scaphiopus couchi*, unlike that of most other amphibians, becomes darker rather than lighter in response to epinephrine and norepinephrine. F. C. G. van de Veerdonk suggested that intermedin liberates a catecholamine in the skin of *Xenopus* which in turn (or with intermedin) stimulates synthesis of cyclic AMP, which then initiates darkening. In *Scaphiopus*, however, the melanin-dispersing action of intermedin does not appear to involve catecholamines. The β-adrenergic blocker, propranolol, inhibits the response to intermedin in *Xenopus* but not in *Scaphiopus*, but it does block darkening of *Scaphiopus* by catecholamines. It appears then that through α-adrenergic receptors catecholamines cause lightening of the skin, whereas through β-adrenergic receptors they cause darkening.

Reptiles. Reptiles show a broad spectrum of first-messenger chromatophore regulation ranging from hormones exclusively to what appears to be nervous control alone, with the intermediate condition of a combination of nervous and hormonal control. The lizard *Anolis carolinensis*, whose chromatophores are noninnervated, requires intermedin to darken its skin when it is on a black background. In addition to background responses, under stress *Anolis* exhibits excitement pallor followed by excitement darkening. These excitement reactions occur in response to adrenal catecholamines. M. E. Hadley and J. M. Goldman found that most melanophores of *Anolis* possess both α- and β-adrenergic receptors; some melanophores, however, have β-receptors alone. Excitement pallor is regulated through α-adrenergic receptors, whereas β-adrenergic receptors are involved in excitement darkening. In *Anolis* also the intracellular level of cyclic AMP may be responsible for the control of pigment migration in the melanophores.

For background information *see* CHROMATOPHORE in the McGraw-Hill Encyclopedia of Science and Technology.

[MILTON FINGERMAN]

Bibliography: M. Fingerman, *Ann. Rev. Physiol.*, 32:345–372, 1970; E. Florey, *Amer. Zool.*, 9:429–442, 1969; M. E. Hadley and J. M. Goldman, *Amer. Zool.*, 9:489–504, 1969; R. R. Novales and R. Fujii, *J. Cell. Physiol.*, 75:133–135, 1970; K. R. Rao, *Z. Vergl. Physiol.*, 65:233–241, 1969.

Circuit (electronics)

With the advent of the space age, the development of highly sophisticated electronic products has increased at a rapid rate. Thus the need to develop suitable test equipment to evaluate rigorous electronic requirements of electronic devices and systems during production has been accentuated. In addition, lengthy manual test times have compounded and, in some cases, actually impeded equipment production. Consequently, the emphasis in production-line testing has been shifting from manual testing to automatic check-out of electronic equipments. Similar problems also exist in field maintenance shops, where lengthy manual test times prevent rapid repair of equipments and result in lengthy waiting lines. In the past few years automatic test equipment has been found to be eminently successful in evaluating electronic devices and systems during production testing.

Among the chief advantages of automatic test equipment are rapid and comprehensive testing of highly complex electronic equipments; improved test repeatability; highly accurate test results; and printed records of test results. Automatic test equipment can rigorously test large electronic systems in a matter of minutes, whereas performing the same tests manually could take hours. Similarly, the testing of smaller electronic devices (that is, printed-circuit boards) can be completed in seconds using automatic test equipment; manual testing could take several minutes to perform the same tests. In addition, the tests are performed under the instruction of a digital computer, which minimizes false human judgments when evaluating test results.

General characteristics. Automatic test equipment can be easily adapted to meet specific user applications. The basic concept of an integrated hardware and software system can be used over a broad spectrum of applications such as electrical, mechanical, pneumatic, and electrooptical testing.

When combined, the hardware and software elements perform the appropriate test functions required by the testing application.

System operation. Figure 1 is a pictorial representation of a typical automatic test system. The basic hardware components of automatic test equipment are illustrated in Fig. 2. It can be seen that programmable stimuli and sensor equipments can provide the appropriate measurement functions as required by the testing application (that is, power supply, signal generator, and digital voltmeter). Usually the inputs and outputs of the stimuli and sensor equipment are channeled to the interface of the unit under test by means of internal programmable switching trees and matrices. Control of the stimuli and measurement devices is provided by a digital electronic controller which facilitates the interchange of information between the stimuli and sensor equipments and the control computer.

The function of the control computer is to furnish central control and data evaluation for the automatic test system. Peripheral equipments which could facilitate automatic test equipment usage include entry of information by the operator and provision of mass-memory devices used for storage of numerous test programs. These mass-memory devices could consist of disk-storage, magnetic-tape, or drum-storage devices. The control computer, which operates under the direction of a previously loaded test program, provides the digital control signals to the stimulus and measurement subsystem. In addition, on the basis of measurement information, it processes and evaluates information that it receives from the sensor equipments. Final comparisons which evaluate test results are made in the computer by comparing the sensor signals with prestored test limits.

System software. Originally, automatic test equipment was designed by engineering personnel who had a minimum of experience and background in the use and application of digital computers. Automatic tests were envisioned as consisting mostly of measurement and comparison against fixed limit operations; consequently, programming language requirements were minimized. It was necessary to utilize a particular programming language to generate test program instructions. As experience with automatic test equipment grew, sophistication of the on-line test operations increased, placing a larger emphasis on language flexibility and use. Consequently, automatic test languages have been developed which are used to write automatic test procedures which are unique in some ways compared to languages used with conventional computer systems. Since automatic testing can be considered a form of process control, a language designed for this purpose must be able to direct the testing processes by causing stimuli to be generated and response data to be gathered and evaluated as well as to be called for action by the system operator.

Problem-oriented languages developed for automatic test equipment are designed around the problems intended to be solved using these languages. Specific machine functions then become invisible or below the level of awareness of the programmer. He does not have to concern himself with registers, memory location, or channel assignments but can focus his efforts on solving the problem at hand. Consequently, problem-oriented languages are said to be machine-independent (that is, the same program and problem will give the same results regardless of the make or model of the computer used).

Automatic test equipment utilizes internal com-

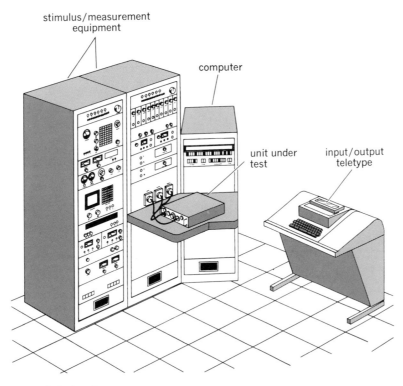

Fig. 1. Typical automatic test system.

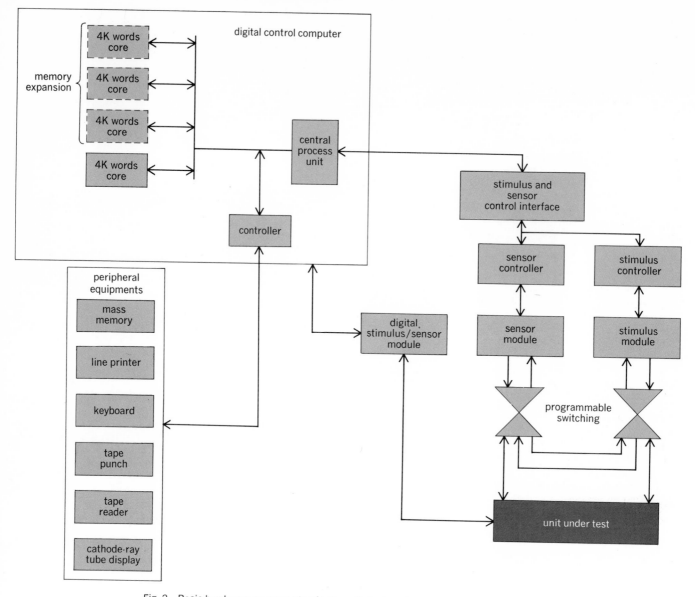

Fig. 2. Basic hardware components of automatic test equipment.

puter processes. In addition, response data are used to perform arithmetic and logic manipulations to arrive at a test decision. Therefore it is desirable to express these internal computer operations in a problem-oriented manner as opposed to addressing specific computer functions. Consequently, programming languages have been developed for automatic test equipment which enable the control of all equipment required to arrive at a solution and which have a sufficient syntax to express the logical steps necessary to control the hardware.

Once the test program has been written using a problem-oriented language, it is necessary to convert the program statement into instructions which will command the automatic test equipment to perform accordingly. To accomplish this task, use is made of a compiler which is itself a series of programs and which converts test programs into machine-oriented instructions (Fig. 3). The instructions are then used by the control computer to fur-

nish central control and data evaluation for the automatic test equipment.

Recent developments. Recent developments in automatic test equipment have been concerned with improving present system performance by conventional approaches. Typically, modern automatic test equipments can compensate for signal discrepancies that might exist between the unit under test and the actual test function module that generates or monitors the signal; these discrepancies can be common in radio-frequency testing because of signal path characteristics. More advanced automatic test systems, such as the U.S. Navy's VAST (versatile avionic shop test) system, which is used for maintenance of the avionics systems of carrier-based aircraft, utilize a priori signal information with the aid of a computer to compensate for signal modifications inherent in automatic test equipment.

An additional method to further improve system performance is to utilize repetitive measurement-sampling techniques to minimize the effects of both externally and internally generated noise. Typically, measurements made on a signal can be stored in the computer while additional iterative measurements can be made on the same signal. The repetitive measurements are then processed in the computer using averaging techniques. In so doing, more accurate measurements of the response signals of the unit under test can be obtained with the aid of the computer.

Among the problem-oriented languages for automatic test equipment is VITAL (VAST interface test application language), which was derived from the ATLAS (abbreviated test language for avionics systems) language. The use of such universal test-equipment languages has freed the test engineer from becoming deeply involved in understanding the specific mechanics of operation of the automatic test equipment. Instead, he can concentrate on generating the testing requirements for the unit under test without being concerned with individual characteristics of the automatic test equipment utilized.

Future trends. Advanced information-processing techniques developed in recent years lend themselves to automatic test equipment usage. Thus higher system performance and more accurate measurements of signals will be readily attainable. These techniques, such as Kalman filtering, which can predict signal trends based upon previous measurements, have found application in optimal control theory, and can be used for failure-prediction analysis as well as for applications in inertial system testing.

Also, advanced testing techniques such as one which applies an impulse input to a unit under test and evaluates the resultant spectral components of the output signal from the unit under test could be utilized. Such a technique would enable an evaluation to be made of the transfer functions of the unit under test to determine whether the unit is operational. It should be noted that this type of testing will make greater use of the computational capability inherent in digital computers rather than depending on numerous sensor devices.

It is apparent that in the near future increasing use will be made of closed-loop testing techniques. The use of these techniques can automatically compensate for signal discrepancies that might appear between the unit under test and the sensor-stimuli/input-output, thereby further optimizing system performance. In addition, closed-loop testing will enable greater latitude on the part of individual sensor equipment accuracy requirements and will still maintain extremely accurate test capability because of the dynamic characteristics inherent in the loop gain.

Finally, the broader development and use of unique stimuli and sensor equipments such as infrared scanning devices, which scan a unit under test in a raster fashion (similar to a television-picture scan), will provide an even broader base to expand the testing applications of automatic test equipment from the electronics equipment production line to that of other expanded areas of testing.

For background information *see* CIRCUIT (ELECTRONICS); DIGITAL COMPUTER; PRINTED CIRCUIT;

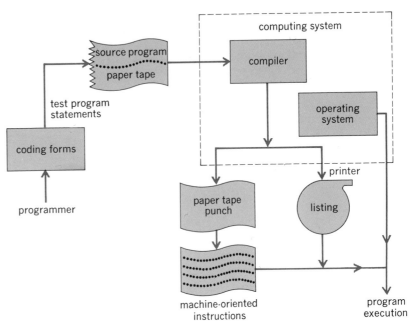

Fig. 3. Conversion of test programs to machine-oriented instructions.

RELIABILITY OF EQUIPMENT in the McGraw-Hill Encyclopedia of Science and Technology.

[THOMAS H. O'BRIEN]

Bibliography: R. G. Loughlin, The AN/USM-247 versatile avionic shop test (VAST) system, in D. M. Goodman (ed.), *Automation in Electronic Test Equipment*, vol. 6, 1969; P. M. Lowitt, *Principles of Automatic Test Equipment*, lecture for course "New Manufacturing Techniques for the Electronics Industry," State University of New York, Stony Brook, December, 1969; *Proceedings of the Automatic Support System Symposium for Advanced Maintainability*, IEEE Catalog no. 69-C-57, suppl., Nov. 3, 4, and 5, 1969.

Collagen

In recent years intensive studies in laboratories all over the world have led to striking advances in the knowledge of collagen. There is now a much more detailed picture of the chemical and physical nature of the individual building blocks (tropocollagen molecules), the way they aggregate into fibrils by a self-assembly process, their maturation by covalent cross-linking, and their association with other connective tissue components. This article describes the more important developments.

Collagen is the principal structural protein of the body and is widespread throughout the animal kingdom. It is a fibrous protein and is laid down as very long thin fibrils showing a characteristic periodic banding (Fig. 1). In tissues these fibrils are surrounded by a glycosaminoglycans-rich ground substance. The relative proportions of fibrils and ground substance (and other components) vary widely from one supporting tissue to another, depending on the mechanical functions it has to perform. Tendon is very largely composed of collagen, with the high-tensile-strength fibrils arranged roughly parallel to the direction of pull. In skin and basement membrane the fibrils are in random ar-

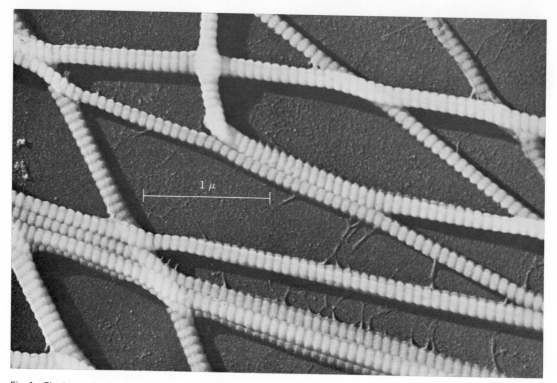

Fig. 1. Electron micrograph of collagen fibrils from human achilles tendon.

rangement; in the cornea there is a remarkably regular array of fibrils; and in bone the collagenous tissue undergoes mineralization by the deposition of calcium hydroxyapatite, initially in or on the collagen fibrils. Thus collagen also forms the basic scaffolding on which a variety of biological structures can build.

Synthesis. Collagen protein is synthesized in fibroblasts (osteoblasts in bone, chondroblasts in cartilage) and is then secreted from these cells, apparently in monomeric form. Subsequent stages in the assembly of monomeric collagen molecules (tropocollagen) into the characteristic fibrillar form occur extracellularly. The intracellular synthetic process occurs on ribosomes associated with an abundant endoplasmic reticular system and resembles that occurring in other cells that synthesize protein for subsequent secretion. The precise mechanism by which the large tropocollagen molecules leave the cell is obscure.

Structure of tropocollagen. Physical studies show that the collagen molecule is a very long, semirigid rod, almost 3000 A (300 nm) long and about 14 A wide. Its semirigidity can be ascribed to its construction from three helically wound polypeptide chains, held together for most of their lengths by a series of interchain noncovalent interactions. Each polypeptide chain has a molecular weight of approximately 100,000 and comprises about 1000 amino acid residues extending over the full length of the molecule. In some animals all three chains are different, but in others, two of the three chains appear to be identical.

The lengths of these chains make the determination of amino acid sequences difficult and laborious. The most rewarding studies have been those in which the chains are broken into fragments by cyanogen bromide and certain enzymes. For the most part every third residue is glycine (gly); the imino acids proline (pro) and hydroxyproline are common, constituting together roughly one-quarter of all residues. The sequence of the N-terminal region of chick and rat skin collagen has now been established, and it is likely that the complete sequence will be elucidated during the next few years.

Like amino acids are not all distributed uniformly along the molecule. Polar regions, rich in polar amino acids, alternate with nonpolar regions, in which imino acids and nonpolar amino acids predominate. This alternation is manifest in electron micrographs of stained collagen as a series of alternating dark and light striations, which are due to the uptake of metallic cations or anions by the acid or basic residues of the polar regions (Fig. 2).

The main structural features of the molecule were established in 1955 by x-ray diffraction studies. The characteristic high angle pattern, which may be used as a hallmark for the identification of collagen, can be interpreted in terms of a triple-helical arrangement of polypeptide chains based on a nonintegral screw axis with equivalent units related by an axial translation of 2.9 A and a rotation of 108°. The three chains are held together by interchain hydrogen bonds. Recent x-ray studies of synthetic polytripeptide and polyhexapeptide models resembling collagen have clarified the structure and have shown that a diversity of tripeptide sequences of the form gly.pro.pro, gly.pro.ala, gly.ala.pro, and so on, can give the same molecular conformation, in all cases with glycine (the smallest amino acid) near the central axis of the triple helix and one NH . . . O interchain hydrogen bond per tripeptide. Nuclear magnetic resonance measurements support this structure. In naturally occurring collagen it is

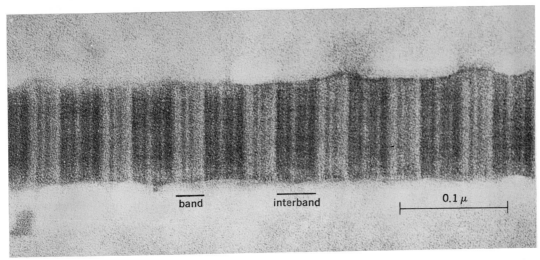

Fig. 2. Electron micrograph of negatively stained collagen, showing the characteristic band-interband structure. The darker transverse striations are regions rich in polar amino acids.

probable that such a structure relates only to the nonpolar regions. Far less information is available about the structure of the polar regions.

Short regions at the N-terminal ends of the three chains do not have glycine in every third position, and this fact and other data suggest that this end region of the molecule is nonhelical. A stabilizing intramolecular covalent cross-link can occur between lysyl residues of adjacent chains in this region. An extrahelical region may also be present at the C-terminal end. Recent evidence indicates that these extrahelical regions (telopeptides) may be important in the assembly of molecules into fibrils by determining molecular orientation and restricting lateral growth.

Assembly into fibrils. The typical banding of collagen fibrils, shown in Figs. 1 and 2, has a periodicity of about 680 A. Fibril diameters range from 200 A (or less in growing tissue) to 2000 A, with each fibril tending to be uniform in diameter along its length (apart from the band-interband corrugations).

The extracellular location of the assembly process and the observation that monomeric collagen can be made to assemble into fibrils in the test tube show that the process is one of self-assembly; that is, it can occur spontaneously without outside direction. The tropocollagen molecules come together in ordered parallel or near-parallel array as a result of specific interactions between them, these interactions involving a large number of weak noncovalent bonds. The main driving energy for the process is provided by entropy changes which result from the exclusion of water molecules from the structure. The resulting association corresponds to a pronounced minimun in the free energy of the system and requires correct pairing between appropriate polar regions. The occurrence of axial periodicity is the natural outcome of self-assembly, as only in this way can correct pairing everywhere be achieved.

The band-interband structure arises from the nonintegral relationship between the length of the tropocollagen molecule and the periodicity. This ratio is 4.4 for collagen, so that for 0.4 of a period molecules overlap, and in the remaining 0.6 of

each period, gaps occur (Fig. 3). Initial deposition of calcium hydroxyapatite in bone mineralization probably occurs in these gaps.

Low-angle x-ray scattering of wet specimens shows that the lateral packing of the tropocollagen units in assembled fibrils can be ordered, in part at least. It is possible that the restricted lateral growth of collagen fibrils and the uniformity of diameter may depend upon this ordering, with the rodlike units deforming on assembly and requiring greater deformation with increasing fibril diameter.

Maturation. In order to fulfill the mechanical requirements of connective tissues, further stabilization of the assembled fibrils is provided by the covalent cross-linking of neighboring chains, intermolecularly as well as intramolecularly. Much work has been devoted to the elucidation of the chemical nature and distribution of these crosslinks. Other studies have been concerned with the interaction between collagen and the glycosaminoglycans of connective tissues.

Denaturation. The stability of the triple-helical structure is maintained in solution under favorable conditions but can be destroyed by mild heat treatment or by certain enzymes. The helical part of the molecule resists attack by the usual proteolytic enzymes, but collagenase, which is specific for the collagen type of structure, can bring about cleavage of the molecule. Absorption of collagen, as in the postparturition uterus or in the remodeling of

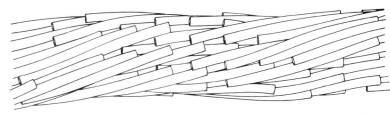

Fig. 3. Rough sketch showing how self-assembly of rodlike molecules of tropocollagen can give rise to a periodic-structured fibril with bands (overlap regions) alternating with interbands (gap regions). Interunit interactions will be such that adjacent rods can be staggered only by 680 A, or integral multiples thereof. In this sketch rods are grossly foreshortened in comparison with tropocollagen molecules.

bone, probably involves the action of collagenolytic enzymes. Collagenase has recently been shown to be present in synovial tissue from patients with rheumatoid arthritis, and it may well play a part in the disease process.

For background information *see* BONE; COLLAGEN; MUSCLE; PROTEIN in the McGraw-Hill Encyclopedia of Science and Technology.

[JOHN A. CHAPMAN]

Bibliography: E. A. Balazs (ed.), *Proceedings of the NATO Advanced Study Institute: The Chemistry and Molecular Biology of the Intercellular Matrix*, May 16–28, 1969, Santa Margherita (Ligure), Italy, 1970; K. A. Piez et al., *Brookhaven Symposia in Biology*, 21:345, Brookhaven National Laboratory, Upton, N.Y., 1969; G. N. Ramachandran (ed.), *Treatise on Collagen*, vol. 1, 1967, and B. S. Gould (ed.), vol. 2, 1968.

Communication cables

Most new communication cables being placed in service today are buried in the ground. Since water can find its way into the cables and impair service, methods have been devised recently to fill the cables with a substance to prevent the entry of water. The filling material used most extensively and the one which provides the best balance between cost and performance is petrolatum (petroleum jelly). Various compounds have been formed by adding materials to the petrolatum base to raise the melting point and to offset greasiness. Many developments are in progress on this new cable design.

In the last several years legislators have introduced ordinances and laws which require utility lines to be placed "out of sight." The usual treatment is to bury the cables and wires in the ground in lightly populated areas or to place the cables and wires in underground conduit with manholes spaced 1000 ft apart or less in heavily populated areas.

Until recently cables with plastic insulated conductors and water-resistant sheathing have been used for direct burial in the earth. However, the sheath is frequently damaged by construction activity, rodents, or lightning, and these sheath ruptures allow water to enter the core of the cable and to run along the conductors. If the water travels to a splice in the conductors or to an open spot in the plastic insulation, corrosion takes place and the copper conductor is eaten away in a matter of weeks. In 1968 the Bell System experienced 50 service-affecting troubles per 100 mi of buried cable. Even if this disruption of continuity (and service) does not occur, the transmission quality of the cable is impaired when the air in the core is displaced by water because the capacitance increases by as much as twice its nominal value.

The need has been established therefore for a multipair communication cable which can be buried in the ground and suffer sheath ruptures without causing impairment to service. This basic need seems to have been met by the design of filled cables which have the desired capacitance when filled, which precludes the entrance or conduction of water within the sheath and which still permits the normal placing (plowing or trenching) and splicing operations under the variety of temperature and other environmental conditions encountered in the United States. The sheath structure is greatly simplified, since it no longer bears the burden of excluding water but serves only to protect the core during placing and to shield the pairs from external electrical influences.

About 5 years ago the British Post Office, which operates the telephone system in Great Britain, pioneered in the manufacture and use of buried communication cables filled with petrolatum. Since that time many other countries have adopted this fundamental cable design. In the United States there are two main departures from the British design in that the filling material either is compounded of a blend of petrolatum and high-density polyethylene or is composed entirely of low-density polyethylene. These filling substances are superior to plain petroleum because of their higher melting points. That of the petrolatum and polyethylene blend is raised from 125 to 250°F by the addition of high-density polyethylene, and the melting point of the low-density polyethylene is 150°F.

Some differences between these two filling compounds are: The polyethylene-petrolatum compound is not compatible with polyethylene insulation; thus a change in manufacture to polypropylene insulation is required. The filling compound, however, is less greasy than others, which is an advantage to the craftsman who must work with the cable. The low-density polyethylene filling substance is compatible with polyethylene insulation; therefore no change in manufacture is required. However, this substance is sticky and harder to work with than are the other fillers.

Many improvements are expected in the manufacture of jelly-filled cables, and the most optimistic projections are that these cables will be manufactured to sell for only a small premium above nonfilled cables (because of the simpler sheath requirement) and that they will be economical to use in all buried-cable applications because the potential maintenance savings will offset the small increase in price.

For background information *see* COMMUNICATION CABLES in the McGraw-Hill Encyclopedia of Science and Technology.   [WILLIAM K. GRAUER]

Bibliography: M. C. Biskeborn and D. P. Dobbin, *Bell Lab. Rec.*, March, 1969; N. S. Dean, *Trans. 18th Int. Wire Cable Symp.*, December, 1969.

Communications, electrical

As in most fields of technology, computers have in the past few years begun to play a markedly increasing role in communications. Several reasons account for this: (1) The emphasis in communications has shifted dramatically to digital systems. Such systems, with numerical quantities transmitted (binary digits are the most common example), lend themselves very naturally to all-digital or computer processing. (2) The widespread use and availability of electronic integrated circuits have in turn helped motivate the increased emphasis on digital systems. In many cases the use of digital integrated technology results in substantial reductions in cost. (3) The use of computers and high-speed digital processing has enabled more sophisticated and complex communication systems to be designed, with a resultant capacity and performance far exceeding that deemed possible in the not too distant past. (4) The flexibility and improved performance made possible through the use of computers and digital processing have in turn led to applications of communications technology to fields previously considered outside the traditional range of communications—geophysical and oceanographic exploration, seismic signal processing, environmental sensing, automatic character and pattern recognition, biological and medical signal processing, and so on.

By computer or digital processing is meant the carrying out of numerical calculations according to prescribed procedures or algorithms. The calculations may be simple enough to require only a few logical circuits for their implementation, or they may, in some cases, require the use of the largest general-purpose computers available. The types of computers thus run the gamut from hard-wired digital processors, with the calculations carried out by built-in circuits, to special-purpose digital computers, with a combination of hard-wired and program-controlled (software) logical operations, to general-purpose computers, with most of the processing done under the control of external programs. There may thus be many different processors of all types carrying out different operations in a typical system. A currently debatable point is whether to continue to separate different operations and carry out each in a special-purpose machine or combine as many of them as possible, using a larger computer to perform the various tasks either simultaneously or sequentially (time-shared).

Three major tasks are typically carried out by computers in a communication system: (1) switching and routing of messages for most efficient signal transmission, as well as the necessary book-keeping operations involved in message transmission; (2) coding of signals and the necessary decoding on reception to enhance communication in the presence of noise, interfering signals, and perturbing effects of the medium over which communication takes place; and (3) digital filtering and various signal-processing functions required to extract the pertinent data from the signals transmitted. In some systems, these various tasks may be combined in one operation; in others, one or more of these tasks may be missing. In some systems, different computers perform each task; in others, one computer may share tasks. The focus in this article will be primarily on the use of computers in the filtering and processing operations.

Consider as an example a highly idealized picture of a digital communications system as shown in Fig. 1. In this case the necessary switching and message routing operations are not indicated. Transmission from one fixed point to another only is assumed for the sake of simplicity. The box labeled source is, as the name indicates, the point from which the message to be communicated originates. The message may be a voice signal or group of voice signals in the case of telephony; it may be a television picture or facsimile picture or a group of such messages; or it may be important data to be transmitted in the case of a telemetry system. The types of data commonly transmitted are too many to enumerate.

In all communication systems the data must first be processed before being transmitted. Different

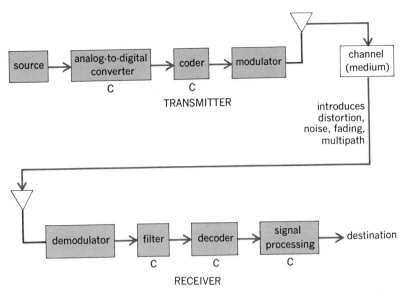

Fig. 1. Simplified digital communications system. Digital processing and computing are carried out in the boxes labeled C.

types of data or independent messages may be combined for more efficient use of the system and the composite message then further processed or coded to prepare it for more effective transmission over the channel. These operations are shown carried out in the analog-to-digital converter and coder. At the receiver the desired signals must be separated from others being received at the same point. Distortion and noise introduced during transmission must be minimized and compensated for. This is all commonly done by devices called filters, as shown in Fig. 1. The decoder then restores the original message from the transmitted coded version. Any further signal processing (separation of individual messages from those transmitted in combined form, signal enhancement techniques, extraction of most significant data, and so on) is then carried out in the box labeled signal processing.

Two of the subsystems noted above that have undergone particular development in the past few years are discussed in more detail in the next sections.

Digital filter. A typical form for such a filter is shown in Fig. 2. Each delay element delays the input number by one time unit. The a and b coefficients are multipliers. Tracing through the example shown, it is apparent that the output y_n at the nth time instant is related to the input x_n by the pair of algebraic equations labeled (1). The vari-

$$y_n = a_0 w_n + a_1 w_{n-1} + a_2 w_{n-2}$$
$$w_n = x_n - b_1 w_{n-1} - b_2 w_{n-2} \tag{1}$$

able w_n is an auxiliary variable that may be eliminated to give Eq. (2), the equivalent single equation relating y_n to x_n. The subscript $(n-1)$ implies a delay of one unit from time n.

$$y_n + b_1 y_{n-1} + b_2 y_{n-2} = a_0 x_n + a_1 x_{n-1} + a_2 x_{n-2} \tag{2}$$

For a given sequence of numbers $x_n, n = 0,1,2$, etc., Eq. (1) and (2) generate a sequence $y_n, n = 0,1,2$, etc. By adjusting the a and b coefficients appropriately, the relation between the input and output sequence can be made to obey a

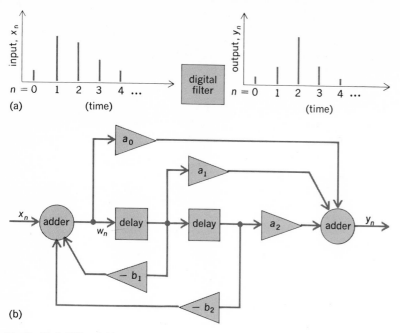

Fig. 2. Digital filter. (a) Input-output relations. (b) Configuration.

prescribed pattern. This is the filtering or processing operation. Equations (1) and (2) are referred to as difference equations in mathematics.

Filters have always been used in communications systems. It is their implementation in terms of digital operations that is newly developing. Other types of configurations that perform the same processing, satisfying the same equations as above, may be derived. Engineering and economic considerations dictate which to use. Much more complicated structures involving more addition, multiplication, and delay elements may also be derived.

The design of digital filters (that is, choice of configuration, complexity, choice of coefficients, and means for synthesizing the various elements with digital operations) is currently an extremely

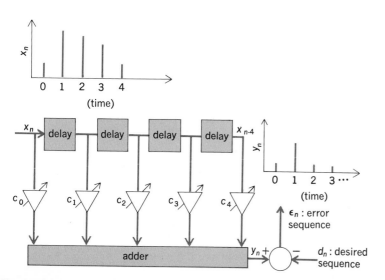

Fig. 3. Adaptive equalizer. The c coefficients are automatically adjusted by the sequence ϵ_n.

active area in the communications field. Digital filters have been designed for various special signal-processing problems as well; they have been used for processing (filtering) acoustic signals, for medical and biological signal processing (for example, electrocardiograph records), in radar signal processing, and so on.

Combinations of three basic operations are called for in digital filters: delay, addition (or subtraction), and multiplication. Simple digital filter structures are commonly hard-wired, in which case shift registers are used to realize the delay units. The adders and multipliers comprising the arithmetic unit may be implemented by using serial two's-complement adders.

Adaptive equalizer. A particular class of digital filters that has been receiving a great deal of attention in the past few years and has recently begun to be marketed commercially is the so-called adaptive equalizer. This device attempts to overcome distortion and noise introduced during transmission of digital signals over channels or media such as telephone lines or the ionosphere (shortwave broadcasting of digital signals). The effects of the channel are usually unknown beforehand and in fact may vary randomly during signal transmission. A filter or processor designed to compensate for these unknown changes must thus be capable of automatically adjusting its characteristics. The adaptive equalizer that performs this automatic signal correction is an example of a nonrecursive digital filter (the digital filters discussed previously were of the recursive type). Figure 3 is a simplified diagram of an adaptive equalizer.

The algebraic equation satisfied by this device is given by Eq. (3). It is called nonrecursive because

$$y_n = c_0 x_n + c_1 x_{n-1} + c_2 x_{n-2} + c_3 x_{n-3} + c_4 x_{n-4} \quad (3)$$

there is no dependence on previous values of the output y_n. The c coefficients here are shown adjustable since their settings will depend on the characteristics of both the signals (data) and the channel over which they are transmitted. Various specific schemes have been developed and proposed for automatically adjusting these coefficients. Generally they all derive some form of error information by comparing the equalizer output y_n at each time interval ($n = 0,1,2, \ldots$) with some known or derived time sequence $d_n(n = 0,1,2, \ldots)$. This sequence of error numbers $\epsilon_n = y_n - d_n(n = 0,1,2, \ldots)$ is used to adjust the coefficients.

In practice, the equalizers operate in two modes. In the first, or training, mode, initiated prior to the transmission of actual data, a known sequence of data symbols is transmitted. This known sequence is simultaneously switched in at the receiver as well. The effect of the channel is then easily measurable and the c coefficients may be iteratively adjusted until the error sequence ϵ_n drops to within tolerable limits. In the second, or tracking, mode the system is assumed operating reasonably well with the channel fairly well compensated for. It is then possible to derive error information from the output data symbols y_n themselves. Should the channel characteristics begin to change, with a corresponding increase in the number of errors detected, the c coefficients would automatically change to reduce the errors.

The performance of these equalizers is determined by their ability to minimize errors and the time taken to adjust the coefficients to the derived positions. For some channels, such as the telephone plant, in which the channel characteristics may change slowly (times measured in seconds), it may not be critical to adjust the coefficients rapidly. For some channels, such as the ionosphere, the channel characteristics change continuously and randomly in milliseconds. The rapidity of iteration is then highly significant.

More rapidly converging equalizers obviously require more complex strategies (algorithms) and hence become more sophisticated special-purpose computers. The equalizers currently being marketed are all hard-wired. It is possible to conceive of more complex devices with rather large memory as well as arithmetic requirements in which both hard-wired and software programs would be used. As this happens it may become advantageous to combine the automatic equalization function with other necessary filtering, synchronization, decoding, and signal-processing functions carried out elsewhere in the communication system. A larger computer could then time-share the necessary memory and computational requirements among these different functions. Such a computer would require much more software programming.

For background information *see* COMMUNICATIONS, ELECTRICAL; DATA COMMUNICATIONS; INTERPOLATION; MODULATION; SWITCHING CIRCUIT in McGraw-Hill Encyclopedia of Science and Technology.

[MISCHA SCHWARTZ]

Bibliography: Computer Processing in Communications: MRI Symposium Proceedings, vol. 19, 1970; B. Gold and C. Rader, *Digital Processing of Signals*, 1969; R. W. Lucky, J. Salz, and E. J. Weldon, *Principles of Data Communication*, 1968; M. Schwartz, *Information Transmission, Modulation, and Noise*, 1970.

Communications satellite

One of the most technically challenging tasks of communications in the 1970s is the development of communications satellites suitable for providing television broadcasts directly to individual homes or to small communities. This capability could be applied to implement an educational-instructional television system in a developing country at a lower cost than that required to create an equivalent system using conventional techniques. Similarly, direct-broadcast satellites could provide supplementary television services in developed regions. Recent studies have shown conclusively the technical feasibility of constructing this type of communications satellite, and a pilot project has been initiated by NASA and India to demonstrate such a capability in 1973–1974. The project will use two satellites in the advanced technology satellite (ATS) series, *ATS-F* and *ATS-G*.

System concept. Communications satellite development over the past few years has centered on the geostationary orbit where the satellite is visible from approximately 40% of the Earth, and appears stationary in the sky to an observer on the Earth. Such satellites are in wide commercial use by the International Telecommunications Satellite Consortium (INTELSAT) for relaying communications, including television, between large and electronically sensitive Earth stations located in various countries. The relayed television is then transmitted to home viewers by conventional local television broadcast stations.

Direct-broadcast types of communications satellites would relay television signals directly to the home or to a community viewing facility, thereby eliminating the need for local television broadcast stations (Fig. 1). It is important to note that a conventional television broadcast station provides service within a radius of approximately 50 mi from its transmitting tower. Consequently, to service a 1,000,000-mi² area, more than 100 broadcast stations, as contrasted to a single direct-broadcast satellite, would be required. Additionally, each conventional broadcast station must be connected to the source of the television program. This is often accomplished by terrestrial radio-relay facilities.

Recently completed studies have concentrated on three aspects of satellite broadcasting: system optimization, economic comparisons, and satellite technology. Some of the system models involve providing service to areas where no or few conventional television broadcast stations exist (for example, India, Alaska, and Brazil). In many of these areas the population is not homogeneously distributed, and a large portion lives in small villages. There has been extensive analysis of one form of this type of direct-broadcast satellite, in which the Earth station is a small, simple installation located at a community center or school where residents may view the television broadcasts. The broadcasts can be extended beyond the community center to a few locations by cable or, if many villagers have television sets, by a small rebroadcast installation. Other system models involve areas where many conventional television broadcast stations exist and where most of the homes contain television receiving sets. In these cases, the satellite would broadcast the television signal directly to the home receiving set with a suitable adapter.

Technical considerations. The most significant technical parameter involved with direct broadcasting is the amount of effective radiated power (ERP) required from the satellite. The community form of direct broadcast requires the least ERP, 45–55 dBW (decibel wattage) per television channel being typical. This form will be demonstrated in the India-NASA pilot project. The direct home form may require about 65–75 dBW per television channel. Current satellites provide approximately 37 dBW ERP per channel. The actual amount of satellite ERP required for direct broadcast is pri-

Satellite system parameter values

Parameter	Assumed values
Radio frequency	800; 2500; 12,000 MHz
Transmission	Vestigial-sideband amplitude modulation; frequency modulation
Earth stations	Antenna diameter, 3–15 ft; noise temperature, 300–1100°K
Picture quality	TASO* 1–3 (45–30 dB)

*Television Allocations Study Organization picture-rating system.

marily dependent on the assumptions made for operating frequency, type of modulation, sensitivity of the Earth station or home receiving installation, and desired quality of the television picture. The table shows the ranges in assumptions which have been used in various studies.

The ability to have more satellite ERP, holding all other satellite parameter values constant, permits reducing the sensitivity of the Earth station (or television receiver adapter) and, consequently, the cost. Since there may be many tens of millions of such stations for a direct home broadcast system, the reduction of the cost of the television receiver adapter by even $20–40 could save in total over a billion dollars. Satellite ERP is the product of the transmitter output power and the antenna beam directivity or gain. The former is set by the available prime power in the satellite, and the latter by the Earth coverage area desired. Consequently, a significant increase in satellite ERP for a given coverage area requires a larger prime power system on the satellite. This results in a larger, heavier, and more expensive satellite and launcher. The cost of a typical launch vehicle for minimum types of direct television broadcast service to individual homes is $16,000,000.

Current restraints. Further development of communications satellite technology is required for direct television broadcast service. The major areas are in satellite prime power systems (Fig. 2 shows a lightweight, roll-out solar array providing a threefold improvement in prime power per pound), thermal control, transmitter efficiency, and antenna stabilization and pointing. Likewise, for direct home service, the cost of the television receiver adapter must be reduced substantially below $100 in quantity production.

A second restraint is that no radio frequencies are internationally allocated for providing such services. Proposals for allocations will be presented at the World Radio Administrative Conference in June, 1971, at Geneva. Also, fears have been expressed that direct broadcast services may be used for dissemination of propaganda.

Lastly, the economic potentials must be more clearly assessed. Although direct-broadcast satellite systems may be a more economical way of providing television than another method, large investment costs (that is, greater than $50,000,000) will be required for the most minimal operational type of system. The sources of such funds or their equivalents in revenues must be determined.

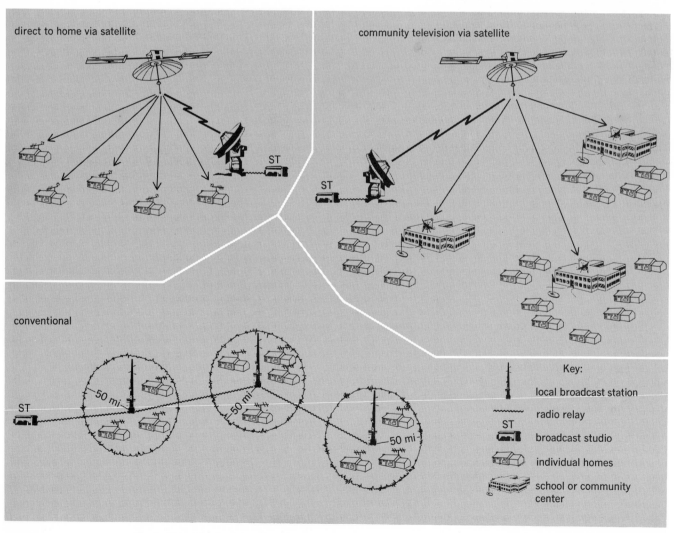

Fig. 1. Television broadcasting systems.

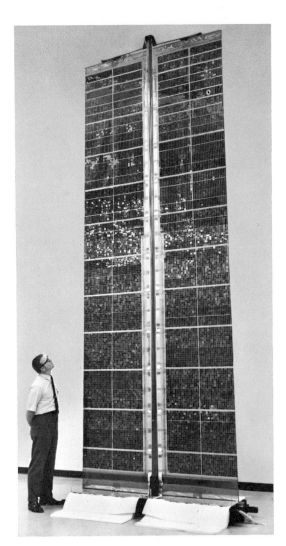

Fig. 2. Developmental roll-out solar cell array for satellite prime power system. (*Courtesy of Lockheed Missile & Space Co.*)

For background information *see* COMMUNICATIONS SATELLITE; RADIO SPECTRUM ALLOCATIONS; TELEVISION TRANSMITTER in the McGraw-Hill Encyclopedia of Science and Technology.

[ROBERT D. BRISKMAN]

Bibliography: R. W. Hesselbacher, Space broadcasting: Where is the technology? What can it do? *1970 IEEE Int. Conv. Dig.*, 70C 15-IEEE, March, 1970; *Television Broadcast Satellite Study*, TRW Systems Group – General Electric Co., NASA Contracts 3-9707 and 3-9708, November, 1969; *Useful Applications of Earth Oriented Satellites*, vol. 10: *Broadcasting*, National Academy of Science, 1969.

Complement, serum

Deficiencies of complement proteins (C1 to C9) or of proteins affecting the complement system have been found in man and in animals. Although some of these deficiencies may be associated with symptoms, others surprisingly are not.

C5. Deficiency of C5 inherited as an autosomal recessive trait has been described in inbred mice. Affected animals have no detectable C5 protein.

Whereas initial observations indicated no increased susceptibility to infection in affected animals, more recent evidence suggests a possible deficiency in the clearance of some intravenously administered bacteria.

C6. Rabbits with inherited C6 deficiency have no C6 protein but do not appear to be unduly susceptible to infection.

C2. In man C2 deficiency has been described in several families. Affected persons are homozygous for the trait and have less than 5% of the normal serum concentration of C2 when measured both functionally and as C2 protein. Heterozygotes have 40–60% of the normal level. Nevertheless, no undue susceptibility to infection is associated with C2 deficiency. Hemolytic complement, bactericidal activity for gram-negative bacteria, immune adherence, and the enhancement of phagocytosis of gram-positive bacteria are moderately reduced in heterozygous, and markedly reduced in homozygous, C2-deficient sera in the laboratory.

C3. Individuals heterozygous for C3 deficiency in one family have half-normal C3 concentrations owing to inheritance of one "silent" C3 gene. Hemolytic complement is slightly reduced in partial C3 deficiency, and enhancement of phagocytosis is half-normal. Affected individuals are asymptomatic. However, individuals homozygous for C3 deficiency have not yet been found.

A patient with an entirely different disorder involving C3 has been described. He is a young man with a lifelong history of numerous infections with pyogenic (pus-producing) organisms of the upper and lower respiratory tracts, bloodstream, and other systems. Hemolytic complement and C3 concentration are markedly reduced, and chemotaxis for normal leukocytes, enhancement of phagocytosis of pneumococci, and bactericidal activity for smooth gram-negative bacteria in the laboratory are all grossly deficient or not demonstrable in his serum. All of these defects are correctable by the addition of a 5-6S beta pseudoglobulin (but not purified C3) from normal serum to his serum in the test tube.

Evidence for continuous inactivation of C3 in the body of this patient was obtained, and it is presumed that he is congenitally lacking an inhibitor of some protease which has C3 as a substrate. However, neither the inhibitor nor the protease has been identified. His serum lacks several normal serum proteins in addition to the one that restores complement-mediated functions. The human protein which complexes with a protein in cobra venom to form an enzyme capable of inactivating C3 is not detectable in his serum. Similarly, the C3 inactivator, an enzyme capable of destroying the activity of cell-bound C3, is missing from his serum. Finally, a recently isolated beta globulin of unknown function but for which a monospecific antibody is available, is also absent.

The infusion of 500 ml of normal plasma into this patient resulted in a correction, complete or partial, of all complement-mediated functions for as long as 17 days. In addition, C3 concentration rose almost tenfold (owing to decreased inactivation), and all of the missing proteins appeared in his serum. It is presumed that the normal plasma supplied the primarily deficient inhibitor, whereas the other proteins, perhaps because they, too, had been acted upon by the hypothetical protease, appeared because of diminished "consumption."

Hereditary angioneurotic edema. In hereditary angioneurotic edema, a disease inherited as an autosomal dominant trait, patients have recurrent episodes of localized noninflammatory edema of the skin and the linings of the gastrointestinal and respiratory tracts. A deficiency of a serum protein capable of inhibiting the protease and esterase activity of the first component of complement (both C1r and C1s), plasmin, PF/dil, and kallikrein has been found in all patients. During attacks, free C1 esterase activity can be found in patients' plasma. The serum concentrations of functionally active C4 and C2, the natural substrates of C1 esterase, are markedly reduced or not detectable in serum during attacks and are usually somewhat reduced between episodes. A vasoactive peptide, perhaps responsible for the local swellings, has been isolated from plasma from patients during attacks and may be identical with a vasoactive peptide derived from C2 by the action of C1 esterase.

All patients with hereditary angioneurotic edema are deficient in serum C1 esterase inhibition. In 85% of kindred the concentration of the C1 esterase inhibitor measured immunochemically is markedly reduced to 5–30% of that in normal serum, whereas in the remaining 15% of families, the C1 esterase inhibitor protein concentration is normal or elevated. Thus, in the latter patients the inhibitor is dysfunctional or nonfunctional. Since the proteins in most of these families have different electrophoretic mobilities from those of the normal C1 esterase inhibitor, they are probably the products of mutant structural genes. Since both the low and normal concentration forms of the disease are inherited as autosomal dominant traits, all affected persons are heterozygotes.

C5 dysfunction. A familial defect in C5 associated with increased susceptibility to infection with gram-negative bacteria and *Staphylococcus aureus* in infancy has been found in two unrelated kindred. Although hemolytic complement and C5 protein concentration are normal in affected individuals, there is failure of the normal complement-mediated enhancement of phagocytosis in laboratory tests, which is correctable by small amounts of normal serum or purified C5 but not C5-deficient mouse serum. Normal plasma infusions correct the defect and apparently prevent the frequent infections. C5 isolated from the serum of one affected individual had 10% of the hemolytic efficiency of C5 from normal serum. It thus appears that C5 is structurally abnormal and dysfunctional in these kindred.

Acquired complement deficiencies. Lowered serum concentrations of one or more complement components occur in association with a variety of diseases, such as acute and subacute glomerulonephritis, progressive glomerulonephritis of childhood, and systemic lupus erythematosus. No direct evidence has been found for undue susceptibility to infection or for other symptoms related to complement component deficiencies in these disorders.

The role of complement in host defenses is clearly complex and knowledge is still rudimentary. Nevertheless it has been established that certain disturbances of the complement system can produce increased susceptibility to infection.

For background information see BLOOD; BLOOD GROUPS; COMPLEMENT, SERUM in the McGraw-Hill Encyclopedia of Science and Technology.

[CHESTER A. ALPER]
Bibliography: C. A. Alper et al., *N. Engl. J. Med.*, 282:349, 1970; V. H. Donaldson et al., *J. Clin. Invest.*, 48:642, 1969; M. E. Miller and U. R. Nilsson, *N. Engl. J. Med.*, 282:354, 1970; H. J. Müller-Eberhard, *Annu. Rev. Biochem.*, 38:389, 1969.

Culture, tissue

Tissues derived from a wide variety of species of plants can now be cultured, and it can be confidently predicted that the test-tube propagation of tissues from any species, including monocotyledons, previously recalcitrant in this respect, is now technically feasible. Until recently the tissues cultured had been derived from somatic, as opposed to gametic, cells, but one of the most significant advances of the last few years has been the recognition that pollen grains could also serve as a source of cell cultures. This discovery has considerable implications for both basic and applied studies of the genetics of higher plants and could, if successfully exploited, prove of enormous value in plant breeding programs. Another highlight of plant tissue culture research has been the discovery that plant protoplasts, that is, cells lacking walls, can be fused to form hybrid cells.

Pollen culture. That pollen grains could be induced to forego their normal pathway of development and to form a callus tissue or even embryoids and ultimately plants was recognized several years ago. Anthers of *Datura innoxia* grown on a basic nutrient medium with 2% sucrose and 15% coconut milk developed numerous embryoids. Cytological examination showed these to have arisen from the pollen grains and to have a haploid complement of chromosomes. In this work only mature pollen grains could be induced to develop in this manner. A few years elapsed before any further success was reported, this time with various species of the genus *Nicotiana*. The tetraploid *N. tabacum* and *N. rustica* and the diploid *N. sylvestris* and *N. affinis* were all induced to develop embryoids from pollen grains.

The basic procedure adopted with these species has been to surface-sterilize flower buds at various developmental stages, aseptically remove the stamens (taking care not to damage the anthers), and place the anthers on a nutrient agar medium. The filament should be completely removed prior to plating since it very readily forms a callus which can be confused with that arising from haploid tissue. The environmental conditions have been 28°C by day and 22°C by night, with a day length of 12–14 hr, but the precise conditions necessary have not been defined. The medium is a simple one containing inorganic nutrients, a few vitamins, and sucrose. The presence of auxins at low concentrations has little effect, and no growth hormone appears essential. After 4–6 weeks, plantlets can be seen emerging through a fissure in the wall of some of the anthers (Fig. 1), and up to 40 or 50 per anther have been reported. After they have grown to an adequate size, they can be teased apart, transferred individually to fresh medium, and allowed to grow until they are ready for transplanting into soil. Several hundred such plants have now been grown, and the vast majority, as would be expected, are haploid in the case of the diploid parental species, and diploid in the case of the tetraploids.

Considerable discrepancies exist regarding the precise developmental stage of those pollen grains which can be successfully induced to differentiate into embryoids. In the initial studies, stages were not characterized precisely, but it appeared in one series of experiments with *N. tabacum* that only tetrad stages were of value. In the tetrad stage the four products of the meiotic division of the pollen mother cell are still attached (Fig. 2). In other experiments, neither the very immature tetrad stages nor the very mature pollen, immediately prior to shedding, appeared to develop embryoids. These appeared only when anthers were plated which had pollen in the uninucleate phase and in which starch had not been laid down. There were differences in the constitution of the nutrient agar used in these two experiments which could account for the discrepant results, but it is more likely that they are due to an imprecise characterization of stages. Most anthers have a range of developmental stages within them, and sampling one anther out of the many that are usually found in a given flower can at best serve only as an indication of the stages to be found in the cultured anthers. A further assessment of the situation in *N. tabacum* has shown that the crucial stage for successful development is the time of the first mitotic division in the pollen grain. If slightly earlier stages are plated, they can progress to this crucial stage in culture and still give rise to plantlets.

The two nuclei produced as a result of the mitotic division in the pollen differentiate into quite distinct forms. One, the generative nucleus, becomes small and densely staining, and the other, the vegetative nucleus, much larger and diffuse. The generative nucleus, or its daughter nuclei, is destined to fuse with nuclei within the ovule, whereas the vegetative nucleus has no function other than a role in the organization and growth of the pollen tube. The vegetative nucleus, however, gives rise to the embryoid in *N. tabacum*. Characteristically there are also two types of pollen grains induced within the anthers after plating; one has a cytoplasm which stains intensely with acetocarmine; the other does not show this response and is the type which gives embryoids. The biochemical basis which underlines this difference in staining response of the two types of pollen grains has not been investigated. Subsequent development of the pollen grains to a multicellular state, embryoids, and ultimately plantlets appears to follow the normal pattern of embryos derived from fertilized eggs.

The ease of production of haploids in *Nicotiana* species is striking—up to 50% of anthers have been known to give rise to plants—but this success has not been reproduced elsewhere. Numerous attempts in a large number of laboratories and using a wide range of plant species have so far, with one published exception, failed to repeat the *Nicotiana* story. This situation will almost certainly be a transient one, and the precise conditions necessary for other species will be defined sooner or lat-

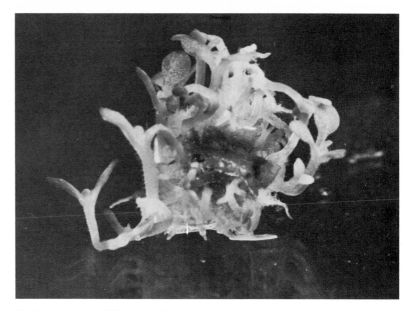

Fig. 1. An anther of *Nicotiana tabacum* showing numerous plantlets emerging from fissures in the wall. (*Courtesy of N. Sunderland*)

er. The one known exception is rice, where a haploid callus has been induced to develop from within anthers. The rate of success was low, 20 calluses being obtained from approximately 3500 anthers: seven haploid plants grew from the subcultured callus tissues. Only anthers with mature pollen grains were used, and this could in part account for the low rate of success. This indirect production of haploid plants through the intermediacy of an undifferentiated haploid callus has also been seen in *Nicotiana*.

Application of pollen culture. Should the technique of producing haploids succeed ultimately in crop plants on the scale that it has with *Nicotiana*, then the plants so produced could be exploited in several ways. The first application will arise from the ready availability of homozygous diploids in those species whose normal production depends on a long and time-consuming process of inbreeding. If the haploids are treated with the drug colchicine, a doubling of the chromosome number occurs following an upset in the normal processes of cell division, and the diploid forms which ensue must necessarily be homozygous and can be of value in several ways in plant breeding. In the first place, genetic studies of many kinds will be facilitated. Also, such homozygotes are the parents of the F_1 hybrids commonly grown as commercial

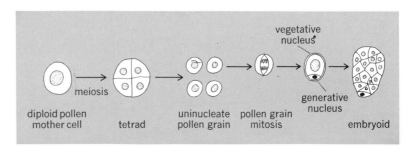

Fig. 2. Developmental sequences in the formation of an embryoid from a pollen grain in *Nicotiana tabacum*.

crops, but whereas it normally takes six to seven generations of selfing or backcrossing to produce a reasonable degree of homozygosis, pollen culture allows its production in one generation. Should the generation time of a plant be long, as it is, for example, in many tree crops, then conceivably the technique of F_1 hybrid production could be contemplated much more readily than has hitherto been possible. If haploid plants were to be exploited in this way, a vast array of different genotypes would have to be grown and induced to form homozygous diploid forms, which would then have to be tested for their combining ability and suitability as parents of F_1 hybrids.

The time-saving element of the technique also enters into plant breeding programs in another way. When two plants of an inbreeding species are crossed in order to recombine the attributes of one with that of the other, it is necessary to wait for several generations to elapse before an accurate assessment can be made of the potential of the recombinants. This period is required for them to become reasonably homozygous again. With a method of producing completely homozygous diploid forms in one generation, however, the potential saving in time and money is enormous.

Again, haploids offer novel possibilities for inducing and recognizing mutations on a scale not hitherto possible in higher plants. Most induced mutations are recessive, and with inbreeding species it is not until the second generation that such mutations occur in the homozygous condition and are recognizable. Since a mutation is in the first place a rare event and since only a small proportion of tissues in a plant having a mutation may be mutant, large populations have to be grown and maintained for two generations in order to have a reasonable chance of success in detecting the mutation. In an outbreeding species, such homozygous recessive mutants will occur very infrequently; they will occur if a comparable mutation has been induced in both male and female parents, an unlikely event, or when, in later generations, heterozygous progenies derived from a common parent and bearing the same mutation intercross. Treatment of haploids with mutagens allows the immediate recognition of mutations in the first generation, and radiation treatment of such forms in *N. tabacum* already has yielded a high number of mutations.

Fusion of plant protoplasts. Plant protoplasts have been induced by a variety of techniques, both mechanical and enzymatic. Once the cell wall is eliminated, the membranes of different cells can be brought into contact and cell fusion can be induced to occur. The essential features of the recently developed technique for inducing fusion are covered by a patent application but, in essence, involve the provision of a suitable ionic environment for the protoplasts; in this environment sodium nitrate appears to play an important role. Fusion at an interspecific, as well as at an intraspecific, level has been observed. Although the demonstration of fusion has in itself been a significant achievement, much remains to be done before the technique can be exploited. Nuclear fusion has not been demonstrated, the protoplasts or fused units have not been shown to be viable, and, equally important, it has not been possible to regenerate cell walls on most protoplasts. One of

the primary problems is that of modifying the enzymatic methods of removing the walls so that undue damage to other cell constituents does not ensue and the viability of the cells is unaffected. Commercial samples of cellulases and pectinases, though effective in many instances, probably contain numerous and deleterious contaminants; on the other hand, extremely pure isolates of single enzymes can be ineffective. The limited success already achieved in this field indicates the value of further effort, since rich rewards could be obtained in terms of novel hybrid plant combinations.

For background information *see* BREEDING (PLANT); CULTURE, TISSUE; FLOWER; PLANT GROWTH in the McGraw-Hill Encyclopedia of Science and Technology.   [D. R. DAVIES]

Bibliography: J. P. Nitsch and C. Nitsch, *Science*, 163:85, 1969; J. B. Power, S. E. Cummins, and E. C. Cocking, *Nature*, 225:1016, 1969; R. U. Schenk and A. C. Hildebrandt, *Crop Sci.*, 19:629, 1969; N. Sunderland and F. M. Wicks, *Nature*, 224:1227, 1969.

Cycadales

New evidence bearing on the ancestry and evolution of the Cycadales has recently been reported from Upper Pennsylvanian and Lower Permian rocks of the southwestern United States. Although it is known that the cycads were highly evolved by Jurassic time, until recently the oldest known putative cycads were of Upper Triassic age. It had been widely assumed that the cycads evolved from Pennsylvanian pteridosperms (seed ferns), but convincing evidence was lacking. These discoveries in the United States extend the known history of the cycads roughly 100,000,000 years, and this lineage becomes one of the longest known among the gymnosperms.

Cycad classification. A few background remarks are essential to an understanding of the significance of the fossils. Classification of living cycads is based largely on morphology of the seed-bearing female cones; taxonomically the male cones are not as important. The female cones consist of numerous spirally arranged megasporophylls (seed-bearing appendages). The cone of *Zamia* is extremely compact, and the much-reduced megasporophylls are small peltate structures with two lateral ovules each. In *Cycas*, however, the cone is not compactly organized but is composed of a loose crown of terminal megasporophylls, borne like ordinary leaves. These each produce basally two lateral rows of several seeds and terminate in leaflike expansions. The leaflike appearance of *Cycas* megasporophylls reflects a foliar origin of these organs; because of this, *Cycas* is regarded as a primitive member of the living cycads. The new fossil megasporophylls are interpreted as ancestral elements in the evolution of a *Cycas*-like megasporophyll.

New fossil evidence. Three types of fossil megasporophylls constitute the new evidence. Geologically the oldest is *Spermopteris*, first described in 1960 by A. A. Cridland and J. E. Morris, from the Upper Pennsylvanian of Douglas County, Kans. *Spermopteris* had a megasporophyll in which unmodified leaves bore two rows of small seeds on their presumed lower surfaces, one row on either side of the midrib (Fig. 1a). Without seeds the leaves of *Spermopteris* would be indistinguishable

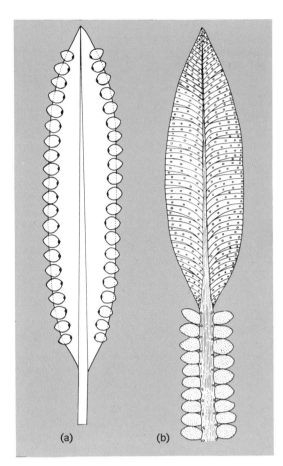

Fig. 1. Hypothetically ancestral and derivative cycadean megasporophylls. (a) Semidiagrammatic representation of *Spermopteris*, with superficially borne seeds; solid black dots indicate points of seed attachment. (b) Reconstruction of new genus A, showing elaminar ovuliferous base and small glands between seeds and veins. (*From S. H. Mamay, Cycads: Fossil evidence of late Paleozoic origin, Science, 164(3877):295–296, 1969*)

from *Taeniopteris*, a common, geographically widespread form-genus of late Paleozoic and Mesozoic foliage characterized by a simple linear outline and dichotomous venation. Cridland and Morris interpreted *Spermopteris* as a pteridosperm but postulated no unusual evolutionary alliances for the genus. However, its leafy form and bilaterally arranged ovules render *Spermopteris* the ideal morphological archetype from which a *Cycas* type of megasporophyll might have been derived. S. H. Mamay has interpreted *Spermopteris* as a precursive cycadean and an evolutionary link with the Pennsylvanian pteridosperms.

A second form, new genus A (Fig. 1b), is based on one specimen from Lower Permian rocks near Elmo, Kans. This is a stout naked axial fragment to which are attached two rows of small seeds in a clearly bilateral arrangement. The seeds have well-preserved cuticles with cycadlike characteristics. Small resinoid globules, evidently of glandular origin, are attached to the axis in positions alternating with those of the seeds; obviously these were a natural biological feature of the fruiting axis. The basic significance of the glands lies in the presence of similar bodies on taeniopterid leaves found in the same beds with new genus A. Because of the similarity of glands, the known ovuliferous habit of *Taeniopteris* (*Spermopteris*), and other indirect evidence, new genus A has been interpreted and reconstructed as the fertile basal part of a primitive cycadean megasporophyll with a taeniopterid distal laminar expansion. A spermopterid Pennsylvanian form was probably ancestral to new genus A, and new genus A is thought to have evolved through phylogenetic restriction of the ovules to the sporophyllar base, followed by reduction of the lamina in the fertile area. The result would be a megasporophyll of fundamentally modern cycadean morphology. Geologic relationships are consistent with this hypothesis, because *Spermopteris* is significantly the older of the two forms.

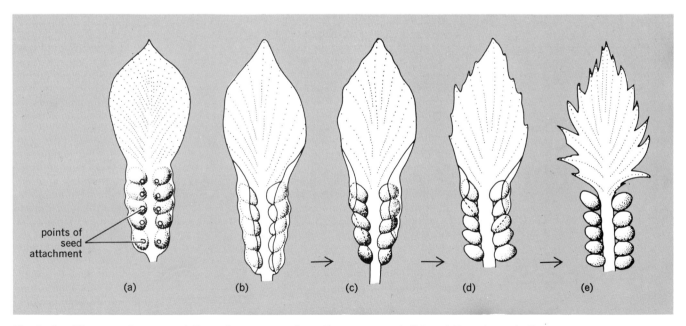

Fig. 2. Semidiagrammatic representations of new genus B and supposed derivatives. (a) Top view of genus B; (b) lower view of a; (c,d) hypothetical intermediate stages in proposed evolutionary series producing (e) a *Cycas*-like megasporophyll from b through a reduction of basal lamina. (*From S. H. Mamay, Cycads: Fossil evidence of late Paleozoic origin, Science, 164(3877):295–296, 1969*)

The third form, new genus B, is from the Lower Permian of Baylor County, in northern Texas, and is approximately the same age as new genus A. Several specimens occurred at two localities. New genus B is a small megasporophyll with a sterile terminal lamina and 4–6 pairs of basal ovules, which are closely appressed against each other and attached to the basal part of the lamina a short distance from the midrib (Fig. 2a). The lamina in the fertile basal area is slightly modified and extends around the tips of the attached ovules; the lower side of a megasporophyll has not been seen, but it is assumed that the ovules were not completely enclosed (Fig. 2b). Because of the presence of its basal lamina, new genus B seems to be an important morphological intermediate, standing between *Spermopteris*, with its undifferentiated lamina, and the modern *Cycas* type of megasporophyll, with its naked stalklike seed-bearing base (Fig. 2e). It is not difficult to envision the *Cycas* type of megasporophyll evolving from new genus B through phylogenetic reduction of the basal lamina of the latter (Fig. 2b–e).

Interrelationships of fossils. The ancestral nature of *Spermopteris* appears clear enough, but relationships between new genera A and B are questionable because the two were essentially contemporaneous. Probably they were both derived from a common spermopterid ancestor and may represent early divergences of two cyadean lineages. Their consistent association with *Taeniopteris* foliage is important; together with the fact that *Taeniopteris* is common to all Permian and Triassic floras that contain cycadlike fossils, this association lends strength to a favored hypothesis that all cycads were derived from taeniopterid elements in the Pennsylvanian pteridosperm complex.

Precise taxonomic assessments of the early cycads would be premature at this time; such assessments will depend on knowledge of many other morphological details of the fossils. It is reasonable to predict, however, that much important information on early cycadean evolution will eventually be discovered in Lower Permian strata, particularly those representing the Wolfcampian time interval.

For background information *see* CYCADALES; PALEOBOTANY in the McGraw-Hill Encyclopedia of Science and Technology. [SERGIUS H. MAMAY]

Bibliography: A. A. Cridland and J. E. Morris, *Amer. J. Bot.*, 47(10):855–859, 1960; S. H. Mamay, Cyacads: Fossil evidence of late Paleozoic origin, *Science*, 164(3877):295–296, 1969.

Cytology

Much new information has been obtained in the field of cytology, the study of the formation and function of cells, since the improvement of the technique of microirradiation, which was pioneered by S. Tchakhotine in 1912 with an ultraviolet microbeam.

Development of the laser provided a source of highly chromatic, coherent light and made it possible to produce beams of tremendous energy flux. A microirradiation instrument with a ruby laser as the source was developed in 1962. As the potential of this apparatus was explored and as advances in laser technology made new sources of coherent

light available, the capabilities of this unique instrument have been broadened and a wider variety of biological and physical investigations have become possible.

Laser microirradiation. The principal devices used are (1) the ruby laser ($\lambda = 6943$ A) and the neodymium laser ($\lambda = 10,600$ A) with doubled ($\lambda = 5300$ A) or quadrupled ($\lambda = 2650$ A) frequency giving pulses of various duration (from 50 nsec to 300 μsec); (2) the argon laser ($\lambda = 4880$ and 5145 A), continuous or with 60 pulses per second; and (3) the conventional (noncoherent) ultraviolet.

Some organelles are affected by microirradiation because they contain natural chromophore, compounds that absorb electromagnetic radiation. For example, nucleic acids absorb strong ultraviolet light at 2650 A. Some organelles which do not contain such chromophores can be stained artificially with vital stains. It has been shown that the laser beam, because of its tremendous energy, can destroy organelles which have absorbed very small, nontoxic quantities of these specific stains (such as Janus Green for mitochondria).

Physical parameters. Laser microirradiation was used to study the uptake of low levels of stain (Janus Green B) by the mitochondria of tissue-culture cells and the selective lesions induced by the irradiation. A comparison of the thermal effects produced in the cells and in a stained model system such as albumin permitted the calculation of the degree of concentration of stain in the mitochondria from the staining media. It was found to be approximately 500 times more concentrated in the mitochondria than in the surrounding medium. Evaluation of the various parameters of laser irradiation, particularly thermal events, resulted in optimal irradiation times for the measurement of local stain concentration as well as for the selective destruction of the stained mitochondria. For the studies of the latter phenomenon, a lower limit for irradiation duration is set by optical saturation and dielectric breakdown.

Ultraviolet laser microbeam. The lesions produced by laser microirradiation at 2650 A are very different from those produced by a conventional ultraviolet source of the same wavelength. The functional difference between the two types of damage is interpreted to be primarily a function of the duration of the irradiation. With an ordinary ultraviolet source, exposure times are of the order of a minute, and it is possible to study the resulting phenomena under isothermic conditions. With irradiations lasting only several nanoseconds, however, adiabatic approximations (a gain or loss of heat to surroundings is assumed not to take place) are permitted. The classical photobiological reactions that are found at a wavelength of 2650 A are thought to be masked by attendant phenomena produced by thermal denaturation of the biological material in the irradiated zone.

Argon laser microbeam. An argon-gas laser microbeam has been used to induce lesions in mitotic chromosomes and nucleoli of salamander lung cells in the test tube. These cells were photosensitized by 5-min pretreatment incubation with a dilute solution of acridine orange or quinacrine.

Necrotaxis. Necrotaxis is the phenomenon involving the movement of leukocytes toward another leukocyte or the destruction of a red cell by laser irradiation. Investigations aimed at analyzing this phenomenon have been initiated. Thus far the effects of varying the cellular environment and changing the irradiation target have been studied. Tentative calculations have been made on the rate of diffusion of necrotactic substances under varying conditions, such as in synthetic media and under alterations of viscosity, different pHs, and different temperatures.

For background information *see* CELL (BIOLOGY); LASER in the McGraw-Hill Encyclopedia of Science and Technology.   [M. C. BESSIS]

Bibliography: M. W. Berns, R. S. Olson, and D. E. Rounds, *J. Cell Biol.*, 43:621, 1969; G. Moreno, *C. R. Acad. Sci.*, 270:988, 1970; G. Moreno, M. Lutz, and M. Bessis, *Int. Rev. Exp. Pathol.*, 7:99, 1969; C. Salet, M. Lutz, and F. S. Barnes, *Photochem. Photobiol.*, 11:193, 1970; C. L. Smith and P. P. Dendy, *Cell Tissue Kinet.*, 1:225, 1968; S. Tchakhotine, *Biol. Centralbl.*, 32:623, 1912.

Deoxyribonucleic acid (DNA)

It has been known for about 15 years that cells have the ability to repair defects in DNA, the primary genetic material of living systems. This repair requires enzymes, and its study promises to provide important insights into the nature of DNA-protein interactions and into such outstanding genetic problems as the replication and mutation of DNA.

The impetus for these studies came with the identification of pyrimidine dimers as the major inactivating lesion in cells irradiated with ultraviolet light of wavelengths in the range 250–270 nm. Dimers are formed by covalent cyclobutane linkage between adjacent pyrimidine bases of a single strand of the DNA helix. Acid hydrolysis of irradiated DNA followed by chromatography allows the separation of dimers from DNA bases. This assay and the use of mutants that are defective in different stages of repair have led to the identification of three processes by which organisms overcome the inactivating effects of lesions in DNA.

Photoreactivation. Enzymatic photoreactivation involves monomerization of dimers in place by an enzyme that does not function without light. The reaction has been demonstrated to occur in two steps. The first step, binding of the enzyme to a dimer, is independent of light. The second step, monomerization of the dimer, requires light of wavelength 310–500 nm, depending on the organism. The rate-limiting step of the overall reaction appears to be the slow diffusion of DNA and enzyme together. The photoreactivating enzyme from yeast has been extensively purified and has an estimated molecular weight of 60,000 to 90,000. The enzyme is found throughout the animal and plant kingdoms, with the interesting major exceptions of some bacteria and the placental mammals. Although dimers are the only known substrate for the enzyme, the presence of the enzyme in internal organs such as the heart and liver not reached by light suggests it may have other, unknown functions.

Excision repair. A second repair mechanism (Fig. 1) involves the physical removal of dimers and their replacement by the correct base sequence. Some or all of the steps in this process

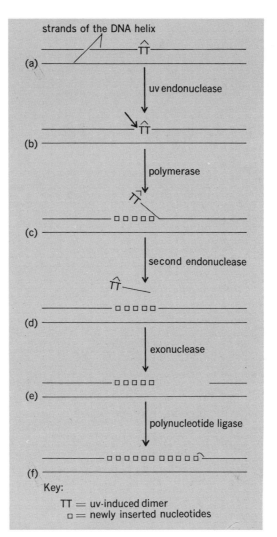

Fig. 1. Diagram of steps in excision repair of deoxyribonucleic acid (DNA). The precise sequence is not known, nor is it clear whether separate enzymes are responsible for steps a–d. (a) Dimers produced by ultraviolet light between adjacent pyrimidines in a single strand of the DNA helix. (b) Dimer is recognized by a repair endonuclease which makes a nick in the phosphodiester backbone. (c) Dimer is displaced from DNA duplex by activity of a polymerase which synthesizes DNA using the complementary strand as template. (d) Dimer is freed from DNA helix, as part of a short oligonucleotide by a second endonucleolytic nick. (e) The gap may be enlarged by an exonuclease, causing the release of mononucleotides. Subsequent polymerase activity fills in the gap. (f) Repair synthesis is completed by a sealing of the phosphodiester backbone.

appear to be involved in the repair not only of dimers but also of damage due to ionizing radiation and errors produced by certain chemical mutagens. As with photoreactivation, excision repair has been found in all species examined, with the exception of murine rodents.

Recombination repair. Among radiation-sensitive mutants of the bacterium *Escherichia coli*, the rec⁻ class also show a marked reduction in the ability to undergo genetic recombination. This observation prompted the proposal that DNA is also repaired by a mechanism which bypasses lesions by recombining with homologous DNA that

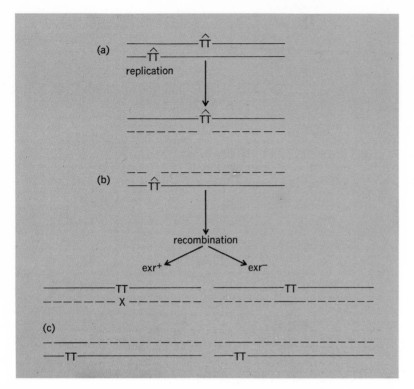

Fig. 2. Diagram of recombination repair mechanism which is intended only to indicate the few steps in recombination repair that have been identified, and not to describe the mechanism of recombination. (*a*) Ultraviolet irradiation produces dimers in DNA. (*b*) If the dimers are not removed by photoreactivation or excision repair, replication of DNA may result in discontinuities in the replicated DNA because of the failure of dimers to act as templates for the newly synthesized strand. (*c*) Following replication, recombination events initiated at the gaps restore the continuity of the newly synthesized DNA. Recombination may introduce errors (X) in exr⁺ cells (left) or be error-free as in exr⁻ mutants (right).

has the correct information sequence. Recently, it has been shown that, if replication of ultraviolet-irradiated bacterial DNA occurs, discontinuities in the newly replicated strand occur at about the same frequency as nonexcised dimers in the parental strands, implying that gaps occur opposite the dimers. Recombination events initiated at the gaps may lead to reconstitution of a chromosome free from dimers, as indicated schematically in Fig. 2. In accord with this model, there is evidence indicating that recombinational repair of ultraviolet-irradiated phage λ is efficient when DNA is replicating, and that it complements excision repair, which can occur in the absence of DNA replication.

Error-correcting mechanisms in mutagenesis. Ultraviolet-induced mutations in surviving cells of *E. coli* are only rarely produced as a result of errors in the photoreactivation and excision repair processes. However, the presence of an unrepaired dimer is correlated with a high probability of mutation induction, apparently because recombinational repair of the discontinuities in newly replicated DNA can introduce errors into DNA. The exr⁻ class of radiation-sensitive mutants may have an altered recombinational repair mechanism which does not permit the introduction of error into DNA, since these mutants, which show a modest reduction in recombination ability, do not permit ultraviolet induction of mutations.

When *E. coli* mutants are treated with 4-nitroquinoline-1-oxide, they behave in the same way as do ultraviolet-treated cells, suggesting that mutation induced by this agent may also be produced by recombination repair. The involvement of repair enzymes in certain types of diseases, particularly cancers, has been strongly implied by the finding that this compound is also a carcinogen. Further, xeroderma pigmentosum, a rare form of skin cancer induced by exposure to ultraviolet light, is due to a genetic defect in a step in excision repair.

Excision repair enzymes. A number of laboratories have succeeded in partially purifying an endonuclease which recognizes dimers in ultraviolet-irradiated DNA and which provides the function missing in cell unable to perform the first step in the excision reaction. Polynucleotide ligase is able to restore the integrity of the phosphodiester backbone of DNA, and probably performs the final step of excision repair, since cells defective in this enzyme are radiation-sensitive.

Recently it was demonstrated that DNA polymerase, the enzyme thought to be responsible for normal semiconservative replication of DNA, can perform all of the intermediate steps in the test tube. The polymerase and exonuclease functions of this large molecule (molecular weight 109,000) can be separated by proteolytic cleavage of the molecule. It seems possible that a number of exonuclease preparations which can excise dimers, but which have no polymerase activity, may have resulted from such cleavage during the extraction procedure. The excision properties of the polymerase raise the question as to what complex functions the DNA polymerase ("Kornberg's enzyme") may play in the living organism.

For background information *see* BACTERIAL GENETICS; GENETIC CODE; MOLECULAR BIOLOGY in the McGraw-Hill Encyclopedia of Science and Technology.   [JOHN M. BOYLE]

Bibliography: J. S. Cook, in A. C. Giese (ed.), *Photophysiology*, 5:191–233, 1970; C. C. Richardson, *Ann. Rev. Biochem.*, 38:795, 1969; B. Strauss, *Curr. Top. Microbiol. Immunol.*, 44:1, 1969; E. M. Witkin, *Proc. 12th Int. Congr. Genet.*, 3:225, 1969.

Diffraction grating

The use of optical gratings in spectroscopy has increased a hundredfold in recent years; this increase results from new processes for replicating gratings, from an increase in the number and quality of engines available to rule masters for replication, and from a general improvement in grating permanence and performance. World sales of gratings have exceeded $20,000,000 since 1952, compared to roughly $350,000 previously.

The introduction of ruling engines that are controlled interferometrically with monochromatic light waves has made possible increases in grating size, resolving power, and efficiency and freedom from false lines and scattered light.

Most gratings now used are of the reflection type, and about three times as many plane gratings as concave are being produced. The advantages of the concave grating are now offset by improvements in the efficiency of mirror coatings and by the lower astigmatism, larger blaze angles (angle of maximum reflection from the groove sides), and greater speed and compactness obtainable with plane grating spectrographs.

Replication of gratings. In the basic system of W. Frazer and J. White the master grating is first coated with a thin parting layer; on this a layer of aluminum is deposited to take the form of the grooves. The aluminum layer is then backed with one of epoxy resin, which fills the space between the aluminum and a thick plate of glass or fused silica. After curing, the replica is separated from the master at the parting layer.

Although grating masters are sometimes spoiled during replication, under good conditions it has been possible to make 50 or more replicas from a single master. Today most commercial gratings are products of replication, and a wide variety of sizes, shapes, and groove spacings has become available. A properly made replica may show higher efficiency than its master, should have equal resolving power and freedom from scatter, and is likely to be less delicate.

Choice of groove spacing. It has been widely assumed that the resolving power of a grating increases with the number of grooves it contains and that its dispersion decreases with increasing groove spacing. Both statements are incorrect, and a false premium has been put on ruling many grooves per inch.

All gratings, regardless of groove spacing or number, give the same intrinsic dispersion at the same wavelength and angles of use. Assuming optical perfection, resolving power, in turn, depends only on the ruled width of the grating, on the sum of the sines of the angles of illumination and reflection, and on the wavelength. The figure shows the relative variations of resolving power and dispersion of any perfect grating with angle of use in autocollimation, for any selected wavelength. The tangent of the angle of incidence and diffraction is represented by the symbol r.

The property of a grating affected by groove spacing is its free spectral range, or the separation of its orders of interference. Once overlapping orders have been separated by a device such as crossed dispersion with a weak prism or grating, large increases in resolution and dispersion can be achieved by working at high angles of incidence and reflection. Blazing plane gratings at large angles for use in autocollimation led to the development of the echelle, now coming into wide use in both spectrographs and monochromators.

Echelle gratings. The echelle, a plane grating successor to the powerful reflection echelon, became practicable only with the perfection of the interferometrically controlled ruling engine. Gratings ruled on purely mechanical engines are seldom useful at high angles because most ruling defects produce results which increase as the square of the sine of the angle of use. Echelles of high quality have been produced which greatly increase effective dispersion, have ghosts and satellite intensities weaker than 10^{-4}, and give resolution in excess of 1,000,000 in the visible.

Grating ruling engines. At present the supply of gratings is produced from masters ruled on about 20 engines, most of which are limited to ruling widths of 6 in. or less. Almost all are based on H. A. Rowland's design, the larger sizes being under interferometric control.

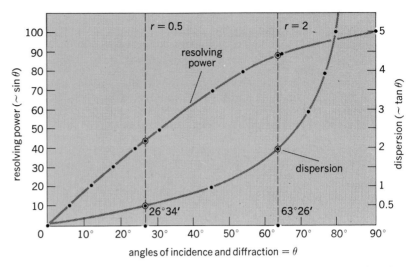

Curves showing the variation of resolving power and dispersion with angle of use in autocollimation of a diffraction grating. (*G. R. Harrison and G. W. Stroke, Attainment of high resolution with diffraction graftings and echelles, J Opt. Soc. Amer., 50:1154, 1960*)

G. R. Harrison and colleagues at the Massachusetts Institute of Technology developed a system of interferometric control of groove ruling, spacing, and orientation which they have used on three ruling engines of 10-, 18-, and 24-in. capacity respectively. In these engines the coated blank is moved at constant speed during the ruling process in order to reduce elastic and frictional deformations and to simplify arrangements for keeping blank and diamond advance properly synchronized. This is done photoelectrically, using stabilized laser light and closed-loop servo systems. Several hundred large gratings, up to 12×15 in. in size, have been produced on these engines. Echelles have been produced up to 8×16 in. in size, blazed at angles of from 63 to 79° and giving resolving powers in excess of 10^6.

H. W. Babcock of Mount Wilson Observatory has applied stop-start interferometric control to a 10-in. machine which represented a peak of development of the purely mechanical ruling engine. W. R. Horsfield uses interferometers to control a hydraulic blank-advancing mechanism which takes the place of a screw in a 6-in. engine. Bausch and Lomb, of Rochester, N.Y., Jarrell-Ash Co., of Waltham, Mass., and Diffraction Products, Inc., of Chicago, Ill., have applied interferometric control to engines of intermediate sizes, as have Hitachi, Ltd., and others in Japan and Jobin et Yvon in France. F. M. Gerasimov, of the Soviet Union, has described a ruling engine in which a master grating corrects the excursion of the blank. D. A. Davies and G. M. Stiff, of the Commonwealth Scientific and Industrial Research Organization (CSIRO) of Australia, and J. Dyson, of the National Physical Laboratory in England, are building controlled engines, as are G. W. Stroke and A. E. Johnson in the United States. The Stroke-Johnson engine is designed to rule very large blanks while they remain at rest, the ruling diamond being reciprocated in the y direction while slowly advancing in the x direction.

Gratings have been produced directly by means of light waves using holography, a process which is still in the experimental stage. It will probably be found most useful for gratings of fine groove spacing.

Fourier spectroscopy, used with a Fabry-Perot etalon interferometer, gives a very powerful supplement to grating spectroscopy. The method has been most successful in the infrared, but gratings, which give data that can be reduced more simply, are still to be preferred for spectroscopic operations which can be carried out with them.

For background information *see* DIFFRACTION GRATING; HOLOGRAPHY; INTERFEROMETRY in the McGraw-Hill Encyclopedia of Science and Technology.   [GEORGE R. HARRISON]

Bibliography: G. R. Harrison and S. W. Thompson, *J. Opt. Soc. Amer.*, vol. 60, May, 1970; G. W. Stroke, *Encyclopedia of Physics*, vol. 29, 1967.

Direction-finding equipment

Two important advances in the radiopositioning field have been made recently. The principles involved have been known for many years, but the equipment necessary for practical field applications has just recently become available. These developments now allow a fuller utilization of two systems in worldwide use, loran C and shoran.

Loran C. Loran C is strategically placed in various high-density marine traffic areas of the world. The system uses three land-based stations in a master-slave relationship to produce hyperbolic lines of position in an overlapping grid. Although the system is excellent as a navigational aid, two adverse features limit the system accuracy, thereby limiting its use as a radiopositioning system for oceanography and similar applied sciences. (1) The hyperbolic configuration produces areas of poor geometric control in the vicinity of the base line extensions. (2) The distance to the farthest of the three stations sometimes causes the received signal from that station to be of poor quality. Both these shortcomings are overcome by using the system in a range-range mode. Since ranges from only two of the three stations are required to define a

position, the most distant station need not be used. Further, the range-range geometric coverage gives superior accuracy to the hyperbolic geometry.

The loran C rho-rho system employs the technique of measuring ranges at the mobile vessel from the very precise time-stable transmitters in the system. The distance from each station is measured by comparing the arrival time of the received signal against an ultraprecise frequency standard contained on the vessel. The transit time of the received pulse is a precise measure of range since the velocity of propagation of the radio energy is constant. One such frequency standard of comparison is a rubidium standard whose frequency is maintained by the natural frequency of the atom of a rubidium isotope (hence the name atomic clock).

The figure shows the elements of the system. The master and slave stations in the loran C chain transmit in bursts of energy on identical frequencies (100.000 kHz) but are uniquely coded to allow identification of each station. The gated clock associated with receiver channel 1 allows a small portion of a particular cycle of each master station burst to arrive at time-interval counter 1. The same gated clock allows a small portion of a particular repetitive cycle of the atomic clock output to arrive also at time-interval counter 1. Since the portion of the cycle (or pulse) which originates at the atomic clock arrives at the counter without delay, whereas the pulse which originates at the master station arrives at the counter with a delay equivalent to the transit time from master station to mobile vessel, the counter changes as a direct function of range to the master station. Similarly, time-interval counter 2 changes as a direct function of range to the slave station. In order to have the counters record actual range from the two stations, it is necessary to calibrate the equipment on the vessel at a known point. This calibration can be done in port or by other means such as position derived from satellite navigation signals. Results of tests conducted in the Gulf of Mexico indicate a consistent return to position capability within 100 ft.

Shoran. Shoran is a range-range system which operates on an interrogate-transpond principle. The system is capable of measuring ranges with an accuracy to within 75 ft. Unlike phase comparison systems, the readings are free of ambiguity. The operating frequencies are in the vhf-uhf range of from 230 to 310 MHz. Until recently the system was limited to line-of-sight ranges, which seemed characteristic of this portion of the frequency spectrum. As a matter of fact, path losses alone limited the operating range to the radio horizon. (For sea-level base station installations and vessel stations, the range was 35–40 mi.)

Studies made under United States government grants indicated that the energy was being propagated well beyond the radio horizon, and the problem became one of developing hardware to increase the overall system gain from the existing 115-dB level to a 185-dB level. The hardware became available in 1968 and, with the use of high-gain antennas and solid-state preamplifiers of low noise design, the system operating range has been extended well beyond 200 mi from sea-level installations. Extensive checks made at ranges up to 250 mi indicate that the accuracy of range measure-

Loran C in range-range configuration.

ment has not been degraded to any significant amount.

For background information *see* DIRECTION-FINDING EQUIPMENT; LORAN; NAVIGATION in the McGraw-Hill Encyclopedia of Science and Technology. [JOSEPH DE LERNO]

Bibliography: O. J. Baltser, Navigation by means of VLF radio transmissions, *1963 National Winter Convention on Military Electronics*, 1963; B. B. Barrow et al., Indirect atmospheric measurements utilizing rake tropospheric scatter techniques, *Proc. IEEE*, 57:537–551, 1969; J. DeLerno and A. Marchal, Offshore radiopositioning systems, *Offshore Technology Conference*, April, 1969; *The LORAN C System of Navigation*, Jansky and Bailey, Inc., Division of Atlantic Research Corp., Washington, D.C., 1962.

Doppler radar

A system of navigation, Doppler-difference radar, intended for use by low-flying planes whose small size and low cost preclude the use of expensive on-board hardware is currently being developed by Ohio University's Avionics Research Group. In this class of vehicle the burden of costly precision equipment must be borne by ground-based installations.

When an airplane or other moving vehicle carrying a radio receiver is approaching a stationary transmitter, a received frequency is shifted by a factor equal to the closing rate divided by the velocity of the radio waves (very close to the speed of light). In the conventional Doppler radar, the frequency of the transmitter is available for direct comparison with that of the return signal, since the transmitter is on board and may be assumed not to change or to change by a known amount during the short time required by the two-way trip from transmitter to target and return. There is also a doubling of the Doppler shift, since the target sees and returns a shifted frequency which is again shifted at the receiver and in the same sense.

In a passive Doppler scheme in which the transmission is from the ground or from a quasi stationary space vehicle to a moving receiver, there must be an on-board clock synchronized to the transmitter clock to permit measurement of the Doppler shift. Since this shift is the difference between two very large numbers, the precision requirements are extremely severe and the equipment is generally delicate and costly.

To accomplish these objectives, the Doppler-difference system has been proposed and is being studied. To make available a stable reference at the receiver to facilitate accurate determination of the Doppler effect, two transmitters are established at different, known locations and synchronized to the same clock. These constitute a "station pair" or "transmitter pair." In general, the signal from each transmitter of a pair will be shifted differently as received by the plane, which will have different closing rates for the two stations of the pair. The difference in these frequencies is measured, and is called the Doppler difference, hence the name of the system.

Design of system. In its most elementary form the Doppler-difference system requires four station pairs at five known locations as shown in the figure. Pairs are indicated by letters subscripted 1

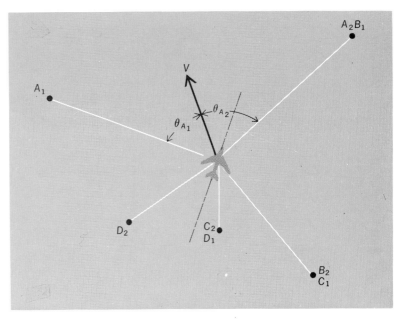

Geometry of the Doppler-difference radar system.

and 2; thus A_1 and A_2 transmit a single coherent frequency f_A; B_1 and B_2 transmit an exactly coherent frequency f_B; and so on.

Four tuned radio-frequency (trf) receivers in the aircraft are tuned to receive these four frequencies. If the receivers were fixed in space with respect to the transmitters, these unshifted continuous-wave (cw) signals would produce no outputs. However, a moving receiver will receive Doppler-shifted frequencies which are different for the two stations of each transmitter pair (except in the case in which the closing rate is the same for both stations of a pair). These four Doppler differences are functions of the position coordinates and velocity components of the plane with respect to the station locations. If altitude and rate of climb are disregarded, four numbers are required to specify the location of the plane in position and velocity. The four beat frequencies available from the trf receivers supply four equations involving these numbers, and are shown below. Here v is the

$$\Delta f_A = f_{A_1} - f_{A_2} = f_A (v/c) (\cos \theta_{A_1} - \cos \theta_{A_2})$$
$$\Delta f_B = f_{B_1} - f_{B_2} = f_B (v/c) (\cos \theta_{B_1} - \cos \theta_{B_2})$$
$$\Delta f_C = f_{C_1} - f_{C_2} = f_C (v/c) (\cos \theta_{C_1} - \cos \theta_{C_2})$$
$$\Delta f_D = f_{D_1} - f_{D_2} = f_D (v/c) (\cos \theta_{D_1} - \cos \theta_{D_2})$$

ground speed of the plane, c is the speed of propagation of the electromagnetic waves, and the θs are angles measured between the plane's track and the line between the plane's location and the transmitter corresponding to the subscript.

If these equations can be solved simultaneously by a special- or general-purpose on-board computer, the navigation problem is solved. Because of the complexity and nonlinearity of this system of equations, it has not been solved by a computer capable of being placed on board.

Operating modes. Other modes of operation are possible which do not require as much computer capability and may prove useful. For instance, a position may be known with good accuracy, as when over a radio marker or some visual check point. In this case, two of the station pairs may be used to solve for the velocity components, and since the equations are linear in these unknowns, the solution is simple and rapid. Using the calculated value of velocity and the known position, a new position can be calculated for extrapolation in time and the velocity measured at the end of the interval of extrapolation. How long or how far this process of augmented dead reckoning can be carried on needs to be carefully studied.

The above mode of operation requires only two station pairs. The remaining two pairs may be used alternately to check position by entering the extrapolated position and making redundant measurements of velocity.

Another use of the two extra station-pair measurements is to calculate position using the measured velocities. However, this procedure requires solution of two nonlinear equations, and although more direct in approach, it may require considerably more computational equipment.

Advantages and disadvantages. The following advantages are to be expected from the Doppler-difference system. (1) It is nonsaturating, that is, it permits any number of users. (2) It is passive, requiring no high-powered transmitter for the plane. (3) One-way transmissions require less power, even on the ground. (4) No on-board clock is required. (5) Only frequency differences have to be measured, involving simple beat frequency. (6) The radio-frequency (rf) equipment is simple and inexpensive.

The following disadvantages are foreseen. (1) Straightforward simultaneous solution of the four equations awaits the development of analytical methods or of faster, more powerful on-board computers. (2) In its simplest form, the method requires line-of-sight frequencies in order to produce high enough Doppler differences to allow sufficiently rapid and sensitive measurement of the beat frequency. Higher plane speeds produce higher beat frequencies but also require faster measurement because of their more rapidly changing position. (3) Much of the simplicity obtained for the rf sections may be offset by increased complexity and cost of the signal processing of beat frequencies and of the computer. This complexity is expected to change favorably with advances in computer technology.

For background information *see* DOPPLER EFFECT; DOPPLER RADAR; RADAR in the McGraw-Hill Encyclopedia of Science and Technology.

[GUSTAVUS E. SMITH]

Bibliography: R. H. McFarland, *All-Weather, Low-Level Navigation*, ECOM-0084-1, Ohio University, 1969; R. H. McFarland, *All-Weather, Low-Level Navigation, Project Themis*, ECOM-0084-2, Ohio University, 1970; G. E. Smith, A Doppler-difference navigation system, *National Aerospace Electronics Conference*, Dayton, Ohio, 1969.

Drilling, deep-sea

The drilling vessel *Glomar Challenger* has completed a successful 18-month drilling program in both the Atlantic and Pacific oceans and will continue for an additional 30 months. Results of the initial program include substantial support for the concept of continental drift or sea-floor spreading, recovery of oil from salt diapirs in about 12,000 ft

of water, and recovery of the oldest sediment (Jurassic) ever taken from the deep oceans. The vessel has drilled in 20,146 ft of water, penetrated 3231 ft below the ocean floor, and suspended a total drill string of 20,760 ft. A hole reentry system, which will provide the capability of recovering the drill string, replacing a worn-out bit and reentering the same hole, has been adapted for use by the *Glomar Challenger* in oceanic depths of over 20,-000 ft.

Organizational program. Marine geologists have long wished to investigate the sediments below the ocean floor. Previous investigations using piston cores were limited to the near-surface sediments, which represent the accumulation of only the past few hundred thousand years. In some areas geologists were fortunate to find older sediments exposed on the ocean floor, where they could be sampled, but in general the older sedimentary record remained untouched.

Project Mohole demonstrated that it was feasible to drill in very deep water utilizing the technology of the petroleum industry, and paved the way for a program of drilling in the deep sea for the purpose of recovering old and deeply buried sediment. In May, 1964, Lamont-Doherty Geological Observatory, Rosenstiel Institute of Marine and Atmospheric Sciences of the University of Miami, Scripps Institution of Oceanography, and Woods Hole Oceanographic Institution formed a consortium called the Joint Oceanographic Institutions for Deep Earth Sampling (JOIDES). The primary function of JOIDES is to foster programs to investigate the deep-ocean basins by drilling and coring. The University of Washington became a member of JOIDES in 1968.

The first deep-sea drilling program encouraged by JOIDES was undertaken on the Blake Plateau off the eastern coast of Florida by the drilling vessel *Caldrill* in depths of up to 1000 m. E. T. Bunce and coworkers reported on this program in 1965. The program, supported by the National Science Foundation and operated by the Lamont-Doherty Geological Observatory, was very successful and encouraged planning for a more ambitious and extensive program of drilling in abyssal depths of the main ocean basins.

In response, in 1966 the National Science Foundation awarded a contract to the Scripps Institution of Oceanography for an 18-month drilling program. A subcontract was given to Global Marine, Inc., to supply a vessel capable of drilling in up to 20,000 ft of water with a penetration of 2500 ft below the ocean floor. On Mar. 23, 1968, the drilling vessel *Glomar Challenger*, built especially for this project, was launched, and began the first of nine cruises in August, 1968. The success of the 18-month drilling program prompted the National Science Foundation to extend the program for an additional 30 months.

Vessel. The 400-ft, 10,500-ton *Glomar Challenger* is the only vessel capable of drilling in the deep-ocean basins. The feature that makes the vessel unique is the dynamic positioning system (Fig. 1) which enables her to maintain position to within a 100-ft radius while drilling in about 20,000 ft of water. Positioning is accomplished by reference to a sonar beacon on the ocean floor. Pulses from the beacon, received by a series of hydrophones mounted in the hull, are processed by a computer which calculates the position of the ship relative to the beacon. The computer, which has direct control over the ship's propulsion system, maintains position by automatically commanding power to the main screws or to one or more side thrusters or tunnel thrusters. The tunnel thrusters are large tubes or tunnels mounted in the hull and extending from one side of the ship to the other. Propellers in the tunnels provide thrust at an angle of 90° to the thrust of the main screws. The combination of the main screws and the four thrusters, two in the bow and two in the stern, enables the *Glomar Challenger* to maneuver in any direction, or even to rotate 360° in its own length.

Mounted amidships is a 142-ft drilling derrick which towers almost 200 ft above the water. The derrick has a hook-load capacity of 500 tons. Up to 23,000 ft of pipe can be laid horizontally in 90-ft lengths on an automatic pipe racker mounted forward of the derrick.

Sediment samples are taken in a 30-ft core barrel lined with a plastic sleeve. The barrel is dropped through the hollow drill pipe and locks in place over a 2.5-in. hole in the drill bit. As the bit cuts through the sediment, it leaves a central cylinder which is pushed into the core barrel. A retrieving tool attached to a cable is lowered through the drill pipe and latched onto the core barrel. The core barrel is then pulled up through the drill pipe and onto the derrick floor. The sediment is extruded from the core barrel in its plastic tube and is examined on board by sedimentologists, paleontologists, and chemists. The average time for the entire operation is about 2 hr.

Scientific program. The primary objective of the deep-sea drilling program is to determine the geologic history of the ocean basins. To achieve this ambitious goal, JOIDES established site selection panels for both the Atlantic and Pacific oceans. These panels, which represent the diverse interests of the entire marine scientific community, were charged with the responsibility of establishing a drilling program that would yield the maximum scientific returns. They recommended sites to sample the oldest sediments, to test the hypothesis of sea-floor spreading or continental drift, and to recover complete sedimentary sections for paleontologic and stratigraphic studies. The proposed program is shown in Fig. 2 by the completed track of the *Glomar Challenger* at the end of the initial program.

Old sediment. The program has been successful in retrieving the oldest sediments (Jurassic age, 140,000,000 years) ever recovered from the deep-ocean basin. The report published in 1969 by M. Ewing and coworkers discussed this accomplishment. These ancient sediments have now been recovered from both the Atlantic and Pacific oceans, but it is probable that somewhat older sediments may be available in other areas. The drilling results, combined with data gathered from many years of geophysical surveying in the oceans, indicate that sediment significantly older than that already recovered probably does not exist. It has been concluded that the ocean basins are relatively young features in comparison with the continents (3,500,000,000 years) or the Earth itself (4,500,000,000 years).

Continental drift. Prior to the drilling, geological and geophysical data were mounting in support of

Fig. 1.   Illustration of how a system of pulses from acoustic beacons on the ocean floor, which are picked up by a ship-mounted hydrophone array, are fed into a computer and translated into corrective action by propulsion units (tunnel thrusters and ship propellers) which automatically keep the *Glomar Challenger* precisely on station. (*Deep Sea Drilling Project*)

the concept that the continents were once part of one great land mass and that about 200,000,000 years ago they were separated and are still moving apart. It was noted that in all the major oceans there exists a large, continuous mountain system called the mid-ocean ridge (locally referred to as the Mid-Atlantic Ridge, Mid-Indian Ridge, East Pacific Rise, and so on). The geophysical data indicated that the crest of the ridge is a zone where new ocean crust is being formed as the flanks of the ridge separate. The farther the crust is from the ridge crest, the older it is. According to theory, the crust is being reassimilated where deep trenches have formed, such as around most of the perimeter of the Pacific Ocean. In areas where few or no trenches have formed (Atlantic Ocean), the continents are coupled to and moving with the sea floor.

In order to test this concept, a series of holes were drilled across the ridge in the North and South Atlantic and in the equatorial Pacific. The age of the oldest sediment overlying the ocean

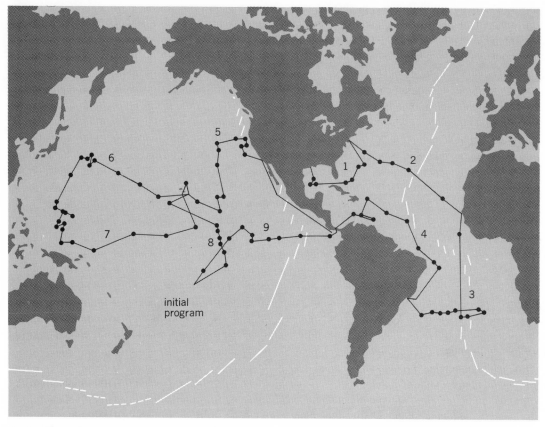

Fig. 2. Chart showing sites drilled by the *Glomar Challenger* during the initial 18-month program of drilling in the deep ocean. The vessel was built especially for this project. (*Deep Sea Drilling Project*)

crust at each site demonstrated without exception that the age of the crust increased with increasing distance from the ridge crest. Figure 3 is a compilation of results from three cruises of the *Glomar Challenger* demonstrating the relationship between age of the oldest sediment and distance from the ridge crest and providing substantial evidence in support of sea-floor spreading.

Chert. The recovery of the old sediments from the oceans has been hampered by the unanticipated presence of extensive layers or lenses of chert. In the western North Atlantic, the chert appears to form an extensive interval of up to 140 ft thick of interbedded chert and soft sediments. The upper layer appears to be a fairly synchronous surface, as reported by M. N. A. Peterson and coworkers in 1970, of lower to middle Eocene (about 50,000,000 years). Cherts were also recovered in the western North Atlantic.

In the Pacific Ocean cherts were found in sediments of almost all ages older than Miocene (18,000,000 years). Instead of forming a discrete synchronous horizon, as it appears to be in the western North Atlantic, the chert appears to be nodular, lenticular, and diachronous on a regional level. Certain sediment types such as volcanic ash appear to silicify preferentially over typical deep-sea clays.

The chert is composed primarily of christobalite (silicia) and appears to be the result of secondary silification of radiolarian ooze, deep-sea sands, volcanic ash, biogenic calcium carbonate, and in some cases deep-sea clay. No satisfactory explana-

tion has been proposed to account for the vast extent of the silification or the synchroneity of the upper surface of the Eocene chert in the North Atlantic.

The recovery of complete sedimentary sequences for paleontologic and biostratigraphic studies has been very successful in the Cenozoic (the past 70,000,000 years) sediments. The relatively well-established faunal zonation of the foraminifers was correlated with the less precise and, in some cases, incomplete zonations of the Radiolaria and nannofossils. The Cenozoic-Mesozoic boundary was cored at several locations in both the Atlantic and Pacific oceans. The Mesozoic (70,000,000–200,000,000 years) stratigraphically remains relatively obscure owing to the difficulty encountered in penetrating the chert horizons, the common occurrence of unfossiliferous clays, and the low core recovery in stiff or consolidated older sediment.

Salt domes and petroleum. In 1962 J. Ewing and coworkers obtained seismic reflection data in the Gulf of Mexico. They reported a number of diapiric or intrusive structures and interpreted them as salt domes, which may be associated with petroleum accumulations. The *Glomar Challenger* encountered oil- and gas-saturated cap rock after drilling about 400 ft into one of these structures. This was also discussed in the 1969 report of M. Ewing and coinvestigators.

The major mineral components of the core are calcite, sulfur, and gypsum; minor constituents are pyrite, quartz, tourmaline, and dolomite. These

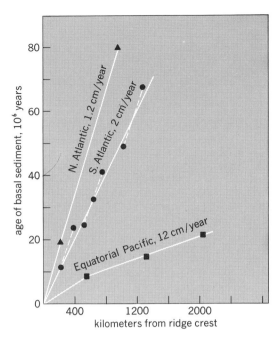

Fig. 3. Graph of distance from the ridge crest versus age of the basal sediments. The linear plots are consistent with the concept of sea-floor spreading. The different slopes of the lines reflect the different spreading rates in the various ocean basins. (*Deep Sea Drilling Project*)

components are identical with those of the transition zone of cap rock of salt domes surrounding the Gulf of Mexico. Cap rock is considered to be composed of an aggregate of relatively insoluble components of a salt diapir after the salt of the upper part of the structure has been removed by circulating water. The presence of cap rock implies that the structure is a salt diapir.

Igneous rock. Igneous rocks have been recovered at numerous sites in both the Atlantic and Pacific oceans, and preliminary studies indicate that they are typical oceanic tholeiite basalts. Laths of plagioclase are common, and in some cases vesicles have been reported.

The sediment-basalt contact is generally characterized by baked sediment and basaltic glass, and at a few sites pieces of baked sediment have been found between basalt layers, as reported by Peterson and coworkers. It is not certain, however, whether the sediments found within the basalt are fragments trapped in the basalt or are part of a continuous layer. The baked sediment at the contact indicates that the basalt was intruded into or plowed through the thin soft sediment rather than spread over the ocean floor as a flow. Concern has been expressed that a considerable amount of sediment may be buried below the basalt and that consequently the dating of the basal sediments in the test of sea-floor spreading may be invalid. However, as noted in Fig. 3, the relationship of sediment age to distance from the ridge axis is surprisingly consistent and is considered strong evidence that no significant thickness of sediment is buried beneath the upper basalt.

Program extension. The initial 18-month drilling program has been extended by the National Science Foundation for an additional 30 months.

The advisory panels for both the Atlantic and Pacific oceans provided new programs based on additional geophysical data and the results of the earlier drilling program. During this 30 month extension, plans called for 6 months of drilling in the Indian Ocean, and a panel was assembled to devise a suitable program.

A subpanel of the Indian Ocean panel was assembled to provide a program of 2 months in the Mediterranean Sea.

The emphasis on the extension will shift from the spreading ridges to the older parts of the ocean basins and continental margins. In order to successfully recover the older sediments, a reentry system is being developed which will enable the engineers to replace a worn-out bit and put the drill string back in the same hole while operating in about 20,000 ft of water. The proposed system will consist of a base plate and 16-ft-diameter funnel worked into the soft sediments on the ocean floor and a rotating beacon located in the drill bit. The base plate will be located by the sonar beacon, and the drill string will be maneuvered over the funnel by hydraulic jets mounted above the bit. The jets are actuated by pumping water from the drilling vessel down the drill string. The success of this system will enable the *Glomar Challenger* to achieve many of the objectives that the scientific community has proposed.

For background information *see* MARINE GEOLOGY; MARINE SEDIMENTS; SEDIMENTATION (GEOLOGY) in the McGraw-Hill Encyclopedia of Science and Technology.

[N. TERENCE EDGAR; M. N. A. PETERSON]

Bibliography: E. T. Bunce et al., *Science*, 150:709–716, 1965; J. Ewing et al., *J. Geophys. Res.*, 67:2509–2527, 1965; M. Ewing et al., *Initial Reports of the Deep Sea Drilling Project*, vol. 1, 1969; M. Peterson et al., *Initial Reports of the Deep Sea Drilling Project*, vol. 2, 1970.

Drug addiction

Various mechanisms of action for the addictive drugs have been proposed. It is currently considered that these drugs induce addiction by a direct action on the basic conditioning mechanisms of the limbic system so as to either block transmitters that reduce reward or increase punishment or to potentiate transmitters that have the opposite effect. In addition, addictive drugs induce withdrawal effects, but this appears to play a less important part in the establishment of the craving for the drug. The interference with the chemical mechanisms of conditioning causes the patterns of behavior that led to taking the drugs to become so powerfully reinforced, and the behavior-directing mechanism of the brain originating in the limbic circuits to become so altered, that these patterns become too powerful to be broken by an act of "will," that is, by cortically originated directives to the altered limbic system. It is well known that monkeys become more quickly addicted to morphine if they make the injection themselves; that is, there must be some behavior emitted that can be reinforced.

There is now considerable evidence that the brain mechanisms mediating reinforcement are chemically coded. Positive reinforcement would appear to be linked to norepinephrine (NE) release. Likewise, negative reinforcement (pain) would appear to be in part a function of excessive release of acetylcholine (ACh). Thus addictive drugs should be those which potentiate central NE mechanisms and depress central ACh release.

Addictive drugs of the amphetamine and cocaine class potentiate NE action by several mechanisms: release of NE, antagonism of monamineoxidase (MAO), blockade of uptake from the synaptic cleft (the major mechanism of inactivation of catecholamines), or some direct action on the receptor. Single injections of morphine lead to decreased levels of brain NE, and repeated injections lead to increased levels. Morphine may also induce an increased rate of synthesis.

During repeated injections of morphine, the urinary excretion of NE also rises. However, because morphine blocks release, all these effects may be secondary to the main effect of morphine, which is blockade of ACh release. There appears to be some balance in transmitter amine functions in the brain so that blockade of one transmitter may lead to an unbalanced overactivity of one of the other transmitters. In the case of morphine, depression of ACh release may lead to an unbalanced relative overactivity of the NE system. The result is a shift in the reinforcement mechanisms in the direction of increased positive and decreased negative reinforcement so that addiction and craving develop. Likewise, in certain psychoses, the knocking out of hydroxytryptamine (5HT) mechanisms (for example, by d-LSD) leads to an unbalanced activity of central adrenergic—mainly dopaminergic—activity that may contribute to the symptomatology and that may in turn be reduced by phenothiazines. However, the hallucinogenic drugs do not induce a compulsion to take them. The repeated use of lysergic acid diethylamide (LSD) and similar drugs results mainly from the pressures of a drug-orientated subculture and from the "willful" choice of the user, rather than from any subcortically determined compulsion.

Thus the most dangerous addictive drugs, such as the amphetamine-like drugs, opiates, and cocaine, seem to act on the chemical reinforcement mechanism in the brain so as to reinforce powerfully the drug-taking pattern of behavior. A disruption of these mechanisms eventually occurs, followed by the inertia and apathy of the chronic addict.

Another class of truly addictive drugs may act directly on ACh receptors. The action of chronically administered barbitone sodium in rats has been investigated. After 40 days there was no change in brain ACh levels, in brain cholinesterase or choline acetyltransferase activity, or in the ability of brain slices to synthesize or store ACh. However, the thermal response to acetylcholine-like drugs was altered. Normally these drugs cause a fall in body temperature (if given by intraventricular injection), an effect blocked by atropine. However, in the barbiturate-treated rats, pilocarpine induced prolonged hyperthermia. It is of interest that morphine decreases the incorporation of P^{32} into phospholipids in the brain, whereas ACh has the opposite effect. This indicates the depressant effect of morphine on ACh functions, probably release of the neurotransmitter.

Problems of tolerance. The fact that addictive drugs produce tolerance certainly aggravates the addiction problem, but tolerance may depend on different mechanisms. One widely held theory for this effect is that the drugs induce the formation of larger amounts of the enzymes that destroy them. This appears to be the case for the barbiturates but not for morphine. No clear relation has been established between hepatic demethylase activity and the development of tolerance. Studies with radioactivity-labeled morphine have shown that neither the distribution nor the metabolism of morphine is concerned with the development of tolerance.

Various theories have been proposed to explain the development of tolerance to morphine.

1. The surfeit theory proposes that, because morphine blocks only the release of the transmitter, the ongoing synthesis of ACh builds up ACh levels in the axon terminal so that some ACh eventually trickles over the dam and tolerance to the blocker develops. Likewise, when the morphine is removed, the excess ACh now being released causes the withdrawal symptoms, which are a mixture of sympathetic and parasympathetic discharge. This hypothesis can be amended slightly to include the notion of induction of ACh synthetic enzymes by negative feedback from the starved receptor to the genome of the cholinergic neurone.

2. The supersensitivity theory suggests that, as is the case following any denervation, chemical or physical, the postsynaptic neurone becomes supersensitive to the transmitter by some modification of the receptor site or the receptor membrane—for example, the development of new receptors, the synthesis of a new protein that alters the response, or the repression of synthesis of a normal protein that stabilizes the membrane. For instance, botulinum toxin is known to act by preventing ACh release. One to two weeks after the administration of a sublethal dose of the toxin, new receptors appear on the muscle. Thus the effect of the morphine is overcome and tolerance develops. When the morphine is removed, there are now too many receptors and the withdrawal symptoms result.

3. There is now considerable evidence that morphine tolerance is dependent on protein synthesis in the brain. It is prevented by protein synthesis inhibitors of various kinds, such as actinomycin, which inhibits ribonucleic acid polymerase, and cycloheximide and puromycin, which act at different sites in the protein synthetic mechanisms. These inhibitors have no effect on the analgesic response, only on the development of tolerance. The effects of morphine can be detected by suitable techniques for as long as a year following a single large dose, an effect that could be mediated by the modification of protein synthesis or by some immune mechanism. The synthesis of new protein could be concerned with the feedback control of ACh synthetic enzymes, with the construction of new receptors or some other modification of the synaptic membrane as discussed above, or with some basic brain mechanism concerned with conditioning or memory, in which protein synthesis has been shown to play an important role.

Of course, these mechanisms are not mutually

exclusive and two or more may operate over different time courses.

A link between morphine dependence and 5HT has been suggested by the finding that the rate of synthesis of 5HT increases as tolerance to morphine develops and that inhibition of this synthesis by *p*-chlorphenylalanine markedly decreases tolerance to and the development of physical dependence on morphine.

Sleep. Ian Oswald and coworkers have shown that the common drugs of addiction at first suppress rapid-eye-movement (REM) sleep; following their withdrawal, a rebound excess occurs, which may last for many weeks. This occurs with barbiturates, amphetamines, and opiates. Drugs such as amitriptyline and chlorpromazine, as well as electroshock treatment (ECT), also suppress REM sleep, but there is no rebound following their termination and these drugs are not addictive. Reserpine, which induces dysphoria rather than euphoria, enhances REM sleep, as does LSD. The fact that sleep is controlled in part by the pattern of activity of the brain amines provides a further link in the chain of evidence connecting addiction and brain amine function.

Psychosocial factors. No account of drug addiction can possibly be given solely in physiological terms. The first "fix" is determined by social and cultural factors and, in the final analysis, drug addiction needs to be controlled by attention to these factors as much as to judicial measures or to medicinal procedures. The present epidemic of drug addiction arises not from the increased efficiency of drug pushers but from deep-seated sociocultural factors.

Recent studies have determined that most addicts have shown evidence of social deviance or of very unpatterned social behavior before they started on drugs. Such delinquent behavior stems in part from a mismatch of the schedules of reinforcement provided by the environment and those needed by that particular person for his proper personality development. In the language of operant conditioning, these people have lived in decayed and fragmented social subcultures, have received little in the way of stable schedules of reinforcement, and have often been subjected to conflicting stimulus control. Thus the brain mechanisms determining their adult patterns of consumatory (impulse gratification) behavior have been wrongly programmed, so that impulse control remains poor and the ability to avoid drugs giving immediate impulse gratification is diminished. Thus the ghetto environment tends to produce people with damaged personalities, and the damage makes living in the reduced environment of the ghetto even more difficult, since it erodes the capacity to plan, save, and endure that is the classical puritan way of leaving the ghetto.

However, since many addicts come from comfortable homes, one must here be witnessing a wider breakdown of a culture. Parental neglect, indifference, excess consumption of alcohol, preoccupation with purely materialistic values, as well as foolish overpermissive child-rearing practices can erode personality in the suburb as powerfully as does the slide into apathy and oblivion that characterizes the ghetto.

Many people show in their behavior a need to live their lives in the context of some system (in technical anthropological language, a myth) that provides a pattern of meaning in life. Since the widespread decay of the Christian system there has been a fierce competition of surrogates, for example, Hitlerian neopaganism, Marxist reductionism, and Mao-Castroite romanticism. That the psychology of fanaticism has been much neglected by Western scientists is curious, since, as Arthur Koestler has pointed out, a most powerful — and potentially destructive — human trait is the drive to sacrifice oneself for "the cause." The philosophical impulse behind Western civilization derives from a blend of the classical philosophies of the 17th and 18th centuries (in particular, John Locke) and the Greco-Judaic tradition. The recent attempts to replace it have hardly led to any improvements and do not reach down to any extent to illumine the life of the heroin addict in the ghetto. The addicts, however, do form a subculture which approves of their way of life and allows peer reinforcement of "successes," such as a good "bag" (robbery). This secondary psychosocial reinforcement is a powerful factor in maintaining addiction. But the origins of this subculture extend deeply into Western culture and are results of an internal deculturation. Normally a culture suffers deculturation only on the impact of a more powerful foreign culture, as the culture of the American Indians was destroyed by Western culture in its expansive stage. Western culture is unique in that it creates its own forces leading to internal deculturation, largely by its devotion to "progress," science, and technology to the virtual exclusion of any other consideration. A brave attempt to believe only what is "true," that is, scientifically established, has led to the problems of living in a progressively more Darwinian world of which the victims are not only the massacred but also the sensitive, the deprived, and the weak, who are then liable to become drug addicts. The academic proponents of positivistic philosophies bear a heavy burden of responsibility, for, as Koestler has again shown, there are no more horrendous determinates of human behavior than abstract philosophical systems. Until Western civilization solves this problem, drug addiction will remain but one of the many symptoms of the failure to provide ordinary people with any account, satisfactory to them, of the meaning of their existence.

For background information *see* BARBITURATES; COCAINE; MORPHINE; NARCOTIC; PSYCHOTOMIMETIC DRUG in the McGraw-Hill Encyclopedia of Science and Technology.   [JOHN R. SMYTHIES]

Bibliography: H. Steinberg (ed.), *Scientific Basis of Drug Dependence*, 1969; World Health Organization, *Biochemistry of Mental Disorders*, WHO Tech. Rep. Ser. no. 427, 1968.

Earth, heat flow in

Recent studies of heat flow in the Earth have been concerned with finding the average value of the heat flow for continents, oceans, and island areas and with determining the distribution of heat flow on the Earth's surface.

Terrestrial heat flow Q is the thermal energy discharged per unit area and unit time by conduction from the Earth's interior through the Earth's solid surface. It is the product of the geothermal gradient (the rate of temperature increase with depth Z near the top of the Earth's crust) $\Delta T/\Delta Z$ and the thermal conductivity K of the Earth's material, $Q = K(\Delta T/\Delta Z)$.

To measure the geothermal gradient on land, any facility through which the Earth's internal temperature is accessible (mine, coal field, tunnel, oil well, and so on) can be utilized. To obtain the best estimate of undisturbed geothermal gradient, correction must be applied for the effect of such near-surface thermal processes as flow of underground water, diurnal and seasonal variation of atmospheric temperature, long-period temperature fluctuations during the ice ages, topographic relief, evolution of topography, and heterogeneous distribution of thermal conductivity. In the oceans, the deep-sea water is usually free from short-period temperature fluctuations, and it is possible to measure a reliable geothermal gradient in the upper few meters of oceanic sediment by inserting a probe in the soft and homogeneous layer of the sediment. *See* DRILLING, DEEP-SEA.

Heat-flow values. The heat flow of 2.81 μcal/cm^2 sec, measured in the Mohole experiment, drilled through 180 m of the ocean sediment to the basaltic basement near Guadalupe Island, is nearly equal to the heat flow (2.83 μcal/cm^2 sec) measured by the probe in the surface layer of the sediment near the drilling site, an observation that suggests the heat flow is constant from the top of the sediment to the underlying basement. However, A. H. Lachenbruch and B. V. Marshall reported in 1968 that the temperature of the deep-ocean current through the Denmark Strait northwest of Iceland fluctuated more than 0.1°C over a few weeks. Thus the thermal gradient determined from temperature measurements in the top few

meters of sediment is not representative of the steady-state gradient. Clearly, careful interpretation is necessary when heat-flow measurements are made in variable environments.

If geothermal areas (volcanoes, hot springs, crests of the oceanic ridges) are excluded, the thermal gradient ranges from 0.08 to 0.6°C/10 m on land and from 0.1 to 2.5°C/10 m in the sea. In most cases, thermal conductivity is determined in the laboratory on rocks and sediments recovered from the site where the geothermal gradient was measured. Recently, measurements of thermal conductivity in ocean sediments in place were made by J. G. Sclater, C. E. Corry, and V. Vacquier. The thermal conductivity of rock ranges from 2 to 15 mcal/cm sec °C, whereas the thermal conductivity of ocean sediment is more uniform, ranging from 1.8 to 2.4 mcal/cm sec °C.

Since 1939 more than 3000 heat-flow values have been reported (Fig. 1). Most of the determinations have been made in North America and Europe, in the Pacific, Atlantic, and Indian oceans, and in the area around Japan. The amount of data is increasing rapidly in the United States, the Soviet Union, Czechoslovakia, and India, as well as in the oceans. Despite few determinations in many very large areas (Asia, South America, Africa, and the Antarctic), the following conclusions drawn from the existing data seem to be valid and most important: (1) The average heat flow for the entire Earth is 1.5 μcal/cm^2 sec. (2) Heat flow is correlated with the age and history of the Earth's crust. (3) The average heat flow for the continents is nearly equal to that for the oceans.

The Earth's thermal energy is due mostly to the decay of radioactive nuclei. The total output of thermal energy by terrestrial heat flow is 2.5×10^{20} cal/year. This estimate is quite close to the rate of heat generation, 2.3×10^{20} cal/year, of an Earth

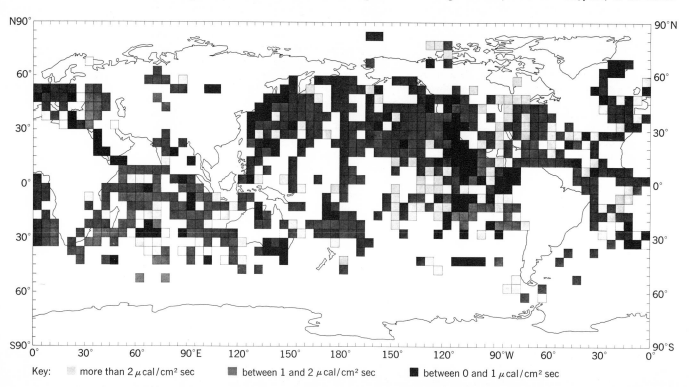

Key: ▨ more than 2 μcal/cm^2 sec   ▦ between 1 and 2 μcal/cm^2 sec   ■ between 0 and 1 μcal/cm^2 sec

Fig. 1. Distribution of heat flow as averages for 5 × 5 degree (longitude and latitude) areas.

assumed to have the average concentration of the radioactive elements of chondritic meteorites: uranium, 1.1×10^{-8} g $(U^{235} + U^{238})$/g: thorium, 4.1×10^{-8} g Th^{232}/g; and potassium, 8.5×10^{-4} g K^{40}/g. This observation, called the chondritic coincidence, has provoked debates about the gross chemical composition and the origin of the Earth. If thermal conduction is assumed as the mechanism of energy transportation for the whole Earth, the heat of the lower mantle and of the core is not transmitted to the surface within 4.5×10^9 years, the age of the Earth. Therefore, for the chondritic coincidence to be of any significance, these radioactive elements must be concentrated in the outer layer of the Earth, and the heat currently being produced is reaching the Earth's surface.

Distribution of heat sources. Recently, evidence concerning the distribution of heat sources in the Earth's crust was obtained by R. F. Roy, D. D. Blackwell, and F. Birch and by Lachenbruch. In several provinces on the North American continent where the Earth's crust has been formed by the intrusion of granitic plutons, the heat flow Q was found to be correlated linearly with surface radioactivity A by the relationship $Q = DA + Q_0$. According to this relation, represented by Fig. 2a, the observed heat flow can be decomposed into two terms: one, DA, represents a component which varies in proportion to the surface radioactivity, and the other, Q_0, is a constant throughout the province. Areas have undergone different degrees of erosion z, but surface radioactivity $A(z)$ varies linearly with heat flow Q. The difference between Q and Q_0 is equal to the rate of heat production in the crust integrated with respect to depth; that is, $Q - Q_0 = \int_z A(z')dz'$. As shown in Fig. 2b, the first term, DA, requires that the concentration of radioactive elements be highest near the surface and decrease exponentially downward. The second term, Q_0, is independent of the radioactive heat generation concentrated in the upper layer of the crust and must represent the heat flowing from the mantle into the crust; it provides useful information on the distribution of temperature and heat sources in the mantle. The different values of mantle heat flow characterizing the individual provinces, 0.40 μcal/cm^2 sec for the Sierra Nevada, 0.84 μcal/cm^2 sec for the New England area, and 1.40 μcal/cm^2 sec for the Basin and Range province, may imply that large-scale horizontal heterogeneities exist in the upper mantle under the continents. *See* ENERGY SOURCES.

The general idea that most of the radioactive elements under the continents are concentrated near the surface is consistent with the statistical analysis of heat-flow data. The lowest average heat flow (1.0 μcal/cm^2 sec) is observed in the Precambrian shields, where the Earth's crust was formed more than 600,000,000 years ago; the top layer of the crust, where most of the radioactive elements had been concentrated, has been eroded away. Moderate average heat flow (1.4–1.6 μcal/cm^2 sec) is observed in areas where the crust was formed during the Paleozoic and Mesozoic, and the highest flux (2.0 μcal/cm^2 sec) exists in Cenozoic orogenic areas, where the crust is in the process of formation.

In oceanic areas where the radioactive granitic layer is absent, the location of heat sources has

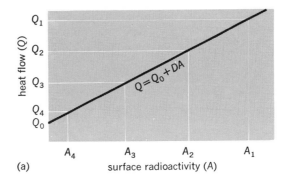

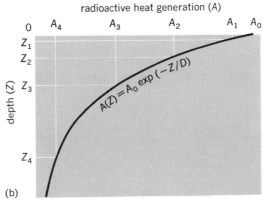

Fig. 2. Correlations of heat flow. (a) Linear relationship between heat flow Q and radioactivity $A(Q = Q_0 + DA)$. The term Q_0 represents the component of heat flow which is independent of surface radioactivity. (b) The relation $Q - Q_0 = DA$, that is, $\int_z A(z')dz' = DA(z)$, implies that the distribution of radioactive heat sources is given by $A(z) = A_0 \exp(-z/D)$.

puzzled many geophysicists, since the equality of heat flow between oceans and continents was established as early as 1956. It was expected that the mantle material underneath the oceanic crust would be more radioactive than that under the continents. However, recent examinations of the radioactivity of eclogitic and peridotitic nodules, which supposedly originated in the mantle, did not reveal any systematic tendency for oceanic mantle materials to be more radioactive than continental ones, as shown by H. Wakita, H. Hasegawa, S. Uyeda, and H. Kuno. The origin of oceanic heat flow still remains obscure.

In oceanic areas the correlation of heat flow with large-scale topography is remarkable. The oceanic ridge system is characterized by areas of heat flow as high as 8 μcal/cm^2 sec. Heat flow is moderate, and usually quite uniform, on the ocean floor away from the ridge system, and is less than 1 μcal/cm^2 sec toward marginal areas which are associated with oceanic trenches. On the circum-Pacific island arcs, heat flow is more than 2 μcal/cm^2 sec in the marginal seas (Sea of Okhotsk, Sea of Japan, and the Fiji basin) inside the arc, in contrast to the area of heat flow of less than 1 μcal/cm^2 sec lying outside the arc. *See* VOLCANO.

According to the sea-floor spreading hypothesis, the oceanic crust is created at ridge crests, transported across the ocean floor, and annihilated beneath the trenches. Seismic and magnetic studies of the ocean floor revealed that the thickness of the lithosphere (the oceanic crust and subjacent upper mantle, which behave as a rigid plate) is about 70 km and that the rate of spreading is 1–6 cm/year. Calculations on the basis of plate tectonics, a model of crust-mantle dynamics of sea-floor spreading, showed that these values are quite consistent with the general pattern of heat flow, gravity, and topography of the oceanic ridge-trench system, as shown by D. P. McKenzie, by E. R. Oxburgh and D. L. Turcotte, and by N. Sleep. A satisfactory explanation has not yet been given of the origin of high heat flow in the marginal seas inside the arc. However, a convective overturn of the mantle material, rising under the oceanic ridge and sinking under the island arc, may transport significant amounts of heat from the sources buried in the deep interior of the mantle, and may thus offer a partial solution to the enigma of continental-oceanic heat-flow equality.

For background information *see* CONTINENT FORMATION; EARTH; EARTH, HEAT FLOW IN; MARINE GEOLOGY in the McGraw-Hill Encyclopedia of Science and Technology.   [KI-ITI HORAI]

Bibliography: A. H. Lachenbruch, *J. Geophys. Res.*, 73:6977–6989, 1968; A. H. Lachenbruch and B. V. Marshall, *J. Geophys. Res.*, 73:5829–5842, 1968; D. P. McKenzie, *J. Geophys. Res.*, 72:6261–6273, 1967; E. R. Oxburgh and D. L. Turcotte, *J. Geophys. Res.*, 73:2643–2661, 1968; R. F. Roy, D. D. Blackwell, and F. Birch, *Earth Planet. Sci. Lett.*, 5:1–12, 1968; J. G. Sclater, C. E. Corry, and V. Vacquier, *J. Geophys. Res.*, 74:1070–1081, 1969; N. Sleep, *J. Geophys. Res.*, 74:542–549, 1969; H. Wakita et al., *Geochem. J.*, 1:183–198, 1967.

Earth tides

Progress in obtaining and analyzing longer, more accurate series of Earth tide observations has made clear that ocean tide effects, both loading and attraction, are significant and must be considered in the geophysical interpretation of the data. Therefore many recent research studies have used the available ocean cotidal and corange charts for estimating the deformation of the Earth due to tidal loading and for calculating the varying gravitational attraction of the nearby water mass at different stages of the tide. The effects of ocean tides on Earth tide observations can best be observed by a network of stations, varying in distance to the coast, which systematically monitor the ocean loading or attraction contributions to the observations.

Western Europe tiltmeter networks. Two tiltmeter networks in Western Europe have been established by the Institut de Physique du Globe de Paris and the Observatoire Royal de Belgique, in Normandy and in the Massif of the Ardennes, respectively. Both networks are aiming at the determination of (1) the deformation of the Earth's surface due to ocean tidal loads in the Atlantic Ocean, the English Channel, and the North Sea, and possibly (2) the regional variations due to heterogeneities in the crust or in the mantle, and even the anelastic properties of the mantle. That the effects of ocean tidal loading on the tilt of the Earth's surface are greater than those of the Earth's tidal tilt in Western Europe makes the separation of various effects on the tilt measurements, particularly the geological influences, very difficult, if not impossi-

ble. The data so far obtained still lack a general consistency.

United States tidal gravity profile. Most recent results, from a United States transcontinental tidal gravity profile established by the Solid Earth Tides Group of Columbia University, for the principal tidal constituents M_2 and O_1 have shed direct light on the long-standing problem of the influence of the ocean tides on solid Earth tides.

The network of the transcontinental profile consisted of nine semipermanent (6-month observation) stations around latitude 39–41°N, extending from New York City to Point Arena, Calif.

The observed relative values of the gravimetric factor and the phase do indeed follow a definite logarithmical pattern with respect to distance from the Atlantic and Pacific oceans. Figures 1 and 2 show the relative differences of gravimetric factor $\Delta\delta$ in percent and of phase κ in degrees for M_2 and O_1 (solid circles). The maximum difference of gravimetric factors for M_2 between New York City and Point Arena amounts to approximately 8%. There is a lag of about 4° for the phases of M_2 in both New York City and Point Arena; these phases decay toward midcontinent, where the phase lag is less than 1°. The gravimetric factors for O_1 are nearly equal for all stations east of Kansas, and the value of $\Delta\delta$ increases at a rate of about 0.14% per 100 km toward the West Coast. The phases for O_1 are equal, with a nearly constant lag of about 1° across the continental United States. They gradually begin to lead west of Ephraim, Utah, and the lead increases very rapidly to 5° in Point Arena. These results demonstrate the different tidal regimes in the nearby Atlantic and Pacific oceans. In the Pacific, both diurnal and semidiurnal tides are important, and therefore the ocean loading due to both O_1 and M_2 is important. In the Atlantic, only the semidiurnal tides are important. This is borne out by the small changes in gravimetric factor and phase due to O_1 near the Atlantic end of the network.

Quantitative calculations were made for each station by taking into account the effects of ocean tides on tidal gravity, namely, a variation in the height of the point of observation, a distortion of the tidal potential, and an addition to the variation of the vertical component of acceleration of gravity due to the water mass of the ocean tide, making use of the cotidal and corange information for the M_2 and O_1 ocean tidal constituents. The observed values of $\Delta\delta$ and κ as shown in Fig. 1 agree remarkably well with those calculated (triangles) for the M_2 constituent. For the O_1 constituent the agreement is not as good (Fig. 2), but the general trend for the values of $\Delta\delta$ and κ indicates that, even though the cotidal and corange information for the O_1 ocean tidal constituent is inferior to that of the M_2 ocean tidal constituent, the O_1 ocean constituent is a primary influence on the O_1 Earth tidal constituent. Nevertheless, there is a considerable degree of uncertainty about the ocean tides on open oceans. The agreement between the observed deviations of the gravimetric factors and the phases and the calculated deviations due to the influence of ocean tides merely substantiates the fact that the influence of ocean tides on tidal gravity is of primary importance.

Although the question of the influence of geological structure on tidal gravity has drawn much at-

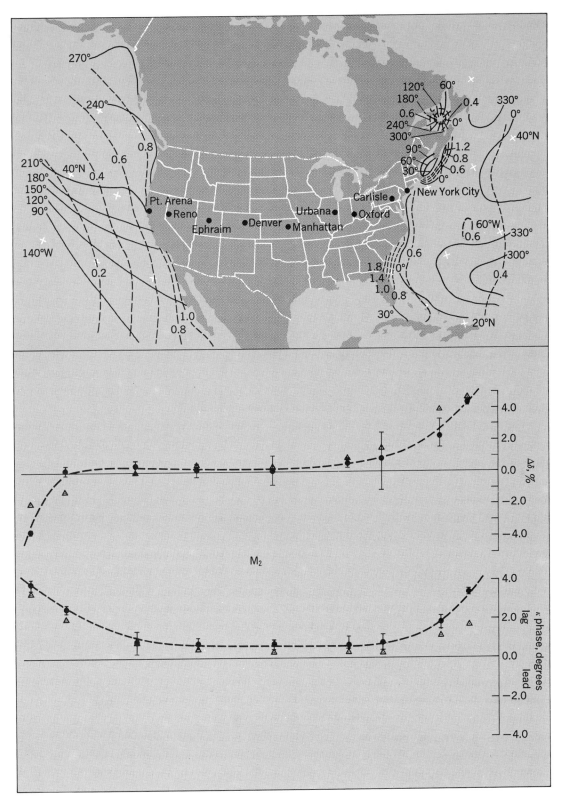

Fig. 1. Observed values of Δδ (in percent) and κ (in degrees) for M_2 tidal constituent (solid circles) compared with calculated values of Δδ and κ (triangles). Cophase (in degrees) and coamplitude (in meters) of M_2 ocean tidal constituent are also shown by solid and broken lines, respectively.

tention in the past, the theoretically calculated gravimetric factors for several Earth models involving drastic differences in the crustal and upper mantle structures are nearly constant. The small residual deviations of both the gravimetric factors and the phases, after subtracting the effects of ocean tides on tidal gravity, do not seem to correlate with the major different geological provinces such as the Interior Plains and the Rocky Mountains.

The Earth models deduced from seismic body waves, for example, the Jeffreys-Bullen and the

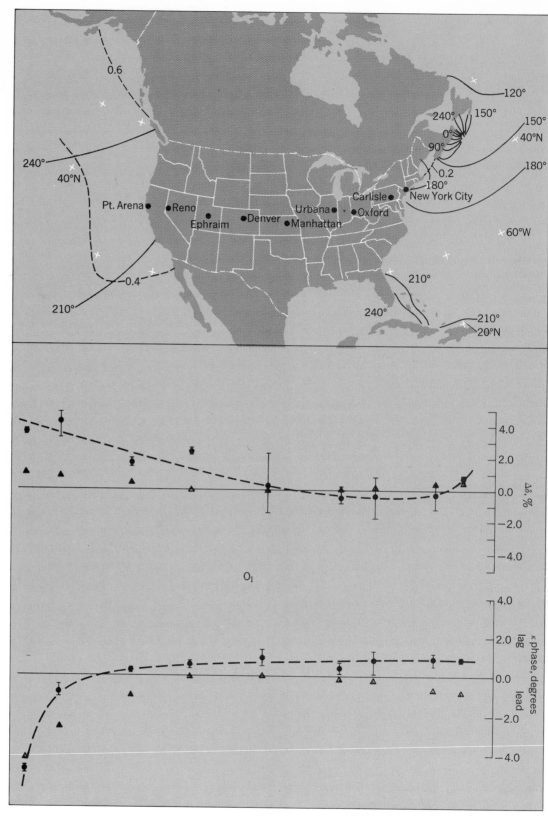

Fig. 2.   Observed values of $\Delta\delta$ (in percent) and κ (in degrees) for O_1 tidal constituent (solid circles) compared with calculated values of $\Delta\delta$ and κ (triangles). Cophase (in degrees) and coamplitude (in meters) of O_1 ocean tidal constituent are also shown by solid and broken lines, respectively.

Gutenberg-Bullen models, have long been regarded as close approximations to the actual Earth. The observed areal strain of solid Earth tides with improvements of extensometer measurements at Ogdensburg, N.J., which is nearly independent of the load strain caused by the ocean tidal loading off the Atlantic coast based on the Boussinesq solution, confirms the theoretical tidal strain based on the Gutenberg-Bullen Earth model with a New York–Pennsylvania crustal structure, which has the characteristic Love numbers $h = 0.621$ and $l = 0.084$. These Earth models have been further supported by good agreement between the theoretical and experimental values of the periods of the free oscillations of the Earth, although refinements are still needed.

It is now evident that it is fruitless to try to verify the theory of Earth tides by using the inferred or theoretically calculated cotidal and corange charts to make corrections for the indirect effects of ocean tides on solid Earth tides, as has been done in the past. If measurements can be made to an accuracy of 1% or better, it is more appropriate to consider the possibility of mapping ocean tides on the open oceans by means of extended Earth tidal gravity measurements on the adjacent lands supplemented by a few ocean-bottom stations of ocean tides measurements and by the tidal information of nearshore tidal stations.

International Symposium on Earth Tides. The Sixth International Symposium on Earth Tides was held in France at the University of Strasbourg, Sept. 15–20, 1969. It was organized by the Permanent Commission on Earth Tides, International Association of Geodesy, IUGG.

Many aspects of the meeting were comparable to those of previous Earth tide symposia in that the papers dealt primarily with instrumentation and data analysis. With regard to instrumentation, considerably greater emphasis is being placed on possible contributions by various potential sources of error, on calibration, on the interchange of instruments between stations, and on simultaneous observations using multiple sensors at the same site. With this increased sophistication has come an awareness that the geophysical interpretation of the observations may hinge on a high degree of accuracy.

A similar change has come about in data analysis in that there is now an active program to use longer series to achieve both more accurate results and a high degree of resolution. Thus the fixed 29-day analysis is no longer the ultimate end product, and least-squares analysis of data that may not be continuous and equally spaced in time is accepted as a basic tool. With this has come an interest in other frequency bands, not only the long-period tides but also high-frequency tides. There is considerable interest in M_3, and frequencies as high as 8 cycles per day are being investigated.

The various resolutions adopted at the symposium included (1) the recommendation that long-term observational series be planned as regular observatory-type programs in a number of different regions of the Earth; (2) the recommendation that complete Earth tide stations be established in the equatorial regions and in the Southern Hemisphere, particularly in Africa, South America, and Australia, and at Schiltach, Black Forest, Germany, to fill the gap in the net between France, Belgium, and Luxembourg on the one hand and the stations at Berchtesgaden (Germany), Czechoslovakia, and Austria on the other hand; and (3) the endorsement of efforts now under way in the oceanographic community to develop more accurate cotidal and corange charts by means of deep-sea observations and hydrodynamic model studies.

For background information *see* EARTH; EARTH TIDES; TERRESTRIAL GRAVITATION in the McGraw-Hill Encyclopedia of Science and Technology.

[BERNARD D. ZETLER; JOHN T. KUO]

Bibliography: J. T. Kuo, *J. Geophys. Res.*, 74: 1635–1644, 1969; J. T. Kuo et al., *Science*, in press; P. Melchior and A. Venedikov, *Phys. Earth Planet: Interiors*, 1:363–372, 1968; J. C. Usandivaras and B. Ducarme, *Bull. Acad. Roy. Belg. Cl. Sci.*, June 7, 1969.

Earthquake

Reports of recent research on earthquakes have been concerned with earthquake prediction and control and with earthquakes resulting from underground nuclear explosions.

Prediction and control. Earthquake prediction, long a hope of scientists in earthquake-prone countries, is advancing toward realization. In the United States, as in Japan, the approach to earthquake prediction is largely empirical: attempting to find some earthquake forerunner in the sequence of changes that continually take place in an active fault zone that may signal the approach of an earthquake. Earthquake control, beyond hope until recently, will soon be tested with the first deliberate experiments to stop and then restart small earthquakes in a zone of water injection in a Colorado oilfield.

Earthquake mechanics. In the United States, research on earthquakes is now concentrated mainly on improving understanding of the mechanics of earthquake generation. This improved understanding forms the essential background for applied experiments in earthquake prediction and control. In California, dense clusters of seismographs that telemeter seismic signals over telephone lines to the United States Geological Survey's National Center for Earthquake Research in Menlo Park have been installed along the central part of the San Andreas fault zone (Fig. 1). From these telemetered clusters, supplemented by networks of portable seismographs to record aftershock sequences, it has been shown that the earthquakes are occurring along active faults at depths no greater than 15 km. This suggests that the storage of elastic energy is taking place mainly in the upper part of the Earth's crust.

In some segments of the San Andreas fault, this energy is released almost continuously in the form of fault creep—the quiet, steady-to-episodic slippage along the fault—accompanied by small to moderate earthquakes. This "safety valve" of fault creep and small earthquakes may inhibit the storage of really large amounts of elastic energy. In other segments of the fault, however, such as the San Francisco Peninsula, where the great 1906 earthquake did so much damage, the safety valve seems to be missing: Few small earthquakes are recorded there, and no fault creep has been observed. Elastic energy is presumably being steadily stored in the upper part of the Earth's crust near San Francisco, someday to be released in another severe earthquake. If such an earthquake could be predicted, lives could be saved, injuries could be prevented, and vital public services could be preserved.

Earthquake prediction. There are encouraging signs that earthquake prediction may be possible in the fairly near future. The rate of fault creep

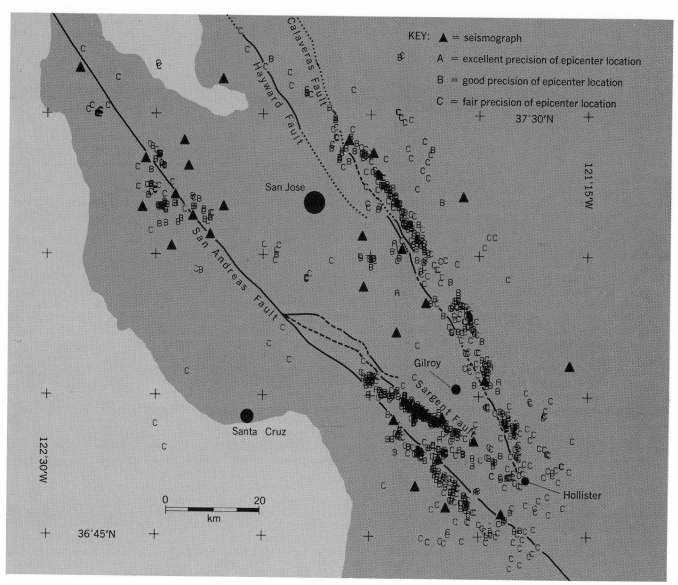

KEY: ▲ = seismograph

A = excellent precision of epicenter location

B = good precision of epicenter location

C = fair precision of epicenter location

37°30'N

121°15'W

122°30'W

36°45'N

0   20
km

San Jose

Santa Cruz

Gilroy

Hollister

Hayward Fault

Calaveras Fault

San Andreas Fault

Sargent Fault

Fig. 1. Microearthquakes along the San Andreas and related faults between San Francisco Bay and Hollister from March, 1968, to April, 1969. (*National Center for Earthquake Research, USGS*)

along the San Andreas fault appears to vary with time. Rapid changes in the rate of creep (speeding up in some places, slowing down in others) have been observed to occur prior to small earthquakes, and trial predictions based on fault creep have been attempted, some successfully. Variations in the tilt of the Earth's surface as measured on sensitive tiltmeters have also been observed prior to earthquakes, and these may have predictive value. Rapid but tiny changes in the strength of the Earth's magnetic field have been detected prior to several fault-creep episodes in California. These are but pieces of a pattern of fault-zone behavior that seems to be emerging, but they suggest that short-range prediction of earthquakes (on the order of hours or days) may be achieved through continuous monitoring of ground tilt, strain, seismic activity, and magnetic-field fluctuations.

Ultimately, prediction may be based on accurate determination of the internal forces (stresses) that are acting in the Earth's crust. If these stresses are large enough to overcome the strength or frictional resistance of rocks in a fault zone, an earthquake will occur. Direct measurement of rock stress in drill holes is difficult and expensive, but it can be done in shallow holes. It may also be possible to measure changes in stress by measuring changes in the electrical resistivity and seismic velocity of rocks in fault zones. Laboratory experiments have showed that such resistivity and velocity changes are caused by changes in rock stress.

Earthquake control. The possibility of earthquake control is suggested by the man-made earthquakes caused by injection of fluids into the stressed rocks of the Earth's crust. At the Rocky Mountain Arsenal near Denver, Colo., the Earth's crust was strong enough to resist the large stresses acting there until waste fluids were injected into ancient Precambrian rocks at the 12,000-ft-deep Arsenal well. The waste fluid, injected at high pressure, reduced the frictional forces which tend to oppose sliding along fractures in the rock. The fracture strength of the rock was reduced to a value below the rock stress and caused a series of earthquakes which continued long after the injection of fluid was stopped (Fig. 2). The correlation in Fig. 2 and the location of the center of the earthquake zone near the well demonstrate that earthquakes resulted from fluid injection.

A similar occurrence of earthquakes was discovered at the Rangely oilfield in northwestern Colorado. At Rangely, water is injected at high pressures into the edges of the oilfield to increase oil recovery. The Rangely earthquakes, located by a dense seismic network, occur along a preexisting fault which is active only in the part of the field where the fluid pressure is abnormally high, thus suggesting that they were brought on by fluid injection. An attempt to stop the earthquakes by reducing the pressure in the earthquake zone is being planned. If the earthquakes can be stopped, an attempt will be made to start them again by raising the pressure.

If the Rangely and related experiments are successful, the first steps toward earthquake control will have been taken. Ultimately, such techniques might be applied to the San Andreas fault to convert stored elastic energy to harmless fault creep or small-to-moderate earthquakes or both, thus eliminating the hazard of destructive and death-dealing earthquakes.   [LOUIS C. PAKISER]

Underground nuclear explosions. Large underground nuclear explosions in Nevada cause movements on geologic faults that can be measured in feet and initiate earthquake sequences that last for weeks. Although these effects are impressive, they are confined to within 15 km of the explosions, and the largest aftershocks touched off are much weaker than the explosions. Of great interest to scientists is the convincing evidence that the explosions cause the release of natural forces within the Earth's crust. An explanation of how this occurs would be an important step toward understanding the mechanism of natural earthquakes and their aftershock sequences.

Evidence for explosive effects. The first clues that explosions could affect geologic processes were found in the seismic waves that radiated to distant points on the globe. An explosion alone should not generate a horizontally polarized shear wave; however, such horizontal motion, transverse to the direction of wave propagation, was observed from explosions at the Atomic Energy Commission's Nevada Test Site and reported in the early 1960s. Waves from a seismic source that travel along the Earth's surface also provided a clue. An explosive source should emit waves that are approximately the same amplitude in all directions. The waves observed from some underground nuclear explosions, however, had an asymmetric pattern, one that contained an earthquakelike component.

Fault movements in the vicinity of many explosions provided further evidence concerning the nature of the explosion effects. A fault running the length of Yucca Flat at the Nevada Test Site was repeatedly displaced by explosions. In general, larger explosions have greater effects. To date the largest explosions in Nevada, with a yield of about 1 megaton (roughly equivalent to an earthquake of magnitude $6\frac{1}{2}$), have triggered faulting up to 8 km in length. Movement of one side of a fault with respect to the other side has reached 1 m in verti-

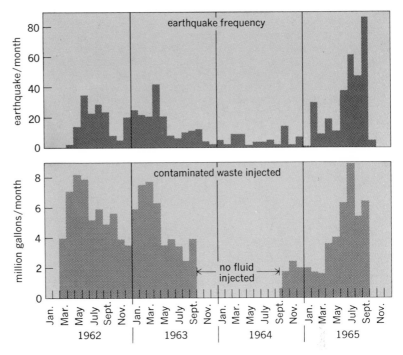

Fig. 2. Correlation of number of earthquakes each month during 1962–1965 near the Rocky Mountain Arsenal near Denver, Colo., with millions of gallons of fluid injected each month in the Arsenal waste-disposal well. (*Courtesy David M. Evans, Colorado School of Mines*)

cal offset and 15 cm in horizontal offset. Most of the explosion-induced fault movements occur on faults that were recognized as geologic boundaries and mapped as such before detonation. Moreover, the relative displacement on the faults has always been in the same direction as that of the most recent recognizable natural movement.

Reports of earthquake activity associated with underground nuclear explosions also began to appear in the early 1960s. They failed to attract very much attention, however, probably because the earthquakes observed were very small and occurred close to the explosion cavity. When explosions with a yield of about 1 megaton were detonated in Nevada in 1968, the resulting aftershock sequences attracted considerable attention. Some of these aftershocks reached a magnitude of 4 on the Richter scale, which is large enough to be felt strongly in the test-site vicinity.

The recognition of strong aftershock excitation by the 1-megaton explosions led to a number of investigations of the phenomenon. One study showed that all explosions with a yield equivalent to magnitude 5 or larger are followed by seismic activity for at least 1 day. Much of this activity is very near the explosion site and is probably the result of deterioration and collapse of the cavity produced by the explosion. The evidence indicated, however, that in many cases earthquakes occurred as far as 20 km from the explosion, and in one case, 40 km.

Detailed monitoring. These findings, coupled with suggestions that the nuclear tests could be producing a cumulative effect on the seismicity at even greater distances, prompted detailed monitoring of the explosion given the code name BENHAM, a 1.1-megaton detonation in December,

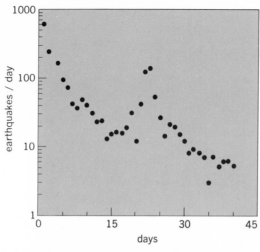

Fig. 3.   Number of earthquakes per day with magnitude greater than 1.3 versus number of days after the BENHAM underground nuclear explosion.

1968. Twenty-seven seismographs were deployed within 30 km of that shot. The explosion touched off a remarkable aftershock sequence. During the first day after detonation, earthquakes more than 1.3 in magnitude were occurring at the rate of 1000 per day. As is usual in aftershock sequences, this rate decreased exponentially during the next 3 weeks to 10 events per day (Fig. 3). It then dramatically increased to 100 shocks per day during the fourth week, before continuing its decline to fewer than five shocks per day after 6 weeks. Most of the

aftershocks occurred within 5 km of the BENHAM site. The flurry of earthquakes during the fourth week was associated with a growth of the main zone of seismic activity to a distance of 8 km. Scattered earthquakes were detected to 13 km from the shot. The rock movements responsible for the aftershocks, the nature of which was inferred from the pattern of initial ground movement from the aftershock waves, were similar to those of long-term fault movements in the Nevada region (Fig. 4). This was further evidence that natural processes can be excited by an explosion.

Subsequent large underground explosions at the Nevada Test Site reactivated and enlarged the BENHAM aftershock zone and created new seismically active areas nearby. The effects of the later explosions were markedly less than those of BENHAM, which is not surprising, since the aftershocks of BENHAM probably drained much of the earthquake potential from the immediate region.

Extrapolation of present results. The seismic effects of the Nevada explosions observed to date do not appear to constitute a hazard to life or property. The largest earthquakes caused are very much smaller than the explosions themselves, and have occurred close to the detonation point.

What the seismic effect of explosions with a yield of several megatons or larger will be is unknown. Without an understanding of the mechanism by which an explosion excites the seismic regime, it is difficult to extrapolate present results. This is a problem the Atomic Energy Commission faces in conducting a test of a larger explosion at Amchitka, a remote island in the Aleutian chain.

Fig. 4.   Fault movement caused by a large underground nuclear explosion, 2000 ft from ground zero, in Nevada.

Explosions can excite such natural processes. (*Courtesy F. A. McKeown, USGS*)

The Aleutian Islands lie above one of the Earth's major earthquake belts. Consequently, when the testing program there was announced, fear was expressed that a major earthquake might be touched off, which in turn might generate a tsunami (a giant water wave). A calibration shot, named MILROW, was therefore detonated in October, 1969. Its yield, about 1 megaton, was comparable to the largest yields yet released in Nevada. The aftershocks were few in number compared with the sequence that followed BENHAM, and no large earthquake or tsunami was generated.

For background information *see* EARTHQUAKE; SEISMOGRAPH; SEISMOLOGY in the McGraw-Hill Encyclopedia of Science and Technology.

[ROBERT M. HAMILTON]

Bibliography: *Bull. Seismol. Soc. Amer.*, vol. 59, no. 6, December, 1969; R. M. Hamilton, F. A. McKeown, and J. H. Healy, *Science*, 166:601–604, Oct. 31, 1969; J. H. Healy et al., *Science*, 161: 1301–1310, 1968; L. C. Pakiser et al., *Science*, 166:1467–1474, 1969.

Eclipse, astronomical

A total eclipse is a rarity at any geographical location. The umbral path of the Mar. 7, 1970, solar eclipse, the only total eclipse to occur over the continental United States in the 20th century, conveniently passed close to two major rocket-launching facilities, Eglin Air Force Base, Fla., and the NASA site at Wallops Island, Va. Eclipse studies performed in the United States were coordinated through the offices of the National Science Foundation. Ground-based optical investigations were best afforded in Mexico. Locations used by the Air Force Cambridge Research Laboratories, one of the major groups investigating primarily aeronomic eclipse effects, are shown in Fig. 1, which is

a display of the path of totality over North America in universal time. Aeronomy is concerned with that part of the upper atmosphere where dissociation and ionization processes are important, and includes the ionosphere. This region is influenced considerably during a solar eclipse.

Ozone. Ultraviolet radiation of wavelength less than about 3100 A is strongly absorbed by ozone and results in its dissociation. The obscuration of this radiation in the course of an eclipse likely perturbs the ozone concentrations in the mesosphere (50–85 km) and upper stratosphere (30–50 km). Photodissociation of ozone yields atomic and molecular oxygen, mainly in excited states. These species play a role in ionospheric chemistry, as does ozone. Both react with O_2^-, the negative ion initially formed in the lower ionosphere, detaching its excess electron; O_3^- is created when ozone reacts with O_2^- by means of the charge-transfer process $O_2^- + O_3 \rightarrow O_3^- + O_2$. Further reactions with minor atmospheric gases result in the production of negative ions less subject to detachment. Photodetachment also occurs, and is greater for O_2^- than for any other negative ion species expected.

Ozone reacts with atomic hydrogen to yield molecular oxygen and vibrationally excited OH* molecules. Observations of the Meinel bands emitted by OH* can provide some information about O_3 and H. Atomic hydrogen is a minor constituent in the mesosphere, where OH* radiates, but is the predominant atmospheric component above about 1200 km. The Meinel bands and the 1.27-μ line of excited molecular oxygen were measured from an aircraft (Fig. 1). Rockets fired from Florida during totality and during third contact carried instruments for detecting ozone and the total atmospheric density. Similar ozone observations were carried out from Wallops Island. The density

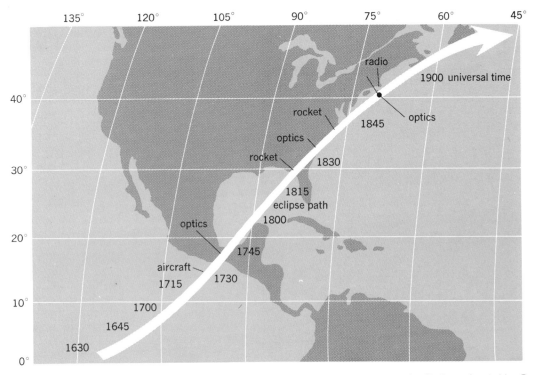

Fig. 1. Path of totality for the Mar. 7, 1970, solar eclipse over North America in universal time showing field sites used for investigations by the Air Force Cambridge Research Laboratories.

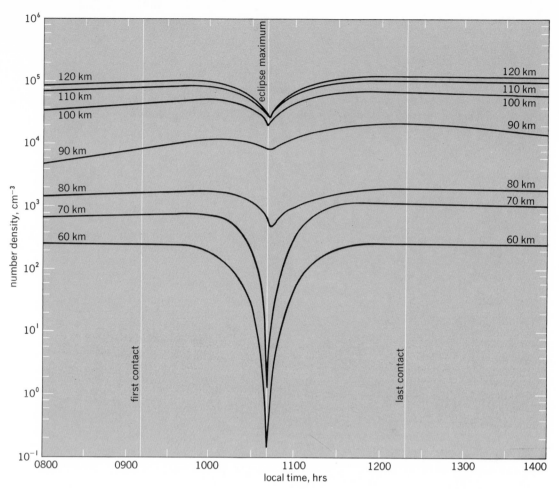

Fig. 2. Model of electron concentration versus time for selected E- and D-region altitudes during the Nov. 12, 1966, solar eclipse over Brazil. Calculations are computerized and are typical for a midday eclipse.

measurements were made in the stratosphere, mesosphere, and lower thermosphere. No major density effect was anticipated. *See* AIR SAMPLERS, ROCKET-BORNE.

Electron concentrations. Figure 2 is an estimate of the electron concentration patterns expected at various altitudes in the lower ionosphere for the Nov. 12, 1966, solar eclipse. These results are typical of a midday eclipse. The calculations are from a computer code developed by T. J. Keneshea which combines the known chemistry at these altitudes with a uniform obscuration of the solar flux, except for an x-ray limb effect which allows an x-ray flux of 15% at totality. The problem is complicated by the fact that x-rays apparently have a nonuniform distribution over the solar disk. Solar x-rays are major ionizing agents in the E region (85–150 km), especially the lower half, but are of minor significance in the D region (60–85 km) when the Sun is "quiet."

Although only the electrons are responsible for radio-wave absorption and are of interest in other radio-communication effects, knowledge of the various types of ions is required to understand the physical-chemical processes controlling the electron concentration. Mass spectrometric measurements of positive and negative ions were obtained during the Mar. 7, 1970, eclipse with rockets launched from Wallops Island. Observations of electron concentrations were made simultaneously by such instruments as Langmuir probes.

The total positive ion concentration, which is equal to the total concentration of negative particles, is proportional to the square root of the ionization production rate in the D and E regions under normal daytime conditions. Negative ions are prominent only in the lowest portion of the ionosphere, the D region, where the number of neutral

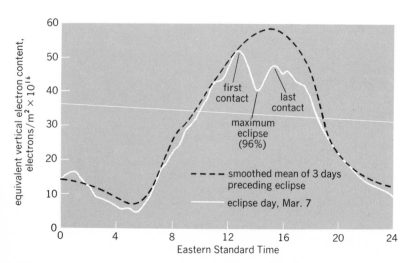

Fig. 3. Total electron content versus eastern standard time for the Mar. 7, 1970, solar eclipse.

particles is sufficiently large for three body processes, such as $e + O_2 + O_2 \rightarrow O_2^- + O_2$, to be significant. The much larger depletion of electrons with decreasing altitude in the D region as compared to that in the E region (Fig. 2) is a result of a strong increase in the number of negative ions at the expense of the electrons because of the interruption of photodetachment and (ozone) photodissociation processes during the eclipse. The concurrent changes in the positive ion concentrations in the D region are far less pronounced. *See* IONOSPHERE.

The influence of a solar eclipse on the F region is much more complex. This is the largest portion of the ionosphere and has a peak concentration near 300 km at mid-latitudes. The high vacuum in this region permits considerable particle motion. This includes diffusion along the Earth's magnetic-field lines and electromagnetic drifts which allow the positive ions, mainly O^+, and electrons to cross field lines. These movements, or storm effects, are sometimes so intense that they dominate the eclipse effects.

The total electron content (TEC) of the ionosphere per square meter, at least 95% of which is contributed by the F region, is shown in Fig. 3 for approximately mid-latitudes on Mar. 7, 1970. The TEC measurements were obtained from observations made at Hamilton, Mass., of the amount of polarization twist of linear, attitude-stable, VHF radio waves transmitted from the geostationary satellites *Early Bird*, *ATS 3*, and *ATS 5*.

The ionization production rate in the mid-F region is much less than that at E-region altitudes. The electron concentration is larger in the former region, however, because of the much slower loss processes. This leads to a slower response of the F region in the absence of motions. Furthermore, daytime electron concentrations in the mid-F region are proportional to the ionization rate. This fact, plus a slow response, makes the F region also sensitive to the much greater partial eclipse path besides the path of totality. A decline in the E-region ionization production rate to about 10% at totality decreases the electron density by about the square root of this factor about 30%. The finite loss rate keeps the decrease near 50%. Since the F-region decay is slower, its peak concentration, which is controlled partly by diffusion, does not decline to 10% during totality. On the other hand, some longer time appears required before the eclipse effects are entirely mitigated, provided that motions are minimal. This may be the case for the results in Fig. 3.

For background information *see* AERONOMY; ECLIPSE, ASTRONOMICAL; IONOSPHERE in the McGraw-Hill Encyclopedia of Science and Technology.   [WILLIAM SWIDER]

Bibliography: H. N. Ballard et al., *J. Geophys. Res.*, 74(2):711, 1969; J. A. Klobuchar and C. Malik, *Nature*, 226(5251):1113–1114, 1970; E. A. Mechtly et al., *Radio Sci.*, 4(4):371, 1969; H. Rishbeth, *Space Sci. Rev.*, 8:543, 1968.

Ecosystem

The planetary ecosystem is composed of vast arrays of smaller ecosystems of varying scopes; the internal functional relations (the "structure") and external interrelations of these ecosystems have changed greatly during geological time. In order to

Ecological hierarchy		Level
Descriptive units	Functional units	
Planetary biota	Biosphere (ecosystem)	High
Province	Provincial system (ecosystem)	
Community	Community system (ecosystem)	
Population	Population system (niche)	
Individual	Ontogenetic system	Low

Fig. 1. The parts of the ecological hierarchy.

evaluate the modes and paths of their evolution in an orderly manner, it has recently been proposed that the planetary ecological structure be considered as a hierarchy (Fig. 1) and that evolutionary processes operating upon an ecological unit of any hierarchic level be defined in terms of changes in the components of that unit that represent the next lower level. Since many evolutionary processes are ecologically regulated, the evolution of ecosystems involves much feedback.

Populations. The smallest ecological unit that can be considered to evolve is the population, which is composed of all the individuals in a given place and time that have drawn their gametes from a gene pool within which gene flow may occur. Since populations are subsystems of species, they form units in the taxonomic, as well as the ecological, hierarchy and have received much more study than have units at higher ecological levels. The evolution of a population, when regarded as a change in the taxonomic hierarchy, is commonly defined as a change in gene frequency within its gene pool. Components of this change that are of special interest for ecosystem evolution are (1) change in gene quality, based on a change in nucleotide sequence owing to a mutation, crossover, or similar event; (2) change in gene proportion, based on the preferential survival of individuals bearing certain genes owing to selection, drift, or other processes; and (3) change in gene diversity, that is, the rise or fall in the total number of different genes in the pool.

The evolution of a population also changes the state of the ecological hierarchy, and it is possible to define the evolution of populations (as ecological units rather than taxonomic units) in terms of the changing frequency of individuals that have particular functional properties. By analogy with the genetic changes, these changes may be in (1) quality, arising from the appearance or disappearance of certain phenotypes; (2) proportions, arising from the differential success of certain phenotypes; and (3) diversity, in this case simply changes in population size. Of course, these phenotypic changes mostly have genetic bases or effects or both. Thus individuals are the primary subsystems of populations, which are in turn primary subsystems of communities (Fig. 1). Changes in the functional aspects of the individuals cause changes in the

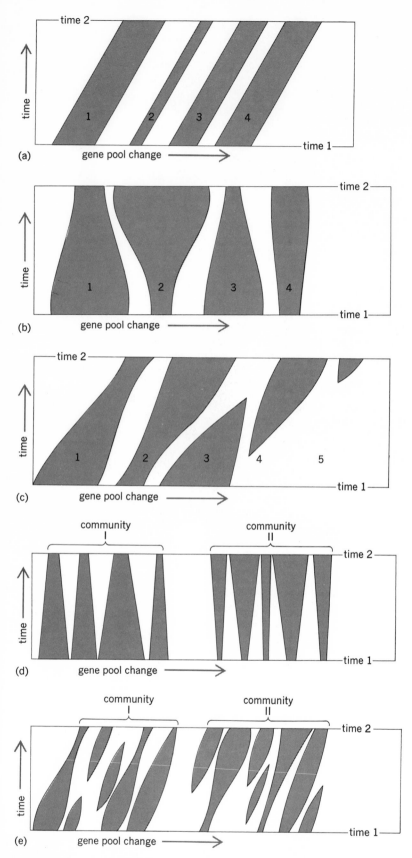

(a)

time

gene pool change ⟶

(b)

time

gene pool change ⟶

(c)

time

gene pool change ⟶

(d)

time

gene pool change ⟶

(e)

time

gene pool change ⟶

Fig. 2. Some components of the evolution of ecological units. Shaded areas represent populations; their horizontal dimensions indicate population size. (a) Populations evolving in concert owing to qualitative genetic changes. (b) Population quality constant but population proportions change, so that community evolution occurs. (c) A community evolves in quality, proportion, and diversity of populations. (d) Two communities change in proportions, so that provincial evolution occurs. (e) A province evolves in quality and proportion of communities.

functional aspect of the population, which is usually termed a niche.

Communities. At the next highest level in the ecological hierarchy, the community is composed of a characteristic assemblage of populations living in a given place and time. Functionally, communities are ecosystems that involve energy flow and nutrient cycling. Their evolution may be defined in terms of change in the quality, proportion, or diversity of their component populations. Changes in population quality (Fig. 2a) have been described above. Changes in population proportions involve relative changes in population sizes and thus quantitative changes in the energy flow among the populations forming the community (Fig. 2b). These changes are commonly due to competition, predation, and physical environmental fluctuations. Finally, changes in diversity within communities involve increase or decrease in the number of species represented by populations (Fig. 2c).

Theoretical considerations suggest that such diversity changes may come about chiefly because of (1) changes in the spatial heterogeneity of the environment, resulting in more or fewer habitats for the accommodation of different species populations; (2) shrinkage in average niche size, permitting more species to be associated in the same range of environment, usually accompanied by a trend toward smaller, more efficient populations and favored during stable environmental regimes; and (3) the expansion of niches and a concomitant decline in diversity, accompanied by a trend toward larger, less efficient populations and favored under fluctuating environmental regimes. Intracommunity diversity is thus a function of the type of adaptive strategy (especially niche breadth and population size) employed by many populations under given environmental conditions. These suggestions are supported by much evidence but are not completely verified.

Provinces. Biogeographic provinces form the next highest ecological level. They are composed of communities that live in a certain region at a given time. Provincial biotas are distinctive for either or both of two reasons. First, different provinces may be based upon distinctive environmental regimes and therefore contain distinctive taxa and have distinctive structural styles of communities. Second, provinces may be separated by barriers to biotic interchange for a significant length of time and thus have evolved in isolation.

Provincial evolution may be defined in terms of changes among the communities that are analogous to changes within ecological units at lower levels. Qualitative community changes have been described above. Changes in community proportion usually involve changes in the environment which proportionately alter the area occupied by different biotopes (Fig. 2d). Changes in community diversity may be due to changes in spatial heterogeneity on the scale of the biotope, permitting the appearance or causing the exclusion of communities, or they may be due to changes in the functional range of the community ecosystem owing to narrowing or broadening of population niches, which can lead to a restriction of a community to only a portion of a biotope previously occupied or to the extension of a community into habitats from which it was previously absent (Fig. 2e).

Total planetary biota. The highest level in the ecological hierarchy contains all the organisms on Earth and forms an ecosystem that is usually termed the biosphere. Provinces form the primary subsystems of this level. At any time the state of the biosphere depends upon the quality, proportion, and diversity of the provinces, and it may evolve by changes in these properties. Changes in provincial quality are described above. Provincial proportions depend upon the relative geographic extents of provincial environmental regimes. Provincial diversity depends upon (1) diversity of environmental regimes, (2) spatial replication of similar regimes by isolating barriers, and (3) the functional ranges of the biotas, which, if narrow, tend to be sensitive to relatively small changes in the environment, leading to greater provinciality.

The factors of quality, proportion, and diversity, which have been employed herein to discuss the states and changes of ecological structures, are somewhat interrelated and normally change in concert. In general, for a given percent of internal change in an ecological unit, those structures on lower levels have a higher frequency of change than those on higher levels. However, each change in the functional character of a population, which may be based upon a minor change in some of the phenotypes of which it is composed, causes a change in the community and thus in the province and in the biosphere. On the other hand, most such changes, which occur constantly and in profusion, do not form steps in biosphere evolution, but are quickly canceled by other events. Never-

theless it is a summation of such small changes over long periods of time in trends mediated by environmental change that leads from one configuration of the biosphere structure to another.

For background information *see* ECOLOGY; EVOLUTION, ORGANIC in the McGraw-Hill Encyclopedia of Science and Technology.

[JAMES W. VALENTINE]

Bibliography: R. Margalef, *Perspectives in Ecological Theory*, 1968; J. W. Valentine, *J. Paleontol.*, 42:253–267, 1968.

Electric organ (biology)

Recent studies have shown that many nonelectric fishes generate extremely weak electric signals that can be remotely received with electrode antenna and solid-state amplifiers.

Some background is helpful in placing these results in context.

Electric fishes. Fishes that produce electrical discharges have intrigued man since the ancient fishermen of the Nile painfully learned that handling a certain catfish resulted in excruciating sensations, and the physiology and anatomy of the few fishes that produce strong discharges of several hundred volts, such as the electric catfish and electric eel, have been studied for centuries. This type of electric fish generates short, rapid bursts of pulses that may be used in defense or predation. One marine fish and some rays and skates emit similar signals, but only of a few volts.

Still weaker pulses, of the order of tenths or hundredths of a volt, are generated by four fami-

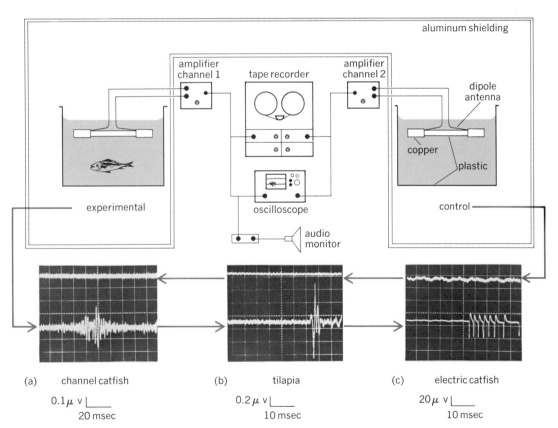

(a)   channel catfish   (b)   tilapia   (c)   electric catfish

0.1μ v |___   0.2μ v |___   20μ v |___
   20 msec   10 msec   10 msec

Fig. 1. Experimental equipment and representative signals received from fishes. When goaded with a plastic stick, both (a) the channel catfish and (c) the electric catfish produce alternating current bursts, but those of

the electric catfish are much stronger and more regular. (b) *Tilapia* signals are spikelike and are frequently generated spontaneously.

lies of fishes that dwell in turbid tropical rivers. Their pulses are emitted either in long bursts or continuously; their rate of emission and intensity vary with the species and the activity of the fish. Although these fishes have long been known, an explanation of their function was not forthcoming until Hans Lissman's ingenious experiments about 20 years ago. Using bits of food as a reward, he showed that one such fish, *Gymnarchus niloticus*, could discriminate between objects of different electrical conductivity. As suggested earlier by C. W. Coates, the results strongly indicated that the electrical production of the fish maintains an electrical field around its body and that disruptions in this field could be sensed. Thus the fish possesses a radarlike system that is possibly used to locate underwater objects.

In all known electric fishes these pulses are generated by special organs that, with one exception, consist of modified muscle cells called electroplaques. Because of these modifications, the electric species must represent the ends of long lines of evolutionary development. However, the nine families of fishes which have electric species are widely separated taxonomically and, in some cases, only one or two species of a group are known to be electric. This apparent lack of intermediate forms has long puzzled biologists.

Nonelectric fishes. A few years ago news items in the press and in articles in popular magazines and trade journals told of newly discovered signals generated by nonelectric fishes that were transmitted over long distances. Because the signals were nonacoustic and because an electric field propagated in water is known to decay by the cube of the distance, it was believed that a unique form of energy was involved.

Recent studies by a group of U.S. Navy researchers cast some doubt on this hypothesis. All the fishes and the one amphibian tested in aquarium tanks and in controlled experiments did produce distinctive signals but, with one possible exception, only when the fishes were stimulated into rapid movements. A small electric catfish produced signals several hundreds of times stronger, but the signals could be received with the same equipment. Furthermore, the signals generated by all the fishes weakened rapidly with distance, as would be expected for electric fields in water. When displayed on an oscilloscope, the signals from the nonelectric fishes were strikingly similar to those recorded directly by electromyograph from isolated vertebrate muscles. Figure 1 shows the equipment used and typical signals recorded in the controlled experiments from two nonelectric fishes and the strongly electric catfish.

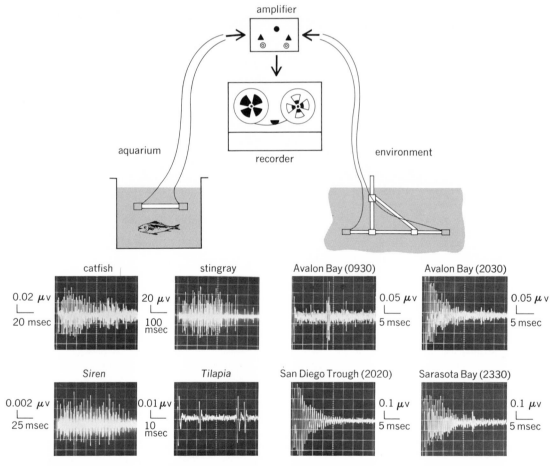

Fig. 2. Equipment used to record electric signals from fishes in a public aquarium and in various marine environments. Electrodes are separated by 2 ft in the aquarium antenna and by about 10 ft in the environmental antenna. The relatively long signal bursts of the catfish and stingray and of the salamander (*Siren*), an amphibian, are typical for rapid movements of nonelectric fishes. The environmental signals are basically of two types: a descending ramp burst, and several sharp spikes. Their audio frequencies sound, respectively, like a richochet and a click.

Futile attempts were then made to record signals from 50 *Tilapia* (one of the species previously tested) held captive in a copper-screened cage suspended in a lake. In the same experiment, a multitude of signals were received from a control antenna unprotected by copper screening 60 ft from the experimental enclosure. Although these signals were similar to those generated by fishes in the laboratory tests, they also resembled the well-known "spherics" normally received with radio antennas. These staticlike spikes and bursts originate from numerous physical causes, and the copper screen would effectively shield their reception from within the cage. The same spheric types of signals were later recorded in tests made in fishless swimming pools and small ponds and a number of marine environments. Figure 2 shows typical signals generated by fishes and an amphibian and the similar types that can be recorded in marine environments.

Scattered throughout the extensive literature on the biology of electric fishes are brief notes by Lissman and others on electric signals received from nonelectric fishes. On the basis of all of these bits of evidence, the Navy workers concluded that the fish-generated signals were very probably produced by white-fiber muscle contraction and that caution must be exercised in attributing the source of electric signals recorded in natural environments to fishes.

Possible applications. Although these conclusions regarding fishes producing a new energy form are essentially negative, several appealing prospects emerge. The ability of solid-state electronic components to remotely sense extremely weak electric signals (of the order of tenths of a microvolt) generated by small animals submerged in water suggests that they could be an important physiological and medical diagnostic tool. Apparently, atmospheric signals, ground currents, and other magnetic radiations become electrically coupled into water, the water thereby acting as an immense and effective antenna that can be easily tapped with electrodes. This process seems to offer possibilities in the study of radio physics.

Perhaps it is time to reinvestigate the problem of evolutionary intermediates that lead to the highly specialized modifications represented by the electric fishes. It is difficult to conceive of complex animals living in a conductive medium whose movements produce electric potentials that have not been put to use by the animals. Modern electronic techniques now provide the tools for a renewed search.

For background information *see* ELECTRIC ORGAN (BIOLOGY) in the McGraw-Hill Encyclopedia of Science and Technology.   [ERIC G. BARHAM]

Bibliography: E. G. Barham et al., *Science*, 164: 965–968, 1969; T. H. Bullock, *Brain Behav. Evol.*, 2:85–148, 1969; R. D. Keynes, *Physiology of Fishes*, vol. 2, chap. 5, 1957; H. W. Lissman, *Sci. Amer.*, March, 1963.

Electric shock protection

The ground fault interrupter (GFI) is the most successful device for eliminating the hazard from low-voltage electric shocks in the home, on the farm, and in industry. It is the newest of four recognized means for reducing the danger from electric shock (isolation and guarding, insulation, grounding, and

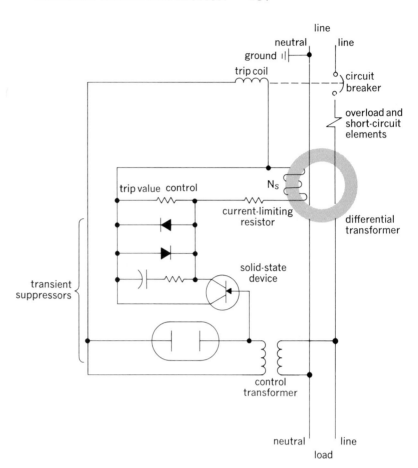

Fig. 1.   Schematic diagram of a ground fault interrupter.

Operating characteristics. The GFI is a device that interrupts an electric circuit when the fault current to ground exceeds a preselected value less than that required to operate the overcurrent devices of the circuit. Such apparatus have been used to protect high-voltage power lines since the 1920s, and they were set to operate at 10–20% of the minimum operating current, or trip value, of the associated overcurrent devices. Thus a power circuit breaker having an overload trip value of 100 amp could be set to trip on ground faults of only 10–20 amp, which was considered a great achievement of the day. In the 1930s the importance of protecting against low-voltage burndowns in industrial equipment was recognized in Europe, and devices were developed having a line-to-ground trip value of about 500 milliamperes (ma). More recently, low-voltage two-wire GFIs having a trip value of 25–30 ma have been developed in Europe.

Figure 1 is a schematic diagram of a modern GFI used in the United States. The device is composed of the customary molded low-voltage circuit breaker with its overload and short-circuit trip elements, in addition to a highly sensitive trip device which actuates the circuit breaker only upon the occurrence of current flowing to ground, which in the case of accident would be the current through the human body. Both the Underwriters' Laboratories, Inc., and the Canadian Standards Association require that the ground-current trip value be not in excess of 5 ma, a current safe for human beings,

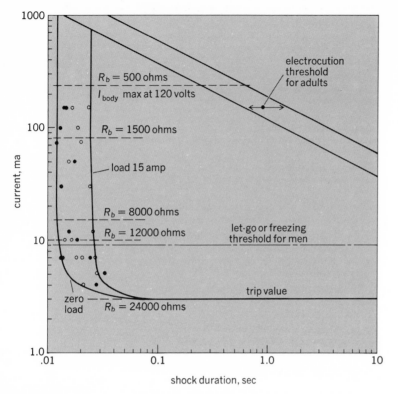

Fig. 2. Operating characteristic curves of a commercial two-wire 15-amp 120-volt ground fault interrupter.

peres, the sparks produced by 120- or 240-volt circuits are feeble and are very unlikely to start a fire.

Although the ground fault interrupter is no panacea for all low-voltage electrical hazards, it does offer a tremendous increase in protection against fires of electrical origin and serious electric shock accidents due to line-to-ground contacts.

For background information *see* CIRCUIT (ELECTRONICS); GROUNDING, ELECTRICAL; GROUNDING, ELECTRONIC-EQUIPMENT; LIGHTNING AND SURGE PROTECTION in the McGraw-Hill Encyclopedia of Science and Technology. [CHARLES F. DALZIEL]

Bibliography: C. F. Dalziel, Ground fault interrupter increases safety, *IAEI News*, 41(4):26–39, July, 1969; C. F. Dalziel, *IEEE Spectrum*, 7(1):55–62, January, 1970; C. F. Dalziel, *IEEE Trans.*, pt. 3, pp. 978–983, February, 1963; C. F. Dalziel and W. R. Lee, Discussion, *IEEE Trans. Ind. Gen. Appl.*, IGA-4(6):676–677, 1968; C. F. Dalziel and W. R. Lee, Re-evaluation of lethal electric currents, *IEEE Trans. Ind. Gen. Appl.*, IGA-4(5):467–476, 1968.

Electrical utility industry

During 1970 the electrical utility industry in the United States completed 25,900,000 kw of new generating facilities and the transmission and distribution to deliver their power output to ultimate consumers. Its capital expenditures for the 1970 portion of this expansion program exeeded $12,500,000. This record expansion raised the industry's total power capability to 338,512,000 kw by year end, thus maintaining by a wide margin the nation's world leadership in the generation and use of electrical energy (see table).

Power shortage. Despite this massive expansion program, however, several regions were critically short of power. More specifically, the available power resources, although sufficient to serve normal power demands and predicted peaks, provided too little reserve for the occasional break-

and that the device safely interrupt low-impedance short circuits up to 5000 amp. Few, if any, other mechanisms have ever been required to operate over such a range of currents, namely 5,000,000:5.

The high ground-current sensitivity is accomplished by means of the differential transformer and the solid-state amplifier. Protection of the secondary circuit against high currents is provided by the saturable core of the differential transformer, the current-limiting resistor, and the back-to-back diodes. The diodes, together with the resistance-capacitor network, protect against false tripping due to high-speed transients such as induced lightning strokes and electric hash due to appliances, especially appliances with commutator motors.

Figure 2 shows the operating characteristic curves of a commercial two-wire 15-amp 120-volt GFI. The electrocution threshold for all adults and the let-go threshold for men are included to give proper perspective. The horizontal lines indicate body current for variously assumed body-circuit resistances. It is generally accepted that the minimum likely body resistance R_b in low-voltage accidents is 500 ohms, and for the perspiring hands of a technician, 1500 ohms. Corresponding resistances for dry hands or casual contacts are too variable to warrant mentioning precise figures; however, experience shows that an accident victim either lets go a defective appliance immediately or, if he freezes to the circuit, the current increases rapidly and he is saved by operation of the device.

Fires. Although many fires are caused by short circuits, electric arcing, or overheated conductors, little is known regarding the minimum current necessary to start a fire. Laboratory tests indicate that when the current is limited to a few milliam-

United States electric power industry statistics for 1970*

Factor	Amounts in 1970	Increases over 1969, %
Generating capability, installed kw (10³)		
Hydro	54,368	5.2
Fossil-fueled steam	268,080	6.5
Nuclear-fueled steam	6,232	55.9
Gas-turbine and diesel	9,832	135.0
Total	338,512	8.3
Energy production, kwhr (10⁶)	1,539,500	6.5
Energy sales, kwhr (10⁶)		
Residential	442,000	8.8
Commercial	309,900	8.5
Industrial	574,400	3.4
Other	59,700	8.1
Total	1,386,000	6.4
Total revenue (10⁶)	$21,658	7.6
Total capital expenditures (10⁶)	$12,522	18.7
Customers (10³)		
Residential	63,400	1.6
Total	72,050	1.5
Residential usage, kwhr (average)	7,020	6.8
Residential bill, ¢/kwhr (average)	2.09	0.0

*SOURCE: *Electrical World*, Feb. 2, Mar. 2, and Sept. 15, 1970, with extrapolations.

downs that are inevitable in such extensive and complex systems. Early in the year the Federal Power Commission, after a survey of facilities, alerted the public to potential power shortages in certain areas of the nation.

This alert proved prophetic in the Northeast when, during a summer heat wave, the largest generator of Consolidated Edison Co. suffered a winding failure that took it out of service for the rest of the year. The loss of this 1,000,000-kw generator, occurring while the company's 275,000-kw Indian Point nuclear plant was down for extensive repairs, completely wiped out the reserves. The company immediately dropped its system voltage 5% to reduce its customers' power demands as much as possible without disrupting their service, and called on other utilities for assistance. These utilities responded as fast as their generating stations could handle additional loads and, in remarkably short order, were pushing power into New York City to the capability limits of the transmission grid. Assistance came in this manner from points as remote as Quebec and Ontario, and from a group of utilities in the Ohio Valley whose capacity had been unloaded by a voluntary cutback in consumption by the Atomic Energy Commission. Their assistance enabled Consolidated Edison Co. to carry the summer peaks.

Criticism, official as well as public, bombarded Consolidated Edison Co., asserting that the company had failed to meet its obligations as a public utility, that it had not provided enough capacity to serve its customers, and that it had not planned far enough ahead for such an emergency. In addition, a Senate subcommittee launched an investigation to determine whether Consolidated Edison Co., in its all-out effort to maintain electric service to its customers, had added to the air pollution in New York City.

In reply, the utility pointed out that its proposed 2,000,000-kw pumped-storage hydroelectric plant near Cornwall, N.Y., more than sufficient in itself to cope with this emergency, would have been in service had its construction not been blocked for more than 7 years by hearings and court procedures. The utility further observed that a large nuclear-fueled generator, Indian Point no. 2, also would have been available had it not been thrown 2 years behind schedule by interventions and labor trouble. In addition, its heavy tie to the massive 500,000-volt system in New Jersey and Pennsylvania—the tie that has been cited by the Federal Power Commission as the most vitally needed transmission circuit in the United States—had been blocked by similar interventions. And, finally, the utility was still waiting for official sanction of plans for building the only alternative generating facilities that could be completed in time to carry the summer peaks of 1974. This plan would have added 1,600,000 kw of steam units fired by low-sulfur oil at an existing plant site in the city. The sanction, when it finally came, specified that the addition be limited to 800,000 kw and that it burn natural gas whenever possible.

The New York "brownout" is a warning of the importance of finding a solution to the conflicting demands of protecting the environment on the one hand and the need for adequate and reliable electric service on the other. The consequences of

failure to reconcile both sides of this argument are becoming well known, but the recent emergency clearly demonstrates the imminence of the dangers involved in postponing a solution.

A similar emergency could have occurred in the West North Central region but for good fortune in the weather pattern and the behavior of generating units. This area, too, had been cited in the Federal Power Commission warning of deficient reserve capacity. Yet the recently completed nuclear station that had been expected to relieve the shortage could not be placed in service because of litigation over the State of Minnesota's assertion that operation of the plant would be detrimental to the state's citizens. This claim, in total conflict with findings released by the Atomic Energy Commission, is still unresolved.

Fossil fuel shortage. The electrical utility industry is facing still another dilemma—a critical shortage of fossil fuels that could be burned under today's tight restrictions on the gaseous products of combustion. These restrictions bar the utilities from using coal from most of the mines that had supplied them in the past, and forced them to compete with the metallurgical industry for the less plentiful coal it requires. In addition, many mines that could not comply with strict new safety regulations discontinued operations. At about the same time, natural gas suppliers, foreseeing a shortage of this fuel, began to turn down industrial contracts, including those of electric utilities. Many utilities promptly initiated boiler modifications and began to negotiate for low-sulfur fuel oil, most of which comes from Eastern Hemisphere fields; these supplies appear increasingly inadequate as the producing nations tighten their export allowances. At this point, the best hope seems to lie in the early completion of sulfur-extracting facilities that would enable domestic and South American oils to comply with today's restrictions. Whether this can be accomplished in time to resolve the shortage or whether regulatory authorities will suspend their restrictions until the fuel supply can be adjusted remains to be decided by future events.

Gains in technology. Power system technology continued its gains during 1970. In fact, the ability of other utilities to relay power over such long distances to assist power-short areas this past summer was a test of modern technology. Directly involved was the use of transmission circuits operating faultlessly at up to 500,000 volts. Involved, also, was the precise control of power flows through the vast interconnected network without exceeding safe operating limits at any point. Neither of these feats would have been possible with the facilities available as recently as 5 years ago.

Major control centers today coordinate the generation and transmission of power among groups of interconnected utilities in many regions of the United States and Canada. By direct "hot-line" communication channels to adjacent regional control centers, often hundreds of miles away, these centers coordinate the exchange of power among interconnected regions for optimum economy as well as emergency assistance. This type of operation has become feasible only with the development of sophisticated apparatus, often computer-directed, to provide instant information about the status of all key components of the far-flung power systems under centralized control.

Transmission capability. Transmission facilities for today's operating patterns must have massive capability. And the industry has swung into this need with such vigor that most major power utilities in a great arc from New Jersey through Virginia and Tennessee to Louisiana are linked in a vast and still-growing 500,000-volt interconnection. So, too, are the major utilities of the Pacific Coast states, which also have, as of 1970, the additional transmission capability of the western world's longest and highest voltage direct-current line. This line extends from the Columbia River to Los Angeles and is rated at 1,440,000 kw. Both interregional high-capacity systems connect, in turn, to somewhat lower capacity systems, generally operating at 345,000 volts, which extend grid coverage into essentially all of the contiguous 48 states and part of Canada.

Transmission of even greater capability is a reality, today, in about 300 mi of 765,000-volt circuits completed by the American Electric Power Co. Destined by about 1972 to extend from western Virginia to the shore of Lake Michigan, these 765,000-volt lines will have more than double the carrying capacity of 500,000-volt lines which, until recently, has been the largest permitted by existing technology. In fact, although only about a third of the transmission lines added during 1970 will operate at voltages from 345,000 to 765,000, these lines have nearly three times the capability of the remaining two-thirds that will operate at 230,000 volts or below.

Such high-capacity transmission has become a necessity for three important reasons. First, it offers the best solution to the pressing need for heavier interconnections between major power systems to facilitate the interchange of emergency assistance, as well as to enlarge the market for the lowest cost power. Second, such lines make the most efficient use of the land area tied up in transmission rights of way. Third, as generating units are built in ratings upwards of 1,000,000 kw, transmission must have comparable ratings to carry their output to major load areas.

The lack of cable to extend such high-capacity transmission into metropolitan areas somewhat limits their fullest usefulness. But cable already is available for 345,000 volts, and the Westinghouse Electric Corp. testing station at Waltz Mill, Pa., after completing tests on sample lengths of cable designed for 500,000-volt service, is ready for the next step—765,000-volt cable.

Generating capacity. Generating capacity added to United States power systems during 1970 set the all-time record of 25,900,000 kw, according to Edison Electric Institute data. Fossil-fueled steam-electric units, at 16,302,000 kw, accounted for the largest share of this increase; 5,650,000 kw of gas-turbine and diesel units came second; and nuclear-fueled steam plants were a poor third at 2,252,000 kw. Hydroelectric plants, pumped-storage as well as conventional, contributed only 1,696,000 kw.

Completions in all categories but gas turbines, however, were substantially below those scheduled, which, had they been completed, would have added 32,104,000 kw and given the electric utilities a more comfortable reserve during the summer load peak. Reasons for this slippage most frequently were the shortage and unrest of skilled labor, delayed delivery of apparatus stemming from strikes at manufacturers' plants, continued opposition to nuclear units, and time lost in repairing vital components of completed generating units that failed during shakedown operation. Thus the completion of some 6,200,000 kw of generating additions slipped into 1971, which, if it is not also plagued by slippage, could exceed the 1970 record by a substantial margin.

Nuclear power plants. Nuclear-fueled generation seems assured of much greater importance in the future but, as of 1970, it accounts for less than 2% of the 338,512,000 kw of generating capacity in service. With over 70,000,000 kw of nuclear units under construction or planned, however, it is assured of continued growth toward the 130,000,000 kw or more predicted by the Atomic Energy Commission for 1980. If it achieves this goal, nuclear-fueled steam plants will then constitute about 20% of the total generating capacity in the United States and will be generating nearly 40% of the electric power. *See* NUCLEAR POWER; REACTOR, NUCLEAR.

The longer range objective of using breeder reactors in nuclear power stations is still awaiting the construction of demonstration plants. At least four teams of utilities and manufacturers have done enough preliminary work, however, to participate in such plants. They hope to prove that their ability to convert inert but fertile material to fissile fuel (even as they use the fissile isotope of natural uranium or of man-made plutonium to generate the steam) will extend the availability of nuclear fuels indefinitely. At this point, however, it would be futile to anticipate commercial power breeders much before 1985. *See* NUCLEAR FUELS.

Fossil-fueled steam plants. In spite of this rapid nuclear expansion, fossil-fueled steam plants must provide the backbone of capacity additions during most of the decade. They will, however, be increasingly of a modified type, known as cycling, to coordinate more economically with nuclear plants which achieve their lowest power cost in continuous full-load operation because of their very low fuel cost. These cycling fossil-fueled units will be designed for economical operation on a variable-load basis, with frequent shutdowns during light-load periods. Thus they will not require the advanced technology that is applied in base-load stations and their investment costs will be lower. This presupposes, of course, that the problems of air and water pollution will be solved, as otherwise there will be neither sufficient low-sulfur fuel nor enough suitable sites for such an expansion. But the success being reported today for several experimental prototypes for controlling such effluents lends assurance that satisfactory methods will soon be available for commercial use.

Hydroelectric plants. Hydroelectric plants will continue their moderate growth rate, but more in the form of added generating units at existing plants than at totally new sites. The reasons for this are the scarcity of feasible sites that have not already been developed and the growing need for easily controlled peaking generation to complement highly efficient fossil-fueled or nuclear steam plants.

Much of this future hydro development, in fact, will be in the form of pumped storage in which off-peak power is used to pump water into man-made upper reservoirs so that about 65% as much power can be generated during heavy load periods by drawing down the stored water.

Gas-turbine and diesel plants. Gas-turbine and diesel power units have been expanded in recent years to fill the capacity void left by delay in completing other generating plants. Gas turbines are particularly attractive for such use because their investment cost is relatively low and their quick-starting capability assures them a continuing usefulness as peaking and emergency power units long after the initial need for added power capacity has been corrected.

Power sales. Power sales slackened a bit in 1970, growing only 6.4% over the record level of 1969. The slowdown stemmed from the relatively slack industrial sector of the economy, where the reduced activity level was reflected in a very small increase in power requirements. The residential sector, which is the largest contributor to electric utility revenues, grew a healthy 8.8% in energy sales and was followed closely by the commercial sector at 8.5%. Growth in these two sectors reflected, in part, the rapid acceptance of electric space heating, now used in about 4,700,000 dwelling units and a growing number of public and commercial buildings.

Underground distribution. Distribution expansion to serve this growing residential and commercial demand reached a record level in 1970, in spite of the low level of housing construction. A contributing factor, of course, was the growing trend toward serving such customers with underground power systems. The magnitude of this swing to underground is revealed by a recent *Electrical World* survey which showed that 33% of the distribution transformer capacity installed in 1970 was in types used for underground distribution and that 2,791,472 residential customers were slated for underground connection during the 7-year period starting in 1970.

For background information *see* ELECTRIC POWER SYSTEMS; NUCLEAR POWER in the McGraw-Hill Encyclopedia of Science and Technology.

[LEONARD M. OLMSTED]

Bibliography: 1970 Annual statistical report, *Elec. World*, 173(5):45–64, Feb. 2, 1970, and 173 (9):41–72, Mar. 2, 1970; W. C. Hayes, *Elec. World*, 171(18):87–106, May 5, 1969; L. M. Olmsted, *Elec. World*, 174(6):35–50, Sept. 15, 1970; L. M. Olmsted and P. H. Leech, *Elec. World*, 172(8):51–74, Aug. 25, 1969; 16th Steam station cost survey, *Elec. World*, 172(18):41–56, Nov. 3, 1969.

Electrodeposition

Electrodeposition of paint, introduced in the late 1950s, has found acceptance in the past few years for the coating of merchandise ranging in size from building steel trusses and car bodies, through office furniture and appliances, to nuts and bolts. This coating process consists of submerging metal-

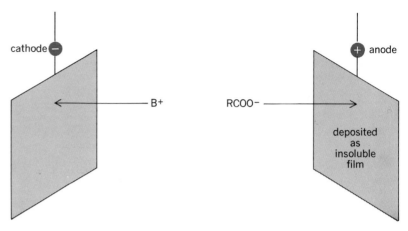

Fig. 1. Schematic diagram for the electrodeposition of certain alkaline anodically depositing resins. Pigmented or unpigmented resins can be used.

lic or other electrically conductive objects in specially formulated water-dispersed paints that are available in practically all colors. The workpiece acts as one of the electrodes in a direct-current field, and the metal tank or metal plates are the counterelectrodes.

The process underlying the electrodeposition of paint is traced back to the deposition of clay (1809), of aluminum powder (Elphal process), and of waxes, asphalts, and rubber (1920s) in aqueous or organic solvent baths and is referred to as electrophoretic deposition, electropainting, and so on. The recent successful adoption of paint deposition is due to the development of the method of applying water-dispersed, synthetic, film-forming macroions. This technique is referred to most frequently as electrocoating.

Advantages. The increased use of electrocoating is a consequence of the extremely even thickness of the deposited film (usually 0.5 to 1.0 mil). This film extends into such recesses as those in automobile doors, joints, and threads, and results in paint and cost savings and quality improvement, virtual absence of fire hazard and pollution from organic solvents, highest uniformity of product, and substantial cost savings.

Basic principles. Most types of film-forming resins (R), such as acrylics and epoxies, can be formulated to contain ionizable groups such as carboxylic acid (—COOH) and amine groups (—NH_2). The solubilization or dispersibility can be symbolized as Eq. (1) or Eq. (2).

$$\begin{array}{ccc} RCOOH & + & BOH & = \\ \text{Acid resin} & & \text{Base} \\ \text{(water-insoluble)} & & \text{(water-soluble)} \end{array}$$

$$RCOO^- \quad + B^+ + H_2O \quad (1)$$
Film-forming anion
(water-dispersible)

$$\begin{array}{ccc} RNH_2 & + & HA & = \\ \text{Basic resin} & & \text{Acid} \\ & & \text{(water-soluble)} \end{array}$$

$$RNH_3^+ \quad + A^- \quad (2)$$
Film-forming
cation
(water-dispersible)

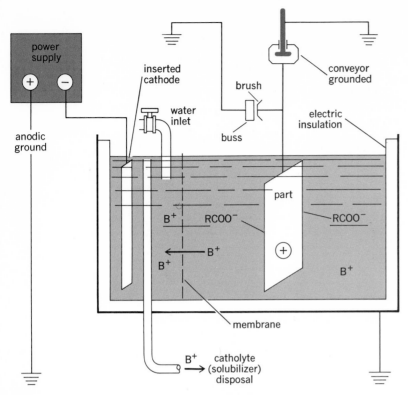

Fig. 2. Solubilizer removal by electrodialysis (completely solubilized feed).

usually provided by alkali, ammonia, or organic amines. Whereas the pigment and film former leave the coating bath as electrodeposits on the workpiece, counterions have to be removed from the bath by vapor phase removal, ion exchange, or electrodialysis. The last process is shown schematically in Fig. 2. Another principle of bath maintenance is the reuse of the counterion for the solubilization of feed (Fig. 3). In any case, the bath is maintained in operating condition whenever all bath components which leave the tank are quantitatively replaced.

Power sources. Rectifiers delivering direct current of less than 10% ripple factor are usually necessary. Various output voltage controls, such as tap switches, induction regulators, and saturable core reactors, are used. Power sources for production installation are equipped with automatic devices to hold a preset voltage, and many are also equipped with maximum amperage settings. In the latter case, at the start of the coating operation a predetermined constant amperage is drawn at comparatively low voltage. As the first deposited coat builds up electrical resistance, the amperage stays constant and the voltage rises, until, after about 10 sec, the preset maximum voltage is reached. The coating process is then completed under constant voltage. The current requirement is usually less than 3 amp/ft². Coating voltages from 50 to 500 volts are usually applied.

Tank and hanging device. From $2\frac{1}{2}$ to 50 gal of bath is provided per square foot of merchandise entering per minute. Electrically grounded steel tanks are widely used as cathodes, with the workpieces hung on a carrier (hook), a section of which is insulated from ground (Fig. 3) and makes contact with a bus bar over the tank. An alternative design uses tanks made of, or lined with, electrically insulating material. Electrodes in properly chosen size and spacing (Fig. 2) provide one terminal of the field, and the grounded hanging device forms the other terminal. In many such installations, however, the hooks are equipped with a contactor and a bus bar to ensure ground connection.

Agitation. Pumps, draft tubes, eductor nozzles, and so forth move the entire tank volume from 2 to 10 times per hour.

Cooling. Practically all the supplied electric energy is converted to heat and removed by devices such as heat exchangers.

Bath composition. The concentration of the paint solids for anodic deposition is usually between 7 and 15 wt %, at a pH from 7 to 9. The concentration of the solubilizer ranges from 50 to 100 milliequivalents of base per 100 g of nonvolatiles. Thus a 10 wt % bath has a concentration of base of about 0.05 to 0.1 N.

Processing and testing. To compete with conventional paints, workpieces are given the customary pretreatment, such as cleaning and phosphating.

As the workpieces emerge from the electrocoating tank, bath droplets are rinsed with water, a procedure which is, however, omitted in some installations. An air blowoff is provided, particularly when a spray paint coat is to be immediately applied. A bake cycle follows, ranging from 5 to 25 min at 250–400°F air temperature.

The chemistry of anodic deposition is schematically represented in Fig. 1. The resin (R) is chosen to give well-adhering deposits which cure into durable paint films. Counterions (A⁻ or B⁺) are chosen to form water-soluble products at the electrode. For cathodic deposition of paint, for instance, acetate ions are used; these ions reconstitute in the vicinity of the anode to give acetic acid. Formate ions, which are also sometimes used, leave the anode as carbon dioxide. Counterions for the more widely practiced anodic deposition are

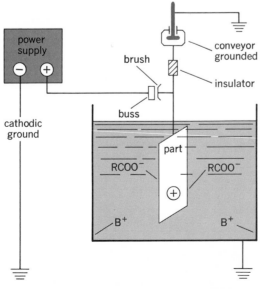

Fig. 3. Solubilizer reuse (solubilizer deficient feed).

Tests for electrocoating bath maintenance and performance have been developed. The finished pieces, when subjected to the customary test procedures, usually meet or exceed conventional specifications.

For background information *see* ELECTROCHEMISTRY; ELECTROLYSIS; ELECTROPLATING OF METALS in the McGraw-Hill Encyclopedia of Science and Technology.   [GEORGE E. F. BREWER]

Bibliography: H. R. Chandler, *Advances in Electrophoretic Painting*, 1969; M. W. Ranney, *Electrodeposition and Radiation Curing*, 1970; R. L. Yeates, *Electropainting*, 1966.

Electron paramagnetic resonance (EPR)

Electron paramagnetic resonance has become, since its discovery about 25 years ago, an important research technique in the laboratory. Supplemented by additional sophistications, such as electron nuclear double resonance (ENDOR) and other dual-resonance techniques, it represents an invaluable tool in the study of transition metal ions, radiation effects, chemical kinetics of molecular fragments produced by radiolysis and photolysis, compounds of biological interest such as metalloproteins and deoxyribonucleic acid (DNA), and various crucial problems in which the validity of both the theory and the computational method is tested.

Transition metal ions. In transition ions the nature of the isotropic hyperfine interaction has received particular attention. Theoretical Hartree-Fock calculations of this term, which is believed to be almost entirely due to polarization of the inner s electrons (core polarization), have been made for the three series of transition metals. The magnetic fields determined at the nucleus per unpaired spin, 130, 375, and 900 kilogauss, respectively, seem to be in good agreement with experimental findings. A. J. Freeman and collaborators performed these studies. Values for the lanthanide ions have also been calculated, comparing well with available experimental data. At the same time, restricted Hartree-Fock parameters of many atoms have been published by C. Froese-Fischer, so that the spin Hamiltonian of paramagnetic resonance can be interpreted with reasonable approximation and the unpaired electron wave function determined with less ambiguity. In general, more powerful theoretical approaches have been applied to specific cases using the molecular orbital, configuration interaction, and crystal field methods. More ambitious efforts to link covalent bonding with magnetic properties have also been made, although the problem is far from being completely solved. Studies by J. Owen and J. H. M. Thornley and by B. R. McGarvey are examples of these efforts. In the years ahead chemists may expect that more elaborate theories will shed additional light on this matter. Paramagnetic resonance has been observed and successfully interpreted in many rare-earth compounds. As expected, it has been found that the effect of the crystal field in removing the orbital degeneracy is significantly smaller than that found in the case of the transition ions. A number of studies of single crystals of fluorite doped with ions of the lanthanide series have been made. This field is particularly attractive because of the relatively simple crystal field, and will certainly be more deeply explored in the future. The experimental results on actinide ions have been interpreted by C. A. Hutchison in terms of the interaction of the crystal field and the spin-orbit coupling with the $5f$ electron levels in a manner similar to that of the rare earths. In some cases substantial agreement of theoretical interpretation and experimental results has provided reliable information about the electronic configurations of these elements.

Radiation effects. There is a vast body of work in the field of inorganic and organic radicals and radical ions in single crystals as well as in amorphous and polycrystalline matrices, and an atlas of paramagnetic resonance spectra has been published by B. H. J. Bielski and J. M. Gebicki to facilitate identification. The value of EPR as a routine analytical tool, however, is still restricted since most radicals exhibit resonance absorption in the same region of the spectrum and a complete analysis is in most cases very difficult. Identification of radiation-produced paramagnetic centers in known substances has become, on the other hand, a standard practice, especially in the case of single crystals. Correlation of EPR and optical absorption data has led to important information covering both primary and secondary effects of irradiation, and represents a major research tool in the field.

Chemical kinetics. With the improvement of experimental techniques, short-lived paramagnetic species are attracting the interest of experimentalists. Thus special procedures are being developed which consist of production of paramagnetic species by pulse radiolysis, storage, and appropriate retrieval of experimental information with an improvement of several orders of magnitude in the signal-to-noise ratio. The application of these special procedures to triplet states such as that of naphthalene-h_8 have permitted, among other things, quantitative determination of the influence of isotopic substitution in their lifetimes. In another series of experiments involving the determination of the rate constant of the energy-transfer process from phenanthrene-d_{10} to naphthalene-d_8 in mixtures of methylcyclohexane and cyclopentane at low temperature, a correlation between the rate of transfer and the viscosity of the medium was found by B. Smaller and collaborators. With appropriate instrumentation, it has been possible to study lifetimes as short as a few microseconds. These studies have proved valuable in the understanding of kinetics of decay and recombination of short-lived species, in particular, the hydrogen atom and the hydrated electron in aqueous solutions.

Biological subjects. In systems of biological interest, after pioneering work on irradiated amino acids by W. Gordy, work has been done in the detection of radiation-induced paramagnetic centers in proteins. EPR has also been successfully used by W. E. Blumberg and collaborators in the study of the biochemistry of metalloproteins. This technique has served the dual purpose of leading to information about certain physical and chemical properties of the metal sites and of partially elucidating the role played by the various paramagnetic ions in the function of these proteins. The studies have covered vanadium, titanium, manganese, cobalt, iron, copper, and molybdenum in com-

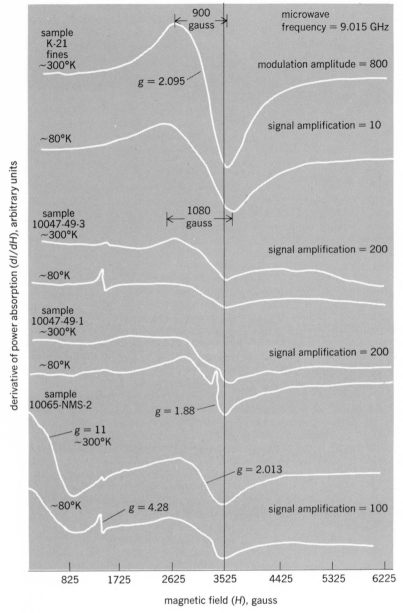

derivative of power absorption (*dI/dH*), arbitrary units

magnetic field (*H*), gauss

Effect of temperature on resonance spectra of several lunar samples. (*From R. A. Weeks et al., Magnetic resonance properties of lunar material, Science, 167:704–707, 1970*)

operators whose coefficients may be regarded as operational parameters. Results have been obtained for transition ions and free radicals in single crystals and in amorphous and polycrystalline samples. This method proved particularly useful for the study of spectra that undergo striking changes as the conditions are modified. As shown by R. Livingston and collaborators, one example is acetoin produced by photolysis of diacetyl in isopropyl alcohol in acid media, which results from an equilibrium between four species whose relative concentration can be changed by adjusting the pH of the solution.

Paramagnetic resonance of lunar fines and rocks collected during the *Apollo 11* mission was observed by R. A. Weeks and collaborators at 9 and 35 GHz and at several temperatures. A sextet of equally spaced lines was detected in several samples and assigned to Mn^{2+}. The width and shape of these lines were similar to those of Mn^{2+} in amorphous compounds such as tektites. The lunar samples exhibited a broad absorption at a Zeeman factor $g = 2$, with a line width of about 950 gauss which was attributed to Fe^{3+} with strong interaction. An absorption at $g = 1.89$ was tentatively assigned to Ti^{3+} on the basis of its thermal behavior. Additional lines were also observed. The figure illustrates some typical spectra. *See* Moon.

This review has covered only a random sampling of results and applications of EPR. Many important contributions have been omitted for lack of space, and the future role of EPR has not been duly evaluated. There is no doubt, however, about the expanding horizons of EPR in physics and chemistry, and any educated prediction, bold as it may sound, may well fall behind reality in the years ahead.

For background information *see* Crystal field theory; Electron paramagnetic resonance (EPR) spectroscopy; Molecular orbital theory in the McGraw-Hill Encyclopedia of Science and Technology.   [JUAN A. MC MILLAN]

Bibliography: P. B. Ayscough, *Electron Spin Resonance in Chemistry*, 1967; B. H. J. Bielski and J. M. Gebicki, *Atlas of Electron Spin Resonance Spectra*, 1967; R. A. Weeks et al., Magnetic resonance properties of some lunar material, *Science*, 167:704–707, 1970; Teh Fu Yen (ed.), *Electron Spin Resonance of Metal Complexes*, 1969.

Electronic music synthesis

Much interest is currently being generated by the subject of electronic music synthesis. Electronic music synthesis is both a means and a process. By means of the synthesizer itself, a series of notes representing a part of a musical composition can be recorded as a separate entity. All the entities can then be combined to form the entire recorded musical composition. Since this combining process cannot be done in real time, the electronic synthesis of music is not a performing type of rendition. Electronic music synthesizers may be classed as manual, programmed, or computer-actuated systems. The way music is written for these synthesizers and the nature of the musical renditions are described in the following sections.

Properties of a musical tone. For purposes of discussing the current interest in the formal synthesis of music by the three types of synthesizers,

pounds of biological interest. Among them, iron, copper, and molybdenum have received particular attention. Quantum-mechanical descriptions have been obtained in the case of copper because of a strong axial *g* anisotropy that permitted a full interpretation of the spectra in amorphous and polycrystalline material, and in the case of iron in single crystals of methemoglobin. Important information has been obtained about biological mechanisms involving the cytochrome *c* oxidase. EPR studies by R. G. Shulman and collaborators of ultraviolet-induced triplet states in DNA have led to the conclusion that the triplet state is localized in the thymine residue.

Special studies. Computer programs have been developed by J. H. Mackey and collaborators to synthesize EPR spectra from a parametric spin Hamiltonian, which is a linear combination of spin

the physical properties of a musical tone should be mentioned. These properties are represented in Fig. 1 and defined below.

1. Frequency is the number of sound waves occurring per unit of time (the second).

2. Intensity (or loudness) of a sound wave is the sound pressure, usually expressed in decibels above 0.0002 microbars.

3. Growth is the time required for a sound to build up to some fraction of the ultimate value.

4. The time required for sound to fall to some fraction of the original value is termed decay.

5. Duration is the length of time that a sound persists without interruption or discontinuity in the output.

6. The length of time during which there is no change in the intensity is called the steady state of a sound.

7. Portamento is a uniform glide in frequency from a sound of one frequency to a sound of another frequency.

8. A complex sound wave is made up of the fundamental and the overtones. The timbre of a tone is expressed in the number, intensity, and phase relations of the components, that is, the fundamental and overtones.

9. Vibrato is modulation in the low frequency of a musical tone. The modulation may be either of, or a combination of, the types known as frequency or amplitude modulation.

10. Deviation is the term applied to a departure from the regular and is one of the characteristics of the beauty and artistry of some types of music.

In all electronic music synthesizer systems, a single series of tones is generated at a time. As a result, all of the characteristics of a tone depicted in Fig. 1 can be applied to each tone in a series of tones. This is, of course, impossible in the case of conventional musical instruments or in the case of the conventional all-electronic musical instruments because, in general, more than one tone is generated at a time. As a matter of fact, the entire composition is produced in one run. Thus electronic music synthesizers provide the most powerful system for the generation of musical sound and the resultant production of recorded music yet invented.

Recording. When a single series of tones is produced by a manual electronic music synthesizer, by a programmed electronic music synthesizer, or by a computer, the recording is usually made with a synchronized multichannel magnetic-tape recorder. In most cases, the magnetic tape is in the form of 35-mm magnetically coated film with sprocket holes. In this way, the magnetic-tape recorder can be synchronized with any picture or program. Some form of synchronized multichannel recorder is employed in all types of electronic music synthesis.

Seven different audio inputs representing seven series of tones are recorded on seven separate tracks (Fig. 2). This magnetic tape is reproduced, and the audio signals in the seven channels are mixed and recorded on a single track of a seven-track recorder (Fig. 3a). When the seven tracks of this recorder are filled with seven different operations, the recorded tape is reproduced and combined to a single track representing 49 tone sequences (Fig. 3b). If the recorder in Fig. 3b is a

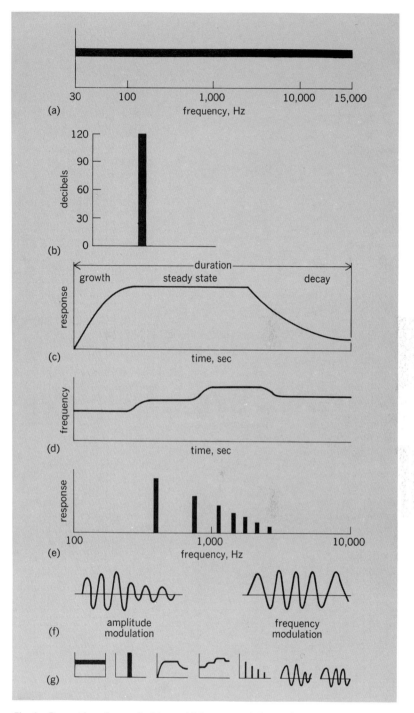

Fig. 1. Properties of a musical tone. (a) Frequency (pitch). (b) Intensity (loudness). (c) Growth, steady state, decay, and duration. (d) Portamento. (e) Timbre. (f) Vibrato. (g) Deviations.

seven-channel recorder and all the tracks are filled, the total final result represents 343 tone sequences. These can be combined to a single track again, and so on.

Manual synthesizer. The manually operated electronic music synthesizer (Fig. 4) is based upon the generation of the properties of a tone to produce the desired final musical tone. In the synthesizer of Fig. 4, only one series of tones is produced at a time. Each series of tones is recorded with a recorder of the type shown in Fig. 2. The various

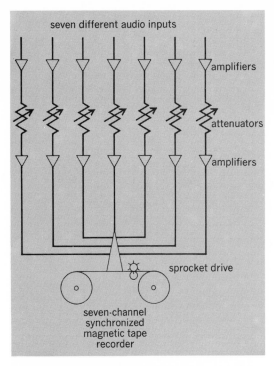

Fig. 2. Schematic diagram of a synchronized seven-channel recorder for recording seven different and separate audio inputs or tone sequences on seven tracks of the magnetic tape.

different series of tones are combined as depicted in Figs. 2 and 3 and as described above. The synthesizer in Fig. 4 can produce a tremendous range of tones with all manner of variation in each series of tones; this cannot be done with the electronic organ. For example, the performer changes the volume, timbre, and some of the other characteristics with one hand, while he plays on the keyboard with the other hand. Therefore great physical dexterity and considerable musical talents are required to produce acceptable music by means of the electronic music synthesizer of Fig. 4. The main reason for this is that each series of tones must possess distinct musical qualities in order to produce musical effects beyond the conventional. In addition, there must be a perfect synchronism of the different series of tones which is also carried out subjectively, that is, "by ear."

Programmed synthesizer. The use of a programmed electronic music synthesizer opens up an entirely new field for the production of recorded music (Fig. 5). To synthesize or produce any musical tone whatsoever, the programmed electronic music synthesizer must provide the facilities for producing all the characteristics of a tone represented in Fig. 1.

The program for each series of musical tones is recorded on the coded paper record, which controls all the functions of the electronic music synthesizer. The information is recorded and stored in the paper record in the form of black ink lines.

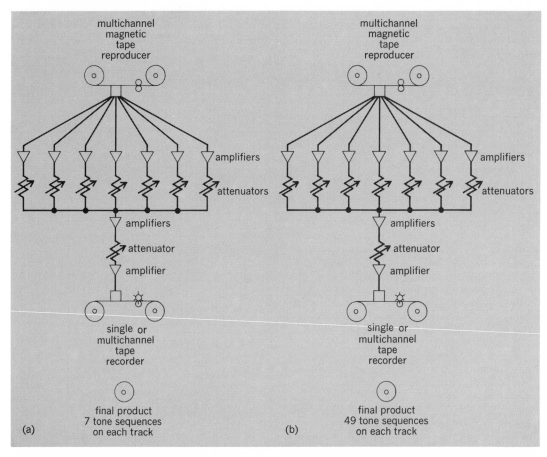

Fig. 3. Combining audio inputs or tone sequences on one track. (a) Combination of the seven recorded in Fig. 2 on the track. (b) Seven tracks of a are combined to one track of the seven-track recorder.

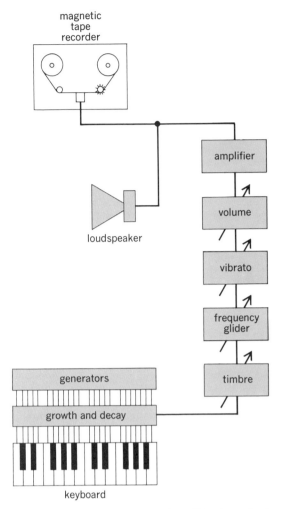

Fig. 4. Diagrammatic representation of the elements of a manual electronic music synthesizer.

When the paper is run through the machine, the sensors scan the paper record.

The paper record of Fig. 6 shows that a binary code system is used. With the use of this system, 2 different functions can be obtained from 1 track, 4 from 2 tracks, 8 from 3 tracks, 16 from 4 tracks, and so forth.

The preparation of the record of Fig. 6 is simple. The musician simply marks the paper with a wide ink pen to obtain the type of sound he desires. The marking of the paper can be done anywhere. After the record is completed, the musician runs the record through the synthesizer and listens to the series of tones by means of the loudspeaker. If he is not satisfied, he can make changes by adding new lines or by marking out existing lines with a white marker. In general, no changes are required for the frequencies because they are taken directly from the music. The length of the line is determined from the note. Therefore, except for errors, the frequency markings are straightforward. As mentioned, changes can be made in the other functions. After the record is satisfactory, the paper record is recorded on magnetic tape as shown in Fig. 2. The recorder is the multichannel recorder of Figs. 2 and 3. This process is carried out for each series of tones until all of the series of tones of the composition have been completed. The

method of combining the series of tones has been described earlier and shown in Figs. 2 and 3.

Entirely new tone complexes and combinations which cannot be achieved with conventional instruments are possible with the programmed synthesizer. With conventional instruments, the musician is limited to the use of lips, mouth, fingers, hands, and feet to perform the different functions. In the programmed electronic music synthesizer, there are no such limitations. In addition, in the case of the manual operation, there is the problem of synchronizing the various series of tones. However, in the programmed electronic music synthesizer, as is evident from the direct drive between the paper record and the magnetic tape recorder, there is no problem of synchronizing the series of tones.

Digital computer. Music may be synthesized by means of a digital computer. The key process involves a digital-to-acoustic converter. A schematic block diagram of the conversion process is shown in Fig. 7. A magnetic tape is prepared by the computer, on which is recorded successive digitized samples of the acoustic output. The numbers are then converted to pulses, the amplitudes of which are proportional to the numbers. The pul-

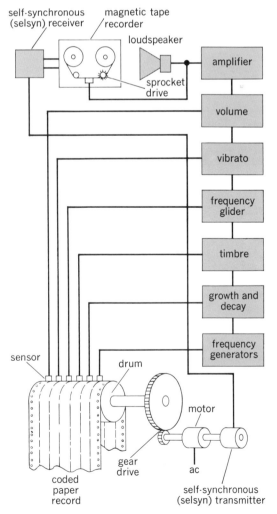

Fig. 5. Diagrammatic representation of the elements of a programmed electronic music synthesizer.

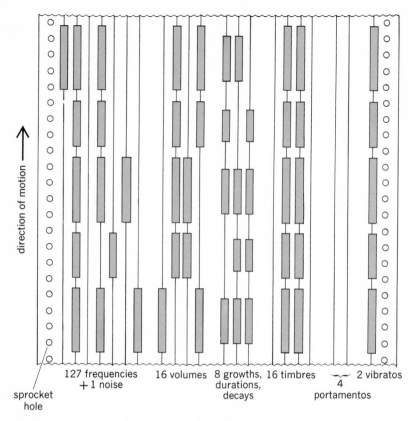

127 frequencies   16 volumes   8 growths,   16 timbres   2 vibratos
+1 noise   durations,   4
                                   decays   portamentos

sprocket
hole

Fig. 6. Coded paper record for programmed electronic music synthesizer.

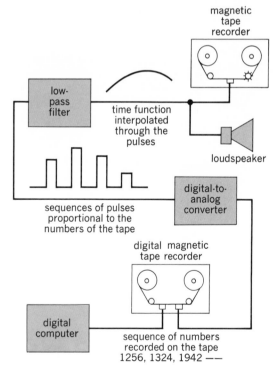

Fig. 7. Digital-to-acoustical converter.

ses are smoothed by a low-pass filter to obtain the input for an audio loudspeaker. If the sampling rate is 20,000 per second, the top frequency from the loudspeaker will be 10,000 Hz. Each sample is produced from a four-decimal number. The signal-

to-noise ratio will be of the order of 60 dB. Within the limits of the frequency range and the signal-to-noise ratio, the converter can produce any sound whatsoever, provided the appropriate sequence of digital samples can be generated.

The basic form of the generating program is a scheme for producing a sequence of sounds representing individual "instruments." The "instruments" are formed by combining a set of basic building blocks termed unit generators. Appropriate combinations of these unit generators can produce sounds of almost any desired complexity.

The first step in the production of a musical selection is to punch a set of cards which specify the "instruments" of the "orchestra." The "instruments" and the computing program are fed into the computer, which punches a card deck that is the music-generating program or "orchestra." Next a series of note cards or score must be prepared. These provide the properties of a tone as shown in Fig. 1.

As in the case of the programmed synthesizer, the process is simplified if the composer listens to one series of notes at a time, makes suitable modifications, and then groups the combination of tones after each one has been approved. Minor modifications can then be made in the group if necessary. The series of tones are then recorded as depicted in Figs. 2 and 3.

The exposition above has described one process for the electronic synthesis of music employing a digital computer. As the programming for computers is simplified, the general procedure will be simplified. However, the composer operating the synthesizer, regardless of the method employed for the electronic synthesis of music, is the final judge of the rendition. The analog and digital synthesizers can produce any tone, regardless of whether it has ever been produced before. Therefore the method which will be selected is the one that provides the composer with the easiest means for translating what he has in mind into the final sound product.

For background information *see* MUSICAL INSTRUMENTS in the McGraw-Hill Encyclopedia of Science and Technology.   [HARRY F. OLSON]

Bibliography: M. V. Mathews, *The Technology of Computer Music*, 1969; H. F. Olson, *Music, Physics, and Engineering*, 1967.

Element 105

Element 105 was synthesized and identified unambiguously for the first time in March, 1970, at the heavy ion linear accelerator (HILAC) in the Lawrence Radiation Laboratory, Berkeley, Calif. The team that made the discovery consisted of A. Ghiorso, M. J. Nurmia, J. A. Harris, K. A. Y. Eskola, and P. Eskola. They suggested that the new element be called hahnium, symbol Ha, in honor of Otto Hahn, the discoverer of nuclear fission.

The hahnium isotope had a half-life of 1.6 sec and decayed by emitting alpha particles with energies 9.06 (55%), 9.10 (25%), and 9.14 (20%) Mev. It was shown to be of mass 260 by identifying lawrencium-256 as its daughter by two different methods.

Characteristic alpha emissions. The method used to produce the new alpha-particle activity of hahnium was similar to the technique of identifying the alpha-particle-emitting isotopes of element

104 (suggested name rutherfordium), except that the new process was considerably more elaborate and sophisticated. The hahnium nuclide was formed by bombarding a 60-microgram (μg) target of californium-249 with 84-Mev nitrogen-15 ions. The transmutation recoil atoms which were knocked out of the 300-μg/cm² target by the intense nitrogen beam (approximately 4×10^{12} ions/sec) were stopped in helium gas at a pressure of 600 torr. They were then carried by the helium stream through a small orifice into a rough vacuum, where they impinged upon the periphery of a 45-cm-diameter magnesium wheel. Most of the transmutation products were adsorbed in a small area on the surface. The wheel was rotated periodically to place these collection spots next to solid-state detectors arranged at high geometry so that the alpha-particle radiations could be examined with good resolution.

Since the rate of production of the element 105 atoms was very low (only about 10 atoms per hour were detected), it was necessary to make many measurements simultaneously. To this end, seven detecting stations were used, and four surface-barrier alpha-particle detectors were employed at each station. As shown in the illustration, these four detectors were arranged in pairs, two crystals alternately being placed (shuttled quickly) next to the flange of the wheel so that they could record the alpha particles emitted by the "mother" element 105 atoms and two crystals fixed in position so that they would be opposite two similar mother crystals and thus aid in the detection of the "daughter" element 103 (lawrencium-256) atoms. The daughter atoms could have reached the mother detectors only as a result of the recoil energy imparted to them by the alpha-particle decay of their mother atoms. The known daughter activity, 30-sec 8.4-Mev lawrencium-256, was thereupon identified as a unique recoil product and shown to be transferred to the mother crystals with the same half-life as that observed for the complex 9.1-Mev peak. This half-life was determined by the relative number of alpha particles of this energy which were detected at each station.

Further proof. An additional proof that Lr²⁵⁶ was the daughter of Ha²⁶⁰ was afforded by an experiment in which there was shown to be a time correlation between the disappearance of Ha²⁶⁰ and the appearance of Lr²⁵⁶ as measured by certain characteristic alpha emissions. The interval from the time of emission of an approximately 9.1-Mev alpha particle to that of an approximately 8.4-Mev alpha particle was shown to correspond with the expected sequence.

A rough excitation function (relative cross section as ordinate, energy in millions of electron volts as abscissa) for the Cf²⁴⁹ (N¹⁵,4n) Ha²⁶⁰ reaction was found to peak at 84 Mev. In addition, the peak production rate of the 9.1-Mev activity was found to be about 1.5 alpha counts per microamperehour, which corresponds to a cross section of 3×10^{-33} cm². Both of these quantities agree with the values predicted for this reaction.

Soviet experiments. Previous work on element 105 was reported in 1968 by G. N. Flerov and colleagues at Dubna Laboratories in the Soviet Union. They claimed to have discovered two isotopes of element 105 produced by the bombardment of Am²⁴³ by Ne²² ions. The transmutation products

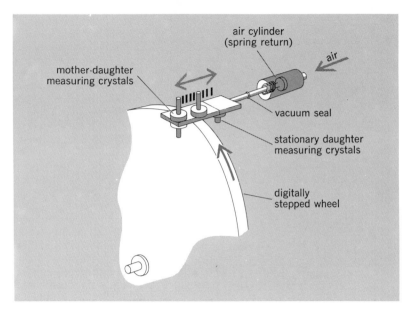
Crystal shuttle configuration.

were carried by a helium gas stream through an annular semiconductor detector which viewed the collecting surface. In the gross spectrum alpha-particle peaks were observed at 8.3, 8.7, 9.0, and 11.6 Mev, which were attributed mostly to a lead impurity in the target. In the high-energy region between 8.8 and 10.3 Mev the Soviet scientists looked for delayed coincidences with alpha particles in the range from 8.35 to 8.6 Mev. They knew that this region was occupied partly by Lr²⁵⁶, and assumed that Lr²⁵⁷ also emitted similar alpha groups with the same half-life. They found "peaks" at 9.4 and 9.7 Mev which seemed to give a statistically valid correlation with the lower-energy alphas and came to the conclusion that they were detecting 105²⁶¹ with $E_\alpha = 9.4 \pm 0.1$ Mev and $0.1 < T_{1/2} < 3$ sec, and 105²⁶⁰ with $E_\alpha = 9.7 \pm 0.1$ Mev and $T_{1/2} > 0.01$ sec. The rate of production of these events was extremely low, with only 10 delayed coincidences observed in 400 μamp-hr of heavy ion bombardment.

The Lawrence Radiation Laboratory work did not confirm the above findings for the following reasons: (1) The isotope 105²⁶⁰ has an energy of only 9.1 Mev; and (2) the daughter of 105²⁶¹ does not have similar decay properties to Lr²⁵⁶, as assumed by the Dubna group, and thus could not give a valid alpha-alpha correlation (the Lawrence group found that Lr²⁵⁷ has an energy of 8.87 Mev and $T_{1/2} = 0.7$ sec).

Early in 1970 the Dubna group reported finding a spontaneous fission activity with $T_{1/2} \cong 2$ sec, also produced by Ne²² bombardment of Am²⁴³, which they think could be due to an isotope of element 105. The cross section to form this nuclide was about 5×10^{-34} cm², and this value is about that which can be calculated for the Ne²²,5n reaction to make this nuclide. However, since the spontaneous fission branching ratio of Ha²⁶⁰ was found by the Lawrence group to be less than 20%, it is very unlikely that the 2-sec activity can be assigned to this isotope. A possible assignment could be to mass 261, if it is due to element 105.

For background information see CRYSTAL COUNTER; ELEMENT 104; NUCLEAR CHEMISTRY in the McGraw-Hill Encyclopedia of Science and Technology.

[ALBERT GHIORSO]

Bibliography: G. N. Flerov, *Proceedings of the International Conference on Nuclear Structure*, Tokyo, 1967; A. Ghiorso, *The Transuranium Elements: The Mendeleev Centennial*, the 13th Robert A. Welch Foundation Conference on Chemical Research, November, 1969, in press; A. Ghiorso et al., *Phys. Rev. Lett.*, 24:1498, 1970; J. Sanada (ed.), *Suppl. J. Phys. Soc. Jap.*, 24:237, 1968.

Elements and nuclides, origin of

The past two have been the most productive years in a decade in man's quest to understand the origin of atomic nuclei. In a series of brilliant papers written in the late 1950s, A. G. W. Cameron, W. A. Fowler, Fred Hoyle, and their many colleagues developed the idea that atomic nuclei have been assembled by nuclear reactions in the interiors of stars. By surveying the systematic properties of nuclei, they discovered a set of thermal environments of various initial compositions that would produce, by thermonuclear reactions, the prominent features of the natural abundance distribution. Recently it has been discovered that the environments yielding the best reproduction of the abundances are not to be found in the slow evolution of stars, during which time there is admittedly an interesting and important sequence of nuclear composition changes, but rather in the dramatic last-second explosions which extensively alter the composition of the star. It now seems virtually certain that all elements and their isotopes with atomic number greater than or equal to $Z = 6$ (carbon) have been synthesized during the explosions of massive stars (roughly 20–40 times the mass of the Sun).

Heavy element production. It has long been realized (since Lord Rutherford) that atomic nuclei cannot always have existed. The fact that radioactive U^{235}, with its moderately short half-life of 700,000,000 years, still constitutes nearly 1% of the much longer lived U^{238} assures scientists that much of the uranium was produced in the few-billion-year period immediately preceding the formation of the solar system because, if the uranium had been produced much earlier, the U^{235} would by now have decayed away.

This line of reasoning has also sharpened significantly in the past 2 years. G. J. Wasserburg, J. C. Huneke, and D. S. Burnett have found a large concentration of neutron-rich isotopes of xenon in the whitlockite grains of the St. Severin chondritic meteorite. They attribute this excess to the fission fragments resulting from the spontaneous fission of Pu^{244}, which requires a Pu^{244} concentration equal to a few percent of U^{238} at the time that the whitlockite cooled to the point of retaining xenon gas.

Inasmuch as the half-life of Pu^{244} is only 82,-000,000 years, this relatively large Pu^{244} concentration seems to require that about 5–10% of the heavy neutron-rich nuclei was synthesized by some events near the solar system roughly 100,-000,000–200,000,000 years before the solar system formed. This short time is only about 1% of the total age of the Milky Way Galaxy, and strongly suggests that explosions of massive stars formed in association with the Sun contaminated the solar-system material with freshly produced nuclei. Unfortunately, it has not yet been experimentally confirmed that the fission of Pu^{244} actually produces neutron-rich isotopes of xenon in the correct relative proportions, so that this exciting development must still be regarded with some caution.

However, the picture is consistent with the one implied by the discovery, by John Reynolds and coworkers, that radioactive I^{129}, with a half-life of only 17,000,000 years, was also present as 1 part in 10^4 of stable I^{127} at the time the meteorites formed. Thus the evidence grows more convincing that the production of new heavy elements has occurred continuously since it began, soon after the metal-deficient globular clusters formed, about 13,-000,000,000 years ago, until immediately before this solar system formed, about 4,700,000,000 years ago. Presumably that nucleosynthesis is still occurring, and D. D. Clayton has recently shown that one may expect to detect the fresh radioactivity by techniques of gamma-ray astronomy.

Explosive burning. The evidence that the elements were synthesized at the moment of explosion rather than in the stable nuclear burning epochs within the life of a star comes from comparing the nuclear abundances produced in the two cases with the actual nuclear abundances observed in nature. The key thermonuclear feature of explosive burning is that the nuclear fuels combust at temperatures considerably higher than the temperatures at which the same fuels burn in a static star in a state of hydrostatic equilibrium. The final abundances are considerably different because of the higher ignition temperature in the explosive case.

The overheating in the explosion may result either from the fact that the advanced fuels, pri-

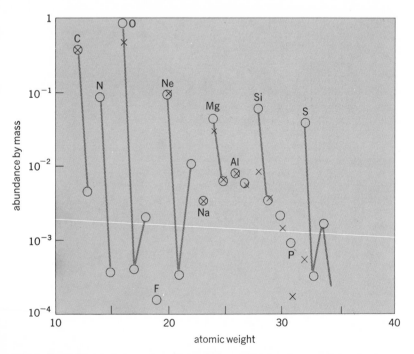

Fig. 1.   Final abundances produced in explosive carbon burning are shown by crosses. For comparison the observed solar-system abundances are shown as open circles, with solid lines connecting all stable isotopes of the same element.

marily C^{12} and O^{16}, first ignite in a degenerate electron gas or, for the noncentral zones, from the compressional heating produced as a strong pressure wave propagates outward from an exploding core. In either case large amounts of thermal energy are liberated in a short time compared to the ability of the star to compensate by hydrodynamic expansion, with the result that the star is given positive energy sufficient to disrupt it.

The first indication of the importance of the dynamics of the expansion on the final nuclear products came in a study of carbon burning. David Arnett and James Truran established a numerical scheme for solving the nuclear reaction network that results when C^{12} nuclei begin to undergo the fusion reactions shown below. A large number of

$$C^{12} + C^{12} \rightarrow Na^{23} + p + 2.238 \text{ Mev}$$
$$\rightarrow Mg^{23} + n - 2.623 \text{ Mev}$$
$$\rightarrow Ne^{20} + \alpha + 4.616 \text{ Mev}$$

coupled reactions become possible as the liberated protons, neutrons, and alpha particles begin to react with all of the nuclear species generated within the gas; in fact, Arnett and Truran used rates for 99 different nuclear reactions in their carbon-burning network. Despite this complexity, the results that they obtained for the final abundances following carbon burning at a constant temperature near $10^9°K$, a value at which carbon would be expected to burn in the core of a star in hydrostatic equilibrium, were very simple. Only Ne^{20}, Na^{23}, and Mg^{24} were produced in substantially significant quantities, and approximately in the ratio of their abundances as observed in the solar system. Although this work confirmed that these three nuclear species almost certainly owe their high natural abundance to carbon burning in stars, the source of the other isotopes in this mass range remains a mystery.

At this time it occurred to Arnett to make an important numerical experiment. He supposed that, for one or a combination of the reasons mentioned earlier, the carbon could be forced to ignite explosively at temperatures near $2 \times 10^9°K$, which is far in excess of temperatures at which carbon naturally burns in a static object, and at initial densities in the range $10^5 < \rho(\text{g-cm}^{-3}) < 10^9$. A typical result of the reaction network in this case is shown in Fig. 1, where the open circles indicate relative solar abundances and the crosses represent the calculated final abundances.

Before the explosion, the gas was taken to be virtually half C^{12} and half O^{16}, as produced in a previous epoch of helium burning, plus 2% O^{18}, which is the result of the earlier conversion within the same star of all of the original carbon, nitrogen, and oxygen nuclei into O^{18} by hydrogen burning and helium burning in turn. In the computation of Fig. 1, the carbon has burned furiously for about one-tenth of a second, at which time reactions are frozen by the falling temperature associated with the vigorous expansion of the gas. Most of the carbon and virtually all of the initial oxygen remain unburned, so that the final ratio C^{12}/Mg^{24} matches the solar ratio. More significantly, the nuclei Ne^{20}, Na^{23}, Mg^{24}, Mg^{25}, Mg^{26}, Al^{27}, Si^{29}, and perhaps Si^{30} and some P^{31} are produced in ratios closely approximating the solar abundances. This dramatic result provides strong evidence that these nuclei

owe their existence to the explosive ejection of partially burned shells of carbon.

If the peak temperature in carbon burning rises as high as $3 \times 10^9°K$, the O^{16} also will ignite explosively, thereby driving the temperature even higher. The primary reaction channels for reactions of $O^{16} + O^{16}$ are $Si^{28} + He^4$, $P^{30} + d$, $P^{31} + p$, and $S^{31} + n$, and the corresponding reaction network builds nuclei at $Ar \geq 28$. A typical result of the reaction network is shown in Fig. 2. Virtually all of the nuclei having $32 \leq Ar \leq 42$ are produced in their observed ratio to Si^{28} in this particular expansion. Only the rare neutron-rich nuclei S^{36} and Ar^{40} are not adequately produced in that mass range; however, these two nuclei are produced by neutron reactions on S, Ar, and Ca seed nuclei present in small amounts in explosive carbon burning. The success of this calculation is so great, it seems that the ideas must be very close to truth.

The mass zones responsible for the nucleosynthesis discussed have not come from extremely dense objects. If the preexplosion density were as great as 10^9 g-cm^{-3}, the temperature would be forced sufficiently high just to remove the electron degeneracy so that C^{12}, O^{16}, and Si^{28} would all be burned into iron-peak nuclei in a time much shorter than the hydrodynamic time scale. To avoid an inconsistent overproduction of iron relative to those nuclei produced in Figs. 1 and 2 seems to

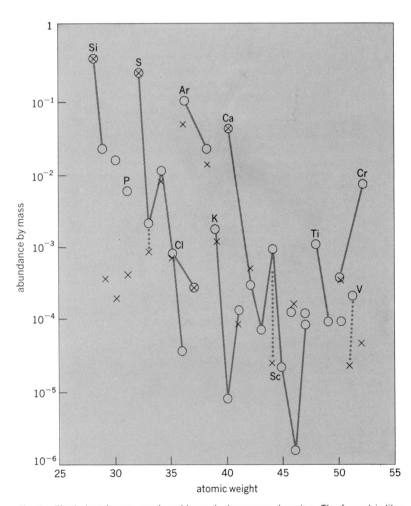

Fig. 2. Final abundances produced in explosive oxygen burning. The format is like that illustrated in Fig. 1.

require explosions in massive stars (20–40 $M_\odot$), for which the densities are lower and the distribution of mass zones most nearly assumes the required relative proportions. It is not clear whether the center can explode, allowing an outgoing pressure wave to ignite overlying shells, or whether implosion of the evolved core, perhaps leading to formation of a neutron-star remnant, must somehow trigger detonation of noncentral shells. The proper dynamic description of the event and the stellar evolution leading up to it remains to be solved quantitatively.

Relative abundance of nuclei. The model of a single event having several concentric mass shells in the differing stages of nuclear burning provides an ingenious solution to an old problem: Why should the relative abundance of nuclei in this mass range remain relatively constant although their absolute concentrations (relative to hydrogen) vary? All that is required is that the relative masses of the explosive carbon-burning, oxygen-burning, silicon-burning, and *e*-process (which is complete silicon burning) shells take on roughly constant proportions in the massive-star explosions responsible for most of the element production. Such a situation may be nothing more than the natural outcome of the physics governing star formation and evolution.

The heavy elements, such as the radioactive species mentioned earlier, are apparently synthesized in the most central regions near the boundary between the collapsed neutron-star remnant and the ejected matter. The calculation of the dynamic details of these mass zones will be very difficult to perform with confidence, although considerable progress has shown that they will indeed be produced in favorable circumstances. This problem will require much more research.

The only nuclei that seem likely not to owe their origins to explosions of stars are the lightest nuclei, H^1, H^2, He^3, He^4, and perhaps Li^7. If this actually is the big-bang universe suggested by the 3°K blackbody radiation apparently pervading it, these nuclei probably are primarily remnants of that original fireball. The truth of that idea will rest on the unresolved question of the He^4 abundance in the oldest stars, although considerable help may come from the He^{3+} microwave line recently detected in HII regions by H. C. Goldwire, R. Predmore, and G. K. Walters. Perhaps one may also hope for the detection of radio lines from molecules containing H^2 nuclei.

For background information *see* CARBON-NITROGEN CYCLE; ELEMENTS AND NUCLIDES, ORIGIN OF; STELLAR EVOLUTION in the McGraw-Hill Encyclopedia of Science and Technology.

[DONALD D. CLAYTON]

Bibliography: W. D. Arnett and D. D. Clayton, Explosive nucleosynthesis in stars, *Nature*, in press; D. D. Clayton, *Phys. Today*, 22:28, 1969; D. D. Clayton, *Principles of Stellar Evolution and Nucleosynthesis*, 1968; W. A. Fowler, New observations and old cosmochronologies, in *George Gamow Memorial Volume*, in press.

Energy sources

Advances have occurred in the past few years in obtaining energy from geothermal sources. World electricity generating capacity from geothermal sources amounted to about 752,000 kilowatts (kw) in 1970; 60% of this was from live steam wells, and the remainder was from hot water from which steam is flashed to supply the geothermal turbine. There are two main problems for geothermal power plants. The first is to select materials which will withstand the corrosion from the geothermal fluids, and the second is the disposal of waste water. Because there is no boiler, the power plant can operate unattended. Condensers are used with the geothermal turbines because the condensing steam is more than adequate to supply the water evaporated in the cooling tower.

The first geothermal power station went into operation at Larderello, Italy, in 1913. By 1970 Italy had a capacity of 362,000 kw. At the same time, the capacity in New Zealand was 192,000 kw; United States, 82,000 kw; Japan, 33,000 kw; Soviet Union, 5000 kw; and Iceland, 2500 kw. In 1970, at Cerro Prieto, Baja California, Mexico, a 75,000-kw plant was under construction for operation in 1971. Other countries which have drilled geothermal wells, but which by 1970 had not installed generating units, are Chile, El Salvador, Guadeloupe, Martinique, Nicaragua, the Philippines, Taiwan, and Turkey.

United States geothermal resources. The development of geothermal resources in the United States has been primarily in California and Nevada. The only commercial generation of electric power in the United States is at The Geysers in Sonoma County, northern California. Geothermal wells have been drilled in 14 other areas in California and Nevada, but by 1970 had not been developed for power.

In the United States geothermal steam has been found only at The Geysers. In the other areas, geothermal energy is in the form of hot water. Steam can be obtained from the hot water by flashing to a lower pressure. An alternative method for extracting the energy from hot water is by heat transfer to a gas which boils at a low temperature. In 1970 Magma Energy Co. was designing such a plant for use at Brady Hot Springs, Nev., using isobutane for powering a turboexpander. This turbine will extract more heat energy than a steam turbine. However, individual circumstances will determine its economic practicality. For water above 400°F a steam turbine is best.

Geothermal power. Steam wells were drilled at The Geysers in a joint venture by Magma Power Co. and Thermal Power Co. in 1955–1957. This steam is used by the Pacific Gas and Electric Co., San Francisco, to supply a 12,000-kw unit placed in commercial operation in September, 1960. As additional wells were proved, unit 2, of 14,000 kw, went into commercial operation in March, 1963; units 3 and 4, each 28,000 kw, were placed in commercial operation in April, 1967, and November, 1968. In June, 1967, Union Oil Co. of California joined Magma-Thermal in developing The Geysers area. In 1970 units 5 and 6, each 55,000 kw capacity, were being constructed, and are scheduled for commercial operation in June and November, 1971. As Union-Magma-Thermal develops additional steam, 400,000 kw additional generating capacity will be installed in 1972–1975.

Steam impurities. Geothermal steam contains rock dust and noncondensable gases. Most of the

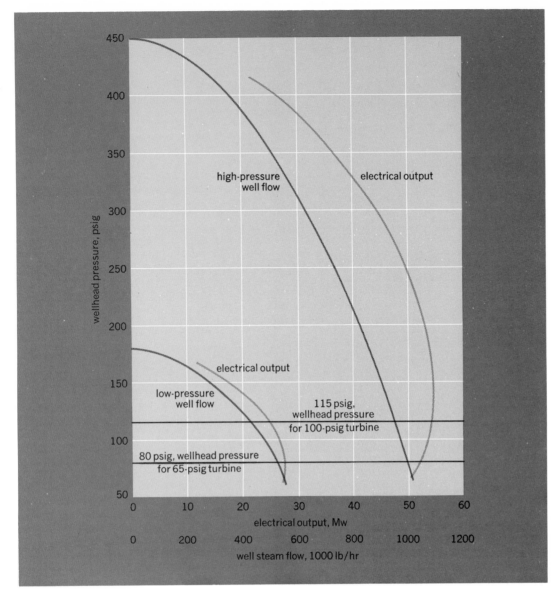

Fig. 1. Relation of well steam flow to wellhead pressure and electrical output. The electrical output curve shows the output possible if the turbine had been designed originally for the pressure and steam flow available at the corresponding wellhead pressure, allowing 15 psi for steam-line pressure drop.

rock dust can be removed by centrifugal dust separators on each well. The remaining fine dust does no harm to the turbine but requires that the control valves and automatic shutoff valves be grease-lubricated to assure that they will not be stuck by the dust.

The noncondensable gas at The Geysers is a small part of the steam, varying from 0.2 to 1.8% by weight for the different wells. As the wells are used, there is a drop in the percentage of gas. The noncondensable gas consists primarily of carbon dioxide. A typical analysis for unit 3 shows that carbon dioxide is 82.5% by weight of the total. The other gases are methane, 6.6%; nitrogen-argon, 1.4%; hydrogen sulfide, 4.5%; and ammonia, 3.8%.

The noncondensable gas not absorbed in the cooling water must be removed from the condensers to maintain a vacuum. Because there is a small percentage of gas at The Geysers, it is more economical to use steam jet gas ejectors than to re-move the gas with vacuum pumps.

The corrosion problem from the noncondensable gas does not add too great an economic burden if the right materials are used initially. Carbon steel can be used for the steam lines, but stainless steel is necessary for condensate lines. Unprotected copper and cadmium plating should not be used, because of hydrogen sulfide corrosion. Turbine materials and design are essentially conventional except that special attention is given to erosion shields and shaft seals, which should be made of stainless steel. Aluminum is better than galvanized steel for switchyard structures. Aluminum also can be used for condensate water pipe and valves where the velocity is too low for erosion to occur. Pumps handling condensate water should be made of stainless steel.

Inlet and exhaust pressure and temperature. The pressure for geothermal steam turbines is 50–100 psig. The optimum pressure is that

Fig. 2. Units 1 and 2 of The Geysers Power Plant of the Pacific Gas and Electric Co., Sonoma County, Calif. Units 3 and 4 are in the center background almost at the skyline. In the middle background is the Thermal Power Co. maintenance shop.

at which the most electric generation is obtained from the steam flow available. This relation is shown in Fig. 1 for live steam fields. For flashed steam from hot water, the turbine inlet steam pressure is 50–75 psig. The temperature of the steam is essentially that of saturated steam. The live steam at The Geysers (Fig. 2) is 355°F at 100 psig and has 17°F superheat. The design exhaust pressure can be 3–4 in. Hg absolute, depending on the cost of the steam supply. The higher exhaust pressure is used with lower steam costs.

Geothermal plant location. A geothermal plant should be located close to the steam wells to reduce the cost of the steam supply lines and to reduce their pressure drop. Because a greater quantity of steam is obtained from the wells when the pressure is lower, it is a disadvantage to use up the available pressure in line drop.

For background information *see* EARTH, HEAT FLOW IN; ENERGY SOURCES; STEAM in the McGraw-Hill Encyclopedia of Science and Technology.   [ALBERT W. BRUCE]

Eye (invertebrate)

During the past several years, significant advances have been made in the understanding of the mechanisms of transduction of light energy into electrical signals in photoreceptor cells. Invertebrate photoreceptors were chosen for many of these studies because of their relatively large size and accessibility, properties which facilitate experimental manipulation.

Depolarizing receptor potentials. The ventral rudimentary eye of the horseshoe crab (*Limulus polyphemus*) has large, accessible photoreceptor cells. R. Millecchia and A. Mauro demonstrated that these photoreceptors maintain a membrane potential of 45–60 mv, inside negative, at rest in the dark. As has been found previously in many other invertebrate eyes, the cells respond to illumination by a change in membrane voltage (the receptor potential), the inside of the cell becoming more positive (depolarization). The receptor potential is multiphasic; there is an initial, larger, transient component which can overshoot zero potential with bright stimuli, and a steady component, keeping the membrane depolarized as long as the illumination is present. Both components of the responses are graded in amplitude with graded changes in stimulus intensity. *See* BIOPOTENTIALS AND ELECTROPHYSIOLOGY.

The mechanism generating the receptor potential was studied by Millecchia and Mauro. They used the voltage-clamp technique to measure the light-induced current (that is, the total membrane current necessary to hold the membrane voltage constant in the dark minus the total membrane current required to keep the membrane at that same voltage in the light). They found that, for the transient component, the plot of light-induced current versus voltage increased monotonically with positive slope for increasingly positive values of voltage. The voltage (that is, the reversal potential) at which illumination induced no change in membrane current was independent of light intensity. Moreover, they found that replacement of the sodium ions by other substances significantly reduced the size of the light response. On the basis of these data, they proposed that light causes an increase in membrane conductance to the sodium ions; consequently, the inflow of positive ions makes the membrane voltage more positive.

Recently, J. E. Brown and coworkers obtained other evidence supporting this hypothesis. The reversal potential for the transient component was determined with the current-clamp technique. Here the reversal potential is that membrane voltage at which illumination elicits no further change in membrane voltage. At voltages more negative than reversal potential, the light response is in the positive direction; at voltages more positive, the light response is in the negative direction (Fig. 1a).

If the transient component arises from a sodium ion conductance change, the reversal potential of the transient component should approximately follow the Nernst relation. In Fig. 1b the reversal potential of the transient component is plotted versus extracellular sodium ion concentration. For sodium ion concentration near normal values, the data can be approximated by the Nernst relation (dashed line). In experiments in which the extracellular concentrations of the other common ions were changed, the reversal potential is relatively unaffected. These experiments thus support the hypothesis that light causes an increase in membrane conductance to sodium ions.

Previously, T. G. Smith, W. K. Stell, J. E. Brown, and colleagues proposed that a change in the activity of an "electrogenic sodium pump" might underlie the steady component of the receptor potential. Recent experiments by J. E. Brown have shown this to be unlikely. In Fig. 2 the cell is seen to be capable of generating a large light response, having both transient and steady components, when held at 2°C. The membrane does depolarize when cooled, and this is interpreted as evidence for the reduction of activity of an electrogenic sodium pump. Hence, the cell is capable of generating large light responses when the activity of the presumed pump must be small.

Another large photoreceptor cell, that of the barnacle (*Balanus eburneus*), has been studied by H. M. Brown, S. Hagiwara, and colleagues. This photoreceptor produces a receptor potential which is similar to that described for the *Limulus* ventral eye photoreceptor. H. M. Brown, Hagiwara, and colleagues performed voltage-clamp experiments on these cells and found that, as in the *Limulus* photoreceptor, the plot of light-induced current increased monotonically, with positive slope, for increasingly positive voltages. They also found that, at a fixed membrane voltage, the inward, light-induced current was linearly related to the extracellular concentration of sodium ions. However, the reversal potential plotted against extracellular sodium ion concentration had a slope half as steep as that which would be predicted by the Nernst relation. From these data they proposed that the light response is generated by an increase in membrane conductance, principally to sodium ions.

Thus, at present, the best general hypothesis for the generation of depolarizing receptor potentials is that the absorption of light leads to an increase in membrane conductance to a specific cation (or cations); this cation flows down its electrochemical gradient across the membrane, producing the depolarizing receptor potential. The magnitude of the conductance increase varies in time, thereby producing the multiphasic appearance of the receptor

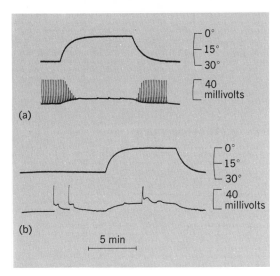

Fig. 2. Receptor potentials recorded from *Limulus* ventral eye photoreceptors in the cold. The upper trace of each pair is a record of bath temperature; the lower record is membrane voltage. (*a*) The cell responded with a receptor potential to repetitive, brief flashes (about 1 per 10 sec). When cooled, the membrane voltage becomes more positive and the amplitude of the light response is markedly reduced. (*b*) Responses to flashes of 30 sec duration. The first two flashes (at about 27°C) elicit large, multiphasic receptor potentials. At about 2°C, the first flash elicits a large receptor potential; the second flash, about 1 min later, elicits only a small, steady depolarization. As in *a*, the membrane voltage becomes more positive in the cold.

potential. The details of the coupling between the absorption of light by a photopigment and the subsequent changes in membrane conductance await further study.

Hyperpolarizing receptor potentials. Both the *Limulus* and barnacle photoreceptor cells, as well as those of most invertebrate photoreceptors, have large regions of closely packed microvilli (the so-called rhabdomere structure) when viewed in thin sections by electron microscopy. A few invertebrate photoreceptors have membrane specializations derived from cilia-like structures. Two such photoreceptors are found in the distal retinas of the mantle eyes of the pelecypod mollusks *Pecten* and *Lima*. The electrophysiological responses of these photoreceptors were recorded first by J. Toyoda and R. Shapley and later by A. Gorman and J. McReynolds in *Pecten*, and by G. Mpitsos in *Lima*. The membrane voltage in these photoreceptors is found to become more negative (to hyperpolarize) upon illumination. This surprising result has also been found in vertebrate photoreceptors which also have cilia-derived structures (fish, amphibian, and reptilian cones); however, unlike the vertebrate hyperpolarizing receptor potential, that of the invertebrates is accompanied by an increase in membrane conductance.

A hyperpolarizing component of the receptor potential has also been recorded from a rhabdomeric-type photoreceptor by J. Nolte and colleagues in the median ocellus of *Limulus*. In this ocellus there are photoreceptors which generate depolarizing receptor potentials to stimuli of near-ultraviolet wavelengths. When blue or green illumination is superimposed on steady ultraviolet illumination, a hyperpolarization is elicited. This

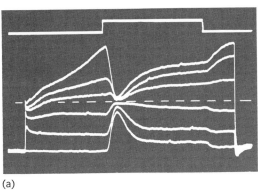

(a)

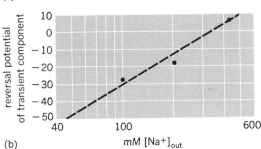

(b)

Fig. 1. *Limulus* ventral eye photoreceptor. (*a*) Measurement of the reversal potential of the light response. (*b*) Reversal potential of the transient component of the receptor potential of a ventral eye photoreceptor as a function of extracellular sodium ion concentration.

hyperpolarization is accompanied by a decrease in membrane conductance.

The detailed mechanisms underlying these hyperpolarizing receptor potentials are not known.

For background information *see* BIOPOTENTIALS AND ELECTROPHYSIOLOGY; NERVOUS SYSTEM (VERTEBRATE); NERVOUS SYSTEM (INVERTEBRATE) in the McGraw-Hill Encyclopedia of Science and Technology.   [JOEL E. BROWN]

Bibliography: H. M. Brown et al., *J. Physiol.*, 208:385, 1970; J. S. McReynolds and A. L. F. Gorman, *J. Gen. Physiol.*, in press; R. Millecchia and A. Mauro, *J. Gen. Physiol.*, 54:331–351; J. Nolte and J. E. Brown, *J. Gen. Physiol.*, 54:636–649, 1969.

Filter, electric

During the past 3 years, research on surface acoustic waves in solids has led to a versatile microminiature technology for electric filtering in the frequency range $10^7 – 10^9$ Hz. Widespread application of these filters seems assured in color television, low-cost solid-state radars, air-traffic control systems, recirculating memories, and numerous forms of frequency- and time-domain signal processing.

Background. Surface acoustic waves, which contain both compressional and shear components in phase quadrature, propagating along and bound to solid surfaces were discovered by Lord Rayleigh in the 1880s. As an example, earthquakes furnish sources for propagating these waves on the Earth's surface. It is of importance for electronic applications that if the solid is a piezoelectric material, the surface acoustic energy is complemented by a small amount of electric energy. This electric energy provides the physical mechanism for the coupling between conventional electromagnetic signals and propagating surface acoustic waves. The coupling is attained by means of transducers having built-in acoustic filtering properties which appear, partially modified, at the electrical terminals giving the filter, to be electric.

Transducers for frequency filters. The basic bandpass filter is shown in Fig. 1. A piezoelectric bar has a polished upper surface on which two transducers, denoted by T, are deposited. The left-hand input transducer is connected, via thermocompression-bonded leads, to the electric source through an electrical matching network. The right-hand output transducer drives the load R_L, usually 50 ohms, through another electrical matching network. R_g and E_g represent the resistance and voltage of the generator. Because these transducers are bidirectional, they lead to filters with at least 6 dB loss even in the pass band. The unwanted acoustic waves are absorbed by black wax terminations at the ends of the piezoelectric bar. At the symmetry plane C-C', a metal baffle serves to isolate electromagnetically the two transducers. This requirement necessitates delay exceeding 100 nanoseconds in the filter.

The transducers originally demonstrated by R. M. White and F. W. Voltmer consist of a set of metal interdigital electrodes, each a few thousand angstroms thick, fed from two bus-bars (Fig. 2a). For this simple transducer arrangement the period p of the interdigital electrode structure is constant and equals one surface acoustic wavelength at the center frequency f_0 of the filter. The width of the metal electrodes is typically $p/4$, being 100 microns (μ) at 10^7 and 1 μ at 10^9 Hz. The 100-μ electrodes are readily fabricated by using techniques standard to the semiconductor integrated-circuit industry of metallization, photoresist, masking, and chemical etching. The 1-μ electrodes require more sophisticated techniques, such as sputter etching or scanning electron microscopy as developed by A. N. Broers. The constant electrode overlap distance w defines the collimated acoustic beam width and is typically 40 wavelengths. The number of periods N defines the total acoustic bandwidth Δf (Fig. 2a).

The acoustic response, both in amplitude and phase, is calculated by regarding the transducer as an end-fire array antenna. The electrical matching networks are arranged to minimize filter loss without overriding the acoustic response. A disadvantage of constant-p, constant-w transducers is that the minimum out-of-band rejection is only 26 dB. A solution lies in amplitude weighting within the transducers, such as varying w according to sin u/u (Fig. 2b), where u equals $2\pi x/Np$ and N is the number of interdigital periods in the main spatial lobe. This sharpens the skirt selectivity at the expense of introducing ripple in the main pass band. Further, the minimum out-of-band rejection is improved to 40 dB. Such filters are finding application in color television.

Both arrangements shown in Fig. 2a and b have a phase response ϕ which is a linear function of frequency. Graded periodicity (Fig. 2c), with constant w, allows phase weighting as typified by a conventional log antenna. Synthesis of a quadratic phase characteristic can result, providing the transducers are deposited so that reflection symmetry exists about plane C-C' (Fig. 1). Reversing the terminals B-B' recovers the linear phase characteristic without affecting the amplitude response. The importance of the graded periodicity technique is that filters of wide bandwidths result at moderate loss.

Currently, surface acoustic wave filters for the frequency domain can be routinely designed with bandwidths between 3 and 50% of the center frequency. For a given bandwidth, filter loss is dependent on the strength of piezoelectric coupling. Piezoceramics are favored at 10^7 Hz. For high-fre-

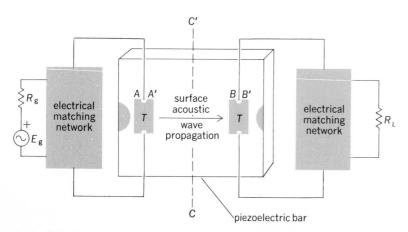

Fig. 1. Basic arrangement of surface acoustic wave bandpass filter.

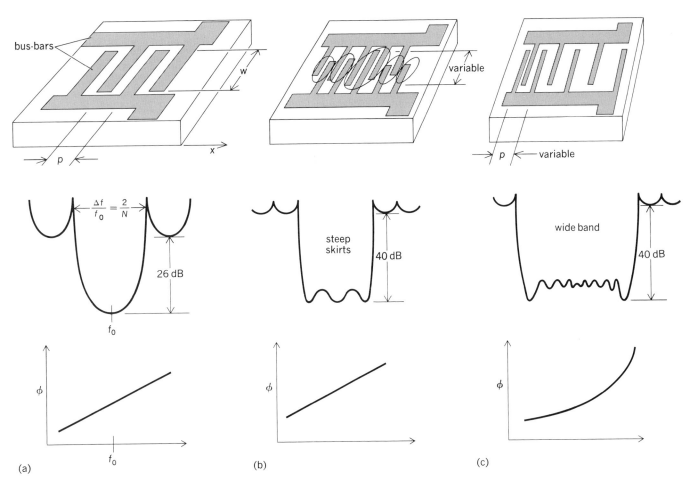

Fig. 2. Transducer configurations and corresponding insertion loss and phase (ϕ) characteristics of electric filter. (a) Interdigital construction. (b) Amplitude weighting technique. (c) Graded periodicity technique.

quency filters, single-crystal materials are dictated from considerations of acoustic propagation loss. Lithium niobate and bismuth germanium oxide are then generally superior to crystalline quartz although more expensive.

Time-domain filters. The goal in time-domain filters is to output an energy pulse of narrow time duration, with minimal spurious response, when the input of the filter is fed with a coded wave train having a substantially longer time duration. The ratio of the time duration of the input wave train to that of the output pulse is defined as the processing gain which is equivalent to the improvement in signal-to-noise ratio. The response of a given filter in the frequency and time domains is related. For example, a filter having a quadratic phase-frequency response (Fig. 2c) exhibits a time delay which is a linear function of frequency. Such time-domain filters find application in high-resolution radar systems where the transmitter radiates energy pulses of duration ΔT which are linearly frequency-modulated over Δf. The time-domain filter is employed in the receiver and has a processing gain of $\Delta T \times \Delta f$. Surface acoustic wave versions of pulse-compression filters are finding engineering application in upgrading the resolution capabilities of existing airborne radars. Processing gains of 100, with spurious response below 40 dB have been realized by W. L. Bongianni and J. E. Dickerman.

Communication systems on a constant carrier frequency f_0 are also enhanced by processing gain. The code of duration MT, containing phase reversals at intervals of T, is propagated down an array of M collinear transducers (Fig. 2a) acting as weak taps, each separated by a distance corresponding to T. The transducer bus-bars are interconnected precisely as the phase reversals within the code. Surface acoustic wave realization with processing gain M has been demonstrated by S. T. Costanza, P. J. Hagon, and L. A. MacNevin. Tapped delay lines of significant versatility are predicted by means of arranging arbitrary interconnections of the M taps utilizing microelectronic switching networks.

Recirculating memories can be based on delay lines possessing the property that a sharp pulse input yields a sharp pulse output. Figure 3 shows three electrode transducers for this purpose constructed by H. J. Whitehouse and G. F. Lindsay. Each element of the code extends over an acoustic wavelength or time duration $1/f_0$. Connection of alternate electrodes to the positive bus-bar and the ground electrode defines the negative code element. Connections to the negative bus-bar and the ground electrode define the positive code element. In this way an arbitrary negative-positive series, or equivalently a binary phase-reversal code, can be established within a single transducer. The output transducer is identical to the input transducer but

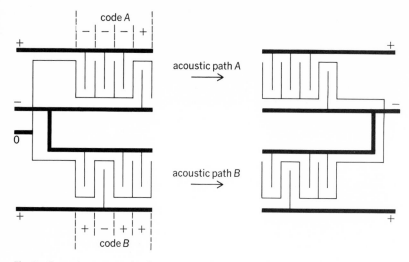

Fig. 3. Transducer array for four-element Golay complementary series.

translated down the piezoelectric substrate. J. M. Speiser recognized that the complementary series of M. J. E. Golay could be encoded by utilizing two parallel acoustic paths A and B, with suitable interconnections, as illustrated for the four-element series in Fig. 3. These special code pairs have the property that an input pulse of one-element length $1/f_0$ gives a tripolar output pulse of duration $3/f_0$ with no time side lobes.

For background information see FILTER, ELECTRIC; PIEZOELECTRIC CRYSTAL; PIEZOELECTRICITY; TRANSDUCER in the McGraw-Hill Encyclopedia of Science and Technology.

[J. H. COLLINS]

Bibliography: W. L. Bongianni and J. E. Dickerman, G-MTT International Microwave Symposium, Newport Beach, Calif., May, 1970; J. H. Collins and P. J. Hagon, Electronics, 42:94–103, Nov. 10, 1969, and 42:102–111, Dec. 8, 1969, and 43:110–122, Jan. 19, 1970; MTT-17, IEEE Trans. Microwave Theory Tech., Special Issue on Microwave Acoustics, November, 1969; Microwave J., March, 1970.

Galaxy, infrared

The nuclei of galaxies and quasars are observed to emit enormous amounts of infrared radiation, in some cases more than 1000 times the output of the entire Milky Way Galaxy at all wavelengths. The energy released is so great that the existence of an entirely new energy source is implied, and the radiative mechanism must be extraordinarily efficient. Attempts to explain these observational requirements have led to views of how galaxies are formed and how they evolve which differ radically from the widely accepted views based only on earlier information.

It is possible that the infrared galaxy phenomenon, now thought to exist in the nuclei of all galaxies, may not be understandable in terms of present physical laws.

The bright quasar 3C273 was the first extragalactic source detected in the infrared. Most of the enormous output of radiation from this distant object is emitted in the infrared. In early 1970, 12 galaxies, including the Milky Way, had been shown to possess powerful sources of infrared radiation in their nuclei.

Characteristic spectrum. Within the uncertainties of the present data, the spectrum, which rises sharply to a peak at about 70 microns (μ), is the same for all galaxies, suggesting a common physical mechanism. Most of the data were obtained with ground-based infrared telescopes. These observations are restricted to the wavelength range $1-25 \mu$ with filters matched to the wavelengths of maximum atmospheric transparency. Beyond 25 μ the atmosphere is rendered opaque by the rotational and vibrational absorption bands of water vapor, and the next useful window, at 1000 μ, is well beyond the peak at which most of the power is radiated. To bridge this gap in the infrared spectrum, F. J. Low and collaborators at Rice University have built an infrared telescope which is operated at altitudes above 15 km in a Lear jet aircraft. The observations have been carried out at the NASA Ames Research Laboratory. In the stratosphere the attenuation at 100 μ is nearly negligible, and accurate relative flux values have been obtained for a number of bright sources: the planets Mars, Jupiter, and Saturn; the infrared nebula in Orion; and, most significantly, the nuclei of the Milky Way Galaxy and of NGC 1068.

Center of the Milky Way Galaxy. E. Becklin and G. Neugebauer reported the discovery of infrared radiation from the direction of the galactic center. Thus it appears that the same physical mechanism is at work in the nuclei of all galaxies. The measured power output from the infrared nucleus of the Milky Way Galaxy is 10^8 times the solar output, or about 0.01 the luminosity of the galaxy as a whole.

Variability and source diameter. Only in the case of the Milky Way Galaxy can the diameter of the infrared source be measured directly. But rapid variations in the infrared output of several galaxies indicate that the dimensions of the regions emitting the enormous infrared fluxes are of similar sizes in all galaxies, no larger than about 1 light-year in radius. The largest variations are seen in the most luminous sources.

Energy requirement. Luminosities as high as 3×10^{46} erg/sec are observed for galaxies whose distances and nuclear masses are well determined (the distance and mass of quasars such as 3C273 are still under debate). All galaxies probably go through the infrared stage of evolution, where they emit at this high level. If this occurs during only 1% of the galaxies' lifetime of 10^{10} years, a serious problem arises: Almost all the mass in the nucleus is lost in the form of far-infrared radiation. The infrared radiative mechanism must be extremely efficient; otherwise the problem cannot be solved by conventional physical means. It is for this reason that speculation has arisen concerning spontaneous creation of matter in the nuclei of galaxies. In 1932 Sir James Jeans postulated such a mechanism to explain the spiral form of galaxies which appear to grow out of their nuclei. The problem of spontaneously creating all the mass of a galaxy, 10^{11} suns, in the nucleus is comparable to the problem of producing the infrared radiation which is now seen to be emitted by the nuclei of galaxies.

Irtrons. To satisfy theoretical and observational

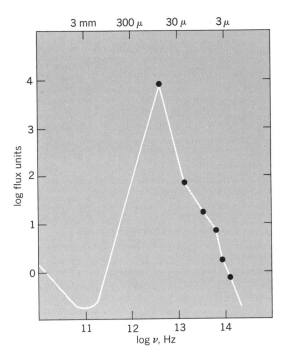

Characteristic spectrum of an infrared galaxy.

constraints imposed by the shape of the spectrum shown in the figure, the size of the emitting region in the galactic center must be made much smaller than the observed diameter. Thus it is necessary to break up the source into many smaller sources. These sources must be nearly identical in all their physical properties, and have been named irtrons because of their characteristic infrared spectrum.

For background information *see* GALAXY; GALAXY, EXTERNAL; QUASARS in the McGraw-Hill Encyclopedia of Science and Technology.

[FRANK J. LOW]

Bibliography: H. H. Aumann and F. J. Low, *Astrophys. J. Lett.*, 159:L159, 1970; E. E. Becklin and G. Neugebauer, *Astrophys. J. Lett.*, 157:L31, 1969; D. E. Kleinmann and F. J. Low, *Astrophys. J. Lett.*, 159:L165, 1970; F. J. Low, *Astrophys. J. Lett.*, 159:L173, 1970.

Gasoline

At no time in the history of gasoline production have there been such extensive and far-reaching changes in its composition, manufacture, distribution, sale, and acceptance. The greatest of these changes is the initiation of an apparently nationwide resolve in the United States to remove the antiknock additives tetraethyllead and tetramethyllead from all gasolines in stepwise fashion during the 1970s. Resultant gasolines would have lower octane numbers. New cars with lower compression ratio engines are being manufactured to operate on the lower-octane gasolines. In order to supply nonleaded and low-lead gasolines without a large drop in octane number, oil refineries are replacing low-octane components with high-octane isoparaffins and aromatics.

During the transition period from leaded to nonleaded gasolines, oil companies will provide gasolines to satisfy the performance requirements of three general, though not distinct, populations of

motor vehicles; namely, the diminishing number of older cars now using either regular or premium gasolines, and the increasing number of new cars using lower octane gasoline. This transition will be a costly one for the oil companies, for the automobile industry, and for the consumer. Costs have been estimated at several billion dollars. This is part of the price of cleaner air in cities.

Air-pollution regulations. The 105,000,000 motor vehicles now on the road in the United States have been identified as a major source of air pollutants. In Fig. 1, which places equal weight on all types of pollutants, it can be seen that motor vehicles produce at least 60% of all air pollution in the United States. However, a large portion of this total is carbon monoxide, which is not a contributor to "aesthetic" pollution. In some large metropolitan areas, however, the motor vehicle contribution is even larger, varying between 70 and 90%. It is not surprising, therefore, that air-pollution regulations have great impact on mechanical changes in the motor vehicle and on the gasoline which it uses.

Although proposals to ban the internal combustion engine have appeared, they do not stand up under any realistic examination of the alternatives, namely, an automobile powered by a battery, fuel cell, or external combustion engine. It is generally agreed at this time that there is no serious competitor for the automobile internal combustion engine in the near future. It is also agreed that existing technology can produce an internal combustion engine which is essentially pollution-free.

Present air-pollution regulations control emissions of carbon monoxide, hydrocarbons, and nitrogen oxides, and new legislation to control emission of particulate matter is proposed. Three types

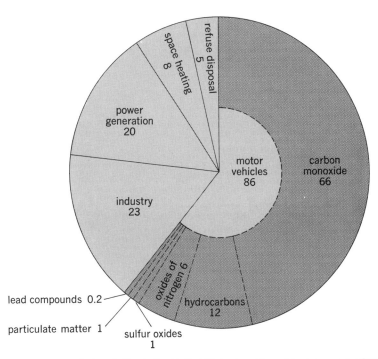

Fig. 1. Sources of air pollution in the United States in million of tons per year. (*From Public Health Service Publ. no. 1548, 1967*)

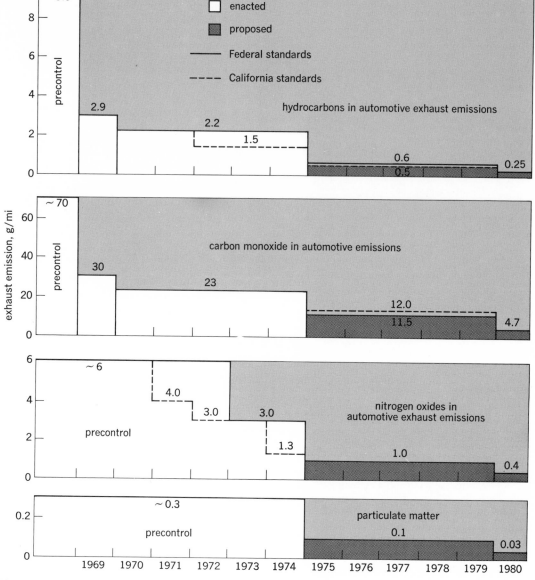

Fig. 2. Federal and California regulations on automotive exhaust emissions.

of regulations have been enacted: (1) control of crankcase emissions, (2) control of the total quantity of hydrocarbons which may be evaporated from the carburetor and fuel tank of a motor vehicle, and (3) control of the total amount of carbon monoxide, hydrocarbons, and nitrogen oxides which may be emitted from the motor vehicle exhaust.

Crankcase emissions. About 20–25% of the hydrocarbon losses from uncontrolled vehicles are attributable to crankcase venting, but crankcase emissions are now satisfactorily controlled. Crankcase emission-control systems were required by law on new cars sold in California in 1963, but were actually installed voluntarily by automobile manufacturers in most new cars sold in California as early as 1961 and throughout the United States soon after. Some used cars also must be equipped in California.

Evaporative emissions. Since about 15 or 20% of all emissions from a motor vehicle without controls enter the air by evaporation, it was not surprising

that California in 1970, and the Federal government in 1971, legislated restrictions on all new automobiles and limited evaporative emissions. Although there are no controls on evaporative emissions from used automobiles, there are proposals which would regulate the composition of gasoline. These proposals (1) specify a lower vapor pressure, which would result in the use of fewer light hydrocarbons in gasolines, and (2) restrict the olefin content of the volatile portion of the gasoline since olefins are the most photochemically reactive constituents evaporated into the atmosphere. Los Angeles County has had an olefin limit on all gasoline sold there for a number of years.

Exhaust emissions. Figure 2 summarizes the regulations on automotive exhaust emissions which have become law in the United States by implementation of the Federal Air Quality Act of 1967 and the California Pure Air Act of 1968. The shaded portion of each bar graph shows the proposals for legislation in 1975 and 1980. For each type of exhaust emission shown in Fig. 2, the level of emission without controls is compared to the subsequent low levels of emissions as exhaust controls are installed to meet the California and Federal standards. Automobile manufacturers have met these lower exhaust emission levels since 1966 by installing better choke control, better control and timing of spark advance, improvement of carburetion to provide leaner fuel-air mixtures to the engine at higher temperatures, and providing for air injection into the exhaust manifold for more complete combustion.

Detergent additives. Although reductions in exhaust emissions have been achieved almost entirely by modifications of the motor vehicle, there also has been some effort to reduce emissions with detergent additives. Such additives are effective in keeping clean those portions of the engine induction system which regulate the metering of both gasoline and air. Fuel passages in the carburetor and inlet manifold surfaces and valves must be kept free of deposits if air-to-fuel ratios are to be maintained at the levels required for optimum exhaust emission control. Gasoline additives of the detergent type include phosphate esters, succinimides, and polybutene amines. Recently these have been employed with high concentrations of mineral oils and have been claimed to be very effective in removing deposits from dirty engines resulting from extremely severe operation or from use of poor-quality gasoline, with a concomitant reduction in hydrocarbon and carbon monoxide in the exhaust.

Nonleaded gasoline. Although pollution authorities have discussed for some years the advantages and disadvantages of removing alkyl lead antiknock additives from gasoline, no concerted move in this direction took place until early in 1970. Although medical evidence seems to be lacking that lead levels presently prevailing in urban atmospheres are related to human health, there is general concern about increasing atmospheric lead much beyond current levels. Lead removal from gasoline also would reduce the particulate lead compounds emitted into the air from automobile exhaust. Such particles may contribute to reduced visibility in the atmosphere. It also has been postulated for some time that lead should probably be removed from gasoline if catalytic mufflers, easily poisoned by lead compounds, are required on motor vehicles to meet exhaust emission regulations. The removal of lead from gasoline would also reduce corrosion (from lead halides) of exhaust systems and plugging of exhaust recirculation systems.

The complete removal of lead antiknock agents from gasoline is not without problems. One of these is accelerated exhaust-valve recession in the absence of the lubricating effects of lead compounds on the exhaust valves. Valve redesign or valve seats of wear-resistant materials may be used in new cars. It appears probable that additives other than lead will control this problem in existing cars.

Another potential problem could be a possible increase in the photochemical reactivity of exhaust emissions if the aromatic content of gasolines is increased to compensate, in part, for the octane numbers lost by removing tetraethyllead. This problem will not arise, however, if the octane number of the nonleaded gasoline remains at the 91 level. Furthermore, as exhaust emission standards become more restrictive, reactors will be employed to reduce hydrocarbon and carbon monoxide emissions from all fuels to acceptable levels.

In a dramatic series of announcements in early 1970, certain automobile manufacturers announced that lead ultimately would have to be eliminated from gasoline and that many 1971 automobiles would be able to operate on gasoline of lower octane quality. Shortly thereafter a number of oil companies announced that they would market a low-octane, nonleaded gasoline to meet the requirements of the 1971 cars. During this time President Nixon sent a message to Congress calling for a lead-free gasoline, and the Secretary of the Department of Health, Education, and Welfare sent a letter to 108 oil companies proposing a timetable for the removal of tetraethyllead from gasoline.

In California, legislation has been proposed which establishes a timetable for removing lead from gasoline. This timetable provides for a low-lead gasoline (maximum 0.5 g of lead per gallon) by July 1, 1971, and a gasoline of zero lead content by July 1, 1974. These would be low-octane grades of minimum 90 Research octane number. A progressive reduction of lead levels in other grades of gasoline would also be required.

Automobile industry program. For 1971 cars the automobile manufacturers are modifying carburetor and distributor systems and are changing cylinder heads or pistons or both to lower the compression ratio of the engine. Many other engine modifications may be required, but operation on 91 octane gasoline without knock is the current design objective.

Exhaust recirculation systems may be required to meet the 1972 exhaust-emission control standards. To meet the more severe exhaust-emission standards in 1974, a thermal or catalytic reactor may be required and may cost as much as $300. The development of reactors for both leaded and unleaded gasolines has been announced. One catalytic system reduces nitrogen oxides, hydrocarbons, and carbon monoxide by 90%, and is claimed to have an almost unlimited life and to allow only small increases in emissions after 50,000 mi of operation on nonleaded gasoline. The same system is reported to produce only about a 50% reduction in exhaust emissions after operating for 50,000 mi on leaded gasoline. Thus periodic replacement during the life of the vehicle would be required. Thermal reactors have also met the 1974 standards over extended periods of operation on leaded or nonleaded gasoline. At present, the problems of effectiveness, cost, and reliability have caused automobile manufacturers to favor devices operating on nonleaded gasoline.

Although all of the above modifications and plans apply only to new automobiles, there is a desire to provide a pollution-control system which can be added to the greater number of older cars already on the road. It has been estimated that cars older than 1966 models emit 85% of the total automotive pollutants. Pollution-control kits for older cars are available at a cost of about $10.00 plus installation. These kits are claimed to reduce exhaust emissions by 30–50%.

One additional factor which must be considered is the requirement to keep new cars, and old cars equipped with pollution-control kits, in excellent operating condition. The criticality of maintenance cannot be overlooked; frequent tune-up will be necessary for best control of emissions. If the car owner does not assume this responsibility, it may be imposed upon him as a prerequisite to annual reregistration.

Oil refiners' program. Removal of tetraethyllead from gasoline poses a complex and costly set of problems for the gasoline manufacturer. Most marketers of gasoline now sell two grades and a few offer three grades. With the 1971 cars using an essentially nonleaded gasoline of 91 octane number, older cars of lower performance requiring a 94 octane regular gasoline, and older high-performance cars requiring a 100 octane premium gasoline, each two-grade gasoline marketer is facing a decision as to whether to try to serve these three car populations with two grades of gasoline or to add a third grade. Either choice leads to manufacturing, storage, and distribution problems which change with the car populations using these gasolines. The rate of transition from leaded gasolines to nonleaded gasolines is determined by legislation and by the rate at which the older cars now on the road are scrapped or are forced by legislation to use nonleaded gasoline.

Another problem for those oil refiners whose average octane number without added lead is below that of the new nonleaded gasoline is the production of the necessary quantities of nonleaded gasoline. This means production of additional quantities of high-octane components.

The two most desired types of gasoline components for nonleaded gasolines are highly branched paraffins and the common aromatic components such as benzene, toluene, and xylene. The highly branched paraffins are manufactured by alkylation of olefins with isoparaffins, whereas aromatics are manufactured by catalytic reforming and purified by extraction. These processes are inherently the most expensive gasoline manufacturing processes. The increased demand for aromatics requires an increase in catalytic reforming capacity in many refineries, and a concomitant increase in demand for the platinum catalysts used in such reforming operations.

All refiners will have to increase production of high-octane components if the lead content is to be reduced in gasolines for existing cars. A study supported by the American Petroleum Institute was made by Bonner and Moore Associates to estimate the cost of manufacturing unleaded gasoline in the United States. The average increase in refinery costs according to Bonner and Moore Associates would be about 2.2 cents per gallon, and the total capital investment in United States refineries would be about $4,000,000,000 to convert all gasoline to an unleaded basis without reduction of octane number in either the regular or premium grades. Other estimates range from $3,000,000,000 to $7,000,000,000.

Effect on consumer. The cost of manufacturing and selling nonleaded gasoline must ultimately be recovered from the consumer. Higher costs are incurred in the manufacture of low-lead and non-leaded gasolines compared to leaded regular gaso-line, since the production of the higher octane components results in a reduction in yield of gasoline under the more severe refinery processing conditions. However, the specific price of any grade or brand will always be determined by the prevailing competitive situation. The consumer may also experience a loss in efficiency of his car from compression ratio reduction and from emission-control devices. The total increase in automobile operating cost from all causes may reach as high as 15%. This is a reasonable price to pay if automobile emissions are reduced to a level acceptable to the public. The automobile manufacturers and oil companies are working to make it so, but whether they will be successful remains to be seen.

For background information *see* GASOLINE; PETROLEUM PROCESSING; REFORMING IN PETROLEUM REFINING in the McGraw-Hill Encyclopedia of Science and Technology.   [F. F. FARLEY]

Bibliography: Air Resources Board, *SAE J.*, 78(3):28–32, 1970; *Chem. Week*, 106(10):12–15, 1970; G. T. Kinney, *Oil Gas J.*, 68(11):91–96, 1970; Post-1974 auto emissions, *Environ. Sci. Technol.*, 4(4):288–294, 1970.

Gene

The material entity of a single structural gene has been isolated in pure form for the first time. The gene specifying the structure of the metabolic enzyme β-galactosidase in the bacterium *Escherichia coli* was isolated, complete with its two genetic control regions, the promoter and operator sites, as a deoxyribonucleic acid (DNA) molecule 1.4 micron (μ) long, having a molecular weight of approximately 2,700,000. This technical achievement, the result of a cooperative effort by a group of eight scientists and a technician at Harvard Medical School, serves to illustrate the current rapidly advancing state of technical sophistication in bacterial genetics.

As an important short-term consequence, the availability of a gene in purified form will make possible laboratory investigations of the control mechanisms which govern gene expression. These control mechanisms can, in principle, exert their restraining or facilitating effects at either of two sequential stages in the process by which a gene is expressed: the stage of transcription, when the message of the gene is copied many times in the form of messenger ribonucleic acid (m-RNA) molecules, or the stage of translation, when the messenger molecules direct the synthesis of a specific protein, the gene product. Controlling actions at either stage can explain equally well the phenomena of genetic control as observed in intact cells. However, clear distinctions could be made in a test-tube system in which specified components were used to reconstruct the living processes. Such a test-tube system had been incomplete because of the lack of a purified DNA molecule containing a known, functionally intact, unit of genetic information.

The long-term significance of this work lies in the fact that the techniques used to isolate this particular gene can in principle be used to isolate any other gene from *E. coli*. By extension to a much higher level of complexity, it is possible to envision the eventual applicability of similar methods to the modification of genetic constitution in higher organisms, including man.

Description. The gene proper consists of a linear chain of about 3700 nucleotide residues whose sequence codes for the sequence of the approximately 1230 amino acids which make up the β-galactosidase molecule. This chain, called the sense chain, is paired with its complementary antisense chain in the well-known double-helical configuration to make a length of about 1.28 μ of duplex DNA. The chromosomes of bacteria and of most DNA viruses are simply single molecules of this same duplex structure but of much greater length. Although the complex structure of chromosomes of higher organisms contains many other components, duplex DNA still carries the genetic message.

The promoter and operator sites are adjacent parts of the same DNA duplex. The promoter is the site to which the transcribing enzyme (RNA polymerase) binds in order to start transcription, and the operator is the site to which the specific repressor protein binds to stop gene expression (possibly by preventing transcription).

The β-galactosidase gene in *E. coli* is the first of a group of three functionally related structural genes which together make up what is known as the lactose operon. The three enzymes coded by these genes are involved in the metabolism of the sugar lactose, and the expression of all three genes is controlled by the one set of promoter and operator sites. This "lac" operon is possibly the best-known example of a genetic control system, having been the object of the studies by François Jacob and Jacques Monod in the late 1950s which laid the foundations of current concepts of genetic control. The lac repressor substance postulated by Jacob and Monod has recently been isolated and purified by Walter Gilbert and B. Müller-Hill.

Isolation. The device used to obtain the purified gene was based on methods developed by Jon Beckwith and collaborators: by taking advantage of certain rare genetic accidents, they transferred the gene from the bacterial chromosome into the much smaller chromosome of a "temperate" virus. Such a temperate virus can incorporate its chromosome into a specific site in the chromosome of its host, from which certain stimuli can make it detach again and multiply to produce many identical progeny viruses. Occasional accidents occur during this detachment process: thus once every 10,000 times or so, the viral chromosome carries away with it an adjacent part of the bacterial chromosome, which is thereafter replicated as an integral part of the viral chromosomal DNA molecule. This is how the β-galactosidase gene was transferred into the viral chromosome. First, however, it had to be moved from its usual location in the *E. coli* chromosome to a position next to the virus site. The agent used for this transfer was an episome, a small extrachromosomal genetic unit which can be passed from one bacterium to another. Episomes too, like temperate virus chromosomes, can go into the bacterial chromosome and out again, but at many different sites rather than at one special one. Episomes can thus accidentally pick up a variety of bacterial genes and reincorporate them into other locations in the chromosome of the same, or of a different, host bacterium.

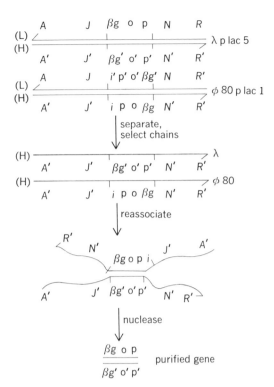

Key: βg = galactosidase   p = promotor site
        structural gene   i = lac i gene (structural
     o = operator site   gene for lac repressor)

Fig. 1. Scheme of isolation of the *Escherichia coli* β-galactosidase gene from the two viral chromosomes into which it had first been inserted in opposite orientations; βg = the β-galactosidase structural gene; o = operator site; p = promoter site; i = the lac i gene (the structural gene for lac repressor); A, J, N, and R are viral genes, shown for orientation.

Thus naturally occurring agents were used to perform the specific gene transfers for this work. However, the crucial part of these experiments was the devising of methods whereby the result of a desired rare accident could be selected for propagation. These methods were in general of the type which ensures that the desired product will be the only one possessing a certain combination of survival characters, such as stability at high temperature, resistance to killing by a certain virus or antibiotic, or the ability to grow on a single food substance. Thus, for example, a virus particle carrying the β-galactosidase gene was selected by its ability to confer on a host cell, which was otherwise incapable of it, the capacity to live on lactose. After the desired products of a gene translocation were obtained, it was of course necessary to use various tests of functional capacity and genetic mapping to find out in detail the genetic content and configuration of the translocated chromosomal segments.

The device that was used to purify the gene actually made use of two different but closely related viruses, λ p lac 5 and φ 80 p lac 1, in which the β-galactosidase gene was carried in opposite orientations, as determined by genetic mapping. In this family of viruses, the two chains of the chromosome can be isolated from each other on the basis of density, yielding a "heavy" (H) chain prepara-

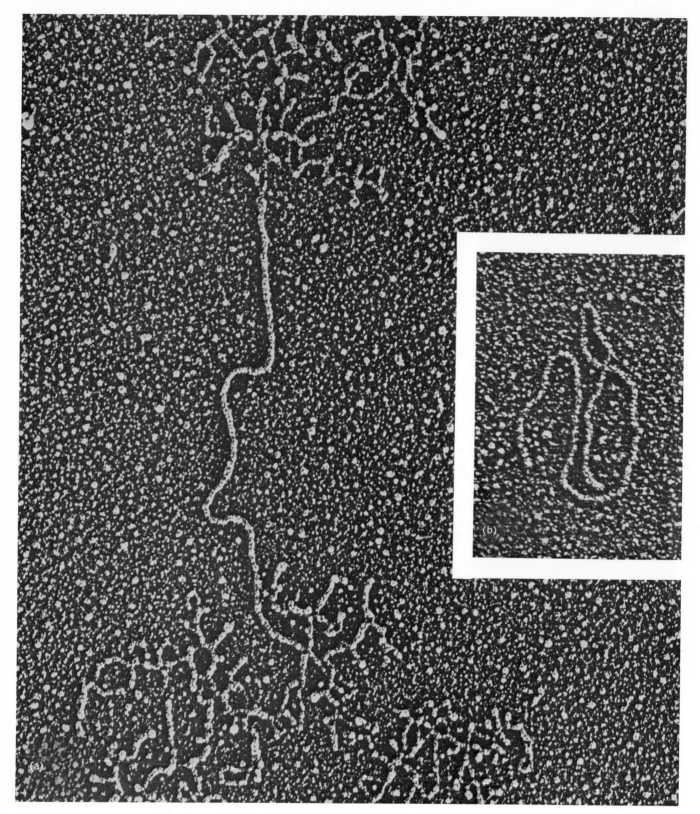

Fig. 2. Electron micrographs of the isolated β-galactosidase gene preparation. (a) Viral single chains are still attached as in Fig. 1 (before the use of nuclease). Single chains bunch up to form "bushes," while duplex chain lies stretched out. (b) Final product after removal of single chains by nuclease.

tion and a "light" (L) chain preparation. It is to be expected, then, that reversal of the orientation of a gene should interchange its sense and antisense chains such that, if in λ p lac 5 the sense chain of the β-galactosidase gene is part of the viral H chain, in φ 80 p lac 1 it should be part of the L chain. Therefore the H chains from the two viruses were mixed and subjected to conditions favoring formation of duplex from single chains. Since nucleotide sequence complementarity between the two chains is a strict requirement for double-helix formation, the product of this reaction is a short length of duplex DNA representing only the bacterial sequences common to the two viruses (Fig. 1). The rest of the viral chains, being identical rather than complementary, remain as single-chain appendages to the duplex segment. This expected product was confirmed under the electron microscope (Fig. 2a), and the single-chain appendages were then removed by incubation with an enzyme (*Neurospora* conidial endonuclease) which specifically degrades single chains but has no effect on double-helical DNA.

The final product was examined by electron microscopy (Fig. 2b), sedimentation behavior, and hybridization studies with lac and viral messenger RNAs, yielding the information that it is a duplex DNA molecule 1.4 μ long, is of molecular weight 2,500,000 – 3,100,000, and contains lac operon sequences but no viral sequences.

Because of the accidental nature of the means by which the gene segment was removed from the bacterium, the isolated product probably contains some bits of its two neighboring genes in *E. coli*, the i and y genes. However, since both are members of the well-characterized lac system, tests showed that not more than an inactive fragment of each was present in λ p lac 5. Moreover, when the observed length of the product DNA is compared with the calculated length of the β-galactosidase structural gene, it appears that the fragments of the i and y genes, together with the promoter and operator sites, occupy no more than 0.14 μ of duplex DNA. In addition to establishing a limit of 10% for the proportion of the i and y material in the product, this value provides a useful upper-limit measurement for the size of the promoter and operator sites, which have not been measured. They therefore can be said to occupy less than 400 base pairs of the DNA molecule.

For background information *see* BACTERIAL GENETICS; GENE in the McGraw-Hill Encyclopedia of Science and Technology.

[LORNE A. MAC HATTIE]

Bibliography: J. R. Beckwith, E. Signer, and W. Epstein, *Cold Spring Harbor Symp. Quant. Biol.*, 31:393, 1966; S. Gottesman and J. R. Beckwith, *J. Mol. Biol.*, 44:117, 1969; J. Shapiro et al., *Nature*, 224:768, 1969.

Hadron

The strong interaction between elementary particles is a manifestation of the most intense force known in nature. Fundamental particles which interact through this strong force are known as hadrons, and include the familiar nucleons and mesons as well as a host of less familiar, more recently identified states of matter. Although the strong interaction provides the force which holds the nucleus together — and is responsible for the energetics of nuclear reactions — it has proved necessary to study interactions at much greater energies than those that characterize nuclear reactions in order to understand the interaction. In the course of these studies, a large number of new states of matter have been discovered, as have a number of regularities and general features. This article attempts to strike some of the high points in the recent research in this area.

In 1948 only two hadrons were known: the nucleon (proton and neutron) and the pion. During the past 22 years physicists have discovered that the strongly interacting particles comprise an extensive spectroscopy, entirely as elegant and rich in physical content as atomic (electron) or nuclear spectra. A recent review article tabulates over 25 mesons and over 50 baryons (not including the different charge states of each or their antiparticles), and the list is still growing. In addition to the spin (angular momentum) J, parity P, and isotopic spin I, the states of hadronic matter are also specified by baryon number B and by a strangeness quantum number S or, equivalently, hypercharge, defined as $Y = B + S$.

In the late 1950s the states of hadronic matter were each referred to as "elementary particles," and each new discovery was hailed as adding to this list of particles. Physicists now recognize that there are a relatively small number of particle groupings or families, each characterized by a hypercharge, a baryon number, and an isotopic spin. Different members of each family may have different rest masses (for example, different energy states), corresponding to differing values of spin, electric charge, and so on, corresponding to different spectroscopic states of the same basic entity. Alternatively, the states may be grouped according to spin and parity, whereby members of these groups have different hypercharge and electric charge. Such groupings of 8 and 10 states arise out of the unitary symmetry group SU(3) and result in a very systematic organization of the elementary particles. A particularly dramatic discovery was the identification of the Ω^- baryon with charge (−1), strangeness (−3), and mass (1672 Mev) after it had been predicted by the SU(3) scheme.

The various hadron states generally decay to lower energy states through emissions of particles (for example, pions or other hadrons) or electromagnetic radiation. When such decays are allowed (corresponding to no change in strangeness or baryon number), these decays proceed very rapidly, with characteristic times of 10^{-22} to 10^{-23} sec. Strangeness-changing decays and decays of mesons into leptons proceed through the weak interaction with characteristic lifetimes of 10^{-8} to 10^{-10} sec, or more than 10^{12} times more slowly than the strong decays. As a consequence of the absolute conservation of baryon number, the only stable ground state of hadronic matter is the nucleon (proton and neutron).

Quarks. Currently, knowledge of the hadronic spectrum is much greater than the understanding of the system which gives rise to it. The situation is analogous to knowing the Balmer formula for the hydrogen spectrum before the discovery of the electron. Murray Gell-Mann and George Zweig, in an attempt to find a physical basis for the SU(3) groupings of particles, postulated that the subunits of hadrons might be very massive particles with fractional values of electric charge, hypercharge, and baryon number. Gell-Mann has referred to these hypothetical building blocks as quarks. According to this model, all baryons might be made up of three quarks and all mesons of a quark-antiquark pair. It may be, however, that quarks do not exist as separate physical entities. They may be only a mathematically convenient parameter in the model, or they may be disallowed by some law of nature yet unknown. Alternatively, they may exist

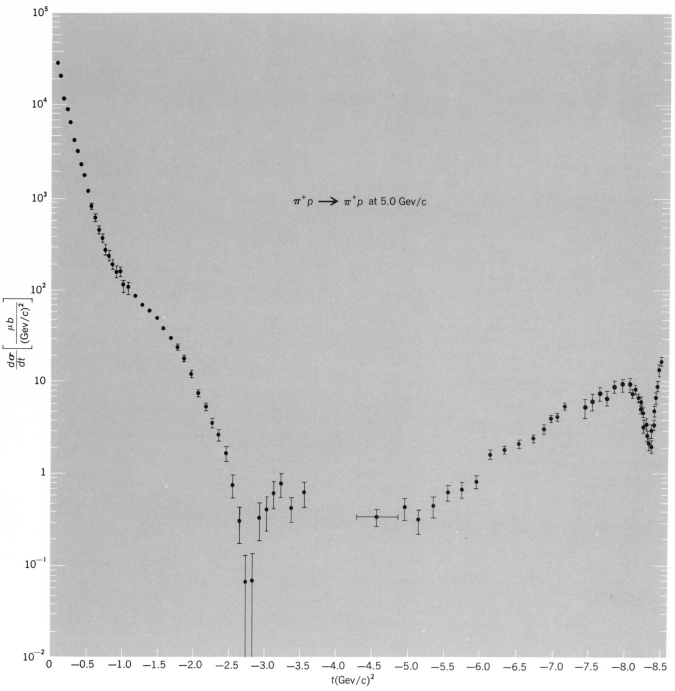

Fig. 1. The differential elastic scattering cross section for $\pi^+ p$ at 5.0 Gev/c from the University of Michigan–Argonne experiment. The range of t covered corresponds to the angular range from 0 to 180°. Note that the ordinate scale is logarithmic.

but be so massive that experimenters with accelerators and with cosmic rays have not yet succeeded in finding them. Many experiments have been carried out, and some reports of quark events from cosmic-ray experiments have appeared in the literature. It would seem prudent at this time to reserve judgment, since these "sightings" may be disproved. In any event, the search for physical quarks will continue as new accelerators and new techniques extend the accessible energy threshold and the lower limits to possible production cross sections. *See* QUARK.

Total cross sections. From electron scattering experiments and from a large body of data from lower-energy nuclear physics, it is well established that a nucleon (proton or neutron) has a physical size corresponding to a radius of about 1.2 fermi (1.2×10^{-13} cm). In interactions of hadrons at higher energy this size again emerges as a characteristic dimension, both for the nucleon and for other hadrons. At energies above 3 Gev, the total cross section of neutrons on nuclei is well fitted by ascribing to the nucleus a radius of $1.27\,A^{1/3}$ fermi (where A is the atomic number) and a mean free path of the neutron in nuclear matter of 3.0 fermis. Recent experiments at the 76-Gev accelerator at Serpukhov in the Soviet Union have given cross sections of negative pions, kaons, and antiprotons which indicate a behavior similar to that of the nucleon except that the pions and kaons exhibit a greater mean free path in nuclear matter and the antiproton a smaller mean free path than the corresponding values for the nucleon.

The elementary proton-proton and neutron-proton total cross sections have been measured at up to 28 Gev, at which point they are nearly the same and equal to about $38-39$ millibarns (mb), or 38×10^{-27} to 39×10^{-27} cm²). From these data it is not clear whether the cross sections have reached an asymptotically constant value or are continuing to fall slowly. The pion and kaon cross sections on protons had appeared to be slowly falling at $20-25$ Gev; however, at least for these negative particles, they now appear essentially constant up to 60 Gev from the Serpukhov data at $24-25$ mb ($\pi^- p$) and about 21 mb ($K^- p$).

Recent experiments at the Deutsches Elektronen Synchrotron (DESY), Hamburg, German Federal Republic, at the Stanford Linear Accelerator (SLAC) of Stanford University, and at the 10-Gev synchrotron of Cornell University have shown that the photoproduction of ρ-mesons on various nuclei may be interpreted to provide a measure of the ρ-nucleon cross section. Although some uncertainty remains in comparing results from the different groups, a value of $25-30$ mb is found for this cross section. Since the ρ-meson decays in about 5×10^{-24} sec into two π-mesons, it was first considered that the ρ-meson was only a resonant state in the scattering of two π-mesons. Since this cross section is comparable to the single pion-nucleon cross section, it now appears that the ρ-meson has an independent (however short) existence of its own.

Elastic scattering. The simplest interaction between elementary particles is the elastic scattering of one particle by another. The magnitude and angular distribution of the scattering of one particle by another is the high-energy equivalent of the subjective experience of "seeing" a macroscopic object scattering light photons from it. The elastic scattering of pions, kaons, and nucleons on protons at energies of several billion electron volts is generally characterized by an exponential falloff of the form $d\sigma/dt \propto \exp(Bt)$, where t is the invariant four-momentum transfer and σ is the scattering cross section. The four-momentum transfer is given by $t = 2p^2 (\cos \theta - 1)$, where p and θ are the momentum and scattering angle in the center of mass. At high energies, $t \cong -p^2\theta^2$, where p and θ are values taken together either in the laboratory or in the center-of-mass coordinate systems. The coefficient B can be related to the radius of interaction r through a Fourier transform. For example, a Gaussian scattering spatial distribution is related to an exponential diffraction scattering by $r \approx \sqrt{5B}$, where r is in fermis and B is in units of $(\text{Gev}/c)^{-2}$. Values of B observed experimentally lead to values of r of $1.0-1.2$ fermis.

Many details of the scattering are very striking and of considerable interest. For example, a University of Michigan–Argonne group has recently established that, in the $\pi^+ p$ elastic scattering, a second, very pronounced minimum occurs in the differential scattering cross section (Fig. 1) which may be connected with a diffraction process, although no such minima occur in the proton-proton scattering at comparable energies. Indeed, it is not clear to what extent elastic scattering can be interpreted in terms of the quantum wave mechanical equivalent of optical diffraction scattering and to what extent it must arise from particle exchange processes. Most probably the truth will involve both points of view.

Interactions. The greatest body of recent data and interpretive effort has gone into the study of inelastic reactions of the form $a+b \rightarrow c+d$, where a, b, c, and d are different states of hadronic matter. Elegant theories have developed from the starting point of H. Yukawa's interpretation of the pion as the quantum of the strong interaction field and the analogy with electromagnetic scattering of two charged particles as the exchange of a virtual photon. Hence the general process $a+b \rightarrow c+d$ is thought of as the diagram in Fig. 2, where x is an exchanged hadron (pion, kaon, nucleon, and so on). The conserved quantum numbers of the strong interaction constrain the nature of the possible processes at the two vertices.

This picture has been carried far and is almost universally adopted as being fundamentally correct. Unfortunately it has not made possible quantitatively correct predictions of new processes. Perhaps it is more nearly correct to say that experiments have continually forced unexpected embellishments and modifications of the theories.

Fig. 2. Diagram representing the process $a+b \rightarrow c+d$.

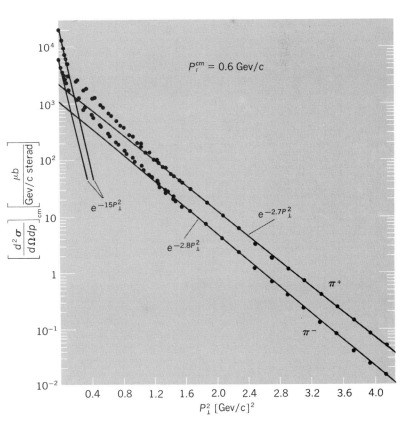

Fig. 3. Data from the reaction $p+p \rightarrow \pi^+ +$ anything from an experiment by the University of Michigan–Argonne group of Alan Krisch. The data are plotted as a function of $p_\perp^2$, the square of the transverse momentum, at a fixed value of center-of-mass longitudinal momentum.

Future. As processes at higher energies are brought under quantitative scrutiny, more complex reaction final states become prevalent and more important for the understanding of the interaction processes. Hence reactions such as $a + b \rightarrow w + x + y + z + \ldots$ become important. Alan Krisch has pioneered in a class of experiments studying reactions of the form $p + p \rightarrow \pi +$ anything, and quantitative studies of the momentum distribution of such pions promises insight into the nature of these reactions. Figure 3 illustrates the Krisch investigations. Other experimenters are exploring the momentum and angular distribution of secondaries from different subsets of reaction channel groupings at up to 25 Gev. Generally, the distribution of reaction products as one proceeds to very high energy seems characterized by a transverse momentum distribution not unlike that seen in elastic scattering, falling exponentially with $p_\perp$. The longitudinal momentum seems to be distributed as E^{-1}, corresponding to a peaking about $p_L = 0$ in the center of mass.

Theorists are now searching for insights into the nature of these high-energy "catastrophic" collisions as a key to the nature of the ultimate structure of hadronic matter. Richard P. Feynman, for example, has argued that the data are consistent with the notion that each hadron is made up of many small granules or "partons," which when rearranged and combined form the observed mesons and baryons. Rolf Hagedorn has proposed a thermodynamic model wherein the substance of hadronic matter is "heated" to a temperature of 150 Mev in a collision, from which it cools by the radiation of mesons.

The internal structure of hadrons, whether quarks, partons, or a mesonic soup, is entirely unknown. It is hoped that quantitative experiments at much higher energies hold the key to this problem, although physicists must be prepared to receive many more surprises before a solution is found.

For background information *see* ELEMENTARY PARTICLE; HADRON; SCATTERING EXPERIMENTS, NUCLEAR in the McGraw-Hill Encyclopedia of Science and Technology.

[LAWRENCE W. JONES]

Bibliography: A. Barbaro-Galtieri et al., *Rev. Mod. Phys.*, 42:87, 1970; G. von Dardel (ed.), *Proceedings of the Lund International Conference on Elementary Particles*, 1969; J. Prentki and J. Steinberger (eds.), *Proceedings of the 14th International Conference on High Energy Physics*, Vienna, Austria, 1968.

Heat transfer

This brief account deals only with the subject of liquid-metal heat transfer, and in fact with only a very small segment of it. Work in this field has increased rapidly in the past few years, owing chiefly to the fact that liquid metals have found an important use as coolants for nuclear power reactors. For example, the liquid-metal fast-breeder reactor (LMFBR), the type of reactor on which the United States is pinning its hopes for the long-range future, is cooled with liquid sodium. In these reactors the heat is generated in fuel rods that are about 1/4 in. in diameter and several feet long. There are generally thousands of such rods in the

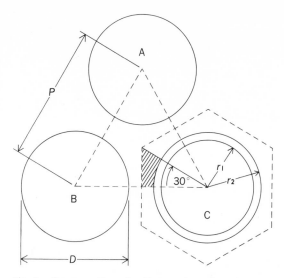

Fig. 1. Cross section of a three-rod portion of a rod bundle with equilateral triangular spacing. The cross-hatched area represents an elemental coolant flow area.

core of a single reactor, grouped in so-called bundles or subassemblies of a few hundred each.

Subassembly arrangement. The rods are arranged in an equilateral triangular pattern, and the spacing between them may vary from reactor to reactor, depending upon the particular type of power system in which the reactor is used. The spacing is generally given in terms of the distance between rod centers, or pitch (P); the most impor-

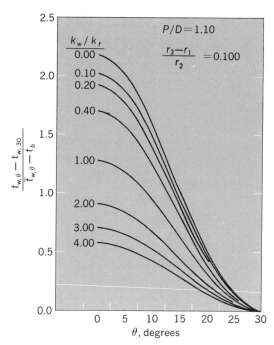

Fig. 2. Typical curves showing the effect of the thermal conductivity of the cladding on the relative circumferential variation of the temperature on the outer wall of the cladding. (*From O. E. Dwyer and H. C. Berry, Effects of cladding thickness and thermal conductivity on heat transfer to liquid metals flowing in-line through bundles of closely spaced reactor fuel rods, Nucl. Sci. Eng., 40: 317–330, 1970*)

tant geometrical factor affecting the heat-transfer behavior of the system is the pitch-diameter ratio (P/D). Structurally, the fuel rods consist basically of a ceramic core (UO_2 or some mixture of UO_2 and PuO_2) encased in a metallic cladding whose thickness is roughly 1/10 the radius of the rod.

It is thus not surprising that most of the single-phase liquid-metal heat-transfer research done in the last few years has dealt with the problem of turbulent-flow heat transfer to liquid metals flowing longitudinally through rod bundles. There is considerable information available on heat transfer to ordinary fluids (such as water) flowing through rod bundles, but this information is not applicable to liquid metals because the heat-transfer behavior of liquid metals in turbulent flow is quite different from that of ordinary fluids. This is because of the high thermal conductivities of liquid metals. When ordinary fluids are heated, the transverse temperature drop in the fluid occurs for the most part in the thin laminar layer adjacent to the heating surface. However, when a liquid metal is heated, the transverse temperature drop is more or less spread across the whole liquid stream. This characteristic of liquid-metal heat transfer makes it a more complicated engineering operation than heat transfer in ordinary fluids. In other words, the heat-transfer correlating equations used in engineering design vary with channel geometry and thermal boundary conditions, whereas with ordinary fluids they generally do not. Also, thermal-entry (that is, near channel inlets) heat transfer for liquid metals can be quite different from that for an ordinary fluid such as water.

Figure 1 is a schematic cross-sectional representation of a three-rod portion of an equilaterally triangularly spaced rod bundle. The area between rod C and the surrounding hexagon represents the flow area for that portion of the coolant stream that receives the heat from that particular rod and transports it away axially. The small cross-hatched section represents an elemental coolant flow area, 12 of which surround each rod. Thus, if the heat-transfer behavior of an elemental 30° wedge is determined, because of symmetry the heat-trans-

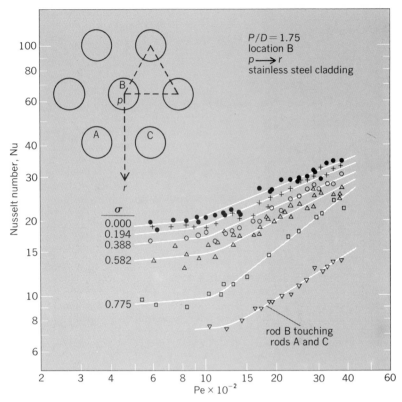

Fig. 3. Experimental results for heat transfer to mercury flowing in-line through an unbaffled rod bundle under conditions of fully developed turbulent flow and heat transfer. The effect of displacing rod B along the path $p \rightarrow r$ on its overall heat transfer capability is shown as a function of relative displacement (σ) and flow rate (Pe). The data were taken under a thermal condition very closely approaching that of uniform heat flux on the inner wall of the cladding.

fer behavior for the whole rod is known.

Because of the irregular shape of the elemental coolant-flow area, the heat-transfer behavior varies circumferentially throughout the 30° arc, passing through a complete cycle in 60°. Thus, there are six such cycles around the whole circumference of a rod. Figure 2 illustrates the circumferential variation of the outer wall temperature of a rod for

Heat-transfer nomenclature

Symbol	Definition
C_p	= specific heat of coolant, Btu/(lb$_m$)(°F)
D	= $2r_2$ = diameter of rod, ft
D_e	= $4m$ = equivalent diameter of rod bundle for in-line flow, ft
$\bar{h}$	= $\bar{q}/(\bar{t}_{w,\theta} - t_b)$ = average turbulent-flow heat-transfer coefficient, Btu/(hr)(ft²)(°F)
k_f	= thermal conductivity of coolant, Btu/(hr)(ft)(°F)
k_w	= thermal conductivity of cladding, Btu/(hr)(ft)(°F)
m	= cross-sectional flow area/wetted perimeter, ft
P	= pitch, or distance between rod centers, ft
$\bar{q}$	= circumferential average outer wall heat flux, Btu/(hr)(ft²)
r_1	= inner radius of cladding, ft
r_2	= outer radius of cladding, ft
t_b	= bulk temperature of coolant, °F
$t_{w,\theta}$	= outer wall temperature at angle θ, °F
$\bar{t}_{w,\theta}$	= circumferential average value of $t_{w,\theta}$, °F
$\underline{t}_{w,\theta}$	= outer wall temperature at angle $\theta = 30°$, °F
v_a	= average linear velocity of coolant, ft/hr
θ	= circumferential angle (Fig. 1) measured from straight line connecting the centers of two adjacent rods, degrees, or radians
ρ	= density of coolant, lb$_m$/ft³
σ	= (actual displacement distance)/(maximum possible displacement distance), dimensionless

slug flow, for the thermal boundary condition of uniform heat flux on the inner wall of the cladding, and for a typical set of design conditions. Actually, slug-flow heat transfer is not greatly different from turbulent-flow heat transfer, in the case of liquid metals, and for the flow rates and rod bundles designs generally used.

Heat-transfer coefficient. An obviously important factor in the design of fuel-rod subassemblies is the average heat-transfer coefficient. This is usually expressed in the dimensionless form hD_e/k_f, called the Nusselt number (Nu). The heat-transfer coefficient itself goes through a rather sharp maximum as the P/D ratio is increased above its minimum value of 1.0. For turbulent flow, this maximum occurs at $P/D \approx 1.15$. At higher P/D ratios, radial heat flow in the coolant dominates, while at lower P/D ratios, circumferential heat flow dominates.

Figure 3 shows some recent experimental results obtained with mercury that illustrate the very deleterious effect of rod displacement (or bowing) on heat-transfer capability with liquid metals. The outside diameter of the rods was 1/2 in., and the cladding thickness was 0.050 in. The Peclet number (Pe), defined as $D_e v_a \rho C_p/k_f$, can in this case be considered a direct measure of the flow rate. The curves in Fig. 3 show the dogleg shape that is typical of channel-flow, liquid-metal heat transfer. The lower, more horizontal portion of each curve represents molecular conduction only (that is, eddy conduction is negligible), and the upper, steeper portion represents the combination of molecular and eddy conduction.

Figures 2 and 3 illustrate the fact that liquid-metal heat transfer is very different from that of ordinary fluids. The differences in the various curves in the figures would practically disappear if the coolant were water, for example. The curves illustrated in Fig. 2 and 3 are for "open" rod bundles; that is, the effect of any spacing devices is negligible.

The table summarizes nomenclature used in studies of heat transfer.

For background information *see* HEAT TRANSFER; NUCLEAR FUELS; REACTOR, NUCLEAR; REACTOR, NUCLEAR (CLASSIFICATION) in the McGraw-Hill Encyclopedia of Science and Technology.
[O. E. DWYER]

Bibliography: D. J. Bender and P. M. Magee, *Turbulent Heat Transfer in a Rod Bundle With Liquid Metal Coolant*, Rep. no. GEAP-10052, General Electric Co., July, 1969; O. E. Dwyer, *Nucl. Eng. Des.*, 10:3–20, 1969; O. E. Dwyer and H. C. Berry, *Nucl. Sci. Eng.*, 40:317–330, 1970.

Hominidae

There is continued interest in the early fossil record of the Hominidae, the zoological family of man. This article discusses new additions to the fossil record of *Australopithecus*, a hominid form associated with the Pliocene-Pleistocene, and recent information on *Ramapithecus*, a form associated with the Pliocene and considered by some paleontologists to be the earliest fossil representative of the Hominidae.

Australopithecus. In the past 5 years there have been important new additions to the fossil record of early Hominidae. These discoveries have resulted from field investigations of new or of previously little-known geological deposits of Pliocene-Pleistocene age in eastern Africa. As a consequence, fossil Hominidae are now known for the first time from radiometrically dated deposits from between 5,000,000 and 2,000,000 years ago. These finds afford evidence of the nature and diversity of such early Hominidae, and at the same time raise important questions in regard to the matter of hominid origins.

All these new discoveries of Hominidae of Pliocene-Pleistocene age have been made in geological formations related to the Baringo Basin and the Rudolf Basin, both portions of the eastern Rift Valley in Africa. The absolute age of these fossil-bearing deposits has been determined by potassium-argon (K/Ar) measurements on associated volcanic rocks and ashes. A succession of local vertebrate faunas of known ages has thus been established. The relative age of other, undated fossil occurrences can now be determined by comparisons with the known temporal range of certain key species (particularly elephants and suids) in that succession. The temporal relationships of these Pliocene-Pleistocene deposits is shown schematically in Fig. 1.

Olduvai Gorge. The youngest such occurrences are in the lower fluviolacustrine sediments and volcanics which form Bed I at Olduvai Gorge, Tanzania. A robust australopithecine skull, assigned to *A. boisei*, was found first, in 1959. It bears close resemblance to *A. robustus*, a species known previously from deposits of broadly comparable geologic age in two caves in the Transvaal, Republic of South Africa. Subsequently, at Olduvai, remains of another hominid species, here termed *A. habilis* but originally designated *Homo habilis*, were recovered from several other horizons in these deposits. The species bears some resemblances, especially in jaws and teeth, to the small australopithecine *A. africanus* from southern Africa, but its brain size was substantially larger. At Olduvai both species were found associated with substantial accumulations of stone artifactual materials and vertebrate food debris. Cultural materials of comparable as well as of somewhat younger age are known in North Africa, particularly in Algeria (Ain Hanech) and in Morocco, but skeletal remains of Hominidae of such antiquity are still unknown there.

Omo Basin. Field studies have been carried out since 1966 by the international Omo Research Expedition in the lower Omo Basin, north of Lake Rudolf, in southern Ethiopia. Richly fossiliferous sediments with associated volcanics occur there and are now known to range in age from somewhat less than 2,000,000 years to slightly older than 4,000,000 years. Remains of Hominidae that have been recovered from most of this time range date back to about 3,500,000 years. Six mandibles, over 100 isolated and associated teeth, a partial skull, and a portion of femur have been recovered (Fig. 2). Fossil vertebrates occur in vast quantities in these deposits.

A robust australopithecine, with dental features very suggestive of those of *A. boisei*, occurs throughout six principal fossil zones, the oldest occurrence dating to nearly 3,500,000 years. A second australopithecine-like hominid, with dental

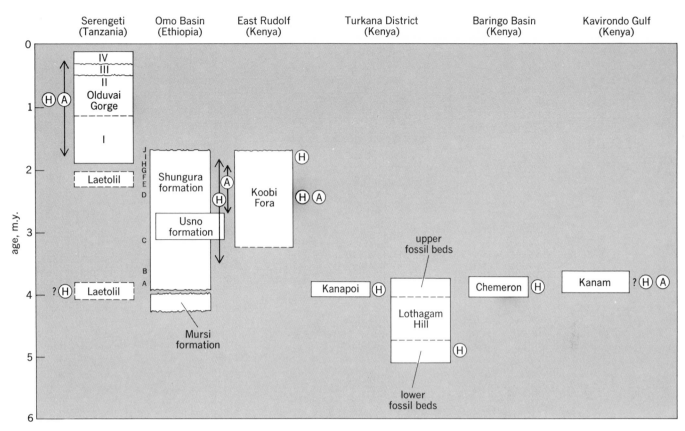

Fig. 1. East African Pliocene-Pleistocene deposits yielding fossil Hominidae (H) or artifactual materials (A), or both.

resemblances to *A. africanus*, also occurs throughout these deposits, and sometimes at the same localities which have yielded the hominid of *A. boisei* affinity. Hence in the Omo succession, as in Bed I at Olduvai at a later time, there is evidence for the coexistence in the same area at the same time of two hominid taxa, presumably at least specifically distinct, and reasonably assignable to the extinct genus *Australopithecus*. Most important has been the acquisition of evidence which demonstrates an age twice that known previously for the very distinctive, robust form of *Australopithecus*.

Rudolf Basin. Additional remains of *A. boisei* were recovered in 1968, 1969, and 1970 in relation to fluviolacustrine and volcanic deposits in the northeastern quadrant of the Rudolf Basin, Kenya. These include portions of mandibles, a palate, a massive mandible with most of the teeth, several skulls, and several bones of the postcranial skeleton. A very well preserved skull is strikingly like the type specimen of *A. boisei* from Olduvai, but is perhaps nearly 1,000,000 years older. Its discovery, and that of the other specimens which represent the same type of hominid, offers splendid confirmation of the significance of the less complete remains referable to this creature from the succession of deposits in the lower Omo Basin.

The discovery in 1968 of a mandible fragment referable to a robust australopithecine, and in 1970 of a small mandible not unlike the one from Olduvai Gorge that was referred to *H. habilis*, from younger deposits in this succession clearly

demonstrates again the coexistence of these different species of early Hominidae. The 1970 discovery will undoubtedly have an important bearing on the issue of whether such creatures should be referred to the genus *Australopithecus* or to the genus *Homo*.

Occurrences of stone artifacts from this range of time are just becoming known. The exposures in the lower Omo Basin afford some artifact occurrences at 1,950,000 and 2,100,000 years, with others suspected (but still uninvestigated in detail) dating to about 2,600,000 years. The eastern Rudolf area has recently afforded artifacts in place in a volcanic ash horizon of 2,600,000 years. This is the oldest such radiometrically dated occurrence discovered anywhere thus far.

Other localities of Pliocene-Pleistocene age in eastern Africa have afforded only fragmentary remains of Hominidae. However, on the basis of evidence supplied by the associated mammalian species, these remains probably antedate the aforementioned specimens and are 4,000,000 to 5,000,000 years old.

A fragment of a distal humerus from the Kanapoi locality, west of the Kerio River, in the southwestern sector of the Rudolf Basin, has characteristics which indicate its hominid nature. It is quite large and could conceivably represent some form of robust australopithecine. However, it is too incomplete a specimen to permit precise determination of affinity, particularly because this portion of the australopithecine skeleton is poorly known. Its age is about 4,000,000 years.

The Lothagam Hill locality, north of Kanapoi,

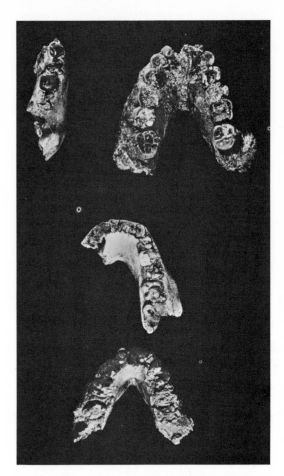

Fig. 2. Mandibles of Hominidae from the Pliocene-Pleistocene deposits of the lower Omo Basin, Ethiopia. The oldest mandible (about 2,600,000 years) is at the bottom of the photograph, and the youngest mandibles (about 1,900,000 years) are at the top.

exposes an older and a younger series of fluvio-deltaic sediments. The older series apparently antedate a basaltic intrusion radiometrically dated at 3,700,000 years, and on faunal evidence may well date back to about 5,000,000 years. They have yielded a small fragment of a hominid mandible with a worn lower molar. Its size and morphological features preclude its assignment to a robust australopithecine such as *A. boisei*, and suggest a smaller creature more like *A. africanus*. However, because of its incomplete state, its exact affinities are difficult to ascertain.

Baringo Basin. A single fragment of a hominid skull, a temporal bone, recovered from a portion of the Chemeron Beds, in the southwestern sector of the Baringo Basin, Kenya, is of an age perhaps comparable with that of the Kanapoi specimen. Most of the characters are compatible with an australopithecine grade of morphology. However, some characters are clearly different from the distinctive features found in the robust form, *A. boisei*. Therefore the Baringo specimen could well represent another species of that genus, perhaps *A. africanus* or a species hitherto unknown.

A specimen of partial cranium, including some of the palate and teeth, discovered at the Cheso-wanja locality in 1970 demonstrates again the existence of a robust type of australopithecine in the range of time represented by Olduvai Bed II.

Kanam. Several fragmentary specimens found some years ago in eastern Africa may also fall within this range of time. These could include a front portion of a hominid mandible recovered in 1932 from fossiliferous deposits at Kanam, near the shore of the Kavirondo Gulf of Lake Victoria, Kenya. It is still not certain whether the specimen indeed derived from these deposits, rather than from stratigraphically higher and younger deposits (as suggested by its chemical composition). At any rate, its incompleteness, poor preservation, and pathological condition almost preclude determination of its affinities.

The situation is different in regard to hominid teeth recovered in 1939 on the surface of, and presumably derived from, the Laetolil Beds, exposures of fossiliferous ashes and sediments along streams draining to the Eyasi trough south of Olduvai Gorge. There are two upper premolar teeth (in a fragment of maxilla). Until recently the age of the deposits and the possible affinities of the teeth were uncertain. Now the relative age of the beds can be compared with radiometrically dated deposits of Pliocene-Pleistocene age elsewhere through their contained mammalian fauna. Faunal elements of two distinct ages are probably represented, and the color and preservation of the hominid teeth suggest their association with the older fauna, which have an estimated age close to 4,000,000 years. The teeth are not large, and their narrowness, relative sizes, and root system diverge from the condition found in either robust or gracile species of australopithecine. Whatever their affinity may ultimately prove to be, they do not appear to represent any known species of *Australopithecus*. They may well belong to another species of Pliocene-Pleistocene Hominidae.

These discoveries of the past 5 years demonstrate the substantial antiquity of Hominidae of the genus *Australopithecus*. They afford evidence for the existence of at least two species of that genus between 2,000,000 and 4,000,000 years ago. Evidence for tool-making behavior by Hominidae has been extended back to about 2,600,000 years. There are hints of a small form of australopithecine, perhaps similar to *A. africanus*, as long ago as about 5,000,000 years. If these indications are confirmed by subsequent investigations and additional material, it is clear that speciation within the australopithecine radiation had occurred by mid-Pliocene time at least, and that the origin of the Hominidae is to be sought at a time no later than the middle to lower Pliocene and perhaps still earlier.   [F. CLARK HOWELL]

Ramapithecus. It is now generally recognized that the earliest fossil form which may with reasonable certainty be regarded as belonging to the Hominidae is the genus *Ramapithecus*. The first remains of *Ramapithecus* to be discovered were not, however, recognized as being hominid but were assigned by their finder, Guy Pilgrim, to a new species of the ape genus *Dryopithecus*, *D. punjabicus*. These finds consisted of parts of lower and upper jaws, came from geological deposits in northern India of early Pliocene age, and were therefore about 12,000,000 years old. The name

Ramapithecus was coined about two decades later, in 1934, by G. Edward Lewis, who applied it to part of an upper jaw found in the same deposits and to which he gave the species name *R. brevirostris*. In his initial description of this form Lewis noted that it bore many features in common with those of Hominidae, but it was only in his Ph.D. dissertation, which unfortunately was never published, that he stated a belief that *Ramapithecus* was ancestral to later hominids. Thus it was not until 1961, when Elwyn L. Simons restudied the specimen, that *Ramapithecus* began to be generally recognized as a hominid. The fact that Pilgrim's material belonged to the same species as that of Lewis was realized shortly thereafter, and because the species name *punjabicus* has priority in time over Lewis's name *brevirostris*, the correct species name of this animal is *R. punjabicus*.

In 1962 Louis Leakey described a fragment of upper jaw from Fort Ternan, a site in Kenya dated at 14,000,000 years, as belonging to a new hominid genus and species, *Kenyapithecus wickeri*. In fact, this specimen does not differ sufficiently from the Indian material to warrant its own genus, and it possibly does not deserve specific distinction either, despite the long time gap involved. Another group of fossils, about 19,000,000 years old and also from Kenya, has recently been described by Leakey as belonging to a species ancestral to the Fort Ternan form, *K. africanus*, but close examination of this material indicates that its affinities lie with one or more species of archaic apes, and that any attempt to regard it as an ancestral hominid cannot be sustained. It now seems overwhelmingly probable that *Ramapithecus* had its origins from a pongid (ape) stock not very long before it is first found in the fossil record, some 14,000,000 years ago.

Dental evidence. Apes differ dentally from hominids in a variety of ways. The hominid tooth row is short from front to back, whereas the reverse is true of the apes. The canine and incisor teeth of hominids are small, whereas the anterior dentition (notably the canines) of the pongids is large. In order to accommodate the long upper canine while the teeth are in occlusion, the front lower premolars of apes are sectorial, that is, elongated and narrow.

On all these points, *Ramapithecus* clearly shows hominid adaptations. Its cheek teeth, in contrast to those of contemporaneous apes, are steep-sided and short from front to back, with relatively little relief on their occlusal surfaces. They are closely packed together, as in later hominids, and as a result become increasingly shortened during life because of attrition between contacting surfaces of adjacent teeth. Both premolars are broad, and a canine from Fort Ternan is greatly reduced in size, although it is morphologically reminiscent of those of early apes (which is not surprising if *Ramapithecus* had been derived from a pongid form in the not too distant past). Although the incisors of *Ramapithecus* are not definitely known, the evidence of roots and sockets indicates that they were small. The structure of the upper and lower jaws indicates that *Ramapithecus* possessed a short face, this being largely a reflection of the short tooth rows.

Behavior and function. What do these resemblances of *Ramapithecus* to later hominids mean in terms of behavior and function? The short face is mechanically advantageous in that it reduces the load arm of the jaw; that is, biting force at the front of the jaw is increased for every unit of force exerted by the muscles of mastication. The origin of these muscles is also moved forward, because the anterior root of the zygomatic arch, where the dominant masticatory muscle attaches, is positioned above the first molar rather than farther back, as in the apes. This increases the power arm of this muscle around the jaw joint, and permits more powerful chewing. Increased masticatory force requires more efficient dissipation of these forces, and the position of the anterior root of the zygomatic arch above the midpoint of the tooth row suggests that all the cheek teeth were of importance in grinding. Powerful chewing is also suggested by the close packing of the teeth; this phenomenon is due to the forward movement of teeth caused by strong stresses along the tooth row. The lower jaw of *Ramapithecus*, insofar as it is known, shows similar adaptations; it is short, and at the front it is deep and strongly buttressed to enable it to resolve masticatory stresses.

What is the adaptive significance of this powerful chewing mechanism? Clifford Jolly has recently proposed a model of hominid divergence, based on the behavior and adaptation of the gelada baboon, which accords very well with these features of *Ramapithecus*. The gelada dental mechanism incorporates many of the traits that have just been discussed, and is related to a very specific feeding behavior, termed graminivorous by Jolly. Geladas feed on large quantities of small tough morsels such as grass seeds and rhizomes. Jolly's hypothesis is that the hominid line diverged when a pongid stock forsook an arboreal existence and a diet of soft fruits, and took to the ground and a graminivorous diet. This hypothesis is suggested to correlate with a shift from a life in the forests to the occupation of more open country. What is known at present of the ecology of *Ramapithecus*, based on faunal and sedimentological analyses in northern India, suggests that the deposits there in which *Ramapithecus* is found represent for the most part a forested environment, well-watered and crossed by numerous streams and rivers. The later part of the time period covered by these deposits, however, appears to have been characterized by a drying trend. *See* PRIMATES, FOSSIL.

Thus the most reasonable hypothesis at present seems to be that, although *Ramapithecus* was probably to some extent arboreal, at least in the initial stages, it was adapted to a terrestrial diet and was presumably spending increasing amounts of time on the ground. It would then have been admirably adapted to occupy a more or less completely terrestrial niche with the onset of aridity and consequent shrinking of the forests at the end of the early Pliocene. It seems unlikely that *Ramapithecus* would have been committed to a bipedal form of locomotion, but almost certainly it would have been facultatively bipedal; if it was descended from a relatively generalized pongid stock, *Ramapithecus* might well have spent much of its existence in the trees with its body in an erect position. If this is so, *Ramapithecus* would have been admirably preadapted to an erect bipedal form of locomotion on the ground.

Relationship with man. It is generally accepted that there are two basic criteria fulfilled by modern man, and indeed by all hominids whose postcranial skeleton is known. The first of these lies in the dental complex, which has been discussed, and the second concerns adaptation for erect bipedal locomotion. *Ramapithecus* had crossed the first of these thresholds, possessing a thoroughly hominid dental mechanism which is ideally suited as a condition ancestral to that seen in *Australopithecus* of the late Pliocene and early Pleistocene and of later hominids. But it is by no means certain, especially in the absence of direct fossil evidence, that it had crossed the other threshold, and there are paleontologists who consider that both requirements should be satisfied before a form can be regarded as hominid.

However, if the ancestors of hominids are best classified as pongids, as seems at present to be the case, it is logical that any form which is ancestral to, or close to the ancestry of, later hominids and which has crossed at least one of the diagnostic thresholds should properly be regarded as hominid. In this view, the assignment of *Ramapithecus* to Hominidae is inevitable.

For background information *see* AUSTRALOPITHECUS; RAMAPITHECUS; ROCK, AGE DETERMINATION OF in the McGraw-Hill Encyclopedia of Science and Technology.   [IAN M. TATTERSALL]

Bibliography: F. C. Howell, *Nature*, 223:1234–1939, 1969; R. E. F. Leakey et al., *Nature*, 226:223–230, 1970; V. J. Maglio, *Nature*, 225:328–332, 1970; J. Martyn and P. V. Tobias, *Nature*, 215:476–480, 1967; B. Patterson and W. W. Howells, *Science*, 156:64–66, 1967.

Hormone

Reports have appeared of recent research in two hormones, thyrotropin-releasing hormone and ovine lactogenic hormone. These reports have indicated that (1) the chemical structure of thyrotropin-releasing hormone may be the same in pigs, sheep, and man and that synthetic thyrotropin-releasing hormone is active in man; and (2) the primary structure of the lactogenic hormone has been elucidated.

Thyrotropin-releasing hormone (TRH). This is the first hypothalamic releasing hormone to be isolated and to have its structure elucidated and synthesis accomplished. TRH is the simple substituted tripeptide L-pyroglutamyl-L-histidyl-L-proline amide shown in Fig. 1. It stimulates release of thyrotropin (thyroid-stimulating hormone, TSH) from the pituitary and thereby controls function of the thyroid gland. Independent studies by two groups of investigators indicate that this compound is the chemical structure of TRH of pigs and sheep. Recent evidence indicates that Pyroglu-His-Pro-NH$_2$ also may be the chemical structure of TRH of man. Furthermore, synthetic TRH is active in man. These results are exciting not only because this is the first definition of the chemical structure of one of the elusive hypothalamic releasing hormones but also because it again represents unequivocal evidence that pituitary hormonal secretion is controlled by specific chemicals secreted from the hypothalamus.

In late 1930, G. W. Harris formulated the neurovascular control theory of adenohypophysial function from his and others' anatomical and physiological findings. In essence, it has been clearly demonstrated that the hypothalamus secretes regulatory substances, called hypothalamic releasing hormones or factors, into the portal system connecting the hypothalamus and anterior pituitary gland, which regulate the release and possibly the synthesis of the anterior pituitary hormones. These substances appear to be liberated from hypothalamic nerve fibers terminating in the capillaries of the hypophysial portal vessels in the median eminence. Chemical and physiological evidence indicates that there are at least seven adenohypophysial hormones and seven companion hypothalamic releasing hormones which regulate their release.

Purification and isolation. TRH was isolated in 1966 from fragments of tissue dissected from the ventral portion of the hypothalamus of pigs. The fragments consisted mainly of the median eminence area and pituitary stalk; after lyophilization, the fragments weighed 10–35 mg each. The lyophilized hypothalamic fragments were pulverized in batches of 10,000–40,000 on dry ice before extraction with acetic acid. Purification was accomplished in essentially six steps: filtration on Sephadex G-25, phenol extraction, ion-exchange chromatography on CM-cellulose (carboxymethylcellulose) columns, countercurrent distribution, free-flow electrophoresis, and partition chromatography on Sephadex G-25. During the purification, the presence of TRH activity was determined by bioassay in mice. In some instances, after adsorption chromatography on a charcoal column, TRH was repurified by analytical scale gel filtration of Sephadex G-25 followed by descending chromatography on Whatman filter paper.

Ovine TRH was isolated in essentially the same way, except that CM–Sephadex G-50 was used instead of the phenol extraction. TRH from bovine and human hypothalami behaved essentially the same as porcine TRH when purified by the former scheme.

The purification methods used for porcine TRH led to no detectable inactivation, and more than 90% of the total TRH activity applied was recovered after each step. Because the material was purified about 570,000 times, the specific biological activity increased proportionally. From two batches of 100,000 and 165,000 porcine hypothalami, 2.4 and 4.4 mg of TRH were isolated.

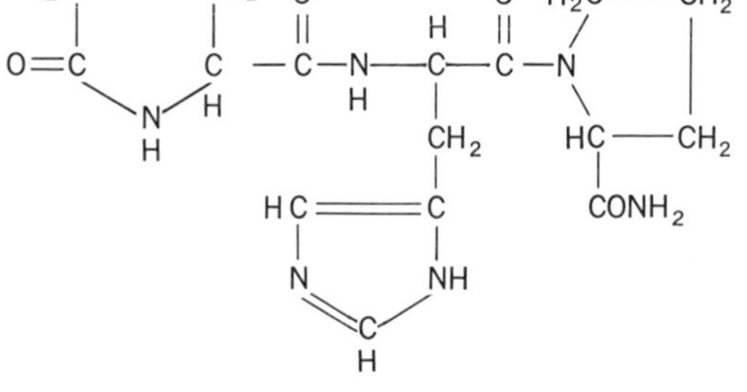

L-Pyroglutamyl-L-Histidyl-L-Proline amide

Fig. 1. Structural formula of TRH.

Chemical structure. After thin-layer and paper chromatography and electrophoresis, the porcine TRH obtained appeared homogeneous. Furthermore, the isolated TRH had only 5–10% of organic impurities detectable by charred spots developed on chromatograms. In the above systems, biological activity could be found only in areas adjacent to a spot which appeared after spraying with diazotized sulfanilic acid (Pauly's reagent), charring, or exposure to iodine vapor. The biologically active areas were ninhydrin-negative but slightly positive to chlorine-*o*-tolidine reagent. Later it was found that the Pauly reaction indicated an unsubstituted imidazole ring of histidine and that the chlorine-*o*-tolidine reaction indicated the presence of some peptide bonds. However, it was found that a number of proteolytic enzymes failed to affect the activity of TRH, whereas 80% of the biological activity was destroyed by the Pauly reagent. After acid hydrolysis, porcine TRH was found to contain histidine, proline, and glutamic acid in equimolar amounts.

During 1969 not only were studies on the further elucidation of the chemical structure of TRH performed but also an intensive approach in the organic synthesis of compounds which may have TRH activity was carried out. Both approaches were valuable and mutually complementary.

The amino acid sequence was determined to be Glu-His-Pro, and only compounds with this sequence prepared by organic synthesis had TRH activity. Dansyl amino acids were not detected after reaction of TRH with dansyl chloride (1-dimethyl aminonaphthalene-5-sulfonyl chloride), a typical aromatic sulfonyl chloride which reacts with primary and secondary amines and also with phenolic hydroxyl groups. Thus free NH_2-terminal or phenolic-OH groups did not appear to be present, nor could a free COOH terminus be detected by the Nedkov-Genov method of hydrazinolysis.

Oxidative nonenzymatic cleavage of TRH with *N*-bromosuccinimide (NBS) was performed, since this method is very selective for peptides containing histidine if tryptophan or tyrosine is not present. This brominating reagent has been found to cleave C-peptide bonds of amino acids which have a double bond in the γ-δ position relative to the carboxyl group of the peptide bond. After NBS cleavage of TRH, followed by dansylation of the reaction products, dansyl-proline was identified by thin-layer chromatography on silica gel. On the other hand, when the reaction products obtained after NBS cleavage were first reacted with phenyl isothiocyanate and then hydrolyzed with trifluoroacetic acid before dansylation, no dansyl amino acids were found. These results indicated that proline was at the C-terminus.

Pyroglutamyl hydrazide, but not other acyl hydrazides, was identified after hydrazinolysis of TRH and chromatography on silica gel. Pyroglutamyl, or 2-pyrrolidone-5-carboxylic acid, is the cyclized form of glutamic acid and, interestingly, has been found to be the N-terminus of gastrin, eledoisin, and fibrinopeptides. The pyroglutamyl N-terminus remains intact under the influence of anhydrous hydrazine, whereas its α-carboxyl end, which is involved in the peptide link, opens with formation of the α-hydrazide, which was revealed as one of the reaction products. Furthermore,

treatment of TRH with 1 *M* NaOH at 23°C, conditions specified for opening the pyrrolidone ring, followed by dansylation, hydrolysis, and identification by thin-layer chromatography, revealed dansyl–glutamic acid. Thus the cyclized form of glutamic acid, pyroglutamyl, appeared to be the N-terminus of TRH.

Mass spectrometry. Mass spectral studies of porcine and ovine TRH were helpful in confirming the amino acid sequence and the structure. A parent molecular ion was not obtained when free TRH was subjected to mass spectrometry at 245°C, and the fragmentation pattern did not proceed in an orderly fashion. However, intense peaks characteristic of pyroglutamyl, histidyl, prolyl, and their degradation products were recorded. Permethylation of TRH increased its volatility, and a parent molecule ion, 432, was recorded at 175°C, indicating a pentamethyl derivative. An orderly fission of permethylated TRH occurred. Peaks characteristic of prolyl, methyl- and dimethylhistidine, Pyroglu, $N(CH_3)_2$, and the methyl group were present in the mass spectrum. The $N(CH_3)_2$ fragment suggested a $CON(CH_3)_2$ grouping at the carboxyl end of proline and was probably derived from prolylamide at the C-terminus. When permethylation was performed after methylation, a peak indicating the prolyl methyl ester appeared in the mass spectrum. This indicated that the amide group of proline was hydrolyzed and that proline was methylated. Thus prolylamide again appeared to be the C-terminus.

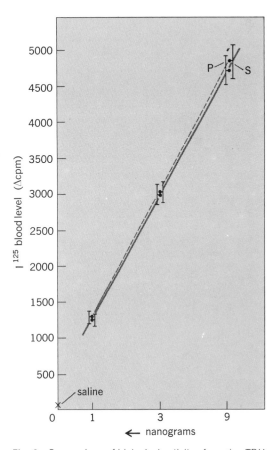

Fig. 2. Comparison of biological activity of porcine TRH and (pyro)Glu-His-Pro(NH_2) in T_3-TRH assay mice.

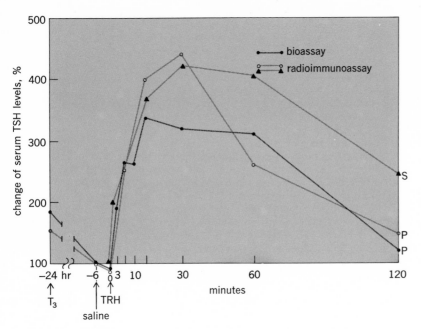

Fig. 3. Porcine TRH administered to cretin pretreated with single dose of T₃ and synthetic TRH administered to normal adult male.

Organic synthesis. Some of the chemical results cited above, as well as the three amino acids and their sequence, were the basis for the organic synthesis steps performed. One of the first active compounds prepared with TRH activity followed treatment of the simple unsubstituted tripeptide Glu-His-Pro with acetic anhydride. Following the acetylation procedure, performed to block the N-terminus, the major product was shown to be Pyroglu-His-Pro-OH. It had 1/5000 the activity of TRH.

In order to block both the N- and C-termini of Glu-His-Pro, this tripeptide was first treated with anhydrous methanol containing hydrogen chloride and then with anhydrous methanol saturated with ammonia. By these chemical steps, TRH was synthesized. Some studies suggest that the latter conditions may be expected to give predominantly, but not exclusively, Pyroglu-His-Pro-NH₂. The methanol-HCl procedure was performed to methylate the free carboxyl groups of proline and glutamic acid, forming the dimethyl ester of Glu-His-Pro. Methanol-NH₃ treatment of this dimethyl ester was performed to convert the methyl ester of proline into prolylamide and to cyclize the half-methyl ester of glutamic acid to a pyrrolidone ring. Even though ammonation of the half-methyl ester of glutamic acid might be expected to result in the synthesis of glutamine, other workers found, under strict anhydrous conditions, that the main product was always L-pyrrolidone carboxylic acid in 70–90% yield. When these same synthetic reactions were carried out on the tripeptides Glu-Pro-His, Pro-His-Glu, and Pro-Glu-His, none of the resulting preparations showed TRH activity.

The structure of L-pyroglutamyl-L-histidyl-L-proline amide, which has a molecular weight of 362, has all the known chemical and hormonal properties of TRH isolated from porcine hypothalami. The R_f values of synthetic Pyroglu-His-Pro-NH₂

and porcine TRH were indistinguishable in 17 diversified chromatography systems. To critically and quantitatively compare the biological activity of synthetic and porcine TRH, a method for quantitating small amounts of this hormone was developed. As little as 0.5 μg of these preparations could be estimated (± 10% error) with a densitometric method by photometry of the charred spots developed after chromatography on glass paper impregnated with silica gel. After chromatography in chloroform/methanol/concentrated ammonia (60:5:0.5), the chromatogram was sprayed with H_2SO_4–dichromate solution, and the charred spots were developed by heating. The regression for synthetic and porcine TRH by this method was linear, between 0.5 and 1.5 μg.

Biological studies. As recorded in Fig. 2, synthetic and porcine TRH were equipotent when assayed by the T_3-TRH method in mice (T_3 represents triiodothyronine). In this assay, the biological response is quantitatively measured by the change in counts per minute (Δcpm) of I^{125} blood level before and 2 hr after the intravenous injection of the sample to be assayed. The I^{125} blood levels are proportional to the amount of the induced release of TSH from the pituitary.

Other biological responses characteristic of both synthetic Pyroglu-His-Pro-NH₂ and porcine TRH were stimulation of TSH release from rat anterior pituitary glands in the test tube; inhibition of the living-body and the test-tube responses by triiodothyronine; elevation of plasma TSH levels in rats and mice as well as man; inactivation when incubated in human or rat serum; and both were active when given intraperitoneally, intravenously, and subcutaneously. When synthetic Pyroglu-His-Pro-NH₂ was given to mice orally, it was also active.

In 1966 C. Y. Bowers and coworkers found that a highly purified preparation of porcine TRH elevated plasma TSH levels in man. Because of the difficulties of isolating TRH from natural sources and the ready availability of synthetic TRH, it is probable that no other studies with natural TRH will be performed in man. As recorded in Fig. 3, administration of porcine TRH or synthetic Pyroglu-His-Pro-NH₂ to man elicited essentially the same pattern of serum TSH response. A quick single intravenous injection of 800 μg of synthetic Pyroglu-His-Pro-NH₂ into a normal male subject elevated serum levels of TSH within 2 min. The peak elevation of TSH occurred after 30 min, and was 370% higher than the level at zero time; after 3 hr the level decreased to normal. The plasma levels of growth hormone, luteinizing hormone, and follicle-stimulating hormone did not significantly change, and the plasma level of cortisol rose only slightly. These studies show the hormonal activity and specificity of synthetic TRH to release TSH in man and indicate its future use for studying the pituitary-thyroid system of man in health and disease.   [CYRIL Y. BOWERS]

Ovine lactogenic hormone (LTH). The primary structure of ovine lactogenic hormone was disclosed in November, 1969. This accomplishment opened the way for the possibility of a synthetic product possessing mammary gland–stimulating activity. In addition, knowledge of the primary structure may provide an explanation of the intrinsic lactogenic activity in the human growth hormone (HGH) molecule.

The first indication that a hormone of the anterior pituitary might effect lactation in mammals came from the experiments of P. Stricker and F. Grueter in 1928. Later, Oscar Riddle and coworkers reported growth of the crop glands of the pigeon about 3 days after the administration of pituitary extracts. Subsequent work by Riddle and others indicated that this effect in birds is due to the same hormone that induces lactation in mammals. A number of terms have since been proposed for the lactogenic hormone, including prolactin and mammotropin. After the discovery by H. M. Evans, E. B. Astwood, and colleagues that the hormone is capable of maintaining luteal function in rats, the term luteotropin was also introduced. For convenience, ovine lactogenic hormone is abbreviated LTH.

Lactogenic activity has been reported to be present in the pituitary extracts of various species, including humans, sheep, cattle, pigs, horses, whales, rabbits, cats, rats, guinea pigs, mice, fish, amphibians, reptiles, and birds. Highly purified preparations of lactogenic hormone have been obtained from sheep and cattle pituitary glands by W. R. Lyons, A. White, C. H. Li, and coworkers. It is not possible to obtain from extracts of human pituitaries a distinct preparation which chemically differs from HGH and possesses only lactogenic activity.

Biological activity. The table presents a listing of some of the reported influences of LTH on various vertebrate organs and functions. The role of LTH in mammary-gland development and function has long been recognized. LTH acts as a synergist with steroidal hormones in the sequential development of ductal and lobuloalveolar components of the mammary tree (mammogenic actions). Secretory activity is also dependent upon the action of LTH in synergism with adrenal hormones to initiate milk production (lactogenic action) and to maintain and augment it (galactopoietic action).

In birds, two of the most striking actions of LTH are stimulation of crop "milk" formation in pigeons and doves and induction of brood-patch development in a variety of species. LTH is shown to promote lean body growth in lizards (*Anolis*), as well as to stimulate regeneration of the tail. LTH also caused increased body weight and increased tail length in *Rana catesbeiana* tadpoles.

The hormone is essential for fresh-water fish to survive osmotic stress. In efts, injection of LTH induced the land-living stage of this species to return prematurely to water. The water-drive activity of LTH is also seen in salamanders. Apparently, the hormone is responsive for the survival of these animals in returning to water (second metamorphosis).

From the listing in the table and the above discussion, it is evident that LTH is remarkable in its broad spectrum of biological actions among the vertebrates. The chemical nature of this extraordi-

NH$_2$—Thr—Pro—Val—Cys—Pro—Asn—Gly—Pro—Gly—Asp—Cys—Gln—Val—Ser—Leu—Arg—Asp—Leu—Phe—Asp—Arg—Ala—Val—Met—
1   5   10   15   20

Val—Ser—His—Tyr—Ile—His—Asn—Leu—Ser—Ser—Glu—Met—Phe—Asn—Glu—Phe—Asp—Lys—Arg—Tyr—Ala—Gln—Gly—Lys—
25   30   35   40   45

Gly—Phe—Ile—Thr—Met—Ala—Leu—Asn—Ser—Cys—His—Thr—Ser—Ser—Leu—Pro—Thr—Pro—Glu—Asp—Lys—Glu—Gln—
50   55   60   65   70

Ala—Gln—Gln—Thr—His—His—Glu—Val—Leu—Met—Ser—Leu—Ile—Leu—Gly—Leu—Arg—Ser—Trp—Asn—Asp—Pro—Leu—Tyr—
75   80   85   90   95

His—Leu—Val—Thr—Glu—Val—Arg—Gly—Met—Lys—Gly—Val—Pro—Asp—Ala—Ile—Leu—Ser—Arg—Ala—Ile—Glu—Ile—Glu—Glu—
100   105   110   115   120

Glu—Asn—Lys—Arg—Leu—Leu—Glu—Gly—Met—Glu—Met—Ile—Phe—Gly—Gln—Val—Ile—Pro—Gly—Ala—Lys—Glu—Thr—Glu—
125   130   135   140

Pro—Tyr—Pro—Val—Trp—Ser—Gly—Leu—Pro—Ser—Leu—Gln—Thr—Lys—Asp—Glu—Asp—Ala—Arg—His—Ser—Ala—Phe—
145   150   155   160   165

Tyr—Asn—Leu—Leu—His—Cys—Leu—Arg—Arg—Asp—Ser—Ser—Lys—Ile—Asp—Thr—Tyr—Leu—Lys—Leu—Leu—Asn—
170   175   180   185

Cys—Arg—Ile—Ile—Tyr—Asn—Asn—Asn—Cys—COOH
190   195   198

Fig. 4.   The amino acid sequence of LTH.

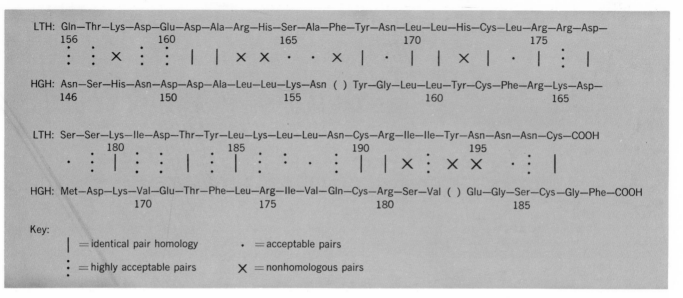

Fig. 5. Comparison of the structures of HGH and LTH.

narily versatile molecule is discussed below.

Structure. LTH is a protein consisting of 198 amino acids, with a molecular weight of 23,000. The exact arrangement of these amino acids in definite sequence has been worked out by scientists at the Hormone Research Laboratory of the University of California (Fig. 4).

It may be recalled that human pituitary glands produce the growth-hormone molecule which possesses both growth-promoting and lactogenic activities, whereas nonprimate pituitaries have separate protein entities for these two activities. Do the two hormones, namely HGH and LTH, have certain structural features in common?

When primary structures of these two hormones were compared, a number of interesting similarities were detected; in particular, the carboxyl terminal portions of the molecules were found to be almost identical (Fig. 5), a result suggesting that the structure present in this area of the molecules is of primary importance for lactogenic activity. There are now attempts to determine, through a variety of chemical studies, whether this is so. It is hoped that the entire structures of HGH and LTH will not be necessary for biological activity. If certain portions of the molecules can be eliminated without destruction of activity, chemical synthesis may provide a practical means of supplying pure hormones for biological and clinical investigations.

For background information *see* HORMONE in the McGraw-Hill Encyclopedia of Science and Technology.   [CHOH HAO LI]

Bibliography: H. A. Bern, *Science*, 158:455, 1967; C. Y. Bowers et al., *Endocrinology*, 86:1143, 1970; R. Burgus et al., *Endocrinology*, 86:573, 1970; K. Folkers et al., *Biochem. Biophys. Res. Commun.*, 37:123, 1969; C. H. Li et al., *Nature*, 224:695, 1969; W. R. Lyons, *Proc. Roy. Soc.*, 149:303, 1958; R. M. G. Nair et al., *Biochemistry*, 9:1103, 1970; A. V. Schally et al., *J. Biol. Chem.*, 244:4077, 1969.

Human-factors engineering

Industrial engineers are concerned with the development of effective production systems through the optimal utilization of man, materials, and manufacturing plant. Human-factors engineering, as applied to industry, is a subbranch of industrial engineering, concerned with the adaptation of the physical working environment as well as with the design of equipment to the needs of man. The aim is to create conditions conducive to working efficiency, assuring at the same time the highest possible level of physical and emotional well-being of the worker. Sometimes the term ergonomics is used as an approximate synonym to human-factors engineering, especially by European authors.

This article focuses upon recent advances in

Some biological actions of LTH

Animal	Biological activity
Mammals	1. Stimulation of mammary gland development and lactation 2. Stimulation of growth 3. Increased fertility of dwarf mice 4. Synergism with androgen in male sex accessory growth 5. Maintenance and secretion of corpus luteum in mouse, rat, and ferret
Birds	1. Secretion of crop milk 2. Formation of brood patch 3. Stimulation of feather growth 4. Feeding of young 5. Synergism with steroids on female reproductive tract
Teleost fish	1. Growth and secretion of seminal vesicles 2. Nest building and fin fanning 3. Melanogenesis and proliferation of melanocytes 4. Lipid deposition 5. Resistance to high-temperature structure
Amphibians	1. Stimulation of eft water drive 2. Stimulation of larval growth 3. Stimulation of limb regeneration 4. Secretion of oviducal jelly 5. Proliferation of melanophores
Reptiles	1. Stimulation of somatic growth 2. Stimulation of tail regeneration 3. Reduction in lipid deposition 4. Hyperphagia 5. Epidermal sloughing

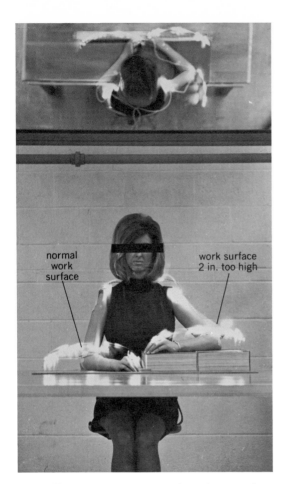

Fig. 1. Chronocyclegram tests. Wrist and arm motions at normal work-surface height and when work surface is 2 in. too high (to the right in both photographs). (*From E. R. Tichauer, J. Ind. Eng., 19(2):96–104, 1968*)

Range, velocity, and precision of specific work-related movements are measured and recorded by kinesiometers. These devices often embody dynamometers to record strength of motion or torques, or both, that are exerted on the joints of a worker's limbs (Fig. 2). Kinesiometers produce kinesiograms, which are continuous records of the more important biomechanical parameters of human performance throughout a complete work cycle.

Electromyography is used in industrial biomechanics whenever the strength and duration of muscular activity needed to produce performance of a specific task are evaluated. Electromyographs, similar to electrocardiographs in construction, amplify and record electric potentials arising in muscles during physical activity. For this purpose, electrodes are fastened with adhesive tape to the skin overlying muscles or muscle groups under study. Needle electrodes directly inserted into muscles require a medical environment and sterility for their application and are not suitable for use under industrial conditions.

Tracings of kinesiograms and electromyograms, when produced simultaneously by multichannel recorders, constitute biomechanical profiles (Fig. 3). In Fig. 3 the higher the myogram tracing, the greater the effort used in rotating the tool. The higher the rotation tracing, the greater the tool rotations. A comparison of the tracings in Fig. 3 indicates that the straight wrist can rotate the tool more and with less effort than if the wrist were deviated. Biomechanical analysis employing the body of knowledge of functional anatomy alone, or in combination with conclusions derived from biomechanical profiles, can be profitably applied to the design, evaluation, and improvement of handtools, machinery, and materials-handling.

industrial biomechanics, a subdiscipline of human-factors engineering, which are of significant usefulness in the adaptation of manufacturing processes to the needs of today's rapidly changing work force and the demands of an even more rapidly changing technology.

Biomechanical methodology. Specialists in biomechanics apply anatomical knowledge to the design of equipment and work situations. This permits the development of environmental conditions conducive to high efficiency, while avoiding conditions stressful to vulnerable structures of the human body.

The geometry of motion patterns demanded by workplace and equipment is studied by the chronocyclegraph (Fig. 1), which provides a visual record of motion pathways obtained from small light-bulbs fastened to anatomical reference points (such as limbs and fingertips) of individuals studied. In Fig. 1 the displacement of the wrist can be performed by simple rotation of the upper arm without elbow movement. However, when the work surface is 2 in. too high, a full and fatiguing shoulder-elbow swing must be performed to produce the same wrist movement. The chronocyclegraph permits detection of erratic, unnecessary, and stressful movements and their elimination by improved workplace engineering.

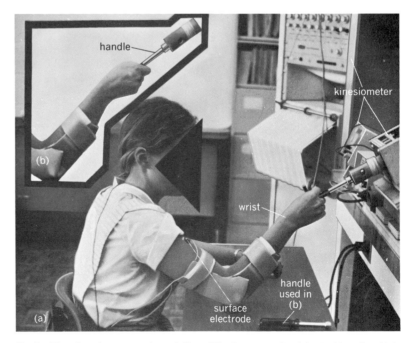

Fig. 2. Kinesiometer measuring rotation of the forearm using (*a*) a tool handle which permits the wrist to be kept straight. A surface electrode simultaneously picks up a myogram of the biceps, which is one of the muscles rotating the forearm, and a potentiometer attached to the end of the tool shaft measures rotation of tool. (*b*) A tool handle which forces the wrist into deviation, which is uncomfortable and fatiguing. (*Adapted from E. R. Tichauer, Electromyography, vol. 8, suppl. 1, p. 200, 1968*)

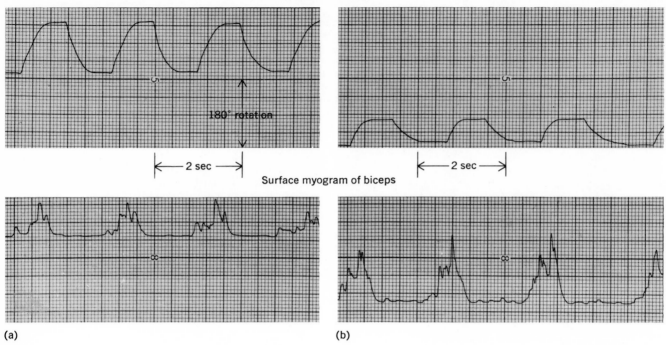

range of rotation

180° rotation

2 sec   2 sec

Surface myogram of biceps

(a)   (b)

Fig. 3.   Biomechanical profile of forearm rotation using tracing of tool rotation and surface myogram of biceps.

(a) Wrist straight. (b) Wrist deviated. (*Adapted from E. R. Tichauer, Electromyography, vol. 8, suppl. 1, p. 201, 1968*)

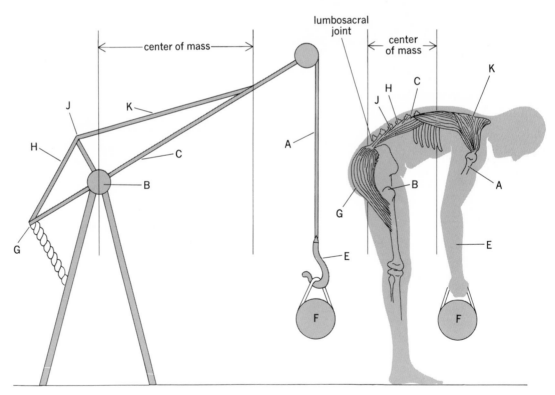

Fig. 4.   Man-crane analogy. The product of the combined weight of torso and load multiplied by the distance of its center of mass from the lumbosacral joint expresses true lifting stress in terms of inch-pounds of torque.

Letters identify similar stress points on crane and man. (*From E. R. Tichauer, in Biomechanics Monograph, American Society of Mechanical Engineers, 1967*)

Analytical biomechanics. Known biomechanical cause-and-effect relationships often permit the application of deductive reasoning as an effective substitute for lengthy experimentation. This is especially desirable in stressful work situations such as lifting, where experimentation could be hazardous to the subject population. Stress conditions within the human body during lifting are analogous to those of a crane (Fig. 4). Therefore lifting stresses acting on the vertebral column and the muscles of the back are not determined solely by the weight of the object lifted. On the basis of analytical reasoning, the magnitude of lifting stresses must be expressed in inch-pounds of torque acting on a risk-exposed anatomical structure, for example, the lumbosacral joint. This is done by adding the estimated weight of the body segment involved in the maneuver to the weight of the object handled and multiplying the aggregate weight by the distance of its center of mass from the lumbosacral joint.

When individuals in upright posture handle objects, simple multiplication of the weight of an article by the distance of its center of mass from the back of the individual lifting yields an approximation, adequate for practical purposes, of the biomechanical lifting equivalent, expressed in inchpounds (Fig. 5). This procedure, which takes the bulk-weight ratio of an object into consideration, is of great importance to safety in modern industry, where lightweight miniaturized components are often packaged in bulky protective containers and where strains of the back are the most frequent cause of disability.

Work tolerance. Industrial engineers aim at the design of man-task systems of high efficiency. Work tolerance is defined as the ability of individuals within a working population to maintain expected levels of efficiency without detriment to health and well-being. A high degree of work tolerance is often achieved through the maintenance of biomechanically sound working postures brought about by the correct relationship between chair height and workbench height (Fig. 6).

A working chair only 2–3 in. too low will induce workers to raise their elbows excessively. This increases forces acting on the shoulder joint and thus causes early fatigue. As a result, performance levels drop off. Furthermore, the raised arm produces tension in those muscles which run from the front of the rib cage to the armpit. Such strained posture can cause vague discomfort and light pain over the chest and shoulders after a few hours. "Heart-conscious" individuals, such as middle-aged overweight males, may then have unfounded fears of developing heart disease. These fears can cause emotional suffering, resulting in greatly reduced work tolerance.

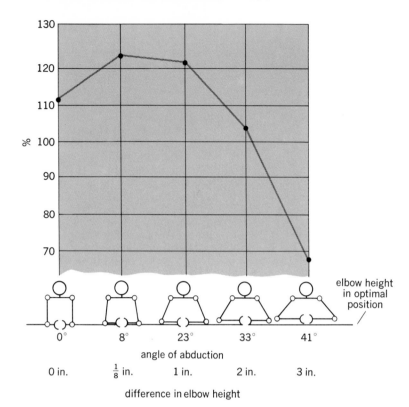

Fig. 6. Work tolerance. Graph shows how performance efficiency (in percent) decreases when the elbows are raised during work. Simultaneously, muscles of the chest and shoulder become stretched and may cause discomfort. (*Adapted from E. R. Tichauer, Proceedings of the Professional Conference, American Society of Safety Engineers, 1968*)

Biomechanics in modern industry. The composition of the work force has changed drastically in the course of the last two decades. Prior to World War II, the industrial worker normally entered a particular branch of manufacturing when he was still in adolescence. Over the years the need for adjustment, within a competitive climate, to an environment of machinery and mechanical processes inculcated efficient and safe working habits into many individuals through experience.

Today, substantial portions of the work force are recruited from population groups of adult age often unaccustomed to manufacturing situations (such as housewives seeking assembly work, white-collar workers experiencing technological obsolescence, and hard-core unemployed and ghetto populations). Here biomechanics assists in promoting easier acclimatization to the industrial environment through identification of potential failure points in man-task systems. This enables human-factors engineers to take preventive action while work methods or training programs are still in the design stage. Thus not only are employees protected, but economic and social disappointments to labor and management alike are avoided.

For background information *see* HUMAN-FACTORS ENGINEERING in the McGraw-Hill Encyclopedia of Science and Technology.

[ERWIN R. TICHAUER]

Bibliography: O. G. Edholm, *The Biology of Work*, 1967; K. F. H. Murrell, *Ergonomics*, 1969; E. R. Tichauer, *Electromyography*, vol. 8, suppl. 1, pp. 197–211, 1968; E. R. Tichauer, *J. Ind. Eng.*, 19(2):96–104, 1968.

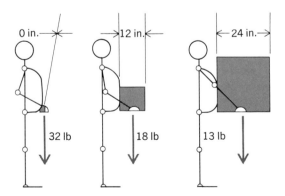

Fig. 5. Biomechanical lifting equivalent. All the boxes shown, in spite of their weight differences, exert identical torques on the lumbar spine and therefore represent lifting tasks of equal severity. (*From E. R. Tichauer, Proceedings of the Professional Conference, American Society of Safety Engineers, 1968*)

Hurricane

Results of modification experiments on Hurricane Debbie, August, 1969, suggest that man may someday use energy stored within these destructive tropical cyclones to blunt their power and reduce the damage caused by them. The experiments were conducted by Project STORMFURY, a cooperative program of the U.S. Departments of Commerce (ESSA) and Defense (Navy) with the assistance of other government agencies and several universities. In addition to the field modification experiments, the program includes a number of theoretical and experimental investigations of the structure and energy processes in hurricanes. Results from some of these projects also suggest that hurricanes can be beneficially modified.

Hurricane damage to property in the United States has been steadily increasing. Hurricanes caused an average annual damage in the United States of $13,000,000 between 1915 and 1924. By the period 1960–1969, this figure had soared to $432,000,000. Even after adjusting these damage statistics to the 1957–1959 base of the cost of construction index, the increase in the last 50 years is 650%. This trend should continue because structures being built in areas vulnerable to hurricanes are increasingly expensive. Hurricanes Betsy (1965) and Camille (1969) each caused more than $1,400,000,000 of damage.

History. The current systematic research on modification of hurricanes extends back to 1961, when R. H. Simpson suggested that hurricanes might have their winds reduced by introducing freezing nuclei into the massive wall clouds around the eye, or center, of the storm. About the same time, Pierre St. Amand and associates in the Navy developed pyrotechnic silver iodide generators which made it practical to deposit very large

quantities of freezing nuclei (in this case, silver iodide crystals) into clouds within a few minutes.

Hurricane Esther was seeded one time on each of two days in September, 1961, and Hurricane Beulah received similar treatment in August, 1963. Results of these experiments were encouraging but not conclusive, because changes in the winds and structure of the hurricanes were of about the same magnitude as changes which often occur naturally in storms.

A multiseeding experiment was designed prior to the 1965 hurricane season, but a suitable hurricane did not occur in the experimental areas until 1969. During these years research continued and resulted in changes of the plan. Improved silver iodide generators also became available before the 1969 season. Therefore the frustration of waiting 4 years for opportunities may not have been in vain. The succession of apparently minor changes to improve the design of the seeding experiment may have made the difference between success and failure for the Debbie experiments.

Modification technique. The modification technique seeks to make the hurricane work against itself. Hurricane clouds contain large quantities of water still in the liquid state at temperatures lower than the freezing temperature. If these supercooled drops can be made to freeze, they will release the amount of heat required to melt an equivalent amount of ice, thereby possibly providing a means of modifying a hurricane. Two questions are suggested by this proposed method: (1) How can the supercooled water be made to freeze on command? (2) Where in the hurricane would introduction of additional heat cause a reduction in the intensity of the storm?

Supercooled water drops will freeze when they contact freezing nuclei. Ice crystals are very effective natural freezing nuclei, and silver iodide crystals have proved effective as artificial freezing nuclei. Researchers in ESSA's Research Flight Facility and National Hurricane Research Laboratory have collected data which suggest that, in most hurricane clouds, there are relatively few ice crystals at temperatures as low as −8°C. This means that adding silver iodide crystals might indeed modify the hurricane clouds and cause addition of heat.

The answer to the question of where to add the extra heat to the hurricane has been sought by theoretical investigations. Several models of hurricanes have been developed in recent years. The one developed by Stanley L. Rosenthal has been used at the National Hurricane Research Laboratory to simulate seeding experiments at several different radial distances from the center of the hurricane. The results from these experiments, supplied by a computer using the theoretical hurricane model, suggest that the greatest reduction of maximum winds can be achieved by adding the extra heat to the supercooled water clouds at distances from the storm center greater than the radius of maximum winds, which is also outside the mass of relatively warm air in the core of the hurricane.

These concepts are illustrated in Fig. 1, which represents schematically the distribution along a radius of: (1) temperatures in the middle and upper layers of the storm, (2) pressure at sea level, and (3)

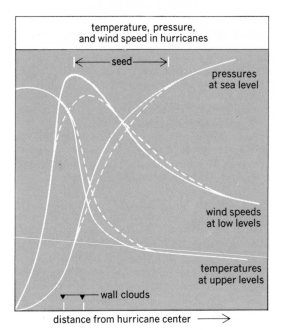

temperature, pressure, and wind speed in hurricanes

seed

pressures at sea level

wind speeds at low levels

temperatures at upper levels

wall clouds

distance from hurricane center ⟶

Fig. 1. Schematic representation of temperatures, pressures, and wind speeds in a hurricane. The broken lines indicate changes that might be caused by the seeding of supercooled clouds along radial band marked seed.

winds near the surface. Note that the air is warm-est near the hurricane center. If heat is added in the band marked seed, the temperature distribu-tion should change so that the difference between the temperatures in the core of the storm and those at the same elevation a few miles farther away should become less (see broken line). Pres-sure at sea level represents the weight of the air above. Changing the temperatures, therefore, should result in changes of the pressure distribu-tion (see broken lines). The wind speeds vary with the rate at which the pressure changes across the storm. Thus, changing the temperatures changes the pressure distribution and eventually the winds. This explanation oversimplifies the events that take place, but the simulated seeding experiment with the theoretical model does indicate that a reduction of maximum winds by a few percent might be achieved. The model simulates many fea-tures of a hurricane quite well, but it is still not perfected. Hence, the modification hypotheses must be tested on a real hurricane.

Debbie experiment. The track of Hurricane Debbie and the areas where it was seeded are shown in Fig. 2. The STORMFURY group seeded the storm five times on Aug. 18 and five times on Aug. 20. Thirteen aircraft participated: nine from the Navy, two from ESSA, and two from the Air Force. Of these, five seeded the storm and the remainder monitored the storm for changes.

Each seeded aircraft flew across the eye of the hurricane at 33,000 ft, entered the wall cloud on the north-northeast side, and dropped 208 silver iodide generators along the seed band indicated in Fig. 1.

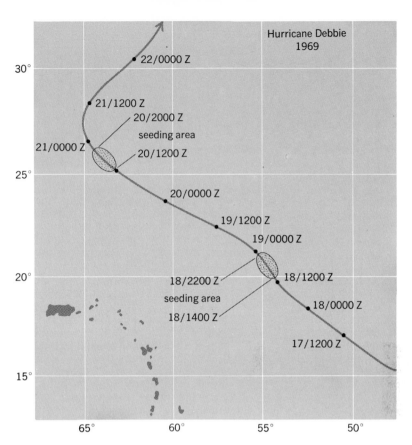

Fig. 2.   Track of Hurricane Debbie, August, 1969. Seeding areas on Aug. 18 and 20 are indicated on the track.

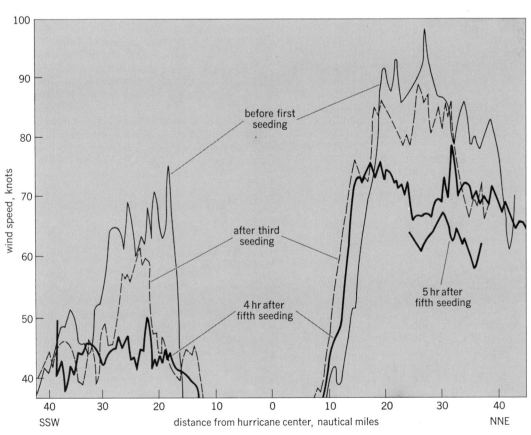

Fig. 3.   Changes with time of wind speeds at 12,000 ft in Hurricane Debbie on Aug. 18, 1969.

The smoke from the generators produced a curtain of freezing nuclei about 20 mi long and 20,000 ft deep. The winds in the hurricane should have carried the nuclei around the storm and gradually distributed them by turbulent action. Each generator had 190 g of silver iodide, and each gram could produce 10^{14} freezing nuclei that would be active at temperatures encountered in hurricanes. The five seeding runs were at intervals of 2 hr.

The monitoring of the storm started a few hours before the first seeding and continued until 6 hr after the fifth seeding. The most complete set of monitoring data was collected at 12,000 ft. The winds, temperature, pressure, moisture, and other meteorological parameters were measured once per second, and the data were recorded automatically on magnetic tape. Of these meteorological parameters, the most interesting data were the winds measured at this 12,000-ft level.

The maximum winds at 12,000 ft in Hurricane Debbie were 113 mph before the first seeding on Aug. 18 and decreased to 78 mph by 5 hr after the last seeding (Fig. 3). The winds were measured by aircraft flying across the storm from south-southwest to north-northeast and in the reverse direction. On Aug. 19 the storm reintensified. On Aug. 20 the maximum winds at 12,000 ft were 114 mph before the first seeding and decreased to 97 mph by 6 hr after the final seeding. These changes represent reductions in maximum winds of 31 and 15%, respectively. Since the force of the winds varies with the square of the wind speed or the kinetic energy, it is interesting that the kinetic energy at the belt of maximum winds decreased 52 and 28%, respectively, on the two experimental days. The winds at 12,000 ft in previous mature hurricanes such as Debbie have been about 95% as strong as the winds near the surface.

Significance. That Hurricane Debbie decreased in intensity following multiple seedings on Aug. 18 and 20, 1969, is well established. What is not known is whether the decrease was caused by the seeding or by natural forces. From analyses of past storms, however, one can conclude that the changes of wind speeds like those in Debbie have not occurred more often than 1 in 40 cases for nonmodified storms. This is based on considerations of the reduction of 31% on Aug. 18, no seeding and a regaining of intensity on Aug. 19, and a reduction of 15% on Aug. 20.

Thus the results of the Debbie experiments strongly suggest that some degree of beneficial modification was attained. The changes that occurred, however, could have been due to natural causes rather than the seeding experiment. One must therefore await a repetition of the experiments before deciding that man has finally found the means of making a hurricane work against itself to reduce its intensity.

For background information *see* HURRICANE in the McGraw-Hill Encyclopedia of Science and Technology.

[R. CECIL GENTRY]

Bibliography: R. C. Gentry, *Bull. Amer. Meteorol. Soc.*, 50:404–409, 1969; R. C. Gentry, *Science*, 168:473–475, 1970; S. L. Rosenthal, *ESSA Technical Memorandum*, ERLTMNHRL 82, U.S. Department of Commerce, January, 1969; R. H. Simpson and J. S. Malkus, *Sci. Amer.*, 211:27–37, 1964.

Hydrodynamics

A recent development in ship hydrodynamics has been the design and application of large protruding bulbous bows to ship hulls.

The hull of a ship can in general be described as being composed of smooth curved surfaces, with a sharp bow and a rounded or squared-off stern, but in recent years many new vessels have been constructed with a large elliptical or cylindrical swelling of the underwater hull form protruding forward of the bow. These bulbous bows extend ahead of what would otherwise be the stem of the ship. Figure 1 is a photograph of a modern fast containership, the *Hawaiian Enterprise*, which has a typical modern bulbous bow. Figure 2 is a photograph of a ship with an even more pronounced bulb.

The primary purpose of a bulbous bow is to reduce the drag, or hydrodynamic resistance, of a ship and hence reduce the horsepower required to drive it at a given speed. There are also effects on ship motions in waves and on steering, but these are minor compared to the power reduction.

Wave drag. Surface waves are created by any ship as it moves through the water. These waves spread out and finally disappear far astern of the ship. The wave system generated by a ship is shown in Fig. 3. The waves represent a continuing energy loss to the ship system, and a substantial portion of the ship's power is used in creating the waves. It follows that a ship form which produces minimum waves will in general require less power to drive it at a given speed. The action of the bulbous bow, if properly designed, is to greatly reduce or cancel the wave system generated at the bow of the ship. This phenomenon occurs when the bulb creates a surface wave system out of phase with the normal bow system.

The history of bulbous bows for ships is documented in technical literature as long ago as the 1860s, when it was reported that a bulblike protuberance was fitted to an armored cruiser. In the early 20th century ram bows were common on warships; however, it is believed that these protruding devices were intended primarily for the use indicated by their name even though they may have had beneficial effects on ship resistance. A theoretical explanation of the effects on wave drag

Fig. 1. *Hawaiian Enterprise*, one of the world's largest containerships with overall length of 719 ft and displacement of 34,700 long tons, features a protruding bulbous bow. (*Matson Lines*)

bulb size could be found. Many ships were built with bulbs of this type until large bulbs having much greater effect were introduced.

Construction of bulbs. In 1961 a 3000-ton Japanese passenger-ferry vessel, the *Kurenai Maru*,

Fig. 2. *Mormacsky,* a roll off – roll on ship, with a bulbous bow 83 ft long and 17 ft in diameter. (*Moore-McCormack Lines, Inc.*)

Fig. 3. Sketch of characteristic bow wave train.

was not then available. During the 1920s and 1930s, the basic mathematical theories were developed which described the wave systems produced by the ship alone, by the bulb, and by the mutual interference of the two wave systems. But practical applications of the theories were not realized until high-speed electronic digital computers could be used to help solve the complex mathematical equations involved.

In the meantime, in the 1940s and 1950s, ships' bows incorporated bulbous shapes which were largely confined to the normal stem profile. These were designed mainly by means of systematic tests in ship model towing tanks. Hull forms were altered by redistributing volume from the forward shoulder of the hull into the bulb so that an optimum, or least, horsepower requirement versus

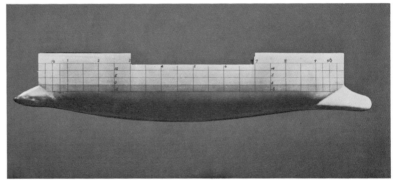

Fig. 4. Model of the idealized hull form of T. Inui.

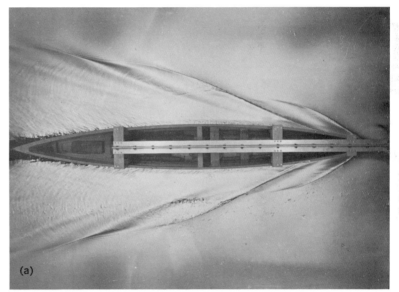

(a)

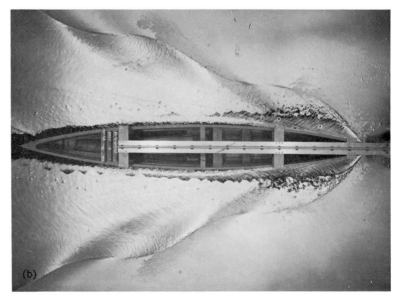

(b)

Fig. 5. Wave patterns produced in a ship model towing tank. (*a*) Patterns generated by a model without bulbs. (*b*) Patterns generated by a model with a bow bulb.

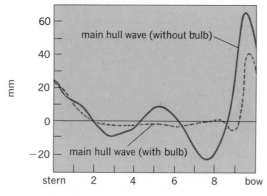

Fig. 6. Measured wave profiles on model hull side.

was fitted with a large protruding bulbous bow and tested competitively with an identical sister ship, the Murasaki Maru. The difference in their speed at the same power output was about one-half a knot, corresponding to a 13% saving in power by the *Kurenai Maru* for the same speed. Aerial photographs of the wave systems produced by the two ships clearly indicated the relative attenuation of waves after fitting of the bulb. The success of the experiment was due to research carried out by scientist – naval architect T. Inui of Tokyo University. Worldwide attention was given to Inui's work. He had applied the theory developed earlier, his own improvements to that theory, and experimentation in a ship model towing tank to produce highly efficient bulb forms. Inui showed that, with an idealized hull form, over a certain limited speed range he could virtually eliminate all waves with a bulb placed at both the bow and stern of the hull. Figure 4 shows Inui's idealized hull form, with both bow and stern bulbs; bow is to the right. Figure 5, which is a photograph taken in the ship model towing tank of the University of Tokyo, clearly indicates the wave patterns generated by the model without bulbs and the model with a bow

bulb. The surface of the water in the tank has been covered with aluminum powder to obtain this photographic effect. A profile of the wave systems at the hull side was also measured (Fig. 6).

Since the wave system created by the bow contributes the greatest amount to the total wave drag and since the fitting of a bulb at the stern is in conflict with the normal arrangement of propellers and rudders, only bow bulbs are normally fitted.

After publication of Inui's work, scientists in many countries began to work on the same problem. Bulbs were incorporated in many types of vessels, including warships, cargo ships, and research vessels. These were all relatively high-speed ships in which the drag due to wave making is a large portion of the total drag and the effects of the bulb could be expected to be substantial.

Another important type of ship is the bulk carrier, of which the oil tanker is the most numerous. These ships normally have very full hull forms and are designed for slower speeds where wave drag is not as significant as it is for high-speed vessels. It had been the belief of naval architects that large bulbs for this class of ship would be detrimental and would probably increase total drag. However, model tank tests were conducted in the early 1960s in the towing tanks at the University of Michigan and Hamburg, Germany, on full-bodied hull forms with large bulbs in order to determine whether potential power savings might be obtained. Because oil tankers normally operate fully laden from oil fields to the refineries and at a lighter draft or ballast condition on the return trip, it was necessary for the model tests to be conducted for both of these conditions. As the theory had indicated, the effect of bulbs in the fully laden condition was relatively small. The dramatic power reductions in the ballast condition were a surprise to naval architects, who had not previously been concerned with optimizing hull forms for ships at light draft. Horsepower reductions at a given speed of as much as 30% were reported. A typical tanker without bulb averages a 1-knot speed increase in both conditions after a bulb is installed, and may be able to make an additional round trip annually.

The large improvements effected for these slow-speed bulk carriers is greater than the amount which can be attributed to wave drag reduction only. It is believed that some of the decrease may be due to other effects, such as a reduction of viscous eddy resistance. However, at present there is a lack of understanding of this phenomenon, and additional research is needed to fully explain the reasons for it.

Many hundreds of oil tankers are now equipped with bulbous bows, either through new construction or retrofitting. One oil company alone reports that all of its ships, more than 200, now have bulbs of this type. Figure 7 shows a new medium-sized tanker with a typical bulbous bow. In order to simplify the construction and to reduce the costs of installation on existing ships, bulb forms consisting of a cylinder with a hemispherical nose have been fitted onto many vessels.

For background information *see* FLUID-FLOW PROPERTIES; HYDRODYNAMICS; SHIP DESIGN in the McGraw-Hill Encyclopedia of Science and Technology.

[RICHARD B. COUCH; JAMES L. MOSS]
Bibliography: J. P. Comstock (ed.), *Principles of*

Fig. 7. *ESSO San Francisco*, with typical bulbous bow. (*Standard Oil Co., New Jersey*)

Naval Architecture, SNAME, 1967; R. B. Couch and J. L. Moss, Application of large protruding bulbs to ships of high block coefficient, *SNAME Trans.*, 1966; T. Inui, Wave-making resistance of ships, *SNAME Trans.*, 1962.

Hydrology

Reports of some of the recent research in hydrology have been concerned with the progress of the International Hydrological Decade program, new analytical techniques in snow and ice hydrology, and water balance in North America.

International Hydrological Decade. The International Hydrological Decade, now in its seventh year (1971), has accelerated progress in the collection, analysis, and use of scientific water information and has advanced hydrological education through a wide variety of postgraduate courses, seminars, and symposia. Perhaps its principal achievement will be the impetus it will have given to demonstrating the usefulness of international cooperation in hydrology, hydrological education, and water resources activities and the need for such cooperation on a permanent basis after the Decade is over, at the end of 1974.

Background. The concept of a long-term international program of scientific studies encompassing the whole hydrological cycle was first advanced in the early 1960s, and the International Hydrological Decade was launched by the United Nations Educational, Scientific, and Cultural Organization (UNESCO) on Jan. 1, 1965. The Decade's main objective was to improve the ability of all countries to cope with individual and mutual water problems. This objective was to be attained by strengthening the scientific base for water use and by stimulating education in hydrology. At its midpoint, 105 countries were participating in the Decade in cooperation with UN agencies, such as the World Meteorological Organization (WMO), Food and Agriculture Organization (FAO), World Health Organization (WHO), and International Atomic Energy Agency (IAEA), and international scientific groups, particularly the International Association of Scientific Hydrology.

Program. The scientific core of the Decade program is the improvement of knowledge of the world's finite water supplies and their distribution in time and space. This central effort is supplemented by a program of about 60 projects involving almost all aspects of the quantity and quality of water in all of its phases. On the education side, emphasis is placed on the training of hydrologists, technicians, and educators and on the exchange of information through symposia and publications. Corollary aspects include the standardization and comparison of instruments and techniques and the encouragement of regional cooperation.

Organization. Although UNESCO provides administrative support for an official intergovernmental coordinating mechanism, each nation is responsible for selecting and carrying out its own program. Thus each country is able to participate in the Decade to the extent of its interest and within its own understanding of its capabilities. The Decade is a program of many countries collaborating to advance their knowledge, individually and collectively, to meet the rapidly enlarging problems of water planning, development, and management.

Each member nation has a national committee to guide and coordinate its participation, and the international leadership is provided by a Coordinating Council. During the first 5 years, the Council was composed of representatives of 21 nations, most of whom were rotated every 2 years. Representatives of international agencies may attend Council sessions and discussions, but they have no vote in its decisions.

Scientific activities. The Decade has placed considerable emphasis on the expansion of data-collecting networks, and many hydrological services have enlarged national networks and institutional capabilities during the past 5 years. World-wide networks of special Decade stations were selected to provide indices of regional and world-wide conditions. These networks now include 1462 river-gaging stations in 75 countries, 259 lake stations in 32 countries, 633 pan-evaporation and 132 lysimeter stations in 57 countries, and 345 groundwater stations in 28 countries. A descriptive list is available from UNESCO. Also, more than 550 representative basins and 220 experimental basins form a network of watershed areas that provides a means for monitoring and studying the effects of natural and man-made changes on water-bearing characteristics. A guide to conducting research on these basins and a preliminary inventory have been published.

An updated summary of global water resources is being prepared. Supplementing this are investigations of snow and ice, which comprise three-fourths of the world's fresh-water supply. The Decade also has been concerned with the development of hydrological mapping and in summarizing scientific research in special fields. One such report is the *Guidebook on Nuclear Techniques in Hydrology*, prepared and published in cooperation with the IAEA. Other reports are in various stages of preparation.

Educational activities. Seventeen symposia have been convened within the Decade program through 1969. Topics have ranged from snow physics and computer techniques to management of aquifers and land subsidence. The symposia, already held in 14 countries and on five continents, will continue through the Decade. Educational seminars were given in Iran, Brazil, Ethiopia, and Czechoslovakia through 1969, and negotiations are under way for five more through 1971. Special courses are being held for hydrology teachers and professors in the Netherlands, United States, and Soviet Union. Technical training courses have been given in Mali, Tunisia, and Kenya, and more are planned. UNESCO-supported international postgraduate training courses in hydrology are offered in Hungary, Netherlands, Spain, Italy, and Czechoslovakia. In addition, fellowships and assistantships in hydrology are offered by Austria, Canada, Czechoslovakia, France, Hungary, Israel, Italy, Netherlands, Norway, Spain, United Kingdom, United States, Soviet Union, and Venezuela. The Decade also is preparing compilations of curricula and syllabi and lists of hydrological textbooks, films, and teaching aids.

Regional cooperation. Perhaps one of the outstanding contributions of the Decade to date has been its stimulation of bilateral and multilateral investigations and programs. Outstanding are the WMO-sponsored project for developing a coordinated network of hydrological and hydrometeorological stations in Costa Rica, El Salvador, Guatemala, Honduras, Nicaragua, and Panama; the inventory of hydrological and climatological data in the La Plata Basin of South America, conducted by the Organization of American States (OAS) in cooperation with Argentina, Bolivia, Brazil, Paraguay, and Uruguay; the International Field Year for the Great Lakes, a joint United States–Canadian study of the hydrology of Lake Ontario and its basin; the Nordic Council, which coordinates hydrological activities among Denmark, Finland, Iceland, Norway, and Sweden; and the cooperation of Mauritania, Upper Volta, and Mali in evaluating the groundwater resources of the Taoudeni Basin. Other regional cooperative arrangements are being considered and developed in Africa, Europe, and Asia.

Mid-Decade Conference. In December, 1969, 71 nations participated in the Mid-Decade Conference, which was convened to review work accomplished, to make recommendations for the latter half of the program, and to exchange views regarding international hydrology after the Decade is over. The consensus was that progress had been made under the Decade in education and in the extent of networks and research, but that the progress was not so great as had been hoped. Shortcomings were laid largely to inadequacies in funding and technical assistance, particularly in education. As a result, the Conference recommended that increased emphasis during the last 5 years of the Decade be placed on practical rather than scientific results; that is, on the application of existing science and technology to urgent problems of public supply and health.

The Conference also agreed that the Decade had demonstrated the benefits of nations coming together in a common forum to discuss their individual and mutual water problems. In its resolutions the Conference asked the Coordinating Council to outline a program of long-term activity in international hydrology, to be discussed by the 1972 UNESCO General Conference, and asked that the international agencies concerned plan suitable machinery, to be ready in 1975, for implementing the long-term program.   [L. A. HEINDL]

Snow and ice. Major problems in snow and ice hydrology are the temporal-spatial distribution of snow and ice water equivalent, the release rate of this water, and how man can influence either the distribution of snow and ice or the release rate of water therefrom to obtain more usable water.

Current attacks on these problems involve development of new instrumentation, including the application of remote-sensing and isotope techniques, energy-balance studies over snow and ice, and management of snow-covered timberlands to increase water yields.

Remote-sensing technique. Determination of snow and ice distribution is an obvious application of remote sensing, yet it has received only limited attention. In the United States, thermal infrared, passive microwave, and K-band radar have been tried in studies of snow and ice. The older techniques of ground and air photography have been refined and successfully used in several studies of snow distribution and depth. Satellite imagery has given interesting information on general snow cover distribution but has been of little help in the practical problems of snow and ice hydrology. *See* SATELLITES, APPLICATIONS.

The problems of remote sensing of snow and ice become increasingly complex as more detailed questions concerning occurrence, depth, water equivalent, and profile characteristics are asked.

The communication of meterological information from remote stations via satellite is particularly appealing because of the frequent remoteness of snow- and ice-covered lands. Such relay of information has been done experimentally. Furthermore, instruments exist for measuring the pressure of snow on the ground which could be used in such a remote system. These pressure pillows have been successfully used in a wide variety of conditions, and range in design from 12-ft-diameter rubber bags filled with alcohol to smaller metal pressure plates. They are readily adapted for converting the pressure of the snow, and hence the water equivalent, to a radio signal. Of course the older, standard weighing precipitation gages could also be used in such a remote relay system.

Isotope technique. Promising results in utilizing the attenuation of natural gamma radiation from the Earth by snow to measure snow water equivalent have been reported by workers in Norway and the Soviet Union. The technique is based on a comparison of measurements made without and with snow cover. Gamma sources for these studies are the naturally occurring potassium-40 as well as the uranium-radium and thorium families. Hand-carried instruments and instruments mounted in low-flying aircraft have been tried. Problems including variable soil conditions and exposed rocks may cause large errors in some situations.

An interesting development in the use of isotope techniques for snow hydrology is the profiling snow gage utilizing gamma attenuation between twin source and detector probes moved vertically through the snowpack. These instruments give very detailed information on the density distribution within the snowpack. From this information inferences can be made about snowpack conditions, including the amount and timing of meltwater release, important factors in flood forecasting.

Other recent applications of isotopes in studies of snow and ice include the use of tritium, oxygen-18, and deuterium as natural labelers of mass transfer within the seasonal snow cover or glaciers and of the role of meltwater in the overall hydrology of an area.

Energy budget. Research continues on the energy budget of glaciers and of the seasonal snow cover. Ablation (snow water decrease) and meltwater have been measured, as have the components of the energy balance of areas ranging in size from glacier surfaces to those that can be measured by small lysimeters. Techniques to isolate meltwater have included artificial construction of a small watershed on a glacier surface and the development of small plastic lysimeters that can be built up or taken down according to the natural accumulation and ablation of the snow cover. Although measurement of radiative transfers are now relatively refined, direct measurement of latent and sensible heat transfer at the snow or ice surface is still a major problem. The development of eddy-correlation techniques and the associated fast-response instrumentation offer hope for improvement of this situation.

Mass flux of snow. The complicated effects of the various roughness elements of topography and vegetation on the wind flow and hence on heat exchange and mass flux of snow are a serious challenge for future research. An advance in the measurement of snow mass flux has been the development of instruments using light attenuation. One such instrument has been used to study snowfall rate in conjunction with cloud-seeding experiments, and another has been used as a particle counter to study the mass flux of snow in alpine tundra.

Snow-cycle modeling. Results from individual process studies of accumulation and ablation of snow and ice eventually must be combined into a working model for utility. One such digital model is being developed and tested by the Hydrologic Research and Development Laboratory of the U.S. Weather Bureau. A major problem in such models is the routing of snowmelt once it has been determined. The isotope-label studies previously mentioned should contribute significantly toward the solution of this problem.

Management for water yield. Hydrologists are interested not only in natural processes of water yield from snow and ice but also in how they might alter these processes to obtain more usable water. The application of various darkening materials to accelerate melt has long been known. The problems now are concerned more with the economics and environmental impact of such measures. Snow fences have been tested in a variety of conditions ranging from use in conjunction with natural accumulation areas above the timberline to the possible trapping of more snow on windy plains areas. Again, the problem is not physical capability but economic efficiency.

Perhaps one of the more promising potentials for managing snow for water yield is in relation to forest management. Over the years a number of studies have demonstrated that forests influence the accumulation and disposition of snow such that forest stand characteristics can be modified to enhance the yield of water from snow from forest land. The most notable demonstration of this potential is the U.S. Forest Service's Fraser Experimental Forest in Colorado, where after 13 years annual water yield increases of more than 25% have resulted from a designed timber harvest pattern of block cuts. This increase is believed to have resulted from a combination of reduced transpiration and increased deposition of snow in the openings. The challenge in managing forests in the snowpack zone to produce more water involves not only understanding the hydrologic processes involved but also the harmonious combination of management for water, timber, wildlife, recreation, and other resource values.

Much attention is currently given to weather modification to increase snowfall and consequent water yields. Less public attention has been given to date to the possibilities of increasing the efficiency of water production from snow-covered lands. Perhaps the efforts at weather management will stimulate interest in land management. It is possible that a synergistic effect could result

from a combination of the two approaches.

[JAMES R. MEIMAN]

Water balance in North America. Demand for water has customarily been balanced against the supply on a predominantly local basis. However, in recent years water shortages have occurred in many localities, including some which have had abundant supplies in the past, and, as the population grows, more areas can be expected to experience such shortages. At present neither the North American continent as a whole nor any of its component countries has an overall water shortage. It would seem possible, therefore, to balance the supply and the demand on a larger scale so that local water shortages could be prevented or eliminated. An examination of the availability of water and the factors involved in the increasing demand for it is given below, followed by a discussion of the various means of achieving a balance between supply and demand.

Availability of water. The availability of water in any area is determined by the balance of certain elements in the hydrologic cycle. Principally, these are precipitation, evaporation, transpiration, groundwater, lake and glacial storage, and stream runoff. The International Hydrological Decade has undertaken the task of studying the balance and the long-term changes among these elements throughout the world. For the purposes of this discussion, however, the long-term average runoff to the oceans is a satisfactory measure of the surplus water available for man's needs in North America.

Table 1 shows estimated water availability in terms of the long-term average runoff from the North American continent as a whole and from some selected subdivisions. The last column gives the availability on a per capita basis. The figures in Table 1 are approximate; better data should be available when the work of the Decade has been completed. The values given are long-term averages of quantities which can vary widely from day to day and from season to season, but somewhat less from year to year. The variability is greater in some areas than in others because natural and artificial storage tends to regulate discharge.

Table 1. Estimated water availability for North America

Area	Average runoff, millions of cubic feet per second*	Population, millions*	Water availability, gallons per capita per day*
NORTH AMERICA	7.3	263	18,000
Canada	3.6	20	115,000
Conterminous United States	1.8	200	5,800
Mexico	0.4	43	6,000
Eastern seaboard from Virginia to New York	0.16	47	2,200
Texas and New Mexico	0.08	12	4,300
Southern California	0.003	10	200
Mexico, central and southern plateau	0.015	16	600
Mexico, between plateau and Gulf Coast	0.1	6.5	10,000

*Rounded values.

Table 2. Estimated water withdrawal in the United States

Use	Billions of gallons per day	Gallons per capita per day*	Annual increase, %
Public water supplies	26	130	} 3.0
Rural domestic supplies	7	35	
Irrigation	155	775	1.8
Steam electric cooling	127	635	2.7
Industrial and other	81	405	2.8
Total	396†	1980	2.4

*Based on 200,000,000 people.
†Includes 70,000,000,000 gal/day of groundwater.

Interestingly, the average per capita availability of water is greater in Texas and New Mexico, areas which are generally associated with water shortage, than it is on the heavily populated east coast of the United States, which is usually thought of as adequately supplied. Table 1 also shows that a large per capita water supply exists when the average is taken over the entire continent or even over each individual country. This is because of the large runoff in relatively lightly populated areas such as Alaska, northern Canada, northwestern United States, and southern Mexico.

Demand for water. The demand for water is not based on man's minimal needs alone; it is a complicated function of need, availability, and custom. Demand breaks down into two categories—withdrawal uses and nonwithdrawal uses. Water withdrawals for the United States are summarized in Table 2, and these are probably not much larger, on a per capita basis, than would be the corresponding statistics for Canada and Mexico. Only a small part of the water withdrawn is lost, a little less than 10%, principally through evaporation and transpiration in connection with irrigation. The rest is returned to the lakes and streams, frequently degraded in quality. Nevertheless, much of it can be and is used again many times over, and some of the withdrawals given in Table 2 include reused water. These withdrawal data can serve as a measure of demand in the absence of a better indicator.

It is worth noting that the average water withdrawal rate per capita exceeds the availability in some areas of the United States. However, because water is reused, only a few areas have real shortages. For example, along the Eastern seaboard of the United States, where there is heavy industrial use of water, treatment of waste water generally yields enough water for reuse to satisfy all demands. On the other hand, areas with heavy demand for water for irrigation have little water for reuse, both because of losses sustained from evaporation and transpiration and because of the increased salinity of the recovered water.

Nonwithdrawal uses include hydroelectric power generation, navigation, transport of waste products, storage of floodwaters, recreation, and enhancement of aesthetic features of the landscape. In general, these uses do not degrade the quality or affect the quantity of the water in serious measure; however, some evaporation losses and quality changes may be associated with impoundments, and quality changes certainly accompany transport of waste products. There is no general way to measure the demand for nonwithdrawal uses be-

cause the demand depends so heavily on such factors as how much waste of what type will be transported, how deep the navigation channels will be, and how much water must flow in a river to make it aesthetically pleasing. These factors determine how much of the available water passes unused through a given river basin. Thus, even though water availability in Table 1 exceeds withdrawal use in Table 2, there still may be water shortages.

Although the availability of water to a given area is fairly well fixed when averaged over long time periods, the demand is not. The last column in Table 2 presents estimated increases in withdrawals. These increases are a product of the increased standard of living. Thus, whereas availability per capita is decreasing, demand per capita is increasing, and sooner or later some water-rich areas will become water-poor.

When the demand factors are considered and weighed against the availability of water, the water-short areas of North America can be identified as parts of the Canadian prairie provinces, the northern plains and the southwestern part of the United States, and northern Mexico including the central and southern plateau. Other areas, such as the heavily populated part of the Eastern seaboard of the United States, may fall into this category in the foreseeable future.

Meeting the demand. Where demand threatens to outreach available water supplies, several courses of action are possible. Such courses, listed in their most probable order of application on the North American continent, include taking measures to conserve water, controlling industrial growth and population, introducing new sources of supply, and transferring water from one basin to another.

Possible conservation measures include the metering of, and adequate charge for, all water used, detection and sealing of leaks in conduits and reservoirs, control of plant growth to retard transpiration, more efficient application of water in irrigation and in industrial processes, and recycling of waste waters from industrial plants and public water supplies. (In the United States the cost of preparing water for highest reuse capability may approach 75 cents per 1000 gal.)

It may become desirable to plan the abandonment of certain industries in some water-short areas in favor of others that require less water. For example, it has been suggested that irrigation be abandoned in certain parts of the southwestern United States. At the same time, shortages of water may motivate certain industries which cannot accommodate themselves to conservation measures to move to areas with water surpluses, and population shifts would accompany these industrial shifts.

Possible methods of adding to the water supply include mining groundwater (as opposed to simply removing it from storage during one season or during dry years and replacing it at other times), desalinating sea water (currently estimated to cost from 75 cents to $1 per 1000 gal) or brackish water, and mining glaciers or towing icebergs to water-short coastal areas. Controlling precipitation by seeding clouds could also be mentioned in connection with increasing the water supply; however, this process might also be characterized as a method of interbasin transfer. Indeed, the question of whether water so produced is really new water or simply water removed from another basin may present insurmountable political and legal obstacles to the practical use of precipitation control.

The data in Table 1 point toward the possibility of transferring water from areas of high per capita availability to those of low availability. Several grand schemes for such interbasin transfer of water have been proposed. Some smaller transfer projects, such as New York City's Delaware River supply and the Colorado–Big Thompson project across the Continental Divide, are already in operation. The California Water Project, which is nearing completion, and the Texas Water Plan, now stalled, are examples of larger projects. The grand schemes, however, visualize the transport of water from the northern parts of Canada, where there is a large surplus, to the central and southwestern parts of the United States and into northern Mexico. It is likely to be many years before any of these large-scale projects reaches fruition, since reasonable cost figures will need to be settled and many international political and social problems overcome. However, lesser interbasin transfer projects will probably continue to develop, and it may be through sequential interconnection of these that one of the grand schemes will eventually be realized.

For background information *see* HYDROLOGY in the McGraw-Hill Encyclopedia of Science and Technology.   [EDWARD SILBERMAN]

Bibliography: R. del Arenal C., Water resources of Mexico, *Water Resour. Bull.*, 5(1):19–38, 1969; L. A. Heindl, IHD bulletin 13. *Trans. Amer. Geophys. Union*, vol. 5, no. 5, 1970; A. H. Laycock (ed.), *Proceedings of the Symposium on Water Balance in North America*, American Water Resources Association, 1969; W. G. McGinnies and B. J. Goldman (eds.), *Arid Lands in Perspective* (including AAAS Papers on Water Importation into Arid Lands), 1969; M. F. Meier, *J. Amer. Water Works Ass.*, 61(1):8–12, 1969; J. R. Meiman (ed.), *Proceedings of the Workshop on Snow and Ice Hydrology*, Colorado State University, 1969; R. L. Nace, *Trans. Amer. Geophys. Union*, 45(3):413–421, 1964; *Symposium on the Hydrology of Glaciers, Cambridge, England, Sept. 7–13, 1969*, The Glaciological Society, in press; *Symposium on the Use of Isotopes in Hydrology, Vienna, Mar. 9–13, 1970*, International Atomic Energy Agency, in press; C. Toebes and V. Ouryvaev, *Representative and Experimental Basins: An International Guide for Research and Practice*, 1970; M. Tribus, Physical view of cloud seeding, *Science*, 168:201–211, 1970; UNESCO, *List of International Hydrological Decade Stations of the World*, 1969; A. Wilson and K. T. Iseri, *River Discharge to the Sea from the Shores of the Conterminous United States, Alaska, and Puerto Rico*, USGS Atlas no. HA-282, rev. ed., 1969; G. Young, Dry lands and desalted water, *Science*, 167:339–343, 1970.

Hypersensitivity

The presence of reaginic antibody in the sera of atopic patients (patients characterized by immediate hypersensitivity to allergens) was first demonstrated by C. Prausnitz and H. Küstner in 1921. The skin sites receiving serum of an atopic patient gave an erythema-wheal reaction upon challenge with the specific antigen to which the donor was sensitive. The observation definitely showed that atopic individuals have antibodies responsible for allergic skin reactions, and suggested the role of such antibodies in allergic diseases. However, the nature of the reaginic antibody remained unknown for almost a half-century. Recent progress in the structure of immunoglobulins and the development of sensitive immunochemical methods have provided a new approach for studies of the antibody and have thus led to the identification of a unique immunoglobulin, immunoglobulin E (IgE), as a carrier of reaginic activity. Characterization of the immunoglobulin has provided much evidence for the idea that reaginic antibodies are one of the causes of allergic diseases. Using the IgE system, information on the mechanisms of reaginic hypersensitivity reactions has also been obtained.

Characterization of IgE. IgE was first detected in a reagin-rich fraction of the sera of atopic patients by K. Ishizaka and T. Ishizaka. Since the results showed that the protein has a different antigenic structure from that of the four known immunoglobulins (IgG, IgA, IgM, and IgD) and that it has antibody activity, it was tentatively called γE. Extensive studies on the correlation between γE antibodies and reaginic activity indicated that γE is the major, or perhaps the only, human immunoglobulin capable of mediating the Prausnitz-Küstner (P-K) reaction. An atypical myeloma protein subsequently described by S. G. O. Johansson and H. Bennich was proved to have the same antigenic structure as γE. Presence of E myeloma proteins confirmed that γE represented a distinct immunoglobulin class, and thus was designated IgE.

The physicochemical properties of IgE as compared with those of other immunoglobulins are summarized in the table. The protein is a glycoprotein with γ_1 electrophoretic mobility, the molecular weight being approximately 200,000. The structure of IgE has been studied with the use of E myeloma proteins. Like other immunoglobulins such as IgG, IgA, and IgD, the protein is composed of four polypeptide chains, two heavy (ϵ) and two light chains. Papain digestion of the protein produced two fragments which correspond to the Fc and Fab fragments of IgG. The antigenic determinants characteristic for IgE were present in Fc fragments but not in Fab fragments which contained light chains. After pepsin digestion of the myeloma protein, a large fragment was obtained whose sedimentation coefficient was 6 S. The fragment was composed of light chains and a portion of heavy (ϵ) chains, and corresponded to $F(ab')_2$ fragments. Amino acid analysis of the fragments suggested that the Fc portion of the molecules has a characteristic structure. *See* IMMUNOLOGY.

Formation and metabolism. IgE is a very minor component of serum protein. The concentration of the protein in normal human sera is in the range of 0.1 to 0.7 μg/ml. Statistically, a significantly raised level of the protein was found in the sera of patients with allergic asthma, hay fever, and especially atopic dermatitis. It was also found that the IgE level was increased by parasite infestation.

Physicochemical properties and biologic activities of human immunoglobulins

Properties	Class				
	IgG	IgA	IgM	IgD	IgE
Physicochemical					
Electrophoretic mobility	$\gamma_1 - \gamma_2$	γ_1	γ_1	γ_1	γ_1
Sedimentation coefficient $(S^o_{20,W})$	6.8	7	18	6	7.9
Molecular weight $(\times 10^3)$	150	180	900	—	200
Carbohydrate, %	2.9	7.5	11.8	—	10.7
Heavy chain	γ	α	μ	δ	ϵ
Light chain	κ, λ	κ, λ	κ, λ	κ, λ	κ, λ
Biological					
Serum concentration, mg %	800–1680	140–420	50–190	0.3–40	0.01–0.07
Antibody activity	+	+	+	?	+
Complement fixation	+	—	+	—	—
Skin sensitization					
Human	—	—	—	—	+
Monkey	—	—	—	—	+
Guinea pig	+	—	—	—	—

An intravenous injection of E myeloma protein in normal individuals showed that the half-life of the myeloma protein was approximately 2–3 days, compared with 25 days for IgG. A short half-life for IgE suggests that IgE reaginic antibodies are being formed continuously in atopic patients and that the production is more than that expected from the concentration in the serum.

Immunoglobulins are synthesized in plasma cells. As expected, IgE is formed in lymphoid tissues. However, distribution of IgE-forming plasma cells among various lymphoid tissues is different from that of IgG-forming cells. Among the lymphoid tissues, tonsils and adenoids removed surgically after recurrent infection possess the greatest number of IgE-forming plasma cells. Bronchial and peritoneal lymph nodes also contain IgE-forming plasma cells. By contrast, very few IgE plasma cells have been found in spleen and subcutaneous lymph nodes. IgE plasma cells are present in respiratory and gastrointestinal mucosa as well, and IgE has been detected in respiratory secretions. It appears that IgE-forming cells are predominantly in the respiratory and gastrointestinal tract. Such distribution suggests that IgE antibodies may be formed locally in the organs and may participate in causing allergic diseases.

Immunochemical properties. Human reaginic antibodies were believed to have only one combining site for antigen per molecule. However, IgE antibodies against ragweed allergen agglutinated red cells coated with the antigen, indicating that the number of combining sites in a molecule is multiple. In view of the structure of the molecule, IgE antibodies appear to have two combining sites. IgE antibodies do not fix-complement when they react with antigen. Lack of complement fixation by IgE antibodies indicates that complement is not involved in reaginic hypersensitivity reactions.

The most characteristic immunochemical property of IgE antibody is its ability to passively sensitize human and monkey skin for allergic reactions. The minimal concentration of IgE antibodies required for giving a positive Prausnitz-Küstner reaction in normal human skin is in the order of 2×10^{-4} to 3×10^{-4} μg/ml. The antibodies sensitize the other tissues of homologous species. Human leukocytes sensitized with IgE antibodies release histamine upon exposure to allergen, and normal monkey lung tissues sensitized with the antibody release histamine and the slow-reacting substance anaphylaxis (SRS-A) upon challenge with allergen. None of the other immunoglobulins sensitize lung tissues for antigen-induced release of the chemical mediators. The fact that IgE sensitizes lung tissues and mediates the antigen-induced release of both histamine and SRS-A, which are chemical mediators implicated in certain forms of anaphylactic bronchospasm in man, strongly suggests participation of the immunoglobulin in respiratory allergy.

Molecular basis of hypersensitivity. Sensitization of homologous species with IgE is probably the result of affinity of the protein for target cells. Recent studies on human leukocytes revealed that IgE molecules are present in basophil leukocytes which contain much histamine. Furthermore, incubation of isolated leukocytes with E myeloma protein resulted in binding of the molecules with basophils but not with the other leukocytes.

The binding with target cells is a characteristic property of IgE. When nonantibody IgE was mixed with reaginic serum and injected into normal human skin, the immunoglobulin competed with reaginic antibody for fixation to target cells and prevented passive sensitization. Blocking of passive sensitization was not observed when reaginic serum was mixed with the other immunoglobulins. The affinity is based on structures in the IgE molecules. The blocking of passive sensitization was achieved with the Fc fragments but not with either Fab or F(ab')$_2$ fragments of E myeloma proteins. Evidence was obtained that Fc fragments actually bound with human leukocytes and monkey lung tissues. It appears that the structures essential for sensitization are present in the Fc portion of IgE molecules.

Studies of immune mechanisms of reaginic hypersensitivity reactions indicated that bridging of cell-bound IgE molecules by antigen initiates the hypersensitivity reaction. It was also found that IgE antibody-allergen complexes could be formed in the test tube, and that chemically aggregated IgE and aggregated Fc fragments were capable of inducing hypersensitivity reactions in normal primate tissues. Monomer IgE or Fc fragments did not do this. The results suggest that possible structural changes in the Fc portion of the IgE molecules may induce enzymatic sequences leading to the release of chemical mediators.

For background information *see* ALLERGY, ATOPIC; HYPERSENSITIVITY; IMMUNOLOGY in the McGraw-Hill Encyclopedia of Science and Technology.   [KIMISHIGE ISHIZAKA]

Bibliography: K. Ishizaka and T. Ishizaka, *Clin. Exp. Immunol.*, 6:25, 1970; K. Ishizaka and T. Ishizaka, *J. Allergy*, 42:330, 1968; K. Ishizaka, T. Ishizaka, and M. M. Hornbrook, *J. Immunol.*, 97:840, 1966; S. G. O. Johansson and H. Bennich, *Immunology*, 13:381, 1967.

Immunology

The complex series of events which occur as a result of the specific interaction of antigen and immunologically competent lymphocytes and antigens constitute the immune response. Recent studies have contributed new information in both of the two broad categories of immune responses. The first category constitutes humoral responses, the production of humoral antibodies (immunoglobulin molecules) synthesized and secreted generally by plasma cells in tissues such as lymph nodes and spleen and released in the serum. The second category concerns cellular responses. Studies have elucidated somewhat these cellular responses, which are mediated by specifically altered or "sensitized" lymphocytes and are responsible for delayed hypersensitivity reactions, resistance to certain infections, and rejection of foreign tissue grafts. The cellular responses are thymus-dependent and can occur independently of serum antibody production. The humoral responses are generally thymus-independent.

Although lymphocytes are morphologically similar, they form part of a very heterogeneous collection of cells which differ in origin, life-span, distribution, and immunological task.

Primary lymphoid organs. The ultimate origin of all lymphocytes is from some stem cell, a pool of which is found first in the yolk sac during embryogenesis, then in the liver in fetal life, and finally in the bone marrow.

As shown in the illustration, stem cells migrate to different sites, such as the primary lymphoid organs, thymus, and bursa, or bursa equivalent, where their differentiation is channeled exclusively along lymphoid pathways. Thymus-derived cells (pathway I) eventually join the pool of recirculating small lymphocytes and populate certain areas of lymphoid tissues. Non-thymus-derived cells (pathway II) populate the follicles of secondary lymphoid tissues, such as lymph node and spleen.

Stem cell differentiation is dictated by inductive influences which prevail at the various sites after they migrate. Nothing is known concerning the mechanism of action of such inductive influences. Lymphocytes are derived from stem cells that have migrated to the "primary lymphoid organs." These are the thymus in all vertebrates and the bursa of Fabricius in birds, or some equivalent

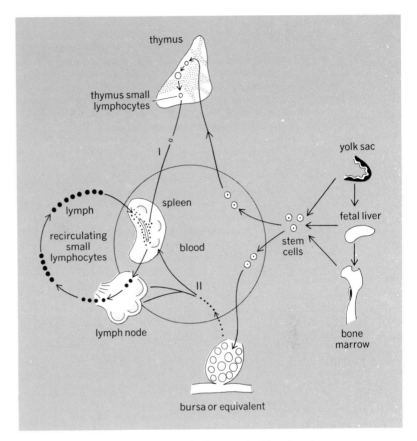

Diagram showing origin of immunologically competent lymphocytes.

organ, tissue, or collection of tissues (as yet unidentified) in other vertebrates. Lymphoid cells which differentiate in the thymus eventually play a role in cellular immune responses, whereas cells which differentiate in the bursa or bursa equivalent can eventually direct the production of humoral antibodies. Antigen plays no role whatsoever in the differentiation and proliferation of lymphocytes from stem cells in the primary lymphoid organs. Extirpation of the thymus early in development impairs the capacity to undertake cellular immune responses, and removal of the bursa from avian embryos or at hatching depresses immunoglobulin and antibody production. Immunological deficiency syndromes akin to experimentally induced defects occur in man, such as hypogammaglobulinemia of Bruton and thymic deficiency associated with alymphoplasia (such as di George's syndrome).

Secondary lymphoid organs. Once lymphocytes have been generated in primary lymphoid organs, some migrate out and either settle in other tissues—the "secondary lymphoid organs"—or recirculate from the blood through lymphoid tissues and back to the blood. Thymus-derived small lymphocytes become part of a complex of long-lived cells which recirculate and also populate certain areas of the secondary lymphoid organs—the "thymus-dependent areas" (such as the paracortical areas in lymph nodes and the periarteriolar lymphocyte sheaths in the spleen). The bursa-derived (or bursa-equivalent-derived) cells occupy the "bursa-dependent areas" of the lymphoid tissues, which include the lymphoid follicles and areas, such as medullary cords in lymph nodes, where plasma cells are found. In contrast to the situation in primary lymphoid organs, lymphocyte proliferation and differentiation is antigen-dependent in secondary lymphoid organs.

Antigen recognition. Specific antigen-binding receptors exist on the outer cell membrane of unstimulated lymphocytes. Thus, for instance, an antigen labeled with a radioactive isotope can specifically react with a small proportion ($< 0.1\%$) of lymphocytes from normal, unimmunized animals. Furthermore, passage of lymphocytes through columns of beads coated with a particular antigen specifically deprives the population of cells having reactivity to that antigen but not to other antigens. Unstimulated lymphocytes also have immunoglobulin molecules on their surface. These can be demonstrated by the binding of a specific antiimmunoglobulin antibody identifiable by conjugation to a fluorescent or radioactive label. The presence of immunoglobulin receptors on lymphocytes can also be inferred from the fact that antiimmunoglobulin sera induce transformation of small lymphocytes to large blast cells, a reaction which mimics the first steps of an immune response.

At least some of the immunoglobulin molecules on resting lymphocytes may be receptors or recognition units for antigen. This is because antiimmunoglobulin reagents can prevent binding of radioactively labeled antigen, protect antigen-reactive lymphocytes from absorption onto antigen-coated columns, and inhibit stimulation by antigen, in tissue culture, of cells taken from animals already immunized to that antigen. It is thus highly probable that antigen recognition is mediated by antigen-binding receptors that are identical to immunoglobulin molecules and are situated on the surface of lymphocytes. Presumably the immunoglobulin receptor is an accurate sample of the product that the cell, or its progeny, will eventually secrete following stimulation with the corresponding antigen.

Antigen stimulation. This leads to cell proliferation and differentiation. Following antigen binding by lymphocyes of the bursa-dependent type, plasma cells are produced which secrete the known classes of immunoglobulins (IgG, IgM, IgA, IgE, or IgD). Each plasma cell is highly specialized and produces antibody of only one immunoglobulin class and subclass and one specificity and affinity. *See* HYPERSENSITIVITY.

Some antigens are presented in such a way that they stimulate thymus-derived lymphocytes to produce, not plasma cells, but large pyroninophilic cells which proliferate and differentiate further to a progeny of "sensitized" small lymphocytes involved in cellular immune responses. In these situations, it has been postulated that a cell-bound antibody is responsible for antigen recognition, an antibody that is not secreted in the serum and that is conveniently referred to as IgX. Antigen activation of such lymphocytes triggers the production of a number of soluble products (not immunoglobulins), or "lymphokines"; among these are inflammation-inducing, mitogenic, cytopathic, and macrophage migration-inhibitory factors. The activity of such factors contributes to the development of delayed hypersensitivity lesions and graft rejection.

Immunological memory. In addition to stimulating the production of "executive cells" (that is, plasma cells and "sensitized" lymphocytes), antigen is responsible for the generation of clones of "memory" cells. Upon reencountering the antigen, these cells are particularly effective in producing an enhanced immune response, for example, a rapid and copious production of high-affinity antibodies. There is evidence that some memory cells are long-lived small lymphocytes which continuously recirculate from blood to lymph.

For background information *see* ANTIBODY; ANTIGEN-ANTIBODY REACTION; IMMUNOLOGY in the McGraw-Hill Encyclopedia of Science and Technology. [J. F. A. P. MILLER]

Bibliography: J. F. A. P. Miller and G. F. Mitchell, *Transplant. Rev.*, 1:3–42, 1969; N. A. Mitchison, in K. B. Warren (ed.), *Differentiation and Immunology*, 1968; G. J. V. Nossal, *Antibodies in Immunity*, 1969; World Health Organization (WHO), *Factors Regulating the Immune Response*, 1970.

Industrial engineering

The expansion of industrial engineering in banking closely parallels the growth of manpower and facilities in the industry. With the advent of new services for both corporate and consumer accounts, the banking industry has acquired a keener understanding of the major factors influencing the cost of providing their services and the manpower required to serve their customers more effectively. The basic industrial engineering tools have been successfully adapted to the banking industry in the areas of manpower and control, cost analysis, and methods improvement.

The myriad of services, reaching into the hundreds in number, range from the traditional areas of demand deposit, saving, and lending to the specialized fields of tax, travel, trust, and computer-oriented accounting systems. A review of the operating statement of most banks reveals the reason for the natural success of industrial engineering expertise in banking. The controllable cost of providing service centers largely around the efficient utilization of the time and talent of manpower. Certainly, from the customer viewpoint, the need for a prompt and well-trained staff is paramount. The logistics of banking revolves heavily around the strategic assignment of personnel and the supporting facilities to serve customer requirements. The steady growth of manpower and banking locations underlines this point vividly (see illustration).

The logistics of both commercial and retail banking require constant and close scheduling of time as well as strict discipline over the procedures to ensure that each account is handled accurately. The vast majority of a bank's staff is assigned to accomplish a series of specific paperwork steps under precise qualitative and quantitative security control. This basic requirement lends itself to scientific and statistical analysis. Although there has been no one pattern as to the areas of banking first approached by the industrial engineers in the field, the initial involvement was concentrated in the areas with high manpower usage, such as operations, branch banking, and credit card operations. The introduction of software and computer hardware programs added a new dimension to be met outside of the normal techniques of manpower analysis. In all cases, public contact or behind the scenes, the opportunity for an engineered approach has proved valuable.

Manpower data development. Heavy initial emphasis was placed upon the use of known applied work-measurement tools adapted to the language of the banking field. The entire spectrum of time and methods analysis techniques has been applied. The degree of refinement varies, but some form of basic predetermined data is most heavily used throughout the industry as a basis for establishing the time values for each assignment. The reasons for the emphasis in the work-measurement area were that most banks had no formal experience in establishing guidelines for manpower control until the beginning of the 1960s, and their expansion demanded a set of information on which to base their planning decisions.

The initial outcome of this effort again varies from bank to bank, but in the majority, two major areas of accomplishment were found. First, the program provided reliable time data to plan and control the number of people required to complete the work loads at different levels of activity. This step is especially important to banks to ensure proper levels of customer service in balance with the cost of providing the service. For example, in branch offices the staff was now able to be reassigned on the basis of the combination of standard time values and the anticipated customer traffic patterns. This also helped the internal paperwork processing by eliminating bottlenecks or delays in completing work caused by inadequate staffing during peak periods. This phase positively contributed to the decision-making process in planning for expansion of the highly related areas of manpower, equipment, and facilities.

Second, the cost analysis phase had a firm foundation for allocating the cost or added value of each service. The work measurement data were tied together in a flow-chart format to arrive at a total standard time for each increment of the cost of the service. Based upon the elements of cost related to the time of the manpower, a cost per labor-hour was established. The combination of the time required multiplied by the cost per hour provided a major share of the direct operating cost of the service. The sum of direct labor, material, and equipment cost provided a dollars-and-cents figure that was useful in evaluating the profitability of a service. The common denominator of standard cost also is widely used as a basis for responsibility accounting and budgeting.

The information resulting from these two phases establishes only the beginning point for most planning and control programs. Considerable selling and training about the principles and use of the facts are required in most banking firms. The supervisory and managerial groups hold the key to the successful and practical application of the continuing review of the results. The language of time and cost often was new or different to the normal banking vernacular. As a result, the purely technical presentation of the data to management had to be supplemented with a great deal of patient explanation and aggressive follow-up. Many programs overlooked this point because the long-standing use of industrial engineering in industry made it appear almost second nature; this was not the case

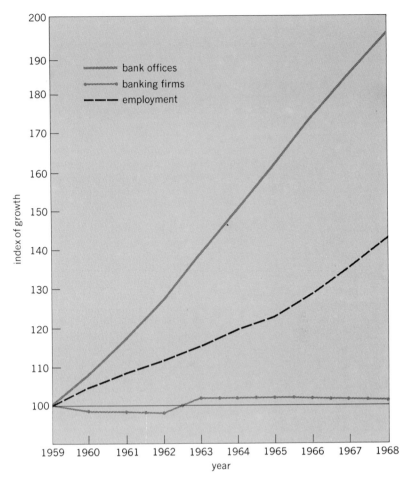

Rate of growth in banking firms and bank offices in the United States during 1959–1968; 1959= 100. (*Annual Report of Federal Reserve System, U.S. Bureau of Labor Statistics, 1969*)

in banking initially. For those who had worked in the industrial engineering field, presenting their methods to the banking industry was often a new experience. Today most methods and costs programs include as a regular part of the study the orientation of all staff members about the development and implementation of the data in order to improve their results-oriented use by all management levels. This vital phase should always be given high priority to improve the return on the investment in the program.

Technology growth. The future use of industrial engineering tools in banking depends heavily on their ability to contribute to solutions in the widening scope of bank activities. The introduction of automated equipment in the computer and communications areas creates a noticeable change from sole reliance on manpower measurement. The analyst is rapidly learning to use a composite of systems, methods, and cost skills. The logic of this combination lies in the fact that the hardware and software programs are heavily dictating the format of accomplishment. The contribution by the industrial engineer will be more limited unless he integrates his efforts into the systems area. This pattern is already evident in industry, where tooling and equipment have long been the paramount concern, and banking is now at this point in its needs.

Significantly, professionally trained engineers

have provided a better insight for bank management in the fields of manpower and cost control. In banking the opportunity to expand the use of these tools in line with their business growth is still outstanding. The greater use of management reporting, the need for operations research, and a keener competition on a broader base of services all contribute to an ever-growing need for the more sophisticated and skilled contribution by industrial engineering in the future.

For background information *see* INDUSTRIAL ENGINEERING in the McGraw-Hill Encyclopedia of Science and Technology. [RICHARD SCHWEITZER]

Bibliography: H. B. Maynard (ed.), *Industrial Engineering Handbook*, 2d ed., 1963.

Interstellar matter

Radio studies of interstellar gas have shown the existence of polyatomic molecules, whose presence in the gas was previously unsuspected. In several cases the physical processes that establish the population of the energy levels yield nonthermal distributions which produce anomalous radio properties.

General comments. In the past 2 years astrophysicists have had to revise considerably their views about the molecular constituents of the interstellar medium. It had long been thought that interstellar space was filled with a very tenuous gas composed principally of hydrogen atoms, averaging about 1 atom/cm³, and some simple, diatomic molecules whose abundance was at least 10^6 times smaller than the atom density, that is, 10^{-6} molecule/cm³. Molecules composed of more than two atoms were considered highly unlikely because of the difficulty of such molecules being formed in such a tenuous gas. Another recognized constituent of the interstellar medium was dust grains, small particles with a typical size of 10^{-5} cm and composed predominantly of carbon, oxygen, and nitrogen, which could play a fundamental role in the formation of complex molecules.

The techniques of radio astronomy have proved very powerful when applied recently to the study of the neutral constituents of the interstellar medium, such as atoms and molecules, because: (1) Electromagnetic radiation at radio wavelengths can travel galactic distances without being absorbed to any appreciable degree, whereas optical and infrared wavelengths are significantly absorbed by the interstellar dust, and (2) many molecules emit radio energy under the typical conditions of interstellar excitation but do not have sufficient excitation to emit energy at infrared or optical wavelengths. When these facts are coupled with the very high sensitivity of modern radio instruments, it is understandable why radio astronomy has had a dominant position in recent years in the discovery of new interstellar molecules.

Ammonia. The first detection of a polyatomic interstellar molecule was by A. C. Cheung, D. M. Rank, C. H. Townes, D. D. Thorton, and W. J. Welch at the University of California, Berkeley. They detected microwave emission from ammonia (NH_3) in the direction of the galactic center at the frequency of the NH_3 inversion line at 23,694.48 MHz, corresponding to a wavelength of 1.25 cm. The total received power has approximately 10^{-17} watt, a value quite typical of current radio astronomy capabilities and indicative of the sensitivity of radio methods. To verify that the received emission was due to NH_3, a second line was detected at a slightly different frequency, 23,723 MHz. Subsequently, the same scientists have detected three other NH_3 lines at nearby frequencies. Because several different lines have been studied, it is possible to determine the kinetic temperature of the radiating molecules because the relative intensities of the lines are a measure of the populations of the molecular energy levels, and this, in turn, is related to the kinetic temperature, assuming that collisions are the effective mechanism for populating the energy levels and that equilibrium is established. Kinetic temperatures determined in this manner range from 20 to 100°K and depend on which pair of observed lines is used in the computation. Since identical temperatures are not obtained for all possible pairs of lines, it must be concluded that the NH_3 molecules are not in equilibrium and that a more complex analysis is required. In spite of the uncertainty in the temperature of the NH_3 molecules, the observations provide an estimate of the total number of NH_3 molecules in a column of 1-cm² cross section in the direction of the galactic center. The value of 7×10^{15} molecules/cm² is typical and can be compared with approximately 10^{22} hydrogen atoms/cm² in the same direction. These numbers lend support to the general statement that molecular abundances are about 10^{-6} less than atomic abundances in interstellar gas.

Water vapor. The same University of California scientists were successful in discovering interstellar water vapor (H_2O) shortly after their detection of NH_3. Water vapor has a microwave transition at 22,235 MHz, that is, a wavelength of 1.35 cm, which was observed to originate from several H-II regions, the region of hot, ionized gas surrounding luminous stars, for example, the Orion Nebula. However, the microwave properties of H_2O are strange indeed. The molecule is an asymmetric rotor with most of its rotational transitions at in-

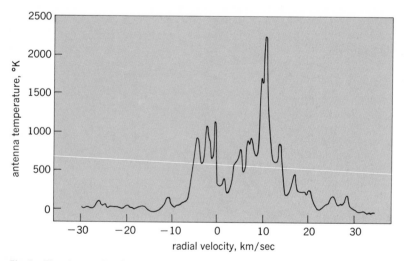

Fig. 1.   The observed H_2O emission spectrum from the H-11 region W49. The vertical axis is in units of antenna temperature, which is proportional to the power received on Earth. The horizontal axis is frequency, but it is expressed in terms of radial velocity using the Doppler formula. Here 1 km/sec equals 74 kHz. The many individual features in the spectrum may arise from spatially separated clouds, each having its own characteristic radial velocity.

frared wavelengths, but two levels, approximately 500 cm^{-1} above the ground state, are close enough together to give rise to the 22,235-MHz microwave line. Transitions from these levels to lower rotational levels can occur in 1 sec or so, with the emission of an infrared photon; therefore the detection of the microwave transition implies a very effective mechanism for keeping the upper rotational levels of H$_2$O populated. Whether this mechanism is due to collisions or radiation is not clear.

Subsequent studies of the H$_2$O emission have shown that the emission is confined to regions whose angular size is less than 0.003 second of arc. This angle corresponds to the angle subtended by a 25-cent coin when viewed from a distance of 1000 mi. For some H$_2$O sources, such as the Orion Nebula, the distance to the source is known; thus the measured angular sizes of the H$_2$O sources can be converted to a linear dimension. The linear sizes are less than 1.5 times the Earth-Sun distance. The small angular sizes of the emitting sources imply a very intense emission in order to give the intensity received on Earth. A sample spectra from the H-II region, W49, is shown in Fig. 1. The intensity of radio sources is customarily expressed in terms of the brightness temperature, which is the temperature that an ideal thermal emitter would have to acquire to radiate the same amount of energy. For the H$_2$O emission this number is typically 10^{13}°K, or higher. The magnitude of this temperature is so great that it precludes interpreting the temperature as being related to the kinetic energy of the radiating molecules. Instead, it suggests that the radiation must be generated by some nonthermal mechanism; therefore the concept of temperature is of limited usefulness.

Another property of the H$_2$O microwave emission, which was recognized from the outset, is that the intensity changes with time, often in dramatic fashion. Many sources show a complex spectra consisting of several features, and it is presumed that each feature originates from a separate cloud in the source. The relative amplitudes of the features from a source are observed to change during a period of several days or a week, and in some cases rather strong features have been seen to disappear altogether and recur at some later time. The origin of these variations is not clear, and will probably remain obscure until the basic radiation mechanism is understood, but intensity fluctuations on a short time scale are consistent with the small linear sizes, as determined from other observations.

The characteristics of the H$_2$O emission strongly suggest that the radiation is generated by some coherent, or cooperative, process, such as maser amplification. The proposal that maser amplification might be responsible for the observed H$_2$O microwave emission was a natural one because similar proposals have been made to explain the intense emission from interstellar hydroxyl (OH) radicals, which have many of the same properties. Maser amplification will occur in neutral atomic or molecular systems whenever the population of the higher energy level of a transition exceeds that of the lower level. When this occurs, the ensemble of atoms is able to add more energy to the radiation field than it subtracts, thus leading to amplification. The medium may be thought of

quite correctly as one which produces "negative absorption." To produce a maser, it is necessary to have a mechanism which can "invert" the normal population distribution of the system. Various radiative mechanisms involving ultraviolet and infrared radiation have been proposed, as have nonradiative processes such as collisions and chemical processes related to the formation of H$_2$O. No inversion process has yet gained wide acceptance, but the maser amplification concept has not been seriously challenged.

Formaldehyde. Scientists' ideas of the interstellar medium took on an added degree of complexity in 1969, when formaldehyde (CH$_2$O) was discovered to be an interstellar ingredient. In 1964 A. E. Douglas and G. A. Elliott studied many molecules and suggested that CH$_2$O might be detectable as an interstellar molecule by virtue of a microwave transition at 4830 MHz, corresponding to a wavelength of 6.2 cm, between low-lying levels. The suggestion appears to have been largely ignored on the grounds that CH$_2$O was too complex a molecule to be expected to exist in the interstellar medium to any appreciable degree. However, the discovery of NH$_3$ and H$_2$O as interstellar constituents forced a reexamination of the question and prompted attempts to detect many complex molecules. The search for CH$_2$O by L. E. Snyder, D. Buhl, B. Zuckerman, and P. Palmer, using the 140-ft radio telescope of the National Radio Astronomy Observatory, Green Bank, W.Va., was immediately successful. The method of observation differed from those used for NH$_3$ and H$_2$O, which were detected by their microwave emission. In the case of CH$_2$O, the radio telescope was directed toward a radio source and the source intensity was measured at the frequency of CH$_2$O and at adjacent frequencies. The presence of CH$_2$O was detected by noting a minimum in the source intensity at the CH$_2$O frequency; that is, the gas was detected by virtue of its absorption of some of the radio energy from the source. This method was used

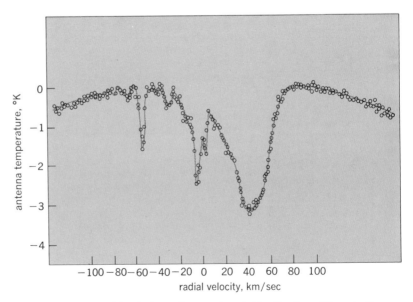

Fig. 2.   The observed CH$_2$O absorption spectrum in the direction of the galactic center. The axes are labeled as in Fig. 1, except that 1 km/sec equals 16 kHz. The various absorption features correspond to different clouds along the line of sight. (*From L. E. Snyder et al., Phys. Rev. Lett., 22:679, 1969*)

because smaller amounts of gas can be detected in this manner. A sample of an actual record showing the CH_2O absorption is given in Fig. 2.

Initially the identification of the absorbing gas as CH_2O rested solely on the coincidence of the frequency of absorption with the known frequency of CH_2O, but any doubt that the absorber was CH_2O was subsequently removed by the detection of absorption by the isotopic species $C^{13}H_2O$ at the proper frequency. The first observations of CH_2O were sufficient to reveal an unexpected result: The CH_2O was surprisingly widespread throughout the Milky Way Galaxy because it could be seen in absorption for almost half of all sources observed. Further observations of the southern hemisphere by J. B. Whiteoak and F. F. Gardner in Australia have revealed CH_2O in 31 of 34 sources. More recent observations have shown that CH_2O appears in absorption when viewed against the general cosmic background radiation, whose brightness temperature is only approximately 3°K. This was also an unexpected result because it implies that the excitation temperature of CH_2O is less than 3°K. Such a result is difficult to envision when the molecules are subject to collisions with other atoms and molecules which tend to produce an equilibrium at about 50°K. It seems as if some mechanism is at work which "antiinverts" the population of CH_2O in its lowest levels and thus makes the gas radiate at a temperature less than 3°K. It is difficult to make estimates of the CH_2O abundance because the population distribution and the mechanism establishing it are uncertain, but the observations appear to be consistent with approximately 10^{14} molecules/cm² and an abundance about 10^{-8} of the hydrogen atom density.

Other molecules. Not all radio studies of the interstellar medium have been limited to complex molecules. The diatomic molecules carbon monoxide (CO) and cyanogen (CN) have been detected by means of their microwave spectra near 100 GHz by R. W. Wilson, A. Penzias, and K. B. Jefferts. Since these discoveries were made as recently as April, 1970, full details of the observations have not been revealed, but it is clear that the study of these molecules will provide important information about the physical conditions of the interstellar medium. It should be noted that these molecules do not contain hydrogen as one of their ingredients and that they are the only two molecules of which this can be said. Optical spectra had previously shown evidence of the CN molecule in interstellar gas, but its abundance and spatial distribution are not well known.

The rapid discoveries of new molecules in the interstellar medium promise to open up new areas of research in the field of celestial chemistry and in the physical properties of gas constituents. A greatly expanded knowledge of the origin, composition, and evolution of interstellar matter can be anticipated.

For background information *see* INTERSTELLAR MATTER; INTERSTELLAR SPACE; RADIO SOURCES (ASTRONOMY) in the McGraw-Hill Encyclopedia of Science and Technology.

[ALAN H. BARRETT]
Bibliography: A. C. Cheung et al., *Nature*, 221: 626, 1969; A. C. Cheung et al., *Phys. Rev. Lett.*, 21: 1701, 1968; L. E. Snyder et al., *Phys. Rev. Lett.*, 22: 679, 1969; J. B. Whiteoak and F. F. Gardner, *Astrophys. Lett.*, 5:5, 1970; B. Zuckerman et al., *Astrophys. J.*, 160:485, 1970.

Ionosphere

The solar wind is the more or less steady expansion of the solar outer atmosphere, an expansion which is supersonic even at the distance of Earth from the Sun. Solar cosmic-ray events are infrequent occurrences superimposed upon this background. The protons and alpha particles released in these events are much more energetic than those composing the solar wind, and they penetrate the Earth's atmosphere. The Earth's magnetic field restricts these energetic particles to high latitudes, where they substantially increase the ionization in the lower ionosphere, the D region, about 50 to 80 km. Since this increase in ionization creates a strong absorption of radio waves at these latitudes, the disturbance has been designated a polar cap absorption (PCA) event. The response of the atmosphere during a PCA is of great interest to atmospheric scientists.

Polar cap absorption. The most extensive experimental study of a PCA ever attempted was recently completed by scientists and engineers from, or under contract to, the Air Force Cambridge Research Laboratories and the Ballistic Research Laboratory of the U.S. Army. Thirty-six rockets were launched from Fort Churchill, Canada, into the PCA which began on Nov. 2, 1969. The PCA died out in 2–3 days. Measurements were made of electron concentrations, ion composition, incoming particle flux as a function of energy, optical emissions, atmospheric density, ozone, and radio-wave absorption. The distribution of electrons with altitude was observed by several instruments, including Langmuir probes, and quadrupole mass spectrometers provided information on the ionic composition. The proton and alpha-particle spectra were obtained by several satellites in orbit during the PCA in addition to those measurements of the incoming particle flux by rocket-borne instruments. Apparently, alpha particles were not important in this PCA. See AIR SAMPLERS, ROCKET-BORNE.

Radio-wave absorption may be determined by a riometer, an acronym for relative ionospheric opacity meter. This device monitors the signal

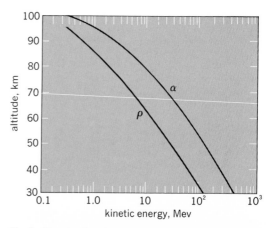

Fig. 1. Penetration depth of vertically incident protons and alpha particles versus their initial kinetic energy.

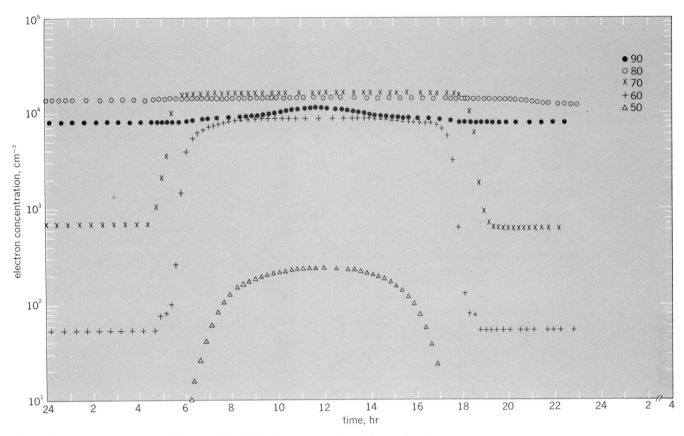

Fig. 2. Diurnal electron concentration model for a PCA with a moderate ionization production rate.

strength of the galactic cosmic radio noise, usually at a frequency of 30 MHz in units of decibels (dB). Absorption occurs when the radio wave excites the free electrons in the atmosphere, which then collide with other particles before the electrons reradiate the wave. The collision frequency of electrons is proportional to the concentration of atmospheric molecules. Thus, although the details are more complicated, the lower an electron lies in the atmosphere, the more effective it is in absorbing radio waves. Ions contribute little to absorption because of their greater inertia. A peak riometer value of abut 11 dB was obtained for the Nov. 2, 1969, PCA a few hours after the onset of the event, which began in the predawn hours.

All solar cosmic-ray events appear to be associated with solar flares. However, a PCA will not occur unless the flare emanates from certain regions of the Sun because of trajectory constraints due to the solar and interplanetary magnetic fields. A PCA begins with enhanced ionization at the geomagnetic poles and first extends to the upper and then to the lower auroral zones. Protons, and sometimes alpha particles, are the energetic species responsible. Their energies range from about one to several hundred million electron volts. Typically, the geomagnetic barrier is depressed, so that at Fort Churchill, for example, 1-Mev protons can enter the atmosphere, whereas the normal geomagnetic field permits only protons of more than about 14 Mev to penetrate into the atmosphere.

The maximum penetration depth of vertically incident protons and alpha particles is shown in Fig. 1 for particles with the indicated initial energy.

The information provided by Fig. 1, coupled with the actual number of incoming particles and their energies, enables the electron-ion pair production rate in the atmosphere to be determined, since it is also known that about 35 ev is expended in creating one electron-ion pair. Thus, a 35-Mev proton or alpha particle will yield 1,000,000 electron-ion pairs distributed over about a 10- to 20-km vertical column with a cross section of 1 cm^{-2}. The total integrated effect of all incoming particles is to produce an ionization rate which changes slowly with altitude and which is typically of the order of $100 - 300$ electron-ion pairs cm^{-3} sec^{-1}. The major ions initially produced are positive ions of oxygen and nitrogen, since these gases are the principal atmospheric constituents. Charge transfer of nitrogen ions with molecular oxygen is so rapid at these altitudes that it is proper to assume that all initially formed ions are O_2^+ ions. The problem rapidly becomes complex beyond this point.

Electron concentration. Figure 2 is a theoretical estimate of how the electron concentration varies with the time of day at Fort Churchill during equinox. The dependence of the electron concentration on time (actually the solar angular position) is due mainly to photochemical effects since the ionization rate has been assumed to be independent of time of day. (There is a slight change at 90 km because of some ionization by the solar x-ray and ultraviolet flux.) Absorption at 30 MHz is calculated to be 8.5 dB at noon and 0.6 dB at midnight for the electron concentrations shown in Fig. 2.

Negative ions are initially formed at these altitudes when electrons collide with O_2 and a third body (O_2 or N_2) to produce O_2^- ions. Production increases rapidly with decreasing altitude, in proportion to the square of the atmospheric molecular concentration. It would be fairly simple to understand a PCA if only electrons, O_2^+ ions, and O_2^- were involved, especially since the number of positive ions is equal to the number of electrons plus negative ions. However, a number of minor atmospheric gases react with O_2^+ ions, and the successive ions formed, to yield a variety of positive ions. The negative ion reaction chain is even more complicated. This chain depends on the various minor gases and on the fact that solar radiation can photodetach the electron from the negative ion, yielding an electron plus a neutral gas molecule. Furthermore, the absorption of ultraviolet radiation by molecular oxygen and ozone in the mesosphere and stratosphere results in the formation of atomic oxygen and electronically excited molecular oxygen. These species react efficiently with O_2^-, causing it to revert to a free electron plus molecular oxygen. The electron concentration at 40 km is calculated to be less than 1 cm^{-3} because, at this altitude, O_2^- production is so great that it overcomes the detachment processes and thus enables more stable negative ions to be formed.

The multiplicity of ions created and the dependence of important processes on the angular elevation of the Sun requires extensive analysis. The data given in Fig. 2 were derived by a computer code developed by T. J. Keneshea which contained approximately 100 reactions for 29 charged and neutral species. The detail that such a code can provide in conjunction with extensive measurements made during Nov. 2, 1969, will enable scientists to better understand the behavior of the atmosphere during a PCA.

For background information *see* AERONOMY; IONOSPHERE; SOLAR WIND in the McGraw-Hill Encyclopedia of Science and Technology.

[WILLIAM SWIDER]

Bibliography: G. C. Reid, *Planet. Space Sci.*, 17:731–736, 1969; J. C. Ulwick et al., in *Space Research X*, 1970; P. Velinov, *J. Atmos. Terr. Phys.*, 32:139–147, 1970.

Isozymes

In the last decade, with the continued development of methods for the effective separation of structurally similar enzymes, there have been numerous reports describing multimolecular forms of enzymes that catalyze identical chemical reactions. Although excellent progress has been made in regard to the physicochemical basis and genetic control of isozymes, the functional significance of only a few isozymic systems has been understood.

Regulatory isozymes. Studies of the cellular regulation of various metabolic activities have led to the discovery of isozymes whose functions are clearly concerned with regulatory processes. In general, it appears that isozymes with control functions are elaborated when a given enzymic reaction is required for more than one metabolic process. The existence of these isozymic types was first recognized by H. E. Umbarger and coworkers in their studies on feedback regulation of isoleucine and valine biosynthesis.

In both these biosynthetic pathways the first step under feedback control is a reaction that,

under certain conditions, is involved also in catabolism.

Under these circumstances, if only a single enzyme catalyzed the critical common step, its feedback control by excesses of the final end product could lead to a deficiency in energy metabolism. In such instances, when the critical step is needed for energy metabolism, the organisms elaborate another, inducible isozyme that catalyzes the same reaction, but which is insensitive to feedback control by the biosynthetic end product. These findings led Umbarger to propose that, whenever an enzyme is under rigid end-product inhibition and repression, another isozyme (or isozymes) will be needed by the cell if there is some other essential role for that enzyme. This principle found support from numerous studies.

E. R. Stadtman demonstrated the first branched biosynthetic pathway shown to be under isozymic control. The pathway concerned the biosynthesis of lysine, methionine, and threonine. He showed that these amino acids are all derived from aspartate by a branched pathway in which the phosphorylation of aspartate (catalyzed by aspartokinase) is the first reaction step. As shown in the figure, there are three independent aspartokinases in *Escherichia coli*, and they are differentially inhibited by the three end products. Moreover, there are only two separate homoserine dehydrogenases catalyzing the first step uniquely involved in the formation of methionine and threonine, and these two dehydrogenases are differentially controlled by methionine and threonine. In addition, the first reaction step of each branch of the pathway is under specific end-product control by the ultimate product of the divergent pathway. A consequence of this kind of control circuitry is that all aspartokinase activity is inhibited when all three end products are in excess. In essence, the production of multiple enzymes catalyzing the first common step in a branched pathway provides for flexibility in the control process; by avoiding the drastic consequences that would arise from unrestrained end-product control of a single enzyme, isozymes confer obvious selective advantage on the organism.

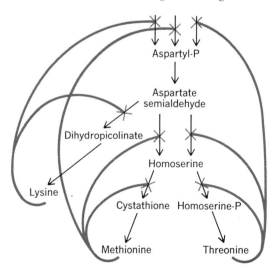

Feedback regulation of lysine, threonine, and methionine biosynthesis in *Escherichia coli*. Number of isozymes of aspartokinase and of enzymes at each branch in metabolic map is indicated by the number of arrows. Gray lines indicate steps subject to feedback control.

Isozymes in thermal acclimation. Isozymes play analogous regulatory functions during thermal acclimation in cold-blooded animals. P. W. Hochachka and G. N. Somero found that, at physiological substrate concentrations, the rates of enzymic reactions may be highly independent of temperature. In many poikilothermic enzymes, active sites are tailored so that a decrease in thermal energy is compensated for by an increase in enzyme-substrate affinity. However, at temperatures below a critical minimum, which depends upon species and isozymes considered, enzyme-substrate affinity drops; that is, the Michaelis-Menten constant (K_m) rises dramatically. This means that, in the case of enzymes from warm-acclimated animals, the K_m values of substrate may be many times higher than substrate concentrations in the animal. Under such conditions these enzymes become highly inefficient and possibly entirely inactive at low temperatures. To circumvent this problem, new isozymes are elaborated during low-temperature acclimation which differ kinetically from those of warm-acclimated animals in having higher absolute affinities for substrate or in having maximal affinities for substrate at lower temperatures or both.

The primary functional and selective advantage of employing "better" isozymes in thermal acclimation, as opposed to producing altered quantities of single enzyme species, is the production of enzymes with K_m values in a range likely to be optimal for regulation of catalytic activity. Thus, at low temperatures, small changes in substrate concentration or small changes in enzyme-substrate affinity can lead to large changes in the activities of "cold" forms of these enzymes, a condition which is admirably suited to controlling reaction rates. In the case of "warm" variants of these enzymes at low temperatures, very large changes in substrate concentration or in enzyme-substrate affinities are required to yield small changes in reaction rates. This condition is clearly not one which allows efficient control of reaction rates. In evolutionary terms it appears that, during thermal acclimation, there is a strong selection for the biosynthesis of new isozymes which are optimally suited for catalysis and control of catalysis at the given acclimation temperature. This is reflected in the patterns of new isozymes of lactate dehydrogenases, pyruvate kinases, aldolases, phosphofructokinases, citrate synthases, acetylcholinesterases, and presumably many other enzymes during acclimation and during evolutionary adaptation to temperature.

For background information *see* ENZYME in the McGraw-Hill Encyclopedia of Science and Technology.   [PETER W. HOCHACHKA]

Bibliography: P. W. Hochachka and G. N. Somero, in W. S. Hoar and D. J. Randall (eds.), *Fish Physiology*, 1970; C. L. Markert, *Ann. N.Y. Acad. Sci.*, 151:14–40, 1968; E. R. Stadtman, *Ann. N.Y. Acad. Sci.*, 151:516–530, 1968; H. E. Umbarger, in D. M. Bonner (ed.), *Control Mechanisms in Cellular Processes*, 1961.

Leaf (botany)

Of the three major vegetative organs of the higher plant—leaf, stem, and root—leaves are by far the most varied structurally, and their morphological diversity in turn accounts in part for the great variety of shoot form exhibited by flowering plants.

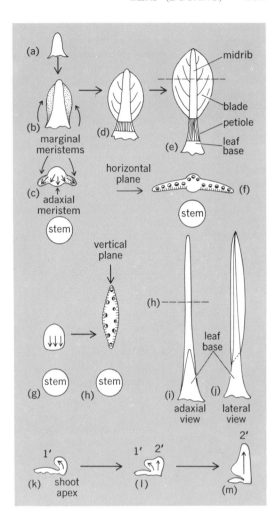

Fig. 1. Leaf morphogenesis. (*a–f*) Course of development and mature form of a conventional dorsiventral leaf. (*g–j*) Development and morphology of ensiform types of leaves in monocotyledonous flowering plants. (*k–m*) Stages in the early morphogenesis of an ensiform leaf, interpreted according to the sympodial theory, in which the primary leaf apex (1′) is succeeded by a secondary (2′) apex.

Although leaves are usually considered to be relatively stereotyped dorsiventral appendages, they actually range in structure from radially symmetrical organs, almost indistinguishable from stems, to elaborately subdivided compound leaves, as well as the bizarre tubular leaves of insectivorous and epiphytic tropical species. Recent investigations of the growth and differentiation of some of the more divergent and problematic types of foliar appendages in flowering plants suggest that there is a basic pattern of development common to leaves in general and that even the most radical differences in form are due simply to variations on this morphogenetic theme rather than to fundamental differences in ontogeny. These findings have important implications for experimental studies of plant morphogenesis and suggest that investigations of comparative development can also aid in the solution of problems of plant phylogeny.

General features of leaf development. Leaf morphogenesis is characterized by particular patterns of meristematic activity (cell division) responsible for progressive delimitation of the major morphological regions, namely, leaf base, petiole,

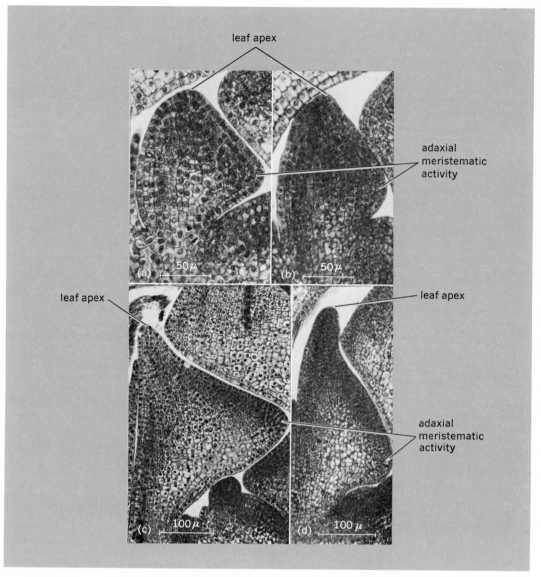

Fig. 2. *Acorus* leaves. (*a*) Median longitudinal section of a leaf primordium from a Wisconsin population of *A. calamus* with marked adaxial meristematic activity but little extension of the apex. (*b*) Leaf primordium of comparable length from an Iowa population of *A. calamus* with less adaxial but more apical meristematic activity. (*c, d*) Comparison of older leaf primordia showing differing degrees of adaxial versus apical growth in leaf primordia of *c* the Wisconsin population and *d* the Iowa population of *A. calamus*. (*From D. R. Kaplan, Comparative foliar histogenesis in Acorus calamus and its bearing on the phyllode theory of monocotyledonous leaves, Amer. J. Bot., 57:331–361, 1970*)

and blade (Fig. 1*e*). Following initiation from the apical meristem of the shoot, the leaf primordium usually undergoes a period of apical and intercalary growth in length, establishing the petiole-midrib axis (Fig. 1*a*). The blade in turn is initiated as a pair of winglike outgrowths from the lateral margins of this axis, and blade formation and early extension in breadth are accomplished by characteristic marginal meristems (Fig. 1*b* and *c*). In addition, activity of an adaxial meristem, resembling a cambium in histology and localized on the upper (adaxial) surface of the primordium, is responsible for radial thickening of the petiole-midrib axis and its protuberance as a rib in the adult blade (Fig. 1*c* and *e*). Apical, marginal, and adaxial meristematic activities are of varying but limited duration, and as these regions mature, cell division becomes restricted to the central and basal sectors of the pri-

mordium; ultimately cell division ceases in the leaf as a whole, and subsequent expansion is due to cell enlargement only. The blade and leaf base are the first regions to be differentiated, and the petiole is intercalated between these two zones in the later stages of development (Fig. 1*d* and *e*); rapid elongation of the petiole is usually responsible for emergence of the leaf from the bud.

Two examples from recent studies of leaf development illustrate how simple changes in the degree of expression and timing of the basic patterns of meristematic activity outlined above can result in profound changes in mature leaf form.

Ensiform leaves in monocotyledons. Ensiform, or sword-shaped, leaves are widespread among monocotyledonous flowering plants, particularly the iris, lily, amaryllis, and sedge families. Appendages of this type differ from conventional dorsiventral leaves in (1) the lack of a true blade region; (2) the vertical rather than horizontal plane of flattening (Fig. 1h and j), which is due to a change in direction of extension rather than secondary twisting; and (3) the radial rather than bifacial distribution of vascular and photosynthetic tissues (compare Fig. 1h with f). Recent developmental investigations of the ensiform leaf of *Acorus calamus* in the family Araceae by Donald R. Kaplan have shown that its divergent orientation is due to a simple change in growth emphasis: Adaxial meristematic activity, normally responsible for limited thickening of the leaf axis (Fig. 1c), is accentuated, and results in the marked radial extension of this leaf (Fig. 1g and h), whereas marginal meristematic activity, normally responsible for blade initiation, is suppressed (compare Fig. 1g with c). Adaxial meristematic activity is expressed very early in development and causes the ventral surface of the leaf primordium to protrude markedly (Fig. 2a). Although the radial alignment of cell files at the adaxial surface is identical to leaf-thickening meristems in general (Fig. 2a), some morphologists in the past have interpreted this exuberant radial growth as resulting from a primary leaf apex (1' in Fig. 1k) rather than from an adaxial meristem. Furthermore, it was suggested that ensiform leaves developed by a sympodial pattern rather than by a conventional course of leaf morphogenesis. According to the sympodial interpretation, radial growth proceeds from the primary apex (1' in Fig. 1k) which, however, soon ceases its growth and is succeeded by a secondary apex (2' in Fig. 1k and l) from the abaxial (dorsal) side of the primordium; the secondary apex is thus considered to be responsible for the longitudinal extension of the leaf (Fig. 1m).

Histogenetic comparisons between different populations of *A. calamus*, however, have disclosed reciprocal variations in the degrees of apical versus adaxial meristematic activity which have helped to harmonize the sympodial interpretation with standard patterns of leaf meristematic activity. For example, leaf primordia in a population of *Acorus* from Wisconsin exhibit a predominantly basal distribution of growth, and show, correspondingly, a very active adaxial meristem but a relatively late extension of the leaf apex (Fig. 2a and c), whereas primordia from an Iowa population of the same species show the inverse relationship of meristematic activity, that is, cell division is initially apical in distribution, and therefore extension of the leaf apex occurs earlier in development and adaxial meristematic activity is reduced by comparison (compare Fig. 2b and d with a and c). Leaf development in the Wisconsin population appears sympodial only because adaxial meristematic activity precedes apical growth, whereas leaves of the Iowa collection seem more conventional because apical growth is expressed before radial growth, as in most dorsiventral leaves.

Interpopulational developmental comparisons of the above kind have thus helped to clarify what in the past seemed to be an aberrant course of leaf development. Comparison of minor variations in meristematic activity has been relatively unexplored in comparative morphology and has great potential as an aid to understanding plant development.

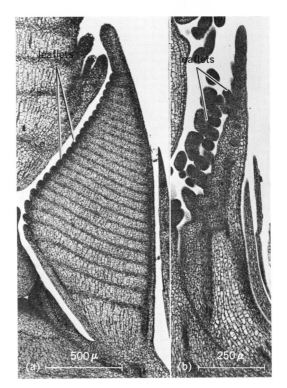

Fig. 3. Comparison of primordia of (a) rachis leaf of *Oxypolis greenmanii* with those of (b) pinnatifid leaf of *Carum carvi*. In the rachis leaf leaflets are initiated but remain small and the leaf axis is thickened, whereas in the pinnatifid leaf the pinnae are extended but the leaf axis is unthickened by comparison.

Rachis leaves in dicotyledons. A second example of clarification of a problematic leaf type by the study of comparative development is the so-called rachis leaf of certain genera in the parsley family (Umbelliferae). Although the majority of leaves in Umbelliferae are elaborate compound appendages with the blades dissected into numerous lateral leaflets or pinnae (Fig. 4a), a few species possess linear, radially symmetrical leaves that resemble the axis of a compound leaf that has been stripped of its leaflets (Fig. 4b), hence the term rachis leaf coined by the German plant morphologist Wilhelm Troll. Recent studies of the ontogeny of rachis leaves in *Oxypolis greenmanii* by Kaplan have shown that pinnae are actually initiated and that they arise in the same manner and sequence as leaflets in leaves of other Umbelliferae; however, they soon become arrested in their growth and differentiated as water glands or hydathodes (Fig. 3a). Because growth in girth of the leaf axis is accentuated and extension of the leaflets is suppressed, the pinnae in *Oxypolis* ultimately become dwarfed and inconspicuous in the mature leaf (Figs. 3a and 4b). By contrast, leaves of Umbelliferae that have well-developed leaflets show the inverse relationship: Extension of lateral pinnae is accentuated and thickening of the rachis axis is reduced (compare Fig. 3b with a).

As in the preceding example, simple changes in the degree of emphasis of basic patterns of meristematic activity account for the distinctive morphology of rachis leaves. However, before further generalizations can be made on the mechanisms of leaf development in flowering plants as a whole, a

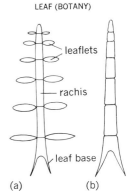

leaflets

rachis

leaf base

(a)   (b)

Fig. 4. Umbelliferae leaves. (a) Pinnately compound leaf typical of many Umbelliferae. (b) Rachis leaf of some Umbelliferae.

greater number of studies of a range of leaf types will be necessary. Of equal importance to such developmental inventories will be experimental studies designed to probe the causal factors that control these divergences in leaf morphology. Both approaches are goals of current research on leaf development.

For background information *see* LEAF (BOTANY) in the McGraw-Hill Encyclopedia of Science and Technology.   [DONALD R. KAPLAN]

Bibliography: K. Esau, *Plant Anatomy*, 1965; D. R. Kaplan, *Amer. J. Bot.*, 57:331–361, 1970; D. R. Kaplan, *Suppl. Bot. J. Linn. Soc.*, in press; W. Troll, *Vergleichende Morphologie der höheren Pflanzen*, 1939.

Leukemia

Two lines of evidence have recently suggested that human leukemic cells may carry antigens which could be considered tumor-associated transplantation antigens (TATAs). Such antigens, associated with tumor cells but not with normal tissue, have been demonstrated in several animal species. These antigens were so named because they were demonstrated by transplantation techniques. In order to detect antigens associated with tumor cells in man, one is restricted to immunological methods other than direct transplantation techniques. Consequently, evidence for the existence of TATAs is much more limited in man. In addition to presenting an interesting problem to the biologist, TATAs play a prominent role in research on immunotherapy of neoplasia. Currently, the presence of TATAs on tumors has encouraged investigation aimed at either enhancing an individual's own immunological response to the foreign TATA, and thereby destroying the tumor cells carrying the antigen, or destroying tumor cells by replacing the patient's own immunological system with a well-matched bone marrow transplant which could react to the TATA but not react to the normal tissue of the host because of the excellent histocompatibility match.

Use of autologous leukemic cells. F. M. Kourilsky and J. F. Dore analyzed sera from patients with leukemia for the presence of antibodies directed against the TATA. On the supposition that a patient may react to TATA on the leukemic cells by making antibodies to this foreign antigen, they tested the sera with autologous leukemic cells. They used a variety of techniques which demonstrate antigen-antibody reactions: membrane fluorescence, complement fixation, cytotoxic tests, and immune adherence. Depending on the method used, they found that between 10 and 20% of the sera tested contained antibodies which would react specifically with the autologous leukemic leukocytes. These studies suggest that, in at least some cases of human leukemia, the patient himself can respond to foreign antigens on the leukemic cells and produce antibodies directed against those antigens.

Mixed leukocyte culture test. Another approach attempting to demonstrate leukemia-associated antigens has been to use the mixed leukocyte culture test. In this test, peripheral blood leukocytes of one individual (A) are mixed with leukocytes (treated with mitomycin C and designated subscript m) of another individual (B). The untreat-

ed responding cells (A) enlarge and divide in response to foreign histocompatibility antigens present on the treated stimulating cells (B_m). After the cells are cultured for 7 days, incorporation of tritiated thymidine into acid-precipitable material is studied in allogeneic (AB_m) and isogeneic (AA_m) cell mixtures. The enhanced incorporation of tritiated thymidine in normal allogeneic mixtures (stimulation in mixed leukocyte culture test) is a measure of the response of A cells to foreign histocompatibility antigens of B_m cells and reflects the antigenic disparity of the two individuals at HL-A, the major histocompatibility locus in man. If the two test cells are identical at HL-A, the cell mixtures AB_m and BA_m will not incorporate thymidine at a rate significantly higher than their respective controls, AA_m and BB_m (nonstimulation, identity in the mixed leukocyte culture test). If a mixture AB_m shows no stimulation, this result will be accepted as indicating identity only if, in suitable controls in the same experiment, the test cells (A) react as responding cells and the treated cells (B_m) act as stimulating cells.

W. H. Fridman and Kourilsky froze the leukemic cells of patients in relapse. They waited for the patients to undergo a remission and then tested the frozen cells as stimulating cells (treated with mitomycin C) with the "normal" peripheral blood lymphocytes of the patient in remission. In 7 of 10 patients tested, positive stimulation in the mixed leukocyte culture test was obtained. In this case, if the reaction of the responding cells is to a new tumor-associated antigen or antigens on the leukemic cells, as will be discussed below, then it is possible that the patient has become sensitized to that antigen during the course of the disease and is reacting because of that sensitization.

A second study obviated the possibility that the responding cells had been overtly sensitized to the "leukemic antigen" in the living subject. In the human population, HL-A is a highly polymorphic locus, and thus individuals are generally heterozygous at this locus; two unrelated individuals will usually be heterozygous for different alleles. Approximately 25% of sibling pairs should be identical at HL-A. Two such identical siblings will be heterozygous for the same alleles. Any siblings with different alleles will be reciprocally HL-A disparate. Thus no sibling pairs should stimulate in one direction but not in the other when both parents are unlike heterozygotes. Consonant with these considerations, 28.2% of several hundred sibling pairs studied failed to stimulate in mixed leukocyte culture tests, and all of the nonstimulatory cell mixtures were nonstimulatory in both directions. No nonreciprocal stimulating pairs were found.

Marilyn Bach, Patricia Joo, and Fritz H. Bach examined 36 leukemic patients. Of these families, 30 were studied when the patient was in remission; that is, no leukemic cells appeared in a smear of the peripheral blood. In these families the investigators found sibling pairs identical in the mixed leukocyte culture test in a frequency not significantly different from that found in normal sibling pairs. In all of the sibling pairs in these families that were identical in the mixed leukocyte culture test, there was no exception to the rule that the siblings were nonstimulatory in both direc-

Experimental study of family of a 34-year-old male with acute lymphoblastic leukemia*

Cell mixture	Concentration of stimulating cells, number of cells × 10⁻⁶/ml	Activity, counts/min	p†
A		593	
AB$_m$	0.5	572	>0.5
AB$_m$	1.0	696	>0.5
AB$_m$	1.5	796	>0.5
B		442	
BA$_m$	0.5	2,832	<0.05
BA$_m$	1.0	53,793	<0.02
AC$_m$	0.5	14,672	<0.02
X		472	
XB$_m$	0.5	3,816	<0.05

*A represents patient; B, sister; C, another sibling; and X, unrelated individual.

†Probability that a given allogeneic mixture is different from the isogeneic control by chance alone.

tions. Six families with a leukemic member were studied at a time when the patient was in relapse; that is, 30–100% of the leukocytes in a smear of peripheral blood were leukemic blast cells. In three of these families there was reciprocal stimulation in all patient-sibling pairs tested; this is consistent with findings in the normal population, indicating that the patients had no siblings identical in the mixed leukocyte culture test.

In each of the remaining three families, cells of the patient did not respond to the cells of one sibling. In one case, the patient had a sibling, identical in the mixed leukocyte culture test, who showed reciprocal nonstimulation as in normal individuals. In the other two cases, the cells of the patient showed nonreciprocal stimulation with the cells of one sibling. In both of these unusual cases, cells of the leukemic patient were stimulatory to cells of one of his siblings, but cells of the sibling were not able to stimulate the cells of the patient. Appropriate controls with cells from unrelated individuals demonstrated that the leukemic test cells could both stimulate and respond to allogeneic cells.

A representative experiment from one of the two exceptional cases is given in the table. In this experiment, cells of the patient cultured alone incorporated 593 cycles per minute of tritiated thymidine. In a mixture with stimulatory cells (treated with mitomycin C) of the well sibling at three concentrations of stimulating cells, between 572 and 796 counts/min were incorporated; that is, there was no significant stimulation. Several concentrations of stimulating cells were used.

Responding cell concentrations are 0.3 × 10⁶ mononuclear cells per milliliter for healthy siblings and unrelated individuals, and 0.75 × 10⁶ mononuclear cells per milliliter for patient's cells. Stimulating cell concentrations presented refer to total leukocytes per milliliter. Leukocytes from the test individuals are incubated for 7 days at 37°C, and on the seventh day, 2 μcuries of tritiated thymidine is added to each culture for 5 to 5.5 hr. Counts-per-minute measurement refers to the average for washed acid precipitates of replicate cultures.

Mixture AC$_m$ verified that the patient's cells can respond to the stimulating cells of another sibling (14,672 counts/min as compared with a control of 593 counts/min); mixture XB$_m$ verified that the

cells treated with mitomycin C (B$_m$) of the sister can stimulate allogeneic cells (3816 counts/min incorporated, compared with 472 counts/min for the control of X cells cultured alone). When the cells of the sister (B) were tested for their response to the stimulating cells (A$_m$) of the patient, there was clear stimulation.

Similar experimental results were obtained in a study of the second patient. This nonreciprocal stimulation between siblings in two of three families with a leukemic member, wherein cells of the patient failed to respond to cells of his siblings, is in sharp contrast to results obtained in families with no leukemic members. In the latter group, of 291 sibling pairs tested, cells of 75 pairs were reciprocally nonstimulatory, and no sibling pairs showed nonreciprocal stimulation. The probability of finding these two cases in the three families with a leukemic member tested, by chance alone, with no cases in the normal families, is 0.008, establishing the significance of these findings in relation to these patients. The probability, calculated by the Fisher exact test, is a conservative estimate, because of the statistical test used and because only the 75 pairs of reciprocal nonstimulators are considered.

There are at least three possible interpretations of these findings that invoke special antigenic properties of leukemic cells. (1) The cells of the leukemic patients may possess leukemia-associated antigens which are stimulatory in the mixed leukocyte culture test. (2) Leukemic cells may express their minor loci antigens (non-HL-A) more strongly and, in these cases, disparity at minor loci can result in stimulation in mixed culture. (3) Blast cells may show their own unique antigens.

It is thus not totally clear whether the responses obtained in the mixed leukocyte culture test are measuring a leukemia-associated antigen. Nonetheless, they do suggest this possibility and, with the serological evidence quoted above, provide some reason to believe that, in human leukemia, there may be leukemia-associated antigens.

For background information see HUMAN GENETICS; LEUKEMIA in the McGraw-Hill Encyclopedia of Science and Technology.

[FRITZ H. BACH]

Bibliography: M. L. Bach, F. H. Bach, and P. Joo, *Science*, 166:1520, 1969; W. H. Fridman and F. M. Kourilsky, *Nature*, 225:277–278, Oct. 18, 1969; G. Klein, *Cancer Res.*, 29:1741–1746, September, 1969.

Lightning

Lightning, the most violent form of electrical disturbance which occurs in the Earth's lower atmosphere, has been the subject of intensive studies. Three basic questions may be asked: What is the nature of lightning? What mechanisms are responsible for the production of lightning? Can lightning be controlled or prevented? Answers to the first question can now be given, often with a high degree of precision, though much remains to be done; answers to the second question are contentious, but qualitative patterns are emerging; and answers to the third question are purely speculative, though a number of possibilities are open to investigation.

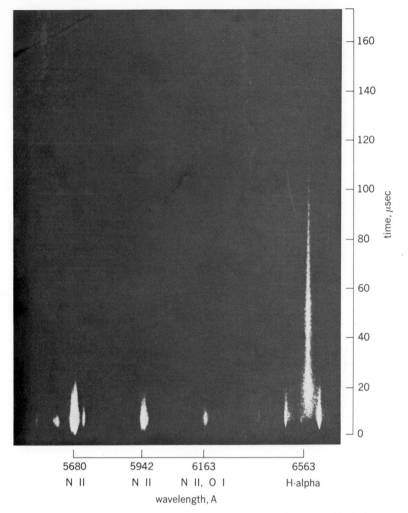

5680 N II 5942 N II 6163 N II, O I 6563 H-alpha

time, μsec

wavelength, A

High-speed, time-resolved spectrum of lightning stroke. (*Courtesy of R. E. Orville*)

Nature of lightning. Studies of the nature of lightning may be conveniently divided into two sections, one dealing with electrical properties and the other with properties revealed by optical techniques.

Early optical studies were concerned with the identification of the luminous channels down which the electrical charges passed and of successive luminous components of a lightning flash; such techniques are now routine. In the early 1960s attention turned to spectrographic studies using slit spectrographs, and in the mid-1960s, slitless spectrographs (those with a slit often recorded only indirect light) were used. Time-resolved spectra were also obtained, but the limit of resolution of 20 milliseconds was rather long in comparison to the length of the transients which occur during a flash. Recently R. Orville has improved the resolution of slitless spectrographs to between 2 and 5 μsec, enabling a detailed history of the degree of ionization, electron temperature and electron density, and other quantities to be calculated for both leader and return strokes.

A high-speed, time-resolved spectrum of a lightning stroke is shown in the illustration. A 10-m section of a lightning return stroke has been isolated and the spectral emissions streaked in time. The time resolution is 5 μsec. Singly ionized atoms emit first, followed by the continuum radiation, which in turn is followed by neutral emissions (H-alpha).

Much of the definitive work on the electrical properties of lightning was done in the late 1950s and early 1960s. The electrical components of a flash were isolated, the position of charge centers within clouds determined, and the quantities of electrical charge involved in various phenomena evaluated. Such properties are still being measured, since each cloud or cloud system responsible for the production of lightning is unique. In view of the statistical nature of cloud formation and separation of electric charge, one of the pitfalls to be avoided is the attribution of average values to the various parameters which culminate in a lightning flash. Recent work on the electrical nature of lightning has been almost exclusively concerned with recalculation of earlier work or its reappraisal using new concepts; included in such studies are calculations of the magnetic field of a lightning return stroke, calculations of the radiation field of a lightning return stroke, calculations of the radiation field and current of the lightning stepped leader, and recalculations of charge distribution in thunderstorm clouds.

Production of lightning. At least five mechanisms which probably occur within clouds have been proposed to explain the separation of charge in sufficient quantities to cause lightning; each mechanism is thought to be self-sufficient. Many other mechanisms of unknown potential almost certainly operate simultaneously and compound the experimental difficulties of trying to assess the efficacy of the mechanisms thought to be the most important.

Some of the processes are conducive to simulation on a laboratory scale, though the correspondence of laboratory conditions to "in-cloud" conditions has not always been good (artificial hailstones stuck to rigid supports in a wind tunnel, for example). Many experiments are now being reperformed in vertical wind tunnels which permit freefall conditions to prevail. On the basis of such experiments four principal theories of thunderstorm electrification require further investigation in actual clouds. They are: (1) The Reynolds-Brook process, involving the collision of ice crystals with soft-hail pellets in the presence of supercooled water droplets; (2) the Workman-Reynolds process, involving the collision of water drops or ice crystals with hail pellets; (3) the Latham-Mason process of drop splintering during collision and freezing; and (4) the Sartor process of induction in polarizing electric fields. A fifth mechanism proposed by B. Vonnegut involves large-scale circulation of air currents carrying space charge and is not amenable to laboratory investigation. Airborne studies performed by C. D. Stow tentatively support mechanisms 1 and 4, but contributions from 2 and 4 could not be evaluated. Investigations confirm the importance of treating each cloud as a completely independent system. At times, apparently identical clouds adjacent to each other exhibit completely different electrical characteristics. The most recent investigations employ Doppler radar and other techniques in an attempt to provide much more detailed information about individual cloud systems; it is becoming clear that, without proper incorporation of the influence of cloud dynamics into calculations on cloud electrification and the production of lightning, progress in solving these problems will be very limited.

Prevention of lightning. In its most general terms, the problem of preventing lightning must be considered in three phases: Prevent the thundercloud from forming; if it has formed, prevent it from separating charge; and if it is separating charge, short it out internally. Methods of short-circuiting large convective clouds have been tried. One method, known as chaff seeding, involves the distribution from an aircraft of large quantities of metallic needles which by going into corona are supposed to provide an internal conducting path, thereby minimizing the separation of large quantities of electric charge. Experiments have been inconclusive, and some workers consider the principle of this method to be based on erroneous assumptions. Experiments in which ice crystals are nucleated in order to act as "chaff" have on occasion appeared to enhance the production of lightning, an explanation for which can readily be found in terms of at least two of the mechanisms for charge generation mentioned earlier.

Stow has recently proposed methods by which mechanisms 1–4 might be influenced. The Reynolds-Brook and Latham-Mason processes require the existence of supercooled water drops. Effective seeding should eliminate major contributions to charging from either process. The Workman-Reynolds process, being electrochemical, is susceptible to trace impurities so that the introduction of the ammonium radical into water of which the cloud constituents are formed should annul or even reverse the electrical charge separation which would normally take place. The Sartor process of induction works most efficiently with ice crystals with a wide dispersion in their sizes, since this facilitates collision and gravitational separation. Early seeding of a cold cloud with large quantities of nucleating material tends to produce a monodispersion of crystal sizes, thereby reducing the chance of collision and separation under gravity. Similarly, it is possible by seeding techniques to stunt the growth of a convective cloud before it has a chance to become mature; this would seem to be the most appropriate method but implies anticipation of a development which may never take place.

It is the statistical nature of almost all meteorological phenomena that makes assessment of lightning prevention methods difficult.

For background information *see* ATMOSPHERIC ELECTRICITY; LIGHTNING; LIGHTNING AND SURGE PROTECTION in the McGraw-Hill Encyclopedia of Science and Technology.   [C. D. STOW]

Bibliography: R. E. Orville, *J. Atmos. Sci.*, 25: 827–856, 1968; C. D. Stow, *Bull. Amer. Meteorol. Soc.*, 50:514–520, 1969; C. D. Stow, *Rep. Progr. Phys.*, 32:1–67, 1969; M. A. Uman, *Lightning*, 1969.

Liquid crystal

Liquid crystals may be identified as a state of matter existing between the boundaries of the usual crystalline solid and the isotropic liquid. In recent years the study of this state of matter has become particularly fruitful. Recent research has revealed new information on molecular ordering in thermotropic and lyotropic liquid crystals and on their optical, electric, magnetic, and solvent properties.

Historical background. Historically the states of matter have been sharply divided into solid, liquid, and gaseous. In the solid state the atoms of the molecules can show small vibrations about fixed lattice positions but cannot rotate (except under special circumstances). The solid state has long-range order which gives it physical and optical properties quite different from those of liquids and gases. One normally thinks of liquids as having short-range order, but long-range order can occur in some liquids, giving rise to anisotropies in certain physical and optical properties. Such "anisotropic" liquids are liquid crystals. They are liquids in the sense that they do not support shear deformation and that they flow; however, they have some optical properties similar to those of solids. The fluidlike behavior of liquid crystals arises because of loss of translational order. Rotational or orientational long-range order may persist, however, giving rise to anisotropy.

Friedrich Reinitzer is credited with the discovery of liquid crystals. He prepared cholesteryl benzoate and, on heating the compound, made two observations. He found that there was not a sharp melting point from the solid to the isotropic liquid but an intermediate system that was turbid. He also found that the turbid system changed color as the temperature changed.

Liquid crystals can be prepared by heating certain organic crystals (thermotropic liquid crystals) or by mixing two or more components (lyotropic liquid crystals). Molecular shape is important in liquid crystallinity. Those molecules which show liquid crystallinity are elongated and often contain "flat" segments, for example, benzene rings. They are "rigid" along their long axes, with one or more double bonds commonly present. A guide to determine whether a molecule will form a liquid crystal is to consider the ratio R of the length to the width of the molecular frame. If $R \gg 1$ and the molecule is relatively rigid, there is a strong possibility that the compound will show liquid crystallinity.

The liquid crystalline state results from significant intermolecular interactions which preserve extensive solidlike order in the fluid. The molecular interactions that lead to liquid crystalline formation may be separated into (1) dipole-dipole interactions, (2) dipole-induced dipole interactions, (3) dispersion forces, and (4) hydrogen bonding.

Molecular ordering. A measure of the degree of internal order (or molecular alignment) in the liquid crystal is given by the expression below. Here

$$S = 1/2 \langle 3 \cos^2 \theta - 1 \rangle$$

θ is the angle between the long axis of a molecule and the axis of preferred orientation, which coincides with the symmetry axis in uniformly oriented nematic liquid crystals. The angular brackets indicate that the average value is used. For perfect parallel orientation of molecules $S = 1$, and for the isotropic liquid $S = 0$. In nematic liquid crystals the range of S values lies between 0.3 and 0.8.

Thermotropic liquid crystals. Thermotropic liquid crystals exist in two classes, namely nematic

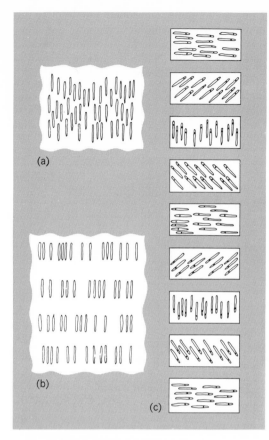

Fig. 1. Schematic diagrams of molecular arrangements. (a) In uniformly oriented nematic liquid crystal. (b) In a smectic A liquid crystal. (c) Molecular packing in cholesteric liquid crystal. Note the helical design.

(threadlike) and smectic (greasy). A special modification of nematic liquids which has received considerable recognition is the cholesteric liquid crystal, a twisted nematic.

The smectic texture has higher symmetry than that of the other classes of liquid crystals and ex-

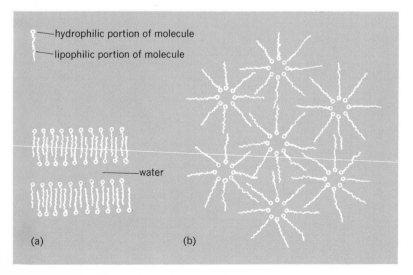

Fig. 2. Schematic representation of structures of lyotropic liquid crystals. (a) Lamellar packing. (b) Complex hexagonal packing. View is through end of rod. Water is in center of rod.

ists at lower temperatures. Molecules in smectic liquid crystals are arranged in layers, which gives them a stratified structure. The layers are not rigid but form flexible two-dimensional strata that may slide past one another. However, it is difficult to have movement in any direction other than tangential to the layer surfaces. In smectic liquid crystals there is translational long-range order in one dimension, whereas in nematic and cholesteric liquid crystals all translational order is lost. Molecular motion in smectic liquid crystals is rather slow, and they are generally quite viscous. Optically, smectic liquids resemble a uniaxial, birefringent crystal. The velocity of light transmitted parallel to the molecular layers is greater than that transmitted perpendicularly. There are at least five different smectic textures known. They differ from one another primarily in the arrangement of the molecules within the strata. The smectic A arrangement is represented schematically in Fig. 1b.

Molecules in nematic liquid crystals in the confines of their container are spontaneously oriented along their long axes (Fig. 1 a). The nematic mesophase is more mobile than the smectic. Molecules in nematic liquids may be substantially oriented and their mobility reduced by adhesion to supporting surfaces. The surface effect is so persistent that the system may be converted to a cholesteric structure by twisting the cover glass relative to the surface on which the nematic liquid crystal has been laid down. Nematic liquid crystals are very responsive to electric and magnetic fields. Potential uses of nematic liquid crystals include electronic clocks with no moving parts, electronic window shades, numeric readouts, display devices, and storage cells. Liquid crystals may in the future yield a practical thin-screen television display. The low voltage and power consumption of these displays make them attractive.

The cholesteric structure is a twisted nematic structure (Fig. 1c). The distinguishing features of cholesteric liquid crystals are their iridescent color and their large optical rotatory power. The molecules in the cholesteric structure are parallel along their long axes and packed in layers, but superimposed on the parallel molecular arrangement is a spontaneous and continuous twist. The effect is cumulative, and an overall helicoidal architecture results. An optically active molecule is necessary to form a stable cholesteric structure. The twisted molecular layers are responsible for the diffusion of reflected white light into colors and for the rotation of the polarization of transmitted light. The cholesteric structure functions as a grating for visible light. The color observed depends on the angle of the incident light, the nature of the cholesteric material, and the temperature. The color can also be affected by mechanical stress and by intrusion of traces of foreign vapors. If a planar cholesteric structure with a uniform twist is prepared in thin layers, it will reflect circularly polarized light selectively. This differs from optical rotation in optically active molecules, which originates in the selective absorption of one circularly polarized component of the light. In cholesteric liquid crystals the wavelength of the reflection band is determined by the pitch of the helix, while in the optically active molecule the absorption band is determined by the compound's chemical constitution.

Lyotropic liquid crystals. Liquid crystalline systems in great numbers can be prepared by the combination of two or more components. These systems occur in both the inanimate and animate world. The role of liquid crystals in living systems appears to be an important one. Phospholipids are amphiphiles, and amphiphiles are known to form lyotropic liquid crystals with water. Amphiphilic compounds are characterized by having in the same molecule two groups which exhibit two greatly different solubility properties. These groups are (1) a hydrophilic portion of the molecule, which tends to be water-soluble, and (2) a lipophilic group, which tends to be water-insoluble but soluble in organic solvents. This property of amphiphilic molecules gives interesting structures, for example, when they form lyotropic liquid crystals with a second component. Common structures are identified as lamellar, cubic, hexagonal, or complex hexagonal packing (Fig. 2). The water-containing mesophases may form around a core of hydrophilic groups with water in the core; the organic portion of the molecule projects outward. The water-containing mesophase may also be formed by the organic portion of the molecules being projected toward the center of the rodlike or spherical-shaped particles, with the hydrophilic portion of the molecule projected outward and surrounded by water.

For background information *see* BIREFRINGENCE; OPTIC ACTIVITY; STERIC EFFECT (CHEMISTRY) in the McGraw-Hill Encyclopedia of Science and Technology.   [GLENN H. BROWN]

Bibliography: G. H Brown, *Anal. Chem.*, vol. 41, no. 26A, 1969; G. H. Brown (ed.), *Liquid Crystals*, vol. 2, pts. 1 and 2, 1969; G. H. Brown and W. G. Shaw., *Chem. Rev.*, 57:1049, 1957; A. Saupe, *Angew. Chem.*, 7:97, int. ed., 1968.

Magnet

In the last 3 years, significant improvements have been made in the properties of permanent-magnet materials. This progress has been due to the use of new hexagonal, intermetallic compounds having the formula Co_5RE, where RE means rare-earth element. These permanent magnets have a high combination of magnetic anisotropy and saturation which are important factors in determining their properties.

Magnetic dipole. The elementary magnet is equivalent to a tiny loop of wire (of about atomic dimensions) carrying current. This is called a magnetic dipole. Permanent-magnet materials are composed of many small regions of parallelly oriented magnetic dipoles. These regions are called ferromagnetic domains. In modern permanent magnets these domains are extremely small, usually of the order of several hundred angstroms. In general, the purpose of a magnet is to supply a magnetic field in an external air gap. Its properties are best illustrated by the hysteresis loop (Fig. 1). The quantity B_s is known as the saturation induction; B_r is the residual or remanent induction; H is the applied field; and H_c is the coercive force which is the reverse field necessary to reduce B to zero. For permanent magnets it is important that this field (H_c) be large enough to resist external demagnetizing forces that tend to reduce the flux density B to zero. Sometimes the intrinsic coercive

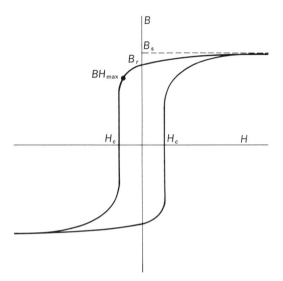

Fig. 1. Ferromagnetic hysteresis loop.

force H_{ci} is used, which is the field necessary to reduce (B-H) to zero. The maximum-energy product BH_{max} indicated on the demagnetization curve (second quadrant) is a figure of merit for permanent magnets. This represents the operating point of maximum efficiency, where a given amount of flux will be carried by the smallest amount of material. However, other criteria have to be considered that depend upon the application of the magnet. Sometimes coercivity, ductility, or raw materials cost is of paramount importance.

Production techniques. The new permanent magnets can be made by casting or powder metallurgy techniques. When they are cast, some of the cobalt is replaced with copper, which results in solid material having substantial permanent-magnet properties. An outstanding feature of these cast alloys is their high value of intrinsic coercive force and their stability after heating to tempera-

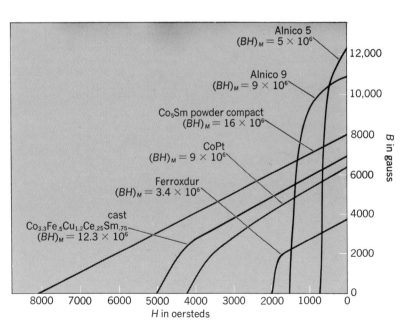

Fig. 2. Comparison of demagnetization curves of some new and old permanent-magnet materials.

tures above room temperature. High-energy product alloys were found in the $Co_{5-x}Cu_xSm$ and $Co_{5-x}Cu_xCe$ systems, as well as in mixtures of these two systems, with and without iron additions. An alloy of $Co_{3.3}Cu_{1.2}Fe_{0.5}Ce_{0.25}Sm_{0.75}$ exhibited $B_r = 7070$ gauss, $H_c = 5000$ oersteds, and $(BH)_M = 12.3 \times 10^6$ gauss-oersteds (g-oe). The Co_5RE compounds with or without copper additions can also be made into magnets by powder metallurgy methods. In general, this process consists of grinding the compound to a fine powder and pressing in a strong magnetic field to orient the particles. Sometimes the material is sintered after pressing. Powders made without copper have to be more finely ground and are less stable; however, their maximum-energy products are higher.

Figure 2 shows a comparison of the figure of merit and second quadrant hysteresis loop for some of the older, more available alloys with those for the new permanent-magnet materials. In general, the latter have higher values of coercivity and maximum-energy products. Since their values of B and H are more nearly equal, the demagnetization curve tends to become more linear than in the older materials. In industry, magnets are used as components in telephone receivers, loudspeakers, automobiles, electric meters, direct-current motors, memory systems, computers, and scientific apparatus.

For background information see FERROMAGNETISM; HYSTERESIS, MAGNETIC; MAGNETIZATION in the McGraw-Hill Encyclopedia of Science and Technology.   [ETHAN A. NESBITT]

Bibliography: J. J. Becker, *IEEE Trans. Mag.*, 4: 239, 1968; K. H. J. Buschow et al., *Philips Tech. Rev.*, 29:336, 1968; D. K. Das, *IEEE Trans. Mag.*, 5:214, 1969; E. A. Nesbitt et al., *Appl. Phys. Lett.*, April, 1970; K. J. Strnat, *IEEE Trans. Mag.*, 4:255, 1968.

Marijuana

Recent investigations have cast light on the major constituents and physiological action of the marijuana plant (*Cannabis sativa*). To permit a better understanding of the direction of recent studies, some results of the early research are summarized below.

Initial studies. About 1895, T. B. Wood, W. T. N. Spivey, and T. H. Easterfield isolated cannabinol from cannabis extracts. Cannabinol was not isolated again until 1932, when R. S. Cahn determined the structure of the molecule in all but the orientation of the substituent groups. He also cast doubt on the euphorigenic activity of cannabinol.

Not long after Cahn's work, University of Illinois chemists led by R. Adams isolated cannabinol by a method different from Cahn's; as a result of their procedure, they isolated a new constituent, cannabidiol. In collaboration with pharmacologists at Cornell Medical College, they reported both substances to be inactive. Their study of cannabidiol structure, however, led to determination of cannabinol structure (1-hydroxy-6,6,9-trimethyl-3-pentyldibenzopyran) (Fig. 1), which in turn led to the euphorigenic tetrahydrocannabinols. Heating cannabidiol with an acid catalyst converted it into a mixture of stereoisomers and double-bond isomers of tetrahydrocannabinol. The investigators suspected that such a mixture was in natural cannabis.

A. R. Todd and coworkers arrived at the structure of cannabinol at about the same time that the group led by Adams did. Todd's synthesis of cannabinol included tetrahydrocannabinol as an intermediate (Fig. 2).

The group led by Adams went on to synthesize compounds with activity similar to that of natural tetrahydrocannabinol. For example, one isomer

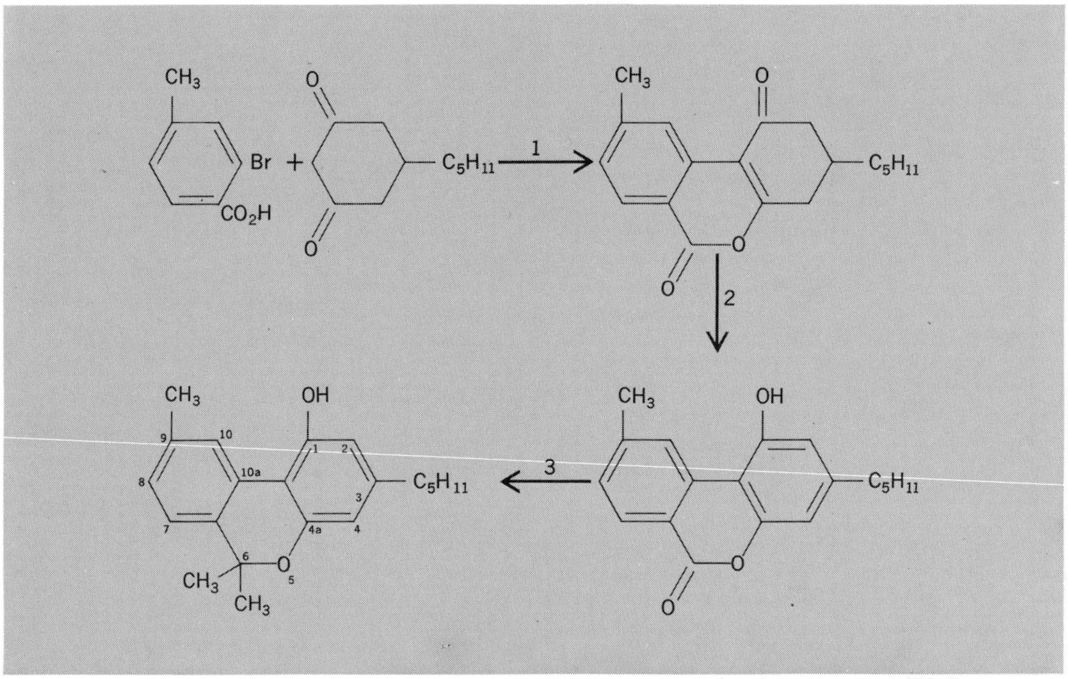

Fig. 1. Adams's synthesis of cannabinol. Reagents are (1) $Cu(OAc)_2 \cdot NaOEt$, (2) S, and (3) MeMgI.

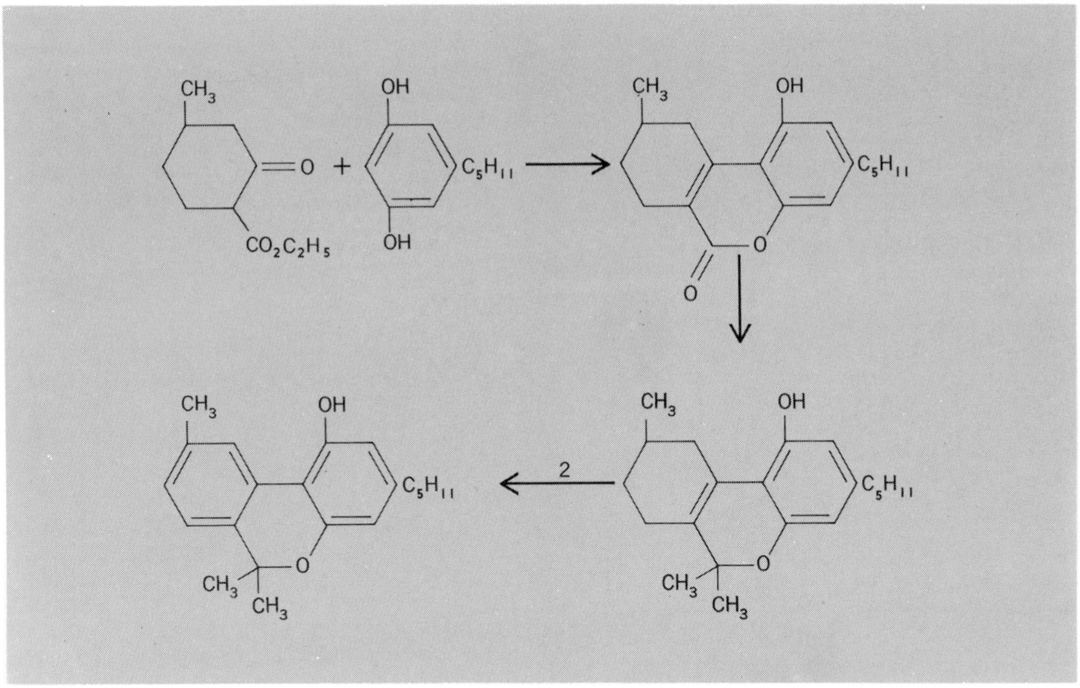

Fig. 2. Todd's synthesis of cannabinol. Reagents are (1) MeMgI, and (2) Pd or Se.

with the double bond conjugated to the benzene ring proved to have marijuana activity of about 1/10 the natural form. In 1940 Adams and his group also tried to synthesize the isomer with a 8,9 double bond. Although this attempt failed, E. C. Taylor and E. J. Strojny in 1960 successfully adapted the Adams method to the preparation of tetrahydrocannabinol model compounds (Fig. 3).

A large number of the tetrahydrocannabinol homologs have been prepared, especially compounds such as in Fig. 4. Such compounds usually have been tested by one or both of the two tests used in the pharmacological study of marijuana. These are the dog ataxia test and the Gayer corneal anesthesia test. In the series $R = n$-alkyl (Fig. 4), the 1-hexyl compound is the most potent in animal tests. Its activity is surpassed by $R = 1$-methylheptyl and by the $1',2'$-dimethylheptyl compound. The compound represented in Fig. 4 by $R =$ hexyl is known as Parahexyl or Synhexyl and is one of the few members of the series tested for euphorigenic activity in man. Its effective dose was found to be 5–15 mg in man. S. Loewe in 1944, however, reported the activity threshold at 200 mg in man.

Research in the past decade. Investigations during the 1960s have established the biogenetic precursors of cannabidiol and have identified and synthesized two tetrahydrocannabinol isomers in natural cannabis. Y. Gaoni and R. Mechoulam in 1965, using methods such as nuclear magnetic resonance and infrared spectroscopy, reported that the Δ^1-3,4-trans isomer of tetrahydrocannabinol is the major active constituent of natural cannabis. They synthesized that isomer and in 1966 also identified a new active constituent, cannabichromene. Earlier, in 1964, they had isolated cannabigerol, yet another new constituent.

The other naturally occurring isomer of tetrahydrocannabinol is Δ^6-3,4-trans. It is similar to the Δ^1 form in marijuana activity. E. C. Taylor and coworkers in 1966 produced it and two additional isomers, dl-Δ^1-3,4-cis-tetrahydrocannabinol and dl-Δ^6-cis-tetrahydrocannabinol. Their one-step synthesis gives the Δ^6 trans isomer in 20% yield under very mild conditions. They have also used nuclear magnetic resonance to distinguish between Δ^1 and Δ^6 trans isomers by a difference in the locations of the olefinic double bond. Both Taylor and R. L. Hively note that because of easy thermal isomerization, euphorigenic activity due to smoking marijuana may actually come from the Δ^6 isomer rather than from the Δ^1 isomer. It might be possible to link this thermal isomerization to the difference in effect between ingested and smoked marijuana, although the major differences reported have been in time of incidence and duration of effects rather than in the quality of the effects themselves.

L. Grlic and coworkers in the period 1961–1964 reported on variation in chemical composition and potency of cannabis of different age and provenance. Variation results mainly because some of the constituents are unstable and change form. Cannabidiolic acid is regarded as a genuine natural precursor of cannabidiol. With age, cannabidiolic acid converts by decarboxylation to inactive cannabidiol, which in turn yields by intramolecular condensation the active tetrahydrocannabinols. Further aging converts tetrahydrocannabinol into inactive cannabinol by spontaneous dehydrogenation. These changes are faster in tropical areas than in temperate areas, so that various samples of a given age can produce varying effects. Ecological and geographical factors also are important in biological effects because the resin of *C. sativa* (in which most of the active constituents are presumed concentrated) is produced in greater quantities by plants growing in hot dry climates.

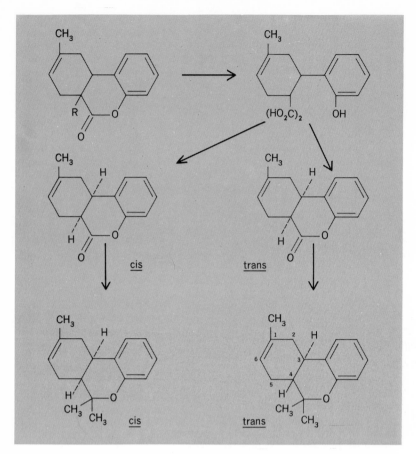

Fig. 3. Synthesis of model tetrahydrocannabinols by E. C. Taylor and E. J. Strojny.

Physiological studies. Numerous studies of the physiological effects of cannabis and its constituents may be found in the literature. Most of these involve experiments with animals and a very few with man. However, the results are highly varied and even confusing. Experiments with whole cannabis or with its resins may vary with the source of the plant and probably with the concomitant varying of constituents. Experiments with isolated tetrahydrocannabinol probably have varying results because the "tetrahydrocannabinol" was in fact a mixture of isomeric compounds. Experiments with individual, clearly identified chemical principles are rare.

In 1925 E. Joël found that no marijuana effect is seen in decerebate cats; only swaying occurs in decorticated cats; whereas swaying, decrease in

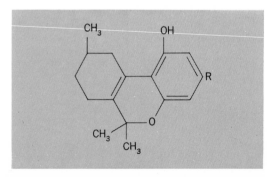

Fig. 4. A model tetrahydrocannabinol compound.

motility, cataleptic perserverance, and general inhibition occur in normal cats. The conclusion is therefore that marijuana acts preponderantly upon the hemispheres. B. C. Bose and coworkers in 1964 made electroencephalograph (EEG) studies in rabbits which indicated initial depression of the parietal area of the cortex with a simultaneous stimulation of the frontal regions, followed by depression of both. Recovery was characterized by increased excitability of neurons.

Concerning gross behavior, S. Gersohn stated that most workers have shown that extracts of marijuana induce stimulation and excitement in animals, followed by general depression, and that other behavioral effects include catalepsy, excitation, aggressiveness, and also sedative and hypnotic effects. Rats given cannabis resin every day for 5 months, by C. J. Miras in 1965, showed depression followed by normal behavior or hyperexcitability, followed again by depression. After 5 months, treated rats showed a lower level of reproductive activity than did the controls. The newborn of treated rats developed normally.

In maze-learning experiments, by E. A. Carlini and C. Kramer in 1965, rats receiving cannabis extract before the trials were better performers. After the sixth trial, their performance was significantly better than that of the control group. This effect was not observed in the rats injected after the trials; their running time was longer than the control running time, but the number of errors was the same. Studies by Carlini and Kramer on tolerance in rats are inconclusive, and no effect has been found on the ribonucleic acid (RNA) in the rat brain.

The studies of H. Isbell and coworkers in 1967 are very important because they are the first demonstration of marijuana activity of tetrahydrocannabinols of known chemical structure in man. Δ^1-*trans*-Tetrahydrocannabinol was given to patients orally and by smoking. By either route of administration this isomer "caused no significant changes in pupillary size, respiratory rate, systolic or diastolic blood pressures, or threshold for elicitation of the knee jerk. Pulse rates at rest were consistently elevated. Patients developed injection of the conjuctivae after larger doses." Furthermore, response to the tetrahydrocannabinol is statistically reproducible if the same patients are used under the same conditions. Δ^1-Tetrahydrocannabinol was three times as potent when smoked as when taken orally; some thermal conversion of Δ^1-tetrahydrocannabinol to Δ^6-tetrahydrocannabinol may have occurred in smoking, but such a change is unproved. The conclusions of this study may be summarized as follows: (1) Δ^1-Tetrahydrocannabinol causes subjective effects similar to or identical with those of marijuana; (2) the effects include alterations in mood, sensory perceptual distortion, and, in high doses, depersonalization, delusions, and hallucinations; (3) the drug in high doses is a psychotomimetic agent; (4) the tetrahydrocannabinols can be assayed quantitatively in man using either the subjective responses or increase in pulse rate; (5) synthetic cannabidiol dimethyl ether and cannabichromene are much less active than Δ^1-tetrahydrocannabinol or have no activity at all in man; (6) the activity of crude extracts of cannabis resin correlates best with their tetrahydro-

Major constituents of cannabis

Name	Formula	Pharmacological properties
Cannabidiolic acid		Sedative and antibacterial
Cannabidiol		No psychotomimetic activity
Δ^1-trans-Tetrahydro-cannabinol		Euphorigenic activity
Δ^6-trans-Tetrahydro-cannabinol		Euphorigenic activity
Cannabinol		No psychotomimetic activity
Cannabigerol		No psychotomimetic activity
Canna-bichromene		Mild or no euphorigenic activity

cannabinol contents.

In summary, the name, chemical structure, and (if known) physiological activity of the major constituents of cannabis are listed in the table. Two systems of numbering are used in the nomenclature of these compounds. Under the system used in Fig. 1, the two tetrahydrocannabinols mentioned in the table, Δ^1 and Δ^6, would be Δ^9 and Δ^8, respectively.

For background information *see* MARIJUANA;

PSYCHOPHARMACOLOGIC DRUGS; PSYCHOTOMIMETIC DRUG in the McGraw-Hill Encyclopedia of Science and Technology.   [LESTER GRINSPOON]

Bibliography: R. Adams, *Harvey Lect.*, 37: 168–197, 1942; D. F. Downing, *Quart. Rev.*, 16:132–162, 1962; N. R. Farnsworth, *Science*, 162:1086–1092, 1968; S. Gersohn, *Behav. Neuropsychiat.*, 1:9–18, 1970; H. Isbell, *Bull. Probl. Drug Dependence*, pp. 4832–4846, 1967; A. R. Todd, *Endeavour*, 2:69–72, 1943.

Measles

The mode of development of measles virus, the structure of the virion, and the relationship of measles virus to the paramyxoviruses have recently been clarified by electron-microscope studies. Measles virus first became susceptible to study by modern virological techniques in 1954, when J. F. Enders and T. C. Peebles showed that it could be propagated in human- and monkey-cell cultures. Nearly all that is known about this virus has been learned since then, much of it recently.

In 1957 Enders and colleagues described the presence of intranuclear inclusions, as well as intracytoplasmic inclusions, in measles-infected cells. The presence of intranuclear inclusions distinguishes measles virus (and the closely related distemper and rinderpest viruses) from paramyxoviruses such as mumps, parainfluenza, and Newcastle disease virus, which are morphologically very similar but produce only cytoplasmic inclusions in infected cells. The nature of the intranuclear inclusions has been investigated in several recent ultrastructural studies, and it has been shown that they consist of aggregates of tubules identical to the nucleoprotein helices that constitute the internal component of the mature measles virus particle, or virion. The cytoplasmic inclusions consist of similar aggregates of tubules. In both types of inclusions the tubules may be associated with a matrix of filamentous material.

The formation of the cytoplasmic tubules clearly represents a stage in the maturation of measles virions which occurs by a process that is common to all paramyxoviruses. The tubules move toward the surface of the cell and become aligned just below the cell membrane. The cell membrane locally undergoes alterations, becoming denser and acquiring on the outer surface an array of fine projections. The membrane forms protrusions, or "buds," which eventually pinch off to form roughly spherical particles consisting of an envelope that is derived from the cell membrane and encloses a bundle of tubules; these particles are mature measles virions. They are rather pleomorphic and vary in size from 300 to 700 nm.

Budding activity resulting in the formation of particles similar to those described but containing no detectable tubules is observed at the surface of some cells. These particles are thought to be defective or aberrant virus particles, similar to the "incomplete" particles which have been recognized for some time in association with influenza and other viruses.

The relationship of the intranuclear aggregates of tubules characteristic of measles virus infection to the maturation process of the virus is not clear. The intranuclear inclusions seem to form quite independently of the cytoplasmic inclusions. In

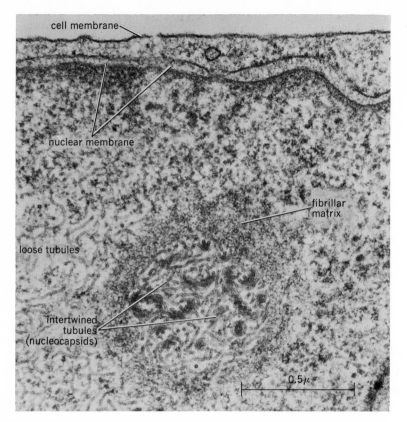

cell membrane

nuclear membrane

fibrillar matrix

loose tubules

intertwined tubules (nucleocapsids)

0.5μ

Fig. 1. Electron micrograph of thin section of measles-infected monkey kidney cell showing intranuclear inclusion.

any one cell, inclusions are seldom present at both sites. Even at an advanced stage of infection, when tubules may occupy almost the entire nucleus, the nuclear membrane remains intact and there is no evidence that intranuclear tubules migrate into the cytoplasm. Thus it appears that tubules of intranuclear origin do not take part in the budding process which results in the formation of mature infectious virus.

Sequence of development. The events associated with measles virus replication in several different cell lines have been studied by the techniques of thin sectioning and negative staining. Details of the infectious process differ according to the type of host cell and the conditions of infection, but, in general, the sequence of events follows a typical pattern. The first indication of virus activity is the formation in the cytoplasm of matrix areas consisting of fibrillar material about 24 hr after infection. The matrix areas resemble the "factory" areas that are found early in the infectious cycle of other viruses, such as vaccinia, and represent areas where viral components collect prior to assembly into mature virions. In measles virus–infected cells the matrix regions appear to be the sites of assembly of the nucleoprotein component of the virus, since tubules are first found in close association with these areas. By 48 hr after infection, aggregates of tubules are present also in the nuclei of some cells. These tubules are identical to the ones observed in the cytoplasm but tend to be more tightly packed and intertwined (Fig. 1). Associated fibrillar matrix material is generally less abundant in the nucleus. Budding activity at the cell surface with formation of mature measles virions is detect-

ed about the same time as the nuclear inclusions appear, and continues until the cells become degenerate (3–4 days).

Correlative observations can be made by using the negative-staining method to examine materials associated with, or released from, infected cells. This method involves disruption of the cells by freeze-thawing or exposure to hypotonic conditions or both, resulting in the loss of cell architecture, but it has the advantage of revealing the structure of the tubules (nucleoprotein helices) with greater clarity. Aggregates of tubules with serrated edges (the herringbone appearance characteristic of the internal component of paramyxoviruses) are readily detected in cell preparations 20 hr and later after infection. Complete virions consisting of coiled tubules enclosed in an envelope covered with fine projections 10 nm long can be detected about 40 hr after infection. The virions cover a range of size similar to that observed in sections, and the content of tubules also varies considerably. Measurement of tubule lengths indicates that the variation is not due to random variation in length of tubules but to the incorporation within the same envelope of different numbers of tubules of fixed length.

Nucleoprotein helices. Nucleoprotein helices (tubules, nucleocapsids) released from infected cells or from virus particles have the appearance shown in Fig. 2 when negatively stained with sodium phosphotungstate. The dimensions of isolated helices are somewhat greater than those of helices enclosed in envelopes or observed in sections; this is because of their greater susceptibility to flattening and stretching during specimen preparation. The distribution of lengths of helices shows a peak between 1.4 and 1.6 μ. When a correction is made for the stretching effect, the mean length is 1.15 μ. Another way of assessing the uniformity of length is to count the number of serrations or helical turns in each tubule. A sharp peak at 204 turns per helix is obtained. When these results are compared with similar measurements on nucleocapsids of various paramyxoviruses (NDV, mumps, Sendai, SV5), a striking similarity is apparent, both in the mean length of the helices and in the number of turns or serrations. Thus the structural similarity of measles and paramyxoviruses extends to the details of the nucleocapsids. This implies similarity also in the molecular weights of the viral ribonucleic acid (RNA). The conformation of the nucleic acid molecule in these viruses is not precisely known but, on the assumption that it is similar to that of the RNA in tobacco mosaic virus, where it is well established, the molecular weight can be estimated to be 6×10^6 to 7×10^6 daltons.

Measles and paramyxoviruses. It has been shown that the virions of measles and paramyxoviruses are essentially identical. On morphological grounds, measles clearly belongs to the myxovirus group, more precisely to the paramyxovirus subgroup, which is distinguished from the influenza subgroup on the basis of the diameter of the nucleocapsid (~ 17 nm) and the overall diameter of the virion (~ 300 nm), which are approximately twice the corresponding values for members of the influenza subgroup. However, measles differs from other members of the paramyxovirus subgroup in two important respects: It does not possess the

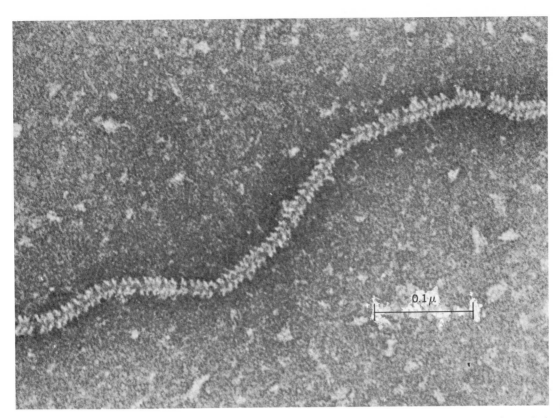

Fig. 2. Part of a measles virus nucleocapsid, negatively stained with sodium phosphotungstate, showing the characteristic herringbone appearance and serrated edges of the tubule.

property of myxophily, that is, the characteristic interaction with mucoproteins which gives the group its name; and, as previously discussed, infection with measles results in the formation of intranuclear inclusions. The latter property may be of importance in relation to the outcome of measles virus infection both in the laboratory and in the patient, as discussed below.

Intranuclear inclusions and virus formation. It has been noted that the number of budding particles containing viral tubules and the yield of infectious virus vary markedly according to the site of formation of tubules in infected cells. Cells in which most of the tubular formation is intranuclear show little budding activity, and very little infectious virus is produced. The tubules seem to be imprisoned in the nucleus so that very few or none succeed in completing the maturation process by acquiring an envelope. These observations are of interest in relation to the disease of the central nervous system usually referred to as subacute sclerosing panencephalitis (SSPE). Brain cells obtained from victims of this disease contain intranuclear inclusions consisting of masses of tubular structures and bearing a striking resemblance to the inclusions of measles-infected cells. This finding, together with immunological evidence, led to the strong suspicion that the disease is a manifestation of measles virus infection. However, attempts to isolate measles virus from infected brain cells met with failure until recently, when the isolation of measles virus from brain biopsies was reported from several laboratories. Perhaps the most convincing evidence for isolation of the virus was obtained by F. E. Payne and colleagues, who "rescued" the virus from brain cells obtained from a patient with SSPE by cocultivating them in tissue culture with cells of a simian line known to be susceptible to infection with measles virus.

The difficulty in obtaining infectious virus from diseased brain is probably related to the fact that the site of formation of tubules in infected brain cells is predominantly intranuclear—a site associated with very low production of infectious virus. Inability of the virus to spread readily from cell to cell would account for the slow progress of the disease. It is interesting to note that SSPE is an example of a disease in which the etiological agent was identified with a fair degree of certainty by electron microscopy before it was possible to isolate and identify the agent by standard virological and immunological techniques.

For background information *see* MEASLES; NUCLEOPROTEIN; PARAMYXOVIRUS; VIRUS in the McGraw-Hill Encyclopedia of Science and Technology.   [A. F. HOWATSON]

Bibliography: M. Matumoto, *Bact. Rev.*, 30:152, 1966; T. Nakai et al., *Virology*, 38:50, 1969; M. Nakai and D. T. Imagawa, *J. Virol.*, 3:187, 1969; F. E. Payne et al., *N. Engl. J. Med.*, 281:585, 1969.

Microscope, electron

The scanning electron microscope (SEM) has been developed during the past few years; it avoids many of the disadvantages of both the light microscope and the transmission electron microscope. Because of its limitations in resolution, depth of field, and effective magnification, the light microscope is not useful in studying very small objects such as surface details on sand grains. The trans-

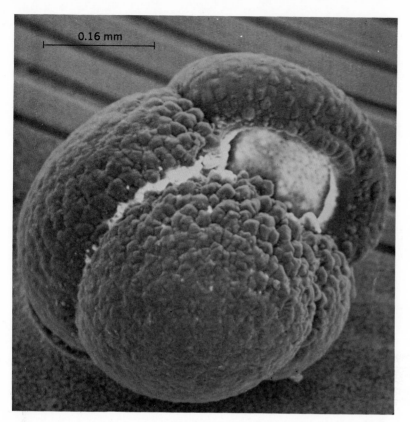

Fig. 1. Scanning electron microphotograph of the foraminiferid *Globigerina inflata* from core V23-14 raised by Lamont-Doherty Geological Observatory's ship *Vema* from the North Atlantic Ocean (depth of water 3177 m, depth in core 88-90 cm). (*Courtesy of L. Cavallero*)

mission electron microscope (TEM) gives better results, but only films less than 200 A thick can be used. It is also difficult to relate the magnified portion to the entire object because most transmission instruments are incapable of magnifying less than 600×. Also, since replication of objects is usually necessary, the original is not viewed directly and fine irregularities may not be discernible. This article will discuss recent studies of geologic specimens with the SEM.

The instrument. The SEM permits objects to be viewed directly at magnifications from 5× to 100,-000×. In operation, an electron beam of 1-50,000 volts is generated, passed through a series of condenser lenses, and focused directly on the surface of the specimen. The electron beam, 100-200 A in diameter, is scanned over the surface. Secondary electrons are generated, picked up by a detector, and displayed as a spot on a cathode-ray screen, with the brightness of the spot corresponding to the intensity of secondary electron emission from the specimen. Differential emission of these electrons produces contrast on the cathode-ray screen.

The image produced resembles dark-field incident illumination in a light microscope. W. Hay and P. Sandbert reported that the great depth of field produces a three-dimensional effect and that stereopairs can be made for actual three-dimensional study. The specimen can be tilted in any direction during observation.

Pollen and spores. Considerably more SEM work has been done with microfossils than with

megafossils. In particular, excellent progress has been made in the study of pollen and spores; the SEM makes possible the detailed study of wall ornamentation which cannot be examined qualitatively or even resolved with the light microscope. With the light microscope, oak pollen is not confused with other genera, but species of oak (*Quercus*) cannot be distinguished from each other. Identification of poorly preserved pollen in soils, floodplain alluvia, or playa lakes likewise requires the superior resolution, tilting ability, and depth of field of the SEM. P. Martin has pioneered these studies.

Foraminiferids. Foraminiferids (microscopic protozoans that float near the surface of the ocean) secrete calcium carbonate shells, which are found in great numbers in the fossil record. In some types of perforate foraminiferids, pore groupings are probably of phyletic importance. In addition, pore size and shape may be related to ecological conditions such as temperature; the widespread availability of foraminiferids has aided in the tracing of geographic and chronologic temperature movements in the oceans. Before the development of the SEM, it had been difficult to study surface topography and pores because of the problems of replicating features of great depth (Fig. 1).

Coccoliths. Some SEM work has been done with coccoliths, unicellular plants that are enclosed in an external skeleton of calcareous plates and average 2-5 μ in diameter. The hard parts of these plants have been examined with the TEM and used in stratigraphic and ecologic studies; but since the specimens can be tilted in the scanning microscope to obtain both plan and side views of the same coccolith, identification is considerably more reliable.

Ostracods and conodonts. Morphological studies have also been made of other microfossils. SEM photographs of ostracods have been used to separate genera on the basis of detailed morphology. A few SEM photographs have been made of conodonts; detailed studies with the SEM should provide new ideas on the mode of secretion and biological affinities of this group. Perhaps future examination will permit the fine subdividing of conodont lineages and the refining of stratigraphic zonation based on these organisms. Tertiary tintinnids have also been studied with the SEM in terms of morphology and systematics.

Echinoderms. Echinoderms are represented in the literature by only a few photographs; they are highly porous and difficult to observe with the light microscope or to replicate for use with the TEM and thus they are excellent candidates for SEM study. Other megafossils such as scleractinians, trilobites, graptolites, alcyonarians, brachiopods, and crinoids have features that could be defined in great detail with the SEM and will probably be studied in the near future.

Apollo 11 Moon samples. Perhaps the most extensive use of the SEM in geology as opposed to paleontology has been the study of the lunar samples returned to Earth on July 24, 1969, by the three-man *Apollo 11* team. More than 500 scientists from nine countries were given portions of the 22 kg of material available. The examination techniques were extremely sophisticated, drawing on

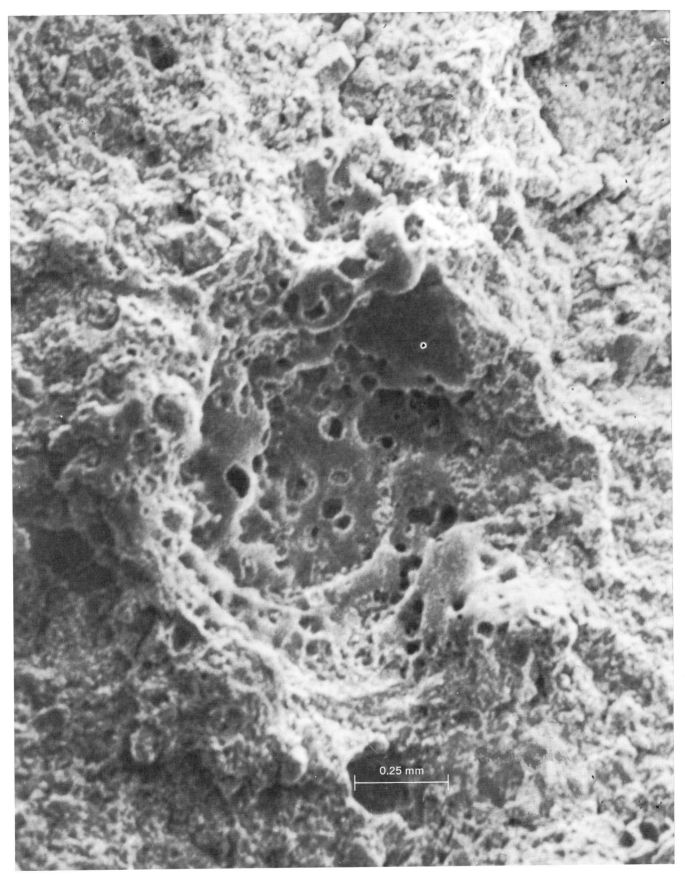

Fig. 2. Scanning electron microphotograph of surface of lunar rock from *Apollo 11.* Crater with glass in the center of photograph is probably the result of cometary or meteoric impact. (*From P. Cloud et al., Micromorphology and surface characteristics of lunar dust and breccia, Science, 167:776–778, 1970*)

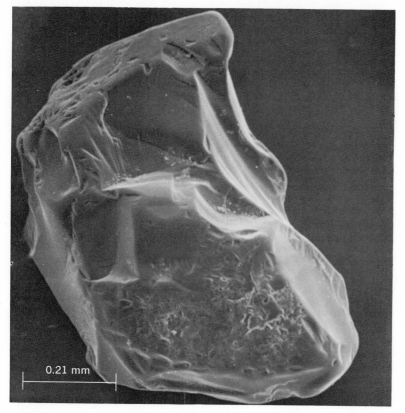

Fig. 3. Scanning electron micrograph of surface of sand grain taken from the Continental Shelf off Argentina (depth of water 73 m, depth in *Vema* core V14-41 99 cm). The irregular and jagged surface features are characteristic of glaciation.

experience in meteorite study, terrestrial petrology, and mineralogy. The SEM was used by many investigators and provided an immediate wealth of information. *See* MOON.

Examination indicated, for example, that the lunar soil and rock contain glassy spheres as small as 100 A whose shape and structure are consistent with the theory of cooling during free flight. It was also found that the soil was modified by impact or shock weathering, that some of the minerals in the soil have small pitted areas containing broken and unbroken glassy bubbles with overhanging edges resulting from outgassing, and that the change in the lunar day-night temperature is probably effective in shattering rocks and crystals by means of contraction and expansion. The SEM was also used to search for evidence of life on the Moon; no such evidence was found (Fig. 2).

Much of the above information could have been obtained by other means, but without the SEM it would have required a much longer time.

Quartz sand grains. The surface morphology of quartz sand grains has been studied for the past 8 years by the TEM to determine the various modes of transportation and the chemical environments through which the grains have passed. It has been possible to distinguish between high-, medium-, and low-energy littoral grains, coastal, periglacial eolian, and desert grains, and glacial and glacial-fluvial grains. In addition, several modes of diagenesis (processes that convert fresh sediments to consolidated rocks) have been discovered. Most of these textures have been duplicated experimental-

ly and used to decipher chemical and mechanical environments in the fossil record. Frequently, as many as five distinct textures may be observed on a single sand grain and the relative time sequence can usually be determined. This permits study of short-term environmental changes in the fossil record.

By using the SEM, new environmental features of the sand grain have been discovered, and under low magnification, environmental relations between various parts of the grain become evident with the simulated three-dimensional view. The simplification of preparatory techniques has increased the number of grains that can be examined in a given time by a factor of 10 (Fig. 3).

Sedimentary rocks. The light microscope has been used to identify the mineral constituents and to study the relationships between grains in coarse-grained thin sections of sedimentary rocks. However, such study is limited to a single plane, and it may be important for interpretation of origin to observe the spatial relationships between grains and their surface textures. This has been done to some extent with the TEM, particularly with fine-grained rocks. However, the SEM is a highly preferable tool. Preparation of sedimentary rocks for the SEM is quite simple; the chip or sample is etched, if desired, coated, and examined.

Other studies with SEM. The SEM has been used to differentiate the texture of hard chalks (composed entirely of interlocking crystals) from that of soft chalks (built up of loosely packed coccoliths). Another study has considered the nature of contacts of fine-grained crystals with one another, and with coarser crystals, particularly in dolomites; also, the deposition of noncarbonates or cementitious material along grain boundaries has been considered.

Researchers are now studying loess particles with the SEM; the technique permits calcium carbonate encrustations and other adherent materials on quartz grain surfaces to be seen clearly for the first time and environmental interpretations to be made. Studies of clay minerals have benefited from SEM techniques; some work has been done comparing Pleistocene and laboratory-flocculated clay sediments.

Experimental rock deformation may benefit from SEM work which will permit detailed study of surface textures before and after deformation. Grain-to-grain relationships and textures in fine-grained igneous and metamorphic rocks will undoubtedly be studied.

At the present time, the SEM is becoming increasingly important in geology and paleontology; as the resolution and ease of operation increase and the price decreases, it will be used routinely in almost every Earth science laboratory.

For background information *see* MICROSCOPE, ELECTRON; PALEOBOTANY; PALEONTOLOGY in the McGraw-Hill Encyclopedia of Science and Technology.   [DAVID KRINSLEY]

Bibliography: P. Abelson (ed.), *Science*, 167: 583–680, 767–778, 1970; J. Gillott, *J. Sedimentol. Petrol.*, 39:90–105, 1969; D. Krinsley and S. Margolis, *Ann. N.Y. Acad. Sci.*, 31:457–477, 1969; P. Sandburg and W. Hay, *Proceedings of the Symposium on the Scanning Electron Microscope*, 1968.

Moon

Experimental results from the two successful manned landings on the Moon have revealed much new information on its seismic activity, on the isotopic abundance in the solar wind, and on the chemical composition of lunar material.

SEISMIC MEASUREMENTS

The Apollo seismic experiments are the principal means for determining the Moon's internal structure; the nature of tectonic processes, such as mountain building and formation of faults, that may be active within the Moon; and the number and masses of meteoroids which strike the lunar surface. Two seismic stations have been installed on the lunar surface as part of the *Apollo 11* and *12* missions. They have revealed that the Moon is a very quiet and stable body relative to the Earth. Small moonquakes, probably induced by the tidal stresses generated within the Moon by the Earth-Moon gravitational attraction, have been detected.

Experiment description. Seismology seeks to understand the structure and dynamics of a body by studying vibrations of the surface of the body. The instrument which measures surface vibrations is a seismometer. The Apollo experiment contains four seismometers mounted within a single framework (Fig. 1). The instrument weighs 11 kg, and is constructed principally of beryllium. It is surrounded by a blanket of insulating material (aluminized mylar) which protects the instrument from the extreme temperature variations which occur at the lunar surface. Three of the seismometers form a matched triaxial set and are sensitive to low-frequency signals. The fourth seismometer is sensitive to vertical motions only and at higher frequencies (up to 20 Hz). The seismometers can detect signals as small as 1 A. Power is supplied by a nuclear battery fueled by plutonium-238. Data from the seismometers are transmitted to a network of receiving stations on Earth. The instrument is operated by a series of 15 commands, sent from Earth, which control such functions as change in sensitivity, calibration and adjustment of the seismometers, and thermal control of the instrument. The *Apollo 11* seismic station functioned for 21 days before a failure in the command-receiving system finally terminated the experiment. The *Apollo 12* seismic station continued to function as of Oct. 1, 1970.

Recorded signals. Early in the *Apollo 11* mission it was discovered that continuous background motions of the lunar surface, analogous to microseisms on Earth, are completely absent on the Moon, or are so small that they cannot be detected by the lunar seismometers. Thus the seismometers are able to operate on the lunar surface at sensitivities 100–1000 times higher than those that can be achieved on Earth. During the 21-day period of *Apollo 11* operation, thousands of signals of a great variety of shapes and sizes were recorded. It soon became clear that the great majority of these signals were produced by the lunar module itself, presumably by venting of gases and by thermal effects on its metal structure. One type of signal, classified as type L, however, shows no regularity

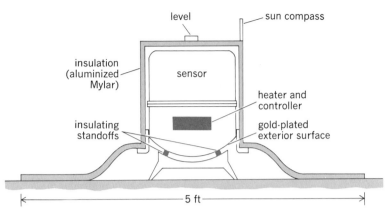

Fig. 1. Schematic diagram of the seismometer that was left on the Moon.

in time of occurrence and covers a wide range of amplitudes and frequencies; 83 type L signals have been detected on the *Apollo 11* records. L signals are complex and of unexpectedly long duration, with gradual increase and decrease in signal amplitude. The familiar pattern of signals corresponding to various types of seismic waves, as normally observed from earthquakes, is not observed in any of the lunar signals.

Since the *Apollo 12* seismic station was installed at a greater distance (130 m) from the lunar module than was the *Apollo 11* station (16.8 m), the number of disturbing signals from the lunar module recorded by the *Apollo 12* seismic station was greatly reduced. L signals, however, were again detected at a rate of about one per day.

Perhaps the most significant signal thus far recorded was generated by the impact of the lunar module ascent stage during the *Apollo 12* mission. The impact took place at a distance of 73 km from the seismometer, with an equivalent energy release of approximately 1 ton of TNT. The recorded signal and two of the largest L signals recorded to date are shown in Fig. 2. The total duration of the impact signal was approximately 55 min. The signal from an equivalent source on Earth would have lasted perhaps 2–3 min. Most important, however, is the fact that the signal character is identical to that of other L events. Thus it is evident that impulsive sources on the Moon, such as impacts, can produce signals of the L type.

Preliminary interpretations. There is good agreement between the numbers and amplitudes of the lunar seismic signals and those expected from meteoroid impacts. From this comparison it appears that the number of meteoroids which collide with the lunar surface is smaller than that which enters the Earth's atmosphere by a factor of approximately 2 (for meteoroids in the mass range from 1 g to 10 kg). This estimate is subject to considerable revision as data accumulate. Some of the recorded seismic events (at least 58 in 8 months) are moonquakes. The moonquakes occur primarily near the time of closest approach between the Earth and the Moon each month (perigee). Thus the moonquakes appear to be triggered by tidal stresses within the Moon which reach maximum levels at perigee. All the moonquakes are small (between 1 and 2 on the Richter magnitude scale).

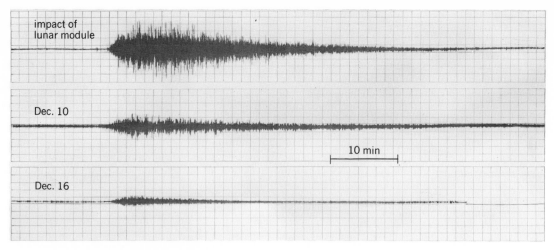

Fig. 2. Record of seismic signals received on the long-period vertical component seismometer from the lunar module impact, which took place at a distance of 73 km from the seismometer, and from two large natural sources on Dec. 10 and Dec. 16, 1969.

Several hypotheses have been advanced to explain the unusual character of the recorded signals. The weight of present evidence favors the hypothesis that seismic energy propagates through the outer shell of the Moon to depths of at least 10–20 km, with extremely low attenuation relative to the surface of the Earth and with a high degree of scattering. Intense scattering of seismic waves would tend to increase the duration of the observed seismic signals and to suppress the appearance of distinct phases within a wave train. A medium which shows low attenuation and high scattering is unlike anything observed within the Earth. Cold blocks of differing composition in welded contact might show these properties. Whatever the mechanism is determined to be, it will provide important evidence on the origin and evolution of the lunar interior.

The fact that no seismic signals with characteristics similar to those typically recorded on the Earth were observed during 4 months of combined recording period for *Apollo 11* and *12* is a major scientific result. The high sensitivity at which the lunar instruments were operated would have resulted in the detection of many such signals if the Moon were seismically as active as the Earth and if it had the same seismic transmission characteristics as the Earth. Thus, although other interpretations are possible, first seismic results suggest that the Moon is an extremely quiet and stable body relative to the Earth, except for the disruptive influence of tidal stresses. The concept of plate tectonics as it applies to the Earth, with formation of mountain chains and great fractures in the outer shell, cannot be applied to the Moon at this point in its history. This conclusion implies the presence of lower specific thermal energy in the lunar interior than is present in the interior of the Earth.

It must be remembered that all results obtained to date apply only to the mare basins, in which the measurements have been made. Observations from the lunar continental regions, planned for future missions, may be quite different.

[GARY V. LATHAM]

SOLAR WIND COMPOSITION

Manned landings on the Moon have made it possible to determine for the first time the isotopic abundances of some rare gases in the solar wind.

Solar wind. The solar atmosphere is so hot (about 1,000,000°C at the inner corona) that even the Sun's high gravitational attraction cannot completely impede its escape. Thus electrons and highly charged ions emanate from the Sun, forming a steady supersonic stream of solar material which flows outward into interplanetary space.

Fig. 3. Deployed SWC unit with astronaut E. E. Aldrin, Jr. (*NASA*)

The existence of such a solar wind was anticipated by L. Biermann 20 years ago as an explanation for the tails of comets. In 1958 E. Parker developed a hydrodynamical model of the expansion of the solar atmosphere. He predicted theoretically the most important features of the solar wind, especially its continuous flow and its supersonic velocity, which is the same for all particles. Because the magnetosphere is shielding the Earth from the undisturbed impact of the charged solar wind particles, only satellites and space probes to distances of more than about 15 earth radii (100,000 km) can serve as measuring platforms. Since 1959 Faraday cups and electrostatic analyzers have been flown out to these distances and were reported to give the following average solar wind data: Flow velocity is approximately 400 km/sec for all ions (0.8 kev/nucleon); flux is 2×10^8 ions/cm² sec; most ions are hydrogen nuclei (protons); and helium (He^4) ions make up 4% of the total flux but their relative abundance, however, varies strongly; oxygen and He^3 ions have been detected a few times. Most of the ion species occuring in the solar wind are either too rare to be detected or have an energy-charge ratio too similar to be separated by the analyzers.

Isotope abundances. No magnetic field or atmosphere prevents the incident particles from striking the lunar surface and from being partially trapped by the soil material. Thus the analysis of lunar material should give a good idea of the elemental and isotopic composition of the solar wind, that is, at least of those components which are frequent on the Sun but very rare in solid material. The noble gases fullfill this requirement well. However, several insufficiently known effects prevent direct determination of the composition of the solar wind by this method. First, the lunar material contains a varying amount of noble gases of other than solar origin (for example, those produced by cosmic radiation or nuclear decay). Second, the trapping probability for the solar wind particles and the saturation effects in the lunar material are little known. Third, thermal diffusion losses of light noble gases are severe. Even when these effects are not present, the measurements would yield only average abundances over long and unknown time periods.

The Solar Wind Composition (SWC) Experiment, in contrast, directly yielded absolute wind fluxes of helium (He^4, He^3) and neon (Ne^{20}, Ne^{21}, Ne^{22}) isotopes at a given time.

SWC experiment. This experiment was proposed and prepared by J. Geiss and P. Eberhardt, of the Physics Institute of the University of Bern, and P. Signer, of the Institute for Crystallography

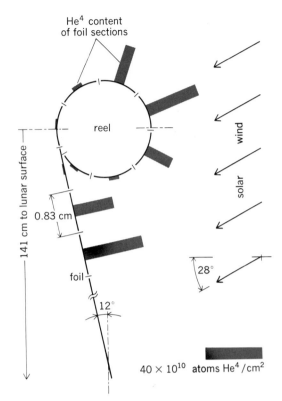

Fig. 4. Upper part of aluminum foil with reel as exposed on the lunar surface during the second landing. The shaded bars represent the measured helium concentrations in each of the small foil pieces.

and Petrography of the Federal Institute of Technology, Zurich. Shortly after the landing, an astronaut extends a piece of aluminum foil 1 ft wide and $4\frac{1}{2}$ ft long which has been rolled on a reel. Then he positions the foil perpendicular to the solar rays (which are almost parallel to the solar wind direction) and pushes the mounting pole, which is fixed to the reel, into the lunar soil (Fig. 3). Previous laboratory experiments have shown that atoms striking the foil surface with solar wind velocities penetrate approximately 10^{-5} cm into the foil. A large and calibrated fraction of the incident atoms is firmly trapped. During the first two landings a solar wind flux was integrated over an area of 4000 cm² and over a time period of 77 min and $18\frac{3}{4}$ hr, respectively. At the end of the crew's extravehicular activity, the foil was rolled up onto the spring-driven reel, and then returned to Earth. Because of the low initial rare gas content of the 0.6-mil foil, the analysis by the laboratories of Bern and Zurich has been restricted to noble gas atoms.

Table 1. Solar wind composition from SWC experiment compared with solar wind trapped in lunar dust, trapped gases in aubritic meteorites, and composition of Earth's atmosphere

Isotopes	Solar wind composition experiment		Lunar dust bulk	Aubritic meteorites	Earth's atmosphere
	Apollo 11	*Apollo 12*			
He^4 flux, ions/cm² sec	$(6.2 \pm 1.2) \times 10^6$	$(8.1 \pm 1.0) \times 10^6$	—	—	—
He^4/He^3	1860 ± 140	2450 ± 100	2550 ± 250	4000 ± 400	730,000
He^4/Ne^{20}	430 ± 90	620 ± 70	90 ± 20	380 ± 70	0.3
Ne^{20}/Ne^{22}	13.5 ± 1.0	13.1 ± 0.6	12.65 ± 0.20	12.2 ± 0.5	9.8
Ne^{22}/Ne^{21}	—	26 ± 12	31.0 ± 1.2		34.5

Fig. 5. Photograph of a freshly broken surface of a sample of fine-grained basalt with spherical vesicles (gas cavities). (*NASA*)

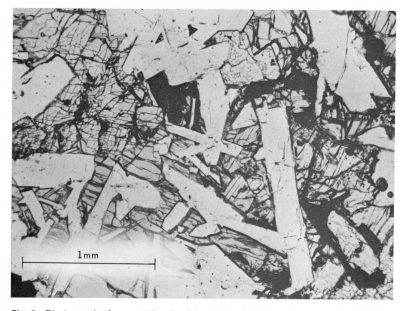

Fig. 6. Photograph of a very thin slice (about 30 μ thick) of a coarse-grained gabbro, as seen under the microscope. The clear, long laths are plagioclase; the black grains are ilmenite; and the light-gray grains which display pronounced cleavage are pyroxene. (*NASA*)

Small sections of the foil were melted in an ultra-high vacuum system, and the noble gases thereby released were analyzed in mass spectrometers. Five special foil pieces, some of which were irradiated prior to flight by a known amount of neon, were attached to the foil at positions shielded from the solar wind. The analysis of these pieces gave information on the interference caused by dust contamination, by ions scattered back from the lunar soil, by chemical etching of the foil surface, by rocket exhaust, and by thermal diffusion losses. None of these effects was found to have a significant influence upon the results.

First results. During the exposure the uppermost part of the foil remained rolled up around the reel for approximately one turn. A long strip of this portion of each foil was cut into small sections, and their individual He[4] contents were measured (Fig. 4). Thus information on the angular distribution of the directions of incidence could be obtained. For both landings, the results are clearly consistent with a highly directional flow of the solar wind along the expected direction. The solar wind arrives essentially unperturbed at the lunar surface, even for low angles of incidence (15–30° above the lunar horizon).

The helium (He[4]) fluxes estimated from the measurements on the fully irradiated parts of the foil (Table 1) agree well with the averages observed by satellites and space probes. The analysis of larger foil sections yielded solar wind abundances of the rarer ions He[3], Ne[20], Ne[21], and Ne[22]. Table 1 shows the results, as available at the date of writing, in comparison to the composition of the trapped rare gas (that is, solar wind) components in the lunar dust from Mare Tranquillitatis and in aubritic meteorites.

Almost all abundance ratios are clearly nonterrestrial. Moreover, the large difference between the helium-neon ratios in lunar dust and in the solar wind indicates heavy losses of helium in lunar material due to thermal diffusion.

The SWC He[4]/He[3] and He[4]/Ne[20] ratios are likely to correspond to true solar wind abundances. Their differences can be interpreted as being due to time variations in the solar wind. These variations could be caused by a fractionation effect at the source of the solar wind. Results of expositions during further landings should give good insight into these processes. Reliable values of the solar surface noble gas composition could then be estimated; these values are very important for every theory of the solar system.   [F. BUHLER]

SAMPLES RETURNED BY APOLLO 11

On July 24, 1969, the first samples from the Moon were returned to Earth for direct scientific study in laboratories around the world. Until this time, knowledge about the chemical composition of lunar material had been derived from data collected by several Surveyor spacecraft equipped with alpha-particle backscatter instrumentation, and indirectly from the study of meteorites, from the study of cosmic rays, and from electromagnetic radiation from stars and planets. The rock samples brought from Tranquillity Base by *Apollo 11* consist of basaltic igneous rocks and of microbreccias, which are compacted mechanical mixtures of lunar soil and small rock fragments. The lunar soil consists of a mixture of crystalline and glass fragments and a very small amount of meteoritic material. The glasses occur as irregular fragments, spheres, dumbbells, and ellipsoidal bodies.

Rocks. As reported by the Lunar Sample Preliminary Examination Team (LSPET), the lunar samples returned by *Apollo 11* weighed about 22 kg, of which 11 kg consisted of rock fragments over 1 cm in diameter and the remainder of finer particulate material. The three major rock types are: fine-grained, vesicular crystalline igneous rocks (basalts) (Fig. 5); coarser-grained, sometimes vuggy, crystalline igneous rocks (gabbros) (Fig. 6); and microbreccias, which are made up of a mechanical mixture of lunar soil and small rock fragments, compacted into coherent rocks (Fig. 7).

The lunar basalts and gabbros, both of which have typically igneous textures, are made up of a mixture of pyroxene (an iron-magnesium-calcium silicate), plagioclase (a calcium-sodium-aluminum silicate), and ilmenite (an iron-titanium oxide) and small amounts of cristobalite (SiO_2), olivine (an iron-magnesium silicate), and troilite (FeS). In these lunar rocks, pyroxene is the most abundant mineral, whereas in terrestrial basalts and gabbros, plagioclase generally exceeds pyroxene or equals it in abundance. The microbreccia and lunar soil are mixtures of the minerals which occur in volcanic rocks plus 10–20% glass. All the min-

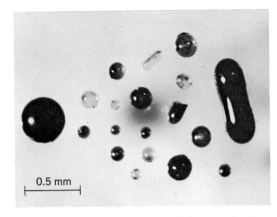

Fig. 8. Glass spherules and dumbbells of varying size, color, and composition, from lunar soil.

erals and glasses in these rocks are unaltered, and many are transparent and clear because of the lack of alteration processes on the lunar surface.

The average chemical compositions of the lunar rock types are given in Table 2 and are compared with the composition of a terrestrial oceanic basalt which is found on the ocean crust, ridges, and rises. These samples are from Mare Tranquillitatis, and their analyses are taken from C. Frondel, C. Klein, J. Ito, and J. C. Drake and those of the oceanic basalt from A. E. J. Engel and C. G. Engel. The compositions of the four lunar materials in Table 2 are quite similar, but they differ considerably from the terrestrial material. The lunar rocks contain no water and lack Fe_2O_3 (ferric iron). The richness of TiO_2 as well as FeO in lunar rocks is a result of the great abundance of ilmenite. Na_2O and K_2O generally are much less abundant in lunar rocks than in oceanic basalts.

Soil (fines). The lunar soil (particle size less than 1 mm) consists of a mixture of the same minerals which occur in the rocks, such as pyroxene, plagioclase, and ilmenite, and about 20% glass. This glass varies widely in shape, grain size, color, and chemical composition. Most of it is dark brown to almost black, and occurs in irregular and rounded forms and sometimes as thin sheets covering rock fragments. The most remarkable, although less common, glass variety occurs as minute spheres and dumbbells and in ellipsoidal shapes. The colors of such glasses vary from clear, transparent, and colorless to light green, light brown, dark brown, and opaque. Figure 8 shows a variety

Fig. 7. Photograph of a very thin slice of microbreccia as seen under the microscope. Angular and broken crystal fragments'of pyroxene and plagioclase are enclosed in a very fine grained, dark gray to black ground mass. At the upper right a glass spherule is visible.

Table 2. Average chemical compositions of lunar rock types compared with terrestrial basalt

Component	Vesicular basalt	Gabbro	Microbreccia	Soil	Terrestrial oceanic basalt
SiO_2	40.60	41.65	41.69	41.77	50.01
TiO_2	11.28	8.79	7.35	7.42	1.37
Al_2O_3	8.05	11.00	13.52	13.68	16.18
Fe_2O_3	–	–	–	–	2.32
FeO	20.09	18.20	16.05	15.98	7.07
MgO	7.90	7.80	8.40	8.38	7.71
CaO	10.57	11.22	11.76	11.68	11.33
Na_2O	0.44	0.41	0.42	0.41	2.79
K_2O	0.24	0.12	0.14	0.13	0.22
P_2O_5	0.15	0.12	0.12	0.12	0.13

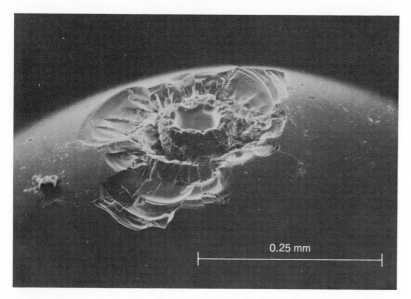

Fig. 9. Example of microcratering, caused by hypervelocity impact of tiny particles, on a dark-brown glass sphere. The diameter of the sphere is approximately 0.75 mm, and the diameter of the inner crater, inside the raised rim, is about 50 μ. This photograph was taken through a scanning electron microscope.

of such glasses. The colorless glasses are similar to plagioclase (Ca-Na-Al silicate) in composition, and dark-brown glasses represent compositions of varying mixtures of pyroxene (Ca-Fe-Mg silicate), ilmenite ($FeTiO_3$), and plagioclase. These glass particles in the lunar soil are the result of local melting of the lunar soil and rock fragments by impact of meteorites and high-velocity particles which have caused melting, splashing, evaporation, and condensation. The surfaces of the glass spheres and rocks show minute craters, which are the result of hypervelocity impacts of tiny particles (Fig. 9).

The chemical composition of the lunar soil is very similar to that of the igneous lunar rocks but differs somewhat in the content of less abundant elements. For example, nickel is more abundant in the soil than in the rocks because of the presence of small meteoritic, iron-nickel fragments in the soil.

Terrestrial material comparison. The three rock types and the soil have no exact terrestrial or meteoritic analogs. They are close in composition to some oceanic basalts and some meteorites, but their high titanium content combined with their highly reduced state (no ferric iron) is not found on Earth. The textures and mineral assemblages of the rocks indicate crystallization from lava flows such as those found on Earth in presently active or long extinct volcanic centers.

For background information *see* MINERALOGY; MOON; PETROLOGY; SEISMOGRAPH; SEISMOLOGY in the McGraw-Hill Encyclopedia of Science and Technology.   [CORNELIS KLEIN, JR.]

Bibliography: W. I. Axford, Observations of the interplanetary plasma, *Space Sci. Rev.*, 8:331–365, 1968; F. Bühler et al., Apollo 11 solar wind composition experiment: First results, *Science*, 166:1502–1503, 1969; A. E. J. Engel and C. G. Engel, Lunar rock compositions and some interpretations, *Science*, 167:527–528, 1970; C. Frondel et

al., Mineralogical and chemical studies of Apollo 11 lunar fines and selected rocks, *Geochim. Cosmochim. Acta*, in press; C. Frondel et al., Mineralogy and composition of lunar fines and selected rocks, *Science*, 167:681–683, 1970; J. Geiss et al., Apollo 11 and 12 solar wind composition experiments: Fluxes of He and Ne isotopes, *J. Geophys. Res.*, vol. 75, no. 31, Nov. 1, 1970; G. V. Latham et al., *Apollo 11 Preliminary Science Report*, sect. 6, NASA SP-214, pp. 143–169, 1969; G. V. Latham et al., *Science*, 165(3890):241–250, July 18, 1969; Lunar Sample Preliminary Examination Team (LSPET), *Science*, 165:1211–1227, 1969; A. McGarr, G. Latham, and D. Gault, *J. Geophys. Res.*, 74(25):5981–5994, 1969; Y. Nakamura and G. Latham, *Bull. Seismol. Soc. Amer.*, 60(1):63–78, 1970.

Muon

The muon (mu-meson) was discovered to be an important constituent of cosmic radiation in 1935, and has been the object of intensive study since, both in cosmic rays and in accelerator beams, which have come to dominate the field since the 1950s. The muon belongs in a special category when compared to the vast expansion in the number of elementary particles known since the advent of the large accelerator. It is known as a lepton along with the electron and the two kinds of neutrinos. In all, there are eight leptons: e^+, e^-, μ^+, μ^-, $\bar{\nu}_e$, ν_e, $\bar{\nu}_\mu$, and ν_μ, with the antiparticle and particle listed for each of the four types. These particles all share several important properties: They all have 1/2 unit of spin, come in pairs, and, most remarkable, have no strong interaction. Thus the muon is able to reside in the lowest (K−) orbit of a heavy-lead nucleus for extremely long times (on a microscopic scale, some 10^{20} orbits) before reacting with the protons in the lead nucleus, notwithstanding the fact that the orbit is entirely inside the nucleus. The properties of the muon are usually classified in terms of static measurements, for example, mass, charge, spin, magnetism, lifetime, and decay properties, and in terms of its interactions with matter. All of these properties have been under intensive investigation in the last few years.

Static properties. The mass of the muon is 207 times that of the electron. In fact, this number, 207.76, is very precisely known and constitutes one of the puzzles of the muon since it is the only number known so far which differentiates the muon from the electron. The magnetism of the muon (its magnetic moment) derives from the fact that it is a charged body possessing a spin angular motion. The measurement may be made with very high precision because of its curious decay properties (discussed below), and this precision confirms a theory of P. A. M. Dirac that the muon has no finite extent and no anomalous interaction with the outside world. The muon is unstable and decays in two-millionths of a second into an electron (of the same charge) and two neutrinos. The failure of the muon to disintegrate into an electron and a photon was a great puzzle until the discovery of the muon neutrino and the recognition that the muon (and its neutrino) possess a property (sometimes called mu-ness) which must be conserved in all interactions involving muons. Thus a muon cannot simply disappear in an interaction without giving rise to "its own" neutrino. *See* ATOMIC CONSTANTS.

Sources. The muon is produced chiefly by the decay of heavier mesons, mainly the pion and the kaon. The sea-level cosmic-ray flux of muons is such that about 10 muons pass through an average person per second. Having no strong interaction, they are very penetrating; muons have been observed 8000 ft underground. The parent pions are generated by collisions of high-energy protons incident upon the Earth from outer space. These protons strike atoms of the Earth's atmosphere, creating showers of pions and kaons which in turn decay into muons. In accelerators, much the same process enables the formation of intense beams of muons, which are used either to study the properties of these particles or as tools in other investigations. *See* HADRON.

Interactions. The muon engages in weak interactions (those responsible for radioactivity) or, by virtue of its electric charge, in electromagnetic interactions. Its weak decay into electron and neutrino has already been mentioned: $\mu^+ \rightarrow e^+ + \bar{\nu}_\mu + \nu_e$. It can also react with protons: $\mu^- + p \rightarrow n + \nu_\mu$, and the latter reactions have been carefully studied in order to compare them with a corresponding reaction engaged in by electrons (K capture). The two reactions appear to be identical. The above reactions are studied with muons at rest. Beams of muons from high-energy accelerators such as the 30-Bev alternating gradient synchrotron (AGS) at Brookhaven Laboratory can reach as high as 20 Bev. They have been used in scattering experiments in an attempt to detect some structure in the muon which would differentiate it from the electron and provide a clue to the origin of the mass difference. It is known that the higher the energy and the larger the deflection of the incident particle, the more intimate is the collision. In this way, muons and electrons have been compared (using protons as the "ruler") and found to be identical down to distances of the order of 10^{-15} cm.

On the other end of the energy scale, the positive muon has been observed to collect a negative electron when coming to rest in pure gases, and the resulting "atom," muonium, has been studied with great precision. Again, the purely electrical forces are fully accounted for by the current theories of electromagnetism.

Muon as tool. When negative muons are stopped in various elements, they eventually are captured into atomiclike orbits which are largely inside the usual electron shells because of the high mass of the muon. Thus a new type of atom is formed that is very much like a hydrogen atom, the nucleus of variable charge and a single muon replacing the customary electron. The transition of the muon from higher to lower energies gives rise to x-rays, which have been measured over a very large number of nuclei and transitions. The detailed analysis of these data is a rich source of information on the properties of the nucleus about which the muon is in orbit.

The decay of the muon was one of the early experiments which established the breakdown of the law of parity conservation, it having been observed that the muon prefers to emit electrons backward relative to its spin. This asymmetry is useful in determing the muon spin direction, and thus, the muon "magnet" can be used to probe for microscopic magnetism in various materials, in nuclei, and in atoms. *See* NUCLEAR STRUCTURE.

The extremely penetrating properties of high-energy muons have recently been used as a technique for studying the detailed electromagnetic structure of protons and neutrons in previously unexplored regions. This is because a positive and negative muon emerging from a high-energy collision (a dimuon) is equivalent to a very massive photon, that is, quantum of electromagnetism. A study of the frequency of emission of such dimuons delineates these properties with great sensitivity. The emission process is rare and would normally be submerged in a large background of competing processes but for the ability of the muons to penetrate a thick iron shield which serves to screen the unwanted particles.

For background information *see* ELEMENTARY PARTICLE; LEPTON; MUONIUM in the McGraw-Hill Encyclopedia of Science and Technology.

[LEON M. LEDERMAN]

Musical instruments

Electronics is playing an increasingly important role in machines related to music. The use of electricity in the production of music began more than 80 years ago. Nevertheless, until recent years the progress in the application of such technology to the art of music has been slow. Tradition has considerable influence upon the development of musical instruments.

G. F. Diekmann was granted a patent in 1887 on means for producing music by combining electrical circuits with tuned strings such as those used in pianos. By 1930 approximately 100 United States patents had been granted on electronic musical instruments of various types. On May 5, 1938, the League of Composers presented a program and demonstration, *Music and Electricity*, at the Town Hall Club in New York City. A number of the instruments in this program used electronics to either amplify or modify or both amplify and modify the tones produced by tuned strings in a piano, in instruments of the violin family, in a guitar, and in an instrument that sounded like a kettledrum. Benjamin F. Miessner, an active pioneer in this field, said at the time that he was convinced that all the traditional instruments would be remade and vastly improved with the modern methods of electronics, that new and enduring instruments would be developed, and that musical horizons would be greatly expanded by their use. Tradition provided considerable inertia, however. Some of Miessner's many inventions did find their way slowly into commercial instruments, but it was about 30 years later that a resurgence occurred in the application of electronics to amplify and modify tones of individual musical instruments.

Guitars. The guitar was the first instrument to make popular use of electronics. Electric guitars use a suitable pickup to transduce the motion of the guitar strings to electrical signals. These sig-

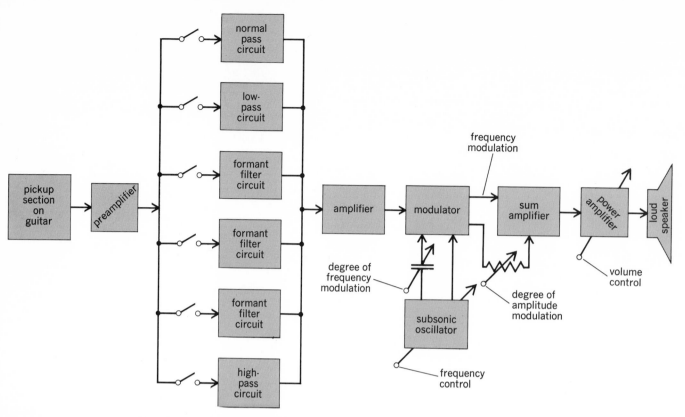

Fig. 1. Electronic system for an electric guitar.

nals may be amplified and fed into one or more loudspeakers, and usually provision is made to modify the character of the signals in several ways. The tone may be modified so that a guitar of a given form and size, with a fixed tone capability without electronics, can be made to sound like many instruments in the guitar family through the use of electronics. In addition, either vibrato or tremolo can be added to tones electronically, and a variety of novelty effects, such as "fuzz tone," can be produced.

One United States patent describes a system for electric guitars and the like as illustrated in Fig. 1 and as modified in Fig. 2.

The pickup section on the guitar may consist of one or more electromagnetic or piezoelectric transducers per string. The output of the pickup section is in electrical form representing the tones, including all partials, generated by the vibrating strings

of the instrument. The design of the pickups and their locations with respect to the strings influence the relative amplitudes of the partials of the tones, but the positions selected for the system of Fig. 1 provide a full complement of partials.

The electrical signals produced by the pickup section in Fig. 1 are amplified by an isolating amplifier and passed to a series of circuits which may be selected individually or in combination by means of switches. This series of circuits may include a circuit which passes all partials without coloration, a circuit which passes low frequencies with less attenuation than given higher frequencies, one or more circuits that accentuate one or more bands of frequencies (called formant filters), and a circuit that favors the passage of high-frequency signals and attenuates low frequencies. This series of circuits enables the musician to select the sound of the acoustic guitar, rock-and-roll guitar, classic guitar, Hawaiian guitar, bass guitar, solo guitar, and other sounds. After the electrical signals have passed through one or more coloration circuits, they are again amplified and passed to a modulator circuit that makes use of a low-frequency (subsonic) oscillator to add a vibrato or tremolo effect or both at a desired degree and rate. The tone signals are again amplified and passed to one or more loudspeakers, with a suitable volume control provided.

The system illustrated in Fig. 2 provides clipper circuits between the pickups and the tone-forming filters. The clipper circuits generally serve to augment the strength and number of partials so as to provide greater possibilities for variety in the tone-forming filters. The tone-forming circuits may be the same as those shown in Fig. 1, or they may be

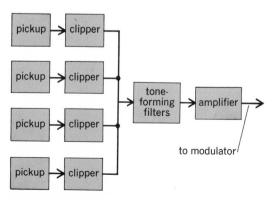

Fig. 2. Modified electronic system for an electric guitar.

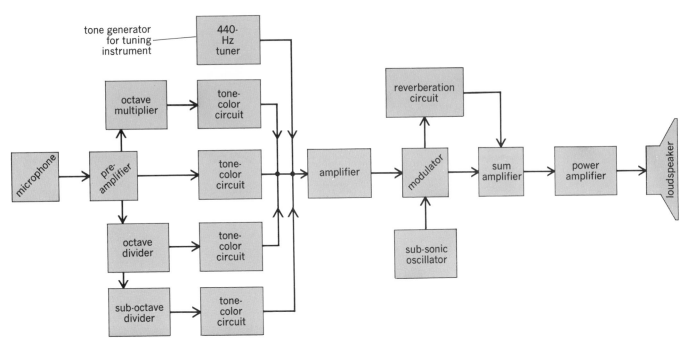

Fig. 3. Electronic modification of wind instruments.

different. The resulting signals may be treated as in Fig. 1.

Wind instruments. The treatment discussed above can be given to the tones of other stringed instruments, including pianos, and to wind instruments also.

In wind instruments only one pickup is generally used, a sound-pressure microphone placed in the mouthpiece or in some other part of the instrument. The electrical signals from the pickup microphone can be treated in ways similar to those illustrated in Fig. 1 or 2. Other variations are common also.

The block diagram in Fig. 3 indicates a simplified composite of functions offered on different commercial products to be used to either amplify or modify or both amplify and modify the tones of individual wind instruments.

In addition to tone modifications in timbre, tremolo, or vibrato embellishment and virtually unlimited sound power, it is possible for one instrument to sound as if it were as many as four octavely related instruments playing simultaneously. This is done by means of circuits that divide or multiply the fundamental frequency of the tone produced by the instrument played by the musician. Also, artificial reverberation or echo effects can be added to the tones. *See* SOUND-REPRODUCING SYSTEMS.

The 440-Hz tuner shown in Fig. 3 is an oscillator that can be used to produce standard note A for tuning the instrument being played.

Commercial power amplification of wind instruments is shown in Fig. 4. Electronic systems to amplify and modify tones of individual musical instruments are being used in marching bands, concerts, motion picture sound tracks, recordings, and many other ways.

For background information *see* MUSICAL ACOUSTICS; MUSICAL INSTRUMENTS; SOUND-REPRODUCING SYSTEMS; TONE (MUSIC AND ACOUS-

Fig. 4. Power amplification of wind instruments in use.

TICS) in the McGraw-Hill Encyclopedia of Science and Technology.

[EARLE L. KENT]

Bibliography: D. W. Elbrecht et al., *Output Systems for Electric Guitars and the Like*, U.S. Patent no. 3,493,669, 1970; E. L. Kent, *Audio Eng. Soc. J.*, vol. 17, no. 3, 1969.

Navigation

Dioscures is the name of a project which is under study by French space and civil aviation agencies. This project will utilize two geostationary satellites per zone to provide an integrated system of radio communications, navigation, and radio location.

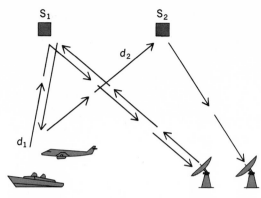

Fig. 1. Measured paths for radiolocation of a plane or ship.

Characteristics of radio-frequency links.

The Dioscures system will offer worldwide coverage. The system has a built-in growth capacity to keep pace with development of air and sea transportation. It is permanently available and its performance, transmission quality, and location accuracy should fully satisfy the requirements of the users.

The system is based on the simultaneous use of two geostationary satellites per zone served and provides for the integration of radiocommunication and radiolocation services. When only the radiocommunication service is required over a zone, one satellite only is used. To keep the cost of the on-board aircraft or ship equipment to a minimum, the calculation of the mobile's position function is performed by the Earth station computer; this procedure eliminates the need for a position computer in each mobile.

For the first generation of satellites, the capacity of the system over the Atlantic zone, where it supplies a complete service, is six telephone circuits or six 1200-baud circuits with complete interchangeability; two telex circuits of 150 bauds; three automatic data-transmission channels giving the following data: heading, velocity, temperature, altitude-position, call-sign, and circuit-busy signals; and 250 calculations of position of mobile every 5 min.

Location. The location is based on distance measurements performed on the ground from measurements of the following paths:

$P_1 = $ Ground – Satellite S_1 – Mobile –
$\qquad\qquad\qquad\qquad$ Satellite S_1 – Ground

$P_2 = $ Ground – Satellite S_1 – Mobile –
$\qquad\qquad\qquad\qquad$ Satellite S_2 – Ground

From the measurements of P_2 and P_1, the satellites-mobiles distances d_1 and d_2 are deduced (Fig. 1). These distances indicate that the mobile lies somewhere on the surfaces of two spheres centered on the satellites. Knowing the mobile's altitude, one can deduce that its position lies at the point of intersection of three spheres. The mobile itself removes the north-south latitude ambiguity.

To supply radiolocation service, one must establish a data transmission link between the ground and the mobile. The Earth station requires the aircraft's altitude j as measured by its pressure altimeter. For a surface vehicle this information is not required. The station calculates the position by taking the geoid as an altitude sphere. The mobile requires its position as calculated by the Earth station.

Telecommunications. In the Earth station – mobile path, the telecommunications are based on digital multiplexing techniques. On a single carrier, three telephone channels, one telex channel at 150 bauds, call signs, and occupation signals are transmitted. At least one telephone channel per satellite may be used for automatic data transmission at 1200 bauds. This information may be utilized for the automatic integrated data system (AIDS).

In the mobile – Earth station path, the system is of the multiple-access type; that is, four mobiles on four different frequencies may transmit to a satellite on one telephone channel which may be multiplexed with a telex channel at 150 bauds. In fact, each satellite will receive five separate carrier frequencies. Four of these will serve the telecommunication function and one is for the location function. These five frequencies will be amplified in a common repeater B. In the satellite itself, a sixth carrier frequency will be used to transmit the telemetry data.

The principle of the multiplexing of data which has been adopted for the ground-mobile path, and the digital coding of these data, allows the setting up of an integrated "location plus telecommunication" system. All the data (including a location signal of a period of 25 Hz) have been gathered together on a same-time multiplex. The transit time of this transmission will be measured on the ground along the ground-aircraft-ground path.

Figure 2 is a diagram of the radio links of the Dioscures system, where VCO is the voltage control oscillator, ampli is the amplifier, VF is voice frequency, F_{T1} to F_{T14} are the carrier frequencies going to or coming from the Earth station, and F_{a1} to F_{a9} are the carrier frequencies coming from mobiles.

Main technical options. Three main technical choices were made to define the Dioscures system. They have a major effect both on the performance (quality of the telephone channel and location accuracy) and on the cost of the project. One of these choices concerned the utilization of frequencies in the L band (1540–1660 MHz) for the satellite-mobile links. These frequencies give reliable propagation characteristics and good location accuracy (better than 1 nm for aircraft and 0.5 nm for ships in 95% of the cases).

The second decision involved the digital coding of all the information (telephone, data) to permit a complete interchangeability of voice and data

Costs of satellites and launchers in millions of dollars

Equipment	Cost
First generation	
First two flight models (240 kg)	40
One additional flight model	5
One Thor Delta M9	5.4
Total for 8.54 launches	118.6
Second generation	
First four flight models	110
One additional flight model	10
One Atlas Centaur	16
Total for 9.7 launches	312.2
Total	430.8

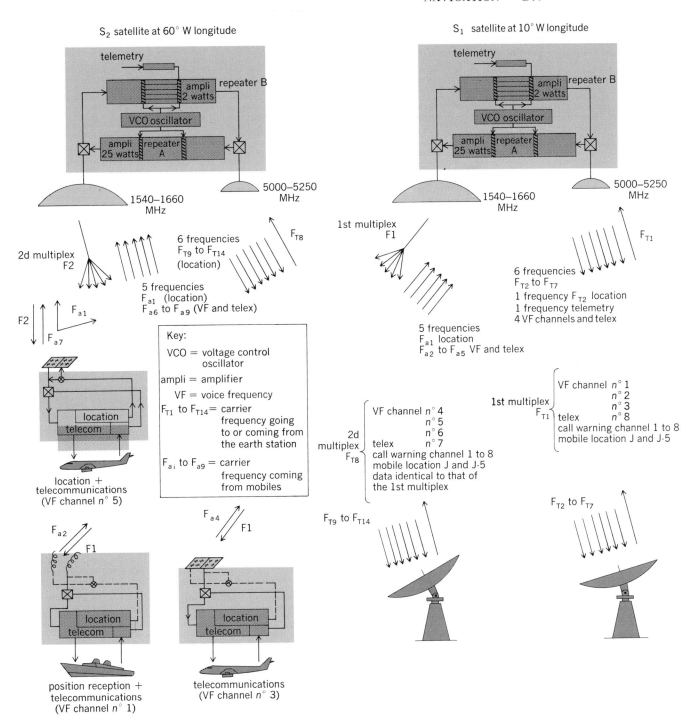

Fig. 2. Diagram of radio links of the Dioscures system.

channels (use of delta coding).

The third decision relates to the important calculation parameters of the radio links and especially the equivalent intrinsic radiated power (EIRP) of the satellites and the characteristics of the aircraft or ship station. This decision made it possible to employ a first-generation satellite weighing less than 250 kg and an aircraft antenna with a nominal gain of 10 dB.

Project costs and economic advantages. The costs of the project have been calculated through studies of the satellites and by the operational simulation of their failures. Prototypes of the aircraft antenna and of the transponder have been manufactured and studies of installation of these units in various types of aircraft have been made. The costs are shown in the table. It should be noted that the simulation of the failures of the launchers and of the four operational satellites has shown that, on the average, 8.54 satellites are required for the first generation (service life of 5 years) and 9.7 for the second generation (service life of 7 years).

Aircraft and ship equipment for a quantity production of more than 300 units would on the average cost (including installation), for aircraft, $58,000 ($13,000 for installation), and for ship, $30,000 ($1,000 for installation).

For 1500 equipped aircraft, it is to be noted that

aeronautical equipment is only of the order of $87,000,000, that is, about 17% of the total cost. This is why a sophisticated aircraft antenna is chosen; its cost is low compared to its economical influence on the complete economical trade-off (cost of space segment plus aeronautical cost).

The economical advantages of such a project cannot be analyzed briefly without distorting the true economic picture. For this reason it is desirable to refer to the article by B. Manuali in the February, 1969, *Telecommunications Journal.*

For background information *see* NAVIGATION; NAVIGATION SYSTEMS, ELECTRONIC; SATELLITES, NAVIGATION BY in the McGraw-Hill Encyclopedia of Science and Technology.   [B. MANUALI]

Bibliography: Centre National d'Études Spatiales – Secrétariat Général de l'Aviation, *Projet Dioscures: 1972–1985,* July, 1969; Intergovernmental Maritime Consultative Organization, *Paper MSC XIX/30 Annexe IX*; B. Manuali, The Dioscures Project, *Telecommun. J.,* vol. 36, February, 1969; B. Manuali and C. Ganier, Economic study of satellite systems for telecommunications, air traffic control and navigation, *Telecommun. J.,* vol. 37, January, 1970; *Philco-Ford Report to Comsat,* TR-DA 1583, December, 1967; Radio Technical Commission for Marine Services, *Paper No. 205-69/SG 57–73.*

Navigation instruments

More than 3000 vessels are involved in collisions each year. This remarkably high collision rate is the cause of much concern among the leading maritime nations. Opposed to the obvious need for improved safety are the economic forces that are direct by-products of the worldwide population-commercial explosion. Shipping densities, already severe in many areas, are becoming still greater. While ships are becoming larger and faster, the competitive pressures to improve the efficiency of ship handling persist.

Against this background, the United States Maritime Administration (MARAD) has undertaken long-term programs designed to improve both the safety and efficiency of the merchant fleet. The work thus far has achieved many encouraging results.

Lookout assist device. One program was initiated with the Sperry Rand Corp. in 1963 to study various problems involved in the conning of modern ships and to undertake certain developments. A recent outcome of this effort is the lookout assist device (LAD). The first purpose of LAD is to detect automatically and warn of the presence of other ships within conning range of modern merchant ships. The second purpose of LAD is to detect, enhance, and relay the audible cues which are transmitted over the wide variety of acoustic signaling devices used at sea. Feasibility tests of LAD have been carried out aboard the Humble Oil Co. ship SS *Bangor* and the Cleveland Cliffs Iron Co. ship *Edward B. Greene.* Both series of tests have been fruitful, achieving all the basic design objectives. Intruding ships were reliably detected to ranges of about 10 mi in all weather conditions, and their bearings and ranges were displayed automatically. Audible cues such as the bearing, number, pitch, and timbre of acoustic signaling devices were automatically detected and relayed to the conning officer with a sensitivity superior to that of unaided hearing. Evaluation tests are still under way aboard the *Edward B. Greene* with the objective of selecting the optimal configuration of LAD for Great Lakes operations.

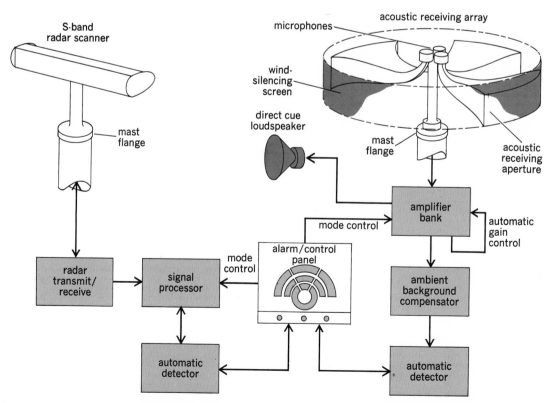

Fig. 1. Functional block diagram of lookout assist device (LAD).

LAD combines a simple radar system and an acoustical reception system to achieve a high probability of detection (Fig. 1). The radar subsystem operates in the S-band (10 cm) and functions completely unattended. Automatic detection is limited to range bands, called guard rings, about a ship. When one ship enters the guard ring of another ship, the radar echo is detected and used to trigger an audiovisual alarm system. The approximate range and bearing are displayed on a lighted alarm panel at the conning station.

The acoustic subsystem of LAD is, in effect, a simple passive in-air sonar, with automatic detection capability. The band of operation (approximately 100–400 Hz) includes all acoustic signals of importance to conning. Both radar and acoustic subsystems have been designed to have very low false-alarm rates. In the acoustic subsystem, the detection thresholds adjust themselves automatically to compensate for the ambient background interference, which is usually high and variable when the ship is under way. Another interesting engineering problem was the gain and phase matching of several high-gain acoustic amplifiers which had to operate over a wide range of frequencies and signal levels.

Other conning problems. Although LAD promises to enhance the probability of detection of threatening ships, the analysis and solution of other problems related to conning continue. Modern disciplines such as operations research, psychophysics, statistical decision theory, and human engineering have been applied in attempts to discover fundamental human limits and to prescribe requirements for the design of collision-avoidance sensors and displays. The density of shipping in many areas of the world's oceans and waterways far exceeds that of former decades and has resulted in new conning problems. These problems will become intolerable unless adequate solutions are discovered. It is not uncommon to observe 20–60 ships on the radar screen at a given time. Unlike automobiles, ships are provided with few road markers and it is seldom practical to direct traffic in an orderly pattern.

The problem is aggravated by serious sources of uncertainty that can be appreciated by referring to Fig. 2, where the various factors involved in a collision-avoidance maneuver are illustrated. Uncertainty enters the maneuver control loop in the form of noise interfering with the displays, environment limiting the effectiveness of sensors, and uncertainty of predicting the movements of other ships. The task of the marine systems engineer is to design sensors and displays which will perform satisfactorily in the environment considering human limitations under stress.

The decisions of conning can be conveniently categorized as involving detection, danger assessment, and maneuver planning. The types and quality of data required in each phase differ radically. For example, simple detection requires only a rough estimate of range and bearing. Danger assessment requires much more accuracy in the data since it involves, among other things, an estimate of the probability of collision. Once the intruder has been assessed as dangerous, the maneuver plan calls for estimates of target speed, heading, class, condition (trawling, towing, under-

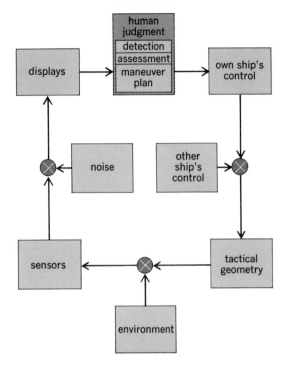

Fig. 2. Collision-avoidance maneuver control loop.

way, and so forth), other ships' intentions, and other variables. Furthermore, some estimate of the errors associated with these variables must be made to ensure safe passage.

Largely as a result of MARAD-sponsored efforts on LAD and related projects, these problems have been studied in sufficient depth to permit the design of remedial hardware for many conning problems. The sensor display systems that are currently under development should result in greatly improved ship safety and yet permit rapid and efficient conning decisions.

For background information *see* NAVIGATION; NAVIGATION INSTRUMENTS; RADAR in the McGraw-Hill Encyclopedia of Science and Technology.   [ROBERT F. RIGGS]

Bibliography: J. Brough and K. D. Jones, An investigation into the use of radar for collision avoidance, *J. Inst. Navig.,* vol. 23, no. 1, January, 1970; R. F. Riggs, LAD: A new family of devices for the avoidance of collisions at sea, *Navigation,* vol. 16, no. 3, Fall, 1969; R. F. Riggs, New conning aids for safe merchant shipping, *Sperry Rand Eng. Rev.,* vol. 21, no. 3, 1968.

Navigation systems, electronic

One of the elements of the existing air-traffic control (ATC) complex is an L-band secondary radar system which furnishes data on the position of aircraft. This system has come to be known as the air-traffic control radar beacon system (ATCRBS). Aircraft transponders are presently required in positive-controlled airspace, and the mandatory areas are being extended. Aircraft are being equipped with new transponders at a rate of over 1500 per month. The ATCRBS is already approaching overload conditions in some of the busiest areas. Simulation studies show that saturation can be expected to occur at several centers. There are several improvements which will extend the life of the current system, but the need for a second-generation data-acquisition system is clearly indicated.

Many types of systems have been proposed, and several have been designed and tested. Four categories of systems are under consideration: (1) systems using a constellation of satellites, (2) trilateration systems using one-way ranging by a network of ground stations (time frequency), (3) trilateration systems using two-way ranging from a network of ground stations, and (4) rho-theta systems based upon the present ATCRBS system.

Rho-theta systems. Analyses have shown that a rho-theta system will best meet requirements of siting, accuracy, reliability, simplicity, and capability for evolutionary growth. It appears feasible to design a second-generation beacon system which can maintain a high degree of compatibility with the present ATCRBS and still meet the requirements of the new air-traffic control system. The following six factors favor the choice of a rho-theta system.

1. Line-of-sight coverage. Coverage is required over the entire airways and approach structure, and down to very low altitudes in the vicinity of airports. In the case of multistatic systems, it may often be difficult to find adequate sites for the multiplicity of ground stations required, whereas a single rho-theta site located on the airport almost always meets the siting requirements.

2. Accuracy. The highest measurement accuracy requirements occur in the highest density areas and where closely spaced parallel paths exist. These areas are usually near the central airports, and the converging angular measurement accuracy of a rho-theta system best meets this requirement.

3. Multipath. Certain sites suffer from code distortion due to reflections from surrounding objects. Analysis and test show that narrow-beam systems are less susceptible to these effects than omnidirectional systems.

4. Aircraft antenna shadowing. Signal dropouts are more easily solvable in a monostatic system either by using multiple aircraft antennas in a reciprocal reply system or by having a redundant ground network.

5. Siting costs. The site, buildings, facilities, and remote maintenance costs will probably make a ground-based multistatic system too expensive.

6. Compatibility. A new data-acquisition system must be phased into the existing system, which consists of rho-theta nav-aids, radars, and beacon interrogators. The phaseover will be easier to accomplish with a compatible second-generation ATCRBS.

Design of the new system. The conceptual design has emerged of a new rho-theta system which will have a high degree of compatibility with the current ATCRBS system, providing that there is adherence to the following premises: (1) All aircraft will carry beacon interrogators and will automatically encode their barometric altitude. (2) The ground interrogator environment will be upgraded in such a manner as to remain compatible with aircraft which are still equipped with standard beacons. (3) The ground ATC system will be automated and the data-acquisition accuracy, traffic capac-

ity, and reliability will have to be adequate to serve this automated system. (4) The beacon system will be the primary data-acquisition system for en route and terminal areas, and upgraded vortac will be the primary navigation system. (5) The Federal Aviation Agency (FAA) will exercise control over the environment to prevent the unrestrained use of interrogators, which would overload the system.

The new data-acquisition system will evolve from the current beacon system. It will resemble the current system in that coverage will be provided by a network of interrogators operating on a common channel. Replies from aircraft will supply azimuth and range information, as well as identity and altitude codes. Most aircraft will be roll-call interrogated and sent ground-derived control information. Some of these aircraft will transmit data-link information back to the ground. Some aircraft will continue to use only standard beacon transponders and voice communications.

The new system, like the present one, will be monostatic, that is, it will measure position from a single site. A computer at each site will make the system capable of continued operation in the event of central computer failure. Redundant coverage will aid system reliability.

Growth phases. Compatibility and evolutionary growth from the current system into the new system are of primary concern. Two phases of growth may be considered.

In phase 1 improved interrogators operating only in modes 3A and C will be added to the present system and in some cases will replace those now in existence. They will provide greater angular measurement accuracy and better angular resolution. Better coverage and better reception will be achieved by the ability to separate the interrogator from the radar. Interrogators can use certain sites more advantageously than can radars. When all aircraft report altitude, three-dimensional information can be delivered to the central computer, which will then be able to deliver beacon overlays which have been "registered" in the coordinates of any radar. By this procedure, many separate interrogators at individual radars will not be needed and can be turned off, thereby controlling system overload problems.

In phase 2 the new interrogators will transmit and receive two-way data-link information which will be interleaved in time and space with the standard modes 3A and C interrogations and responses. The data-link information will be discretely addressed to those aircraft which equip themselves with new transponder equipment. This new equipment, together with the new interrogators, will provide the measurement accuracies necessary in high-density airspace.

Those aircraft still equipped with standard beacons will continue to be accommodated in other airspace and will be controlled by voice instructions, as at present. However, in the densest areas, standard beacon reporting and voice control will not be adequate for two reasons: (1) Voice relay of automatically generated control information will be inadequate because of delays, insufficient radio channels, and the need for too many controllers. (2) The beacon replies would be garbled because aircraft densities would greatly exceed the positional resolution of the beacon system.

Those aircraft flying in mixed airspace which has become too crowded may be required to be equipped with the intermittent positive control (IPC) decoder. This decoder will work in conjunction with the beacon receiver and will be activated by the ATC computer to furnish guidance for navigation or collision avoidance when necessary. In addition, it will be used to trigger discrete replies from the transponder on the aircraft with which it is associated. Since all of these equipped aircraft can be sequentially addressed in a range-ordered manner, there will be no garbling. The fleet must become equipped with the IPC receivers early enough to prevent standard beacon reply garbling.

For background information *see* AIR-TRAFFIC CONTROL; NAVIGATION SYSTEMS, ELECTRONIC; RADAR in the McGraw-Hill Encyclopedia of Science and Technology.   [ROBERT C. RENICK]

Bibliography: Department of Transportation, Federal Aviation Agency (FAA), *The National Aviation System Plan: Ten Year Plan 1971–1980*, March, 1970; Department of Transportation, Federal Aviation Agency (FAA), *Report of DOT Air Traffic Control Advisory Committee*, December, 1969; R. C. Renick, *Proc. IEEE*, pp. 413–422, March, 1970.

Nervous system (invertebrate)

Invertebrate nervous systems range in structural complexity from dispersed, two-dimensional assemblages of nerve cells, termed nerve nets, to elaborately organized ganglia and nerves such as those of cephalopod mollusks, whose neural complexity rivals that of lower vertebrates. From this diversity have come many observations that have been prominent in the development of modern neurophysiological concepts. Present ideas about nervous conduction are based mainly on experiments with giant axons of squids; synapses in squids and crayfish have contributed importantly to an understanding of the mechanisms by which nerve cells interact; receptors in crayfish and horseshoe crabs have revealed basic principles about coding of sensory information; and small aggregates of cells in arthropods and mollusks are now being intensively studied to determine the mechanisms of information processing by nervous systems. The structure and function of invertebrate nervous systems are comprehensively treated in a monograph by T. H. Bullock and G. A. Horridge. The following discussion is concerned with nervous systems in their simplest form, the nerve nets of cnidarians.

Morphology of cnidarian systems. The most simply constructed multicellular animals are the sponges (phylum Porifera). The presence of nerve cells in sponges has been the subject of controversy, but the best available physiological and histological evidence suggests that sponges lack nervous systems. In the evolutionary tree, the first indisputable nervous systems are found in the Phylum Cnidaria (also referred to as Phylum Coelenterata), the group including sea anemones, jellyfish, and hydroids.

Cnidarians are composed essentially of two epithelial layers. With a few exceptions, musculature in cnidarians occurs as epitheliomuscular cells, epithelial cells with elongate, contractile bases. The nerve cells form a dispersed net lying among

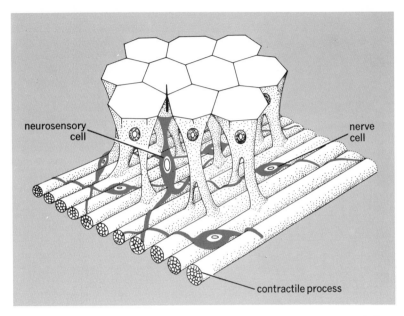

The organization of an epitheliomuscular cell layer. Cnidarians are composed of two such layers. (*After G. O. Mackie and L. M. Passano, Epithelial conduction in hydromedusae, J. Gen. Physiol., 52:600–621, 1968*)

the contractile processes of the epitheliomuscular cells (see illustration) or just below the nonmuscular epithelial cells. The cells of the net have two or three elongate processes from the cell body, and the processes are morphologically similar. Cells thought to be sensory cells lie between the epithelial cells, and contact the net below. The nerve cells interact where they contact one another in crossing. Here the cells come into close proximity, but usually there is not fusion. Electron microscopy reveals that these junctions or synapses are often not polarized; both processes show similar membrane thickenings at the contact point, and there are vesicles in the cytoplasm of both cells at the synapse. At some synapses, however, only one of the two nerve processes contains vesicles; transmission at these structurally polarized synapses may be unidirectional. Nerve nets may be associated with one or both epithelial layers of cnidarians, and in some instances there is evidence that the nerve cells of one epithelial layer form two functional nets, coexisting in the same tissue but capable of independent activity. In medusae there are generally local nerve cell concentrations, marginal nerve rings or marginal ganglia, in addition to dispersed nets.

Nerve net physiology. The relatively simple organization of cnidarian nervous systems has attracted many investigators interested in the evolutionary origin of nervous systems and the role played by primitive nervous systems in controlling animal behavior. Since the nerve cells in cnidarians are small and typically difficult to visualize in living material, direct observations, such as recording electrical activity from single cells, have been difficult. Consequently, many conclusions about nerve net functioning are somewhat tentative.

The following conclusions seem to have been demonstrated with some certainty: (1) Nerve cells, like those in higher animals, conduct all-or-nothing impulses. (2) As might be expected from the dispersed arrangement of nerve cells, conduction in nerve nets is diffuse. This is classically shown by the ability of activity conducted in a nerve net to spread through interdigitating cuts in the tissue containing the net. (3) Conduction velocities in cnidarian nerve nets range from about 0.1 to 1 m/sec. Since measured conduction times include synaptic delays of unknown number and duration, the conduction velocity in the nerve cells themselves must be greater than this. (4) Some nerve nets respond as single, all-or-nothing conducting elements. In other nets there are requirements for facilitation at the junctions between nerve cells; the first impulse arriving at a junction along a nerve cell does not cross to the other cell but temporarily converts the junction from a nontransmissive state to a transmissive one, paving the way for following impulses. A consequence of requirements for facilitation is that varying temporal patterns of impulses originating from one area, for example, from a sensory cell, become expressed as varying spatial patterns of spread, the distance of spread being greater the greater the number and frequency of evoked impulses. This is one of the simplest forms of information processing by nervous systems. (5) Nerve nets often initiate contraction in the musculature underlying them. Here again, since there are often requirements for facilitation at the junction between the nerve cells and the musculature, the contraction evoked is a function of number and frequency of impulses in the net. Activity in nerve nets may also excite or inhibit centers of spontaneous activity (pacemakers), and there are suggestions that nerve cells may in part control firing of nematocysts, the stinging cells of cnidarians.

Conducting epithelial cells. Perhaps the most significant recent advance in cnidarian neurophysiology was the demonstration by G. O. Mackie in 1965 that, in some instances, epithelial cells themselves are electrically excitable and can conduct behaviorally significant signals. Epithelial conduction, like nervous conduction, is diffuse, and the reported conduction velocities range from 0.02 to 0.5 m/sec. Since cells of conducting epithelia are probably joined by low-resistance junctions, electrical current generated at an active area can flow through and excite neighboring cells. Only one surface of the epithelial cells may be excitable; this was shown by the observation of R. K. Josephson and M. Macklin that conduction results in the appearance of a brief, rather large electrical potential difference across the epithelial layer, which would not be the case if both inner and outer cell faces were simultaneously and symmetrically active.

Conducting epithelia are now being intensively investigated, both because they offer a new mechanism of behavioral control which helps to account for the discrepancy between the rather simple organization of cnidarian nervous systems and the often surprisingly complex behavior of the animals themselves and because excitable epithelia seem to be likely evolutionary precursors of nerve cells in cnidarians and higher animals.

For background information see COELENTERATA; NERVOUS SYSTEM (INVERTEBRATE) in the McGraw-Hill Encyclopedia of Science and Technology.
[ROBERT K. JOSEPHSON]

Bibliography: T. H. Bullock and G. A. Horridge, *Structure and Function in the Nervous Systems of Invertebrates*, 1965; R. K. Josephson and M. Macklin, *J. Gen. Physiol.*, 53:638–665, 1969; G. O. Mackie, *Amer. Zool.*, 5:439–453, 1965; G. O. Mackie and L. M. Passano, Epithelial conduction in hydromedusae, *J. Gen. Physiol.*, 52:600–621, 1968.

Neurophysiology

During every era of man's scientific activities, an answer is sought to the problem of how man's mind and body interrelate. Typically, each generation of mechanistically oriented psychologists turns to the currently available technology for a source of ideas and intellectual tools. Thus, over the course of history, models based on fluids, gases, mechanical devices, telephones, and computers have emerged as contemporary mechanical models of mental processes. In the last two decades, the model of interest has had a peculiarly close fit to the psychological world, for it is based upon the data obtained from electrophysiological studies of the brain itself. Although at first glance it may appear that the threshold of a breakthrough in the quest for reductionistic explanations of mental action may have been arrived at, a little thought suggests that, even in this instance, caution should be used, for this approach may be merely another fad. This model is a particularly enticing and seductive one, since quite clearly the brain is the seat of mental life and neurons are the structural components of the brain. Nevertheless, recent evidence from several laboratories suggests that even electric potentials recorded from neural tissue are not necessarily identifiable with psychological processes.

Current problems. One of the most important problems facing psychobiology in the 1970s is to distinguish among three categories of neurophysiological potentials: those that do not correlate with mental phenomena; those that do in a merely concomitant, but irrelevant, fashion; and those that represent true identity processes. The first category of potentials contains those electrophysiological signals which are simply random and spontaneous events and which do not vary in concert with any information process of significance in the nervous system, even though they might reflect some metabolic status of the organism. The second category of potentials may be considered to be signs which, if they are functionally related to stimuli, may convey some useful information to the neurophysiologist, yet still be behaviorally insignificant. The last category—the identity processes—represents the real target of the psychobiological quest, the true codes, as indistinguishable from behavioral processes as the electric motor is from the rotary motion of its shaft.

Two neurophysiological correlates. In the past 20 years, two main varieties of neurophysiological signals have been of particular relevance to this sort of psychobiological reductionism. The first is a body of data which has been accumulated primarily through the use of the intracellular microelectrode. It deals in large part with the processing of afferent, or sensory, information by simple nerve networks. On the basis of interactions among single cells, certain integrative functions, called by such names as feature extraction or feature filtering, are performed in such a way that only the critical information-bearing components of the original stimulus pattern are propagated toward the central nervous system. Some classical observations in psychology were immediately associated with such simple integrative mechanisms when they were first reported, and new searches were instituted for phenomenological events of comparable nature. In making these associations, it was sometimes forgotten that electrophysiological data become explanatory models or microtheories when applied to explain psychological phenomena.

The second major category of electrophysiological recordings which have been associated with behavioral events is based almost exclusively on the pioneering work of G. D. Dawson. Dawson discovered that, by appropriate statistical averaging techniques, the tiny potentials evoked in the human brain by impulsive natural stimuli (such as light flashes and clicks) could be detected by electrodes laid gently upon the scalp. This meant that human brain neurophysiology was opened to inquiry without surgical intervention. The electroencephalogram (EEG), enjoying the same advantage, has the disadvantage that its main feature, the 10-Hz alpha rhythm, is suppressed by intellectual activity; thus any correlations between EEG rhythms and mental processes have to be tenuous indeed. But the evoked brain potential is a quite different phenomenon; it correlates not only with stimulus amplitude and quality but also with the mental state of the subject as measured in highly rigorous experiments. Furthermore, unlike other physiological measures such as the galvanic skin reflex, the heart rate, or breathing dynamics, the evoked scalp potential is derived from the brain itself. Even more striking, unlike the single-cell action potentials referred to above, whose duration is but a millisecond or two, the evoked brain potentials seem to persist for durations which are of the same order of magnitude as the tentatively associated psychological phenomena. A record of evoked brain potential is shown in the illustration.

Some emerging discrepancies. The flowering of psychobiological thinking was one of the great developments in experimental and physiological psychology of the 1960s. Yet it is entirely possible that, in the rush to accept the "objective" physiological data, a few important clues were ignored and the familiar tendency to fall victim to a new fad had reoccurred. It turned out that the evoked brain potential did not correlate perfectly with stimulus characteristics in all experiments. Lateral inhibitory interaction, the reciprocal depressing effect of one neuron on an adjacent one, is among the most often used neurological explanations of perceptual phenomena. Yet when carefully controlled psychophysical experiments are used to explore the problem, it becomes increasingly clear that not all of the features of that neurological mechanism are found in the psychological process. Furthermore, the retroactive masking by a second visual stimulus of an earlier one is now known to depend often upon the shapes of the stimuli being identical or even upon the meaning of printed word patterns. Such cognitive effects seem to be able to swamp out any effects which depend simply upon the features of the geometry of the stimuli, such as

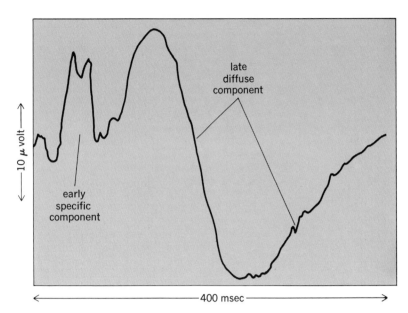

A tracing from a photograph of a computer-analyzed signal showing a typical photically evoked brain potential recorded from the occipital cortex representing the usual early specific and the late diffuse components.

the distance between parallel lines.

The impact of these new data is that psychobiologists must now consider the strong possibility that there has been too much uncritical theorizing without sufficient consideration of some of the possible alternatives. Correlations are not identities, and electrical response and behavior may concur because both are functions of the stimulus rather than because they are identifiable with each other.

This counterargument is posed as an alternative for discussion, suggesting that some of the proposed associations between the specific data of the neurophysiologist and the psychophysicist may have been premature. Nevertheless, it is also quite clear that these associations represent the vanguard in a new sort of physiological reductionism. It is not necessary to reject the monist notion that the brain and the mind are inseparable to call attention to the disparities which occur in many of the overly simplistic microtheories so far presented.

For background information see BIOPOTENTIALS AND ELECTROPHYSIOLOGY; NEUROPHYSIOLOGY in the McGraw-Hill Encyclopedia of Science and Technology.   [WILLIAM R. UTTAL]

Bibliography: D. L. Clark, R. A. Butler, and B. S. Rosner, *J. Comp. Physiol. Psych.*, 68:315, 1969; S. Sutton, in E. Donchin and D. B. Lindsley (eds.), *Average Evoked Potentials: Methods, Results and Evaluations*, NASA SP-191, 1969; W. R. Uttal, *Perception and Psychophysics*, 1970; W. R. Uttal, *Perspect. Biol. Med.*, 10:627, 1967.

Nobel prizes

The Nobel prizes for 1970 were awarded to two United States scientists and an economist, two Swedish scientists, a French scientist, a Soviet writer, a British scientist, and an Argentinian scientist.

Physiology or medicine. Three scientists from the United States, Sweden, and Britain shared the prize for "their discoveries concerning the humoral transmitters in the nerve terminals and the mechanisms for their storage, release and inactivation." They were Julius Axelrod, chief of the section of pharmacology at the National Institute of Mental Health, Ulf von Euler, president of the Nobel Foundation, and Sir Bernard Katz, professor and head of the department of physiology at University College, London.

Chemistry. Luis F. Leloir, director of the Institute of Biochemical Research – Campomar Foundation in Buenos Aires, was honored for his discovery of sugar nucleotides and their role in the biosynthesis of carbohydrates.

Physics. This prize was awarded to Louis Néel, professor at the University of Grenoble, and Professor Hannes Alfvén, who teaches at the University of California in San Diego and the Royal Institute of Technology in Stockholm. Néel was cited for work on antiferromagnetism and ferrimagnetism, and Alfvén for work on magnetohydrodynamics.

Literature. Aleksandr I. Solzhenitsyn was honored for his novels.

Peace. Norman E. Borlaug, director of the International Maize and Wheat Improvement Center of Mexico, won this prize for research into improved strains of wheat and rice.

Economic science. This award was given to Paul A. Samuelson, institute professor at the Massachusetts Institute of Technology, for his efforts to "raise the level of scientific analysis in economic theory."

Noise control

Noise, often defined as unwanted sound, is a significant pollutant of man's living and working environment. It can startle and awaken people, interfere with the hearing of speech or other wanted auditory signals, and under some conditions cause a degree of deafness because of damage to the neural receptors in the ear. However, this pollutant has somewhat different aspects from those of other environmental pollutants currently receiving attention.

First of all, noise is dissipated immediately into harmless, minute amounts of heat, whereas some pollutants may accumulate in the atmosphere, soil, or water over periods of time. Second, in man's general environment, noise by and large causes only psychological problems and, except when it causes deafness, does not physically harm people or animals.

Measurement of noise. Before discussing the current studies of effects of noise on people, a few words about the physical measurement of noise are necessary. Ideally, the average level of noise, following some special differential weighting of the frequency components, should be measured every 1/2 sec. Two ways have been developed for applying differential weighting to the different parts of the frequency content of a noise in order to make the sound measurements agree with subjective judgments. One, and generally the more accurate, is to filter the noise into narrow frequency bands of energy and then apply a differential weighting to each band. The result of this band-weighting process, plus a few other mathematical manipulations, is a unit for perceived noisiness and loudness, the PNdB. The other method is to use a sound-level meter that automatically applies the frequency weighting to the noise spectrum. These sound-level meter readings are called dB(A), dB(B), dB(C), or dB(D), depending on which of several weightings are used. Generally the favored weightings for this purpose are dB(A) and dB(D).

Even when a noise is measured every 1/2 sec during its occurrence, there is still the problem of how most simply to report the value to be assigned to a total noise, say, that of a truck driving by. In the past, engineers have usually measured and reported only the PNdB or dB(A), (B), (C), or (D) value occurring during the 1/2-sec interval when the noise is at its maximum level. This is called the maximum perceived noise level (Max PNL).

Common experience and experimental data show that not only the maximum level but also the duration of noise contribute to its perceived noisiness. For example, the noise from an aircraft engine run-up that lasts for 1 min is more bothersome than the same noise that lasts only 5 sec, even though both noises reach the same maximum level for a given 1/2-sec interval. To account for the duration factor, it is becoming standard practice to integrate the levels that are present during each 1/2-sec interval of a noise and let the sum, divided by some common reference duration, represent the total value of the noise. The result of this summation is called the effective perceived noise level (EPNL) of a noise, as distinct from the Max PNL.

There is one more step to noise measurement before an adequate and useful description of a noise environment is achieved. That step is incorporated into the composite noise rating (CNR). The CNR is the sum of the EPNLs of the individual noises that may occur during a typical 24-hr day in a neighborhood, except that noises that occur at night are given more weight (a matter of 10 dB) than those occurring primarily during the daytime.

Table 1 summarizes units of noise measurement that, according to the present state of the art, should or can be used for the measurement of environmental noise for the purposes of estimating the general acceptability of the noise to people

Table 1. Physical measurements for estimating the subjective perceived noise level (PNL) of sounds

Category and method	Basic unit of measurement	
	Band spectra	Sound-level meter
PNL, Max PNL: The level in 1/2-sec intervals (the highest value is called Max PNL)	PNdB Max PNdB	dB (D), (A) Max dB(D), (A)
EPNL: The 1/2-sec PNLs summed over the duration of a noise	EPNdB	EdB(D), (A)
CNR: EPNLs of different noises summed over 24-hr periods – 12 (a penalty of 10 is added to nighttime occurrences)	EPNdB	EdB(D), (A)

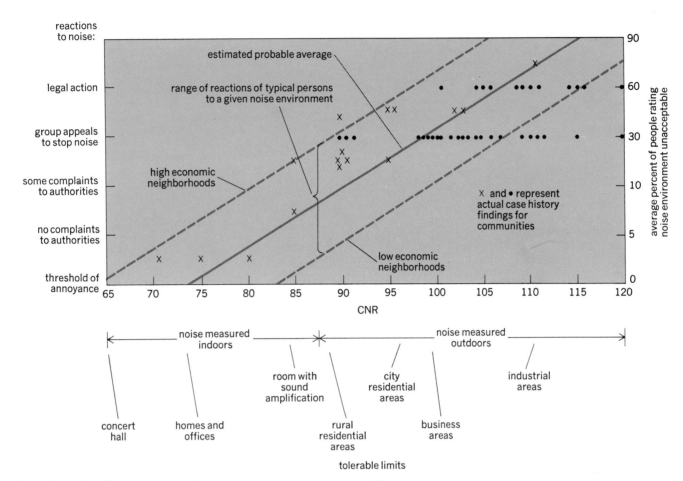

Fig. 1. General relations between reactions of people to environmental noise and suggested tolerable limits that might be set for different rooms and areas. Present legislation against noise is not adequate.

Table 2. Suggested tolerable maximum levels in various rooms for more or less continuous noise from 7 A.M. to 10 P.M.*

	Max PNL			EPNL† EPNdB EdB(D') EdB(A')	
Type of space	dB(A)	dB(D)	PNdB		CNR
Broadcast studios	28	35	41	78	66
Concert halls	28	35	41	78	66
Legitimate theaters (500 seats, no amplification)	33	40	46	83	71
Music rooms	35	42	48	85	73
Schoolrooms (no amplification)	35	42	48	85	73
Apartments and hotels	38	45	51	88	76
Assembly halls	38	45	51	88	76
Homes	40	47	53	90	78
Motion-picture theaters	40	47	53	90	78
Hospitals	40	47	53	90	78
Churches	40	47	53	90	78
Courtrooms	40	47	53	90	78
Libraries	40	47	53	90	78
Offices					
Executive	35	42	48	85	73
Secretarial (mostly typing)	50	57	63	100	88
Drafting	45	52	58	95	83
Meeting rooms (sound amplification)	45	52	58	95	83
Retail stores	47	64	60	97	85
Restaurants	55	62	68	105	93

*The noise levels outdoors from sources located outdoors (aircraft, road traffic, and so on) would be typically about 20 dB greater for the average house and 30 dB for masonry or well-sound-insulated buildings than the levels given in this table.

†dB(A') − 13 = dB(A); dB(D') − 6 = dB(D).

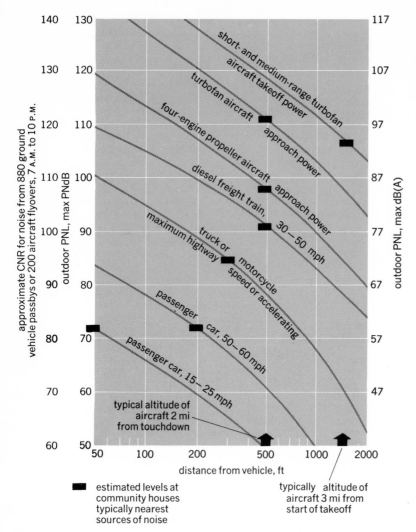

Fig. 2. Maximum perceived noise level (Max PNL) from various transportation vehicles and composite noise rating (CNR) for a specific number of aircraft flyover noises.

do manual work. Second, the sense of values of the people involved, as well as economic and social factors, dictates to some extent how much annoyance will be tolerated. Sociological studies show that in general it is the better-educated, psychologically adjusted, and economically advantaged individual who complains and mounts legal action against noise. Figure 1 shows data regarding the reactions of people and communities to different amounts of environmental noise. In Fig. 1 and Table 2 different tolerable limits are suggested for different rooms and areas.

In Table 2 equal Max PNLs for different noises are comparable to each other only when the noises have a broadband spectrum and do not contain any strong pure-tone or line spectrum components; noises or noise environments of equal EPNL or equal CNR values are presumably equal in their effects on people regardless of the spectral or temporal complexities of the noise.

Environmental noise in and around the home and office is seldom sufficiently strong to cause, even after years of exposure, significant amounts of deafness, although some of the reduction in hearing acuity as a function of aging (presbycusis) is possibly due to noise in such environments in industrialized countries.

Although it is possible that noise can eventually harm a person or animal as the result of emotional and related physiological reactions, such as increased blood pressure, there is evidence that in most people and animals such reactions cease with repeated exposures to the noise. Such adaptation apparently tends to prevent harmful effects of noise on the body except to the ear.

Present legislation and practices do not appear to protect people adequately from noise pollution. One reason is that general environmental noise mostly creates psychological rather than demonstrable physiological problems; laws forbidding the creation of a general (psychological) nuisance are often not applicable because the noise, such as from public transportation, may be considered by courts as a nuisance necessary for the well-being of a community and therefore legal. However, some success in obtaining relief through present legal proceedings lies in the claim of a taking of

and the setting of tolerable limits.

Environmental noise and behavior. Certain effects of noise on man can be reasonably predicted regardless, within limits, of the source of the noise, that is, whether it be from a motorcycle, airplane, washing machine, and so on. This is approximately true only when the noises are an expected, regular part of the environment and do not carry any unusual information with emotional connotations; for example, the buzz of a mosquito or sound of an approaching siren evokes responses that are peculiar to those sources and not to the physical contents of the sound itself. It should be clearly understood that the noise measurement and evaluation techniques used for noise-control purposes are approximately valid only when the noises in question are either of equal meaning or have inconsequential meaning to the listeners. Many significant environmental noise pollutants, such as from transportation vehicles, fall in this category, and only to the extent that they do can the management and control of environmental noise be a practical engineering matter.

First, how much annoyance is considered tolerable depends upon what one is expected to do in a given situation—to converse, listen to music, rest,

Table 3. Permissible noise exposures specified by U.S. Department of Labor and estimated amount of hearing loss incurred from these exposures

Permissible noise exposures		Estimated amount of noise-induced permanent threshold shift equaled or exceeded in 25% of people after 20 years of almost daily exposure	
		Average hearing level at 1000, 2000, and 3000 Hz Re ISO*	Average hearing level at 500, 1000, and 3000 Hz Re ISO*
hr	dB(A)		
8	90	35 dB	25 dB
6	92	35 dB	25 dB
4	95	35 dB	25 dB
2	97	30 dB	20 dB
1–1/2	100	30 dB	20 dB
1	105	30 dB	20 dB
1/4 or less	115	30 dB	20 dB

*International Standards Organization Recommendation R389-1964, a standard reference zero for the calibration of pure-tone audiometers. American National Standards Institute, New York.

property by the noise, that is, the property loses some value as a residence because the noise interferes with normal use (sleeping, speech communications, and so forth). Figure 2 illustrates typical noise conditions that might be found in the general environment from various vehicles.

Noise in industry. Noise inside an industrial structure or in or near a noisy machine such as a tractor, truck, or aircraft can have two primary deleterious effects: It can interfere with the performance of work that involves speech or other auditory communications, and it can cause permanent deafness. The effects of noise on work performance that does not require hearing are usually negligible, sometimes slightly increasing and sometimes slightly decreasing productivity. Presumably because of adaptation, as mentioned earlier, harmful emotional or physiological effects of noise in industry, other than to the ear, have not been scientifically demonstrated, although they possibly occur to some extent.

Much is known about the quantitative relations between amount of noise-induced deafness and continuous (up to 8 hr) or intermittent exposure to so-called steady-state (nonimpulsive) noise. Information as to the effects of impulsive noise, such as gunfire, is not so precisely known. Amounts of deafness are measured in terms of the number of decibels a pure tone must be increased above a normal hearing level (HL) in order for it to be heard by the noise-deafened person when in the quiet and after several weeks away from exposure to intense noise.

Table 3 shows permissible industrial noise exposures as specified by the U.S. Department of Labor. Also tabulated are estimates of the amount of permanent hearing loss found in 25% of the people exposed for 20 years to different levels and durations of workday, nonimpulsive noise. It is estimated that exposure to impulsive noise, such as gunfire, should not exceed 140 dB(A).

For background information *see* NOISE, ACOUSTIC; NOISE CONTROL; NOISE CONTROL IN BUILDINGS in the McGraw-Hill Encyclopedia of Science and Technology.   [KARL D. KRYTER]

Bibliography: W. Burns, *Noise and Man*, 1969; Department of Labor, Safety and Health Standards, *Federal Register*, vol. 34, no. 96, May 20, 1969; K. D. Kryter, *The Effects of Noise on Man*, 1970.

Nuclear aircraft

Nuclear energy offers the possibility of an aircraft that could fly anywhere on the surface of the Earth or remain aloft for weeks at a time without refueling. The major obstacle to this accomplishment has been that aircraft have not been large enough to carry a safe heavy nuclear power plant. This was one of the basic reasons that the United States aircraft nuclear propulsion (ANP) program was abandoned a decade ago. In the past few years, however, the development and introduction into military and commercial service of the Lockheed C-5 and the Boeing 747 aircraft have shown that aircraft in the 1,000,000-lb weight category are not only feasible but practical, desirable, and profitable.

Problems and attractions. Aircraft with gross weights of at least 1,000,000 lb are necessary to

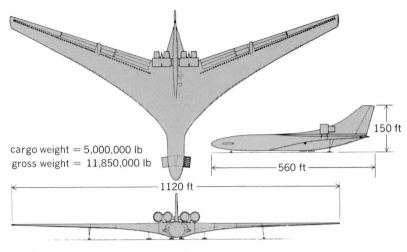

cargo weight = 5,000,000 lb
gross weight = 11,850,000 lb

150 ft

560 ft

1120 ft

Fig. 1. Large nuclear-powered aircraft.

make nuclear aircraft feasible. A practical nuclear aircraft would have complete shielding so that neither the flight and ground crews nor the passengers would receive radiation doses significantly greater than those normally received on Earth. Also, it would have safety provisions designed to prevent the release of radioactive material in any accidents.

In addition to virtually unlimited range and duration, nuclear aircraft may have an economic attraction. Because energy from nuclear fuel costs only a fraction of that from kerosine, nuclear-powered aircraft could significantly reduce the cost of air transportation. This factor, in addition to the potential economy of construction and operation of very large aircraft, could make air transportation competitive with transportation by truck, rail, and general cargo ships. Inland cities built around large airports could become new world trade centers.

The increasing demand for air transportation will require larger and larger aircraft. Aircraft weighing many thousand tons may be required (Fig. 1) to handle the traffic. The larger the aircraft, the more attractive nuclear power becomes.

Safety features. In 1964 the National Aeronautics and Space Administration (NASA) initiated a low-level effort to reassess the feasibility of nuclear aircraft. This new effort, with cooperation from the U.S. Air Force, was prompted by the fact that aircraft no longer seemed limited to sizes that ruled out nuclear power plants. The goal is to determine whether it is possible to provide practical and safe nuclear aircraft power plants that have complete shielding and that will not release fission products in the worst possible aircraft accidents.

Prime attention has been focused on the safety problems. Major aircraft accidents involve impact at high speeds. Such impacts are highly destructive unless special design provisions are made to protect parts such as the reactor containment vessel. Its rupture would allow the escape of radioactive fission products. Means of absorbing kinetic energy during crashes to prevent reactor containment vessel rupture are being investigated.

Another problem receiving attention is the rupturing of the containment vessel due to afterheat and reactor meltdown. Afterheat is the heat energy

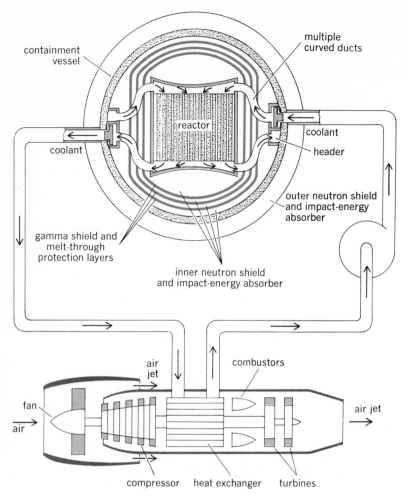

Fig. 2.   Nuclear aircraft power plant.

passages in the hot reactor and picks up the fission-generated heat. The hot helium is then ducted to helium-to-air heat exchangers located forward of the conventional combustors of ordinary turbofan engines. The air that is heated in flowing through the heat exchanger expands through turbines which drive the compressors and fans. Propulsive thrust is provided by the fan airflow. The turbofan engines can be operated by nuclear power or by combustion of kerosine or by both. *See* REACTOR, NUCLEAR.

The reactor is surrounded by various layers of material constituting shielding, containment vessel, impact energy–absorbing material, and melt-through protection material. The gamma shielding consists primarily of multiple layers of heavy material such as lead, uranium, or tungsten. The containment vessel acts as a portion of the gamma shield. The neutron shielding is composed of relatively light materials with high hydrogen atom concentration, for example, water, lithium hydride, or organic solids or liquids. The use of organic materials such as plastic or chemical aircraft fuel would be limited to the outer shield layers, where the radiation levels are sufficiently low that radiation damage would not occur.

During an impact with the Earth, the containment vessel and shield materials are designed to absorb the kinetic energy of the reactor and shield assembly without rupturing the containment vessel. For example, the outer shield material can be made of a plastic honeycomb material that can absorb kinetic energy as it deforms during the impact (Fig. 3). The gamma shield can be made of refractory materials such as tungsten or uranium dioxide to prevent molten materials from melting through the containment vessel. Shield materials thus serve not only as shielding but also as melt-through protection, impact energy absorbers, and containment vessel. Because materials are used to perform multiple functions, substantial weight savings are obtained.

Other requirements. Other features are required to make a nuclear aircraft practical. Among these are a reactor fuel which will permit operation of about 10,000 hr between refuelings. Long-life high-temperature oxidation-resistant heat exchangers are required to heat the air of the turbofan engine. Reliable lightweight long-life pumps

liberated by the radioactive decay of the elements that are produced as a result of fuel fissioning. Heat energy continues to be produced even after the reactor is shut down. It amounts to a few percent of the normal reactor power and reduces with time to less than 1% after a day. In an accident which destroyed all normal reactor cooling systems, this afterheat would cause the reactor to increase in temperature and to melt. The volatile reactor materials and fission products would form vapors which would condense in lower-temperature regions and therefore would tend to move toward the relatively cool containment vessel. In so doing, they would distribute themselves uniformly around and near the inside surface of the containment vessel. In order for the afterheat to be removed through the containment vessel without melting, the containment vessel must be large enough, that is, it must have sufficient surface area, to remove the afterheat by radiation and convection to the surrounding air. Work is under way to demonstrate that this can be accomplished without excessive weight penalties.

Power plant construction. Figure 2 shows a schematic drawing of a nuclear aircraft power plant that incorporates shielding and safety provisions. The fissioning uranium releases energy within the reactor. A heat-transfer medium (coolant), such as high-pressure helium, flows through

Fig. 3.   Honeycomb impact energy absorber.

and valves are required to handle high-temperature heat-transfer mediums that transfer heat to the propulsion engines. *See* HEAT TRANSFER.

The goal of unlimited range, duration, and potentially low-cost air transportation makes it worthwhile to continue to investigate and solve the problems of nuclear aircraft.

For background information *see* AIRCRAFT; NUCLEAR AIRCRAFT PROPULSION; REACTOR, NUCLEAR in the McGraw-Hill Encyclopedia of Science and Technology. [FRANK E. ROM]

Bibliography: M. M. Miller, The nuclear airplane—now, *Ind. Res.*, 12(3):48–50, March, 1970; F. E. Rom, The nuclear powered airplane, *Technol. Rev.*, 72(2):48–56, December, 1969; F. E. Rom, Status of the nuclear powered airplane, *J. Aircraft*, vol. 7, no. 4, July-August, 1970; J. M. Wild, Nuclear propulsion for aircraft, *Astronaut. Aeronaut.*, 6(3):24–30, March, 1968.

Nuclear fuels

A significant development in the United States and other major countries has been the change in emphasis in reactor development programs in the past 5 years, characterized by the systematic shifting of available resources to the development of fast-breeder reactors for wide-scale commercial use. This change has been brought about in the United States by the successful development of light-water reactor plants, as evidenced by the large-scale commitments of the utilities to the construction and operation of such plants. Other important factors, however, have been instrumental in accelerating this change. These include the rapidly increasing demands for energy, the significant advantages of electricity in meeting these demands, and the increasing appreciation of how meeting these demands will affect resources, environment, and the health and safety of the public. *See* ENERGY SOURCES.

The United States, France, West Germany, Japan, the United Kingdom, and the Soviet Union are concentrating large resources on an unprecedented priority basis on the development of the liquid metal—cooled fast-breeder reactor (LMFBR), using sodium as the coolant. In the United States the LMFBR development program is the highest priority effort in the nation's energy programs. The LMFBR holds the greatest promise of successfully achieving the near-term objectives of making available the full potential of nuclear fuels. Its economic potential is best exemplified by the results of a cost-benefit analysis of the United States breeder reactor program. This report notes that, in addition to many other important benefits the kilowatt-hour (kwhr) cost savings ($207,000,000,000 in a 50-year period, 1970 dollars, undiscounted) are paralleled with other quantifiable benefits, including a potential reduction in the demand for U_3O_8 of 1,400,000 tons over a 50-year period and a reduction in the maximum separative work demand of 85 ktonnes in the same 50-year period. These results were based on a 1984 introduction date of the LMFBR, rising uranium costs, a constant fossil fuel cost, and a United States electricity-generation rate of 8,000,000,000,000 kwhr in the year 2000. Naturally these benefits would mount with an increase in fossil fuel costs, as is the current trend.

In a large measure, the ability to produce unprecedented amounts of energy reliably, safely, and economically by the efficient exploitation of nuclear and other natural resources will be a vital component in the ability of the United States to maintain its high standard of living and assure its survival for the foreseeable future. Abundant low-cost nuclear energy is the base for introducing new products, a means of replacing inefficient or less desirable sources of energy, and a means of improving the quality of the environment. The objectives of the reactor development program of the Atomic Energy Commission (AEC) relate directly and strongly to these goals, and their realization in a timely manner will exert a powerful influence on the economic and social well-being of the United States and other countries. *See* ELECTRICAL UTILITY INDUSTRY.

Program. The target date of introducing the LMFBR on a large commercial scale in the 1980s has been predicated on a successful research and development program estimated to cost the AEC $2,000,000,000 during the next 15–20 year period. This includes $270,000,000 research and development support for a proposed demonstration plant and initial commercial plant programs, but does not include the capital or operating costs of the demonstration plants or the research and development support by utilities and reactor manufacturers.

AEC program plans have been issued in a 10-volume series, each volume covering a major LMFBR program element. These plans have compared the ultimate objectives of the program with the current state of the art and have identified the deficiencies and problems associated with each of the elements of the program. Tasks and their priorities and schedules have been delineated, and the resources needed to accomplish the tasks have been identified, including the training of personnel, the establishment of new facilities, and the upgrading of existing facilities. The plans have established interfaces between the various program elements, have provided for the dissemination of developing information, and have provided for a continuous correlation of the implementation of the program, including reviews of the plans on a continual basis. A direct result of these plans has been the increased delineation and focusing of the program elements with a consequent reduction of the diversification which so often in the past has impeded progress. Each contractor, industrial or laboratory, now identifies his work as part of the program plan, and technical leads have been delineated. Program assumptions are clearly outlined to contractors prior to their submission of proposals for continuing or new work.

Plans to proceed with LMFBR demonstration plants are based on the successful implementation of the research and development program outlined in the LMFBR program plans. These plans call for three 300- to 500-MwE plants to be built in the 1970s. Legislative authority has been requested to enter into cooperative arrangements for participation in the research and development, design, construction, and operation of the first LMFBR demonstration plant. Decisions to proceed with a second and third demonstration plant will depend on many considerations, including budgetary limitations in the AEC and the limited availability of resources in the industrial and utility arena.

The most recent legislation request supplements ongoing authority to conduct the first project definition phase (PDP) of the demonstration plant program. Atomics International, General Electric Co., and Westinghouse Electric Corp. are conducting a PDP, each for its particular demonstration plant concept. The PDP will define the demonstration plant project in sufficient detail to provide all participants with an assessment of the extent of future second-phase effort required and an identification of the technical aspects, including risks.

An important preliminary to the PDP was the series of 1000-MwE commercial plant follow-up studies made by Atomics International, Babcock and Wilcox, Combustion Engineering Inc., General Electric Co., and Westinghouse Electric Corp. These studies assembled reference design descriptions, estimated capital and fuel-cycle costs, and determined research and development requirements for five commercial LMFBR plant concepts. An objective was to use this information in LMFBR research and development planning. This information is now being applied to the work being done in the PDP.

Facilities. Many facilities are required for each of the LMFBR program elements. Physics has required the use of extensive computer installations for analysis and digestion of test information; neutron cross section measurement machines; and zero power critical assemblies from ZPR-III, through ZPR-VI and ZPR-IX, to the largest plutonium critical of its kind in the world, the zero power plutonium reactor (ZPPR). The components element requires an entire testing complex at the Liquid Metal Engineering Center in Santa Susana, Calif. Component experience derived from the experimental breeder reactor EBR-II and the Fermi reactor are being factored into component design for the fast flux test facility (FFTF) and the demonstration plants. The FFTF and the demonstration plants will, in turn, provide further experience for extrapolation to the first commercial plants.

Fuels and materials development has required the construction of extensive fuels and materials irradiation fast flux test reactors, such as EBR-II and the FFTF (now under construction), and the use of thermal flux irradiation test reactors, such as the engineering test reactor (ETR), advanced test reactor (ATR), General Electric test reactor (GETR), and Oak Ridge research reactor (ORR).

The FFTF, a 400-megawatts (thermal) (MwT) test facility with an initial flux of about 7×10^{15} neutron flux and with closed-loop capabilities to 1400°F sodium, will be the largest and an extremely important test facility in the LMFBR program, as well as a focal point for other significant engineering efforts in the LMFBR program. With its initial design essentially complete, orders are being placed for components. FFTF operation is scheduled for the mid-1970s.

EBR-II (operating at 50 MwT) has served as the fast flux reactor workhorse of the irradiations program of the United States. It has completed over 35,000 Mw-days of operation, and has irradiated PuO_2-UO_2 fast breeder fuel pins to over 120,000

Mw-days per tonne of heavy metal. EBR-II was scheduled to go to 62.5-MwT operation in the latter half of 1970.

Safety. Safety has required the extensive use of the transient test reactor facility (TREAT), a power burst facility. Plans are under way for an up to 19-fuel-pin failed fuel propagation loop in ETR, results from which will determine the need for more extensive facilities. Safety criteria for LMFBR plant design are being developed, and engineered safety features for the FFTF and the demonstration plants will be prototypical of the large-scale commercial plants.

Core geometry. One important aspect of reactor safety has been the universal approach to what is called unspoiled or regular core geometries as compared to spoiled geometries (such as an extreme pancake, annular, or modular design), which previously appeared to be an attractive means to avoid positive sodium void reactivity coefficients. The sodium reactivity coefficient is only a part of the overall reactivity coefficient, however, and it is this overall coefficient which is decisive. Also, the concern for coolant voiding is reduced by the knowledge that sodium superheat, greater than 400°F for extremely clean, nonflowing sodium, is unquestionably lower than 100°F (and possibly only a few degrees Fahrenheit) for flowing sodium under normal reactor operating conditions. The general consensus of the nuclear industry is that so-called spoiled geometries entail economic penalties with marginal safety benefits. Further, it is evident that the expenditures for the development of engineered safety features for detection, prevention, and mitigation of fuel failure in regular geometry cores have a far better payoff than any other route which could be taken. *See* HEAT TRANSFER.

Voids in stainless steel. A major problem in the use of the now-dominant stainless steel cladding for uranium-plutonium oxide fuel for fast reactors is the void formation in stainless steel under fast neutron irradiation, resulting in swelling of the steel.

This swelling has had implications on burnup limitations, material qualifications, and core design. The resolution of this problem includes the use of designs that minimize swelling effects and steels that are modified to provide more resistance to swelling. Significant progress has been made in developing engineering equations and establishing confidence limits at exposures of 5×10^{22} neutron velocity time (nvt) and clad temperatures to 1000°F. Design equations for fluxes beyond 1×10^{23} nvt probably will be confirmed by late 1972 or early 1973. Modifications to steels include thermomechanical pretreatments and small additions of dispersants such as titanium. Thermomechanical treatment of niobium-stabilized, high-carbon stainless steel produces a stable defect structure resistant to swelling up to 1300°F.

Detecting fuel failure. A third major problem is the development of instrumentation for detecting the potential or onset of fuel failure, or core failure, and the development of other engineered safety features for either preventing or mitigating, or both, fuel and core failures. The thermocouple has been, and will continue to be, the main means for detection, but more sophisticated instrumentation is being developed. This instrumentation includes magnetic or eddy-current sodium flow meters in each subassembly; high-temperature in-core neutron detectors; acoustic detectors for onset of sodium boiling; fission-product detectors, both in sodium and in the cover gas over the sodium; and isotopic tagging of fuel.

Levels of safety. Consistent with the policy followed since the inception of the nuclear energy program in the United States, emphasis on safety is dominant in all engineering and design activities. Three levels of safety are being provided in all LMFBR reactors. The first can be categorized under quality assurance, essentially assuring superior quality in design, construction, and operation. Components and systems must be safe and reliable, and must perform in a predictable manner. Each component and system is analyzed for malfunctions or faulty conditions to ensure that the plant safety is not jeopardized. Inherent safety features are used to advantage. The application of engineering principles and strong quality-assurance measures is backed by high standards on the basis that accidents can be prevented.

The second level of safety, on which priority is placed, is the use of engineered safety features, not only in the core but also for all components and systems. The third level of safety is positive assurance that, for any unpredictable or unlikely event, the public is protected by adequate containment barriers. Even if all other lines of defense fail, public protection will be guaranteed.

These three levels of safety will provide a high level of assurance that safe, reliable, and economic fast-breeder reactors will be capable of incorporation into the large network of electricity supply stations needed for the future economic growth of the United States. In the development of these reactors there is also a dominant trend to environmental protection. Absence of air pollution and the use of high-efficiency machines to reduce thermal effects are part of this trend. *See* NUCLEAR POWER.

A programmatic base has been generated on which the nuclear industry can (1) create a viable competitive nuclear industry; (2) demonstrate the safety, reliability, and economics of fast-breeder reactor power plants by designing, constructing, and operating demonstration plants on a utility network; and (3) introduce fast-breeder reactor power plants on a large scale into the commercial arena.

Environmental technology. The nuclear industry has accepted the challenges imposed by the gigantic task of meeting unprecedented demands for electricity and other forms of energy and for eventually supplementing and controlling water resources and recycling much of the wastes. These challenges, particularly to the young engineer and scientist, are of an unprecedented nature. There are compelling incentives to apply new scientific research and process engineering principles to the achievement of the chief objective of releasing man from a dependence on high-cost and rapidly diminishing energy sources and other valuable natural resources. Constructive efforts in the area of environmental technology are an important part of this challenge. They include the alleviation and amelioration of problems which an energy-based society has created. The challenges, above all, include the continuing need for an overriding concern for the health and safety of the public, which, from the beginning, has characterized the conduct of nuclear power programs.

These demands of the future call for effective leadership and management, increased knowledge and understanding, and concerted action within the industry to reconcile conflicting demands between energy needs and environmental considerations and to apply energy to reduce other types of pollution. This approach provides the basis for a constructive and realistic course of action.

For background information *see* NUCLEAR FUELS; NUCLEAR POWER; REACTOR, NUCLEAR (CLASSIFICATION) in the McGraw-Hill Encyclopedia of Science and Technology.   [MILTON SHAW]

Bibliography: Atomic Energy Commision, *Cost-Benefit Analysis of the U.S. Breeder Reactor Program,* WASH 1126, April, 1969; LMFBR Program Office, Argonne National Laboratory, *LMFBR Program Plan,* prepared for the USAEC, Division of Reactor Development, WASH 1101: *Overall Plan,* WASH 1102: *Plant Design,* WASH 1103: *Components,* WASH 1104: *Instrumentation and Control,* WASH 1105: *Sodium Technology,* WASH 1106: *Core Design,* WASH 1107: *Fuels and Materials,* WASH 1108: *Fuel Recycle,* WASH 1109: *Physics,* WASH 1110: *Safety,* 1968.

Nuclear power

Perhaps the most significant development in nuclear power at the end of the 1960s was a somewhat sudden focus on the environmental effects of nuclear generating stations. This concern, primarily on the part of environmental preservationists, was expressed in regard to effects of both thermal and radioactive discharges, and had the effect of delaying the construction or operation of several nuclear power stations. During this period also, results of studies at several stations were published which indicated that no significant radiological effects could be detected in the vicinity of operating nuclear stations and that there were no calamitous effects of heat additions from these stations. Public interest was sufficiently aroused, however, to generate a review of environmental effects of generating electricity by the Congressional Joint Committee on Atomic Energy in hearings held during the winter of 1969–1970.

Thermal effects. In any steam-electric generating station, less than half the heat liberated from the fuel is used to turn the turbine generator; the remainder is rejected to the environment. The fraction of the heat which is rejected is dependent upon the efficiency of the system, which in turn is a function of the maximum temperature at which the boiler (whether fossil- or nuclear-fueled) can operate. The average thermal efficiency for all fossil plants in the United States in 1968 was about 33.2%, resulting in about 2 kw of heat energy being rejected for every kilowatt of electricity generated. Of this total, about 15% is rejected in stack gases, and the remainder to the condenser cooling water. The most modern fossil plants have thermal efficiencies of up to 40%, with correspondingly lower heat rejection to both stack gases and water. Present light-water reactors (that is, those cooled and moderated with ordinary H_2O) operate with an efficiency of about 32%, slightly less than that of the average fossil-fueled plant, but with a much higher heat rejection to water because no heat is wasted with combustion gases as is the case with fossil-fueled plants. In comparison, present nuclear plants may reject more than 60% more heat to the water than do the most modern fossil-fueled plants of the same size. Thus all steam-electric plants reject heat to their environment; the higher heat-rejection rate for present nuclear plants means only that less nuclear capacity than fossil can be installed at a given location for the same environmental effect. For perspective, however, it has been calculated that for a lake the size of Lake Michigan, heat from the cooling water of 6500 nuclear plants of 1000 Mw each would be required to equal the amount of heat added by the Sun on an average August day.

Means are available for modifying the path of waste heat from the turbine condenser to the atmosphere, which is the eventual heat receiver. These means include cooling towers (being built or planned at more than 15 sites) and cooling lakes (being used at more than five sites). However, each of these alternatives has its disadvantages: Cooling towers evaporate tremendous quantities of water to the atmosphere (a 1000-Mw generating unit may evaporate about 20,000,000 gal of water per day), and in unfavorable climates this additional moisture may create fog, local precipitation, icing on roads, and so on; cooling ponds consume much less water than do towers but require much more land area, approximately 2–3 acres for each electrical megawatt of generating capacity. Thus it is important to evaluate both the environmental and economic impact of heat-rejection alternatives, particularly in the light of data from ecological studies in the vicinity of power stations, which have shown little, if any, widespread damage from heat.

A comprehensive study at one nuclear power station, the Yankee-Haddam Neck plant in Connecticut, has indicated that, although some changes have taken place in the relative abundance of the species existing in the Connecticut River before plant operation began, there has been "no significant deleterious effect on the biology of the river" as a result of plant operation; shad still migrate past the plant, avoiding or passing through the warmer portions of the river. On the basis of the results, the study director, D. Merriman of Yale University, coined the word calefaction to represent "the state of being warmed" to avoid the semantic bias of thermal pollution since no harmful effects had been observed.

Similar studies on the Columbia River section affected by the Hanford operation of the Atomic Energy Commission (AEC) have shown no adverse effects on salmon over a period of about 20 years, despite the simultaneous operation of as many as nine large reactors at that site. Although the results of the Connecticut River and Columbia River studies are not predictors of the effects of heat discharge at every power station, they are another indication that nuclear (or fossil) plants can be operated without the environmental devastation from heat claimed by some.

Radiological effects. The second area of environmental concern is that of radiological effects resulting from discharges of radioactive materials in liquids and gases.

There are three sources of radioactive materials created in nuclear power plants. These are the fission products, radioactive elements produced by the fissioning of uranium; the activation products, resulting from the creation of radioactive elements from stable elements present in the reactor core and coolant; and the corrosion products, small quantities of metals which are corroded from the reactor system by the coolant and made radioactive by the neutron flux.

The only significant source of fission products arises from the existence of minor defects or small holes in the nuclear fuel cladding, which permit the gaseous and volatile fission product elements to migrate from the hot fuel rod to the coolant. In the absence of defects, this source does not exist. The activation products and corrosion products are inescapable results of exposing materials to a neutron flux. For the most part, the activation products are composed of gases which have very short half-lives and hence decay quite rapidly, many in a matter of seconds to a few minutes. Some of these, however, have longer half-lives, notably tritium with a half-life on the order of 12 years. Tritium is produced primarily by activation of boron in the coolant of the pressurized water reactors. The corrosion products are primarily metal oxides and are mostly insoluble. *See* TRITIUM.

Of these three sources, the fission products constitute the major potential source of activity release from a nuclear plant. All operating plants have had some, very small fraction of defective fuel elements, and it is the usual practice to design waste-handling systems which are capable of managing quantities of radioactivity and volumes of gas and water which are much greater than those expected in the lifetime of the plant. All wastes receive efficient processing by filtration, ion-exchange demineralization, or evaporation followed by measurement before release to assure that discharges to the environment are maintained well below the established standards. Discharges from nuclear power stations, as a result, have averaged only a few percent of present AEC standards. The confirmation of this effectiveness has been provided by radiological surveillance programs routinely conducted by plant operators and state agencies in the environs of every nuclear station. The surveillance programs have shown no significant change in environmental radioactivity levels from those existing prior to plant operation from natural sources or nuclear weapon test fallout.

In April, 1970, the AEC issued proposed changes to its regulations which essentially would require that nuclear power plants restrict radioactive discharges to values "as low as practicable." Although this philosophy had in essence been followed in the past, its formalization in regulations led to questions of the definition of "as low as practicable." Considering the undetectably small effect of present operating stations on radiation in their environs, the answer to the question of lowest practicable discharges will play a large role in the future regulation of nuclear power plants but not in their effect on the environment. *See* REACTOR, NUCLEAR.

For background information *see* NUCLEAR POWER in the McGraw-Hill Encyclopedia of Science and Technology.   [MORTON I. GOLDMAN]

Bibliography: *Environmental Effects of Producing Electrical Power*, Hearings before the Joint Committee on Atomic Energy, 91st Congress of the United States, Oct. 28–31 and Nov. 4–7, 1969, Jan. 27–29 and Feb. 24–26, 1970; D. Merriman, *Sci. Amer.*, 225:5, May, 1970.

Nuclear structure

Major progress has been made in the recent past in understanding the structure of atomic nuclei; this progress has flowed not only from greatly improved experimental techniques and devices but also from a much more coherent theoretical approach to the nuclear structure problem than has hitherto been available. There have been a number of surprises. Quite apart from the marked impetus which this increased understanding provides in nuclear physics itself, these results have far-ranging consequences in the sister sciences and in areas as diverse as the examination of the constancy of the fundamental constants over cosmological epochs and the reduction of the cost per kilowatt-hour of electrical power to the citizen.

As a unique bridge in nature between the very few body problems of elementary-particle physics and the extreme many-body problem of solid-state or plasma physics, the nucleus displays a bewildering richness of phenomena, but the very fact that a limited number of nucleons underlie these phenomena holds promise of their eventual understanding in terms of the structure of the nuclei. Because of lack of precise knowledge of the nature of the nuclear forces and because of inadequate mathematical tools, a first-principles calculation of the characteristics of nuclei has been precluded, although, as noted below, impressive progress has been made in recent years. Lacking such a fundamental approach, nuclear physicists have had recourse to models of the nucleus. Again reflecting the wealth of phenomena involved, these models, depending upon the aspect of nuclear behavior on which they focused, were characterized by apparently contradictory hypotheses and assumptions. Outstanding examples were the liquid-drop model developed to reproduce some of the phenomena of fission and the shell model developed to correlate some of the more fine-grained and individual characteristics of nuclei. The former necessitated a nucleon mean free path much less than nuclear dimensions, and the latter, one much greater; both worked remarkably well.

Only very recently have all these models coalesced into a unified understanding of essentially all aspects of nuclear structure. This shall be illustrated in three main regions.

Heavy nuclei. Perhaps the most striking progress has been in the region of heavy nuclei, where adequate instrumentation had previously made precision studies impossible. Pioneering studies by Nelson Stein and his collaborators at Yale University, by Peter von Brentano and his collaborators at Heidelberg University and elsewhere, and by William Parkinson and his collaborators at the University of Michigan have shown that the textbook concepts of single particles and single holes moving in well-defined orbits relative to a core composed of the remaining nucleons have almost precise validity. In totally unexpected fashion these workers have shown that many of the quantum states, even at high excitation energies, in all nuclei in this region have wave functions of great simplicity, involving a particle or a hole, whichever is appropriate, loosely coupled to a core. One family of these states involves simply the core in its stable state of lowest energy—its ground state; other families involve the simple change to a core in an excited state. Striking agreement has been obtained between relatively simple shell-model calculations in this region and a wide variety of new nuclear information on these single-particle and single-hole configurations.

Beyond this, as first emphasized by Aage Bohr, Ben Mottelson, and their experimental collaborators at Copenhagen, there are totally distinct families of quantum states in this region which involve not single nucleons but tightly coupled pairs of particles or holes moving in equally well-defined orbits. This unraveling of very complex spectra of quantum states into such simple configurations was unexpected in these heavy nuclei. Having succeeded this far, researchers are now devoting major attention to whether some of the remaining, yet unclassified quantum states may not indeed involve motion of more complex clusters of three, four, or more nucleons.

Although, as emphasized above, a complete first-principles calculation of heavy nuclear structure and behavior has not been evolved, there have been important breakthroughs. A few years ago Michel Baranger and his collaborators, then at the Carnegie-Mellon University, undertook the formidable task of calculating the properties of the osmium isotopes, having respectively 186, 188, 190, and 192 nucleons, on the basis of certain reasonable assumptions concerning the nature of the force between two individual nucleons as deduced from nucleon scattering experiments and the properties of the deuteron. Experiments to test these calculations have recently been carried out in many laboratories; the theoretical predictions have been shown to be amazingly good, and small discrepancies can be eliminated by quite reasonable and minor modification of the input nucleon-nucleon forces. This success marks the opening of a new level of sophistication in nuclear theory and a new approach to a fundamental understanding of nuclear structure which has been made possible by full utilization of the contemporary power of major computing facilities.

Figure 1 shows spectra demonstrating the striking improvement in the instrumentation for studies on nuclear gamma radiation which has occurred in recent years. These data are part of the Yale University study of the osmium isotopes, which was carried out by bombarding isotopic targets with oxygen ions and subsequently examining the gamma radiation emitted as the nuclei returned to their lowest energy states. The upper curve in Fig. 1 represents the best results obtainable with the previous standard detector, a sodium iodide crystal optically coupled to a photomultiplier and electronic amplification system. The lower curve is a repetition of the same experiment using a new lithium-germanium semiconductor detector. As is obvious, particularly in the region of channel 200, whole series of previously unobserved peaks, each corresponding to a separate nuclear gamma radiation, are disclosed. These new detectors have revolutionized nuclear spectroscopy.

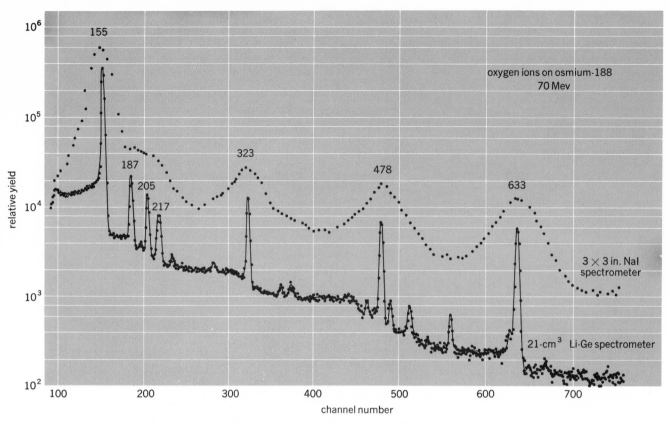

Fig. 1. Spectra obtained from studies of nuclear gamma radiation.

Parallel fundamental model studies, but of a slightly more phenomenological nature, have been carried out, and are being continued, by the collaborating groups at Oak Ridge and at the University of Rochester led by Edith Halbert and Bruce French, respectively. An important word of caution must be inserted here, however: Work by M. H. Macfarlane and his collaborators at the Argonne National Laboratory has shown that the fact that theoretical model calculations can reproduce or predict any given set of nuclear phenomena is not in itself a guarantee that the model wave functions provide any direct description of physical reality as it might be evidenced in further data. This remains an important ambiguity.

Shapes of nuclei. Considerable progress has been made recently in understanding the shapes of nuclei and how these shapes vary throughout the periodic table. David Hendrie at Berkeley and Anthony Aponick at Yale, with their respective experimental collaborators, have shown conclusively that the scattering of alpha particles from these nuclei can be interpreted, using a theory due to Norman Glendenning of Berkeley, to show that the nuclear shapes are not the simple footballs previously assumed, but that there are significant polar and equatorial depressions and bumps which vary systematically as additional nucleons are added to the system; this implies that not only quadrupole but also hexadecapole and higher moments are present in the nuclear shape.

In light nuclei, extensive work in laboratories around the world has finally delineated what happens to the nuclear shape in going through a major shell, in particular, from oxygen to calcium. As additional neutrons and protons are added, the initially spherical nucleus assumes a football shape, which is very pronounced in some of the neon and sodium isotopes; this shape gradually moves back toward sphericity and then at silicon plunges from a football to a doorknob shape; this then returns toward sphericity, which it reaches at calcium after a slight overshoot toward football shape at argon. Measurements sensitive to the higher moments, that is, finer topographical features, have not been made in this region.

In medium-mass nuclei in the region of tin and in selected other regions, it has been shown that the nuclei behave much as though they were superconducting. The nucleons in their ground states are paired off coherently, as are the electrons in a superconductor. This phenomenon is being subjected to detailed tests in many laboratories using reactions such as (p,t) or (t,p) wherein an already paired set of neutrons is subtracted or added, respectively, to the target state. This is simply another illustration of the unity of physics and an opportunity to obtain information complementary to that available from, for example, solid-state studies on superconductors.

Systematic analyses of an extensive body of data of many varieties have led John Schiffer of the Argonne National Laboratory, and others, to the conclusion that all nuclei have a neutron halo —that the neutrons extend farther out in the nuclear stratosphere than do the protons—in direct contrast to naive arguments involving mutual electrostatic repulsion of the protons.

Perhaps the major development in recent years, however, has been the understanding of all of the above phenomena as the results of microscopic individual nucleon motions in a generalized shell-model framework. Many groups have contributed to the theoretical work involved; in addition to those already noted, the major groups at MIT and at the Weizmann Institute deserve mention.

Nor has this new understanding been limited to the motion of nucleons in the so-called valence cloud, that is, nucleons outside the core. With the newer, higher-energy probes, it becomes possible to search ever deeper in the nuclear interior. High-energy electron scattering data from Robert Hofstader's group at Stanford a few years ago showed structure suggestive of shells remaining intact within the nuclear interior. Recent studies by A. N. James and his collaborators at Liverpool have used high-energy protons and the $(p.2p)$ reaction, wherein a proton is knocked loose from its binding in the nucleus, to study the interior structure. They have obtained remarkable results which show that the shell structure is maintained throughout the nuclear cores. Similar and complementary information is beginning to become available from $(e,e'p)$ experiments carried out by a group at Frascati. The completion of the high-energy meson factories at Los Alamos, Vancouver, and Zurich will greatly accelerate these important studies and will focus on the momentum distributions of nucleons in nuclei and the correlations which may exist. Study of π^- meson capture provides direct access to the degree of two-nucleon correlation, for example, because energy and momentum cannot be conserved unless the pion is captured by a closely correlated nucleon pair. K^- capture, already studied in preliminary fashion by groups at CERN and at Berkeley, shows promise of determining the degree to which the nu-cleons in the fringes of the matter distribution remain free or cluster into larger aggregations, that is, whether the nuclear surface is smooth or lumpy.

Studies on mesonic atoms, using all the new high-precision technologies, permit unprecedented study of the nuclear electromagnetic moments, shape, and structure. These will be of particular interest when extended to transuranic species.

Extension of periodic table. Work during the past two decades, primarily at Berkeley and more recently at Dubna, has extended the periodic table from $Z=93$ to $Z=104$ and perhaps even $Z=105$. Attention is now focused on the supertransuranic regions which, with the new understanding of the structure of existing nuclei, can be predicted at $Z=114$, $Z=126$, and perhaps $Z=164$, all with $N=184$. In contrast to the species in the $Z=93-105$ region, those of the supertransuranic region, if even quasi-stable, have no known natural production mechanism; hence the fact that they are not found in nature is not an indication that they are unstable. The calculations leading to these predictions have been international in scope, involving, among many others, V. M. Strutinski, a Russian; S. G. Nilsson, a Swede; W. Swiatecki, a Pole; and J. R. Nix, an American. The experimental race toward these new supertransuranic regions involves predominantly the group at Dubna led by G. N. Flerov and that at University of California at Berkeley led by A. Ghiorso, although many other laboratories are attempting to obtain suitable experimental facilities. *See* ELEMENT 105.

However, this extension to new elements is not the only frontier which the nuclear valley of stability presents. In the entire history of nuclear physics only about 1600 nuclear species, both stable and

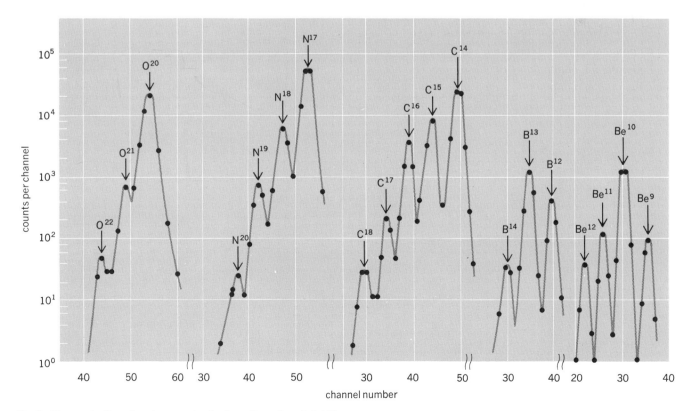

Fig. 2.   The production of nuclear species far from the valley of stability.

radioactive, have been studied; yet it can be shown that 2-Gev uranium ion bombardment of a uranium target might be expected to lead directly to the production of about 5000 different nuclear species far off on the walls of, and flatlands beyond, the valley. Study of these highly neutron- and proton-rich species will provide an extremely stringent test of the understanding of nuclear structure. Although they have never been observed, stable tin isotopes ($Z=50$) are believed to exist with mass numbers all the way from 92 to 176; this is typical of the unexplored range of nuclei awaiting study.

The data illustrated by Fig. 2 were obtained by V. V. Volkov and his collaborators at the Laboratory for Nuclear Reactions at Dubna using heavy-ion bombardment of heavy nuclei. Each peak in the spectrum corresponds to a different nuclear species, and each group corresponds to a different element—oxygen, nitrogen, carbon, boron, and beryllium—reading from left to right. The most stable isotopes of these elements have 16, 14, 12, 11, and 9 nucleons respectively, so that in the case of O^{22}, for example, six additional neutrons have been added to the normal oxygen nucleus; the same is true in the case of N^{20} and of C^{18}. This marks the first time that such neutron-rich isotopes of light nuclei have been observed.

Most striking, however, has been the demonstration by V. M. Strutinsky of Moscow, by S. G. Nilsson of Lund, by G. E. Brown of Stony Brook and Copenhagen, and by many others that all the phenomena discussed above, as well as many others, can be simply understood, on a microscopic level, in terms of the well-defined and well-understood motions of individual nucleons. In the past, qualitative reproduction of nuclear structure phenomena was often considered a major triumph; scientists have reached, and passed, the threshold from which one can now look with confidence to quantitative understanding. Only with this grasp can the potential of the nucleus be exploited as a tool for the fundamental understanding of the operation of the natural forces, the fundamental symmetries, and the conservation laws of all physics. Also, such practical consequences as more economic designs of nuclear energy-generating systems can be anticipated.

For background information see NUCLEAR STRUCTURE; SCATTERING EXPERIMENTS, NUCLEAR; SELECTION RULES (PHYSICS) in the McGraw-Hill Encyclopedia of Science and Technology.

[D. ALLAN BROMLEY]

Bibliography: A. Bohr and B. R. Mottelson, *Nuclear Structure*, 1969; M. Harvey et al. (eds.), *Proceedings of the International Conference on the Properties of Nuclear States, Montreal, 1969*, 1969; E. Hering et al. (eds.), *Proceedings of the 5th International Conference on Heavy Ion Physics, Heidelberg, 1969*, 1970; E. Vogt and M. Baranger (eds.), *Advances in Nuclear Physics*, annual, from 1967.

Nutrition

The role of magnesium (Mg) in protein-calorie malnutrition (PCM) has been investigated in several research laboratories in the tropics and subtropics since 1960, when J. Metcoff and associates in Mexico and R. D. Montgomery in Jamaica reported Mg depletion in PCM. The biochemical lesions in PCM are still incompletely understood but are of obvious importance. The syndrome has been called the greatest killer of infants and young children and the major cause of retarded child growth and development in today's world. The incidence was estimated to be 276,000,000 in 29 technically developing countries in 1968, and it is increasing (*Report of Committee on International Child Health, American Academy of Pediatrics*).

Etiology of Mg deficiency in PCM. A newly weaned child in economically depressed areas of the tropics may develop gastroenteritis and be fed a high-starch, low-protein diet of cassava, polished rice, or corn starch, which are all magnesium-poor foods. Some children are starved and purged to control diarrhea and to eradicate suspected parasites (Fig. 1). Upon hospitalization, such a child usually receives medical and nutritional therapy that, unless supplemented with Mg, may increase the Mg deficit. Renal excretion of Mg may follow prolonged intravenous infusions and the diuresis of edema fluid. Further distortion of the electrolyte balance occurs when electrolytes other than Mg are repleted and the Mg deficit remains uncorrected. The PCM formulas usually contain skimmed milk, casein, or calcium caseinate; skimmed milk contains little Mg, and the latter two contain essentially none. The high protein, calcium (Ca), and phosphorus (P) content of these feeds increases the metabolic requirement for Mg partly through increased anabolism. Finally, serum Mg is reduced by repeated transfusions of citrated blood.

Biochemical role of Mg. Numerous enzyme systems require Mg, including many involved in adenosinetriphosphate (ATP) reactions. W. E. C. Wacker reasons that since ATP is necessary for most of the energy-requiring biological systems, the function of Mg probably also extends to these systems, which include membrane transport; nerve impulse generation and transmission; muscle contraction; oxidative phosphorylation; and the synthesis of protein, nucleic acids, fat, and certain vitamin coenzymes such as thiamine pyrophosphate.

Experimental Mg deficiency. Experimental Mg deficiency in animals leads to profound electrolyte derangement, cachexia, neurological disorders, peripheral edema, and a series of degenerative lesions, especially in skeletal muscle, heart, and kidney.

Although most of the above findings occur in PCM, they cannot be attributed to Mg depletion alone. For example, G. A. O. Alleyne reports a greater potassium (K) than Mg deficit in muscle. Also, protein depletion affects many tissues.

Clinical manifestations. The initial symptoms in Mg-depleted PCM patients are nonspecific. They include weakness, lethargy, irritability, sleeplessness, and anorexia. The heart rate is rapid, and rhythm disturbances may occur. Stools may contain undigested food. Later changes may include staring, nystagmus, oculogyric crises (Fig. 2), tetany, or convulsions. The boy in Fig. 2 weighed 6.4 kg and had a history of diarrhea and vomiting. He was treated with intravenous fluids and tube feeds of vitamin B–enriched milk-protein formula. Oculogyric crises developed and were relieved by injected Mg. The plasma Mg was 1.1 meq/liter, and the muscle Mg was 12.3 meq/kg wet weight. Electrocardiograph changes of the se-

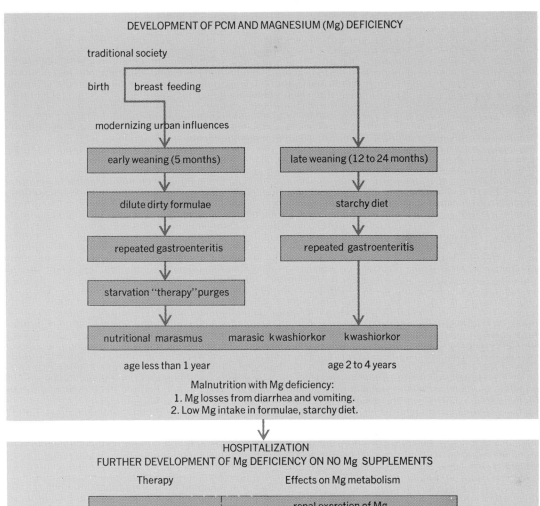

DEVELOPMENT OF PCM AND MAGNESIUM (Mg) DEFICIENCY

traditional society

birth   breast feeding

modernizing urban influences

early weaning (5 months)	late weaning (12 to 24 months)
dilute dirty formulae	starchy diet
repeated gastroenteritis	repeated gastroenteritis
starvation "therapy" purges	

nutritional marasmus   marasic kwashiorkor   kwashiorkor

age less than 1 year   age 2 to 4 years

Malnutrition with Mg deficiency:
1. Mg losses from diarrhea and vomiting.
2. Low Mg intake in formulae, starchy diet.

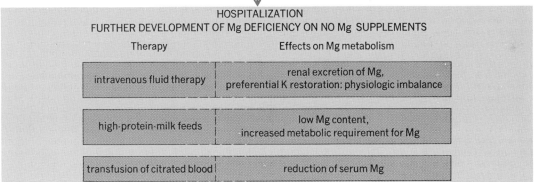

HOSPITALIZATION
FURTHER DEVELOPMENT OF Mg DEFICIENCY ON NO Mg SUPPLEMENTS

Therapy   Effects on Mg metabolism

intravenous fluid therapy	renal excretion of Mg, preferential K restoration: physiologic imbalance
high-protein-milk feeds	low Mg content, increased metabolic requirement for Mg
transfusion of citrated blood	reduction of serum Mg

Fig. 1. The development of Mg depletion and PCM in a child is shown in the upper part of the diagram. Hospital treatment is shown in the lower part of the diagram. (After D. S. McLaren)

NUTRITION

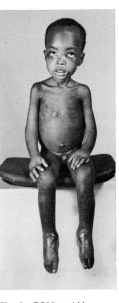

Fig. 2. PCM and Mg depletion in 3-year-old Nigerian boy.

riously ill child included low-amplitude complexes and abnormal repolarization, with a flat or inverted T wave over the left side of the chest. Partially treated children may develop a "recovery pattern," with asymmetrical late peaking of the T wave. Mg administration to patients who have failed to improve after a therapeutic trial without Mg may be followed by rapid improvement in all of the above clinical and electrocardiograph findings. This suggests that Mg was the limiting factor necessary for restoration of normal metabolism, not that the symptoms were due to Mg deficiency alone.

General considerations in diagnosis. Mg deficiency should be considered in any patient who, having sustained major losses of body fluids, remains lethargic and anorexic and fails to thrive after replacement of water, calories, sodium (Na),

and K. A plasma Mg content below 1.4 meq/liter in a normally hydrated or dehydrated patient suggests Mg depletion. However, plasma Mg may be normal in severely depleted patients. A 24-hr urine Mg value of less than 1.0 meq suggests body depletion unless intake is low and absorption poor. In that case, a test dose of Mg is injected; retention of more than 40% indicates Mg depletion. Normal muscle in children contains about 20 meq/kg wet weight.

Mg studies in PCM. On admission of the patient to the hospital, plasma and red cell Mg are often normal, but during the first 2 weeks of Mg-unsupplemented therapy the plasma value may drop to less than 1 meq/liter, and at this time patients may develop neurologic symptoms, which are thought to be related to a critically low ionized fraction of plasma Mg. The 24-hr urine Mg value is low on admission; R. D. Montgomery's Mg-unsupplemented patients had persistent low values for 40 days. P. J. Pretorius and coworkers reported urinary retention of 60% of injected Mg. Muscle Mg values of about one-half normal have been found in Mexico, Jamaica, Nigeria, and Thailand; the abnormal distribution of other cellular constituents in PCM complicates the interpretation of these values.

A mean cumulative retention of over 50 meq Mg has been found during the first 3 weeks of balance studies conducted in Jamaica and South Africa by Montgomery and by G. C. Linder and coworkers. Fifty meq Mg is about one-half of the Mg content of soft tissue in a normal 8-kg child. Bone Mg is also lost. P. J. Pretorius and coworkers reported high fecal losses sometimes leading to negative Mg balance; they attributed this to malabsorption as well as to diarrhea.

Therapy. The aim of therapy is to replete all deficient nutrients at a reasonable rate. Protein, K, and Mg are particularly important for tissue repair, repletion, and growth in the young PCM patient. Mg is readily available and can be given orally or parenterally (intramuscularly or intravenously) in suitable dilutions. In view of the large fecal losses that occur in the acute stage of PCM, parenteral administration is advisable initially. Magnesium sulfate heptahydrate (epsom salt) can be given by any route. The best dietary sources include nuts, legumes, whole grains, and certain seafood. Suitable blends of peanut flour, soya flour, and skimmed milk are well tolerated in PCM and are good sources of protein and K as well as Mg. Studies to determine the optimal dose and duration of therapy are in progress.

For background information *see* NUTRITION in the McGraw-Hill Encyclopedia of Science and Technology.   [JOAN L. CADDELL]

Bibliography: G. A. O. Alleyne, D. J. Millward, and G. H. Scullard, *J. Pediat.*, 76:75, 1970; J. L. Caddell, in E. B. Flink and J. E. Jones (eds.), The pathogenesis and clinical significance of magnesium deficiency (symposium), *Ann. N.Y. Acad. Sci.*, 162:874, 1969; A. W. Wilkinson and I. Harris, *Brit. J. Surg.*, 56:628, 1969.

Observatory, astronomical

In recent years significant advances in observational astronomy have been achieved in space. Electromagnetic radiation that never reaches the Earth because of absorption by the Earth's atmosphere is now being measured with rocket- and satellite-borne instruments. The advent of spacecraft capable of escaping the Earth's atmosphere has opened unprecedented opportunity for measuring astronomical phenomena in previously unexplored regions of the electromagnetic spectrum.

Astronomical observations from sounding rockets have been made for more than two decades. However, experiments conducted in this way are severely limited because the time above the atmosphere is only a few minutes. For this reason the National Aeronautics and Space Administration initiated several programs to provide astronomers with improved research facilities in space. These spacecraft, the observatory series of unmanned satellites, monitor electromagnetic and particle radiation emitted by a variety of astronomical sources. Experiments on the Orbiting Astronomical Observatory (OAO) study stars, galaxies, and other objects located beyond the solar system. Although the first OAO malfunctioned, the second one (launched on Dec. 7, 1968) has been an outstanding success and has produced a wealth of important new astronomical data. The Orbiting Solar Observatories (OSOs), which are designed primarily for solar instrumentation, have also enjoyed outstanding successes and have yielded important new information about the outer solar atmosphere and the dynamic phenomena occurring there. Figure 1 shows an OSO spacecraft which carries 7–9 experiments. *See* ASTRONOMICAL SPECTROSCOPY.

Solar atmosphere. Astronomers and physicists have long been fascinated by the outer solar atmosphere, where the most spectacular forms of transient solar phenomena, such as flares, occur. Flares are energetic explosions lasting from minutes to hours. They can emit large quantities of cosmic and x-ray radiation that affect radio communication on the Earth by producing changes in the terrestrial ionosphere. Flare radiation can also affect men and equipment in spacecraft located above the protective layers of the Earth's atmosphere. To understand and hence to predict these transient phenomena, astronomers must obtain a thorough knowledge about the conditions in the medium in which they occur.

A striking characteristic of the solar atmosphere is the temperature inversion. The mean temperature of the visible surface of the Sun is 5800°K. Motions below the solar surface produce shock waves which deposit large amounts of heat in the layers making up the outer solar atmosphere, the chromosphere, and the corona. This heating raises the temperature in the corona to several million degrees Kelvin and produces the solar wind, which interacts with the terrestrial magnetic field. The Sun is unique because it is the only star sufficiently close to Earth to permit astronomers to study the three-dimensional structure of its atmosphere.

Much of the radiation produced by the chromosphere and corona appears at wavelengths below 3000 A, which cannot penetrate the Earth's atmosphere. Wavelengths of particular interest are those in the extreme ultraviolet (UV) (300–1500 A) and x-ray (1–300 A) portions of the solar spectrum. Although the intensity of solar radiation in the visible is remarkably constant, the short-wavelength radiation shows significant variations which depend on the state of solar activity. To study these variations over long periods of time, it is necessary to use spacecraft such as OSOs that are above the Earth's atmosphere for months or years.

Experiments on the first six OSOs have returned thousands of hours of measurements in x-ray and UV wavelengths. Of special significance for studying the structure of the solar chromosphere and corona are the OSO spacecraft with the capability of taking pictures, that is, the capability of measuring at selected wavelengths the distribution of brightness across the solar surface. Experiments on *OSO 2* (launched on Feb. 3, 1965), *OSO 4* (launched on Oct. 18, 1967), and *OSO 5* (launched on Jan. 22, 1969) have produced thousands of x-ray and UV pictures. Instruments on these satellites

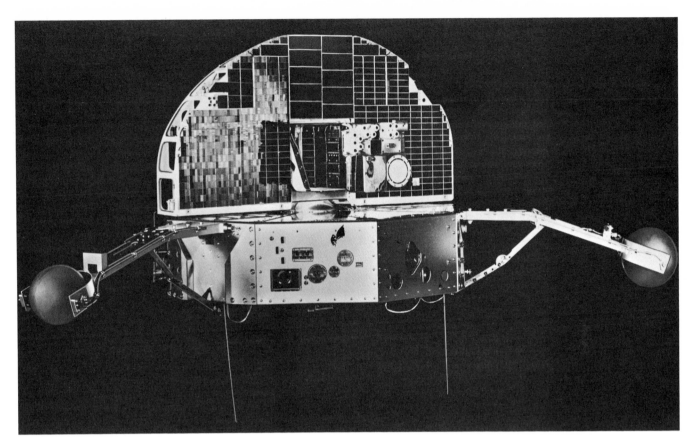

Fig. 1. Orbiting Solar Observatory (OSO) spacecraft.

were able to resolve details on the Sun down to 1 arc minute, or approximately 1/30th of a solar diameter. Pictures were obtained at a rate of one every 5 min.

OSO 6 (launched on Aug. 9, 1969) introduced the ability to take pictures in rapid sequences, one every 30 sec, and the ability to point at a selected feature on the Sun and take an UV spectrum. The rapid picture-taking capability is used by an x-ray experiment and an UV experiment. The spatial resolution of the *OSO 6* spacecraft, 0.5 arc minute, or 1/60th of a solar diameter, is also better than that of the earlier OSOs. Future OSOs, *OSO H* through *K*, will introduce additional refinements.

In addition to picture-taking experiments, each OSO contains instruments that monitor the UV and x-ray radiation emitted by the entire solar disk. These instruments, many of which can observe with time and spectral resolution superior to those of the picture-taking experiments, are particularly valuable for studying transient solar phenomena such as flares with lifetimes of minutes.

Solar flares. X-ray observations of solar flares are among the most interesting observations obtained by OSO instruments. Large flares can increase the total solar x-ray output at some wavelengths by one or more orders of magnitude. Flares are sufficiently complicated that development of an adequate flare theory has not been possible. Satellite measurements of x-ray and UV radiation emitted by flares will undoubtedly play an important role in determining the physical processes occurring during these events. Of particular value

are simultaneous observations of spectral lines produced by ions in different stages of ionization. Such observations, in which different spectral lines often exhibit markedly different behavior, provide measurements of the amount of flare material at different temperatures.

The OSO data provide a rich source of observational material about the structure of the solar atmosphere. An important characteristic of many x-ray and UV spectral lines and continua is that they are emitted by material which is confined to a rather narrow range of temperature. Consequently, a measurement of the intensity of the radiation at a selected wavelength provides a measurement of the amount of gas above the solar surface having some well-defined temperature. Similarly, a picture made at the same wavelength shows how gas at the specified temperature is distributed across the solar surface.

Figure 2 shows examples of several UV pictures made by a Nov. 26, 1967, experiment on *OSO 4*. The pictures (spectroheliograms) were constructed in the following manner. A photometer in the instrument measured the UV brightness at 1920 points covering the solar disk and the area surrounding it. These measurements were then transmitted to the ground by radio and converted into pictures with a computer-driven cathode-ray tube. The darkest areas represent the regions of greatest intensity on the Sun. The mean temperature of the gas emitting the radiation observed at each wavelength is 10,000°K (Lyman continuum 800 A), 100,000°K (N III 991 A), 300,000°K (O VI 1032 A), 800,000°K (Ne VIII 770 A), 1,300,000°K (Mg X 625

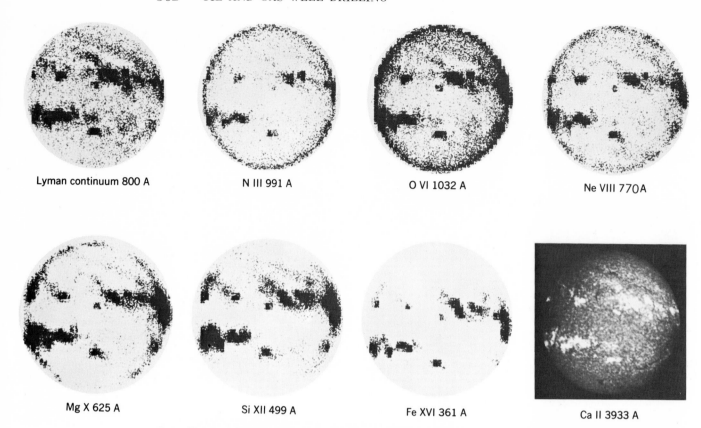

Lyman continuum 800 A N III 991 A O VI 1032 A Ne VIII 770A

Mg X 625 A Si XII 499 A Fe XVI 361 A Ca II 3933 A

Fig. 2. Pictures of the Sun at several extreme ultraviolet (UV) wavelengths (*courtesy of Harvard College Observatory*) and Ca II in the visible spectrum (*courtesy of the McMath-Hulbert Observatory*).

A), 2,500,000°K (Si XII 499 A), and 4,000,000°K (Fe XVI 361 A). The picture in the lower right-hand corner shows how the chromosphere appears in Ca II (3933 A) from the ground. The dark areas in the UV pictures (light areas in Ca II) are active regions where the gas in the solar atmosphere is denser and hotter than in the surrounding quiet regions. As the temperature increases, a greater percentage of the radiation comes from the active regions. Solar flares and other transient phenomena generally occur in or near these regions. L. Goldberg and W. M. Neupert have published results of studies on UV and x-ray pictures.

By studying x-ray and UV pictures and spectra of active regions, astronomers hope to better determine the temperatures and densities there, how these features evolve with time, and why they produce flares. Knowledge of the variation of temperature with height above the solar surface and how this variation with height differs in quiet and active regions is vital to the development of theories which explain how the outer solar atmosphere is heated. Goldberg has described how UV and x-ray measurements are being used for this purpose. Coronal heating theories have been described by M. Kuperus.

The next big advance in solar space observatories will be the Apollo Telescope Mount (ATM), which will be launched as part of the Skylab manned mission. ATM contains three visible-light, two x-ray, and three UV instruments that will be used to study both quiescent and active phenomena in the solar atmosphere with high spatial resolution (5 arc seconds, or 1/360th of a solar diameter) and high time resolution (fractions of a second to minutes). This more advanced space observatory, capable of carrying heavier and more sophisticated instruments than the OSO spacecraft, should significantly increase understanding of solar dynamic phenomena and the structure of the outer solar atmosphere.

For background information *see* OBSERVATORY, ASTRONOMICAL; SPACE PROBE; SUN in the McGraw-Hill Encyclopedia of Science and Technology. [GEORGE L. WITHBROE]

Bibliography: L. Goldberg, *Annu. Rev. Astron. Astrophys.*, 5:279, 1967; L. Goldberg, *Sci. Amer.*, 220(6):92, June, 1969; M. Kuperus, *Space Sci. Rev.*, 9:713, August, 1969; W. M. Neupert, *Annu. Rev. Astron. Astrophys.*, 7:121, 1969.

Oil and gas well drilling

Advances in the oil- and gas-drilling industry in recent years have been effected by improved adaptation of known technology to existing applications. The principal improvements have been made in the areas of drilling fluids, directional drilling, rock- and diamond-bit metallurgy and design, subsurface pressure determination, blowout control, and bit stabilization.

The industry's present problems lie in the continuing decline of domestic rig activity, from 2619 active rigs in 1956 to 932 in April, 1970. Attrition, mergers, and auctions have eliminated all but the largest and hardiest contractors.

The ratio of dry holes to economically successful wells in wildcat exploration in the United States has been shown by M. T. Halbouty to be 45:1 instead of 9:1 as had been believed. Consequently operators have turned to the search for petroleum in practically every country of the free world, concentrating on recent offshore sediments where oil and gas deposits are more prevalent and easier to find. Nevertheless, such exploratory drilling and consequent development require staggering expenditures, leaving little budget money available for drilling activities in the United States.

The awakening awareness of the consequences of pollution, coupled with the fear of blowouts and fires, has caused some operators to limit the number of wells that are drilled from an offshore platform. In this sense, the industry is entering a new era in which controlled, activated technology will play a major role in promotion of greater safety and reduction of hazards.

Drilling rig design. A major development in rig design is represented in the automatic drilling rig developed by Automatic Drilling Machines, Inc. (ADM), of Dallas (Figs. 1 and 2). The ADM rig has combined the rotary table, tongs, and draw works into one unitized assembly. Hydraulically actuated, the rig is electrically controlled. It makes trips into and out of the hole unattended, handling the drill string in either direction. The crew consists of two men instead of the customary five for each 8-hr tour, with one drilling superintendent.

Another principal drilling rig development is the culmination of many years of intensive effort by Roy Cullen Research of Houston. A prototype embodying the Cullen ideas, an electronic rig, is being readied by the Continental-Emsco Co. of Houston. The retractors (Fig. 3) serve as the hoist by which a flexible drill string may be lowered into or removed from the hole. The flexible pipe is coiled into a basket on retrieval. Power is supplied to an electric motor (Reda) through an electric cable that runs inside the drill string. The motor in turn drives the bit through a special planetary gear train.

The advantages of the retractor rig are in automation, control, and savings in trip time when it is necessary to change a bit.

Drilling fluids. In drilling fluids, batteries of anhydrous cyclones are being employed to process the entire mud stream and to remove all except the natural colloidal solids and fine silts. Contractors may now carry low-density muds without resort to tremendous quantities of water for dilution. The result has been improved bit performances, less wear on mud pump fluid ends, and cheaper and better mud systems. On weighted muds, large conveyor-type centrifuges are now almost universally employed to effect a separation between the heavy weighting mineral, barites, and the drilled clays, thereby ridding the system of the undesirable clays and also saving on the barites. The result has been a great improvement in the flow properties of such muds, with large savings in weighting materials and chemicals.

Mud motors. Refinements of mud motors are being employed at depth to obtain positive deflection of the borehole without the assistance of whipstocks, pilot bits, and hole-opening bits. The motor is appended to the drill string at its lowest extremity by a special connection. The connection, commonly called a bent sub, inclines the motor at a slight angle to the axis of the drill string. When drilling commences, the drill string is not permitted to rotate. The body of the motor and the bit

Fig. 1. Automatic drilling rig, developed by Automatic Drilling Machines, Inc., on a Sinclair Oil Co. location in West Texas.

thus function as a chord that governs the degree of vertical curvature desired in deviation. The main drill string follows the path thus created. A mud turbine and a positive-displacement type of mud motor which works on the Moyno pump principle in reverse are the two types of motors.

Rock bits. A threefold change has brought about a drastic improvement in rolling-cutter rock bits. These are the advent of sealed bearings, in which the principal loads are carried by a journal instead

Fig. 2. Prototype of ADM rig as seen from control panel.

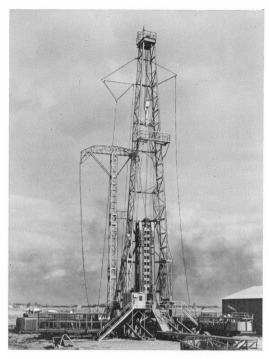

Fig. 3. Continental-Emsco's prototype rig being readied for field trial in yard at Houston.

of the customary rollers; the employment of carbide and segmented journal elements in these bearings; and the use of long carbide inserts instead of milled steel teeth.

Drilling with jets. The most promising new method of drilling involves the employment of jets under pressure differentials approaching 20,000 psi. One way is to drill with pellet-laden drilling fluid so that the velocity imparted to the pellets by acceleration across the jet is sufficient to disintegrate the rock. The second is to drill with mud alone so that the velocity which issues from the bit nozzles is sufficient to cut the rock. A third method is to employ high-velocity jets in combination with rock-bit cutting elements so that the jets impinge against the hole at or close to its outer periphery. Thus the impact force of the jets can be made somewhat additive to the mechanical force applied by the outer or gage teeth of the bit and can aid in destroying the portion of the hole that is hardest to cut. Once the gage or periphery is cut, the inner portion is easily drilled by the inner bit teeth.

Two oil companies are working in the high-velocity area under classified conditions.

For background information *see* BORING AND DRILLING (MINERAL); OIL AND GAS WELL COMPLETION; OIL AND GAS WELL DRILLING in the McGraw-Hill Encyclopedia of Science and Technology.   [ROY A. BOBO]

Bibliography: W. C. Maurer, *Novel Drilling Techniques*, 1968; *Method of Drilling with High Velocity Jet Cutter Rock Bit*, U.S. Patent no. 3,-112,800, Dec. 3, 1963; Modest forecast for 1970, *Oil Gas J.*, Jan. 26, 1970.

Oil field model

The exploration and exploitation of petroleum reservoirs constitute an uncertain and high-risk business. In deciding where to drill, management

needs to be advised on (1) how often petroleum can be found; (2) what the potential reserves will be; and (3) what it will cost to lease the land and to drill the wells. In a statistical sense, the questions can be answered by taking the "how often" and "how much" approach.

The latest applications of modern statistical methods and probability theory have raised the oilmen's capacity to study subsurface geology and engineering parameters as mathematical oil field models, along with the conventional analytical and descriptive subsurface pictures that are generally accepted. During a 4-year period in an area in Kansas the ratio of successful drillings to total holes drilled, that is, success ratio (SR), was held to within reasonable tolerance limits (Figs. 1 and 2).

Decision theory. Geological and engineering data can be analyzed on computers by a number of quantitative decision-making methods that measure the reward-risk alternatives from which an operator can choose courses of action. The problem from the start is one of making a statistical inference, or inferences, or of estimating probabilities that will lead to some sort of reliable forecast from the small amount of data initially available as input.

The process commences by studying only a few limited sources and by making as many inferences

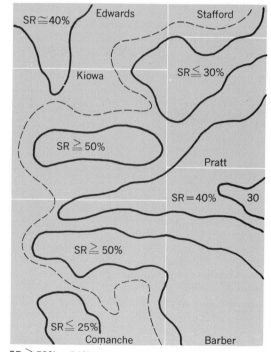

SR ≥ 50% → 54%
SR ≅ 40% → 42%   April, 1964 → April, 1968
SR ≤ 30% → 33%   total holes ≅ 700
SR ≤ 25% → 26%

Fig. 1. Original forecast map of counties in Kansas where predetermined success ratios checked out over a 4-year period. The township and range are 36 mi² with 6 mi to the side. The coordinates in this map are: range 13 west through 20 west and township 25 south through 35 south. (*From J. P. Dowds, In Kansas and Oklahoma mathematical probability approach proves successful, World Oil, 167(7):82–85, December, 1968*)

as is reasonable. At this stage the technologist is unsure, because of inadequate data: The subsurface conditions are unclear, and there is much doubt and uncertainty (high entropy). Even so, predictions and forecasts must be made. Next, older information has to be augmented with new or additional data, and then comparisons made to see whether the original inferences were correct or, percentagewise, how close the forecasts came to being correct. From the new information, a new set of inferences can be projected and the process repeated.

Truth and knowledge are statistical in nature. What was a truth in the past can often be revised to become a new truth as new facts and evidence are accumulated. A series of systematic and sequential examinations causes a steady revision of belief, which in turn leads to better and more accurate forecasts.

By working with Bayes' theorem and the Markov processes, it is possible to construct sequential mathematical models from subsurface data to develop multidimensional pictures as to what could be expected if a number of additional holes were drilled in an area. Lithologies and the consequences of their thicknesses and areal extents can be studied as practical problems in statistical geometry in conjunction with the porosities, hydrocarbon saturations, and so forth that exist in one or more reservoirs.

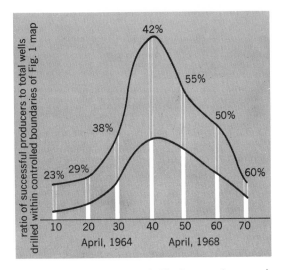

Fig. 2. For the same area as in Fig. 1, over a 4-year period 42% of the wells were successfully completed where a 40% success ratio (SR) had been originally forecast; 55% of wells were completed in areas where a 50% SR had been forecast; and so on.

Bayesian analysis. The powerful Bayesian approach requires two sources of input: (1) deterministic numerical data or quantitative facts that can be observed and measured, and (2) the statistical inferences or subjective attitude (judgment) that

---

Table 1. Bayesian computations in a transition from time$_1$ to time$_2$.*

FROM T_1:
$$A_O = A_\square = A_\triangle = A_\times = 0.25$$
$$A_+ = A_O + A_\square + A_\triangle + A_\times = 100\%$$

TOTAL PROB.
$$P(E_O) + P(E_\square) + P(E_\triangle) + P(E_\times) = 100\%$$
$$.25 + .25 + .25 + .25 = 100\%$$

BY T_2: PREDICTED O's & GOT: 2-O's, 4-$\square$'s, 3-$\triangle$'s, 1-$\times$.
" $\square$'s "   "   : 3-O's, 4-$\square$'s, 2-$\triangle$'s, 1-$\times$.
" $\triangle$'s "   "   : 1-O, 2-$\square$'s, 2-$\triangle$'s, 5-$\times$'s.
" $\times$'s "   "   : 4-O's, 0-$\square$, 3-$\triangle$'s, 3-$\times$'s.

	O	$\square$	$\triangle$	$\times$
O	$P(E\|E_O) = 2/10 = .2$	$P(E\|E_O) = 4/10 = .4$	$P(E\|E_O) = 3/10 = .3$	$P(E\|E_O) = 1/10 = .1$
$\square$	$P(E\|E_\square) = 3/10 = .3$	$P(E\|E_\square) = 4/10 = .4$	$P(E\|E_\square) = 2/10 = .2$	$P(E\|E_\square) = 1/10 = .1$
$\triangle$	$P(E\|E_\triangle) = 1/10 = .1$	$P(E\|E_\triangle) = 2/10 = .2$	$P(E\|E_\triangle) = 2/10 = .2$	$P(E\|E_\triangle) = 5/10 = .5$
$\times$	$P(E\|E_\times) = 4/10 = .4$	$P(E\|E_\times) = 0/10 = 0$	$P(E\|E_\times) = 3/10 = .3$	$P(E\|E_\times) = 3/10 = .3$

FOR O's: $P(E_O|E) = \dfrac{(.25)(.2)}{(.25)(.2) + (.25)(.3) + (.25)(.1) + (.25)(.4)} = 20\%$

FOR $\square$'s: $P(E_\square|E) = \dfrac{(.25)(.4)}{(.25)(.4) + (.25)(.4) + (.25)(.2) + (.25)(0)} = 40\%$

FOR $\triangle$'s: $P(E_\triangle|E) = \dfrac{(.25)(.2)}{(.25)(.3) + (.25)(.2) + (.25)(.2) + (.25)(.3)} = 20\%$

FOR $\times$'s: $P(E_\times|E) = \dfrac{(.25)(.3)}{(.25)(.1) + (.25)(.1) + (.25)(.5) + (.25)(.3)} = 30\%$

*Note that originally the a priori probabilities of $P(E_O)$, and so on, are the actual areas A_O on a map where the events would most probably (predicted) occur, $P(E_O) = A_O$.

Table 2. Bayesian computations from T_2 to T_3 *

FROM T_2:
$$A_O = .4, A_\square = .2, A_\triangle = .1, A_\times = .3$$
$$A_+ = A_O \quad + A_\square \quad + A_\triangle \quad + A_\times = 100\%$$

TOTAL PROB.
$$P(E_O) + P(E_\square) + P(E_\triangle) + P(E_\times) = 100\%$$
$$.40 \quad + .20 \quad + .10 \quad + .30 \quad = 100\%$$

BY T_3: PREDICTED O's & GOT: 10-O's, 0-$\square$'s, 0-$\triangle$'s, 0-$\times$.
  " $\square$'s " " : 5-O's, 5-$\square$'s, 0-$\triangle$'s, 0-$\times$.
  " $\triangle$'s " " : 4-O's, 1-$\square$'s, 5-$\triangle$'s, 0-$\times$.
  " $\times$'s " " : 4-O's, 3-$\square$'s, 2-$\triangle$'s, 1-$\times$.

	O	$\square$	$\triangle$	$\times$
O	$P(E\|E_O) = 10/10 = 1.0$	$P(E\|E_O) = 0/10 = 0$	$P(E\|E_O) = 0/10 = 0$	$P(E\|E_O) = 0/10 = 0$
$\square$	$P(E\|E_\square) = 5/10 = .5$	$P(E\|E_\square) = 5/10 = .5$	$P(E\|E_\square) = 0/10 = 0$	$P(E\|E_\square) = 0/10 = 0$
$\triangle$	$P(E\|E_\triangle) = 4/10 = .4$	$P(E\|E_\triangle) = 1/10 = .1$	$P(E\|E_\triangle) = 5/10 = .5$	$P(E\|E_\triangle) = 0/10 = 0$
$\times$	$P(E\|E_\times) = 4/10 = .4$	$P(E\|E_\times) = 3/10 = .3$	$P(E\|E_\times) = 2/10 = .2$	$P(E\|E_\times) = 1/10 = .1$

FOR O's: $P(E_O|E) = \dfrac{(.4)(1.0)}{(.4)(1.0) + (.2)(.5) + (.1)(.4) + (.3)(.4)} = 60\%$

FOR $\square$'s: $P(E_\square|E) = \dfrac{(.2)(.5)}{(.4)(0) + (.2)(.5) + (.1)(.1) + (.3)(.3)} = 50\%$

FOR $\triangle$'s: $P(E_\triangle|E) = \dfrac{(.1)(.5)}{(.4)(0) + (.2)(0) + (.1)(.5) + (.3)(.2)} = 45\%$

FOR $\times$'s: $P(E_\times|E) = \dfrac{(.3)(.1)}{(.4)(0) + (.2)(0) + (.1)(0) + (.3)(.1)} = 100\%$

*Note that in the area A_O, $A_\square$, and so on, a priori probabilities have been changed because of the recontouring based on additional information.

an individual adds to the system. The technologist must supply or make a priori estimates on the basis of previous work, education, experience, or hunches. Therefore the geologist's or engineer's ideas, combined with the basic facts and downhole data, are needed.

The technique is one whereby the individual supplies some of the a priori probability estimates of the subsurface attributes and parameters. The a priori conditional probabilities developed by Bayesian inference yield results for further examination in terms of a posteriori probabilities. After further interpretation and reestimation, the data are fed back into the Bayesian equations along with new sets of a priori probabilities for the next sequential analysis or iteration.

Step-by-step revisions of belief are computed and mapped in a statistical sense, and knowledge of what to expect becomes more reliable and predictable, provided that the technologist (not the computer) is able to revise and inject reasonable a priori probabilities into the system prior to each sequence run. An infinite number of combinations of subsurface data can be concocted, but it is surprising how rapidly the a priori and a posteriori probabilities approach equilibrium, or a "steady state," as knowledge of the conditions of the subsurface become more consistent. Stationarity is approached and recognized by the fact that the

graph plots and map contours tend to converge from one sequence to the next. The statistical reduction of uncertainties show a direct and corresponding increase in the reward-risk ratio of establishing new petroleum reserves.

Bayes' theorem is defined by Eq. (1), where E_is

$$P(E_i|E) = \frac{P(E_i)P(E|E_i)}{\displaystyle\sum_{r=1}^{n} P(E_r)P(E|E_r)} \quad (1 \le i \le n) \quad (1)$$

are the hypotheses; the probability of one of the four events $P(E_O, E_\square, E_\triangle.$ and $E_\times)$ is the a priori probability that E_i is true or valid. The left side of Eq. (1), $P(E_i|E)$, then, is the a posteriori probability that E_i is true, based on the observed fact that an event E actually occurred or was measurable. The E_rs are the unmeasured values and are expressed as the a priori probabilities, as estimated by the technologist, that an event will occur in the future; these are the $P(E_O)$, $P(E_\square)$, and so forth. Finally, the $P(E|E_i)$ expression in the formula represents the probability that one specific event will occur in place of another event or when another has been forecast. Substituting the event $\square$ into the above equation results in Eq. (2).

$$P(E_\square|E) = \frac{P(E_\square)P(E|E_\square)}{\underset{\text{for } O}{\cdots} + P(E_\square)P(E|E_\square) + \underset{\text{for } \triangle}{\cdots} + \underset{\text{for } \times}{\cdots}} \quad (2)$$

The formula appears formidable to one not familiar with probability/statistical notation; yet, by following through the two sequences, the mathematical manipulations can be readily understood. Table 1 is concerned with four subsurface variables $\bigcirc$, $\square$, $\triangle$, and $\times$, which can be any relative parameters, that is, lithologies, bed thicknesses or areal extent or both, hydrocarbon saturations, types of fossils, or any other factor. Table 1 states that the probability of $\bigcirc$, $\square$, $\triangle$, or $\times$ occurring is 25%, which is the a priori area measurement that the technologist has forecast at $time_1$. However, by $time_2$, where $\bigcirc$s were predicted, it is found that only two $\bigcirc$s occurred, four $\square$s, three $\triangle$s, and one $\times$, and so on also occurred. Next, these values become the conditional probabilities that appear as coefficients in a matrix. Finally, they are substituted into the Bayesian equations, where at the present state of knowledge, the technologist forecaster can reliably forecast the $\bigcirc$s, $\square$s, $\triangle$s, and $\times$s only 20%, 40%, 20%, and 30% of the time, respectively.

Then the forecaster, geologist or engineer, revises the map distribution contours; these revisions change the areas where the four attributes are "most likely" to occur; and these changes in area are again the revised a priori probabilities. As more samples are taken or observations made by acquiring new information, Table 2 can be referred to. From these calculations it is found that the future forecasts for $\bigcirc$, $\square$, $\triangle$, and $\times$ are now 60%, 50%, 45%, and 100%; the uncertainty as to what the predicted outcomes will be for the next sequence has been reduced.

Markov processes and entropy. In the original work by Claude E. Shannon that opened up the field of information theory, he analyzed the statistical structure of the English language by use of Markov chains. In its simplest form, a Markov chain states that the probability of a succeeding event occurring is dependent upon the fact that a preceding event occurred. For example, if the letter Q is known to exist, what is the probability of it being followed by the letter U? Of course, this probability is very high, as would be the probability of the letter H following the letter T. As in the Bayesian statistics, this is in the form of a conditional probability; the chance of something happening when it is known that a certain event has just occurred in time or space. In a geological sense one asks how often will a shale particle lie next to another shale particle, a sand particle occur next to another sand particle, or a sand interval follow a previous sand interval? The distribution of sedimentary rocks and their economic value are a practical example of Markov processes.

If a glass fish bowl in which there are distinct layers of colored marbles is shaken once, vigorously, a few of the marbles will be jostled out of the original strata upward or downward into another zone. If the bowl is shaken again, the mixing increases. The probability distribution as to how much mixing will take place depends on how much took place during the previous shake. This measures the increase in disorder (entropy) as a degenerate Markov chain, and the entropy can be measured by a transition probability matrix. This experiment is allied to the concept of irreversible processes as found in statistical mechanics, whereby the marbles will most probably never be reordered to their original stratigraphic position by continued shaking. To oil geologists and engineers this means that there are reversible processes in nature that give rise to Markov chains. This is how one can study mathematically the development and distribution of sand grains, sand bodies, hydrocarbon saturations, and so on.

The partial transition entropy S_i can be computed from the conditional probability coefficients in the matrix by the use of Eq. (3), where, for each

$$S_i = -K \sum_{i=1}^{\infty} P_i \log P_i \qquad (3)$$

symbol, $S_i = -[P(E|E_i) \log P(E|E_i) + \cdots]$ and where K is any constant selected so that the calculations come out in terms of feet, porosity, hydrocarbon saturations, dollars, and so on. S_i is the entropy or uncertainty measurement in a system in which an event has occurred or, on the basis of previous experience or experiments, can be expected to occur again: $S_{\bigcirc}$ is the entropy (uncertainty) that event $\bigcirc$ (E_0) will occur in a multivariate system where $\bigcirc$s, $\square$s, $\triangle$s, and $\times$s occur.

In Table 3 note that the coefficients in the transition probability matrix for $time_2$ are the same coefficients that were entered into Bayes' formula in time sequence in Table 1. Table 4 demonstrates the calculation of the partial transition entropies from $time_3$.

Results. During the past 15 years several areas covering thousands of square miles have been examined by these methods and others. Figure 1, a map of an area in south-central Kansas where a study was initiated in April, 1964, is an example. By April, 1968, when the SRs were expected to be greater than 50%, producing wells were found 54% of the time. Where 40% SRs should have been obtained, 42% were realized; where an SR of 30% or less was anticipated, the areas drilled out 33%; and where SRs of 25% or less were expected, 26% was actually drilled out.

Figure 2 is a graph of the same area in 10% increments of success by abscissa, demonstrating how the SRs developed over the period of time, and forecasting that operators would have 42% success if they drilled in the 40% area, 55% by

---

Table 3. Partial transition entropies $S_{\bigcirc}$, $S_{\square}$, $S_{\triangle}$, $S_{\times}$ from the matrix at left, which is a measure of the uncertainty that each event will occur in a multidimensional and multivariate model at $time_2$

T_2

	$\bigcirc$	$\square$	$\triangle$	$\times$
$\bigcirc$	.2	.4	.3	.1
$\square$	.3	.4	.2	.1
$\triangle$	.1	.2	.2	.5
$\times$	.4	0	.3	.3

$S_{\bigcirc} = -(.2 \text{ LOG } .2 + .4 \text{ LOG } .4 + .3 \text{ LOG } .3 + .1 \text{ LOG } .1) = .555$

$S_{\square} = -(.3 \text{ LOG } .3 + .4 \text{ LOG } .4 + .2 \text{ LOG } .2 + .1 \text{ LOG } .1) = .555$

$S_{\triangle} = -(.1 \text{ LOG } .1 + .2 \text{ LOG } .2 + .2 \text{ LOG } .2 + .5 \text{ LOG } .5) = .530$

$S_{\times} = -(.4 \text{ LOG } .4 + 0 \text{ LOG } 0 + .3 \text{ LOG } .3 + .3 \text{ LOG } .3) = .460$

Table 4. Entropies at time $_3$

$$T_3$$

	○	□	△	×	
○	1	0	0	0	$S_○ = -(1\ LOG\ 1 + 0\ LOG\ 0 + 0\ LOG\ 0 + 0\ LOG\ 0) = 0$
□	.5	.5	0	0	$S_□ = -(.5\ LOG\ .5 + .5\ LOG\ .5 + 0\ LOG\ 0 + 0\ LOG\ 0) = .301$
△	.4	.1	.5	0	$S_△ = -(.4\ LOG\ .4 + .1\ LOG\ .1 + .5\ LOG\ .5 + 0\ LOG\ 0) = .409$
×	.4	.3	.2	.1	$S_× = -(.4\ LOG\ .4 + .3\ LOG\ .3 + .2\ LOG\ .2 + .1\ LOG\ .1) = .555$

staying in the 50% areas, and further indicating that by staying to the right of the graphic peaks that oil or gas should be found on 50% of the drilling ventures, or more often.

For background information *see* OIL FIELD MODEL; PETROLEUM GEOLOGY; STATISTICS; STOCHASTIC PROCESS in the McGraw-Hill Encyclopedia of Science and Technology. [JOHN P. DOWDS]

Bibliography: J. P. Dowds, *The Colorado School of Mines Symposium on Operations Research and Computer Applications in the Mineral Industries, April, 1968,* 64(3):275–293, July, 1969; J. P. Dowds, In Kansas and Oklahoma mathematical probability approach proves successful, *World Oil,* 167(7):82–85, December, 1968; J. P. Dowds, Statistical geometry of petroleum reservoirs in exploration and exploitation, *Soc. Petrol. Eng. J.,* pp. 841–852, July, 1969.

Optical fibers

Optical fibers are a means of manipulating light that is rapidly finding use in many fields. Some recent developments are their use in chromatography, automatic titrations, and control of chemical reactions. Composed of many small-diameter glass or plastic fibers, fiber optics can be made in a great variety of forms, varying from rigid plates or rods to bundles flexible enough to be tied in knots (Fig. 1). These fibers can conduct light for distances up to 20 ft in a direct path or around corners to inaccessible locations; they can split a light beam one or more times and in any proportion; and they can magnify, cool, shape, or flatten the beam. They can provide localized illumination or retrieve light or an image from a restricted location. Potential applications are so great that new companies are rapidly undertaking optical fiber (fiber optics) fabrication to acquire a share of future sales, expected to climb to $20,000,000–50,000,000 annually within the next 5 years.

Fiber optics have been fabricated from materials that transmit light in the ultraviolet to the infrared region; the frequent dependence of scientific instrumentation on responses in this spectral region should give impetus to using such fibers in the design of this equipment.

Theory. Optical fiber theory is based on the principle of total internal reflectance. It is known that light is transmitted through a smooth transparent cylinder by multiple internal reflections. The same phenomenon occurs when the cylinder is reduced in size to a very small diameter, as in a fiber. However, this light transmission becomes especially efficient (and total internal reflectance becomes possible) when the fiber has a transparent coating or cladding with a refractive index lower than that of its core and when the incident ray is within a certain acceptance angle. Thus, in the typical fiber shown in Fig. 2, $N_2 < N_1$ and θ is the maximum angle (from the longitudinal axis) at which a light ray entering the core can be transmitted through the fiber. The sine of θ is defined as the numerical aperture (*NA*) and is given by the equation below.

$$\text{sine } \theta = NA = \sqrt{N_1{}^2 - N_2{}^2}$$

A light ray entering the core at an angle less than θ is reflected back into the core and travels by a zigzag path of successive reflections to the opposite end, emerging from the fiber at the same angle that it entered. Since the cladding eliminates leakage between the fibers, each fiber of a bundle conducts light independently. For a typical glass fiber, the numerical aperture is 0.63 and θ is 39°.

Absolute transmission through the fiber depends on the transmission area, the finish on the ends of the fiber, reflection losses, and absorption and scattering within the fiber by the transmitting core. Of the light energy striking the end of a fiber bundle within the acceptance angle, about 30% is lost because of reflections (Fresnel losses) at the entrance and exit faces and to the spaces or voids between fibers. Losses also occur as a result of absorption of light energy by the core material, and vary according to wavelength, a typical optical quality glass losing about 10% per foot at 5500 A.

Types. Bundles may be coherent or incoherent. In a coherent bundle the fibers are carefully oriented to permit the transmission of optical images as well as merely light. A bundle is incoherent if the fibers within it are not oriented; these bundles can transmit light only and are sometimes called light pipes.

Coherent and incoherent bundles may be flexible or rigid. Flexible bundles can be bent easily (within limits) and still retain their light-carrying capability. In rigid conductors the fibers are fused together, usually in a parallel arrangement, so that they can transmit images. Diameters of glass fibers range from 2 to 6300 microns (μ), and plastic fibers from 75 to 2500 μ.

The commercial glasses normally used to prepare fiber optics transmit light in the 4000–13,000-A range. For transmission of ultraviolet light, bundles of fibers with quartz cores have recently become available. For transmission in the infrared, arsenic trisulfide fibers are available, and silver chloride is a future possibility.

Glass or quartz fibers are superior to plastic ones in many respects. They are more rugged, can withstand temperatures up to 300°C and higher, are more resistant to chemicals and humidity, transmit more light and therefore can be used for longer lengths, transmit over a broader spectral range (especially in the near-infrared region), and do not show loss of transmission on aging as plas-

OPTICAL FIBERS

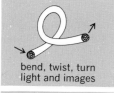

bend, twist, turn light and images

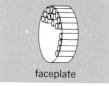

faceplate

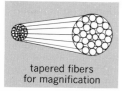

tapered fibers for magnification

Fig. 1. Some basic forms of fiber optics. (*American Chemical Society*)

tics do. Plastic fibers are less costly and more flexible, and their ends are more easily finished than are those of glass fibers; these advantages are not likely to be significant for chemical uses.

Light-conducting bundles are normally encased for protection in a black polyvinyl chloride (PVC) sheath, and heat-shrinkable PVC is used at the terminations. When temperatures exceed 80°C, PVC softens; accordingly, a jacket of flexible stainless steel and close-tolerance terminations replace the PVC sheath for high-temperature uses.

Applications. The development of optical fibers resulted from the attempt to devise a flexible endoscope to examine the stomach by means of an image-carrying light tube introduced through the mouth. Success was achieved when it was found that an assembly of a glass rod within a glass tube of lesser refractive index could be drawn into fibers a few microns in diameter, and that a bundle of these fibers would transmit an image. In a recent version of the device, illumination from an outside source is provided by an outer ring of incoherent fibers, and the image is observed through the inner section composed of coherent fibers. By making it possible to view and photograph many internal cavities and canals of the body with amazing clarity, this tool has facilitated and, in fact, revolutionized medical diagnosis.

It is not difficult to visualize applications of optical fibers to scientific instrumentation, especially if a photosensing device is added to make exact measurements and if different wavelengths of light are used. Such a device was recently reported by M. Beroza, K. R. Hill, and K. H. Norris for the rapid quantitative scanning of thin-layer chromatograms; the scanner (Fig. 3) measures the diffuse reflectance of compounds which appear as discrete spots on a thin-layer plate. The heart of the instrument is an optical-fiber scanning head containing randomly mixed glass fibers; half of the fibers conduct light to a small rectangular area of the plate being scanned, and the other half conduct reflected light to a photosensing cell that has its response recorded. A second fiber optic scans the blank area adjacent to the chromatogram to correct for background variations, and thus provides double-beam operation. Light-absorbing spots appear as peaks on the recorder trace, and the concentration of a chemical is determined from its peak area. Despite the scattering of incident light owing to the uneven surface of the thin-layer plate, a high and constant proportion of the reflected light is captured for measurement because the terminus of the scanning head is very close to the reflecting surface. Use of ultraviolet light (with quartz fibers) and fluorimetry also appears to be feasible with the device.

In another application, by Fisher Scientific Co., two fiber optics facing one another are used for automatic titrations. Monochromatic light passes from one fiber optic through the solution to the other and finally to a photocell. Titrant is automatically added to the stirred solution, and the amount needed to reach a predetermined end point (color) is recorded. The system responds to changes of less than 0.01 absorbance unit.

Other uses of fiber optics are manifold. They transmit or retrieve light or images from vacuum

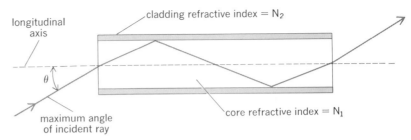

Fig. 2. Schematic of a light path through a fiber.

chambers, from hazardous areas, or from inside equipment (to observe wear or damage). Bundles of coherent fibers can transmit numbers, figures, or lines on a scale from inaccessible locations. Images are readily photographed, and resolution can be excellent since as many as 10,000 image points per square millimeter can be transmitted. A liquid level or a chemical reaction can be controlled by means of a photosensing arrangement. Reliability of lighting can be improved and maintenance reduced by piping light to several locations from a single accessible reliable source; the fibers can be formed in any shape at the exits. A hot source, such as a flame or part of one, is readily viewed because the viewing end is cool. Similarly, cold light can be piped to a microscope stage to illuminate a small heat-sensitive specimen. Expensive mirrors and auxiliary equipment may be replaced with fiber optics. The possibility of manipulating lasers more effectively with them is also being explored.

The faceplate, a flat plate of large area composed of coherent optical fibers, efficiently transfers an image from one side of the plate to the other (Fig. 1). Since thermal and mechanical prop-

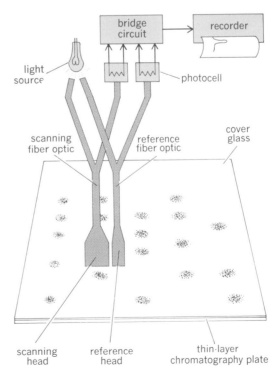

Fig. 3. Thin-layer chromatography (TLC) scanner with dual-beam fiber optic arrangement.

erties of faceplates are nearly identical to those of solid glass, they can be substituted for glass and provide image-transfer capabilities. For example, photographic prints from the image formed on a phosphor within an evacuated cathode-ray tube are poor; with a faceplate front on the tube, however, the image is transmitted point by point to the front, and excellent prints can be made. Such plates will undoubtedly find use on vacuum and other equipment. If tapered fibers are used, images can be magnified or reduced in size depending on the end from which the image is viewed.

The discussion above covers only a few of the potential uses of optical fibers. Undoubtedly they will spur many improvements in scientific instrumentation.

For background information *see* OPTICAL FIBERS in the McGraw-Hill Encyclopedia of Science and Technology. [MORTON BEROZA]

Bibliography: M. Beroza, K. R. Hill, and K. H. Norris, *Anal. Chem.*, 40:1680–1613, 1968; J. K. Crum, *Anal. Chem.*, 41(2):26A–32A, 1969.

Paleoclimatology

In addition to the methods previously applied (the occurrence and isotopic composition of biological fossils and sediments, the morphology of glacier moraines and strand lines, and so forth),

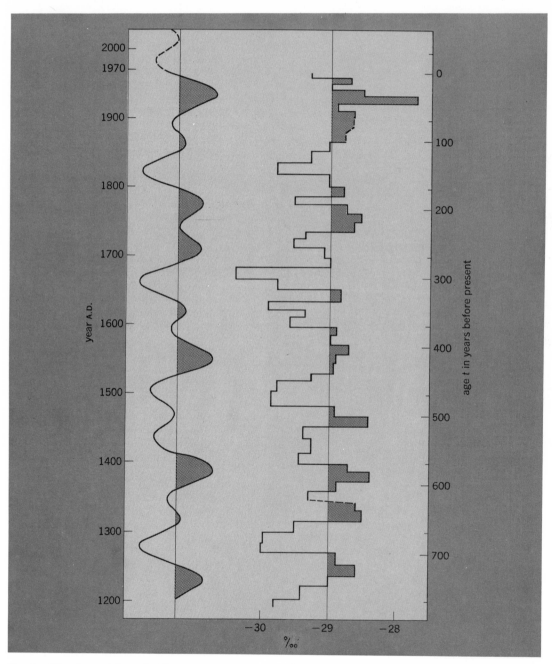

Fig. 1. Climatic variations for the past 800 years. Right, oxygen-18 concentration δ in glacier ice as a function of the time since the deposition of the ice. Left, synthesis of the two dominating harmonics in the step curve, showing main features of climatic changes during the last 800 years. Dashed extrapolation suggests further cooling of the Arctic in the coming one or two decades. (*From S. J. Johnsen, et al., Climatic oscillations 1200 to 2000 A.D., Nature, 227:482–483, 1970*)

a new independent method has been developed to study paleoclimates based on deep ice cores from the ice sheets in Antarctica and Greenland.

As successive layers of snow are deposited on large ice fields, the snow is compressed into solid ice that sinks along with plastic deformation of the layers (horizontal stretching). Thus an ice core penetrating the entire ice sheet contains a continuous sequence of annual layers, the thickness of which decreases downward. In polar snow the concentration of the heavy oxygen isotope O^{18} (expressed by the relative deviation δ from standard mean ocean water) depends mainly on the temperature of condensation in the cloud. Therefore the isotopic composition of increments of the ice core reflects the climatic conditions at the time of deposition of the ice. However, other parameters are not negligible, and a δ profile along the ice core thus reveals the temporal trend of past air temperatures rather than the air temperatures themselves.

Advantages. The method has several advantages: (1) A deep polar ice core represents a continuous sedimentary record far back in time, be-

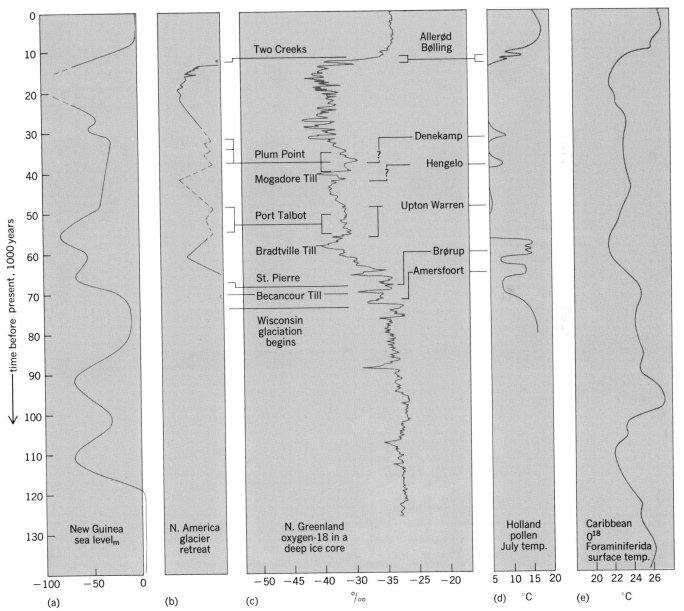

Fig. 2. Comparison between five different climatic records, all spanning the entire Wisconsin glaciation. (*a*) Sea-level changes on New Guinea (*from H. H. Veeh and J. Chappell, Astronomical theory of climatic change: Support from New Guinea, Science, 167:862–865, 1970*). (*b*) Advances and retreats of the Laurentide ice sheet in the Ontario-Erie basins (*from R. P. Goldthwaith et al., in H. E. Wright, Jr., and D. G. Frey, eds., The Quaternary of the United States, Princeton University Press, 1965*). (*c*) Oxygen-18 variations in a deep ice core from Camp Century, Greenland (*from W. Dansgaard et al., in K. K. Turekian, ed.,* *Proceedings of Symposium: The Late Cenozoic Glacial Ages, Yale University, December, 1969, Yale University Press, 1971*). (*d*) European temperature record based on pollen studies (*from T. van der Hammen et al., Stratigraphy, climatic succession and radiocarbon dating of the last glacial in the Netherlands, Geol. Mijnbouwk., 46:79–94, 1967*). (*e*) Generalized paleotemperature or paleoglaciation record (*from C. Emiliani, Paleotemperature analysis of the Caribbean cores P6304-8 and P6404-9 and a generalized temperature curve for the past 425,000 years, J. Geol., 74:109–126, 1966*).

cause no melting or runoff occurs; (2) essentially it is possible to investigate an ice core as far in detail as desirable; and (3) the δ data are more directly connected to past climatic conditions than are primary data obtained by any other method.

Disadvantage. On the other hand, the method has the disadvantage that none of the radioactive ice-dating methods (C^{14}, Si^{32}, Pb^{210}, H^3) can be applied at considerable depths, either because the amount of ice available in a core is too small, or because the radioactive elements have half-lives that are too short to cover more than the upper layers. Instead, a time scale along an ice core may be established mathematically by considering the generally accepted flow pattern in ice sheets and making reasonable assumptions concerning the parameters that influence it. Plotting the measured δs against the calculated ages of the ice then gives the climatic record.

Data. A study comprising about 7000 samples has been carried out on the 1400-m-long top-to-bottom ice core from Camp Century, Greenland, which was recovered in 1966 by United States Cold Region Research and Engineering Laboratory.

Figure 1 shows how δ has varied in the last eight centuries. The shaded areas correspond to relatively warm periods. The climatic optimum that occurred about A.D. 1930 and the so-called little ice age of A.D. 1600–1730 stand out clearly. Fourier spectral analysis shows two dominating oscillations in the δ curve with periods of 78 and 180 years. The synthesis of these oscillations, shown to the left and extrapolated 50 years into the future, results in a prediction of a continued cooling of the Arctic in the coming one or two decades, followed by a slight warming trend (possible climatic effect of the activity of modern man is, of course, neglected).

Figure 2c shows the δ curve for the entire core, except for the deepest 17 m. The time scale has been corrected under the assumption that the period of a persistent δ oscillation, observed in the Fourier spectra, has been constant in the course of time. The δ curve is provided with tentative interpretations for comparison with other studies. All five records essentially agree that the Wisconsin glaciation lasted from approximately 70,000 to 10,000 years B.P. (before present). As might be expected, the details agree most obviously with the North American record (Fig. 2b), but they also agree satisfactorily with the European record (Fig. 2d) in view of the considerable uncertainties involved in C^{14} dating beyond 30,000 years.

Prior to the Wisconsin glaciation, the δ curve must be compared with the lower parts of the records (Fig. 2a and e) that are dated by uranium-thorium methods. All three have minima at approximately 90,000 and 110,000 years B.P., but whereas Fig. 2a indicates two periods of full glacial severity, and Fig. 2e one such period, none of the δ minima can be interpreted as such. The suggestion by C. Emiliani, that partial melting of the Greenland ice sheet during interglacial periods might have left only a thin layer of ice accumulated in the beginning of the preceding glaciation, cannot be ruled out. However, it is difficult to understand why such enormous melting should have left just enough ice to be detected, and should have

done so on two occasions. The question is still open to discussion.

It should be emphasized that, prior to the Wisconsin glaciation, the time scale for the δ curve is encumbered with considerable uncertainty.

For background information *see* GEOLOGIC THERMOMETRY; GLACIOLOGY; PALEOCLIMATOLOGY in the McGraw-Hill Encyclopedia of Science and Technology.   [W. DANSGAARD]

Bibliography: W. Dansgaard et al., in K. K. Turekian (ed.), *Proceedings of Symposium: The Late Cenozoic Glacial Ages, Yale University, December, 1969*, 1971; W. Dansgaard and S. J. Johnsen, A flow model and a time scale for the ice core from Camp Century, Greenland, *J. Glaciol.*, 8:215–223, 1969; W. Dansgaard and H. Tauber, Glacier oxygen-18 content and Pleistocene Ocean temperatures, *Science*, 166:499–502, 1969; C. Emiliani, Interglacial high sea levels and the control of Greenland ice by the precession of the equinoxes, *Science*, 166:1503–1504, 1970; C. Emiliani, *Science*, 154:851, 1966; R. P. Goldthwait et al., in H. E. Wright, Jr., and D. G. Frey (eds.), *The Quaternary of the United States*, 1965; T. van der Hammen et al., Stratigraphy, climatic succession and radiocarbon dating of the last glacial in the Netherlands, *Geol. Mijnbouwk.*, 46:79–94, 1967; S. J. Johnsen et al., Climatic oscillations 1200 to 2000 A.D., *Nature*, 227:482–483, 1970; H. H. Veeh and J. Chappell, Astronomical theory of climatic change: Support from New Guinea, *Science*, 167:862–865, 1970.

Paleozoic fish

The orthodox history of Paleozoic fishes has been based almost entirely on faunas from the western half of the Northern Hemisphere. There is a suspicion, therefore, that this sample of the fossil record is strongly biased, and the hope has been entertained that major gaps in knowledge of the fossil record will be filled by finds from other parts of the world. This hope now promises to be fulfilled to a surprising degree by new finds in Antarctica, China, and Australia. But the presence of North American and European genera in these faunas poses new problems of paleozoogeography.

Gogo fauna. One of the largest and most important of the new faunas has been found in the Fitzroy Trough of Western Australia, near Gogo Station, a cattle farm. The fishes are of early Upper Devonian age and are superbly preserved in calcareous concretions, from which they can be extracted with dilute acetic acid. The bones are uncrushed and the skeletons barely disturbed, making them ideal material for detailed morphological studies. It is arguable that this is the most important fish fauna of its type and age ever discovered. The geology of the Fitzroy Trough is well known from the work of P. E. Playford and D. C. Lowry of the Western Australian Geological Survey. The fossiliferous Gogo shales were deposited between and along the outer margins of stromatoporoid and algal reefs, which grew on a shallow, nearshore shelf, forming what aptly has been termed the Great Devonian Barrier Reef. The Gogo fish fauna thus offers excellent opportunities for studies in sedimentary petrology and paleoecology, fields which are usually ignored by paleoichthyologists. The fauna is rich in both individuals and species, and includes onychodontids, paleonisciforms (ray-finned fishes), dipnoans (lungfishes), and placoderms. The first group is being studied by Mahala Andrews in Edinburgh and the second by Brian Gardiner at London University. Gardiner has already published some preliminary observations. R. S. Miles, responsible for the last two groups, is currently researching the placoderms.

Placoderms. These are jawed fishes with a plate armor in two shields which are connected by a paired hinge joint in the neck region. One shield forms a buckler over the top and sides of the head, and the other a girdle around the anterior part of the trunk. Although they are bony fishes, placoderms seem to be a development of the basal elasmobranchiomorph stock, collateral with the Chondrichthyes (sharks, rays, rabbitfishes). They flourished during the Devonian and became extinct in the Carboniferous following the rise of the chondrichthyans. As far as their adaptive radiation is concerned, placoderms can be regarded in general terms as "armored sharks." There are about 16 species of placoderms in the Gogo fauna, mostly of the order Arthrodira, although the Ptyctodontida and Antiarchi are also represented.

More than two-thirds of all placoderms are arthrodires, a group which can be classified into primitive forms with a long trunk shield (dolichothoracids) and advanced forms with a reduced trunk shield (brachythoracids). The phylogeny of this large order is, however, poorly known, and it has been necessary to divide it into four successive levels of organization to gain some understanding of its evolutionary history (Fig. 1). The characters used to define these levels in Fig. 1 are related to improvements in the locomotor and feeding mechanisms.

As arthrodires evolved, the trunk shield (Fig. 2) became reduced from behind, releasing more of the body muscles for use in swimming, the pectoral fins increased in size and became flexibly mobile hydrofoils, and the fin spines were reduced in size and eventually lost. In the head the neck joints became better developed, and the dorsal gap between the two shields in the neck region enlarged. As the amount of up-and-down movement of the head against the trunk increased, giving a wider gape of the jaws, a wide range of tooth plates was developed and the feeding mechanism became variously modified. As a number of separate lineages approached and attained the pachyosteomorph level of organization in the late Middle and early Upper Devonian, arthrodires underwent a remarkable burst of secondary adaptive radiation, to become the dominant fishes of the time. The Gogo fauna includes several advanced coccosteomorph genera which document the morphological changes that took place during the transition from the coccosteomorph to the pachyosteomorph level. These seem to be mostly native Australian genera, with both crushing and shearing tooth plates and with variously modified snouts. More primitive coccosteomorphs are represented by species very close to *Coccosteus* species from the Middle Devonian of Europe; pachyosteomorphs are represented by three little-modified forms, including *Eastmanosteus*, hitherto known only from the Middle and Upper Devonian of North America and Poland.

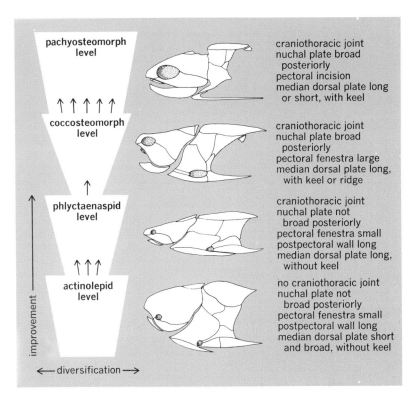

Fig. 1. Successive levels of organization in arthrodires. (*From R. S. Miles, Trans. Roy. Soc. Edinburgh, 68:123–170, 1969*)

A member of the fauna of particular interest is the large arthrodire *Holonema*, which is widely distributed in the Middle and Upper Devonian of the Northern Hemisphere, but always badly preserved. The Gogo specimens solve many of the morphological problems associated with this genus, and because their endoskeletal structures such as the palatoquadrate and opercular cartilage are well preserved, they will make a significant contribution to the knowledge of arthrodire structure. *Holonema* seems to stand somewhere between the phlyctaenaspid and coccosteomorph levels. It shows such an unexpected combination of primitive and advanced characters that it calls into question the whole basis of the present formal classification of arthrodires. The Gogo arthrodires as a whole provide for the first time, in a sufficiently wide range of families, a series of well-preserved structures with many morphological

Fig. 2. Lateral view of the trunk shield of an advanced coccosteomorph-level arthrodire from Gogo, Australia. Note the reduced flank plates and the large fenestra for the articulation of the pectoral fin.

characters (for example, parasphenoids, scapulo-caracoids, pelvic girdles), so that some advance in the analysis of arthrodire phylogeny may be expected.

Antiarchs are represented at Gogo by *Bothriolepis*, a genus with over 40 species and an almost worldwide distribution in the Upper Devonian. The presence of *Bothriolepis* is unexpected, since it is regarded as a fluvial form. The excellent preservation of the Gogo species is helping to solve some of the problems of functional morphology that have long puzzled workers on antiarchs, particularly those associated with the armlike pectoral fins and bony gill covers. Ptyctodonts have been of special interest in recent years as possible ancestors of the living chimeroids (rabbitfishes). Colin Patterson has reviewed this controversial subject. Again, the Gogo ptyctodonts stand to make a unique contribution to the solving of an important problem, because they are so well preserved. They include, for example, ossified neurocrania showing fine details of structure. Two distinct ptyctodonts are present. One is a species of *Ctenurella*, a genus originally described from the Upper Devonian of Germany, and the other has tooth plates which recall those of *Ptyctodus* from the Middle and Upper Devonian of North America.

For background information *see* ANTIARCHI; ARTHRODIRA; PLACODERMI; PTYCTODONTIDA in the McGraw-Hill Encyclopedia of Science and Technology.   [ROGER S. MILES]

Bibliography: C. H. C. Brunton, R. S. Miles, and W. D. I. Rolfe, *Proc. Geol. Soc. London*, 1655:79–83, 1969; R. S. Miles, *Trans. Roy. Soc. Edinburgh*, 68:123–170, 1969.

Particle detector

In physics today, as always, there is a continuing need for precision instruments. No physicist would question the value of precision measurements, or discourage the search for new and precise measuring techniques. It is rare indeed that the introduction of a new technique of measurement does not lead to discoveries of fundamental importance. In high-energy physics in particular, there is a need for an improved instrument to detect γ-rays, the quanta of electromagnetic radiation with energies in the range of millions of electron volts (Mev) and higher. For many years small crystals of inorganic scintillators, such as NaI(Tl) and CsI(Na), have been used with great success by nuclear physicists in the field of γ-ray spectroscopy for γ-rays in the energy range of about 1 Mev. More recently, with the introduction of the so-called solid-state detectors made from single crystals of silicon and germanium, great improvement has been possible in the resolution with which γ-ray spectroscopy can be conducted. The success of these new detectors depends upon the improved ability of their constituent materials totally to absorb any γ-ray incident upon them and at the same time to generate an electrical signal in exact proportion to the energy of the γ-ray absorbed.

At energies much in excess of 1 Mev the interaction of γ-rays with matter takes place in a radically different way, with the result that the detectors appropriate at lower energies become much too small to contain completely the energy of the γ-

ray. One remedy is quite simply to increase the size of the detector by an appropriate amount so that the energy is once again totally absorbed. This is just the approach which has been explored at Stanford University in the past 2 years. In the course of these investigations two attractive properties of this new detector of high-energy γ-rays have come to light. The first is that the energy resolution can be very good, and the second is that the minimum size of the crystal needed to maintain this resolution increases only very slowly with increasing γ-ray energy. Although the application of the total absorption technique to high-energy γ-ray spectroscopy is not a recent idea, having been discussed as early as 1953, it is the recent investigations which have revealed how good the performance can actually be.

γ-Ray detector. Figure 1 shows a schematic drawing of the latest and largest detector of this type to be tested. This device consists of three individually grown single crystals of NaI(Tl), optically cemented together and mounted in a single aluminum case. The overall dimensions of the composite crystal are 16 in. in diameter and 24 in. in length. The γ-rays enter through a thin window at one end, and the scintillation light is viewed by five 5-in. photomultipliers symmetrically mounted at the other end. The signals from these photomultipliers are summed to provide a single output pulse, which is typically recorded in a multichannel pulse-height analyzer.

This type of detector is expected to respond in almost identical ways to γ-rays, electrons, or positrons. All these particles interact in very similar ways with matter, dividing their energy among a vast multiplicity, or cascade, of secondary particles, most of which are rapidly brought to rest in the absorber. For reasons of convenience, since charged particle beams of well-defined energy are much easier to produce than are similar beams of γ-rays, all the tests which have been made so far have employed electrons and not γ-rays.

Figure 2 shows the pulse-height distribution observed when 15-billion-electron-volt (Gev) electrons are incident on the detector. The energy resolution at 15 Gev is 0.9% FWHM (full width at half maximum); at 7.5 Gev it is 1.0%; at 3.8 Gev, 1.5%; and at 1 Gev, 2.0%. Four 2-in. photomultipliers are mounted on the side of the detector, adjacent to two of the individual crystal segments, and are intended to allow some correction to be made for the different intrinsic scintillation yields of the different segments. In this detector these differences are of the order of 10%. This method of correction depends upon the selectivity of the response of the 2-in. photomultipliers to the light directly emitted from the crystal segments to which they are coupled.

Although this technique is certainly effective and improves the resolution by an average of 35%, it is not expected to have the flexibility to compensate exactly for the different scintillation yields. Undoubtedly, total absorption crystal spectrometers in which more carefully matched crystal segments are used will be assembled in the near future. With such instruments the limiting resolution is expected to be set only by fluctuations in the amount of energy escaping from the detector volume, mostly in the form of low-energy γ-rays. This effect can be reliably calculated by Monte Carlo methods since the elementary interactions responsible for the secondary electron-photon cascade are all well understood.

It turns out, for example, that a crystal of the size under discussion should have a limiting resolution, due to leakage, of approximately 0.4% FWHM at 8 Gev. This can be further reduced by simply increasing the crystal volume. Moreover, to maintain this resolution as the incident particle energy increases, the crystal dimensions need only increase in proportion to the logarithm of the incident energy, which is a very slow rate of increase. In addition, the efficiency of the detector is 100%, and the range of particle energies which can be accepted for measurement, the dynamic range of the instrument, is limited only by the range within which the photomultipliers can respond linearly to the light intensity produced.

All the properties of total absorption crystal spectrometers listed above compare very favorably with the alternative methods of electron, positron, and γ-ray detection available at high energy. The energies of electrons and positrons can be determined to a few tenths of 1% by using magnetic spectrometers, but these instruments do not have the dynamic range offered by total absorption devices. They are also usually very large, possess relatively small acceptance apertures, and require considerable amounts of electric power.

In the area of γ-ray detection, however, the properties of total absorption detectors are really unique. No other device offers such good energy resolution in conjunction with 100% efficiency, large acceptance aperture, and large dynamic range. The closest competitor is the pair spectrometer, in which the high-energy γ-ray is converted, with low efficiency, into an electron-positron pair, which can then be analyzed with a magnetic device. The dynamic range is consequently quite poor.

An approximation to the total absorption spectrometer is the ionization calorimeter, which has often been used in high-energy physics. This is a nonhomogeneous structure consisting of alternate layers of a heavy absorber, in which the secondary cascade develops, and radiation detectors to sample the secondary particle flux at various depths in the absorber. The signals from the various detectors are then added to produce an output in proportion to the incident energy. The trouble with this device is that the sampling process introduces large statistical fluctuations into the response, so that the resolution obtained is usually quite poor. The best resolution which has been observed at 8 Gev, for example, is a factor of 10 poorer than that provided by the homogeneous total absorption spectrometer described above.

Neutron detector. High-energy physics today also faces the problem of the detection and measurement of electrically uncharged particles other than the γ-ray, such as the neutron and the K⁰-meson. These particles also deposit all of their energy in an absorber which is sufficiently large and so lend themselves to measurement by the total absorption technique. The nature of their so-called strong interaction with matter is very

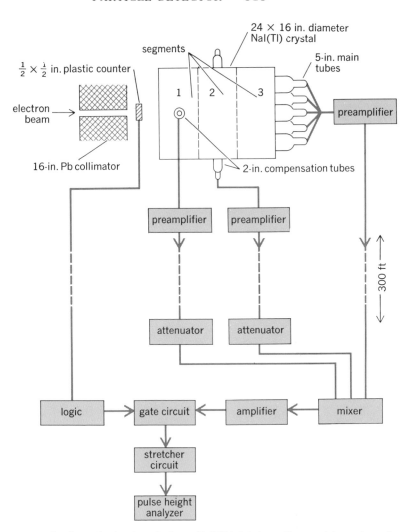

Fig. 1. A schematic drawing of a large NaI(Tl) total absorption crystal spectrometer and the associated experimental apparatus used for test purposes.

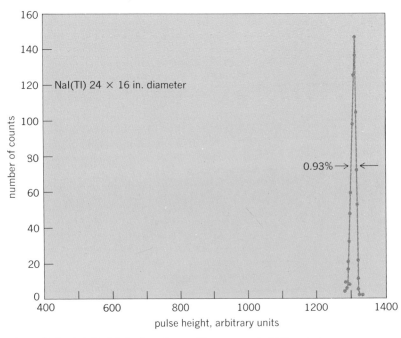

Fig. 2. Pulse-height distribution observed in response to 15-Gev electrons.

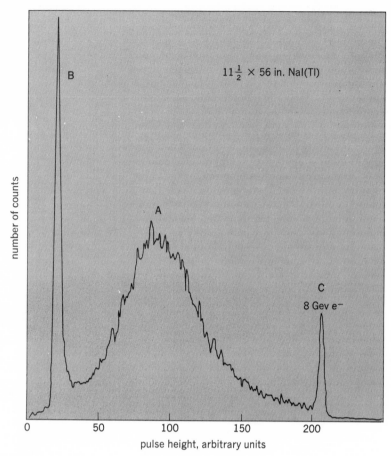

Fig. 3. The pulse-height distribution observed when 8-Gev pions are incident on a relatively small total absorption detector. A is the pion peak, B is produced by muon contamination in the beam, and C is a calibration peak produced by total absorption of 8-Gev electrons.

needed to overcome the process of nuclear binding and in the production of extremely penetrating secondary particles, such as the muon or neutrino, but these are not expected to be important effects.

At the time of writing, crystals of NaI(Tl) have not been assembled in sufficiently large volumes to check the performance of large total absorption detectors for strongly interacting particles. This is not because the component crystals of a suitable size cannot be grown. In fact, the necessary crystals can be grown with no great extension of established crystal-growing methods. Rather, the reason is that large crystals of NaI(Tl), or CsI(Na), are still quite expensive and the accumulation of a sufficient crystal inventory must proceed with deliberate speed.

For this reason, the largest detector which has been tested is a modest 56 in. in length and a quite inadequate $11\frac{1}{2}$ in. in average diameter. This detector was made up of 11 separate NaI(Tl) crystals, each cylindrical in shape, mounted coaxially to simulate one large crystal. For testing, this detector was placed in a beam of high-energy negatively charged pions (also strongly interacting particles) at the Stanford Linear Accelerator Center and its response to pions of precisely defined energy investigated.

Figure 3 shows the response, at peak A, observed at 8 Gev, which is typical of the range 4 – 16 Gev. The sharp peak C at the right-hand side is a calibration peak produced by the total absorption of 8-Gev electrons, and the much wider peak B to the left is produced mostly by a contamination of muons in the beam which pass completely through the crystal, losing energy only by Bethe-Bloch collision-loss processes. It is apparent from this figure that on average only about 50% of the incident energy is deposited in the detector and that the observed fluctuations correspond to an energy resolution of about 65% FWHM. This result agrees very well with the resolution expected for this detector, as calculated by Ranft.

It is quite obvious that serious losses of energy are occurring from the detector volume. Primarily the leakage is in the radial direction, transported mostly by secondary neutrons produced in the detector. Despite this leakage it is important to note that some of the pulses observed in even this small detector represent nearly the full energy of the incident pions. Such pulses confirm that it is possible in principle to contain the whole energy of the incident particle.

Even in the absence of suitably large crystal absorbers it is possible to establish with confidence the absorber dimensions necessary to contain an arbitrarily large fraction of the incident energy. This has been done at Stanford by using a simulation technique in which the average development of a very large number of high-energy cascades in a large block of inert absorber is studied.

The apparatus consists of an assembly of large tin plates whose overall dimensions exceed those of the largest detector that it is required to simulate. The so-called probe plate is pierced by a cylindrical cavity into which a small NaI(Tl) detector can be lowered. With this probe the radial development of the secondary cascade can be measured at any depth in the assembly, and by changing the

different from that of the γ-ray and is such that a considerably larger volume of material is needed to achieve total absorption. At the present time, however, there are no alternative methods of energy measurement from which to choose. Neither the neutron nor the K⁰-meson can be conveniently converted into a pair of charged particles whose energies can be determined with a magnetic spectrometer. On the other hand, all the desirable features of the total absorption method are available if only the volume of the absorber can be made sufficiently large. This has not yet been done. Nevertheless, the expected results can be calculated with a reasonable degree of confidence on the basis of what is generally known about the strong interactions.

This calculation has been done for protons in the range 8 – 100 Gev by J. Ranft, who predicts, for example, that a cylinder of NaI(Tl) 100 in. in length and 30 in. in diameter should be able to define the energy of a 100-Gev neutron to a precision of ~ 10% FWHM. If the diameter of the detector is increased to 60 in., the accuracy of the measurement is expected to improve to 5% FWHM. The available resolution is determined simply by the volume, or shape, of the detector, and no major limitations to the precision of the measurement are expected to arise as the absorber volume is increased. Certainly, limitations are in principle imposed by fluctuations in the amount of energy

position of the probe plate a complete three-dimensional picture of the average cascade development can be obtained. A tin absorber was chosen because of the proximity of tin to cesium and iodine, the usual crystal constituents, in the periodic table.

The experiment was performed with 8-Gev pions, and the results of the measurements compare very well with the expected results calculated by Ranft. Although the measurements do not reveal the fluctuations, and consequently the resolution, which will occur for each absorber size, corresponding to a particular average containment, the agreement with the theoretical result does give added confidence in the resolution which can be predicted theoretically. The calculations suggest that an absorber capable of containing a suitably large fraction of the incident energy will also exhibit an acceptably small degree of fluctuations and therefore will possess an acceptably fine degree of resolution for the measurement of the incident particle energy.

Although at this time it can be concluded that total absorption crystal detectors make very successful detectors for γ-rays at Gev energies, it still remains to establish this method in the field of high-energy neutron and K^0 spectroscopy. It is a very appealing idea, but much work remains to be done. It is clear, however, that in both areas of application the total absorption detector is endowed with many desirable qualities and can be reasonably expected to have a very promising future. In fact, the total absorption technique may prove to be the only practical approach to the precise measurement of the energies of both charged and uncharged particles at superhigh energies.

For background information *see* GAMMA RAYS; PARTICLE DETECTOR; SCINTILLATION COUNTER in the McGraw-Hill Encyclopedia of Science and Technology.

[ROBERT HOFSTADTER; E. B. HUGHES]

Bibliography: R. Hofstadter, *Nucl. Instrum. Methods*, 63:136, 1968; R. Hofstadter et al., *Nature*, 221:228, 1969; E. B. Hughes et al., *Nucl. Instrum. Methods*, 75:130, 1969.

Perception

Attention is neither a simple nor a unified field of study. One sense of attention concerns the degree of alertness which can be maintained. This differs from the sense of attention related to the ability to select some signals while ignoring others. Yet another sense concerns the extent to which conscious attention can be divided between mental operations. Recent experimental analysis has been attempting to separate the components of attention concerned with alertness, selection, and consciousness.

Alertness. This component of attention is known to vary gradually with the time spent performing a repetitive task and over the course of the day. Studies show performance to be quite low at 8 A.M. and to improve during the day in a close relationship to body temperature. Both active and passive repetitive tasks show a decline in efficiency with the length of time they are performed; the decline is manifested in missed signals or delayed responses. Recently the development of alertness has been studied on a microscale (see illustration).

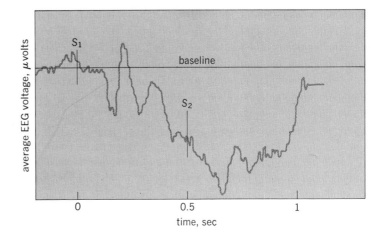

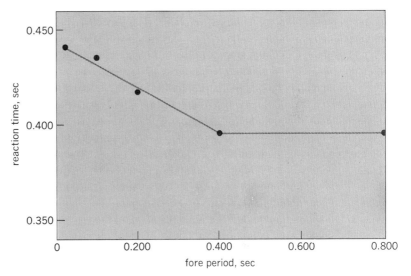

The lower panel shows the development of alertness as measured by improvement in reaction time after a warning. The upper panel shows the negative shift in direct-current potential recorded from the scalp during a 0.5-sec warning interval (S_1 to S_2).

In these experiments, subjects are provided with a warning signal which informs them that some event to which they are to respond will occur after a fixed interval. It takes 0.3–0.5 sec for the warning signal to produce an alertness sufficient for optimal processing of the critical event. The time course is similar whether one makes an overt response to the event (such as pressing a key) or is merely required to detect it. Moreover, this time course of alertness is mirrored by a slow negative shift in the direct-current electrical potential recorded from scalp electrodes. The direct-current shift, first reported by Walter Grey in England, may be similar to the slow potential shifts known to accompany the transition from sleep to waking in animals. The ability to observe the development of alertness over brief time periods may help to determine the relationship of alertness to other components of attention, such as selection.

Selection. To study the selective aspect of attention, psychologists have tried to develop methods which would focus the subjects as fully as possible upon one source of sensory stimulation. In one method, subjects listen to a message presented to one ear and repeat it as rapidly as possible. There is good evidence that, in this situation, retention of information presented to the other ear is nearly zero. However, recent studies suggest that the meaning of information presented to the unattended ear is processed even though subjects seem unable to store or to respond overtly to that meaning. For example, if a synonym of the word presented to the attended ear is simultaneously presented to the unattended channel, the time required to repeat the attended word is significantly increased. Moreover, the interpretation of an ambiguous sentence by the attended ear can be biased by words presented to the unattended channel.

Consciousness. These experiments suggest that selection of information for conscious attention takes place after complex processing has been performed. In this sense one's consciousness is subject to the constraints which are imposed by previously learned ways of classifying information. This conclusion, which seems inescapable from recent studies of attention in normal subjects, is reinforced by the observations of Roger Sperry and Michael Gazzaniga of men and animals whose cerebral hemispheres have been separated by severing the commissure which normally connects them. Although the right hemisphere is able to perform habitual processing, including some reading, the subjects are unable to report this information verbally, and even deny being aware of it. Consciousness, in this sense, appears to require mechanisms related to the speech centers of the left hemisphere. *See* PSYCHOLOGY, PHYSIOLOGICAL AND EXPERIMENTAL.

The idea of a limited capacity implies that signals which require conscious processing tend to interfere with one another. Some recent work suggests that such rivalry extends not only to separate signals but also to different codes of the same signal. In these studies, subjects are shown either a picture or a description of that picture. They are instructed to match their stored representation of the first event against a new stimulus and to respond as rapidly as possible if the two are the same. If both stimuli have the same physical form (for example, if both are pictures), responses are fast and the best speeds are obtained with the briefest intervals between the stimuli. If the physical form of the stimuli changes (for example, a description of a picture and a picture), responses are slow, unless the change is predictable, in which case the speeds improve — but the time interval required to obtain optimal performance is relatively long. These results suggest that subjects can construct visual representations from descriptions just as they can abstract descriptions from the visual forms. They cannot, however, maintain both types of representations in consciousness at the same time. There is a definite rivalry between the two codes for the limited capacity available.

Perhaps these new methods for isolating various components of attention will make it possible to develop more comprehensive theories of the nature of consciousness and particularly of its functional significance.

For background information *see* BIOPOTENTIALS AND ELECTROPHYSIOLOGY; INFORMATION THEORY; PERCEPTION in the McGraw-Hill Encyclopedia of Science and Technology.

[MICHAEL I. POSNER]

Bibliography: L. Karlin, *Psychol. Bull.*, 73:122–136, 1970; N. Moray, *Listening and Attention*, 1969; J. A. Swets and A. B. Kristoferson, Attention, *Annu. Rev. Psychol.*, 21:339–366, 1970; B. Tversky, *Perception and Psychophysics*, 6:225–233, 1969.

Pesticide

An expanding use of pesticidal chemicals in agriculture, public health, forestry, warfare, and home gardens continues to be a source of controversy. For a wide variety of reasons many attempts have been made, particularly in North America, to restrict the use of such chemicals. Considerable scientific evidence has recently been accumulated that documents the deleterious effects on the environment of several of the chlorinated hydrocarbons. On this basis, a number of environmental scientists have attempted to eliminate the use of these compounds, especially DDT. Industry has responded with the considerable resources at its disposal, and the bureaucratic machinery of government has frequently been caught in the middle. Industry has accused the anti-DDT forces of being antitechnology, and environmental scientists have replied that the technology of 1945 is antiquated and that a new technology must be developed that will be compatible with the continued existence of life on Earth. The DDT controversy might therefore be considered a prologue to a much larger controversy of the future that will decide how increasing numbers of people, with decreasing resources, will achieve a technology that will be compatible with both human values and the long-term survival of man.

The term pesticide may be defined as any substance used to kill, or to inhibit the growth and reproduction of, the members of a species considered in a local context as pests. About 800 compounds are now used as pesticides and, with increasing technological sophistication, the list can be expected to grow. Until World War II pesticides consisted of inorganic materials containing sulfur, lead, arsenic, or copper, all of which have biocidal properties, or of organic materials extracted from plants such as pyrethrum, nicotine, and rotenone. During World War II a technological revolution began with the introduction of the synthetic organic biocides. DDT was widely used during World War II to control the vectors of diseases which until that time had been serious health hazards over much of the world. Since then, DDT has been extensively used with spectacular success to reduce the incidence of malaria in the tropics. In the public mind DDT has become almost synonymous with pesticide, yet it is important to point out that DDT is only one of an increasing number of organic biocides. They vary widely in toxicity, in specificity, in persistence, and in the production of undesirable derivatives. The factors that determine how, when, and where a particular pesticide should be used, or whether it should be used at all, are therefore necessarily different for each pesticide. During the recent controversy over pesticide usage, the industry spokesmen have consistently attempted to associate those who believe it is time to ban DDT with those who campaign against the use of all pesticides. The use of chemicals to achieve a measure of control over the environment, however, has become an integral part of technology, and there are no valid arguments to support the proposition that the use of all such chemicals should now cease.

Some take the view that pests include insects, fungi, weeds, and rodents. Many insects, however, particularly the predatory species that prey upon the insects that damage crops, are clearly not pests. Fungi are essential in the processes that convert dead plant and animal material to the primary substances that can be recycled through living systems. In many ecosystems rodents are also an important link in the recycling process. On the other hand, an increasing number of species traditionally considered desirable are becoming "pests" in local contexts. Thus chemicals are used to destroy otherwise useful vegetation along roadsides or in pine plantations, or vegetation that might provide food and shelter to undesirable human populations. The control of any component of the global environment through the use of biocidal chemicals is therefore a problem much more complex than is initially suggested by the connotation of the word pest. The highly toxic chemicals that kill many different species are clearly undesirable in situations in which control over only one or two is sought.

Like any other technological product, pesticides, once used, become waste products and some become pollutants. It can no longer be assumed that pollutants will disappear in the vastness of the sea, since the global environment has suddenly become small. The total amount of waste materials produced by man is now approximately as great as the amount of organic material cycled yearly through nature. Waste products that are readily degraded into elementary materials that can be recycled clearly do not threaten the stability of the environment. In traditional terms they might be compared with the biblical ashes that return to ashes and the dust that returns to dust.

Biodegradable compounds. Of the approximately 800 biocidal compounds used as pesticides, many obey this fundamental ecological law and are degraded to elementary materials that can be recycled. These compounds include the organophosphate insecticides and the carbamates, since no evidence now available suggests that any derivative of these compounds becomes a persistent pollutant. Malathion, diazinon, and parathion are examples of organophosphorous insecticides. Parathion is particularly toxic to many forms of life and has caused the deaths of many persons applying it to crops or who have been accidentally exposed. Other pesticides in this group, however, have been extensively used without causing human fatalities. They are readily metabolized in the mammalian body, and in the environment are degraded to elementary materials. Unwise use of these compounds may result in local fish kills or other loss of wildlife, but with careful use such losses can be minimized. Because of the chemical instability of these biocides they do not travel to areas for removed from the sites of application. Environmental damage, if any, is therefore local. The organophosphorous insecticides inhibit the enzyme cholinesterase, which normally breaks down acetylcholine following the transmission of the nerve impulse across the synapse.

The carbamates are another group of biocides that also inhibit the enzyme cholinesterase. Carbaryl, or sevin, is an example. These biocides are even less persistent than the organophosphates and pose fewer health hazards to man. Indiscriminate use, however, may cause local environmental problems. Carbaryl, for example, is particularly toxic to bees.

Other pest-control measures. Sophisticated methods of pest control are continually being developed. One technique involves the raising of a large number of insect pests in captivity, sterilizing them with radioactivity, and then releasing them into the environment, where they mate with the wild forms. Very few offspring are subsequently produced. Highly specific synthetic insect hormones are being developed. In an increasing number of pest situations, a natural predator of an insect has been introduced, or conditions are maintained that favor the propagation of the predator. The numbers of the potential pest species are thereby maintained below a critical threshold.

Chlorinated hydrocarbons. DDT belongs to a class of biocides that are chlorinated hydrocarbons. The chlorine-carbon bond is rare in nature and is comparatively stable. Few bacteria and fungi are equipped to break it. These biocides are therefore relatively persistent, much more so than are the carbamates and the organophosphates. The chlorinated hydrocarbons are also much less soluble in water but are very soluble in nonpolar media such as lipids. They are therefore readily concentrated by organisms from the environment, and accumulate in fatty tissues. Species higher in the food chains will subsequently accumulate the chlorinated hydrocarbons present in their food. Mobility is another property which makes some of the chlorinated hydrocarbons hazardous to the global environment. DDT and dieldrin become vapors and can be transported anywhere in the world to accumulate in distant ecosystems. Since the chlorinated hydrocarbon biocides are toxic to a wide spectrum of animal life, they have the capacity to cause harm to many nontarget species.

Several of the chlorinated hydrocarbons, however, do not appear to possess the combined properties of intensive use, persistence, and mobility that create the potential for environmental degradation. Lindane is more readily degraded than the others, and chlordane and toxaphene do not appear to possess the mobility of DDT and dieldrin, since they have not so far been detected in marine organisms away from contaminated estuaries.

Of all the pesticides now in use, only five have aroused the concern of environmental scientists who are attempting to consider the long-term effects upon the global environment. Aldrin, dieldrin, endrin, and heptachlor are extremely toxic, persistent, and mobile, and have the capacity to inflict harm upon nontarget species.

The fifth compound, DDT, has yet another property that renders it especially dangerous to the environment. The insecticidal compound, p,p'-DDT, is relatively persistent, but much of it is eventually degraded to another DDT compound, p,p'-DDE. This compound, DDE, is more persistent than the original DDT and appears to be the most abundant of the synthetic pollutants in the global environment. Most of the concern about DDT in the environment is actually about DDE.

Many marine biologists still find it difficult to accept the findings that DDE is now more abundant in marine birds and fish than in many terrestrial organisms. It is not the only pollutant to be so widely dispersed. Polychlorinated biphenyls, another group of chlorinated hydrocarbons that are industrial pollutants, have become similarly dispersed throughout the global ecosystem. The sea is no longer the vast entity that it was traditionally conceived to be. World production figures of many chemicals, including DDT, are now a significant fraction of the amount of organic material that is recycled each year through the world's biomass.

DDE was originally considered to be harmless because of its relative nontoxicity to organisms. Evidence is now rapidly accumulating that DDE has physiological effects that may not kill the organism but which are nevertheless deleterious and may impair the long-term survival of the species. Experiments have shown that DDE may cause several species of birds to lay eggs with abnormal amounts of calcium carbonate. Many species of raptorial and fish-eating birds have experienced low rates of reproductive success over the past 20 years, and in most cases one of the symptoms has been abnormally thin eggshells. In the most spectacular cases, such as the brown pelicans and other sea birds of the West Coast, the shells are so thin that they break during incubation. Increasing amounts of evidence link this phenomenon to DDE.

Physiological effects of DDE that may be associated with thin eggshells and other abnormalities include the inhibition of membrane enzymes that are adenosinetriphosphatases involved with ion transport, the inhibition of the enzyme carbonic anhydrase, and the induction of nonspecific liver enzymes that degrade a wide variety of lipid-soluble compounds, including steroids. All these effects of DDE have been discovered only recently, and more are likely to be discovered in the future.

In 1969 fish caught in the Pacific Ocean were seized and condemned by the U.S. Food and Drug Administration because they had been found to contain high concentrations of the DDT compounds. A number of monitoring programs for pesticides in the environment had previously been initiated in North America, but none had been designed to find the accumulation site of a pesticide pollutant or to find the rate at which it accumulates there. It is now obvious that DDE has been accumulating in the sea, but since there are no data on how fast it is accumulating, it is difficult to know what to expect in the future.

At the present time the environmental science community is convinced that DDE is contributing to the extinction of a number of species of birds. Only 4 years ago the idea that DDE might cause thin eggshells would have been dismissed by most environmental scientists as somewhat preposterous. Moreover, there is no information on what additional effects DDE or another pollutant might have, once a critical threshold level is reached.

The spokesmen for the DDT industry have maintained that DDT has no effects upon man at dosages that might be encountered. No deleterious effects have so far been found in man. A preliminary screening test indicated, however, that p,p'-DDT or a metabolic derivative might be carcinogenic. Almost no work has been done with DDE, the DDT compound that is present in food and in the environment. For reasons that are not clear to an environmental scientist, almost all of the work with DDT has been done with the insecticidal compound, p,p'-DDT, which is applied to fields. Conclusions based upon work with this compound are of doubtful validity if they are meant to include DDE also. Conclusions about the effects of DDE upon man will require many analyses for DDE and other pollutants in human tissues that can be correlated with causes of death, similar to the work that has showed the correlation between cigarette smoking and lung cancer.

Future course of action. The confiscation of the marine fish is an instance of environmental food becoming unsuitable for man because of increasing pollution. The continued release of persistent pollutants into the environment will mean that more and more food will become unavailable.

In considering the expanding populations, the increasing amounts of waste materials produced per capita, and both the known and unknown effects of persistent pollutants, the environmental scientist is forced to assume a very conservative position. Any of the persistent pollutants that are released in amounts comparable to those of the DDT compounds has the capacity to produce an irreversible effect upon the global environment. The chances, therefore, simply are not worth taking if other compounds that will do the same job are available. To preserve the global environment as a fit habitat not only for pelicans and bald eagles but for man himself, a commitment must now be made to a technology consistent with the biblical and ecological law that ashes must return to ashes. Continued use of a substance such as DDT that results in the accumulation in the environment of an undesirable waste product is a violation of this law.

For background information *see* PESTICIDE in the McGraw-Hill Encyclopedia of Science and Technology.   [ROBERT W. RISEBROUGH]

Bibliography: J. W. Gillett (ed.), *The Biological Impact of Pesticides in the Environment*, Oregon State University, August, 1969; J. J. Hickey (ed.), *Peregrine Falcon Populations: Their Biology and Decline*, 1969; M. W. Miller and G. G. Berg (eds.), *Chemical Fallout: Current Research on Persistent Pesticides*, 1969; M. Sobelman, *DDT: Selected Statements from State of Washington DDT Hearings and Other Related Papers*, 1970.

Petroleum products

The demand for petroleum products in the United States continues to grow at a rate of 3–4% per year. Jet fuels and liquefied gases show the largest growth rate, followed by gasoline (Table 1). The overall growth of 3.7% in 1970, however, was less than the 4.4% and 6.8% increases in 1969 and 1968, respectively.

Four of every five households in the United States, or 48×10^6 households, own automobiles. Of these, 16×10^6 own two or more cars, adding about 20×10^6 automobiles. Nonhousehold automobiles bring the total to 87×10^6. Trucks and

Table 1. United States demand for petroleum products in thousands of barrels per day

Product	1969	1970	Increase, %
Motor gasoline	5,503	5,715	3.8
Aviation and jet fuels	1,069	1,135	6.2
Middle distillates	2,719	2,785	2.4
Residual fuel	1,945	2,000	2.8
Liquefied gases	1,184	1,260	6.4
Other	1,595	1,640	2.8
Total	14,015	14,535	3.7

SOURCE: Independent Petroleum Association of America.

buses bring the total number of motor vehicles to 105×10^6. Table 2 shows projections for 144×10^6 motor vehicles in 1985 and a total mileage traveled in that year of 1.5×10^{12}. For the foreseeable future, these vehicles will continue to be propelled primarily by motor gasoline and diesel fuel.

Diesel fuel. Premium diesel fuels are now sold in about 35% of all truck-fueling stations. Three companies, American, Atlantic Richfield, and Union Oil, offer these premium fuels. Union Oil's product contains a detergent additive for maintaining clean injectors and preventing corrosion and rust of fuel system components. Atlantic Richfield's premium diesel fuel contains 0.5% of Lubrizol smoke-suppressing additive, a combustion catalyst. Power output can be increased about 10% for a given smoke threshold level. Skelly Oil Co. is selling at its truck stops a new smoke-reducing additive, Maxflo, which lessens carbon in diesel exhaust reduced by dehydrogenation of the fuel during combustion.

Jet fuel. Airline jet fuel demand is predicted to increase from 12×10^9 gal per year in 1970 to 19×10^9 gal in 1976. These predictions include fuel for the jumbo-jet Boeing 747s now in operation and for the British-French supersonic Concorde (about 1973) but not for the United States supersonic transport expected after 1976.

Greater smoke reduction from jet aircraft can be achieved by using Combustion Improver No. 2 additive of the Ethyl Corp. than by lowering the aromatic content of the fuel. This additive, a manganese compound, aids in oxidation of carbon produced during combustion. Military benefits in smoke-trail reduction and civilian benefits in air-pollution reduction are claimed.

Lubricating oil. The first plant to manufacture refined lubricating oils without solvent extraction comes on-stream in Oil City, Pa., in 1971. Pennzoil United incorporated the Gulf Hydrotreating Process in this 2500-bbl/day plant. This process uses a fixed-bed catalyst to simultaneously hydrocrack, isomerize, and hydrogenate lubricating oil distillates or residues directly to high yields of finished oils of high viscosity index (110–125) and good engine performance.

Table 2. Present and future number of motor vehicles and mileage traveled in the United States

	1970	1985
Motor vehicles	105×10^6	144×10^6
Miles traveled	1×10^{12}	1.5×10^{12}

Asphalt. A thin (1/8 in.) moisture barrier of asphalt inserted about 5 in. below the surface of sandy soil has decreased irrigation water needs by two-thirds and has nearly doubled crop yield of green beans in South Haven, Mich., according to cooperative experiments by American Oil Co. and Michigan State University. Similar barrier experiments at Yuma by the University of Arizona showed that potato crops grown over an asphalt barrier did not increase in yield but required only one-third as much water and only half as much fertilizer as crops grown without an asphalt moisture barrier.

Two applications of asphalt and soil to form building blocks have been reported. One product, developed by Esso Research and Engineering Co. and known as BMX, is competitive with conventional masonry products in compressive, tensile, and flexural strength and has excellent water repellency and freeze-thaw resistance. BMX building blocks can be assembled with conventional mortars, are paintable, and possess a high degree of stability. The essential feature of the BMX building blocks is a curing step which converts the asphalt into a high-strength binder. The other type of building block stabilized by asphalt is a cooperative development of Chevron Research Co. and the International Institute of Housing Technology at Fresno State College. In this instance, asphalt emulsions are used to stabilize and waterproof conventional sun-dried building bricks made of mud, which are common in the undeveloped countries of the world. Such asphalt-stabilized building blocks have been used recently for a variety of construction purposes in the southwestern United States.

Proteins from petroleum. Since only 2% of the world output of crude oil (2×10^9 tons per year) could supply enough protein (25×10^6 to 30×10^6 tons per year) for 2×10^9 people (over half of the present world population), there is considerable incentive to pursue protein synthesis from petroleum. Following considerable protein-synthesis research in Europe and Japan, interest has increased in the United States. Esso Research and Engineering Co. operates a pilot plant in Ohio, and Gulf is planning a demonstration plant at Wasco, Calif., using yeast on a liquid hydrocarbon substrate to produce a product containing 50% protein, most of the amino acids, and many vitamins.

For background information see ASPHALT AND ASPHALTITE; GASOLINE; INDUSTRIAL MICROBIOLOGY; PETROLEUM PRODUCTS in the McGraw-Hill Encyclopedia of Science and Technology

[F. F. FARLEY]

Bibliography: Chem. Week, 105:41, Aug. 2, 1969; T. Durrant et al., SAE J., 77:61, 1969; G. L. Kimmons, R. L. Ferm, and R. Matteson, Ind. Eng. Chem. Prod. Res. Dev., 8:250, 1969; D. T. Rogers and J. C. Munday, Ind. Eng. Chem. Prod. Res. Dev., 8:241, 1969.

Petroleum secondary recovery

Non-Newtonian fluids have recently been found to play an important role in the "exotic" recovery of oil. Non-Newtonian fluids are used in the form of emulsions and thickened water as mobility control fluids and for permeability variation control. Approximately 350,000,000,000 barrels of oil

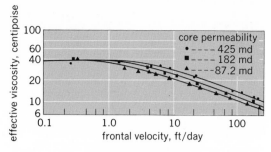

Flow data for a pseudoelastic fluid. Core permeability is absolute permeability in millidarcys (md).

have been discovered in the United States alone. With conventional recovery processes, about 128,000,000,000 barrels will eventually be recovered, leaving about 222,000,000,000 barrels of unrecoverable oil in the ground. Various secondary and tertiary processes have been developed or are in the stage of research and development to recover this "unrecoverable" oil. The newer methods include steam flooding, underground combustion, miscible displacement in the gaseous-to-liquid range, miscible displacement in the oil-water range, and mobility control.

Mobility control. A non-Newtonian fluid may be defined as a fluid whose viscosity is a function of the rate of shear it experiences as it moves through a porous reservoir. A special category is pseudoplastic fluids, which have decreasing viscosity with increasing shear rates.

When hydrocarbons are displaced, such as in water flooding, it can be shown that displacement efficiency and performance are affected by mobility control. For example, if the viscosity of the displacing fluid (water) is less than that of the displaced fluid (oil), the water tends to finger through the oil with reduced efficiency. The same is true if the bank of fluids, miscible with the oil, is injected and followed by a scavenger fluid.

Mobility control is obtained when the viscosity of the displacing fluids is greater (by a factor of 2 to 3) than that of the displaced fluid. If the viscosity is a function of shear rate in the displacing fluid but not in the displaced fluid, a non-Newtonian fluid could result, more or less, in mobility control depending on the rate of displacement.

The illustration shows a typical behavior of the viscosity-to-velocity (and consequently shear rate) relationship of a typical pseudoelastic fluid. Because of varying pore sizes with accompanying varying shear rates, averaged relationships have to be used in flow through porous media. During the range of non-Newtonian behavior the viscosity may be expressed as the equation shown below,

$$\mu' = F(V)^M$$

where μ' is the apparent viscosity, F is a constant for power law relationships, V is the frontal velocity, and M is the slope on the viscosity-velocity curve.

Viscosity control of the mobility-control fluids is accomplished by emulsification or by the use of thickened water. In emulsion and micellar solutions the viscosity is a function mainly of the viscosity of the external phase but also of the proportions of components used in the makeup of the fluids, such as the surfactants.

Polymer use. The commercially most frequently used fluids are thickened waters, which are usually prepared from a solution of high-molecular-weight polymers in water. The viscosity of the polymer solutions is controlled primarily by polymer concentration, polymer molecular weight, and water salinity.

Polymers have to be mixed carefully with the water so as not to break the molecular structure. Breakup of molecular strings results in reduced viscosity and displacement efficiency. Polymers have a tendency to plug pore channels, which may result in a loss in injection efficiency. However, sometimes this has been found to be beneficial in reducing the permeability in thief channels and fingers; hence the application of this type of fluid for permeability control.

Marked differences in the effectiveness of different water-soluble polymers can be found. Apparently partially hydrolyzed polyacrylamides are best. These polymers alter the flow of water without altering the flow of oil.

Polymers retained within the pore structure increase the effectiveness of polymer floods by causing dilatant flow; the viscosity of the solution increases with increasing rates of flow. Dilatant flow behavior is not exhibited in straight capillary tubes. Since polymer solutions do not exhibit dilatancy in bulk flow, it has been suggested that the dilatancy exhibited only in porous media be designated as pseudodilatancy. This phenomenon may occur because the bound polymer molecules in the porous media are "uncoiled" to a greater extent under increased velocity gradients, resulting in apparent increase in viscosity. This theory may support the use of these fluids for permeability control.

Adsorption of polymers on rocks has been demonstrated.

Field results. The application of both categories of non-Newtonian fluids, namely, emulsions and heavy waters, is in the infancy stage of development. Laboratory results look encouraging; however, field tests are inconclusive. The non-Newtonian characteristic is not used to enhance recovery; it is merely an effect which accompanies the use of certain miscible slugs and mobility-control agents.

For background information *see* FLUID-FLOW PRINCIPLES; FLUID-FLOW PROPERTIES; PETROLEUM RESERVOIR ENGINEERING; PETROLEUM SECONDARY RECOVERY in the McGraw-Hill Encyclopedia of Science and Technology.

[HENDRIK K. VAN POOLLEN]

Bibliography: W. H. Barlow, A new oil recovery process, *Proceedings of Southwestern Legal Foundation*, vol. 6: *Exploration and Economics of the Petroleum Industry*, 1968; E. J. Burcik, Pseudo dilatant flow of polyacrylamide solutions on porous media, *Prod. Mon.*, 31(3):27, March, 1967; R. H. Christopher and S. Middleman, Power law flow through a packed tube, *I&EC Fundam.*, 4: 422, November, 1965; J. A. Davis, Jr., et al., Oil recovery using micellar solutions, *API Drilling and Production Practices*, 1968; W. B. Gogarty, Rheological properties of pseudo-plastic fluids in porous media, *Soc. Petrol. Eng. J.*, pp. 149–159, June, 1967; R. M. McKinley et al., Non-Newtonian flow in porous media, *AICE J.*, 12:17, January, 1966; N. Mungan, Know your facts on polymer floods, *Can. Petrol.*, 8(6):25–27, June, 1967; Polymer floods attacking recovery gap, *Oil Gas J.*, pp. 64–67, Jan. 23, 1967; H. K. van Poollen and J. R. Jargon, Steady-state and unsteady-state flow of non-Newtonian fluids through porous media, *Soc. Petrol. Eng. J.*, p. 80, March, 1969; D. J. Pye, Improved secondary recovery by control of water mobility, *J. Petrol. Tech.*, p. 911, August, 1964; B. Sloat, How to flood a tight sand, *J. Petrol. Tech.*, p. 1119, October, 1968; B. Sloat, Polymer treatment boosts production on four floods, *World Oil*, 1969.

Pheromones

Understanding of the behavior of insects has progressed rapidly during the last decade with the isolation of some of the chemical messages that insects employ and with the greater insight of investigators into chemical communication. This sudden acceleration, however, was founded on half a century of careful observations of insects, from which the importance of chemical cues was deduced. Today scientists are in the position to make a similar jump in understanding the behavior of marine invertebrates. There exists a great body of knowledge implicating the importance of chemical communication in the behavior of these animals, and a number of scientists are reporting the initial isolation of some of the chemical messages.

These chemical messages have been termed pheromones, and in insects they have a wide range of functions, from sex attractants and trail markers to alarm pheromones and those which participate in maintaining the social structure of a colony. In the marine environment pheromones have an equally diverse range of functions, and there is probably a direct parallel to all the features of chemical communication found in the insect world except the control of caste structure in the social insect. However, the control of the sex of individuals in a chain of slipper shells (*Crepidula plana*) by a substance diffusing through the water may be analogous.

Epidemic spawning. A widespread phenomenon among sessile marine invertebrates, epidemic spawning has no parallel in the terrestrial environment. Epidemic spawning is the term used to describe the simultaneous release of eggs and sperm by the entire local population of a given species. This phenomenon is of considerable survival value, because it greatly increases the chance for fertilization of the eggs. It has been observed to occur in beds of oysters, mussels, and clams and in populations of abalone, starfish, and sea urchins. Whenever investigated it has been shown to be chemically mediated. In oysters the spawning of the females is triggered by the presence of sperm in the water, with a lag of 6–38 min. The active principle in the sperm is heat-labile and nondialyzable. Reciprocally, the eggs induce instant release of sperm, and the active substance is heat-stable and dialyzable.

Spawning in starfish is controlled by a neurohormone released by the radial nerves. This hormone is a peptide and initiates the enzymatic dissolution of a jelly that cements the eggs in the ovary but does not function as a pheromone. Haruo Kanatani observed the release, during the enzymatic dissolution, of a small molecule which stimulated the final step of meiosis (the reduction in the number

Fig. 1. A male lined shore crab (*Pachygrapsus crassipes*) displaying the typical premating stance with elevated body, extended chelae, and horizontally extended posterior legs.

of chromosomes in the egg). This meiosis-stimulating factor was isolated and identified as a rare purine, 1-methyladenine. Synthetic 1-methyladenine has been shown to induce the release of eggs when injected in sea-water solution. The pheromone activity of 1-methyladenine has apparently not been tested yet. However, this purine would be released during spawning and is of the molecular size and specificity that would be predicted for a pheromone.

Though many efforts to demonstrate a chemotactic response by sperm to eggs have yielded negative results, Richard L. Miller conclusively demonstrated such a response by sperm of the thecate hydroids *Campanularia flexuosa* and *C. calceolifera* to a substance issuing from the aperture of the female gonangium. The response is species-specific. The active principle from *C. calceolifera* is a heat-stable dialyzable polar substance and appears to be a single molecular species. Observations by Japanese workers suggest the activity of a similar substance from the eggs on the sperm of the jellyfish *Spirocodon saltatrix*.

Algae. The male gametes of green and brown algae orient to a substance released by the female gametes. The active substance produced by the female gametes of the brown alga *Ectocarpus siliculosus* is a volatile nonpolar compound. Dieter Müller purified this substance on a gas-liquid chromatograph, but the compound has not been chemically characterized. He found that many small hydrocarbon molecules, as well as some alcohols and aldehydes, will mimic the effect of the natural attractant. Similar observations with other algae indicate that this response is not species-specific.

Crabs. A century ago Louis Agassiz noted that male crabs were attracted to premolt female crabs and exhibited a characteristic behavior. In 1966 Edward P. Ryan implicated pheromones in this response by showing that water from a tank con-

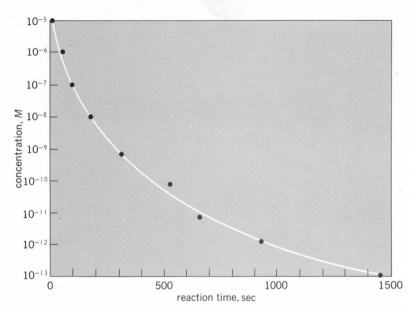

Fig. 2. A determination of the time lapse following the introduction of a solution of crustecdysone into an observation chamber before the male crabs assumed their premating stance. Each point represents the average of six determinations at each concentration.

suspicion that some marine pheromones are only slightly water-soluble and are primarily lipophylic in nature (that is, oillike, or easily dissolved in oils). These properties would be consistent with the efficient perception of pheromones in extremely dilute solutions. It may well be that the most drastic consequence of pollution by petroleum products is the effective masking of the chemoreceptors of marine organisms. Although the effect would not be immediately lethal, it could block reproduction of susceptible organisms at concentrations $10^6 - 10^{12}$ below those exhibiting "lethal" consequences.

For background information *see* ANIMAL COMMUNICATION; CHEMORECEPTION; PHEROMONES in the McGraw-Hill Encyclopedia of Science and Technology.   [JAMES S. KITTREDGE]

Bibliography: H. Kanatani, *Exp. Cell Res.*, 57: 333–337, 1969; R. L. Miller, *J. Exp. Zool.*, 162:23 –44, 1966; D. G. Müller, *Planta*, 81:160–168, 1968; E. P. Ryan, *Science*, 151:340–341, 1966.

Photosynthesis

Recent studies have confirmed that many plants have another carbon-fixation cycle superimposed upon the Calvin cycle. The primary products of photosynthetic CO_2 fixation in such plants are C_4-dicarboxylic acids such as oxaloacetic, malic, and aspartic. The effective enzyme in the carboxylation reaction is phosphoenolpyruvate carboxylase. Plants fixing CO_2 in this manner are called C-4 plants. The cycle that occurs in these plants is in contrast to the Calvin cycle, in which the primary carboxylation product is 3-phosphoglyceric acid (PGA), and ribulose 1,5-diphosphate (RUDP) carboxylase is the effective enzyme. Plants with PGA as the primary product are called C-3 plants.

Sugarcane was the original C-4 plant, and it has since been determined that the panicoid grasses, which are of tropical origin, are C-4 plants, while the festucoid grasses, which are of temperate origin, are C-3 plants. This pathway is also possessed by tropical members of the Cyperaceae, and the dicotyledonous families Amaranthaceae, Chenopodiaceae, Portulacaceae, Euphorbiaceae, and Zygophyllaceae contain species possessing it. It is reasonable to assume that this pathway will be revealed in other families. The taxa possessing the C-4 pathway are tropical in origin, and they tend to be nonsucculent xerophytes.

Physiological features of C-4 plants. The C-4 plants are particularly interesting because they possess a variety of morphological and physiological features which set them apart from C-3 plants. They have very high rates of photosynthesis and are extremely efficient at dry-weight production. Their CO_2 compensation points are very low, and they do not exhibit the light-induced stimulation of CO_2 production which is termed photorespiration.

Morphological features of C-4 plants. All C-4 plants have a readily identifiable leaf anatomy in which the vascular bundles are surrounded by a layer of large bundle-sheath cells containing chloroplasts. The bundle sheath is surrounded by a layer of mesophyll cells. This centric type of leaf anatomy is in strong contrast to the dorsiventral type common to C-3 plants.

Chloroplasts. Dimorphic chloroplasts are found in the bundle-sheath and mesophyll cells, and this

taining an active premolt female crab, when drained into a tank containing male crabs, elicited a behavioral response in the male crabs. By capping the excretory pores of the female crabs and eliminating the response, he determined that the active substance was released in the urine.

J. S. Kittredge, while studying the lined shore crab (*Pachygrapsus crassipes*), found that the male crabs displayed a typical stance in the presence of premolt females (Fig. 1). Using this behavior as a bioassay, the active principle was found to have the chemical characteristics of a polar steroid. The tissues of the female are undergoing changes preparatory for the nubial molt at the time that she is attractive. The chemical observations suggested that the polar steroid molting hormone might be serving a dual function as a sex pheromone. Indeed, dilute solutions of the molting hormone, crustecdysone, excited the male crabs into their premating stance. Under standard conditions the threshold for response was 10^{-13} M, or 0.04 billionths of a gram per liter of sea water (Fig. 2).

This observation of a dual role for the molting hormone sheds light on the conceptual problems of the evolution of pheromones. The improbability of the simultaneous occurrence of the ability to produce a chemical messenger and the appropriate receptor site for sensing its presence has been a puzzle. One can assume that the ancestors of the crabs, having evolved the capability to produce a molting hormone and the receptor sites for this hormone on the target organs, underwent a rather minor evolutionary change resulting in the externalization of the receptor site on the males and the release of the hormone in the urine by the females. The obvious reproductive value, that is, the increased ability of the males to detect nubial females, provided the evolutionary pressure to consolidate this genetic change.

Water pollution. An important by-product of this study of marine pheromones regards pollution of the environment. Initial studies have led to the

dimorphism is the result of differences either in size or in internal membrane morphology in the chloroplasts in these two cell layers. Bundle-sheath-cell chloroplasts also show a greater amount of starch accumulation. Extreme structural dimorphism is illustrated in the chloroplasts of sugarcane (Fig. 1) and related grasses and in the dicotyledon genera *Froelichia* and *Euphorbia*. The stacks of lamellae known as grana are not present in bundle-sheath-cell chloroplasts, while the mesophyll-cell chloroplasts have grana and possess a structure associated with chloroplasts of C-3 plants. Size dimorphism is illustrated by *Amaranthus edulis* (Fig. 2), whose bundle-sheath-cell chloroplasts are much larger than mesophyll-cell chloroplasts but possess a similar structure. Grana are also found in the bundle-sheath-cell chloroplasts of many tropical grasses. The end result of both structural and size dimorphism is an enrichment of stroma in bundle-sheath-cell chloroplasts relative to those in the mesophyll cells. Since the Calvin cycle and starch synthesis enzymes are found in the stroma, bundle-sheath-cell chloroplasts are really functional amyloplasts. Structural dimorphism seems to be the advanced condition, because bundle-sheath-cell chloroplasts of sugarcane possess grana when immature and lose them when mature.

The chloroplasts of all C-4 plants possess a membrane system of anastomosing tubules in the peripheral stroma. This membrane system is contiguous with the inner plastid membrane and is called the peripheral reticulum (Fig. 3). It is the only unique structural feature of chloroplasts of C-4 plants, and it may play a critical role in the photosynthetic pathways of these plants.

Function of chloroplast. The current model for the function of dimorphic chloroplasts states that most of the initial CO_2 fixation takes place in the mesophyll-cell chloroplasts. A primary product, such as malate, is rapidly transported to the bundle-sheath-cell chloroplasts, where transcarboxylation or decarboxylation takes place and PGA is generated. These chloroplasts have the bulk of the RUDP carboxylase and are the most active sites for the Calvin cycle. Since mesophyll-cell chloroplasts can synthesize starch, the greater occurrence of starch in bundle-sheath-cell chloroplasts is probably a result of their location on the carbon-sink gradient. The spatial specialization of the C-4 plant chloroplasts probably accounts for their resistance to photosaturation, which in turn contributes to the high rates of photosynthesis in these plants. The failure of C-4 plants to exhibit photorespiration is undoubtedly due to the reassimilation of CO_2 as a result of the centric leaf anatomy. The high dry-weight yields of C-4 plants can be largely attributed to their preventing the loss of CO_2 produced by photorespiration.

Carbon-fixation pathway in C-4 plants. The carbon-fixation pathway in C-4 plants is essentially the same as in succulent plants, such as the Crassulaceae and Cactaceae, which exhibit the well-known crassulacean acid metabolism. The chloroplasts of the succulent plants fix CO_2 at night into malate and refix CO_2, resulting from the decarboxylation of malate, during the day into PGA; therefore they exhibit a temporal specialization of function very similar to the spatial specialization of function exhibited by the dimorphic plastids of

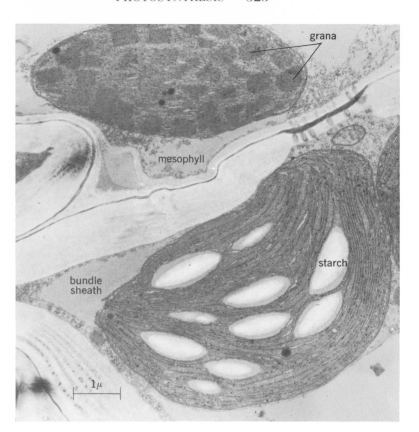

Fig. 1. Electron micrograph of sugarcane chloroplasts. The mesophyll-cell chloroplast has grana; the bundle-sheath-cell chloroplast has starch accumulation but no grana.

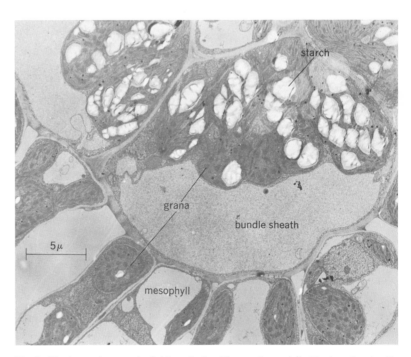

Fig. 2. Electron micrograph of chloroplasts of *Amaranthus edulis*. The bundle-sheath-cell chloroplasts contain grana and large amounts of starch; the mesophyll-cell chloroplasts contain grana and very little starch.

plants with the light-driven C-4 pathway. Both types of C-4 plants represent adaptations to arid environments, where maximum CO_2 fixation capacity must be correlated with minimum water loss.

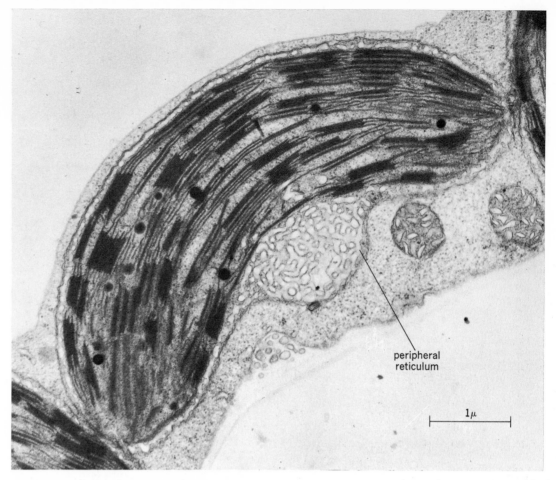

Fig. 3. Electron micrograph of *Portulaca oleracea* showing the peripheral reticulum.

For background information *see* LEAF (BOTANY); PHOTOSYNTHESIS in the McGraw-Hill Encyclopedia of Science and Technology.

[W. M. LAETSCH]

Bibliography: O. Björkman and E. Gauhl, Carboxydismutase activity in plants with and without β-carboxylation photosynthesis, *Planta*, 88(14): 197–203, 1969; M. D. Hatch and C. R. Slack, Photosynthesis by sugar-cane leaves: A new carboxylation reaction and the pathway of sugar formation, *Biochem. J.*, 101:103–111, 1966; W. M. Laetsch, Chloroplast specialization in dicotyledons possessing the C$_4$-dicarboxylic acid pathway of photosynthetic CO$_2$ fixation, *Amer. J. Bot.*, 55(8): 875–883, 1968; W. M. Laetsch, Relationship between chloroplast structure and photosynthetic carbon-fixation pathways, *Sci. Progr. Oxf.*, 57: 323–351, 1969; W. M. Laetsch and I. Price, Development of the dimorphic chloroplasts of sugar cane, *Amer. J. Bot.*, 56(1):77–87, 1969; C. R. Slack, M. D. Hatch, and D. J. Goodchild, Distribution of enzymes in mesophyll and parenchyma-sheath chloroplasts of maize leaves in relation to the C$_4$-dicarboxylic acid pathway of photosynthesis, *Biochem. J.*, 114:489–498, 1969.

Plant, mineral relations of

The significance of the small area at the soil-root interface to the mineral nutrition of plants has been the subject of intensive research recently. Plant-induced changes in this area have been shown to be important in modifying the uptake of nutrients, particularly of phosphorus.

Plants have a higher requirement for phosphorus during early growth than in later growth. Thus increases in the availability of fertilizer phosphorus to plants during this period are of considerable significance to food production.

Studies in 1966 by F. S. Leonce and M. H. Miller, using phosphorus-32, showed an accumulation of phosphorus in or on the root adjacent to the fertilizer source when phosphorus was applied alone. No such accumulation was observed when a small amount of ammonium sulfate, $(NH_4)_2SO_4$, was applied with the phosphorus. However, the phosphorus content of the shoot was greatly increased by the addition of ammonium sulfate.

New techniques such as the use of the soft-particle phosphorus-33 isotope and the electron microprobe analyzer have enabled a more detailed investigation of the soil-root interface area. Using these techniques, Miller's group in Canada have recently shown that phosphorus accumulates on the root surface (Fig. 1) when it is applied as monocalcium phosphate in the absence of $(NH_4)_2SO_4$. Electron microprobe scans of thin root cross sections (Fig. 2) have shown that calcium is associated with the phosphorus on the root surface probably as dicalcium phosphate dihydrate, $CaHPO_4 \cdot 2H_2O$.

Effect of ammonium and nitrate. In an attempt to find explanations for the observed root-surface accumulations, Miller, C. P. Mamaril, and G. J. Blair have shown that the pH of the soil-root interface is lowered when ammonium accompanies the phosphorus fertilizer and increased when nitrate accompanies the phosphorus. It is suggested that, where ammonium is the source of nitrogen, cation uptake by the plant exceeds anion uptake, which results in expulsion of H⁺ ions from within the plant to maintain electrical neutrality. The release of H⁺ ions lowers the root-surface pH. Where nitrate accompanies the phosphorus, there is an excess of anion uptake and a release of HCO_3^- ions from the root to maintain electrical neutrality. This results in an increase in the root-surface pH.

Effect of soil-root interface pH. Other studies, by D. Riley and S. A. Barber at Purdue, have shown marked changes in the soil-root interface pH which modify the mineral nutrition of the plants. They found accumulations of bicarbonate ions associated with nitrate uptake and stress that cause changes in the pH of the soil-root interface which can influence the uptake of ions such as phosphorus and iron and also alter the toxicity of ions such as aluminum in acid soils.

The change in root-surface pH can affect phosphorus availability to plants. The ionic species of phosphorus in solution, whether HPO_4^{--} or $H_2PO_4^-$, is dependent on solution pH with

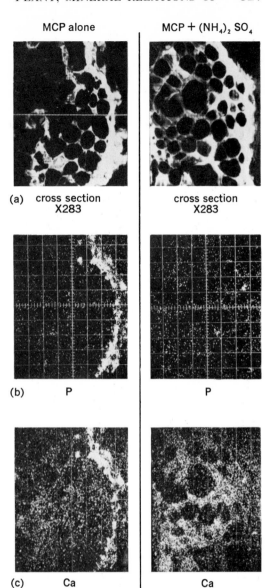

Fig. 2. Electron microprobe scans of root cross sections showing accumulation of phosphorus and calcium on root surface in absence of $(NH_4)_2SO_4$. (a) Electron backscatter photographs of section showing cell walls and outer surface of root (white portions). (b) Phosphorus distribution by Kα x-ray emission. Concentration of white dots indicates phosphorus concentration. (c) Calcium distribution by Kα x-ray emission. Concentration of white dots indicates calcium concentration.

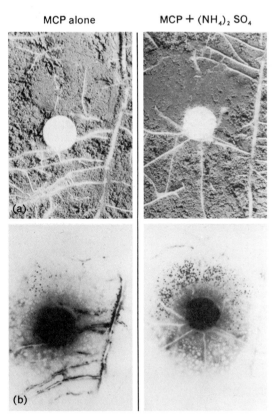

Fig. 1. Photographs and autoradiographs of fertilized zone showing phosphorus accumulation on root surface in absence of ammonium sulfate, $(NH_4)_2SO_4$. (a) Photographs showing fertilizer pellet (white spot) and roots. (b) Autoradiographs showing phosphorus accumulation (black dots). In the MCP (monocalcium phosphate) treatment the phosphorus accumulates on the root surface. No such accumulation is apparent in the MCP + $(NH_4)_2SO_4$ treatment.

$H_2PO_4^-$, the form most readily taken up by plants, predominating under acid conditions. By reducing the root-surface pH, it would be expected that increased uptake of phosphorus would result because of the predominance of the $H_2PO_4^-$ ion. A further reason for the enhanced uptake in the presence of ammonium would be that, with the lower pH, there would be a lower concentration of the HPO_4^{--} ions and thus a lower tendency for compounds such as $CaHPO_4 \cdot 2H_2O$ to precipitate and reduce the availability of the phosphorus.

Further evidence that the change in root-surface pH is responsible for this precipitation comes from solution culture studies by Blair, Miller, and W. A. Mitchell in which changes in root-surface pH were minimized by using fast solution flow. In these studies no root-surface accumulation of phosphorus was found.

Pretreatment of root surface with nitrogen. In addition to the influence of nitrogen on phosphorus uptake through changes in the root-surface pH, evidence is accumulating for a further mechanism within the plant. Investigations by C. V. Cole and coworkers at Colorado and S. J. Thien and W. W. McFee at Purdue have shown that pretreatment of plants with nitrogen, either nitrate or ammonium, results in greater phosphorus uptake. They found no increased uptake of phosphorus when the nitrogen and phosphorus were added together in the uptake solution. They suggest that nitrate or ammonium through themselves are not responsible for this effect but, rather, that a metabolite of nitrogen is connected in some way with the uptake and transport of phosphorus across the root into the xylem.

For background information *see* PLANT, MINERAL NUTRITON OF in the McGraw-Hill Encyclopedia of Science and Technology.

[M. H. MILLER; G. J. BLAIR]

Bibliography: G. J. Blair, M. H. Miller, and W. A. Mitchell, *Agron. J.*, vol. 62, no. 4, 1970; M. H. Miller, C. P. Mamaril, and G. J. Blair, *Argon. J.*, vol. 62, no. 4, 1970; D. Riley and S. A. Barber, *Soil Sci. Soc. Amer. Proc.*, 33:905–908, 1969; S. J. Thien and W. W. McFee, *Soil Sci. Soc. Amer. Proc.*, 34:87–90, 1970.

Plant, water relations of

Under field conditions and similar climates, the daily amount of water evaporated by agricultural plants is nearly the same for most agricultural crops at full cover when soil moisture is not limiting. The amount of energy available to evaporate water is the major controlling factor when actively growing, well-watered agricultural plants fully cover the soil. As a result, mean maximum evaporation rates generally vary between 5 and 6 mm per day in humid moderate climates and between 7.5 and 8.5 mm per day in arid irrigated areas. Daily rates of 10–14 mm have been measured on clear, hot, windy days. The total seasonal amount of water evaporated from an annual crop ranges from 50 to 80% of the amount evaporated from a crop such as alfalfa in a given climate. Small plants that do not shade the soil after planting and mature plants near harvest account for most of this difference. Total seasonal evaporation also depends on the length of the growing season.

When a plant is unable to absorb water from the soil as fast as it can be evaporated from the leaves, the plants may wilt and growth is affected. When this occurs, the plant is said to be stressed. Stress may have detrimental effects before visual symptoms develop. Some crops are more sensitive to stress than are others, and every crop is more sensitive at certain stages of growth. Stress during the period of flowering, for example, can greatly reduce the yield of the marketable portion of a plant that depends on flowering. Studies of these factors and their relations to microclimate and soil moisture are being conducted to develop better water management and cultural practices. The results are essential in understanding plant-water relationships and in the future will play a major role in

plant breeding for higher-producing crops.

Studies are also being conducted on methods of reducing the amount of water lost by evaporation. Evaporation from the soil surface in row crops can be reduced by using plastic films or similar mulching material. Reduction of evaporation through the leaves has been studied, but no practical control mechanism that does not also adversely affect growth has been developed. Studies of practical methods to reduce the heat energy absorbed by the crop by changing the reflectance of the surface are also under way. Large plastic greenhouses are also being explored as a means of controlling the environment and reducing the evaporation loss.

Energy for water consumption. Evaporation from a field of well-watered, actively growing agricultural plants is controlled principally by the heat energy absorbed from the Sun. Additional heat energy is absorbed from the air by irrigated fields in hot, dry climates. Since agricultural plants generally retain less than 1% of the water absorbed from the soil, water consumption by these plants is approximately equal to the water lost by evaporation.

Measurement of water consumption. Water consumption by crops is being measured throughout the world by using tanks of soil (lysimeters) in which plants are grown. The tanks usually are installed within large fields of the same crop, and similar soil-moisture conditions are maintained to minimize unnatural edge environments. The change in tank weight is usually recorded by modern electronic equipment providing daily and even hourly values of evaporation. Another method is to measure the change in soil moisture over a period of 5–10 days or more. This method has been used for many years. Soil samples are removed and their moisture content determined by measuring the difference in weight before and after drying. Within the past decade, a neutron soil-moisture probe has been used for direct measurements of soil moisture in the field. Other methods of direct and indirect measurements are being refined as new theories and instrumentation are developed. These methods require complex instruments and skilled technicians to make the measurements.

Evaporation rates are related to meteorological conditions, soil moisture levels, growth stage, and leaf area per unit land area. Once these relationships are established, these parameters can be used to estimate evaporation for management purposes. They also can be used directly to determine the efficiency of water use (marketable product per unit of water used) for various management practices, new plant varieties, hybrids, and nutrient levels maintained in the soil.

Estimates of evaporation. C. B. Tanner recently reviewed the numerous general methods of measuring and estimating evaporation from plants and soils. Generally, estimates based on energy-balance approximations are the most conservative and reliable. Modern farm management techniques have increased the need for these estimates. M. E. Jensen used modern computer facilities to combine estimates of daily evaporation from agricultural plants to provide managers of irrigated farms with estimates of the current soil-moisture status of each field as well as the predicted date of the next irrigation. This technique is rapidly being adopted and is expected to be in wide use within a few years. E. A. Fitzpatrick and H. A. Nix developed a similar mathematical model for estimating evaporation to simulate the soil-water regime in alternate fallow-crop systems. An estimate of evaporation from soil and plant surfaces is a basic component of the numerous mathematical models being developed to simulate the other aspects of the crop-soil system. Estimated evaporation is also a component of hydrological models of watersheds and river basins.

Plant stress. The degree of plant stress caused by the difficulty the plant has in absorbing water from soil as rapidly as it is evaporated from the leaves influences plant growth and evaporation. When a crop is severely stressed, evaporation decreases because the plants lose turgor and the stomates close. Numerous studies are being conducted to determine the optimum environmental conditions for maximum plant growth and minimum plant stress. The evaporative demand in agricultural fields can be reduced by using modern, automatic sprinkler systems to reduce temperatures and increase humidity. Such cooling may slightly increase the total amount of water evaporated from the field. Plastic field-size greenhouses enable even greater control of the environment.

Some agricultural plants such as alfalfa must be stressed to a certain degree to achieve optimum seed production. A higher plant stress is usually maintained by not irrigating until the soil becomes very dry and then applying only small amounts of water. As a result, alfalfa grown for seed requires less water than alfalfa grown for hay.

The degree of plant stress varies diurnally. Generally stress is lowest during early morning hours and greatest in midafternoon. Techniques for measuring and describing plant stress under field conditions are still being studied. The chemical potential of water may be directly related to the stress to which a plant is subjected. J. W. Cary and H. D. Fisher recently developed a simple, economical freezing-point device that permits rapid measurement of the chemical potential of plant water in the field. Other studies are determining the effects of plant stress on the various biochemical reactions in the plant that are essential to plant growth.

Plant stress generally can be minimized by maintaining high soil-moisture content. However, under field conditions, keeping the soil very wet may produce counteracting adverse effects on plant growth because of lower soil temperatures in the spring or low oxygen levels in the soil. Also, more water may be required to maintain high soil-moisture levels because of increased drainage from a wet soil. The total evaporation from irrigated fields is usually not increased significantly by maintaining very high soil moisture after the plants fully shade the soil.

For background information *see* PLANT, WATER RELATIONS OF in the McGraw-Hill Encyclopedia of Science and Technology.

[MARVIN E. JENSEN]

Bibliography: J. W. Cary and H. D. Fisher, *Agron. J.*, 61(2):302–305, 1969; E. A. Fitzpatrick and H. A. Nix, *Agr. Meteorol.*, 6(5):303–319, 1969; M. E. Jensen, *J. Soil Water Conserv.*, 24(5):193–195, 1969; T. T. Kozlowski (ed.), *Water Deficits and Plant Growth*, vols. 1 and 2, 1968.

Plant anatomy

The use of the scanning electron microscope (SEM) by botanists has been rather limited to date, and its potential is only just being realized. Most botanical studies using the SEM have revealed that the value of the instrument lies not so much in providing information otherwise unobtainable but in providing information with ease and in a form readily understood; that is, the specimen preparation is simple and rapid, and the image obtained is of high resolution and has a distinct three-dimensional quality.

In scanning electron microscopy, which differs considerably from more conventional transmission electron microscopy, a primary beam of electrons is finely focused and made to scan repeatedly in a regular pattern on a specimen surface by deflection coils. Primary electron beam–specimen interaction causes, among other things, the release of lower-energy (secondary) electrons, the number of which varies as the beam sweeps across the specimen surface. The generated secondary electrons are collected by a scintillator-photomultiplier system. The resulting electric signal is then amplified and used to modulate the brightness of a cathode-ray tube which is being scanned in synchrony with the specimen. An image is formed on the face of the cathode-ray tube and it may then be photographed. Figure 1 illustrates these principles schematically.

Palynology. To date the SEM has probably affected the study of pollen structure and taxonomy more than any other area of plant science. Palynologists are continually confronted with the task of examining pollen grains which range from 5 to 250 microns (μ) in diameter. The sculpturing or wall ornamentation of a pollen grain is heavily relied upon for identification purposes. The details of these minute and often extremely complex surface patterns frequently lie beyond the resolution and

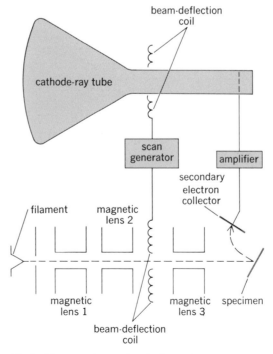

Fig. 1. Diagram of scanning electron microscope.

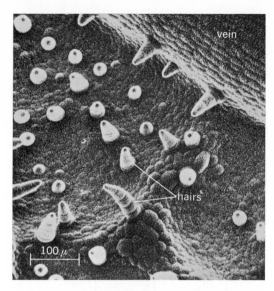

Fig. 2. Scanning electron micrograph of a portion of a young Tropaeolum leaf. Note the granular wax on the cells and the hairs and the very prominent veins.

depth of focus of the light microscope. Configurations reported by conventional light microscopy are often the result of inferences from optical effects such as diffraction.

In the hands of researchers such as Patrick Echlin and P. S. Martin the SEM has produced a more precise knowledge of the surface morphology of pollen and the beginnings of an improved taxonomic system based on this more accurate perception of pollen structure.

Paleobotany. It may seem strange that an electron microscope would be used to study fossil plants; however, it appears that the paleobotanist has needs and problems similar to those of the palynologist. He must be able to observe and accurately record minute surface structures. Thomas Taylor and his colleagues have shown that considerable surface detail remains on fossils from as far

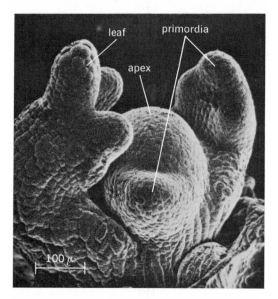

Fig. 3. Scanning electron micrograph of Tropaeolum shoot tip. Apex and a number of leaf primordia are shown.

back as the Pennsylvanian age (about 300,000,000 years ago) and that this surface detail will be useful for taxonomic purposes and perhaps will even provide some insight into fossil plant spore wall development. *See* MICROSCOPE, ELECTRON.

Leaf surfaces. The grosser features of leaf surfaces are readily observable with conventional light microscopy. Indeed, surface features such as hairs, glands, stomatal patterns, and papillae are frequently used by taxonomists as identifying characteristics. More recently physiologists have become interested in these structures and their relationship to such functions as gaseous exchange and water loss. *See* LEAF (BOTANY).

Leaf surfaces have more minute features that are also of interest. The waxy coatings of the cuticle are often beyond the resolution of the light microscope or are too transparent for visualization. Replicas of leaf surfaces have been studied and in some instances have been very useful; however, replicas suffer from a number of drawbacks, not the least of which is the requirement that the surface to be replicated be relatively smooth.

A number of investigators have found that the SEM obviates the problems associated with replica techniques and provides a high-resolution, three-dimensional image of the leaf surface. Yolande Heslop-Harrison has recently shown that wet, unfixed pieces of leaves may be directly examined in the SEM for a brief period of time. Micrographs of material such as this likely represent the native state of the leaf. Figure 2 illustrates the sort of image obtained with wet, unfixed leaf material.

Algae. T. L. Lyon has pointed out some of the advantages of using the SEM in systematic and morphological investigations as related to the algal family Desmidiaceae. The taxonomy of algae below the divisional level is based largely on the morphology of vegetative cells and spores. This being the case, it is expected that the SEM will be used considerably in future phycological studies.

Wood. The study of wood anatomy by conventional methods requires the anatomist to be able to examine a series of essentially two-dimensional sections and then construct a three-dimensional image from them. This is a difficult task that presents a risk of interpretative errors; however, a large body of knowledge of wood anatomy has been accumulated by this method.

So far the SEM has really only confirmed what is already known about wood structure. However, it has done so in a pictorially unique way that allows an immediate three-dimensional visualization of wood structure and requires little if any interpretation. This feature of immediate, uncomplicated visualization coupled with ease of specimen preparation serves to make survey studies more attractive to the wood anatomist and will result in a number of such studies being done in the near future.

Shoot tips. A great deal has been written on the form, structure, and development of shoot apexes and leaves. Usually such information is the result of observations of sectioned shoot tips with the light microscope and involves interpretative problems similar to those in wood anatomy.

The SEM is an excellent instrument for examining the shoot apex itself and the young leaf primordia. Fresh, wet material may be examined briefly. This results in a high-resolution image, such as that seen in Fig. 3, which has a great depth of focus. It is likely that the SEM will be very useful to morphogeneticists in examining the formative responses of meristematic tissues to microsurgery and to the application of growth hormones.

For background information *see* PALEOBOTANY; PALYNOLOGY; PLANT CELL in the McGraw-Hill Encyclopedia of Science and Technology.

[RICHARD H. FALK]

Bibliography: P. Echlin, *J. Roy. Microsc. Soc.*, 88:407, 1969; Y. Heslop-Harrison, *Science*, 167: 172, 1970; T. L. Lyon, *J. Phycol.*, 5:380, 1969; P. S. Martin and C. M. Drew, *J. Ariz. Acad. Sci.*, 5:147, 1969; T. N. Taylor and M. A. Millay, *Proceedings of the 2d Annual Scanning Electron Microscope Symposium*, 1969.

Plant disease

Plant diseases are often considered as static entities by the specialist and layman alike. The specialist knows, on reflection, that as new crop varieties are developed, new races of pathogens evolve and that only rarely does a disease—the interaction of a host, a pathogen, and environment—take on new characteristics and become something different. This article presents recent information on the evolution of different strains of bacterial wilt in bananas.

Evolution of a plant pathogen. In the jungles and banana plantations of Central and South America, a major evolution of a plant disease has been taking place in the 1960s that is unique in the history of recorded plant pathology. The pathogen is the wilt-producing bacterium *Pseudomonas solanacearum*, described by E. F. Smith in 1896, and the hosts include the several species and many varieties of *Musa* known as bananas and plantains.

The organism, earlier considered to be a soil inhabitant and an omnivorous root invader, primarily of solanaceous species such as tobacco, tomato, and potato, has revealed itself as a "species" comprising different races and pathotypes, each with its own specific disease-causing characteristics. Each pathotype has its own ability to survive in soil, to grow within each host's water-conducting vessels and occlude them, to emerge from its host and become available for transmission, and be (or not be) vectored by insects and thus enabled to move rapidly over vast distances and become epidemic.

Although *P. solanacearum* is an omnipresent species in the humid tropics and subtropics with the widest host-range of bacterial plant pathogens, only the evolution and spread of the banana strains are discussed below.

When banana plantations along the Pacific coast of Costa Rica became seriously affected in the mid-1950s by a disease called Moko (Fig. 1), which was found to be bacterial wilt, it was not realized at first that the pathogen differed from the usual soil-inhabiting solanaceous pathogen *P. solanacearum*.

Only later, when reclamation experiments indicated short soil survival and when observations revealed solanaceous weeds commonly wilting in banana plantations where no banana wilt occurred, was it apparent that the banana pathogen must be a different strain or race from the "solana-

Fig. 1. Banana fruit bunch infected with Moko disease, in this case *Pseudomonas solanacearum* pathotype B. Note blackened fruit pulp. (*United Fruit Co.*)

ceous" *P. solanacearum*.

Furthermore, the disease was observed only on a few thousand acres in one area of southwestern Costa Rica, and not on the thousands of well-checked acres of bananas on the Caribbean side of Costa Rica, Panama, Honduras, Guatemala, and Colombia. The then limited geographic occurrence of the disease was an important clue to the evolution of this organism.

A literature search showed absence of the disease in tropical Asia, the homeland of bananas and

Fig. 2. Aerial view of Moko disease, a bacterial wilt, in a banana field in Honduras. (*United Fruit Co.*)

morphology, host-range, and in some physiological characteristics. It was thus apparent that *P. solanacearum* had evolved into several pathotypes in different locations, and in only one was a segment of the population able to bridge the gap and attack the related host banana with enough vigor to come to the attention of man.

The banana host introduced by man from Asia to the Central American jungles presented new, unique opportunities for spread and population increase of this single bacterial type.

Rigorous control measures reduced the disease of banana (and this bacterial population) to insignificance by 1961, when a new epidemic began in bananas in Honduras, which rapidly and inexplicably spread over many square miles.

The bacterium isolated from plants affected in this epidemic appeared different in colony morphology, and it was labeled SFR. The SFR strain was quickly revealed as a highly systemic type with singular ability to ooze abundantly from banana inflorescences, where it was picked up by numerous flying insects that frequent these surfaces (Fig. 3). It was found also to invade fresh flower-dehisced surfaces on the inflorescence. These two characteristics were responsible for the quick spread of the "new" SFR strain throughout 30,000 mi^2 of Central America. All the inflorescence-bearing clumps of the most susceptible "chato" banana that were growing around homes and farmsteads throughout much of Central America became diseased, first in Honduras, then in Guatemala and El Salvador, and finally in Nicaragua. The pathogen crossed dry mountain ranges on the bodies of flies, wasps, and bees, and invaded banana plants in dry regions inhospitable for soil survival or habitation by *P. solanacearum*.

The *P. solanacearum* strains causing bacterial wilt of bananas can be ranked in increasing virulence and field disease-causing ability from D to B to SFR. Their ability to survive in soil and cause disease of the parent host *Heliconia* is ranked in the reverse order. Although it is not evident that these strains evolved directly from one another, it is evident that the banana disease they cause has changed, from one caused by a strain of limited destructiveness and ability to spread to one that is highly destructive, infectious, and transmissible.

The probable origin of the highly insect-transmissible strain SFR has been traced to dry areas of Venezuela, where it is also insect-spread. It was inadvertently brought to Honduras in banana seed pieces in the early 1950s.

The continuing evolution of *P. solanacearum* is dramatized by the discovery in 1965, described by E. R. French and Sequeira, of a disease of plantains in the Amazonian jungles of Peru. The causal strain, labeled A, has characteristics similar to those of SFR, but it differs in several respects, including host-range. This wilt disease was also found to be insect-spread to banana inflorescences and to be advancing at a rate of 15 mi per year up remote tributaries of the Amazon on clumps of banana, which are a main food staple for jungle Indians. It is believed to have come from the llanos area of Colombia.

Laboratory evidence for different strains. Although host-range and reaction remain the most reliable criteria of bacterial identity, laboratory

the entire genus *Musa*. But its earlier presence in British Guiana and Trinidad indicated either two unique evolutionary events or a link between Costa Rica and the far southeastern Caribbean. No link could be found, and efforts to find a local indigenous host for the "banana" bacterium in Costa Rica finally resulted in the discovery by Ivan W. Buddenhagen and Luis Sequeira that various species of *Heliconia*, a New World relative of the banana, were diseased in banana plantations and the adjacent jungle. Further research showed, however, that the vast majority of *Heliconia* bacteria could not cause banana wilt and that most of the few that could were unable to establish a spreading plantation disease in banana fields. Thus, from a bacterial population pool in *Heliconia* (labeled strain D), a strain had been selected in banana plants that could cause a spreading disease in banana plantations (labeled strain B) (Fig. 2).

This highly virulent banana strain was spread on pruning tools and banana seed pieces and threatened many thousands of acres of bananas. It was inadvertently taken to the Caribbean side of Panama, Costa Rica, and Honduras, where it established new foci in banana plantations. An extensive search for diseased heliconias in several countries revealed that heliconias could sometimes be found to be diseased with *P. solanacearum* at many different sites whereas bananas growing nearby were not affected.

Examination of these bacteria indicated that they were of many subtypes—different in colony

Fig. 3. Diseased bud of "chato" banana with bacterial ooze. The insect vector *Trigona* sp. spreads the disease. (*United Fruit Co.*)

the two pathotypes have been observed. The solanaceous strains appeared to induce greater host response with more bacterial entrapment by globular and fibrillar material. The host appears to affect bacterial morphology and cell-wall structure, sometimes resulting in appearance of electron-dense bodies within the bacteria.

Summary. At least three different bacterial banana strains have originated and established five separate centers of spread in very recent times, providing banana diseases all called bacterial wilt, but each with unique characteristics depending on the special characters of the originating strain. These five centers are (1) Venezuela –British Guiana–Trinidad for SFR and its parent forms, (2) a man-created secondary center for SFR in Honduras–El Salvador–Guatemala–Nicaragua, (3) a Colombian center and Peruvian jungle spread for A, (4) a western Costa Rican center for B and D, and (5) spread of B by man to Honduras, eastern Panama, and Costa Rica.

How many other potential banana bacterial wilt strains may exist in jungle areas of Central America and northern South America is unknown. Apparently, small, unrecognized parent populations of bacteria from heliconias (or other hosts) encounter the introduced host banana in different locations, and with rare chance a bacterial clone gets started in a banana plant. From there it is spread to the extent that its unique characteristics of invasiveness, systemicity, aggressiveness, and transmissibility will allow.

For background information *see* BANANA; PLANT DISEASE in the McGraw-Hill Encyclopedia of Science and Technology.   [IVAN W. BUDDENHAGEN]

Bibliography: I. W. Buddenhagen, *Trop. Agr. Trinidad*, 38:107–121, 1961; I. W. Buddenhagen, in K. F. Baker and W. C. Snyder (eds.), *Ecology of Soil-Borne Plant Pathogens*, 1965; I. W. Buddenhagen and T. A. Elsasser, *Nature*, 194:164–165, 1962; E. R. French and L. Sequeira, *Phytopathology*, 60:506–512, 1970; E. S. Kojima and I. W. Buddenhagen, *Phytopathology*, 59:1035, 1969; J. C. Lozano and L. Sequeira, *Phytopathology*, 58:1058, 1968.

Plant disease control

The advent of systemic fungicides in the past 5 years has opened a new era in plant disease control (Fig. 1). Unlike the protective fungicides, which must be applied to the surface of plants and which kill the fungi as they attempt to infect, the systemic fungicides are translocated within the plant and kill the fungi even though infection has already occurred. Each type of fungicide has its attributes and deficiencies, and the systemics will certainly not displace the protectants. They will be useful for many diseases which the protectants have been ineffective in controlling.

The present group of systemics was discovered by the chemical industry as a result of screening thousands of candidate compounds. Once a successful systemic fungicide is discovered, many related compounds mimicking the new structure are developed. For example, Marshall Kulka and Bo von Schmeling found the oxathiin fungicides carboxin and oxycarboxin. Soon thereafter, they developed the related thiazoles. Competitors in Germany and France developed other closely re-

evidence which substantiates the validity of different strains from these different centers is provided by recent serological analysis by Evelyn Kojima and Buddenhagen of many isolates of *P. solanacearum* collected from the various areas. Specific determinants have been found to exist in the SFR strain which make it antigenically different from the B strain and the Peruvian A strain. Other strain-specific antigenic reactions have been observed with certain strains of the tomato race.

A preliminary study of direct DNA-DNA hybridization has also shown that the two banana bacterial strains B and SFR are quite different and no more closely related to each other than either is to the solanaceous pathotype, with about 60% homology in each case.

J. C. Lozano and Sequeira found that the strains which cause wilt of solanaceous and banana plants can also be differentiated by the hypersensitive reaction of tobacco leaves. The banana pathogen produced a rapid hypersensitive reaction and reduction of inoculated population in tobacco tissue; the solanaceous pathogen caused a delayed reaction with marked multiplication and mobility of the bacteria within the host tissue.

The primary interactions of banana tissue inoculated with the banana and solanaceous bacterial strains have been examined at ultrastructural levels (up to 50,000× magnifications) with an electron microscope. Slight differences in interactions with

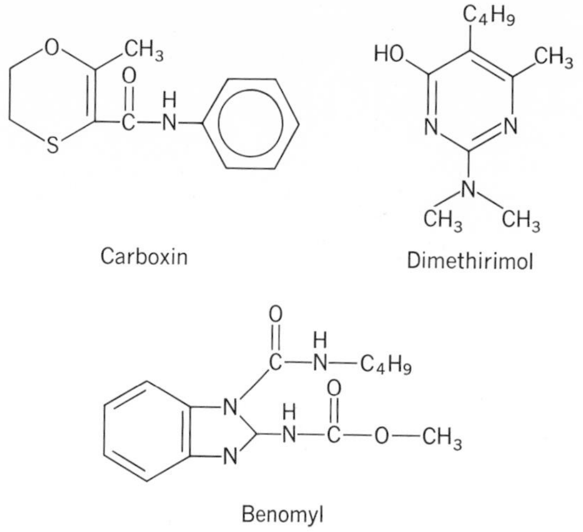

Carboxin

Dimethirimol

Benomyl

Fig. 1. Structures of some systemic fungicides.

lated compounds. Once the chemical industry releases a new fungicide, researchers in universities and in state and Federal governments are stimulated to determine how, when, and where this new chemical will be effective in disease control. For example, the benzimidazole fungicide benomyl was released for research in 1967, and by 1969 scientific journals were replete with reports of successes and failures. Carol Peterson and L. V. Edgington soon discovered that benomyl would move upward in plants but not downward, and to their surprise, that it would also move laterally from the xylem, or sapstream, into the bark.

Fig. 2. Test with bean rust demonstrating systemicity of oxycarboxin in bean leaves. Oxycarboxin applied on the basal half of leaf controls rust over entire leaf (left), but when applied to the apical half of leaf it controls rust only in the apical half (right).

Interestingly, efforts by university and research station scientists which were directed toward finding new fungicides have been rather unsuccessful. Many hours spent evaluating antibiotics and modified growth regulators have proved discouraging. These researchers assumed that antibiotics would have to pass through membranes of the actinomycete which synthesized the antibiotic. Consequently, antibiotics should move through membranes in plants and thus be systemic. Products of this research have been cycloheximide, griseofulvin, and blasticidin S, but they have not been resounding successes. Growth regulators were evaluated by Edgington and others with the thought that a growth regulator like 2,4-D, which is systemic, could be modified slightly in chemical structure to produce a systemic fungicide. This effort has been a failure. However, this research has helped in the evolution of techniques for testing systemic fungicides.

Fungal spectrum. A striking feature of systemics is their toxicity, or lack of it, to certain taxonomic groups of fungi. The oxathiins are toxic to the Basidiomycetes, such as smuts, rusts, mushrooms, and shelf fungi. The pyrimidines dimethirimol and ethirimol are toxic to certain powdery mildews (Ascomycetes). Another new pyrimidine, coded EL-273, is toxic to all powery mildews and also to a few other Ascomycetes. The purine or benzimidiazole fungicides benomyl and thiabendazole are toxic to a broader spectrum of fungi and include most Ascomycetes and Fungi Imperfecti. They are nontoxic to water molds (Phycomycetes) and to certain important Fungi Imperfecti.

Characterstics of translocation. The present systemic fungicides are all translocated upward, but not downward, within plants. When applied to the basal portion of a leaf, they enter and move toward the margins and tip, controlling fungi as shown in Fig. 2. When applied to the apical part of the leaf, they control fungi only in the apical portion. The fungicides appear to be translocated between the living cell protoplasts (not through the cell protoplast), and the movement is governed by the transpiration of the plant through stomates. Since petals of flowers have no stomates, they acquire no fungicide. Similarly, most fruit have no stomates and accumulate only negligible amounts of fungicide, and thus are unprotected from infection by fungi.

Decomposition within plants. The systemic fungicides appear to be more frequently degraded within plants than are protectants, probably because they come into a more intimate contact with enzymes of the plants. For example, carboxin is rapidly oxidized to a nonfungitoxic sulfoxide form. Oxycarboxin lasts longer but ultimately becomes inactivated because of binding to plant polymers such as lignin. The pyrimidines dimethirimol and ethirimol are very rapidly degraded, and a constant supply from the soil is required to obtain long-term disease control. Benomyl is rapidly altered within plants, but fortunately the breakdown product is still fungitoxic.

Practical uses. It is much too early to be sure just how important the systemic fungicides will become. At first, they will probably be used only on ornamentals, where residues are not as critical as with food crops. They are already approved for use in a few instances on food crops. Carboxin is registered for use as a seed treatment of grains for smut fungi borne within and on the seed. Oxycarboxin looks very promising as a control for rust diseases when seed treatment is followed by a foliar spray at midseason. The pyrimidine dimethirimol has been approved in several European countries, where two soil drenches with a protective fungicide replace about 14 sprays for control of powdery mildew of cucumbers, melons, and related crops.

The benzimidazoles may prove to be the most widely used of all systemics. Reports of disease control are remarkably favorable. The decision on registration is yet to be made, although thiabendazole has already been approved for citrus fruit dips.

One feature of many of the new compounds is that fewer sprays are needed than with protectants. This might be because rain cannot wash off the compound after penetration. Another reason may be that systemic movement within a leaf produces very excellent control even though the coverage may have been uneven. The latter factor may prove to be extremely important in determining the commercial success of systemic fungicides.

For background information *see* PLANT DISEASE; PLANT DISEASE CONTROL; PLANT HORMONES in the McGraw-Hill Encyclopedia of Science and Technology.   [LLOYD V. EDGINGTON]

Bibliography: C. J. Delp and H. L. Klopping, *Plant Dis. Rep.*, 52:95–99, 1968; G. J. M. van der Kerk, *Neth. J. Plant. Pathol.*, 75(Suppl. 1):5–20, 1969; C. A. Peterson and L. V. Edgington, *Phytopathology*, 60:475–478, 1970.

Plant movements

Recent research in the plant movement geotropism has been concerned with the roles of the growth hormone auxin and of various cell organelles. This discussion will therefore be limited to the initial hormonal reactions and the sensing mechanism for geotropism, the curving of an organ induced by reorientation with respect to gravity. Geotropic curvature is caused by a lateral difference in cell elongation, an extension which is limited and hence regulated by auxin. When a shoot is placed horizontally it is stimulated by the change in the direction of the gravitational field. More auxin is found in the lower cells of the reoriented organ than in the upper cells. This results in a greater rate of cell extension in the lower half, which, in turn, causes a curvature upward (negative geotropism). Roots, on the other hand, are positively geotropic; they curve toward the direction of the gravitational field.

Primary action of plant hormone. The action of auxin (indoleacetic acid, IAA) on cell elongation occurs rapidly. Michael Evans and Peter Ray showed that the delay between the time of IAA application and its action on cell expansion is only 10 min, a lag later shown to be as short as 2–3 min or less under certain conditions of temperature and hormone concentration. These observations indicate that the initial action of IAA is in the activation of preexisting molecules directly involved in cellular expansion. Evans and Ray further found that treatment of tissues with actinomycin D, an inhibitor of the deoxyribonucleic acid–dependent ribonucleic acid (RNA) synthesis, and with cycloheximide, an inhibitor of protein synthesis, did not

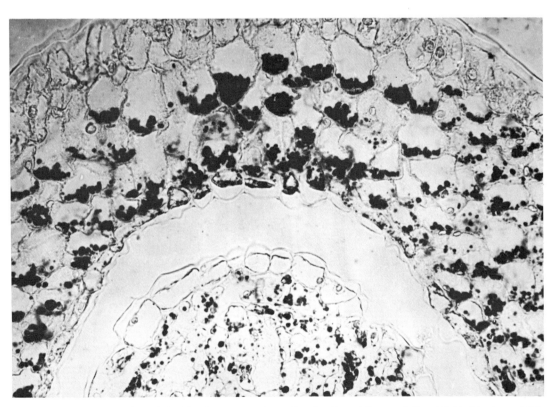

Fig. 1. Longitudinal section of the tip of a young oat shoot showing starch (dark particles) sedimentation in the direction of the gravitational field. (*Courtesy of S. A. Gordon, Argonne National Laboratory*)

extend the latent period of auxin action. These observations contradict the theory that the primary effect of IAA in the promotion of cell extension is by means of a gene-activation or gene-repression mechanism, either by stimulating the synthesis of messenger RNA or by promoting protein synthesis. Yoshio Masuda and Ryoichi Yamamoto had noted that the stimulatory effect of IAA on the synthesis of β-1,3-glucanase (an enzyme that reduces the rigidity of cell walls, which would facilitate cell expansion) was as rapid as 10 min. However, there is no conclusive evidence that glucanase itself promotes cell elongation. Also, earlier work at Masuda's laboratory showed that the action of auxin on RNA synthesis was equally rapid.

Sensor of geotropism. Botanists have speculated for the past 70 years on the identity of the gravity sensor in geotropism. For many years the starch-statolith theory was the accepted explanation. This concept holds that, upon reorientation of an organ, the intracellular starch particles sediment rapidly into the lower portion of the cell (Fig. 1). The position and the rate of movement of these particles within the cell could indicate to the plant the direction and magnitude of the gravitational stimulus. In 1966 Barbara Pickard and Kenneth Thimann showed that geotropic perception was retained in destarched young wheat shoots. Their finding disputes the role of starch grain as the sensor for geoperception. More recently, however, Tor-Henning Iversen presented strong evidence in support of the starch theory. Iversen showed that, although depletion of starch did not affect the growth rate of cress roots, it caused the roots to be unresponsive to gravitational stimulation. This work would have been a more convincing support of the role of starch as a sensor if Iversen had shown that the geotropically stimulated destarched roots would not curve upon reformation of the starch grains. However, Pickard and Thi-

mann's conclusion could be still valid in that starch is required for the physiological implementation of the geotropic stimulus.

Other organelles (the mitochondria, nucleus, and nucleolus) have been also suggested as statoliths for geoperception but without experimental support. Cell organelles that can be directly associated with the expansion of the cell wall are the endoplasmic reticulum and the Golgi apparatus or dictyosomes (Fig. 2) and their vesicles. Material is added to the wall from vesicles, which in some cells appear to be derived directly from the dictyosomes. The rate of vesicle production by the dictyosomes and the time required for a vesicle to traverse the cytoplasm to the cell membrane and wall are compatible with the time required for the appearance of visible curvature in geotropism.

J. Shen-Miller and Ray Hinchman found indications that the Golgi apparatus participates in geotropism. They observed that, in a geotropically stimulated oat shoot, the bottom half of each cell has more dictyosomes than does the top and that these dictyosomes are more active (enhanced vesicle production) than the ones in the top half of the cell. The difference is greater in the lower half of the shoot than in the upper. These observations suggest that the number and localization of activity of the dictyosomes are correlated with the direction of the gravitational field. It remains to be determined whether the kinetics of dictyosome distribution and activation can be correlated with the geotropic response. These findings bear on the function of the dictyosome in geotropism either as a sensor or as an organelle involved in the physiological implementation of the gravitational stimulus. At present the mechanism by which activated geosensors change the distribution of the growth hormone is still obscure.

For background information *see* PLANT MOVEMENTS in the McGraw-Hill Encyclopedia of Science and Technology.   [JANE SHEN-MILLER]

Bibliography: M. L. Evans and P. M. Ray, *J. Gen. Physiol.*, 53:1–20, 1969; T.-H. Iversen, *Physiol. Plant.*, 22:1251–1262, 1969; Y. Masuda and R. Yamamoto, *Develop., Growth, Differentiation*, 11:287–298, 1970; B. G. Pickard and K. V. Thimann, *J. Gen. Physiol.*, 49:1065–1086, 1966; J. Shen-Miller and R. Hinchman, *Plant Physiol.*, suppl., 44:16, 1969.

Plasma physics

Recent fundamental studies of plasma characteristics have led to a better understanding of feedback stabilization, instability threshold measurements, and collective ion acceleration. Experiments with the generation, heating, and control of plasma for fusion reactors have also produced new results. In particular, fruitful investigations have taken place with the closed toroidal magnetic field configuration (Tokamak experiments) and with the method of generating and heating plasma by means of powerful laser pulses.

FUNDAMENTAL STUDIES OF PLASMA

Recent studies on stabilization with negative feedback, instability measurements, and collective ion acceleration are discussed in this section.

Feedback stabilization. It is commonplace to employ negative feedback to stabilize systems with discrete components. More recently, these

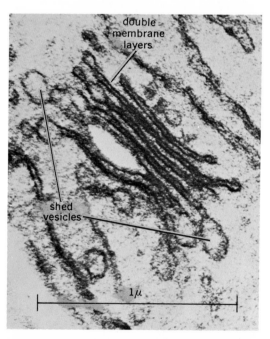

Fig. 2. Electron micrograph of a dictyosome from a young oat shoot. The double membrane layers shed vesicles, which contain proteins and carbohydrates.

ideas have been applied to distributed systems, namely, plasmas suffering an instability which depends upon both temporal and spatial dispersion. Using a hot-cathode reflex discharge in a magnetic field which was subject to an azimuthal drift instability driven by radial electric fields, Ronald Parker and Keith Thomassen stabilized the system using negative feedback. Since the dominant instability ($m=2$) gave rise to a perturbed potential varying as $\cos 2\theta$ with azimuth angle θ, stabilizing fields could be applied at the periphery of the plasma using a pair of plates displaced $90°$ in θ, and driven $180°$ out of electrical phase with respect to one another. The signal applied to the plates was derived from a probe immersed in the plasma, with provision to amplify and phase-shift the probe signal to optimize the negative feedback. Complete suppression of the $m=2$ mode was observed, together with enhancement of the otherwise more weakly unstable $m=1$ mode. This latter result was not unexpected from the theory. Figure 1 shows the spectrum of potential fluctuations as feedback is increased. The $m=2$ mode at 40 kHz is suppressed with increasing gain, while the $m=1$ at 27 kHz is enhanced.

Further experiments by Hans Hendel and coworkers also demonstrated the feedback control principle on a collisional drift wave in a Q-machine. In this case, the dominant instability did not perturb the plasma boundaries greatly, so that feedback control could not be achieved using electrodes outside the plasma. Rather, direct injection of electron current parallel to the magnetic field was provided (with the appropriate phase and amplitude) by inserting two probes into the plasma. One probe sensed the potential fluctuations, and the other injected the current.

Instability threshold measurement. Experiments by the Livermore group under Richard F. Post and collaborative theoretical work by J. Brian Taylor of the Culham Laboratory in the United Kingdom have been reported which typify the high level of sophistication recently achievable in many plasma physics studies. In the experiments, plasma is formed in a magnetic well by injection and stripping of energetic hydrogen atoms. During periods of instability, bursts of radio-frequency (rf) emission at harmonics of the ion cyclotron frequency were observed. The plasma was constituted of hot ions and cooler electrons, the latter having a truncated Maxwellian energy distribution extending up to the ambipolar potential Φ_M. No electron Landau damping was thus expected until the wave's phase velocity fell below $\sqrt{2e\Phi_M/m}$, where e and m are the electronic charge and mass. The observed oscillations are basically electron plasma waves obeying a dispersion relation $\omega = n\omega_{ci} = \omega_{pe}(k_{||}/k_\perp)$, where ω_{ci} is the ion cyclotron frequency, n is the (integer) harmonic number, ω_{pe} is the electron plasma frequency, and $k_{||}$ and $k_\perp$ are the parallel and perpendicular components of the wave number, with $k_{||} \ll k_\perp$. As the injected beam density (and thus ω_{pe}) was lowered, and the rf burst rate dropped and a threshold was identified by extrapolating the burst rate to zero. Threshold densities so determined are shown in Fig. 2 in terms of the dimensionless parameter $\varepsilon \equiv \omega^2_{pi}/\omega^2_{ci}$, which is proportional to nM/B^2, where M is the ion mass and B the magnetic field strength. The independent variable in Fig. 2a is

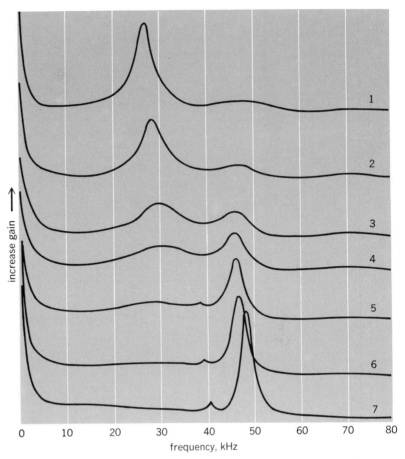

Fig. 1. Suppression of the $m=2$ mode and subsequent appearance of the $m=1$ mode, driven unstable by the feedback network.

$e\Phi_M/W_i$, where W_i is the ion energy. One sees that a linear relationship between ε and $e\Phi_M/W_i$ is suggested. In Fig. 2b, the threshold values of ε are plotted as a function of a parameter which identifies the threshold with the onset of electron Landau damping at the parallel phase velocity $\sqrt{2e\Phi_M/m}$, and one sees the remarkable agreewith experiment.

This work not only shows the level of comparison between experiment and theory achievable in present-day plasma physics work, but also furnishes some practical guidelines in the design of stable high-density mirror-confined plasma.

Collective ion acceleration. At least two schemes have been pursued during the last several years in exploiting the collective plasma feature for acceleration of ions to high energy. The underlying motive is to devise particle accelerators (for high-energy physics research) which might conceivably compete with conventional machines using single-particle acceleration mechanisms.

In 1965 and 1966, experiments were reported by G. Sargent Janes and collaborators on a version of a collective heavy ion plasma accelerator (HIPAC) in which an electrostatic potential well due to a dense electron cloud would accelerate and trap ions. Experiments using this configuration on a large scale have not been reported. More recently, two different schemes for producing collective acceleration have been reported. In one, the electron ring accelerator, a doughnut-shaped cloud of circulating electrons is to be accelerated electri-

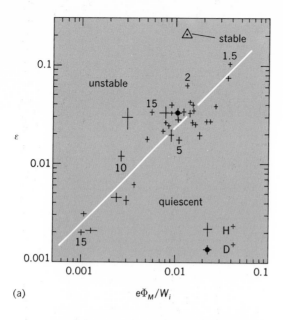

(a) $e\Phi_M/W_i$

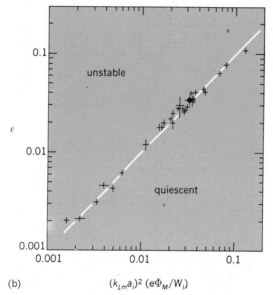

(b) $(k_{\perp m}a_i)^2 (e\Phi_M/W_i)$

Fig. 2. Measurement of threshold densities. (a) Values of ε at threshold for instability as a function of $e\Phi_M/W_i$. Numbers near several points indicate values of W_i in thousands of electron volts. Point $\triangle$ is for a stable plasma decaying by charge exchange after a period of instability. (b) Comparison of measured values of ε and those predicted from theory.

cally or magnetically, carrying along a small number of entrapped positive ions. Production of the electron ring appears to be the major obstacle. Preliminary results in the United States have been published by the Berkeley group in which 3.3-Mev electrons from the Astron facility at Livermore were injected on a 19-cm orbit radius, and compressed to a radius of 3.5 cm by a rising magnetic field. The accompanying betatron action accelerated the electrons to 18 Mev. These rings appear to be stable for as long as the magnetic field pulse lasts, provided that certain precautions are observed which avoid regimes of dangerous instability. The number of electrons trapped in such rings was estimated to be about 4×10^{12}. Entrapment of, say, 10^9 protons in such a ring is then speculated as possible without drastically altering

the dynamical response of the ring to external acceleration. Acceleration of the electrons to a translational kinetic energy of 100 Mev would then presumably bring the protons to 180 Bev.

If not beset by unanticipated instabilities, the electron ring accelerator can apparently be understood in simple classical terms. This state of affairs does not appear to obtain in the other recent instance of collective acceleration reported by J. Rander and colleagues. In these experiments, an intense relativistic electron stream was injected into a gas-filled region as shown in Fig. 3a. Various diagnostic apparatus was employed to analyze distributions in energy of both the injected electrons and of the ions produced by the stream. Figure 3b shows the data. The physical mechanism responsible for the ion acceleration to a momentum value some 40 times that of the electron momentum is not known, although collective interactions such as an inverse Cerenkov effect have been proposed in unpublished work by Ben Eastland and Jonathan Wachtel. [JAY L. HIRSHFIELD]

Tokamak fusion experiments. A closed, toroidal magnetic field configuration is particularly attractive as a controlled fusion reactor design. The plasma ions can collide many times before they are lost, and thus the fuel can be burned efficiently at relatively low temperatures, of order 10 kev.

A basic problem of toroidal confinement, however, is that a simple toroidal magnetic field B_θ cannot provide a plasma equilibrium configuration: The (diamagnetic) plasma would expand along its major radius R, following the negative gradient of the magnetic field strength $\overline{\nabla}B = -B_\phi \overline{R}/R^2$. One way to provide equilibrium is to add a poloidal magnetic field component consisting mainly of B_θ, as in Fig. 4, thus creating helical magnetic field lines. These field lines, in general, will cover magnetic surfaces ergodically, and a well-conducting plasma can then short-circuit the vertical electric field associated with outward motion along R. The Tokamak generates its B_θ by means of an induced toroidal plasma current I.

The transform angle ι traversed by the helical magnetic field lines around the minor torus circumference during one circuit around the major circumference is given by the equation below,

$$\iota = 4\pi IR/10r^2 B_\phi = 2\pi/q$$

where q is called the safety factor. When q is a rational number, each field line on the surface of minor radius r closes on itself after a finite number of circuits. Accordingly, helical charge separations are not ruled out by plasma conductivity, and the associated electric fields can then drive the hydromagnetic "kink" modes (helical perturbations approximately conforming with the field lines). For $q \leqslant 1$, these modes are highly unstable; for large q-values, they become harmless. Typical Tokamak operation has been with $q = 3-4$ on the outer plasma surface $r = a$.

Features of apparatus. A characteristic feature of Tokamak apparatus to date has been the massive slitted copper shell schematically shown in the Fig. 4. Its primary purpose is to act in concert with a weak, externally applied vertical magnetic field to oppose off-centering displacements of the plasma column. A secondary function of the copper shell is to permit induction of helical eddy cur-

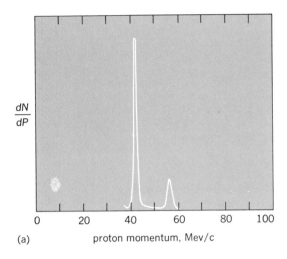

(a)   proton momentum, Mev/c

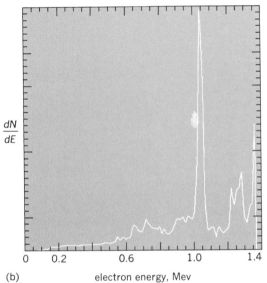

(b)   electron energy, Mev

Fig. 3. Data from ion acceleration experiments. (a) Observed proton momentum spectrum. (b) Typical electron energy spectrum at 1-Mev nominal voltage.

rents that help stabilize the longwave hydromagnetic kink modes (which appear at small q-values). Just inside the copper shell lies a stainless-steel vacuum chamber, usually a thin bellows protected by an annular metal limiter passing the short way around the plasma column.

Plasma confinement experiments with the Tokamak ("toroidal camera magnetic") began at the Kurchatov Institute in Moscow in the 1950s. The results were comparable to those of Ohmic-heated toroidal devices elsewhere: the "stabilized pinch" and the stellarator. With a neutral gas filling of typically 10^{13}–10^{15} atoms/cm³, plasma densities n of the same order, and electron temperatures T_e below 100 ev were produced by the Ohmic heating effect of plasma currents $I \sim 20$ kA. The hydromagnetic equilibrium and stability results were found to conform approximately with theory, but the plasma loss times were short: comparable to the Bohm time $\tau_B = \pi a^2 B_\phi e/cT_e$ observed in stellarators, or even shorter.

Design improvements. In the period 1960–1968, important improvements were made on the Tokamak: notably, the correction of small magnetic field asymmetries and the systematic advance to

larger toruses, with higher discharge currents and optimal current pulse shapes. In the T-3 device ($R = 100$ cm, $a = 15$ cm), currents of about 100 kA in toroidal fields up to 35 kG were used to produce plasma energies about 1 kev per electron-ion pair, as calculated from the measured plasma diamagnetic effect and density. The current rise times were typically 3–10 msec, and the current pulses were extendable to some tens of milliseconds. The energy confinement time τ_E, obtained from the diamagnetic effect and the Ohmic heating input power, was found to increase with plasma energy, apparently reaching values of order 10 msec, or more than 30 times τ_B. The particle confinement times, measured spectroscopically, were found to be even longer. The electron temperature obtained from the plasma resistivity, however, was typically 3–30 times lower than the diamagnetic temperature. This anomaly factor increased strongly with the ratio of the directed electron streaming velocity to the thermal velocity.

Open questions. During 1969 the electron distribution function was measured accurately in T-3 by Thomson scattering of a ruby laser giant pulse and was found to be Maxwellian. The observed T_e values agreed with those deduced earlier from the diamagnetic temperature, and thus the existence of a strong resistivity anomaly was confirmed. Ion temperatures have been inferred mainly from charge-exchange neutral spectra and from neutron yields (when using deuterium). At present, the best results reported on T-3 ($T_e \gtrsim 1$ kev, $T_i \sim 500$ ev, $n \sim 5 \times 10^{13}$ cm⁻³, and $\tau_E \sim 20$ msec $\sim 100\ \tau_B$) lie below the respective parameters for a fusion reactor by only 1–2 orders of magnitude. If the scaling law for τ_E has the a^2 dependence typical for a dif-

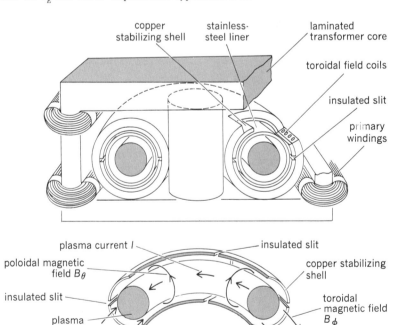

Fig. 4. Diagram of the closed toroidal magnetic field configuration.

Fig. 5. A high-power laser amplifier capable of producing energies greater than 10^3 joules. The amplifier is made up of 12 laser disks, each 15 cm in diameter and 3 cm thick. Amplifiers such as these may be used in parallel to achieve 10^5 joules output.

fusion time (as appears to be the case), then the present diffusion constant is already about right for a toroidal reactor regime, Whether the Tokamak configuration can actually be scaled to a reactor, however, is not clear as yet, particularly since many basic points of physics have not been resolved.

Among the principal open questions are the following: (1) What limits the energy confinement time τ_E? Is it impurity radiation and electron-ion heat transfer, with ion cooling by classical (Galeev-Sagdeev) radial transport? Or, is it anomalous heat transport directly from the electrons? (2) Is the improvement over the Bohm time due mainly to emergence into a plasma regime of long electron mean free paths? Are the shear and minimum-average-B properties of the Tokamak important? Or, is there some intrinsic advantage in the axisymmetric Tokamak geometry relative to stellarators? (3) What is the nature of the anomalous bulk resistivity, and of the still more anomalous skin resistivity during current rise? (4) What is the reason for the empirical peak plasma pressure scaling $\beta_\theta = 8\pi P/B_\theta^2 \sim 1$; is it fortuitous, or related to the Ohmic heating process, or intrinsic in the Tokamak confinement scheme? (5) What is the nature of the gross instability which at present sets an upper limit of about 10^{14} cm^{-3} on the plasma density? (6) Can Ohmic heating be replaced by some more effective plasma heating method, to allow attainment of plasma temperatures in the 10-kev range?

The investigation of these questions and the related endeavor to extend the T-3 parameters toward thermonuclear ignition conditions have at present become the principal objectives of worldwide research on toroidal hot-plasma confinement.

[HAROLD P. FURTH]

LASER-PRODUCED PLASMAS

The development of short-pulse, high-power lasers has enabled experimental work to be carried out on intense vaporization and heating of small dense droplets of thermonuclear fuel. Plasmas produced by heating such droplets have properties which are of significant interest in the field of controlled thermonuclear fusion.

Magnetic confinements. Present-day high-power lasers are capable of delivering 10^3 joules on a time scale ranging from 10^{-10} to 10^{-7} sec. The peak powers in such a pulse may easily be in excess of 10^{13} watts. This burst of coherent electromagnetic radiation, at a wavelength of 1.06μ in the case of a neodymium glass laser system, may be focused down to an area of 2×10^{-4} cm^2, giving rise to local electric fields of 10^{10} volt/cm. A photo of such apparatus is shown by Fig. 5. A plasma heated at such intense field strengths may rapidly rise to temperatures in excess of $10^{7\circ}$K and in some cases $10^{8\circ}$K.

It has been established both theoretically and experimentally that under certain conditions significant absorption of focused laser radiation can be attained in a small sphere of dense plasma. This results in a freely expanding, fully ionized plasma with measured energies greater than $10^{7\circ}$K. Absorption of the incident radiation at modest power density levels (10^{13} watts/cm^2) is predominantly a collisional process leading to a symmetrically expanding spherical plasma whose expansion energy is derived from the thermal energy originally absorbed from the incident radiation. This expanding plasma may be trapped in a suitable magnetic field and thus used to fill a controlled fusion field geometry. Unfortunately, due to the extremely high magnetic fields required and instabilities associated with the field plasma interaction, this approach appears to be limited to ultimate plasma densities of 10^{14} cm^{-3} and temperatures of $10^{7\circ}$K. Some auxiliary method of heating, such as microwave heating, must be employed to raise the temperature almost an order of magnitude to the point where controlled fusion reactions are feasible.

Laser droplet heating. An alternate approach is to achieve significant fusion reactions from a fuel droplet of deuterium-tritium in a time shorter than that associated with the expansion of the plasma into a vacuum background. In this case no magnetic fields are used. Theoretical work indicates that, through the use of an appropriately tailored laser pulse, one can indeed achieve more liberated energy in burning than is deposited in the plasma from the laser beam. This crossover or "breakeven" point occurs at approximately 10^5 joules of incident laser light for fuel droplets a few millimeters diameter.

A reactor based on this concept of laser droplet heating is under study at the Oak Ridge National Laboratory (Fig. 6). Such a reactor producing 50 Mw of electrical power would contain a vortex of liquid lithium formed in a chamber of about 2.4 m in diameter. A fuel droplet would be vaporized once every 10 sec with a laser pulse timed to match the plasma expansion. The neutrons that are produced would be absorbed by the lithium, which in turn would undergo a temperature rise. The lithium would then be circulated, and the heat removed as in conventional generators.

This alternative approach to the problem of controlled thermonuclear fusion is attractive, because it bypasses all of the stability problems associated with magnetic confinement of hot plasmas. However, it is still young and untested. Experimental work in the next few years will go a long way to answer the question of its feasibility.

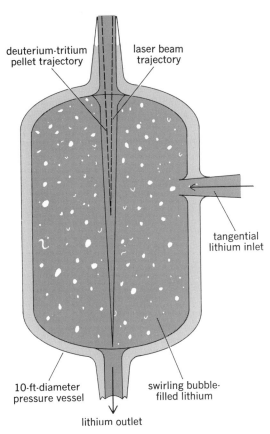

deuterium-tritium
pellet trajectory

laser beam
trajectory

tangential
lithium inlet

10-ft-diameter
pressure vessel

swirling bubble-
filled lithium

lithium outlet

Fig. 6. Schematic drawing of a controlled-fusion reactor utilizing laser plasmas.

For background information *see* FUSION, NUCLEAR; LASER; PLASMA DIAGNOSTICS; PLASMA PHYSICS in the McGraw-Hill Encyclopedia of Science and Technology.

[MOSHE J. LUBIN]

Bibliography: L. A. Artsimovich et al., *JETP Lett.*, 10:82, 1969; L. A. Artsimovich et al., *Nucl. Fusion*, suppl. 17, IAEA, Vienna, 1964; C. C. Damm et al., *Phys. Rev. Lett.*, 24:495, 1970; A. F. Jaught, D. H. Polk, and W. J. Fader, Paper F-7, *Proceedings of 3d Conference on Plasma Physics and Controlled Nuclear Fusion Research, Novosibirsk, U.S.S.R.*, 1968; G. S. Janes et al., *Phys. Rev.*, 145:925, 1966; D. Keefe et al., *Phys. Rev. Lett.*, 22: 558, 1969; M. J. Lubin, H. S. Dunn, and W. Friedman, Paper F-8, *Proceedings of 3d Conference of Plasma Physics and Controlled Nuclear Fusion Research, Novosibirsk, U.S.S.R.*, 1968; R. R. Parker and K. I. Thomassen, *Phys. Rev. Lett.*, 22:1171, 1969; N. J. Peacock et al., *Nature*, 224:488, 1969; J. Rander et al., *Phys. Rev. Lett.*, 24:283, 1970; *Science*, 167(3921):1112–1113, Feb. 20, 1970; T. C. Simonen, T. K. Chu, and H. W. Hendel, *Phys. Rev. Lett.*, 23:568, 1969.

Polarographic analysis

In the last few years the importance of applications of polarography in the trace analysis of inorganic species has somewhat diminished because of the advent of atomic absorption, but polarography and related techniques have become more widely used in the study of organic compounds.

New chemical systems have been shown to be polarographically active; polarographic waves have been applied for functional group analysis; polarography has been used for the study of equilibria, rates, and mechanisms of reactions taking place both in the vicinity of the dropping mercury electrode and in the bulk of solution; relationships between structure and polarographic behavior have been investigated; and polarography has been applied for studies of biological systems, polymers, and pharmaceuticals.

Polarographically active groupings. Electrode processes involving organic compounds at the dropping mercury electrode result in a cleavage or formation of a chemical bond in the investigated organic molecule. Polarographic behavior depends primarily on the nature of the bond involved, even when the molecular environment (that is, the presence and kind of neighboring groups and other substituents, the type of the molecular frame to which the electroactive group is bound, and the spatial arrangement) also affects the electrode process. The reducibility of multiple bonds (for example, $C=O$, $C=N$, $C=C$, and $N=N$) or of strongly polarizable groups (such as NO_2, NO, and C halogen), especially when conjugated, was recognized early. Recently the reducibility of some single bonds such as $C—O$, $C—N$, $C—S$, or $C—P$ when the molecule contains activating groups was recognized. In all classes of compounds a deeper insight into the mechanism of the electrode process, which usually consists of several steps, has been gradually obtained.

Functional group analysis. First, polarography can be applied as a final step to classical functional group analysis. Either the reaction product (for example, semicarbazone in the determination of carbonyl compounds, or *N*-nitrosamine in determinations of secondary amines) is measured, or a decrease in concentration of a reagent (for example, chromic acid in determination of alcohols, or mercuric or silver ions in determination of thiols) follows. Second, the polarographic limiting current can be taken as a measure of the amount of a given electroactive group (for example, quinones, aliphatic nitro compounds, phenazines, or thiols).

Equilibria, rates, and mechanisms. For equilibria established in the bulk of the solution within more than 15 sec, measurement of wave heights of individual components makes possible the evaluation of equilibrium constants. For equilibria that are very rapidly established at the electrode surface, equilibrium constants can be determined from shifts of half-wave potentials. Finally, for some equilibria between these two extremes, which are established in times comparable to the drop time (3 sec), calculation of rate constants is possible (for example, for dehydration of hydrated aldehydes, or for protonation of anions derived from C acids, such as $C_6H_5COCH_2COCH_3$). In this way, rate constants of very fast reactions of the order $10^5 - 10^{10}$ liters mole^{-1} sec^{-1} can be determined.

For slower reactions, rate constants can be found from changes of wave heights with time. Moreover, as some reaction intermediates giving separate waves can be detected, identified, and followed polarographically, and if their half-lives are longer than 15 sec, polarography can prove

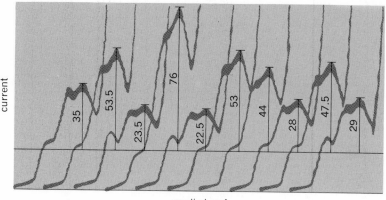

current

applied emf

Polarographic curves for various normal and cancer cases. With the experimental setup, curves above 30 mm correspond to pathological states. Curves starting at −0.8 volt, Hg pool, open to air; full-scale sensitivity 40 μA.

useful in mechanistic studies. Elimination of Mannich bases, hydration of multiple bonds in unsaturated ketones, and aldolization are examples studied.

Polarography and structure. Reduction of most organic systems, particularly in aqueous media, involves steps with a high activation energy and is therefore irreversible. Half-wave potentials of such systems are a function of the rate constant of the electrode process. This heterogeneous rate constant is frequently influenced by structural effects, as are, in a similar manner, rate constants of homogeneous reactions. Therefore, effects of substituents on half-wave potentials in aromatic systems and in aliphatic systems can be treated by the Hammett relations and the Taft substituent constants, respectively. Among steric effects, for example, steric hindrance of coplanarity or effects of cis-trans isomerisms affect polarographic curves. Polarography also enables distinguishing between some epimers, for example, bearing axial or equatorial halogen.

Biological systems. Many naturally occurring substances such as quinones, aldehydes, ketones, ketoacids, sugars, unsaturated acids, peroxides, thiols, and disulfides or some heterocycles are polarographically active and give waves. These waves can be measured for concentrations 10^{-4}–10^{-5} M (with special techniques, even one or two orders of magnitude lower), frequently in the presence of other components of biological systems, and thus rapid analysis of the biological material is possible.

Polarographically active groupings are also present in some vitamins, alkaloids, hormones, terpenoid substances, natural coloring matters, and the like. In particular, determination of ascorbic acid (vitamin C) is simple, rapid, and unaffected by substances interfering in other modes of analysis.

Some proteins, when added to a buffered ammonical solution of cobalt salts, give polarographic waves which correspond to catalytic hydrogen evolution. The detailed mechanism of the catalytic process is not fully understood, but it is known that only some types of protein give this reaction. The protein molecule must contain a suitably situated thiol or disulfide group and must be able to form a complex with cobalt and ammonia. It is assumed

that the proton transferred by this complex undergoes reduction more easily than H_3O^+.

When blood serum is alkali-denaturated and the proteins with largest molecular weight are separated by precipitation with sulfosalicylic acid, the remaining filtrate contains proteins and their fragments, the contents of which differ in the serum of healthy and pathological individuals. Increase of the catalytic wave is observed for cancer cases and inflammatory diseases, decrease of the catalytic wave for hepatitis and other liver diseases. In Europe the polarographic test is used in clinical analysis as a general screening test, in connection with nine other tests as a proof of malignancy, and in controlling cure of cancer and measuring the effectiveness of treatments such as surgery or irradiation. The illustration contains polarographic curves for cancer cases.

Changes in polarographic waves enable the following of structural changes in proteins induced by heat and irradiation, changes in pH, addition of solvents and various chemical reagents, or enzymes. Reduction waves of azo compounds enable the study of the antigen-antibody reaction.

Polarography can be used also for the study of changes in concentration of various reagents, such as metal cations, organomercurials, iodine, or acridine derivatives, when added to protein solution. Some of the methods used in the investigation of proteins can also be used for peptides, such as glutathione, oxytocin, and vasopressin.

Polymers. Polarography enables analysis of monomers and determination of catalysts (initiators and inhibitors) and even of some reactive groupings in polymers. For instance, butylmetacrylate or acrylonitrile was determined in the course of polymerization in polybutyl metacrylate and in copolymers. Phthalanhydride and maleic anhydride were determined in raw materials and oxidation products, such as peroxides and pyruvic acid in monomers or polymers. Among the catalysts that have been determined are benzoylperoxide, hydroquinone, and methyl p-toluylate. Methods for determination of carbonyl, carboxyl, or 1,2-dihydroxy groupings in polymers have been described.

Pharmaceuticals. Polarography is frequently applied to analysis of drugs and pharmaceutical preparations because numerous physiologically active compounds are also polarographically active and give waves. Pharmaceutical preparations are often relatively simple mixtures and are easily prepared for polarographic electrolysis, which enables rapid analysis. The active components are frequently present in small amounts so that use can be made of the sensitivity of polarographic methods. This advantage applies also to determinations of traces of impurities that can possess toxic properties. Determination of vitamins in preparations, of numerous alkaloids and hormones in drugs, and of barbiturates, organomercurials, terpene aldehydes and ketones or peroxides in ether serve as examples.

For background information *see* CLINICAL PATHOLOGY; ELECTROCHEMICAL TECHNIQUES; POLAROGRAPHIC ANALYSIS in the McGraw-Hill Encyclopedia of Science and Technology.   [PETR ZUMAN]

Bibliography: R. N. Adams, *J. Pharm. Sci.*, 58: 1171, 1969; M. Brezina and P. Zuman, *Polarogra-*

phy in Medicine, Biochemistry and Pharmacy, 1958; K. Kustin (ed.), Methods in Enzymology, vol. 16: Fast Reactions, 1969; D. J. Pietrzyk, Anal. Chem., 38:278R, 1966, and 40:194R, 1968; P. Zuman, The Elucidation of Organic Electrode Processes, 1969; P. Zuman, Fortschr. Chem. Forsch., 12:1, 1969; P. Zuman, Topics in Organic Polarography, 1970.

Polymer

A recent advance in the field of polymer science has been the development of colloidal microcrystalline polymers. Colloidal microcrystalline polymers are composed of crystals approximately 25– 10,000 A in maximum dimension, which is the size of viruses and bacteria. Before 1955 single polymer crystals were a laboratory phenomenon studied largely with the electron microscope. By 1968 four new products based on colloidal polymer microcrystals had emerged from the laboratory, and one of them had reached commercial sales around the world. Since then, three more types have been developed in the laboratory, and there is promise that more are to come.

Historical development. In 1955 a scientist at American Viscose Division of FMC Corp. attempted to isolate tiny crystals of cellulose in quantity with the plan of using them to nucleate the viscose solution from which rayon cord is produced. He projected that this would permit greater control of the cellulose structure during the spinning process, enabling the manufacture of a better tire.

In his attempts to produce in quantity cellulose crystals of the desired size, smooth colloidal gels that looked like white petrolatum were accidentally produced. Electron microscopy revealed that the new gels were made of tiny (micro) polymer crystals in the colloidal range (between the size of a flu virus and a wart virus) suspended in water with much larger particles, which were bundles of the single free microcrystals. These gels exhibited unusual viscosity properties. They behaved like butter but would not melt. Furthermore, they became fluid when shaken but recovered thickness quickly when allowed to stand.

Unifying concept. Extensive research with the colloidal suspensions of cellulose microcrystals led to a unifying concept that has been applied to many other polymer precursors.

The starting material for each product must be a high-molecular-weight polymer, a compound composed of very-long-chain molecules. These long molecules, when allowed by nature (natural polymers) or by man (synthetic polymers) to "crystallize" out of solution or from a melt, are the bases upon which all fibers and films are produced. As they pack together, however, some parts of them line up side by side like match sticks into orderly areas (microcrystals), and other parts of the same molecules kink up between the microcrystals to produce areas of imperfection or low order (hinges) (Fig. 1).

The same molecule may pass from one microcrystalline area to several others, depending on the polymer species and on how long the molecules are originally. In any event, preconditions must exist in the polymer to permit some sites or regions of imperfection to become selectively attacked by a chemical means. Naturally, the exact loosening

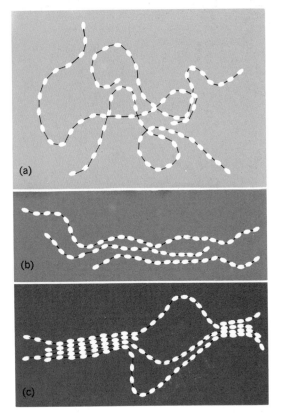

Fig. 1. Schematic arrangements of long-chain molecules. (a) Disorganized. (b) Partially organized. (c) Linear microcrystals forming.

mechanism varies with each polymer precursor, depending on its characteristic morphology.

The prerequisite is to cut, etch, or otherwise weaken such sites without destroying ordered regions in the process. A minimum number of the microcrystals must then be separated in water to produce the new colloidal polymer gels. If too few are freed, a "sand-in-water" separation is effected. The unhinging or etching is usually done by a specific chemical attack which varies for each polymer system. Hydrochloric acid under controlled conditions is used to disconnect the microcrystals embedded in cellulose fiber matrices. Mechanical agitation from a device such as a blender is used to assist in the separation of microcrystals from their unhinged but clustered-together state.

Colloidal polymer microcrystallinity. Within the framework of the foregoing concept, a colloidal polymer microcrystal can be defined as any crystalline or highly ordered or discrete colloidal particle separated from its position within the matrix of a polymer product. These particles possess a maximum dimension (truly colloidal) less than 10,000 A, and more commonly less than 5000 A. Furthermore, the size of the ultimate microcrystal or colloidal particle corresponds closely to the dimension that it possessed in the original polymer matrix prior to the chemical pretreatment and subsequent deaggregation process.

The use of this concept is that when crystalline polymer particles are isolated by the most appropriate means for each polymer precursor in sufficient amounts, new colloidal suspensions having novel properties will be formed.

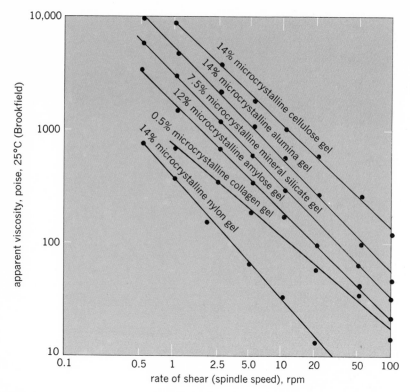

Fig. 2. Viscosities of aqueous microcrystal gels as the function of increasing rate of shear (log scales). Brookfield instrument measures relative viscosities. (O. A. Battista et al., Novel microcrystals of polymers, J. Appl. Polym. Sci., 11:481–498, 1967)

Accordingly, this is not the conventional route of synthesizing long-chain molecules, which is to put them into solution, melt them, or otherwise reconstitute them into fibers, films, or other structural forms. Rather, here one starts with a structure, not unlike a brick wall in some respects, and etches away or loosens the mortar between the bricks under conditions that do not dissolve or swell the bricks. The main difference, of course, is that the bricks that are freed are so tiny that they are invisible to the human eye, and the mortar removed or loosened consists of molecules that intertwine in some way between bricks.

The size of the bricks is predetermined by the history of the polymer precursor. It has been found that these little bricks or microcrystals can vary in size from that of a rabbit papilloma virus up to that of a tobacco mosaic virus. Interestingly, as with the viruses, they may be round or rodlike microcrystals. The shape depends on the molecular geometry within the precursor polymer.

Each single microcrystal consists of hundreds of long-chain or interconnected molecules in a crystalline compact form. Moreover, each crystal behaves as though it were a composite particle of molecular weight well into the millions. This submicron size and the attendant surface phenomena lead to the novel polymer science.

Rheological properties. Certain general rheological characteristics apply to the microcrystal gels, all of which must contain a minimum amount of colloidal submicron particles. They differ from gels, in which the network of chain molecules is held together either by primary bonds or by thermally reversible linkages that are due to secondary attractive forces.

The rheological properties of the microcrystal gels depend mostly on the particle size distribution, particularly on the fraction under 1 μ in maximum dimension. The microcrystal gels are characterized by unusual, extremely high yield values resulting from a three-dimensional network structure. This structure breaks down instantaneously as the elastic limits of the network are surpassed, and will rebuild at velocities different for each microcrystal gel. The flow curves indicate rarely encountered pseudoplastic-rheopectic behavior. The apparent viscosities decrease sharply with increasing rates of shear.

The microcrystals pervade the whole system at low concentrations without settling. Their three-dimensional network has a certain rigidity and, consequently, they should be characterized as gels. Beyond the yield value, increasing rates of shear lead to decreasing apparent viscosities. The gels liquefy reversibly upon shaking and solidify upon standing, the process being time-dependent. Thus all microcrystal gels appear to be thixotropic and pseudoplastic with a yield point. Moreover, rheological studies indicate that they have an irregular, time-dependent, extreme yield point, and portions of their flow curves show rarely encountered pseudoplastic-rheopectic behavior. This property for five microcrystal-gel systems is illustrated in Fig. 2, along with a synthetic microcrystal-gel of alumina.

One common feature of these gels is that the microcrystals recovered from each polymer have on their surface polar groups such as —NH₂, —OH, or —CONH— which are capable of hydrogen bonding. Hence particle-particle interaction or particle-solvent-particle type of bonding has to be assumed as a significant operating force of gelation. It is likely that the small spheres and ellipsoids first form chains of beads which branch off and subsequently pervade the system, whereas the long, rod-shaped microcrystals are probably arranged in the gels like architectural scaffolding or heaped matchsticks. Such arrangements assure the stability of the three-dimensional network.

Commercial developments. Seven types of colloidal microcrystalline polymers are known to exist, five of which are in the commercial, semicommercial, or market development stage. Each takes three forms: gel, powder, and cream. Avicel, the first microcrystalline polymer, which is made of cellulose, has been produced commercially for some time. It can be made into a gel that will not melt. It is used in ice cream, ice milk, canned tuna salads, and hollandaise sauce. Salad dressings and sauces based on Avicel are unique because they can be heat-sterilized, unlike conventional products from which the oils would separate out. In the form of a free-flowing powder, millions of pounds of Avicel are used each year in pharmaceutical tableting alone.

The next product developed was Aviamylose, a microcrystalline starch. It offers the possibilities of uses similar to those of Avicel, to which it is related in structure. Aviamylose, however, has caloric value (4 cal/g).

Avibest, a microcrystalline product of the mineral chrysotile asbestos, is being used in polyester and epoxy resins to control their flow properties. It can also be incorporated into a magnetic clutch

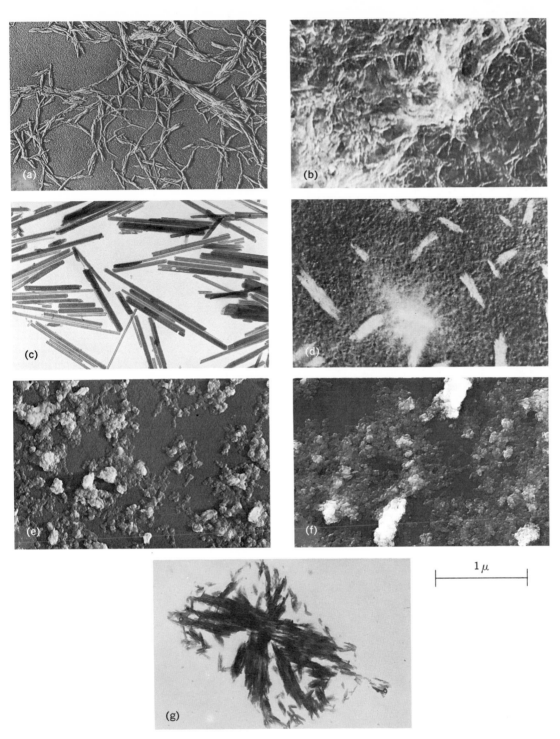

Fig. 3. Electron micrographs of microcrystalline polymer products. (a) Avicel, from wood-pulp cellulose. (b) Aviamylose, from amylose starch. (c) Avibest, from Quebec chrysolite. (d) Avitene, from bovine-hide collagen. (e) Aviamide, from polyamide-6-resin. (f) Aviester, from polyethylene terephthalate polyester resin. (g) Aviolefin, from isotactic polypropylene.

grease capable of serving as a gearless regulator for dampening the movement of mechanical parts such as weigh scales.

Avitene is a microcrystalline form of collagen, a natural polymer that is the chief constituent of connective tissue and bone. It has exciting potential for medical and cosmetic products. One such potential application is in synthetic bone. Crystals of apatite, the chief mineral constituent of bone,

are incorporated in Avitene, leading to compositions that can then be fashioned into bone- and cartilagelike shapes with characteristics almost identical to those of normal bone. More spectacularly, Avitene has been successful in laboratory tests as a novel blood-clotting agent and adhesive for sutureless surgery. Gaping wounds were made in the liver and spleen and many other vital organs, including the heart, stomach, and major blood vessels, of more than 30 dogs. Avitene flour was packed in the wounds; the bleeding stopped rapidly without the use of suturing. This new microcrystalline polymer product has the potential of playing a major role in all types of surgery, especially in controlling the bleeding from small and middle-size vessels. In addition, it could be an important aid in control of bleeding from major vessels where adequate suturing is difficult or impossible because of inaccessible location.

The fifth marketable microcrystalline polymer is Aviamide, a microcrystalline nylon. It is being developed for coating textile glass fibers, thus permitting easier dyeing of fiber glass fabrics, including drapes. Also, it can coat metals and thus may provide a cheaper and quicker method of insulation or wealthering protection. Aviamide offers an aqueous route to producing ultrathin coatings or films of nylon. One special application, for example, would be to use Aviamide to coat supermarket carts and outdoor furniture on which chromium, which is expensive and in short supply, is now being used.

The two newest additions to the microcrystalline polymer family are Aviester, which is microcrystalline polyester, and Aviolefin, which is microcrystalline polypropylene. These are in the laboratory stage of development and are not yet available commercially. However, as with Aviamide, they too could be used to coat metals and glass.

Electron microscopy has clearly revealed that the size and shape of the respective microcrystalline particles are remarkably similar for several different species. An even more striking fact, however, is that, despite the great disparity in chemical composition, colloidal dispersions, creams, and gels of these materials possess very similar rheological, physical, and functional properties. Figure 3 compares individual microcrystals for one grade of product from each of the seven available colloidal microcrystalline polymer products.

For background information see POLYMER; POLYMER PROPERTIES; POLYMERIZATION in the McGraw-Hill Encyclopedia of Science and Technology.

[O. A. BATTISTA]

Polyploidy

Recent studies have revealed the existence of naturally occurring polyploids within several species of South American anurans and, most recently, within a species of North American tree frog.

In a haploid cell, such as a gamete, or in a haploid organism, each type of chromosome is represented only once, the entire set of chromosomes constituting a single complement. Somatic cells of most organisms are normally diploid, a condition in which two complements are present and the chromosomes occur in matching pairs, or homologs. Polyploid organisms have more than two complements of chromosomes—triploids have three sets, tetraploids, four, and so on. Although widespread in plants, polyploidy is rare among animals. This has been attributed to the incompatibility of polyploidy with a highly specialized sex-determining mechanism. In some animals sex is determined by the presence or absence of a Y (or W) chromosome; in others it depends upon the ratio between body, or autosomal, chromosomes and

the X (or Z), or sex, chromosomes. Meiosis in polyploid, separate-sexed animals would produce an imbalance between autosomes and sex chromosomes, thereby resulting in intersexes, in supermales and superfemales, or in largely sterile offspring.

Instances of natural triploidy in animals are often linked with a scarcity or absence of males and propagation by parthenogenesis (activation of an egg without a sperm). This is true of certain lizards, such as species of Cnemidophorus. Studies by T. M. Uzzell, Jr., of the triploid salamanders within the Ambystoma jeffersonianum complex show that the egg is activated by a sperm (of a different species) but that syngamy apparently does not occur.

South American anurans. Two species of South American leptodactylid frogs, Odontophrynus americanus and Ceratophrys ornata, originally reported by F. A. Saez and N. Brum to be diploids with unusually high chromosome counts, were later shown by M. L. Beçak and coworkers and by J. P. Bogart to be polyploids.

Although both O. cultripes and O. occidentalis have a diploid number of 22 chromosomes, at least some populations of O. americanus are tetraploid, with 44 chromosomes in both somatic and germ cells. C. ornata and C. dorsata both are octoploid, each having 104 chromosomes which can be ordered into 13 groups of eight homologs each. Cytophotometric studies of O. cultripes, O. americanus, and C. dorsata by W. Beçak and coworkers have shown that deoxyribonucleic acid (DNA) content occurs in a ratio of 1:2:4, respectively, which thus confirms cytological findings. Further, in a comparison of plasma proteins obtained in starch gel electrophoresis tests, W. Beçak and colleagues found that O. cultripes has only a single albumin-like protein band, whereas O. americanus is polymorphic, having different phenotypes.

In O. americanus there are 11 ring quadrivalents formed during metaphase I of meiosis. Similarly, in C. dorsata, octovalents (as well as hexavalents, quadrivalents, and bivalents) have been observed in spermatocytes I. As Carl P. Swanson pointed out, the presence of multivalents specifies autopolyploidy rather than allopolyploidy as the specific type of polyploid origin involved. Autopolyploids arise through duplication of a single genome (one parental type), whereas allopolyploidy involves duplication of two genomes (through hybridization between two different parental types).

Artificial hybrids obtained by W. Beçak and coworkers in crosses between the diploid O. cultripes and the tetraploid O. americanus are triploids, having 11 groups of three homologs each. These triploids show corresponding DNA values.

Finally, within the family Hylidae M. L. Beçak and coworkers found that Phyllomedusa burmeisteri is a tetraploid, having 52 chromosomes which may be ordered into 13 groups of four homologs each.

North American frogs. It has been recognized for some time that two distinct mating call types, one with a fast pulse rate, the other with a slow pulse rate, exist within what was formerly regarded as a single species of tree frog, Hyla versicolor. Three factors made it evident that each call type actually represented a distinct species: (1) the high degree of genetic incompatibility between individuals of the two call types, as demonstrated by Clifford Johnson; (2) discrimination in favor of their own call type by female frogs, as shown by M. J. Littlejohn and coworkers; and (3) an emerging pattern of sympatry of the two call types. As a result, the slow-pulsed call type retained the specific name of *H. versicolor*, and the fast-pulsed call type was accorded full species rank as *H. chrysoscelis*. Morphologically, the two are indistinguishable, and thus may be considered cryptic species.

The original determination by R. J. Bushnell and coworkers of a diploid number of 24 chromosomes for "*H. versicolor*" (apparently *H. chrysoscelis*) was assumed to apply to both call types as well as to most other species of North American *Hyla*.

Recently, however, A. O. Wasserman discovered that both larval and adult individuals of a Bergen County, N.J., population of *H. versicolor* have 48 chromosomes which can be arranged into 12 groups of 4 each. Further work by J. P. Bogart and Wasserman established that other populations (in Texas and South Carolina) of *H. versicolor* are also tetraploid, whereas *H. chrysoscelis* is diploid ($2n = 24$). A comparison of karyotypes of the two species shows that corresponding chromosomes are essentially similar, with only minor differences in arm length and centromeric position. Quadrivalents may be observed in meiotic figures of *H. versicolor*. Cytological examinations show that sperm and red blood cells of *H. versicolor* are considerably larger than those of *H. chrysoscelis*.

As in the case of *Odontophrynus*, hybrids resulting from the cross between *H. chrysoscelis* and *H. versicolor* are triploids ($3n = 36$). Since triploids are highly sterile because of the random segregation of the three chromosomes in each trivalent, this might explain the high rate of offspring mortality in backcrosses between the *versicolor-chrysoscelis* hybrids as reported by Johnson.

H. versicolor is thus the first known, naturally occurring, sexually reproducing polyploid anuran in North America.

Origin of anuran polyploidy. Precisely what causal factor or factors triggered polyploidy in anurans is unknown; however, extensive work by G. Fankhauser and coworkers has shown that polyploidy in salamanders and frogs may be induced artificially by subjecting eggs to temperature shock. When Bogart and Wasserman exposed *H. versicolor* control (♀ *versicolor* × ♂ *versicolor*) zygotes to cold temperatures, several viable hexaploid larvae resulted. T. M. Uzzell, Jr., and S. M. Goldblatt have suggested that production of unreduced eggs may have played a role in the origin of *Ambystoma* triploids. Bogart and Wasserman believe that this same failure of eggs to become fully reduced in meiosis may have been operating in the origin of anuran polyploids.

Polyploidy and sex determination. Susumu Ohno and others have shown that polyploidy can be sustained in naturally occurring biparental populations in which there are no morphologically recognizable sex chromosomes, or in which they are in a primitive state of differentiation. This is apparently the case in amphibians. An imbalance in the sex determination mechanism is avoided because the W (or Y) chromosome is still genetically equivalent to the Z (or X) chromosome.

Evolutionary implications. Many workers have discounted the importance of the role of polyploidy in animal evolution. Indeed, some have charged that no valid instances of sexually reproducing polyploid animals exist. However, on the basis of comparative DNA value studies, Ohno has stated that various degrees of polyploidization probably occurred early in the evolution of vertebrates, prior to the emergence of terrestrial forms. The fact that the anuran species discussed above have chromosome numbers that are even multiples of those of their diploid relatives, that multivalents occur in meiotic figures, and that DNA values correspond leaves no doubt that they are bonafide polyploids.

All of the instances of anuran polyploidy appear to be relatively recent in origin. Studies of speciation in North American anurans by W. F. Blair clearly establish that the initiation of speciation in the *H. chrysoscelis–H. versicolor* pair is attributable to climatic changes in the Pleistocene. The presence of quadrivalents in meiotic figures and the diploidy of species closely related are cited by M. L. Beçak and associates as factors supporting the view that *O. americanus* is a newly evolved tetraploid. This view is strengthened by the discovery by Bogart of some diploid ($2n = 22$) populations of *O. americanus*. Bogart also has found a diploid ($2n = 26$) individual which has been assigned as a questionable *C. ornata*.

It is likely that polyploidy in anurans is subjected to positive selection pressure in regions of rapid or drastic climatic shifts. Doubling of the genome may confer an additional measure of flexibility on the population.

For background information *see* ANURA; POLYPLOIDY in the McGraw-Hill Encyclopedia of Science and Technology.   [AARON O. WASSERMAN]

Bibliography: M. L. Beçak, W. Beçak, and M. N. Rabello, *Chromosoma*, 22:192, 1967; W. Beçak, M. L. Beçak and F. G. Langlada, *Experientia*, 24:1162, 1968; J. P. Bogart, *Can. J. Gen. Cytol.*, 9:531, 1967; J. P. Bogart and A. O. Wasserman, *Polyploidy in Anuran Amphibians*, in press; S. Ohno, *Sex Chromosomes and Sex-linked Genes*, 1967; T. M. Uzzell, Jr., and S. M. Goldblatt, *Evolution*, 21:345, 1967; A. O. Wasserman, *Science*, 167:385, 1970.

Primates, fossil

Fossil primates are known throughout the entire Cenozoic Era, or Age of Mammals, extending over approximately the last 65,000,000 years. Members of this order are among the oldest distinct groups of placentals, having separated from other mammalian stocks at least by the end of the Cretaceous. Some of the youngest primate fossils are the extinct giant lemurs of Madagascar and related forms, which probably survived into the Christian era on that island, only to fall prey to the early human settlers there. Recent discoveries concerning one specific kind of fossil primate, the giant ape (*Gigantopithecus*), will be considered here. In particular, it has recently been proposed that *Gigantopithecus* resembled men and near-men because it had entered a similar environment.

Distribution. Primates are one of very few groups of placental mammals which are adapted to warm climates in exclusion to most other habitats.

In consequence, their fossils are seldom found in association with fossils which were buried in cooler or drier environments. In mid-Tertiary times, when climates were mild and generally damper than they are at present, primates probably reached their maximum range. They occurred in varying degrees of abundance in North America and Europe as well as in those continents where they still survive. In fact, in the earliest part of their history, primates were distributed in greater abundance in the Northern Hemisphere (ranging from China to Europe and throughout North America) than in southern latitudes. Localities where they occurred abundantly in early Tertiary times, such as the Paris and London regions, as well as Wyoming and adjacent states in the United States, were probably situated considerably farther from the North Pole than they are today. Associated plants and animals in both regions show that the climate in these areas was much warmer than it is today. Primates had reached Africa perhaps in the Eocene Epoch (58,000,000–35,000,000 years ago), and it was apparently there that the higher primates of the Old World, the Catarrhini, differentiated. This group includes the stocks which produced the cercopithecoid monkeys of today and that section of higher primates which includes apes and man.

In the late Miocene, between about 12,000,000 and 16,000,000 years ago, it seems that conditions for migration in and out of Africa were favorable. About the beginning of this period, primates apparently spread out of Africa and investigators know that apes then began to appear in Eurasia. At the very end of the Miocene, monkeys followed the earlier radiation of the apes out of Africa, and at about the same period of time the basic differentiation of manlike forms occurred. These had arisen from earlier apes through the evolutionary acquisition of the characteristic features of the Hominidae. These features, as far as they are to be discerned in fossils, include facial shortening, so that a snout or muzzle does not project out markedly; reduction of canine size and development of large, round, thick-enameled cheek teeth set in jaws constructed to effect powerful grinding, an ability that follows from the more vertically set muscles of mastication. Also, the earliest hominids may have preferred bipedal walking when not in the trees, judging by the locomotor preferences of their descendants. Jaw and facial fragments suggest that the first hominids were smaller than human pygmies and were perhaps no larger than the siamang gibbons or the pygmy chimpanzee; that is, they would have stood about $3\frac{1}{2}$ ft tall.

Dragon teeth. An interesting example of one particular kind of fossil primate is *Gigantopithecus*, the largest primate that ever lived. This animal occurred in southern China probably between 500,000 and 1,000,000 years ago. When this animal was first discovered, only three teeth were found, but this small part of the animal was sufficient to tell that it was a primate, related to man and apes, but larger than any primate previously known. Nevertheless, these teeth were not adequate material to show whether its relationship to man was closer than it was to the apes.

Discovery of *Gigantopithecus* was no less strange than the time and place of its occurrence, far from regions where apes are found today. The Chinese have used fossil bones and teeth as medicine for centuries. These are called *lung ku* (dragon bones) and *lung ch'ih* (dragon teeth), and are considered aphrodisiacs as well as cures for gout, cancer, and tuberculosis. While searching through Chinese drugstore collections of such fossils, Ralph von Koenigswald discovered the teeth in 1935, but for more than two decades no one knew exactly where these *Gigantopithecus* fossils had been recovered. During this time, however, a dozen or so additional teeth were found, all of which reinforced the fact that the animal was truly gigantic for a primate.

In 1955 a cave site in Kwangsi Province in southeastern China yielded a nearly complete lower jaw of this animal which, because of the comparatively small canines, was taken to be a female. Later, two additional jaws, those of a young and an old male, were found, as well as hundreds of separate teeth. Strangely, no vestige of other bones of the skeletons could be located in the Chinese cave site.

Discovery in North India. These jaws and teeth of *Gigantopithecus* posed many problems of interpretation which were resolved more fully by the discovery of a fourth and even more complete jaw of *Gigantopithecus* by a Yale expedition to northern India in April, 1968. Opinions as to whether this animal was more closely related to ape-men and men or to the living great apes had long been debated. The Indian jaw came from a time and place which, judging from other fossils, was a relatively dry, open-country environment. The same, or open woodland, condition seemed to have existed in southeastern China when *Gigantopithecus* lived there.

The jaws have broad, flat, thick-enameled teeth adapted for grinding tougher foodstuff than the usual diet of the more frugivorous apes. Of necessity, because of its large size, *Gigantopithecus* must have spent most of its time on the ground. In fact, if the ratio between cheek-tooth area and body bulk which obtains among living apes applied equally to *Gigantopithecus*, it would have weighted in the vicinity of 600 lb.

Thus it seems that *Gigantopithecus* resembled men and near-men, not because it is closely related to them but because it had entered a similar environment, perhaps foraging for tough foods such as roots, grasses, bamboo, grain, and berries in semi-open country on the ground. There is no evidence that it hunted animal prey.

For background information *see* EVOLUTION, ORGANIC; PRIMATES in the McGraw-Hill Encyclopedia of Science and Technology.

[ELWYN L. SIMONS]

Printing in color

The increasing demand for color printing and improved color-print quality, coupled with pressures for shortened production schedules and cost reduction, has led to the development of new preplate color proofing systems.

The primary use for preplate proofing systems has been in conjunction with offset lithography, which is achieving rapid growth in color printing. The same negative or positive color separations that are used to make the lithographic printing plate from which the production job is printed are also used to make the preplate color proofs. The preplate color proofs show the lithographer the quality achieved in the color separations and serve to simulate the final production printed job. The preplate color proofs are used as an internal product quality-control tool and, in some cases, are used to provide the customer with approval copy.

Until recently, most preplate proofs could be made only from negatives, each color being created on a separate piece of transparent film. The effect of four-color printing is achieved by creating a sandwich of several layers of film, one for each of the colors involved. Systems using positives instead of negatives and creating the proof on a single opaque base are now being introduced. The most notable of these systems is the COLEX 520, which is based on the new dry-powder imaging technology of colography.

Colography. The basic principles of colography were jointly developed by the A. E. Staley Manufacturing Co. and the Columbus, Ohio, laboratories of Battelle Memorial Institute. In its basic forms, colography is both negative-working and positive-working, and can be used to create both continuous-tone and screened images.

Colography is defined as the method of forming pictures or copies of graphic materials in which a photophysically induced latent image is developed by selectively embedding a powder with moderate physical force into a light-altered surface. There are many variations of the basic technology which are utilized to form images for specific applications. The COLEX 520 system utilizes a variation on this technology that is suitable for making multicolor proofs or pictures.

COLEX 520. COLEX 520 preplate color proofing is a positive-working system in which a four-color proof is made on a single white opaque proofing base in approximately 20 min. The proof is made in COLEX 520 dual processing units, using a specialty proofing base stock which has a receptor coating. The five basic steps in the process are coat, expose, develop, intensify, and clear. This basic sequence is repeated for each color to be printed down. Although the system is most commonly used as a four-color proofing system, as many colors as desired can be printed down on the single white opaque proofing base stock. The proof is made in the manner described below.

Coat. The first step in the process is to coat the specialty proofing base stock, which has a factory-applied receptor coating, with proofing sensitizer. The proofing sensitizer is a unique and proprietary formulation of Staley/Graphics and is not a conventional silver or diazo formulation.

In commercial practice the coating is applied by flow-coating techniques in the COLEX 520 coating unit. After the coating is applied, 30–40 sec is required for the sensitizer to dry. The sensitizer can be handled under normal incandescent or fluorescent room lighting conditions without the use of filters.

Expose. The sensitized proofing paper is exposed, emulsion to emulsion, with a screened positive color separation (Fig. 1). Exposures are made in conventional graphic arts plate-making exposure and vacuum frame equipment. Exposure times run from 15 to 90 sec, depending primarily on the type of exposure equipment being used.

Correct exposures are determined by a test strip procedure using an exposure guide furnished by Staley/Graphics.

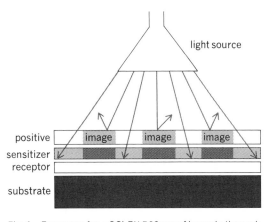

Fig. 1. Exposure for a COLEX 520 proof is made through a screened positive using a conventional graphic arts light source and vacuum frame. (*Staley/Graphics*)

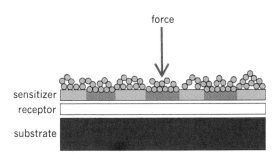

Fig. 2. COLEX 520 image is developed by application of dry powders which are applied to surface of specialized proofing base with moderate force. Powder adheres to image area and is rejected in nonimage area. (*Staley/Graphics*)

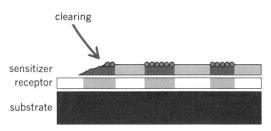

Fig. 3. Sensitizer and powder carrier particles left on the surface of the sheet are removed by the application of a clearing solution. (*Staley/Graphics*)

The exposure step in the process serves to destroy the powder receptivity of the proofing sensitizer in the nonimage area.

Develop. The latent image which was established on the proofing base stock during the exposure step is developed by means of the application of dry powder (Fig. 2). Developing powder of the proper color (magenta, cyan, yellow, or black) is applied to the surface of the proofing stock with an applicator pad. The powder is worked across the image area with a gentle circular motion. Powder particles will adhere to the image area and will be rejected in the nonimage area.

Excess powder is then removed from the surface of the sheet. This is done in the COLEX developing station, first with the aid of an air gun which is used to blow the powder back into the work station. The powder-developing work station contains

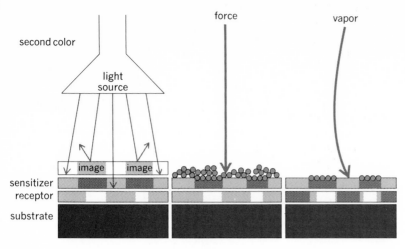

Fig. 4. The second color is put down in the same manner as the first and is overprinted on top of the first color. (*Staley/Graphics*)

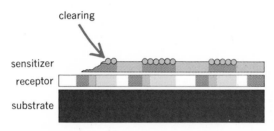

Fig. 5. Sensitizer and powder particles remaining on surface of sheet are removed by application of clearing solution. This step completes the sequence of printing down the second color. (*Staley/Graphics*)

Fig. 6. Two colors have been printed down, and the surface of sheet has been restored to its original condition. The sheet can now be resensitized for the printing down of a third color. (*Staley/Graphics*)

a built-in vacuum and powder-collection system. After the excess powder has been blown off, final traces of powder are removed from the surface of the image area by wiping the proofing base stock gently with a cotton wipe.

The image developed in this step is not permanent and has not yet reached its proper color intensity and hue.

Intensify. The powder image which has been developed is made permanent and brought to its proper color intensity by applying warm moist air to the surface of the image. The application of the warm moist air causes the colorant materials in the powder to migrate from the surface of the sheet into the receptor coat. The image is then actually below the surface of the sheet. The powder particle, without the colorant, and the proofing sensitizer are left on the surface of the sheet.

Clear. The surface of the proofing stock is wiped down with a special clearing solution. This step serves to remove the proofing sensitizer and the powder particles which remain on the surface of the sheet (Fig. 3).

When this step is completed, the sequence of printing down a single color is complete. The five steps can be accomplished in approximately 5 min.

Additional colors are printed down on the same white opaque proofing base stock by repeating the procedure outlined above. An unlimited number of colors can be printed on the same base stock (Figs. 4–6).

Developing powders. The unique method of image development using dry powders permits the COLEX 520 process to be extremely versatile. Colors can be changed simply by changing the color of the developing powder. A variety of process colors and flat colors are available from stock. Custom color powders are formulated and can be matched to almost any conventional printing ink. Therefore with the COLEX system it is possible to get both color variety and color fidelity.

Applications. The COLEX 520 system has been available to the lithographic industry for less than 2 years. Along with other preplate systems, it is beginning to find increasing application in the graphic arts. Research is being conducted by Staley/Graphics to expand the application of colography into proofing for processes other than lithography and for a wide variety of other applications, both within and outside of the graphic arts field.

For background information *see* PRINTING IN COLOR in the McGraw-Hill Encyclopedia of Science and Technology.   [WILLIAM C. LAMPARTER]

Process computer control

The field of process computer control has reached the stage at which well-recognized classes of application have been established. Furthermore, the field has matured to the point that, for each application, the specific requirements of the computer can be detailed accurately. Thus the control and data-handling functions can be performed efficiently. The classes of application and characteristics of the hardware that have become accepted at this time for each task are summarized in this article. In addition, recent progress in hardware design and programming for these tasks is described. Some predictions as to future developments are also covered.

Classes of applications. Some notes and requirements for each type of control computer are covered in the following sections.

Supervisory machine. Initially the supervisory machine was used for data logging, monitoring, and so on. Now it is used to optimize the steady-state operation of a plant or as the middle member of a hierarchy system. This machine is involved in the majority of process-control applications. Its requirements are as follows: (1) All the speed that is available from the manufacturer is needed to utilize present and potential design and programming aids as supplements to the process-control tasks of this machine. (2) The 32K or larger core memories are necessary to take full advantage of the possible executives, compilers, optimization routines, and so forth. (3) Bulk memories of at least 500K and preferably more are required for storing these programs. Speeds of drums or "disks" would be helpful if the above memory can be attained at low enough cost. (4) At least 16-bit word length, plus memory protect and parity, is necessary. More bits are helpful if they serve to improve significantly the instruction repertoire and addressing capability of the machine. (5) Card-handling equipment and line printers are required.

Direct digital control (DDC). DDC was first applied to full-scale plant control in 1962. Third-generation equipment has been influenced greatly by the two classes of requirements for these applications. The first class is for dedicated special control functions, which utilize 16K of core memory of $4\,\mu\text{sec}$ cycle time or less. No drum or line printer is needed. Paper tape equipment for input is sufficient and $14-16$ bit word size are adequate. The second class is for general-purpose control applications. This is similar to the supervisory system except that a 16K core memory may be sufficient. A line printer is needed only for system check-out purposes, not for continuous use.

The dedicated system would be used to completely specify the control requirements ahead of time. The general-purpose system is used to achieve maximum possibility of on-line compilation of control algorithms and flexibility of updating of the complete control package.

Mixed-mode systems. These systems combine the functions of the two classes mentioned above. Generally the mixed-mode system is required when pure DDC will not be profitable or when the overall needs are too small to justify two separate computers. The most common application is for dedicated special control functions, and the requirements are the same.

Hierarchy member. The trend in industry is toward complete plant hierarchies. Requirements for the upper members are as follows: (1) It is necessary to have machines of considerably higher capacity and capability than can possibly be achieved with even the largest versions of the machines considered above; that is, longer word length is needed for wider addressing capability and a larger repertoire of instructions. (2) Magnetic tape, line printer, card-handling, and bulk storage equipment are all necessary for present requirements. Multiprocessing capability is also very important. (3) As before, all possible speed can be utilized.

Special-purpose, dedicated, nonmodified systems. The machines are used for special-purpose tasks such as component testing, system check-out, and analytical instrument monitoring. They are justified for one special task only, and no program modification is planned. Such machines can be much slower than any of those discussed above and can be machine-programmed to conserve memory. Usually $4-8$K core is sufficient, and no bulk storage is required. Since only simple input/output or man-machine interface is required, memory requirements are again decreased.

Systems availability. Currently, the most important factor in the acceptability of process-control computers by company managements, and the major stumbling block in further progress is the problem of overall systems availability. In other words, what percentage of the total time can the computer control system be counted upon to carry out its assigned task? There is, of course, a combination of a very long mean time between failures (MTFB) and a very short average repair time. A figure of 99.95% has been generally accepted by the industry as a desirable goal for overall systems availability.

The second problem in this area is that of rapid and accurate maintenance. This can probably be solved by the use of sophisticated diagnostic techniques by the computer itself and by the "swapping" of major units of the system which have been shown by the diagnostics to be marginal or to have failed. It is interesting that Ferranti Ltd. has already instituted such a procedure for maintenance of its ARGUS 500 systems.

Most of the major hardware problems which were prevalent with the early process-control computer systems have now been solved or the path to their solution is clear. These problems have been replaced, however, by a set of equally difficult ones concerned with the system programming or software.

Programming. In early applications, the size and complexity of the programs required to effect a successful control system were underestimated. In addition, the early computers lacked the programming aids which are now available with most manufacturers' equipments. As a result, large programs had to be squeezed into too small a memory by a programming staff which was also too small. As a consequence, the final installations of the early machines were invariably late and required a very long and frustrating check-out period.

Programming aids such as compilers and special process-control languages have recently eased the programming task greatly. However, this advance has been almost negated by the greatly increased size and complexity of recent overall software packages required for the newer and more sophisticated installations.

The result has been a major shortage of personnel qualified by training, experience, and innovative ability to carry out the programming of today's systems. There is thus a major incentive to reduce this total programming load by developing new and more universal programming aids, by standardizing them for easier personnel training and use, by instituting a ready transfer of successful programs to new computer systems, and by sharing mathematical models and their pertinent programs by different companies and industries.

A standardized set of process-control programs will go a long way toward completing the digital revolution in industrial control that is now well under way.

Software standardization. A set of requirements for achieving standardization is as follows:

1. A high-level programming language should be available. For example, a language which includes USASI standard FORTRAN (X3.9-1966) as a subset would admirably satisfy such a need.

2. The high-level process-control programming language should permit the following industrial control functions to be programmed completely in this language: (a) bit manipulation, (b) bit testing, (c) process interrupt handling, (d) interrupt inhibit and permit, (e) parallel task execution, and (f) file handling.

3. The goal of software design should be to minimize the amount of specific information which a process engineer needs to know about a control computer and its associated software to implement computer control for a known plant.

4. Specialized process-control programming systems should be developed to carry out the following: (a) make the organization and development of a process-control computer system as simple as

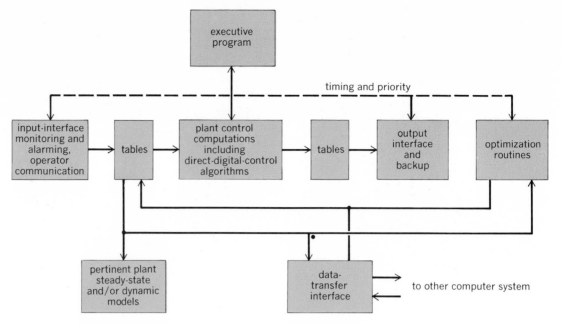

Fig. 1. Block diagram of proposed industrial control program system.

possible; (b) contain an organizational method by which the user can carry out any functions possible with conventional control systems of programming, such as by using a "fill in the blanks" technique; (c) preserve at least as much compatibility between the control practices of different industries as is possible with conventional analog control hardware; and (d) permit on-line addition or deletion of inputs, control computations, and outputs and changes of coefficients without recompilation or reassembly.

5. Terminology should conform to ISA Standard ISA-S5.1, 1968.

6. Software systems should be developed in a modular fashion to the maximum extent possible, the functions of the modules being rigidly defined. These modules should communicate as much as possible through tables. Table format should be such that as newer, more efficient forms are developed substitution of complete blocks and reas-

sembly of the program will be permitted without the necessity of reworking the other blocks. Figure 1 presents one form which such a system might take. It would permit standardized software to be used in a multicomputer system.

Large industrial plants. Because of the wide interest in computer control, it has received even more than its share of this speculation. Some of the resulting proposals concerning the form of future computer-control systems are most intriguing while still being quite technically and economically creditable. One such system has been recently proposed as a result of a major survey of the process control industry.

By making use of the full digital techniques mentioned above, the following concept was developed. Related costs were estimated for a 1250-loop digital-control system for a large industrial plant such as a complete petroleum refinery (Fig. 2). Table 1 outlines the characteristics of the machines involved. Table 2 presents the expected costs for the system; a similar system currently in operation costs $2,500,000−3,000,000.

Note that, for convenience, the 1250 control loops are distributed equally among five separate direct-control computers. The remote multiplexers each have about 25 inputs and transmit information to the control computers digitally. The control computers in turn can transmit information to the supervisory machine. Note also that the crossbar in the large machine, plus the appropriate provision in the small machines, allows all information coming to one of the small computers to be transmitted to the standby computer should the small computer fail for any reason. The large disk of the supervisory machine would contain an image of the program on each small machine for instant transmission to the backup as it was needed.

Supplementation. Should these predictions come to pass by 1975, as expected, they will very nearly meet the requirements which have been made previously for full acceptance of digital process-control systems in the United States.

Table 1. Characteristics of computer systems involved in proposed future digital process-control systems

Function	Value
Remote multiplexers	
Approximate number of inputs	25
Length of analog input leads	<100 ft
Sampling rate	Very fast because of burst mode operation
Direct control computers	
Total inputs each	~250
Memory cycle time	$0.5\ \mu\text{sec}$
Organization	Parallel
MTBF	10,000−20,000 hr
Memory size	32K
Work length	18 bits
Supervisory control computer	
Memory cycle time	$0.4\ \mu\text{sec}$
MTBF	5000−10,000 hr
Memory size	160K
Word size	36 bits
Bulk memory (disk) size	$5\overline{\text{M}}$

Digital control systems will replace analog control systems for essentially all new plant situations, with the following provisions: (1) A fivefold increase in basic machine speed must be achieved over that of third-generation hardware, which now has a $1.75 - 2.0$ μsec memory cycle time. (2) A fourfold increase in reliability must be achieved over that of third-generation equipment, which is now $2000 - 4000$ hr MTBF for all electronic equipment (not including electromechanical equipment). (3) A fivefold decrease in computer system costs must be achieved under those for third-generation equipment, which are now $250,000 - 400,000 for typical application equipment of a single machine with $50 - 250$ loops.

In implementing these larger and more advanced systems, it is expected that the user companies will be doing all the specialized programming required for the system themselves. However, this programming will be much less than that necessary today and will consist mainly of that contained within the model block of Fig. 1. It is hoped that the programming shown in the remaining areas of Fig. 1 will have been standardized and will be available in that form from the vendor. In addition, it is expected that much more use will be made of the modern control theory now being produced by university and research organizations but at present receiving little use in industry. The further research that is now proceeding, as well as the increased capabilities of the machine used, should solve the problems now hindering the wider application of control theory.

Future developments. Some developments in this field that can be expected during the next 10 years are listed below.

1. One, or possibly two, full new generations of process-control machines can be expected. This will mean at least a fivefold increase in raw computing speeds.

2. A greatly increased reliability in terms of a much longer MTBF of the computing hardware will be achieved. Although systems equivalent to those used today will also be maintained more easily, the larger size and more complex nature of the foreseen systems that will be needed to accomplish the much more complex control and data tasks envisioned will more than offset the advantages gained by easier maintenance.

3. An ever increasing use of "natural language" programming for process-control tasks will take place. Present trends indicate that this natural language will probably be an extended FORTRAN with many built-in process-control functions.

4. Much more information on mathematical models of various processes and on applicable

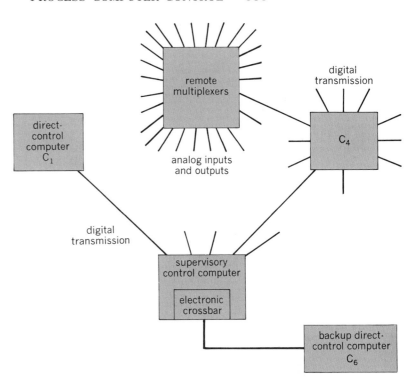

Fig. 2. Possible digital process-control system of the next decade.

control and optimization techniques will be available in the literature. Accurate information on the relative importance of computer control for these processes will also be available. This additional information will not include "justification." The next 10 years should reduce the questions of "if" and "when" to one of "how."

5. Acceptance of digital control as the basic control method and an ever-increasing use of vendor standard routines, if not industry standard routines, along with a burgeoning need for company information systems, will cause application engineers to be increasingly concerned with the overall systems aspects of their installation rather than with the many intimate details of application.

6. One cannot expect any greater compatibility of machine languages through similar instruction sets and codes. Compatibility of programs between different machines, if it is to be achieved, must be at the compiler or natural language level. Thus all of the present symbolic and machine language programs will probably be worthless as soon as the machines for which they have been written become obsolete.

7. The "housekeeping chores" involved in managing the interconnections, data links, and so on in the major hierarchy systems that will be in existence, will require memory allocations and commitments of machine duty cycle far beyond those required today.

8. The capabilities of the envisioned system and the requirements of competition within the industry will demand a sophistication of application far beyond that common today. Although 10 years may not be enough time to achieve it, all of the plant operations must eventually be optimized. Even today, the increase in application requirements occurs much more rapidly than does the normal assimilation of the necessary engineering

Table 2. Predicted costs of a proposed digital process-control system

Item	Cost
Direct digital-control computers	
Each	$ 25,000
Total	$150,000
Supervisory control computer, including peripherals	$150,000
Instrumentation wiring and	
analog-to-digital and digital-to-analog	
distribution	$250,000
Total	$550,000

knowledge and techniques into industry. Thus either a major educational effort beyond any presently dreamed of or a whole new method of application will be required before optimization of plant operations can be accomplished.

9. Large executive programs for management information system hierarchies and subroutine procedures for required plant optimization will be supplied by vendors. The needs for very wide applicability will tend to make these programs overly large and inefficient.

For background information *see* COMPUTER; CONTROL SYSTEMS; PROCESS CONTROL in the McGraw-Hill Encyclopedia of Science and Technology.   [THEODORE J. WILLIAMS]

Bibliography: T. J. Williams, Computer systems for industrial process control: A review of progress, needs, and expected developments, in *Information Processing 68*, pp. 907–916, 1969; T. J. Williams, Interface requirements, transducers, and computers for on-line systems, *Automatica*, 5:845–854, 1969.

Prospecting, petroleum

Emphasis has shifted in recent years in petroleum exploration from structural to stratigraphic trap exploration. Structural traps were discovered with geological and geophysical information in which structural highs, such as anticlines, were the target for exploration. In stratigraphic trap exploration in carbonate rocks, the geologist searches for favorable lithologies which represent environments that at the time of deposition had primary interparticle porosity or in which secondary porosity was prone to develop after deposition. This advance in recognizing criteria for depositional environments in carbonate rocks was a breakthrough in petroleum prospecting. Although regional environmental interpretations of transgressions and regressions, movements of shoreline, and gross aspects of continental and marine sedimentation have been understood for a long time, only during the last 15 years has the science of petroleum geology developed criteria for environmental recognition of specific outcrops, wells, or even hand samples.

Two geological environments are particularly favorable for stratigraphic trap exploration: (1) the high-energy zone in which primary interparticle porosity is formed by deposition of well-sorted calcareous sand under the influence of strong currents or waves (intergranular winnowing), and (2) sebkhas, which are supratidal flats near a shoreline dominated by the local level of the groundwater. In sebkhas, secondary porosity is formed during dolomitization. A description of each of these two environments and criteria for their recognition are given below.

Primary interparticle porosity. During much of the geologic past, shallow seas had depositional slopes of a very low angle, on the order of less than 20 cm/km. Across these slopes there were sharp variations in energy conditons. As a result of these variations in energy conditions, three different sedimentary environments were formed, approximately parallel to the shoreline: the open sea beneath wave base (zone X of M. L. Irwin); a high-energy belt where waves impinge on the sea floor and expand their kinetic energy (a zone roughly coinciding with the shoreline) (zone Y of Irwin); and a broad, shallow depositional zone landward of the high-energy belt in which wave and current action is largely lacking (zone Z of Irwin). The high-energy belt (zone Y) makes a natural stratigraphic trap. This zone can be recognized by a lack of lime mud, which was winnowed out by waves and currents during deposition, and by the fact that the particles making up the rocks of this zone are skeletal fragments and oolites. The zone landward (zone Z) is a low-energy environment in which water depth was extremely shallow, usually less than 1 m in depth, and for a large expanse only centimeters deep. In this environment lime mud is dominant, and the grains are pellets and skeletal fragments, alga-coated particles, or undulating algal laminations (stromatolites). The rocks of this zone are tight and serve as a seal to the reservoir of the high-energy zone. The low-energy zone beneath wave base (zone X), seaward of the high-energy belt, is the site of deposition of fine-grained particles derived from the high-energy belt. This environment is below wave base, and turbulence is at a minimum. Reducing conditions dominate, and abundant organic matter accumulates in this zone, derived from zone Y as an "organic rain" and from indigenous plankton and nekton of zone X itself. This environment, recognized by the presence of lime mud, silt-sized skeletal fragments, and dark coloration resulting from organic matter, serves as a principal source of hydrocarbons for the reservoir forming in zone Y.

Secondary porosity. The rocks formed in the geological past in sebkhas are recognized by finely laminated dolomite or dolomitic limestones, stromatolites, mud cracks, and dearth of fossils, but there are abundant trace fossils and burrow mottles, textures resembling bird's-eye which may result from gases or burrowing, flat pebble conglomerates, bituminous material, and evaporites (gypsum or anhydrite). These structures and mineralogies develop in an arid climate under conditions similar to those of zone Z. They reflect not only low energy but also hypersalinity, upward transpiration of interstitial waters through the porous sediment, and evaporation at or near the sediment-air interface. Dolomite formation results in porosity development; hence the sebkha becomes a reservoir. Evaporites which interfinger with the dolomite serve as a seal, and the abundant organic matter of the algae serves as the principal source of hydrocarbons to the reservoir in the dolomite.

For background information *see* DOLOMITE ROCK; PETROLEUM, ORIGIN OF; PETROLEUM GEOLOGY; PROSPECTING, PETROLEUM in the McGraw-Hill Encyclopedia of Science and Technology.   [GERALD M. FRIEDMAN]

Bibliography: M. Braun and G. M. Friedman, *J. Sediment. Petrology*, 39:113–135, 1969; G. M. Friedman, *Geol. Soc. Amer. N. East. Sect. 4th Annu. Meet.*, pt. 1, p. 20, 1969; G. M. Friedman (ed.), *Soc. Econ. Paleontol. Mineral. Spec. Publ.*, no. 14, 1969; G. M. Friedman and J. E. Sanders, in G. V. Chilingar et al. (eds.), *Carbonate Rocks: Origin, Occurrence, and Classification*, Developments in Sedimentology 9A, 1967; L. V. Illing, G. V. Wood, and J. G. C. M. Fuller, Reservoir rocks and stratigraphic traps in non-reef reservoirs, *7th*

World Petr. Congr. Proc., 2:487–499, 1967; M. L. Irwin, *Amer. Ass. Petr. Geol. Bull.*, 49:445–459, 1965; D. J. J. Kinsman, *Amer. Ass. Petr. Geol. Bull.*, 53:830–840, 1969; L. F. Laporte, *Amer. Ass. Petr. Geol. Bull.*, 51:73–101, 1967.

Psychology, physiological and experimental

The human brain is a bilaterally symmetrical structure which is for the most part richly interconnected by two main bridges of neurons called the corpus callosum and anterior commissure. These structures are surgically sectioned in man in an effort to control the spread of epileptic seizures. Although there is no apparent change in everyday behavior of these patients, dramatic differences in cognitive function can be demonstrated under specialized testing conditions. Because of these studies it now can be said that in normal man these cerebral commissures are largely responsible for the behavioral unity that man enjoys; the neural mechanism keeps the left side of the body up to date with the activities of the right, and vice versa.

Results of tests involving speech. Consider the behavioral responses of patients whose cerebral commissures have been sectioned. In an informal examination, even the experienced interviewer would be hard-pressed to detect change. The patient walks, talks, and behaves in a normal fashion. Only under testing conditions which utilize stimuli that are lateralized exclusively to one hemisphere or the other are dramatic effects observed (see illustration). When visual stimuli are flashed to the

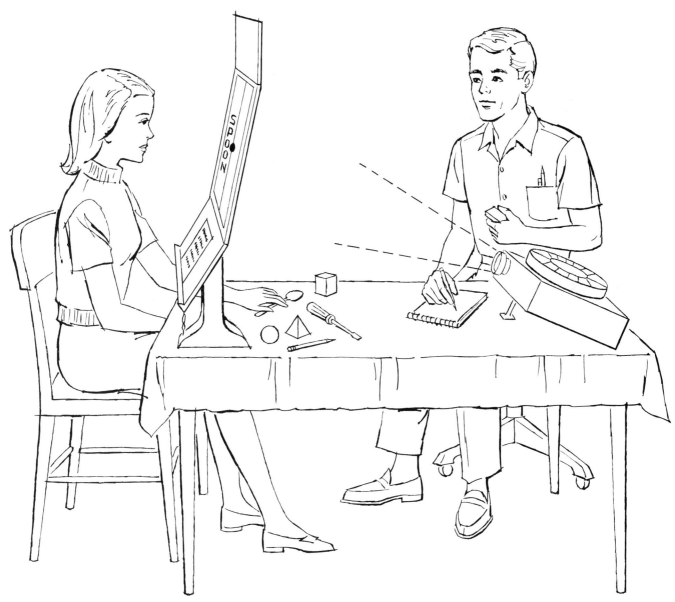

Main features of testing apparatus used in examining split-brain patient. Subject fixates dot in middle of the screen. Visual stimuli are then quickly flashed to either the right or left visual field. Here "spoon" is flashed to the left field (right hemisphere), and the subject is required to make a match by finding the item with the left hand. Because touch information from the left hand is projected to the right hemisphere, the match is easily made. Yet the patient, talking out of the left hemisphere, exclaims that she does not know what the test item is.

left of fixation of the eye, all information is exclusively projected to the right hemisphere, and vice versa. Likewise, when the left hand manipulates objects that are held out of sight, all relevant sensory information projects to the opposite right hemisphere. Similarly, information from objects held in the right hand is relayed exclusively to the left hemisphere. This normal state of affairs permits the separate input-output testing of each half-brain. Using these sensory avenues, remarkable "deconnection" effects are seen in split-brain patients. For example, if a picture of an apple is flashed in the right visual field, the patient describes the object normally. However, if the same picture is flashed in the left visual field, the patient in the early days of postoperative testing denies that the stimulus was presented at all. Gradually, after many test sessions the patient may have the impression that something was flashed, but he is unable to say what. In brief, this disparity of recognition in the two sides of the visual field occurs because the information is projected to the right hemisphere, which is incapable of speech. Because the right hemisphere is now disconnected from the left, information arriving in the right hemisphere cannot be communicated by means of speech.

Results of tests not involving speech. When tests are used which do not require a spoken response, numerous mental abilities are observed to be present in the "disconnected" right hemisphere. If, for example, instead of being asked for a verbal response, the subject is required with the left hand to retrieve from a series of objects the one most closely related to the flashed visual stimulus, good performance is seen. Even though the patient is unable to describe a picture of an orange flashed to the left field, when the left hand searches through a field of objects placed out of view, it correctly retrieves the orange. If asked what the object is, the patient would say he does not know. Here again the left hemisphere controls speech but cannot solve the problem. The right hemisphere solved the problem but cannot elicit speech.

Right hemisphere. Through the use of this kind of test procedure, a variety of other mental abilities were observed to go on in the separated right hemisphere. It can read simple nouns but not verbs. It can recognize some adjectives. It can recognize negative constructions but cannot differentiate constructions in the active or passive voice. It cannot pluralize nouns. In brief, it can manage some aspects of language but by no means all.

Left hemisphere. Despite its linguistic superiority, the left hemisphere does not excel the right in all tasks. Other tests have demonstrated that in some specialized functions, such as arranging blocks to match a pictured design or drawing a cube in three dimensions, the right hemisphere is decidedly superior to the left.

Emotional reactions. In the area of emotional reactions there appears to be equal reactivity in the two hemispheres. Although a brain-bisected patient cannot describe an emotional stimulus presented to the right hemisphere, an emotional response such as laughter is nonetheless elicited. In one instance, when a picture of a nude was flashed to the right hemisphere, the patient said she did not see anything and then chuckled. When asked what the laughing was about, she might say that it was "a funny machine." Here the right hemisphere clearly had enough ability to elicit laughter but not speech. The left hemisphere simply observed the changed emotional state and said that something must be funny, but did not know what.

Summary. The foregoing discussion outlines some of the kinds of changes that can occur in cognition when lesions are made in the nervous system. Behavior does have a neurologic basis, and it is the aim of neuropsychology to come to a better understanding of these processes.

For background information *see* APHASIA; BRAIN (VERTEBRATE); MEMORY; SENSORY LEARNING; VERBAL LEARNING in the McGraw-Hill Encyclopedia of Science and Technology.

[MICHAEL S. GAZZANIGA]

Bibliography: M. S. Gazzaniga, *The Bisected Brain*, 1970; M. S. Gazzaniga, Dyspraxia following division of the cerebral commissures in man, *Arch. Neurol.*, 16:606–612, 1967; M. S. Gazzaniga, The split brain in man, *Sci. Amer.*, 217:24–29, 1967; M. S. Gazzaniga and R. W. Sperry, Language after section of the cerebral commissures, *Brain*, 90: 131–148, 1967.

Psychopharmacological drugs

Reports of recent research in psychopharmacology have been concerned with plant hallucinogens, their use in primitive societies, and the development of analytical techniques for these and other hallucinogenic drugs.

Plant hallucinogens. Psychopharmacological drugs, especially the plant hallucinogens, have been employed by man for millennia, always in a religious or magico-medical context. Only during the past quarter-century have their use, misuse, and abuse constituted a problem in Western society.

Plant hallucinogens have recently attracted ethnobotanical attention, and much new information concerning their use in primitive societies has accrued. The identification of many species hitherto not known to have hallucinogenic status has been matched by spectacular advances in an understanding of their phytochemistry. These studies have been concurrent but in no way allied to the new interest in hallucinogens in sophisticated societies.

It has recently become increasingly obvious that primitive groups employ plants not only to induce visions but also for auditory, tactile, olfactory, and taste hallucinations—often for a combination of these psychotropic effects.

Hallucinogens are concentrated in the plant kingdom among the fungi and the angiosperms. Chemically, they comprise a variety of organic constituents in two broad categories: nonnitrogenous compounds, the most important of which are the phenylpropenes and the dibenzopyrans; and the nitrogenous compounds—alkaloids or related bases, the principal kinds of which are the isoxazoles, tropanes, quinolizidines, phenylethylamines, isoquinolines, tryptamines, β-carbolines, ergolines, and iboga-indoles. The active principles of some hallucinogens are not known.

Hallucinogenic plants are employed almost cosmopolitanly, but many more are used in New World than in Old World societies. The basis of this disparity is probably cultural, for there seems to be no botanical reason to presume that the floras of the two hemispheres differ significantly in genera of plants likely to contain psychoactive constituents.

There are a number of minor or poorly understood hallucinogens, in addition to some that still remain to be discovered or botanically identified. The mescal bean (*Sophora secundiflora*) was formerly the basis of the vision-seeking Red Bean Dance of the American Plains Indians; the beans, containing cytisine, are highly toxic. The red beans of *Erythrina* and the red and black beans of *Rhynchosia* may have been utilized in Mexico as narcotics. In northern Canada, sweet calomel (*Acorus calamus*) may possibly have been valued as an hallucinogen. Another aroid, *Homalomena*, is reputedly used hallucinogenically in Papua; the leaves, mixed with leaves and bark of the himantandraceous *Galbulimina belgraviana*, are employed to induce an intoxication characterized by a sleep in which visions and dreams are experienced. *Galbulimina* contains isoquinoline alkaloids.

A psychotomimetic snuff, *rapé dos indios*, is said to be prepared from fruits of the central Amazonian moraceous tree *Olmedioperebea sclerophylla*, but nothing is known of its chemistry. Tribes of central Asia have long used a mint, *Lagochilus inebrians*, as an intoxicant; recent Soviet investigations point to a polyhydric alcohol as the active principle. In southern Mexico, the Mazatecs of Oaxaca use the leaves of *hojas de la pastora* (*Salvia divinorum*) and of two introduced Old World species of *Coleus* as divinatory agents, but psychoactive constituents have yet to be isolated from these mints.

Among the most recently reported hallucinogens are the composite *Calea zacatechichi*, utilized by the Chontals of Oaxaca as a sacred intoxicant, and two species of puffballs (*Lycoperdon*) employed in divinatory narcotics by the Mixtecs of the same Mexican state. In the southern Andes, the leaves of *tabaco del diablo* (*Lobelia tupa*), which contain lobeline and several of its derivatives, are smoked for their narcotic effect. Two very poorly understood intoxicants, possibly hallucinogenic, are the Chilean endemics *Desfontainia spinosa* var. *hookerii* and *Gomortega keule*, both employed by the Mapuches. The well-known poison *Coriaria thymifolia*, containing picrotoxine-like sesquiterpenes, is reputedly utilized on occasion as a hallucinogen in Andean Ecuador, where it is called *shanshi*; the intoxication is said to be characterized by sensations of flight. Several Andean species of the ericaceous *Pernettya* have fruits that induce mental confusion and permanent insanity. The giant cactus *Trichocereus pachanoi*, known locally as *San Pedro*, contains mescaline and is the basis of a divinatory drink in highland Ecuador and Peru. Species of *Ariocarpus*, *Epithelantha*, and *Pachycereus* are presumably valued as narcotics among the Tarahumare of Mexico.

A still enigmatic hallucinogen is the *kanna* or *channa* of the Hottentots of South Africa, reported more than 225 years ago and usually presumed to be a species of the alkaloid-bearing aizoaceous genus *Mesembryanthemum*; it may well have been another plant, such as *Cannabis* or the anacardiaceous *Sclerocarya caffra*. Undoubtedly, the most curious hallucinogen is *kwashi* of the Bushmen of Botswana, who rub slices of the bulb of *Pancratium* over incisions on the head to induce visual hallucinations.

Agaricaceae. Perhaps the most outstanding recent advances in the study of hallucinogens concern the identification and phytochemical studies of the so-called sacred mushrooms of Mexico, known to the ancient Aztecs as *teonanacatl* (flesh of the gods). Since the 1930s, various species of *Conocybe*, *Panaeolus*, and *Stropharia* have been indicated as the psychotomimetic mushrooms basic to elaborate divinatory rites in Oaxaca. They contain the extraordinarily biodynamic compound psilocybin, a hydroxy indole alkylamine with a phosphorylated side chain (the only indole compound with a phosphoric acid radical known from the plant kingdom) and sometimes the unstable derivative psilocin. Both compounds are novel indoles, having the hydroxy radical substituted in the 4 position. Recent studies based on archeological mushroom stones from the highland Mayan sites in Guatemala dated 1000 B.C. indicate that a sophisticated cult worshiping mushrooms existed 3000 years ago far to the south of its present southernmost range.

The fly agaric (*Amanita muscaria*) has been employed hallucinogenically by certain tribesmen in western and northeastern Siberia. Several mushrooms, taken usually in water, milk, or the juice of *Vaccinium uliginosum* or *Epilobium angustifolium*, induce a complicated, often orgiastic intoxication. These tribesmen, having discovered that the narcotic constituent of the mushroom is excreted with almost undiminished psychoactivity, incorporated a ritual urine-drinking ceremony in order to take full advantage of the biodynamic principle in a region where the mushroom is rare. Although for 100 years muscarine has been considered to represent the active compound of this mushroom, recent research has established that the chief biologically active constituents appear to be muscimole (an unsaturated cyclic hydroxamic acid which is, indeed, excreted almost unchanged in the urine) and ibotenic acid. Other minor compounds may likewise play a role in the intoxication.

This mushroom may represent possibly one of the oldest of the hallucinogens. Recent ethnomycological investigations have indicated the probability that *soma*, the ancient god-narcotic of the Aryan invaders of India 3500 years ago, was *A. muscaria*. The use of *soma* died out 2000 years ago, but more than 1000 hymns to this sacred intoxicant have survived in the *Rig Veda*, in one of which there is a reference which has been interpreted as a description of the urine-drinking custom.

Cannabaceae. Perhaps the most widely used hallucinogenic plant is *Cannabis sativa*, source of marijuana, hashish, *bhang*, *ganja*, *dagga*, and other narcotic preparations. *Cannabis* is a monotypic genus, sometimes included in the Moraceae but now usually separated, together with *Humulus*, genus of the hops plant, into a distinct family. *C. sativa*, native of central Asia, has spread to all parts of the globe except the extremely cold regions and the wet tropical forests and has diversified into many ecotypes or ecological races. A triple-purpose plant, source of hemp fiber, an edible seed oil, and a narcotic resin, it represents one of man's oldest cultivated species.

The active principles, contained in a red oil in the resin, are nonnitrogenous dibenzopyrans present normally in highest concentration in the unripened pistillate inflorescence. They are a mixture of stereoisomers collectively called tetrahydrocannabinols and numerous related compounds. It is believed that Δ^1-tetrahydrocannabinol is the prime active constituent. The first biologically active principle structurally elucidated and synthesized was Δ^1-2-3-*trans*-tetrahydrocannabinol. The widely recognized variations in narcotic effects of *Cannabis* are attributable probably to the instability of certain of the closely related compounds which may be converted by aging, maturation, or ecological conditions of the plant from active to inactive constituents or vice versa. This variation and the use of uncontrolled experiments carried out with crude material of unknown chemical constitution have led to much of the disturbing conflict concerning the effects of the drug, and this state of affairs continues to handicap social and legal controls. *See* MARIJUANA.

Although *Cannabis* in Western societies has assumed a hedonistic role, many cultures of the Old World, especially in India and parts of Africa, have employed the narcotic in a religious context, often as the center of ritualistic cults.

Myristicaceae. In 1954 hallucinogenic snuff called *yakee*, made from the red bark resin of the myristicaceous *Virola calophylla* and *V. calophylloidea*, was reported from various tribes of the Colombian Amazon. More recently, a similar but apparently more potent snuff from tribes in the general area of the Brazilian-Venezuelan frontier has been shown to be elaborated from the resin of *V. theiodora*. Various groups collectively known as Waikás call the snuff *epená* or *nyakwana*. In Colombia only witch doctors employ it, but in Venezuela and Brazil the narcotic is taken by all adult males, either individually at any time or ritually in excess at endocannibalistic ceremonies. The resin is scraped into a pot, boiled down, dried, and pulverized. It is sometimes used unadulterated, but occasionally ashes are mixed with the resin. Sometimes the pulverized leaves of the very aromatic acanthaceous *Justicia pectoralis* var. *stenophylla* are also added.

Contemporary investigations show that the snuff prepared from *V. theiodora* normally contains up to 8% 5-methoxy-*N,N*-dimethyltryptamine, but the resin may contain up to 11%. There is also a lesser concentration of *N,N*-dimethyltryptamine. Two new β-carbolines have likewise been found in this species. There are suspicions that *Justicia* may also contain tryptamines.

Very recently it has been discovered that the Boras, Witotos, and Muinanes of Amazonian Colombia take pellets of the resin of *V. theiodora* orally as an hallucinogen. The tryptamines, normally not effective orally, are activated by the β-carboline bases, which provide the necessary monoamine oxidase inhibitor. It has likewise very recently been learned that certain primitive Makú Indians in the Colombian Vaupés may merely drink the resin without any preparation for its psychotomimetic effect.

Myristica fragrans, source of nutmeg and mace, is thought to have been employed as a narcotic by aborigines in southeastern Asia, where it is native; recently, it has been abused as a dangerous hallucinogen in Europe and North America. Ground nutmeg is taken orally in large doses, usually several teaspoonfuls, when intoxication is sought. The toxicology has not yet been fully elucidated, but the psychotropic effects may be due to elemicine or other constituents of nutmeg oil.

Leguminosae. South American Indians elaborated a number of psychotomimetic snuffs, the best known of which is *yopo*, employed mainly in the Orinoco basin and prepared from the roasted beans of the tree *Anadenanthera peregrina* (*Piptadenia peregrina*), usually with an alkaline ash or lime admixture. This is probably the same snuff encountered by Columbus in 1496 in the West Indies, where it was called *cohoba*. The consumption of *yopo* is excessive and leads to an intoxication marked by fury, bravado, hallucinogenic trances, and eventual stupor. Recent analyses indicate that *A. peregrina* contains five indoles, chief of which are *N,N*-dimethyltryptamine and bufotenine (5-hydroxy-*N,N*-dimethyltryptamine). Contemporary ethnobotanical studies suggest that several narcotic snuffs formerly employed in Bolivia, such as *vilca* or *huilca*, and in northern Argentina, such as *cebil*, were elaborated from beans of the related *A. colubrina*, a species likewise containing these active tryptamines.

In Pernambuco, in eastern Brazil, several Indian tribes formerly took the hallucinogenic drink *vinho de jurema*, prepared from the roots of the leguminous tree *Mimosa hostilis*. The drink, containing *N,N*-dimethyltryptamine, induced glorious visions of the spirit world and was the central element of the ancient *yurema* cult, which exalted bravery and warfare.

Malpighiaceae. The use of *ayahuasca*, *caapi*, and *yajé* in the western half of the Amazon and the Orinoco basins has been known for more than a century. The bark of several malpighiaceous lianas (for example, *Banisteriopsis caapi* and *B. inebrians*) is prepared in either a cold water infusion or boiled decoction, with or without admixtures, to provide a potent hallucinogen. The active principles are β-carboline bases—harmine, harmaline, and *d*-tetrahydroharmine. Only recently has botanical and chemical progress been made toward a clear understanding of the drug. A group of Makú Indians in Amazonian Brazil employ the bark of the related *Tetrapteris methystica* to prepare their hallucinogenic *caapi*, but chemical examination of this species has not been carried out. Sundry plants are occasionally added to the basic preparation of *B. caapi* to heighten and lengthen the visual hallucinations. These additives are presently being investigated, but several have been shown to contain psychotomimetic principles themselves: the solanaceous *Brunfelsia*, with alkaloids (probably tropanes); and leaves of the rubiaceous *Psychotria viridis* and of *Banisteriopsis rusbyana*, from both of which *N,N*-dimethyltryptamine has recently been isolated.

Cactaceae. The peyote cactus (*Lophophora williamsii*), used since long before the conquest of Mexico, has been known botanically for more than a century. Its use has spread since 1880 throughout Indian groups as far north as Canada as a central element in the Native American Church, which claims 250,000 communicants. Peyote—or mescaline, the alkaloid responsible for its extraordinary colored visual hallucinations—continues to interest experimental psychologists. More than 30 alkaloids and their amine derivatives, mostly phenylethylamines and isoquinolines, have been isolated from this species, most of them during the past decade.

Lythraceae. Sinicuichi (Heimia salicifolia), long employed in Mexican folk medicine, constitutes one of the most interesting hallucinogens. One of its five alkaloids, cryogenine, an unusual quinolizidine, has recently been indicated as being responsible for the curious auditory hallucinations characteristic of *sinicuichi* intoxication.

Apocynaceae. Although the Apocynaceae is the plant family richest in alkaloids, apparently only one member is utilized as an hallucinogen: *iboga,* the yellowish root of *Tabernanthe iboga,* a narcotic of great social importance in Gabon and adjacent parts of the Congo in Africa. It is the central element in the powerful Bwiti cult, and sorcerers take it before communing with ancestors or the spirit world seeking guidance. Twelve indole alkaloids have been isolated from iboga, the chief psychoactive constituent being ibogaine, a cholinesterase inhibitor, stimulant, and hallucinogen.

Convolvulaceae. Old reports that the Aztecs utilized morning glories as narcotics have been vindicated by the recent identification of *Rivea corymbosa* and *Ipomoea violacea* as modern Mexican hallucinogens. The Aztec narcotics *ololiuqui* and *tlitliltzen* have been identified as these two species, respectively. Only recently have these morning glories been shown to contain four or five psychotomimetic ergoline alkaloids closely allied to LSD and hitherto known only in ergot (*Claviceps purpurea*), a relatively primitive fungus. These constituents have been found in other convolvulaceous genera and may be widespread in the family.

Solanaceae. Although these narcotics have been known for centuries, new ethnobotanical and phytochemical studies have amplified knowledge concerning the tree *Datura—D. arborea, D. aurea, D. candida, D. dolichocarpa, D. suaveolens, D. sanguinea,* and the recently described *D. vulcanicola.* Significant studies on anomalous cultivars of these species and on a new genus, *Methysticodendron,* allied to *Datura,* have been carried out. The active principles of all these species are tropane alkaloids, and recent studies have shown interesting specific and ecological variation in their chemical composition. Other species of *Datura* have been used hallucinogenically in North America and parts of the Old World.

Botanical and chemical studies of the Chilean endemic, the solanaceous *Latua pubiflora,* known locally as *árbol del brujo,* have been initiated, although the plant has not been employed for more than a century since Araucanian witch doctors malevolently utilized it hallucinogenically to induce permanent insanity.

Another solanaceous genus, *Brunfelsia,* recently found in use in the western Amazon as an additive to other hallucinogens, is at present under taxonomic and phytochemical investigation.

The famous hallucinogens employed in witches' brews and potions in Europe in the Middle Ages and earlier were primarily belladonna (*Atropa*), henbane (*Hyoscyamus*), and mandrake (*Mandragora*), all of which contain tropane alkaloids in high concentrations. [RICHARD EVANS SCHULTES]

Analysis. Most of the work in the analysis of hallucinogenic drugs has naturally centered around marijuana (*Cannabis*) and its ingredients, particularly tetrahydrocannabinol (THC). Many gas chromatographic and thin-layer chromatographic systems have been developed in order to further aid in the identification and quantitation of these components.

Cannabis components. A rapid detection method for the major components of cannabis by J. Karlsen involves the use of microslides for thin-layer chromatography. The plant material is eluted from an aluminum oxide column with acetone. The acetone is evaporated, and the residue is dissolved in benzene and eluted through a silica gel column. The material is applied to micro thin-layer plates of silica gel G. Then 10 ml of benzene is shaken with 1 ml of ammonium hydroxide. After separation of the layers, diethylamine is added, drop by drop, to the benzene layer to get a clear solvent which is then used to develop the plate. After development the slide is dried and redeveloped in the same way. The plate is visualized by using a 0.1% solution of fast blue salt in water. Cannabidiol (orange), cannabinol (violet), THC (red-brown), and cannabichromeme (red-brown) can be separated.

R. Turk and coworkers extracted marijuana with petroleum ether, chilled it in a dry ice—acetone bath to remove other plant principles, and eluted it on a silica gel—silver nitrate—calcium sulfate column. The components are identified by a sensitive thin-layer chromatographic system using silica gel G plates impregnated with silver nitrate and developing the plate with benzene. The components are visualized with 0.15% aqueous solution of napthanil diazo blue B, and the Rf values are cannabidiol, 0.16; THC, 0.36; and cannabinol, 0.63.

Gas chromatography. Many gas chromatographic systems have also been presented for the identification of marijuana components such as 2% OV-17 (a phenyl methyl silicone polymer) on gas chrom Q (a flux-calcined diatomaceous earth which has been acid-washed, base-washed, and silanized) at 210°C; 2% QF 1 (a trifluoropropyl methyl silicone polymer) on gas chrom Q at 170°C; 2% XE 60 (a cyanoethyl methyl silicone polymer) on gas chrom Q at 195°C; 2% SE-30 (a dimethyl silicone polymer) on gas chrom Q at 200°C; 3% SE-30 on chromasorb WHP (flux-calcined diatomite that has been acid-washed and silanized) at 240°C; and 5% SE-30 on chromasorb WHP at 255°C. Excellent results were obtained using 3% OV-17 on chromasorb WHP. This column was used by A. R. Sperling for the separation of equal parts of Δ^1- and Δ^6-tetrahydrocannabinol at 200°C, with relative retention time of 33 min for the Δ^6 isomer and 38 min for the Δ^1 isomer.

Cannabis constituents were also detected in the mouth and on the fingers of smokers by H. Stone and H. Stevens. The presence of condensates on the fingers can be shown for a minimum of an hour after smoking, and in some cases up to 3 hr after smoking. The test is unaffected by the presence of tobacco condensates, other phenolic substances, and food matter. Washing with soap and hot water does not prevent a positive test. The cannabinols are recovered from the fingers by washing with chloroform and from the mouth by an ethanolic saline solution. The solutions undergo cleanup by column chromatography on alumina with subsequent identification by thin-layer and gas chromatography.

Cannabis constituents were also identified in the particulate matter of smoke by F. Fish and W. Wilson. The material was trapped by drawing air through a glass-fiber filter disk. Chromatographic methods indicated the presence of cannabidol, THC, and cannabinol in rooms where cannabis had been smoked. Tobacco smoke did not interfere.

Tryptamines. Sperling analyzed a series of hallucinogenic tryptamines by gas and thin-layer chromatography. With silica gel G plates, a solvent system of ether-acetone–ammonium hydroxide (25:5:0.2) gave the following Rf values: α-methyltryptamine, 0.1; dimethyltryptamine (DMT), 0.19; diethyltryptamine (DET), 0.38; and dipropyltryptamine (DPT), 0.78. The plates were visualized by being sprayed with *p*-dimethylaminobenzaldehyde. The gas chromatographic analysis was performed using a 3% OV-17 on chromasorb WHP column. The temperature was 200°C and the flow rate was 60 ml/min. Retention times were 3 min for DMT, 5 min for DET, and 8 min for DPT.

Lysergic acid diethylamide (LSD). A method for quantitating dosage forms directly has been developed in which the sample is dissolved in methanol and determined by spectrofluorometry without prior cleanup. The excitation wavelength is 320 μ, and the emission wavelength is 400 μ.

The identity of the LSD is subsequently confirmed by thin-layer chromatography.

Mixtures of LSD and 2,5-dimethoxy-4-methyl amphetamine (STP) were separated and analyzed by Sperling, utilizing column chromatography. The substances are eluted through a series of acid-washed purified siliceous earth (Celite 545) columns, the STP being trapped on a 0.1 N HCl column and the LSD being subsequently trapped on a 10% citric acid column. The drugs are then removed from the columns, quantitated by ultraviolet spectrophotometry, and identified by infrared spectrophotometry and thin-layer chromatography, respectively. Recommended systems for the identification of LSD on silica gel G plates are chloroform-methanol (9:1) and chloroform saturated with ammonium hydroxide–methanol (18:1).

D. Anderson differentiated LSD from other indole alkaloids by the ultraviolet degradation products. After ultraviolet irradiation, the solution is applied to thin-layer plates of silica gel G. The plates are developed with chloroform-acetone (1:4) and the compounds are identified by their irradiation spot patterns.

T. Niwaguchi and T. Inoue separated lysergic acid amide and isolysergic acid amide from morning glory seeds by column chromatography utilizing a 0.5% citric acid solution on Celite 545. Thin-layer chromatographic systems used for the identification were acetone-piperidine (9:1), chloroform-methanol (4:1), and acetone–ethyl acetate–dimethylformamide (5:5:1) on silica gel G plates. The total amount of alkaloids in fresh seeds of Heavenly Blue variety was 0.02%, of which 44.8%

Temperature programming for separation of a mixture of hallucinogenic drugs by gas chromatography on a 2% SE-52 column

Compound	Temperature, °C			Retention time, min
	Column	Flash heater	Detector	
Dimethyltryptamine	140	190	170	8.0
Dimethyltryptamine (as the trimethylsilyl derivative)	140	190	170	10.0
Dimethyltryptamine (as the trimethylsilyl derivative)	Programmed	260	260	12.0
Mescaline	140	190	170	5.4
Mescaline (as the trimethylsilyl derivative)	140	190	170	32.8
Mescaline (as the trimethylsilyl derivative)	Programmed	260	260	19.7
Psilocybin	180	235	210	4.6
Psilocybin (as the trimethylsilyl derivative)	180	235	210	21.8
Psilocybin (as the trimethylsilyl derivative)	Programmed	260	260	30.6
Ibogaine	220	280	240	11.7
Ibogaine (as the trimethylsilyl derivative)	220	280	240	12.7
Ibogaine	Programmed	260	260	39.3
Ibogaine (as the trimethylsilyl derivative)	Programmed	260	260	39.9
Lysergic acid diethylamide	250	280	280	14.6
Lysergic acid diethylamide (as the trimethylsilyl derivative)	245	260	280	11.8
Lysergic acid diethylamide (as the trimethylsilyl derivative)	Programmed	260	260	47.6

was lysergic acid amide and 34.2% was isolysergic acid amide.

R. Mesley and W. Evans studied the infrared spectra of LSD and its tartrate salts. Spectra corresponding to amorphous and crystalline forms of LSD base, the neutral tartrate, and two forms of the hydrogen tartrate were obtained. The neutral tartrate is purported to undergo conversion to the hydrogen tartrate on long standing.

M. Lerner separated a mixture of hallucinogenic drugs by gas chromatography on a 2% SE-52 column. SE-52 is a phenyl silicone polymer. Temperature programming was required for best separation. Retention times and temperatures are listed in the table.

Other drugs. K. Genest and D. Hughes used thin-layer and gas chromatography to identify STP, mescaline, DMT, bufotenine, amphetamine, and methamphetamine. Solvent systems consisting of ethyl methyl ketone–dimethylformamide-ammonia solution (13:1.9:0.2) and chloroform-methanol–acetic acid (75:20:5) were used on silica gel G plates, and a system of methanol-chloroform (1:1) was used on alumina G plates. For the gas chromatographic analysis, the two columns used were a 5% SE-30 on chromasorb W and a 3% XE-60 on aeropak 30. For mixtures of these compounds, temperature programming on the 3% XE-60 column gave best results.

Fluorescent techniques. A. Gillespie performed a spectrofluorometric study of LSD, DMT, DET, psilocybin, alpha-methyl-3,4-methylenedioxyphen-ethylamine (MDA), STP, mescaline, benactyzine, phencyclidine, and N-methyl-3-piperidyl benzilate (JB-336). The wavelengths of maximum excitation and maximum fluorescence emission were determined in various solvents and at various pH values. Practical detection limits and optimum solvent systems were determined. The strongly fluorescent compounds were LSD, DMT, DET, psilocybin, MDA, and STP. Mescaline, JB-336, and benactyzine were found to be weakly fluorescent.

E. G. C. Clarke has published an excellent reference book on the identification of drugs, and C. Fulton has written a valuable book on microcrystal tests.

For background information *see* CHROMATOGRAPHY; PSYCHOPHARMACOLOGY; PSYCHOTOMIMETIC DRUGS in the McGraw-Hill Encyclopedia of Science and Technology.   [ALBERT R. SPERLING]

Bibliography: D. Anderson, *J. Chromatogr.*, 41:491, 1969; E. G. C. Clarke, *Isolation and Identification of Drugs in Pharmaceuticals, Body Fluid, and Post Mortem Material*, 1969; F. Fish and W. Wilson, *J. Forensic Sci. Soc.*, 9:37, 1969; C. Fulton, *Modern Microcrystal Tests for Drugs*, 1969; K. Genest and D. Hughes, *Analyst*, 93:485, 1968; A. Gillespie, *Anal. Lett.*, 2:609, 1969; A. Hoffer and H. Osmond, *The Hallucinogens*, 1967; A. Hoffmann, in A. Burger (ed.), *Chemical Constitution and Pharmacodynamic Action*, 1968; J. Karlson et al., *Pharm. Weekbl.*, 104:583, 1969; M. Lerner and M. Katsiaficas, *Bull. Narcotics*, 21:41, 1969; R. Mesley and W. Evans, *J. Pharm. Pharmacol.*, 21:713, 1969; T. Niwaguchi and T. Inoue, *J. Chromatogr.*, 41:491, 1969; R. E. Schultes, *Annu. Rev. Plant Physiol.*, pp. 571–598, 1970; R. E. Schultes, *Science*, 163:245–254, 1969; A. Sperling, *J. Forensic Sci.*, 15:86, 1970; H. Stone and H. Stevens, *J. Forensic Sci.*, 9:31, 1969; R. Turk et al., *J. Forensic Sci.*, 14:385, 1969; H. Wagner, *Rauschgift-Drogen*, 1969.

Pulsar

During June, 1967, an extraordinary new astronomical phenomenon was observed; a group of radio astronomers in Cambridge, England, accidentally discovered certain "stars" which emitted periodic pulses of radio energy. The best-observed object, in the constellation Vulpecula, had a period of exactly 1.33730109 sec. Despite a considerable search, no visual stars were found at the positions of these radio "pulsars." Astronomers were mystified by the precision and short period of the pulsar clocks (variable stars typically have periods measured in days); their nature and origin was completely unknown; and the existing body of information (four objects) was poor.

Since then 46 more pulsars have been discovered, and the number is large enough for astronomers to discuss a "typical period" and "characteristic" location in the Milky Way Galaxy. Two other very important observational discoveries have been made. First, the pulsar clocks do not keep perfect time, as was first thought. In almost all cases, the pulse repetition period is gradually and steadily lengthening. Second, the pulsar with the shortest period (and presumably youngest) has been found in the center of a supernova remnant, the Crab Nebula (Fig. 1). This radio pulsar has a period of only 33/1000 sec.

Pulsar in the Crab Nebula. Chinese and Japanese court astronomers observed a "guest star" in the constellation Taurus that was so bright that it could be seen in daylight. The event occurred in A.D. 1054, at a low ebb of European civilization, and no Western records of the event have been found; centuries later the French astronomer C. Messier noticed a nebulous patch of light in this part of the sky which E. Hubble in 1928 identified with the expanding (velocity 2,000,000 mi/hr) remnants of the stellar explosion seen by the ancient astronomers. Various investigators, in particular F. Zwicky, suggested that during or before the explosion of the outer parts of a star the inner core might implode, leaving a condensed star made of nuclear matter (principally neutrons).

The Crab Nebula was searched for the remnant "neutron" star but, although one fairly good candidate was found (see arrow in Fig. 1), there was no convincing evidence that this star was peculiar in any way. On Nov. 8, 1968, however, D. H. Staelin and E. C. Reifenstein announced that a pulsar had been found in the vicinity of the Crab Nebula. Then in January, 1969, W. J. Cocke, M. J. Disney, and D. J. Taylor, at the University of Arizona, found that there was an object emitting visual light pulses in the vicinity of the Crab Nebula with the same radio period as that of the pulsar discovered in 1968. The final confirmation came when J. Miller and E. J. Wampler at Lick Observatory showed conclusively that the star was indeed flashing on and off about 30 times a second (Fig. 2). Subsequent investigation showed that the period was not constant but increased gradually at the rate of 37/1,000,000,000 sec per day. This very slow rate of increase, if it remained constant, would cause the period to double in about 2000 years. It was

Fig. 1.   The Crab Nebula. Arrow marks the pulsar NP 0531. (*Lick Observatory photograph*)

immediately noted with interest by astronomers that this characteristic time was approximately twice the age of the Crab Nebula, a fact which turned out to be of considerable importance in selecting among possible pulsar theories.

One of the most peculiar properties of the Crab Nebula was found in the pulsar as well. Both emit "light" over an enormous range of frequency, from x-rays through the ultraviolet and optical ranges to infrared, microwave, and finally radio frequencies. The pulsar itself is not an extraordinarily bright object. Its luminosity is about 10 times that of the Sun (emitted mostly as x-rays). The nebula is far brighter, with about 20,000 times the Sun's luminosity, and shines by a process not entirely understood. I. Shklovsky, a Soviet astrophysicist, had shown that the radiation coming from the amorphous central part of the nebula was not simply the radiation of a hot gas, like the light emitted from the outer filamentary network, but was emission from relativistic (that is, very-high-energy) electrons moving in a rather strong magnetic field in the Nebula. Although there had been wide acceptance that the Crab did shine, as suggested, by synchrotron radiation, the origin of the relativistic electrons and strong magnetic fields remained a mystery. The discovery of the pulsar led to the suggestion that it must somehow be the origin of

the continuing activity in the relativistic particles and fields in the nebula.

Many pulsars now known. Of the 50 pulsars that have been found, the Crab object is the only one known to emit visual light. About half of the remaining objects were discovered by radio astronomers in Australia; most of the rest were found by various radio-astronomy groups in the United States; several were discovered in England; and one was discovered in the Soviet Union.

Fifteen of these pulsars have been studied long enough to determine whether their periods are changing. The rate of change of one of these pulsars is undetectable, but the periods of all the others are gradually increasing. The characteristic times range from 20,000 to 100,000,000 years, with a geometric average of 3,200,000 years. If the characteristic time for most pulsars is comparable to the age of the objects (as is the case for the Crab pulsar), then as a group they are not very old (for reference, the Sun is thought to be more than about 6,000,000,000 years old).

Pulsars have been found in all parts of the sky, but lie primarily in the Milky Way near the symmetry plane of the galaxy. Distances to individual objects can be estimated from the measurement of the dispersion of the radio signals, and indicate that the typical pulsar is about 1000 light-years

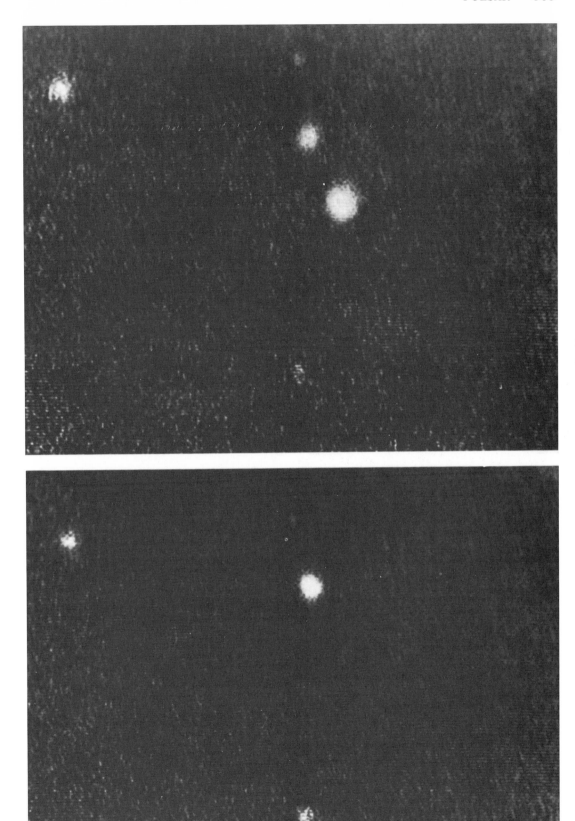

Fig. 2. Detail of four central stars from Fig. 1. Pulsar is "on" in the top frame. In bottom frame (made by a stroboscopic technique) the pulsar is "off." (*J. Miller and E. J. Wampler, Lick Observatory*)

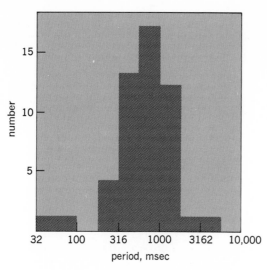

Fig. 3. Distribution of pulsar periods. Ordinate gives number of pulsars found in each period range (total number 50); abscissa, period ranges in milliseconds.

distant from the Earth, but only about 400 light-years from the galactic plane.

The period distribution is rather strange (Fig. 3). There are relatively few objects with very short periods, but this is to be expected; if the periods continuously increase after the birth of the object, it will not maintain a short period for very long. However, there are no objects with periods longer than about 1.3 sec; for some reason the pulsars must "die" when their periods become this long. If it were simply a case of the periods not increasing after about 1.2 sec, many more pulsars with this period, would be seen. When the period reaches about 1–3 sec and the characteristic age a few million years, the objects must become faint and hard to detect; investigators do not yet know why this occurs.

With the information obtained from the present sample of 50 pulsars, scientists can estimate the birth rate of these objects. Dividing the number per unit area near the Sun in the galactic plane by the average age, one obtains a birth rate of one pulsar per square light-year plane of the Milky Way Galaxy over its 10,000,000,000-year history. This rate accords well with the estimates for the rate of supernova events in the Milky Way Galaxy (about 1 per 50 years in the whole galaxy), and seems to confirm the suggestion that many supernova explosions leave pulsars as remnants. In fact, the birth rate of pulsars is so high that most stars in the Milky Way Galaxy having a mass greater than four times that of the Sun must die explosively and leave pulsar remnants.

Theoretical developments. The key to understanding the pulsar phenomenon has been found in the observed period changes. It was already known that the objects involved must be very small. In order for the pulses to be distinct, the sources were required to be smaller than the velocity of light multiplied by the pulse period, which in the most extreme case limited consideration to the common white dwarf stars (radius about 10,000 mi) or the hypothetical neutron stars (radius about 10 mi).

Three mechanisms were considered as a source

of the periodic behavior: The object could pulsate regularly; it could be in a regular orbit about another object; or it could rotate with the observed period. J. P. Ostriker originally suggested that pulsars might be rotating white dwarfs emitting a searchlight beam. White dwarfs were ruled out when the Crab pulsar was discovered, because white dwarfs are not small enough to pulsate, orbit, or rotate as fast as 30 times per second. The objects could not be orbiting neutron stars, because if they were, due to a general-relativistic effect, they would spiral together and the periods would decrease rather than increase. Pulsation was then ruled out because stable neutron stars could not pulsate as slowly as the 2-sec periods of the long-period pulsars. For these and other reasons, T. Gold of Cornell suggested that the pulsars were rotating neutron stars. Actually, as early as 1966, John Wheeler of Princeton had proposed that the remnant of a supernova explosion would be a rotating neutron star, and that one might exist in the Crab Nebula, its stored kinetic energy of rotation providing the energy to keep the Crab Nebula shining. Others (F. Pacini, J. E. Gunn, and Ostriker) pointed out that the neutron stars would be expected to have strong magnetic fields and that a spinning magnet radiates energy away in the form of very-low-frequency electromagnetic waves, and then slows down. From the observed rate of period change in the Crab pulsar and the calculated characteristics of neutron stars, they computed the required magnetic field at the surface of the star. The result, 2×10^{12} gauss (the Earth's surface field is about 1/2 gauss), although incredibly high, was not greater than that expected for a star having an average density of about 5,000,000,000 tons per cubic inch.

Theoretical discussions that followed explored the properties of rotating magnetic neutron stars. Scientists argued that something on the surface of such a star might emit a directional beam—a searchlight—and that the observer would see pulses as the searchlight swept by him. At the California Institute of Technology, P. Goldreich and W. Julian showed that electromagnetic forces near the surface of the proposed pulsar would expel a plasma from its surface, and Gunn and Ostriker showed that when this plasma was in the strong wave field surrounding the pulsar, it would be strongly accelerated away.

According to the various theoretical developments, rotating magnetic neutron stars would tend gradually to slow down and, as they slowed, they would emit enormous quantities of energy in the form of relativistic particles and magnetic fields. Thus the proposed pulsars had exactly the properties required for the energy source in the Crab Nebula. There was quantitative agreement too, since the energy lost by the spinning star (determined from period and period rate of change) was just sufficient to keep the Nebula shining. Finally, according to the theories, the characteristic time discussed above should be approximately twice the age of the object. Here observation agreed with theory for the Crab object, and allowed the other pulsars to be dated.

The origin of the pulses themselves has remained unsolved. Although a great number of theories have been put forward to explain how magnetic neutron stars might emit a searchlight beam, no single theory has gained wide acceptance.

Summary. Although there is general agreement that pulsars are probably rotating, magnetic, condensed stars emitting beamed radiation, the mechanism of the searchlight beam is not understood. In any case, the beamed (pulsed) energy output is small compared to the total output of the objects in high-energy particles and magnetic fields, which is estimated to be $10^{50}-10^{52}$ ergs per object; therefore, in its final throes as a pulsar, a star may emit more energy than it did in its preceding lifetime of burning nuclear fuel. Some of this energy will be emitted as high-energy charged particles and recorded on Earth as cosmic rays. Some, perhaps most, of the energy is expended in accelerating supernova remnants and in providing the energy to keep them luminous. Much of the energy may come out in the initial burst; the dramatic supernova explosion itself may be partly fueled by a pulsar within the expanding supernova shell. These matters and the origins of the pulses are the subjects of continuing investigation.

For background information *see* CRAB NEBULA; PULSAR; RADIO ASTRONOMY in the McGraw-Hill Encyclopedia of Science and Technology.

[J. P. OSTRIKER]

Quark

Very recently the search for quarks has been extended, with negative results, to a higher machine energy as the 70-Gev accelerator in the Soviet Union became operational. An Australian group has made a claim of finding evidence of quarks in very energetic cosmic-ray phenomena, but the interpretation of their results is the subject of a controversy.

The idea that strongly interacting elementary particles (hadrons) might be composed of three fundamental particles (and of the corresponding antiparticle triplet), with electric charges equal to $\pm 1/3$ and $\pm 2/3$ of the electron charge, has been very successful in organizing and simplifying ideas about elementary particles. According to this picture, the proton is an agglomeration of three such fundamental particles, or quarks, and a meson an agglomeration of a quark and an antiquark, the combination having an integral total charge. The first success of this theory was in providing a coherent scheme of classification that predicted the existance of a hitherto unknown particle, the Ω^-, which was discovered at Brookhaven National Laboratory. It has been realized subsequently that static and dynamic properties of hadrons can be calculated if it is assumed that a hadronic property is the sum of the corresponding properties of the quarks which make up the hadron. This quark model of elementary particles has greatly simplified the calculation of both static and dynamic properties such as mass differences, magnetic moments, relative probabilities of various decay schemes, and collision cross sections. The successes of the quark model have greatly strengthened the rationale for the search for free quarks, making it one of the central problems of elementary particle physics. *See* HADRON.

Accelerator searches. One of the first experiments conducted on the newly completed 70-Gev proton synchrotron at Serpukhov, Soviet Union, was a search for quarks. Since the energy of this machine is more than twice that of the Brookhaven alternating-gradient synchrotron and of the CERN proton synchrotron in Geneva, the energy available at Serpukhov for the production of particles in proton-proton collisions is, by the laws of kinematics, 1.5 times greater. Experiments with the lower-energy machines were limited to the maximum mass that the available energy could produce, which was equal to about 3 proton masses. No quarks were found in these lower-energy experiments, and it was assumed that, most probably, if quarks exist they have a greater mass. No quarks have been found at Serpukhov either, and this has established a new lower limit on their mass, about 4.5 proton masses.

Cosmic-ray searches. Cosmic rays extend to much higher energies than do particles accelerated by man, and thus permit the search for quarks of a still higher mass. The intensity of cosmic rays is, however, very low compared with the intensity of particle beams in accelerators; furthermore, cosmic-ray intensity decreases strongly with particle energy. Thus the limits on quark intensities and therefore on their production cross sections become less significant with increasing quark mass. Many cosmic-ray experiments were carried out with negative results until a group of physicists at Sydney, Australia, headed by C. B. S. Mc-Cusker, announced in 1969 that they had found evidence of quarks in extremely energetic cosmic-ray events of about 10^6-Gev energy.

Primary protons and α-particles of such an energy initiate the development of a shower of particles in the atmosphere which, at sea level, attains the number of some 10^5 particles, mainly electrons and muons, spread over a large area. In the center of such an extensive air shower, as the phenomenon is called, in the core region, the particle density amounts to several thousand particles per square meter. In this core region the Sydney group found five particle tracks in cloud chambers which they attributed to quarks with charge $2e/3$. Since a particle ionizes proportionally to the square of its charge, such a quark would produce about half the number of droplets along its track in the chamber that would be produced by a particle with a unit charge moving with the same velocity. The situation is, however, obscured by the fact that the energy and mass of the particles traversing the chamber were not measured, and hence their velocity is unknown. The ionization in gas is a function of the velocity, even for relativistic particles. This fact, in addition to statistical fluctuations in the drop number, could make a slower shower particle with unit charge ionize less than expected on the average. It is most likely that, in the course of the year-long experiment in which about 60,000 particle tracks were photographed, several singly charged particles would ionize lightly enough to simulate $2e/3$ quark tracks.

Conclusions. Physicists are led therefore to conclude that no quarks have so far been found, and the question arises: Have the searches been incomplete, or are quarks only a mathematical convenience to describe some aspect of reality? With investigators' present understanding of elementary-particle interactions, it is reasonable to conclude from the low limits on the quark flux found in the accelerators that quarks do not exist with a mass smaller than 4 proton masses. One can make a similar, albeit weaker, statement up to about 10–15 proton masses from the cosmic-ray results. In the present scheme one would not expect quarks to be more massive than that. At any rate, it would be very difficult to push cosmic-ray measurements much further. Thus the puzzle of strong indirect evidence for quarks from the successes of the theory and the apparent absence of free quarks in nature still exists. Physicists are, however, dealing with a new and unknown domain, and it is possible that an unknown conservation law (of integral charge?) or superstrong interaction which recombines quarks immediately after their production prevents scientists from seeing the free fundamental particles.

For background information *see* ELEMENTARY PARTICLE; QUARKS; SYMMETRY LAWS (PHYSICS) in the McGraw-Hill Encyclopedia of Science and Technology. [HENRY KASHA]

Bibliography: R. K. Adair and H. Kasha, *Phys. Rev. Lett.*, 23:1355, 1969; Y. M. Antipov et al., *Phys. Lett.*, 30B:576, 1969; I. Cairns et al., *Phys. Rev.*, 186:1394, 1969; *Phys. Rev. Lett.*, 23:658, 1969.

Quasar

Recent investigations suggest that quasars, once regarded as uniquely mysterious, have many properties in common with certain galaxies.

Galaxies. A galaxy is a collection of stars and gas whose image on a photographic plate taken with a large telescope appears diffuse. One of the telescopes most frequently used for photographing galaxies is the Mount Palomar 48-in. Schmidt telescope, which can reveal the diffuse character of images as faint as nineteenth magnitude (10^8 times fainter than the brightest star). The limit of resolution of the Palomar Schmidt telescope is about 1 second of arc on the sky (2.8×10^{-4} degrees) when atmospheric conditions are good; similar limits of resolution exist for other Earth-based telescopes.

Galaxies have masses that vary from about 10^5 Sun masses (for the Draco Galaxy in the local group of galaxies) to about 10^{12} Sun masses (for giant elliptical galaxies seen in some clusters of galaxies). The luminosity, or light output, of galaxies varies over a similarly huge range, from about 10^5 times the Sun's luminosity to a few hundred billion solar luminosities. The properties of galaxies have been studied extensively by astronomers since pioneering work (performed in the 1920s) showed that these diffuse objects are really outside the Milky Way Galaxy. Edwin Hubble first established the linear relationship between the distance of a galaxy and the amount by which the emission and absorption lines in its spectrum appear shifted to the red (red shift). Some of the observed absorption lines, such as the H and K lines of once-ionized calcium, are characteristic indicators of the presence of stars. *See* GALAXY, INFRARED.

Quasar characteristics. The puzzling, mysterious characteristics of quasars were immediately apparent when they were first discovered in 1963 by Maarten Schmidt, Jesse Greenstein, Allan Sandage, and colleagues. The quasars appeared as starlike images (that is, unresolved) on photographic plates at the positions of known strong ra-

dio sources (hence the name quasi-stellar radio sources, or quasars). They had large red shifts (shift in wavelength divided by laboratory wavelength) which ranged up to 40%; subsequent discoveries revealed quasars with red shifts that were greater than 2. Their continuum light emission is much bluer than that of ordinary stars, and their emission lines are unusually broad (of the order of 100 A wide). Substantial variations in the intensity of light emitted were detected on practically every time scale on which variations were searched for, from days to years. The combination of these characteristics (starlike appearance, large red shifts, and rapid variations of light) were unexpected in extragalactic objects. In fact, the first quasars discovered had to be more distant, and brighter by a factor of 100 or so, than the brightest and most distant known galaxies if the relation between distance and red shift established by Hubble for galaxies also held for quasars. Yet the volume occupied by a quasar, as indicated by their unresolved images on photographic plates, is at least hundreds of times smaller than that of a normal galaxy (and many billions of times smaller if the time over which the light varied significantly is taken as a measure of their size).

These features of quasars were difficult to understand and led several scientists, especially James Terrell, Fred Hoyle, and G. R. Burbidge, to propose that the Hubble law connecting red shift and distance was not valid for quasars. They suggested that quasars were really much closer than indicated by the conventional interpretation of their red shifts. Most astronomers and physicists did not accept the arguments for the "local" nature of quasars because the characteristics of quasars inferred on the basis of the local hypothesis were at least as peculiar as the characteristics inferred using the more conventional interpretation of red shift as an indicator of distance.

Nevertheless the proponents of the "distant" (or cosmological) origin for the red shifts of quasars were necessarily cautious in their claims because of the unusual nature of the objects with which they were dealing and because of the lack of any direct proof that, for quasars, red shift indicated distance.

Galaxies with quasar characteristics. In recent years a number of galaxies have been studied that have various properties previously regarded as unique to quasars. J. B. Oke investigated galaxies whose continuum emission has the same blue, nonthermal character previously associated only with quasars. In one galaxy, the radio source 3C 371, Oke detected light variations of about one magnitude (a factor of 2.5) over a period of 2 years and a tenth of a magnitude (about 10%) in the course of several days. Light variations in galaxies have also been detected by other workers. A series of galaxies that were selected by F. Zwicky because of their compact appearance on photographic plates were studied by W. L. W. Sargent. Some of these compact galaxies have properties that are intermediate between those of quasars and galaxies. For example, the object Zwicky 0051 + 12 has broad emission lines (of the order of 50 A) and a continuum spectrum similar to that of the famous quasar 3C 273, and is as bright as the fainter quasars if the latter are at the cosmological distances indicated by their red shifts. The bright nuclear region of Zwicky 0051 + 12 (which has a red shift of 6%) might well have been classified as a quasar if it were somewhat more distant, in which case the faint surrounding nebulosity would have been unobservable and the nuclear region unresolved.

Some remarkable observations have also shown that galaxies can have emitting regions that are extraordinarily small. Observations by a group from Princeton University using a 36-in. telescope that was carried in a balloon flying high above much of the Earth's atmosphere have established that much of the light from the Seyfert galaxy NGC 4151 comes from a nuclear region less than 0.2 second of arc in extent. A group of radio astronomers operating telescopes simultaneously in California and Australia have shown that the famous radio galaxy Virgo A has a radio-emitting component that is less than, or of the order of, 0.001 second of arc in size (which corresponds to a size of about 3 light-months at the distance of Virgo A). *See* OBSERVATORY, ASTRONOMICAL.

Quasars associated with galaxies. In 1966 Sandage and William Miller performed a crucial observation. They used an especially sensitive photographic emulsion provided by Eastman Kodak to search for a cluster of galaxies in the directions of the quasars 3C 273 and 3C 48, whose red shifts (16 and 37%, respectively) were among the smallest of any quasars known at the time. Their careful search revealed no obvious cluster of galaxies in the direction of either quasar. The observation was of special importance because both sides in the "local" versus "cosmological" controversy were agreed that one of the very few ways to establish definitely the cosmological nature of quasars would be to find a quasar in the same direction as a cluster of galaxies with the same red shift as the cluster. The effect of such a discovery would be to link the red shift of the quasar to that of a cluster of galaxies whose red shifts had been demonstrated to be an indicator of distance. The negative result of Sandage and Miller discouraged hopes of finding quasars in clusters of galaxies, and subsequent discussions in the standard scientific articles on the subject assumed that quasars were not associated with galaxies.

In the meantime, Sandage had discovered that there are many more quasars with weak or unobservable radio emission than with strong radio emission. He also found a technique for identifying these "radio-quiet" quasars by means of their colors as revealed by photographic studies with optical telescopes. By the spring of 1969, six quasars (including 3C 273) were known with red shifts less than 20%. John Bahcall and James Gunn decided to repeat the observations of Sandage and Miller on the five recently discovered quasars with relatively small red shifts, searching for clusters of galaxies in the direction of these objects. The photographic techniques used were sufficiently powerful that large clusters of galaxies around the quasars could be recognized out to red shifts of about 20%.

Bahcall had an unusual piece of scientific luck in his first night ever using a telescope; he found that the quasar B 264 was in the direction of a previously studied cluster of galaxies. In fact, earlier estimates of the brightness of the galaxies in the

direction of B 264 had led to estimates for their red shifts that were not inconsistent with those of the galaxies having the same red shift as that of the quasar. Moreover, comparison of the known positions on the sky of the quasars with catalogs of previously known clusters of galaxies showed that five of the six small red-shifted quasars (all but 3C 273) lay in the direction of previously known clusters of galaxies. Schmidt obtained spectra for four galaxies in the direction of B 264 that had the same red shifts as that of B 264, thus establishing a physical association between B 264 and the galaxies.

The above result has not settled completely the "local" versus "cosmological" controversy over the nature of quasars despite previous agreement that finding a quasar associated with a cluster of galaxies would end the debate. Two reservations about this agreement are possible: The finding of one quasar in a cluster of galaxies might be by chance and be atypical, and B 264 might have been wrongly classified as a quasar. Both these reservations could be overcome if another object whose quasar credentials were unquestioned could be shown to be in a cluster of galaxies. The most promising case discussed so far is TON 256, a quasar with a red shift of 13% that (as Sandage has emphasized) has a completely stellar appearance and is brighter than the brightest giant elliptical galaxies. Work by Neta Bahcall and John Bahcall has shown that there is a clumping of galaxies in the direction of TON 256 that is unexpected unless there is a physical association between the quasar and the galaxies. Estimates of the apparent brightness of the galaxies lend some support to the possibility that the quasar and galaxy red shifts are the same. It is hoped that some of the difficult measurements of the red shifts of galaxies in the direction of TON 256 will be completed soon.

For background information *see* GALAXY, EXTERNAL; QUASARS; RADIO SOURCES (ASTRONOMY); RED SHIFT in the McGraw-Hill Encyclopedia of Science and Technology.   [JOHN N. BAHCALL]

Bibliography: J. N. Bahcall, M. Schmidt, and J. E. Gunn, *Astrophys. J. Lett.*, 157:L77, 1969; J. B. Oke, *Astrophys. J. Lett.*, 150:L5, 1967; A. Sandage, *Astrophys. J.*, 141:1560, 1965; A. Sandage and W. C. Miller, *Astrophys. J.*, 144:1238, 1966.

Radiation biology

Recent developments in the application of nuclear physics and diagnostic medicine have been in three areas: diagnostic and therapeutic radiology, nuclear medicine, and activation analysis. The applications of nuclear physics in medicine have historically closely followed important discoveries in nuclear physics. The discovery of x-rays of W. Roentgen was followed closely by diagnostic applications. Therapeutic uses of radiation again followed closely behind. More recently, the advent of radioactive isotopes and techniques for applying these to medical problems has led to the field of nuclear medicine, a rapidly growing medical speciality. Analytical techniques employing neutron activation analysis have been developed for medical application, and these are now assuming an increased role in medicine. In particular, activation analysis in the living body, that is, the determination of body composition by neutron activation

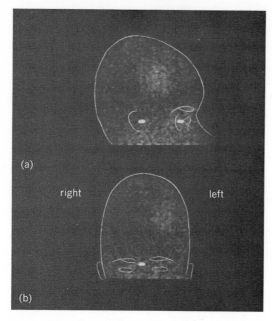

Fig. 1.   Arsenic-74 brain scan. (*a*) Midline lateral view. (*b*) Anterior-posterior view.

analysis, promises to become an important technique in medical research and possibly in medical diagnosis.

Nuclear physics has contributed to these developments both by basic developments in the discipline itself and in the efforts of nuclear physicists working in medical fields. These physicists commonly consider themselves as medical physicists, biophysicists, or medical engineers.

Diagnostic radiology. Diagnostic radiology dates from the discovery of x-rays in 1895. Within a month, Roentgen and others were using x-rays to take radiographs of various objects, including a human hand. The development of diagnostic ra-

diology has progressed continuously since that time. Among other developments, Thomas Edison developed the fluoroscope in 1896, William Coolidge developed the hot-cathode x-ray tube in 1913, and the image intensifier tube was developed in the early 1950s.

More recent developments have been tomography, a technique for examining a given plane in the patient's body, further refinement of the image intensifier, the development of television and radiographic techniques to permit the recording of dynamic function studies, and the use of computers to process x-ray images. In a similar manner, radiation therapy dates back to the discovery of x-rays and the identification of radium by the Curies. Nuclear physics has played an important role in this development in the production of newer and better instruments such as the Van de Graaff accelerator, betatron, and electron linear accelerator and in the development of improved radioisotope sources, such as cobalt-60. This has made it possible to concentrate on ionizing energy more specifically in the area of interest and decrease the dose to surrounding regions. The art of dose calculation or radiation dosimetry has also progressed using techniques developed in large part by nuclear physicists. Computer techniques have recently come into use to more readily calculate dose distribution resulting from various types of radiation to various organs or regions of organs within the patient. Such calculations have made possible improved patient therapy.

Many of the basic phenomena of radiation damage to tissue are still only poorly understood. For example, the oxygen effect is still a poorly understood radiation phenomenon. Almost all tissues have a greater reaction to x-radiation in the presence of oxygen than in the absence of oxygen. This has a deleterious affect in radiation therapy, since the central region of many tumors may be depleted in oxygen. Consequently, attempts to overcome the oxygen effect would be beneficial, and such attempts have been made to increase the oxygen supply to such regions by breathing pure oxygen at elevated pressures. Another interesting approach is the use of radiations that do not exhibit as marked an oxygen effect as x-rays. It has been observed that radiation having a high ionization density, such as heavy charged particles or recoil particles from fast neutrons, has a diminished oxygen effect. Consequently, attempts have been made to use fast neutrons, protons, and other particles for radiation therapy. Californium-252, a transuranic element discovered by G. Seaborg, is a spontaneous neutron emitter and is being studied as a possible internal source of fast neutrons in cancer treatment.

The possible use of negative pions is being pursued at the Donner Laboratories of the University of California. These particles disintegrate light nuclei at the end of their track yielding various heavy particles, including alpha particles, protons, and neutrons. Thus, at the site of the tumor, a large quantity of high specific ionizing radiation is liberated. The principal problem at the present time is the availability of sufficient numbers of negative pions. By 1972, further biological studies in radiation therapy will be possible with the Los Alamos Meson Physics Facility. Other facilities

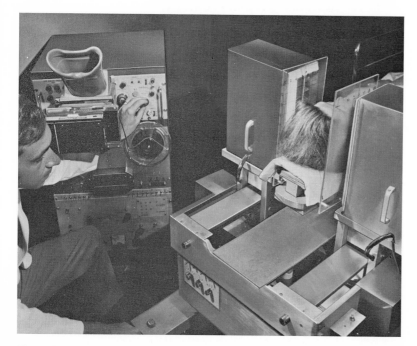

Fig. 2.   Massachusetts General Hospital positron scanner in operation.

are planned in Canada and Switzerland.

Nuclear medicine. Nuclear medicine may be generally divided into the study of biological function within the patient, techniques for visualizing organs or abnormalities, and test-tube studies designed to determine the properties of samples taken from patients. The use of I^{131} to determine the activity of the human thyroid is a classical technique for the study of biological function.

Recently, the imaging of radioisotopes within the body has developed as a result of improvements in isotopically labeled compounds and instrumentation (Fig. 1); this area forms the core of what is now known as nuclear medicine. The final category consists of studies such as the determination of thyroid function by observation of the absorption of I^{125}- or I^{131}-labeled triiodothyronine in human blood cells.

A survey conducted by the Stanford Research Institute showed that approximately 400,000 administrations of radioisotopes to patients for imaging procedures, 300,000 studies of biological function in the living body, and over 700,000 test-tube studies were performed within the United States during 1966. An estimate of the annual growth rate of nuclear medicine was approximately 15%.

Instrumentation in nuclear medicine has seen rapid development in recent years, particularly in the area of radioisotope imaging. One of the most widely used imaging devices at the present time is the Anger camera, developed by Hal Anger of the Donner Laboratories. This device employs a thin large area of sodium iodide scintillator backed by 19 photomultiplier tubes. New isotopes are also coming into use and the 6-hr Tc^{99m} has seen a rapid rise in usage. This isotope is the daughter of the 66-hr Mo^{99} parent and gives rise only to a 140-kev gamma ray.

Recently a small number of cyclotrons have been installed in hospitals. Among the various applications is the production of the three short-lived isotopes, O^{15}, N^{13}, and C^{11}, with half-lives of 2, 10, and 20 min, respectively. The advent of these isotopes, as well as other positron-emitting isotopes, has resulted in recurring interest in positron detection; various devices, including the Massachusetts General Hospital positron camera (Fig. 2), have been or will be applied to studies with these isotopes.

Activation analysis. Activation analysis involves the activation of various radioisotopes within a sample resulting from radiation interaction with the elements present. From the activity of the different isotopes, the composition of the sample can be determined. Biological studies have been aimed largely at determining the role of trace elements in biological material, and particularly in determining the role of trace elements in the etiology of various diseases.

Living-body activation analysis involves the radiation of a patient, or part of a patient, with thermal neutrons or other radiation to induce radioisotopes which can be measured to determine the elemental composition. Perhaps the widest application has been in the determination of total or partial body calcium content in patients with various bone diseases (Fig. 3). With use of small radiation exposures (less than 1 rem) of neutrons, it is possible to determine the total body calcium

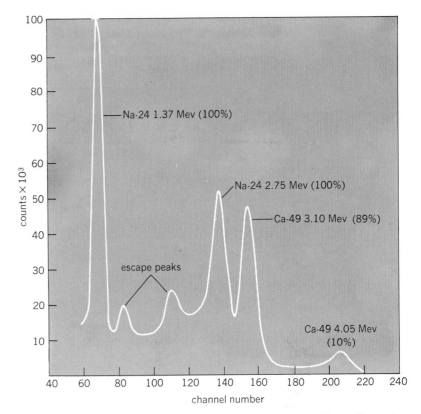

Fig. 3. Gamma spectrum of bone 20 min after irradiation (1 min) in the Massachusetts Institute of Technology reactor on Feb. 19, 1970. Energies are given in millions of electron volts. The escape peaks result from the loss of one or two annihilation quanta (0.51 Mev) from the scintillator.

content with a high degree of precision. By using successive measurements, the effect of various therapy modes can be determined. Living-body activation analysis can be used to determine the amount of other elements, such as sodium, chlorine, and potassium.

For background information *see* ACTIVATION ANALYSIS; RADIATION BIOLOGY; RADIOACTIVITY APPLICATIONS in the McGraw-Hill Encyclopedia of Science and Technology. [GORDON L. BROWNELL]

Bibliography: G. L. Brownell et al., New developments in positron scintigraphy and the application of cyclotron-produced positron emitters, *Med. Radioisotope Scintigraphy*, vol. 1, SM-108/99, 1969; G. L. Brownell, S. Aronow, and G. J. Hine, Radioisotope scanning, in G. J. Hine (ed.), *Instrumentation in Nuclear Medicine*, 1967; E. M. Smith, G. L. Brownell, and W. H. Ellett, Radiation dosimetry, in H. N. Wagner (ed.), *Principles of Nuclear Medicine*, 1968.

Radioactive waste disposal

Radioactive waste disposal or management continues to be a subject of increasing public interest and concern. In general, the results of technical assessments and more than 20 years of operating experience show that radioactive waste-management operations are being carried out in a safe and economical manner, without harmful effect on the public, environment, or resources of the United States. In view of the advanced state of waste treatment and disposal technology, particularly in the solidification of highly radioactive liquid wastes and subsequent storage in selected geologic formations such as bedded salt, the Atomic Energy Commission (AEC) has proposed to establish an industrial requirement for conversion of all high-activity liquid waste to a solid form, suitable for off-site shipment and ultimate disposal in a national repository.

Policy on high-level waste management. During 1968, with indications of substantial industrial interest in the construction of chemical reprocessing plants for the recovery of uranium and plutonium from irradiated fuel, a study was initiated to define the need for an AEC policy regarding the siting of commercial fuel-reprocessing plants and related waste-management facilities. This study, carried out with technical support from major AEC contractors, indicated that there is no need for Federal or state ownership of a reprocessing plant site, nor is there any reason to limit the number of these sites, provided that the site is not proposed for the long-term storage of high-level radioactive waste.

The regional growth of the nuclear power industry through 1980 and by 2000 was an important factor in the study, and the expected influence of this industrial development on the siting and capacity of reprocessing plants and associated high level facilities was analyzed. It appears that, through 1980, nuclear power stations will be concentrated largely in the eastern half of the United States and will consist mainly of light-water reactors. By the year 2000, nuclear power stations will be more uniformly distributed throughout the United States, and will consist of light-water reactors as well as other reactor types. *See* REACTOR, NUCLEAR.

Fuel-reprocessing plants will also initially be located in the eastern portion of the United States. Because of economies achieved by larger plants, proposed fuel-reprocessing designs are tending toward larger capacities (~5 tons/day), even in the early 1970s. Results of the study indicated that 3–6 commercial reprocessing plants will be required by 1985 (~30 tons/day total load). Projected fuel-reprocessing requirements and estimated high-level waste from the civilian nuclear power program through the year 2000 are summarized in Table 1.

With the increasing burnups of reactor fuel, it is evident that larger quantities or inventories of long-lived fission product activity must be handled in an increasing number of chemical processing plants during the coming years. From a regulatory control and long-term safety standpoint, there is a strong incentive to limit the accumulation of long-lived fission products at any one plant. Although many plant sites may be suitable, with properly designed safety and waste-processing systems, each site may not have the required environmental characteristics for assuring the exclusion of this potential radioactive contamination from the biosphere for thousands of years. *See* NUCLEAR FUELS.

In view of the advanced state of solidification and salt-disposal technology, it is considered appropriate to establish the industrial requirement of conversion of all high-activity liquid waste to a liquid form suitable for off-site shipment and ultimate disposal in a national waste repository. The results of the chemical-processing plant siting study are:

1. Reprocessing plants need not be located on Federal land.

2. Reprocessing plants must be designed to facilitate decontamination and removal of wastes from temporary storage.

3. High-level liquid waste inventory must be limited in volume to quantity produced in preceding 5 years.

4. All high-level wastes must be transferred in AEC-approved solid form to a Federal repository no later than 10 years following separation of fission products from irradiated fuel.

5. Repositories will be on land owned and controlled by the Federal government.

6. The Federal government will have physical responsibility, and industry will have financial responsibility, for disposal of high-level wastes.

In the development of this proposed AEC waste-management policy, engineering and economic evaluations were made of high-level waste-management technology. The status of today's technology is summarized in the discussion below.

High-level liquid waste management. High-level liquid wastes originate mainly from the first cycle of solvent extraction in the chemical processing of irradiated fuel and contain more than 99.9% of the nonvolatile fission products. Present practice is to concentrate and store these wastes on an interim basis in specially designed underground carbon- and stainless-steel tanks equipped with devices for removing the decay heat where necessary. More than 20 years' experience with this method of handling highly radioactive liquid waste has shown it to be a practicable means of interim handling. More than 80,000,000 gal of radioactive solutions and sludges have been stored in nearly

Table 1. Estimated high-level waste from the civilian nuclear power industry

Parameters	1970	1980	1990	2000
Installed capacity, 10^3 MwE	6	150	450	940
Spent fuel processed, metric tons/year	55	3000	9000	19,000
Volume of high-level liquid waste				
Annual production, 10^6 gal/year*	0.017	0.97	3.3	5.8
Accumulated, 10^6 gal (if not solidified)	0.4†	4.4	29	77
Volume of high-level waste, if solidified				
Annual production, 10^3 ft³/year‡	–	9.7	33	58
Accumulated, 10^3 ft³	–	44	290	770

*Assumes wastes concentrated to 100 gal/10,000 Mwd (thermal).
†Actual at Nuclear Fuel Services' West Valley Plant.
‡Assumes 1 ft³ of solidified waste per 10,000 Mwd (thermal).

Table 2. Summary of research and development on solidification of high-level waste*

Process and sites*	Time span	Lab scale radioactivity	Pilot plant Radioactivity†	Capacity, liter/hr	Product	Chemical additives	Status of work
Pot calcination							
ORNL	1958–1965	None	None	25	Calcine	Calcium, sulfate	Done
BNW	1959–1962	None	None	10	Calcine	Sulfate	Done
	1962–1970	H	H	20	Calcine	Sulfate, calcium	In progress
Spray							
BNW	1959–1970	H	H	20	Ceramic, glass	Phosphate, borophosphate	In progress
USSR	~1961–1970	?	?	20	Calcine, glass	Borosilicate	In progress
Phosphate glass							
BNL	1960–1970	None	None	20	Glass	Phosphate	In progress
BNW	1964–1970	H	H	20	Glass	Phosphate	In progress
Fluid bed							
ANL	1955–1959	None	L	6	Granules	None	Done
INC	1955–1970	L	I	300	Granules	None	In progress
BNW	1959–1961	No work	None	20	Granules	None	Done
USSR	~1962–1970	?	?	30	Glass, granules	Borosilicate	In progress
Pot glass							
AERE	1959–1966	None	H	6	Glass	Borosilicate	Done
FAR	1962–1970	H	L	20	Glass	Boroalumino-silicate, phospho-silicate	In progress
CPP	1969 startup	No work	H	20	Glass	Boroalumino-silicate	In progress
ORNL	1961–1966	None	None	3	Semiglass	Phosphate, borophos-phate	Done
Rotary kiln							
BNL	1955–1963	None	None	20	Powder	None	Done
FAR	1960–1970	None	None	6	Glass	Phosphosilicate, borosilicate	In progress
Ceramic sponge							
LASL	1959–1964	None	L	4	Ceramic balls	None	Done

*ORNL = Oak Ridge National Laboratory, Oak Ridge, Tenn.
 BNW = Battelle-Northwest, Richland, Wash.
 USSR = Union of Soviet Socialist Republics.
 BNL = Brookhaven National Laboratory, Upton, Long Island, N.Y.
 ANL = Argonne National Laboratory, Argonne, Ill.
 INC = Idaho Nuclear Corporation, Idaho Falls, Idaho.
 AERE = Atomic Energy Research Establishment, Harwell, Berks., England.
 FAR = Center for Nuclear Studies, Fontenay-aux-Roses, France.
 CPP = Center for Plutonium Production, Marcoule, France.
 LASL = Los Alamos Scientific Laboratory, Los Alamos, N.Mex.

†H > 70 Ci/kg of solid.
 I = 0.7 to 70 Ci/kg of solid.
 L < 0.7 Ci/kg of solid.

200 underground tanks throughout the AEC complex. Most of these wastes have concentrations in the range of 1–100 curies/gal; however, some of the high-level wastes have an initial concentration as high as 10,000 curies/gal.

Although current tank storage practices have been successful in preventing significant quantities of radioactive materials from escaping to the environment, these operations require continual surveillance and tank replacement. This need for surveillance, as well as the necessity of transferring liquid waste from tank to tank over periods of hundreds of years, has been a compelling factor for the extensive research and development programs directed at the conversion of high-level liquid waste to a solid form and the ultimate storage of the solid-waste materials in selected geologic environments.

High-level waste solidification. The conversion of high-level liquid wastes to solids as a pretreatment for long-term storage is being developed in several countries that have large-scale nuclear energy programs. Extensive research and development on high-temperature solidification systems for these wastes have been conducted during the past 15 years. Four processes for solidification have been developed in the United States to the point of radioactive demonstration on an engineering scale. These processes are fluidized bed calcination, pot calcination, spray solidification, and phosphate glass solidification.

The pot, spray, and phosphate glass processes have been developed and are being demonstrated at processing rates of 10–20 liters/hr of liquid waste, or the equivalent to the waste obtained from the processing of 1 ton of nuclear fuel per day. The fluidized bed process has been demonstrated at high rates of 300 liters/hr of liquid wastes containing large amounts of aluminum and zirconium salts. In all processes, heat is applied to raise the temperature of the waste to 400–1200°C. At these temperatures, essentially all the volatile constituents (primarily water and oxides of nitrogen) are driven off, leaving a solid or a melt that will cool to a solid. The resulting solids are relatively stable chemically, especially at temperatures lower than those used during processing. The addition of glass-forming materials, such as phosphoric acid, lead oxide, and sodium tetraborate, for the purpose of providing a relatively nonleachable final product has also been intensively studied. Table 2 summarizes the present status of solidification technology for the various processes but does not include small-scale work being done in Canada, Germany, Denmark, India, Japan, and Czechoslovakia.

The full-scale demonstration of the pot, radiant spray, and phosphate glass processes for the solidification of high-level waste from power-reactor fuel reprocessing was planned for completion in 1970 in the Waste Solidification Engineering Prototype at Richland, Wash. As part of this program, the characteristics of the solidified products in various storage environments were determined. The Solids Storage Engineering Test Facility, a facility for the destructive and nondestructive testing of solidified products, has been in operation more than 2 years. The technology developed in this program will provide a reliable basis for the design and operation of industrial waste-solidification plants, and will also provide the various regulatory bodies with information required to assess the safety of such operations.

Disposal in natural salt formations. Although solidification of liquid wastes will provide a large measure of increased safety in the handling of high-activity waste, it does not fully answer the question of how such materials can be kept out of the biosphere. The Committee on Geologic Aspects of Radioactive Waste Disposal of the National Academy of Sciences, adviser to the Commission, recommended that the AEC consider the disposal of such wastes in naturally occurring salt formations. After further study the committee reported that "the most promising method of disposal for high-level waste at the present time seems to be in salt deposits."

Salt has many characteristics that make it particularly attractive for waste disposal. It is widespread and abundant, underlying about 400,000 mi^2 in portions of 24 states in the United States; it has good structural properties, with a compressive strength similar to that of concrete; it is relatively inexpensive to mine; its thermal properties are better than those of most other rock types; and it occurs generally in areas of low seismicity. Most important, salt deposits are free of circulating groundwaters and completely isolated from underground aquifers by essentially impermeable rocks. Furthermore, the structural stability of salt formations tends to be preserved because any fractures which might develop are readily healed by plastic deformation of the salt.

During the 1960s, salt-disposal technology was developed by the AEC's Oak Ridge National Laboratory to the point where confidence can be placed in a system that is, in concept, practical and that will provide the assurance of perpetual isolation of high-level waste from man's biosphere. Project Salt Vault, a demonstration disposal of high-level radioactive solids in a Lyons, Kans., bedded salt mine, using engineering test reactor fuel assemblies from the National Reactor Testing Station (NRTS) in Idaho in place of actual solidified waste, has successfully demonstrated waste-handling equipment and techniques similar to those required in an actual disposal operation. A total of about 4,000,000 curies of fission product activity in 21 containers, each having an average of about 200,000 curies, was transferred to the experimental area in the mine (three separate shipments of seven canisters each on a 6-month schedule) and back to the NRTS in Idaho at the end of the test program. During the period of the field test program from November, 1965, through May, 1967, the salt received a maximum radiation dose approaching 10^9 rads. Results of the field demonstration program can be summarized as follows: The feasibility and safety of handling highly radioactive materials in an underground environment were demonstrated; the stability of salt under the effects of heat and radiation was shown; and data were obtained on the creep and plastic flow characteristics of salt that make possible the design of a safe, long-term disposal facility.

As a result of these studies, the AEC has announced the tentative selection of a site near Lyons, Kans., for an initial salt mine repository for the demonstration of long-term storage of solid high-level and long-lived low-level radioactive wastes. The purpose of the demonstration project will be to provide technical data and experience on operational methods and costs of long-term storage of solidified high-level wastes, and long-term storage of solid materials contaminated with low-level radiation, principally the transuranium elements such as the plutonium now generated at AEC installations.

About 1000 acres of salt space will be required for the repository, which will be capable of handling all solid high-level wastes from commercial fuel reprocessing facilities until the year 2000, in addition to about 1,000,000 ft^3 of low-level, long-lived radioactive waste per year. During 1970–1971 additional geologic and safety studies will be carried out to confirm that all aspects of the operation at this location can be done safely. With funding in 1971, initial operation of the repository could commence in 1974.

Conclusions. Engineering economic studies indicate that waste-management operations carried out in accord with the proposed AEC policy can be accomplished for less than 0.1 mills/kwhr. It is not expected that the costs for waste management, including perpetual storage in a national repository, will be a significant factor in either the fuel cycle or overall costs of nuclear power (waste-management costs are estimated at about 1% of total power-generating cost). The proposed high-level waste-management policy and the establishment of a national salt repository represent a milestone in the AEC's research and development program in effluent control and a prime example of how long-range planning and technological development can be utilized to eliminate pollution conditions before they occur.

For background information *see* RADIOACTIVE WASTE DISPOSAL in the McGraw-Hill Encyclopedia of Science and Technology.

[WALTER G. BELTER]

Bibliography: W. G. Belter et al., *Nucl. News*, 12(11):60–65, 1969; A. G. Blasewitz et al., *Interim*

Status Report on the Waste Solidification Demonstration Program, BNWL-1083, June, 1969; Oak Ridge National Laboratory, *Siting of Fuel Reprocessing Plants and Waste Management Facilities*, ORNL-4451, July, 1970.

Reactor, nuclear

In the past few years major technical advances have occurred in both the high-temperature gas-cooled reactor (HTGR) and the molten-salt reactor (MSR). To date, four experimental HTGR systems have been put into operation. Two experimental MSRs have been built and operated successfully. The technical developments that led to these successes are outlined in this article.

High-temperature gas-cooled reactor (HTGR).
Recent developments have removed all the long-standing major technical barriers to the deployment of the HTGR system. These reactors, which employ graphite to moderate the neutrons and helium as a heat-transfer agent, have appeared attractive from the earliest days of reactor development. The theoretical advantages derive from the fact that this combination of materials yields the lowest parasitic neutron absorption of any set of materials capable of operating at high temperature. When a carbide fuel is used, this system is capable of going to very high temperatures, not only yielding a high thermal efficiency for power conversion but also holding out the hope of being useful as a direct heat source for high-temperature thermal reactions, such as the gasification of coal.

Feasibility questions. Several feasibility questions inhibited the early development of HTGRs. For example, helium is in limited supply, and it was not known for certain whether a high-temperature reactor could be constructed with sufficient leak-tightness to avoid the excessive loss of helium. The relatively low heat-transfer coefficient of a gas at ordinary pressures makes it necessary for reactors using helium as a coolant to operate at high pressures, making the leakage problem somewhat more difficult. Another important question arose from the fact that ordinary graphite is porous and permeable, and it was not known how the fission products generated in the fuel could be kept from circulating in the gas stream, depositing on the heat exchangers and blowers, and rendering their maintenance difficult. As the design studies became more sophisticated, it became apparent that there was also a problem related to the safety of the reactor in case of a large and sudden rupture of the system that would lead to a sharp decrease in the pressure of the gaseous coolant and that would make it difficult to remove the fission product decay heat. *See* HEAT TRANSFER.

Developments of the last few years have allayed fears concerning all these problems. The principal events that led to the present optimistic outlook were the operation of four experimental reactors, which showed that helium can be successfully contained in a high-temperature circulating system; the development of pyrolytically coated fuel particles that retain the fission products and keep radioactivity in the circulating helium system at a low level; and the development of prestressed concrete pressure vessels and the experimental confirmation of the fact that they make the rapid release of the gaseous coolant incredible.

Main features. Table 1 shows the main features of the four helium-cooled reactors that have operated to date. Each of the three larger reactors has achieved excellent availability over periods of 3 months or more. The AVR performance has been particularly impressive, with a 71% overall availability for all of calendar year 1969, including a long shutdown required prior to the start of commercial operation. AVR stands for Arbeitsgemeinshaft Versuchs Reaktor, the name of the organization of German utilities operating it. It is situated at Jülich in West Germany, and was designed by Brown-Bovari-Krupp.

Since these are experimental reactors, in the category of first of a kind, there have been troubles which have kept them from operating continuously over long periods of time. For example, failures of diaphragm compressors in the helium-purification system have troubled each reactor and have been a persistent problem for the AVR. The installation of a new type of nonlubricated piston compressor in the AVR has recently given good service and contributes to the recent high availability. In the Dragon reactor, steam-generator tubes have failed repeatedly because of a combination of abnormal water chemistry, nonuniform heat distribution on the gas side, and severe parallel channel instability leading to intermittent dryout on the steam side. It is believed that these problems are now understood and have been overcome with a revised chemical treatment system for the feedwater and the installation of new and improved steam generators. Dragon was built by a coalition of European countries at Winfrith, England.

The Peach Bottom reactor has been troubled with mechanical damage to fuel elements resulting from the swelling and distortion of fuel-particle coatings. Trouble with these coatings was expected since they represent a very early stage of development, but the excessive swelling of the fuel compacts leading to cracking of the fuel elements

Table 1. Helium-cooled reactors

Reactor	Size		Date first at full power	Maximum core outlet temperature achieved, °F	Best availability periods	
	Mw(thermal)	Mw(electric)			Elapsed time, days	Percent
Peach Bottom	115	40	1967	1380	98	92
AVR	46	15	1968	1600	92	99
Dragon	20	—	1966	1380	98	92
UHTREX	3	—	1969	2400	*	*

*Operated only to test short-term performance before project termination.

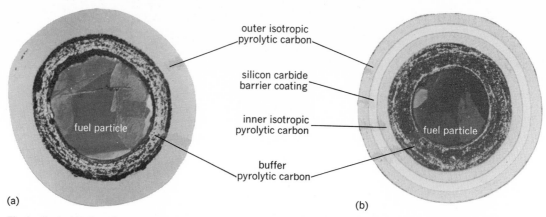

Fig. 1. Coated fuel particles used by Gulf General Atomic in different situations. (a) Biso type. (b) Triso type.

was not anticipated. One fuel element broke, part of it dropping to the bottom of the reactor. Its recovery and replacement gave confidence in the ability to carry out in-core maintenance. Peach Bottom was built by Gulf General Atomic for the Philadelphia Electric Co. in Pennsylvania.

In all three of these reactors it has been possible to show the reasonableness of keeping the helium losses to less than one-tenth of 1% per day, this being the target figure for future large HTGRs. Also, the combination of the use of coated particles and the cleanup systems has held the circuit radioactivity well below permissible levels so that all maintenance operations have been carried out directly.

During 1969 the UHTREX reactor was completed, brought to power, and operated at the remarkable gas outlet temperature of 1310°C (2400°F). UHTREX stands for ultrahigh-temperature reactor experiment, and the reactor was built at the Los Alamos Scientific Laboratory in New Mexico.

Although sustained operation at that temperature is not possible using present fuel, this demonstration showed that fuel damage should not occur if a HTGR is accidentally operated even much above the normal temperature.

Fuel element. Each reactor uses a slightly different kind of fuel element, but they are all based on the use of pyrolytically coated fuel particles. These particles consist of a fuel bead that is a mixture of the oxides or carbides of uranium and thorium, coated with at least two layers of pyrolytically deposited carbon. Generally the inner layer is made of a porous type of carbon which has sufficient volume to accommodate the release of fission product gases and which is thick enough to absorb the recoil fission products and prevent damage to the outer pyrolytic carbon structure. The outer layer of pyrolytic carbon is a higher density material and is impervious to the fission product gases, thus retaining them within the particle. Sometimes a silicon carbide layer is deposited within the outer carbon coating to hold back the strontium and barium fission products, which at high operating temperatures would otherwise migrate through the graphite. The coated particles have overall diameters in the range of 400–800 μ. Figure 1 shows polished sections of typical coated particles. The biso coating uses two layers of iso-

tropic pyrolytic carbon, the inner to absorb the radiation damage and the outer as a container to hold in fission product gases. The triso particles have, in addition, a sandwiched layer of silicon carbide which helps retain the fission products strontium and barium at high temperatures.

In the AVR these particles are incorporated into balls 6 cm in diameter which serve as both the moderator and fuel. In the Peach Bottom and Dragon reactors the coated particles are incorporated in fuel sticks which are encased in graphite.

Gulf General Atomic, the common designation of the General Atomic Division of the Gulf Oil Co., is the only United States vendor of HTGRs.

Pressure vessels. Although none of the presently operating HTGRs employs prestressed concrete pressure vessels, the industry now generally accepts their use as a standard. Much recent research has shown that such vessels do not fail catastrophically, that failure is by relatively slow leakage through cracks, and that the leaks tend to seal on release of pressure. This slow release of pressure is very important for the maintenance of cooling in case a reactor accident should result in failure of the pressure vessel.

For future reactors a pod design such as that shown in Fig. 2 seems to be becoming more or less standard and is, in fact, being used for some carbon dioxide–cooled reactors in the United Kingdom. In the pod design the reactor is in a central cavity, and the steam generators and gas circulators are contained in smaller cylindrical cavities surrounding the reactor; all are within the concrete vessel.

The concrete must of course be protected from the hot gas by an insulating material, and most designers now favor the use of fibrous blankets covered with steel plates as the insulation.

Although the next generation of HTGRs will generate power through conventional steam turbines, studies are being made, particularly in the United Kingdom and Germany, of the direct use of the helium gas coolant to operate gas turbines. The advantages claimed for the use of gas turbines are a lower capital cost for the power plant and perhaps a somewhat higher thermal efficiency.

[H. G. MAC PHERSON]

Molten-salt reactor (MSR). MSRs are being developed for generating low-cost power while extending the United States resources of

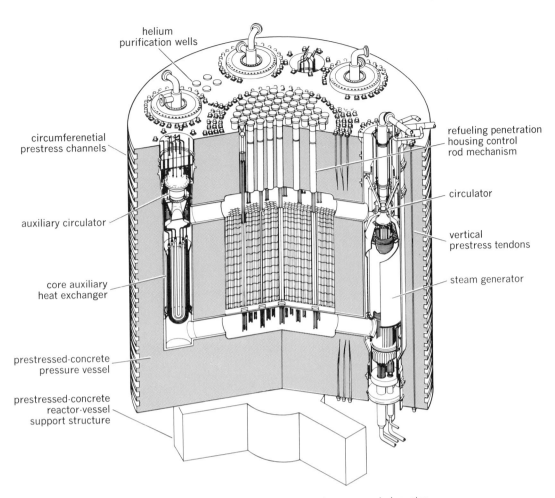

helium
purification wells

circumferenetial
prestress channels

auxiliary circulator

core auxiliary
heat exchanger

prestressed-concrete
pressure vessel

prestressed-concrete
reactor-vessel
support structure

refueling penetration
housing control
rod mechanism

circulator

vertical
prestress tendons

steam generator

Fig. 2. Diagram of the general arrangement of a large high-temperature gas-cooled reactor.

fissionable fuel. The fluid fuel in these reactors, consisting of UF_4 and ThF_4 dissolved in fluorides of beryllium and lithium, is circulated through a reactor core moderated by graphite. The successful operation of the 8 Mw(thermal) molten-salt reactor experiment (MSRE) has demonstrated that operation with this type of fuel is practical, that the salt is stable under reactor conditions, and that corrosion is very low. Processing of the MSRE fuel has demonstrated that MSRs can operate as high-performance converters. New fuel-processing methods under development should permit MSRs to operate as economical breeders. These features, combined with high thermal efficiency (44%) and low primary system pressure, give MSR converters and breeders favorable economic, fuel-utilization, and safety characteristics. Further, these reactors can be initially fueled with U^{233}, U^{235}, or plutonium. The construction cost of a molten-salt breeder reactor (MSBR) power plant is estimated to be about the same as that of light-water reactors, leading to power costs about 0.5–1.0 mill/kwhr less than those for light-water reactors. *See* NUCLEAR FUELS.

Liquid fluorides. A molten mixture of fluoride salts was selected for the fluid in MSRs because of the basic physical chemistry of the fluoride system. Radiation damage is nonexistent in the completely ionic liquid fluorides. In addition, the vapor pressure of the molten salt is extremely low, even at the incandescent temperatures required, and

the salt does not react violently on exposure to air or water. Since a variety of interesting fluorides were known to be stable in contact with some common structural metals, a corrosionless system seemed attainable. High specific heat coupled with reasonable viscosity and thermal conductivity makes these liquids good heat-transfer media. At the same time, although the high melting point of potential fuel mixtures (850–950°F) is not a drawback during power operation, provisions are required for preheating the salt and for keeping it molten during periods when the reactor is shut down.

Two experimental MSRs have been built, and both were operated successfully. The 2 Mw(thermal) aircraft reactor experiment was operated in 1954, but under special conditions different from those for power reactors. The 8 Mw(thermal) MSRE operated from 1966 through 1969, under conditions largely representative of a converter reactor. The molten-fluoride fuel operated at temperatures above 1200°F without corrosive attack on the metal and graphite parts of the system. Further, the reactor equipment operated reliably and the radioactive liquids and gases were contained safely. The fuel was completely stable, with xenon removed rapidly from the salt. When necessary, radioactive equipment was repaired or replaced without overexposing maintenance personnel.

Moderator material. MSBRs are moderated by graphite in direct contact with the fuel salt. Fuel

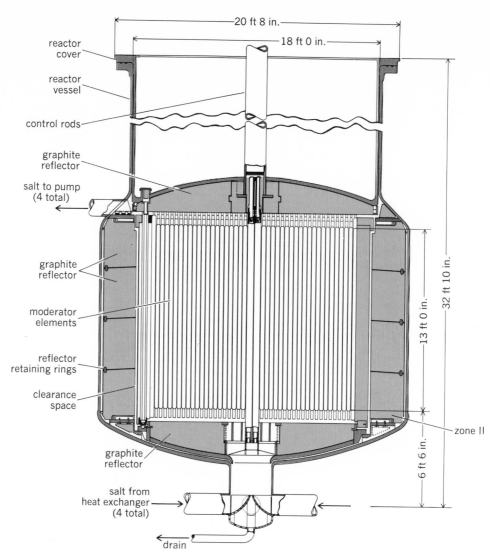

Fig. 3. Cross sectional view of the core of a molten-salt breeder reactor.

salt does not wet graphite and does not penetrate into the pores if material having small pore sizes (<1 μm) is used. Low gas permeability to keep out Xe^{135} can be achieved by pyrolytic deposition of carbon in the pores and on the surface after the graphite parts are fabricated. A problem with graphite concerns the changes in dimension that are caused by fast neutron irradiation. At MSBR temperatures, a number of isotropic graphites first shrink and then begin to expand after continued exposure. The rate of change increases at higher temperatures. Irradiation data have shown that present grades of isotropic graphite can receive up to 3×10^{22} neutrons/cm^2 (>50 kev) at 1300°F before a significant increase in original dimensions occurs.

The core cross section of an MSBR is shown in Fig. 3, and Table 2 lists some of the principal reactor data, based on a 1000 Mw(electric) plant. Optimization studies suggest that a reasonable compromise between low fissile inventory and long graphite life occurs at an average core power density of 22 kw/liter; this provides about 4 years before the graphite in the region of maximum radiation damage receives a fast neutron fluence of 3×10^{22} cm^2/sec and begins to expand beyond its original dimensions.

Coolant salt circuit. Energy generated in the fluid fuel of an MSR is transferred to a coolant salt circuit, which in turn transfers energy to the steam cycle. The fuel salt contains UF_4 and ThF_4 in a mixture of LiF and BeF_2, and is heated from 1050 to 1300°F in passing through the core region; the coolant salt is a eutectic mixture of $NaBF_4$ and NaF (92:8 mole %) having a low cost and a relatively low melting point ($\sim$725°F); it enters the steam generators at 1150°F and leaves at 850°F. A supercritical steam power cycle is used, and steam at 3500 pounds per square inch atmospheric and 1000°F is delivered to the turbine throttle.

One-fluid breeders. Two developments during 1968 established the feasibility of the one-fluid breeder. In the one-fluid breeder the carrier salt contains both the fissile and fertile materials. These developments made MSRE technology and experience more directly applicable to breeder requirements. The first development was the demonstration of the chemical steps in processes which use liquid bismuth to extract protactinium, uranium, and rare-earth fission products selectively from uranium-thorium fuel salt; this led to improved neutron economy. The second development was the recognition that a fertile blanket can be obtained with a salt that contains both ura-

Table 2. Design characteristics of 1000-Mw(electric) MSBR station

Plant	
Useful reactor heat, Mw(thermal)	2,250
Gross electrical generation, Mw(electric)	1,035
Net electrical output of plant, Mw(electric)	1,000
Overall thermal efficiency, %	44
Reactor	
Reactor vessel inside diameter, ft	22
Vessel height at center, ft	20
Vessel wall thickness, in.	2
Vessel head thickness, in.	3
Vessel design pressure, psi	75
Core height, ft	13
Number of core elements	1,412
Volume fraction salt in core, %	13
Volume fraction salt in undermoderated region, %	37
Average core power density, kw/liter	22.2
Maximum thermal neutron flux, neutrons/cm² sec	7.9×10^{14}
Maximum graphite damage flux (>50 kev), neutrons/cm² sec	3.2×10^{14}
Graphite temperature at maximum flux region, °F	1,280
Estimated core graphite life, years*	4
Total weight of graphite in reactor, lb	650,000
Weight of removable core assembly, lb	480,000
Maximum flow velocity in core, ft/sec	8.5
Pressure drop due to salt flow through reactor, psi	18
Total salt volume in primary system, ft³	1,720
Thorium inventory, kg	68,000
Fissile fuel inventory of reactor and processing plant, kg	1,400
Breeding ratio	1.06
Yield, %/year	3.3
Doubling time, compounded continuously, years	21

*Based on operation at 80% plant factor.

nium and thorium by increasing the fuel-to-graphite ratio in the outer regions of the core, the result being decreased neutron leakage losses. Holdup of the protactinium external to the reactor core can be accomplished in an extractive processing system in which the fuel salt is contacted with liquid bismuth containing dissolved active metals. Since thorium has a stronger tendency to become a fluoride than has protactinium, it can reduce protactinium from the salt to the metal. Protactinium, in turn, tends to form a fluoride more readily than does uranium, so that UF_4 can oxidize protactinium from metal back to salt. Based on recent refinements, the protactinium isolation method consists of removing about 95% of the uranium from the fuel salt by fluorination, after which the remaining uranium and the protactinium are removed from the salt by countercurrent contact with bismuth containing lithium and thorium reductant. High concentrations of protactinium can be obtained in the bismuth and retained outside the reactor core until the Pa²³³ decays to U²³³.

The salt leaving the protactinium isolation system, essentially free of uranium and protactinium but containing the rare-earth fission products, is fed to a system in which the rare earths are extracted into a bismuth stream containing lithium and thorium. Contact of this bismuth stream with lithium chloride results in transfer to the salt of a significant fraction of the rare earths but a negligible fraction of the thorium. The final step in the

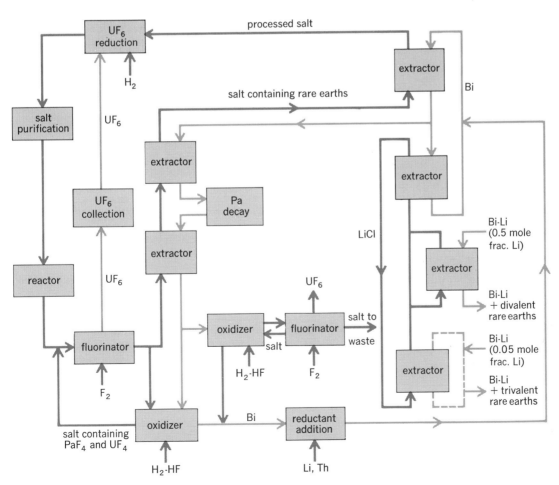

Fig. 4. Processing flow sheet for single-fluid molten-salt breeder reactor.

process consists of removing the rare earths from the lithium chloride by contact with bismuth containing a high lithium concentration. The processing flow sheet for the single-fluid MSBR is shown schematically in Fig. 4.

Outlook. The outlook for MSRs is intimately related to the economic and nuclear performance of such systems. For such systems, the avoidance of fuel fabrication, the ease of processing, and the low fissile inventory result in low fuel-cycle costs. Capital costs benefit from high thermal efficiency, low primary system pressure, and low pumping requirements. Further, converter and breeder reactor plants are basically the same except for the processing system. Thus molten-salt converters can evolve into single-fluid breeders as the processing is improved. This flexibility, generally a result of using a fluid fuel, extends also to the choice of starting up the reactors with U^{235}, U^{233}, or plutonium, depending on which fuel is the most economical at the time.

For background information *see* NUCLEAR FUELS; REACTOR, NUCLEAR; REACTOR, NUCLEAR (CLASSIFICATION) in the McGraw-Hill Encyclopedia of Science and Technology. [PAUL R. KASTEN]

Bibliography: E. S. Bettis and R. C. Robertson, *Nucl. Appl. Technol.*, vol. 8, no. 2, February, 1970; W. R. Grimes, *Nucl. Appl. Technol.*, vol. 8, no. 2, February, 1970; P. N. Haubenreich and J. R. Engel, *Nucl. Appl. Technol.*, vol. 8, no. 2, February, 1970; H. E. McCoy et al., *Nucl. Appl. Technol.*, vol. 8, no. 2, February, 1970; *Proceedings of the Gas-Cooled Reactor Information Meeting, Oak Ridge, Tenn., Apr. 27–30, 1970*, USAEC Rep. CONF-700-401, National Bureau of Standards, 1970; *Proceedings of the Symposium on Sol-Gel Processes and Reactor Fuel Cycles, Gatlinburg, Tenn., May 4–7, 1970*, USAEC Rep. CONF-700-502, National Bureau of Standards, 1970.

Satellites, applications

Recent advances in the design and use of applications satellites have been concerned with furthering knowledge in the fields of geodesy, meteorology, communications, navigation and traffic control, and Earth resource studies. Applications satellites are manned and unmanned spacecraft placed into Earth orbit to obtain information in the five areas cited above. Although spacecraft in the Gemini and Apollo programs do not quite fit into this definition because of their extraterrestrial orbits, pictures of the Earth taken after launching and before returning to Earth have been useful in geodetic studies, and results from these programs are included in this article.

Geodetic studies involve accurately locating and identifying geographical features, regional mapping, and defining the structure of the Earth's gravitational field. Meteorological investigations are for purposes of obtaining a more precise knowledge and understanding of man's environment on Earth and of the forces that control it. The purposes of communications investigations are to provide worldwide communications of all types. Navigation and traffic control investigations are for purposes of providing systems to achieve safe, economical, and swift air, land, and sea transportation. Earth resources studies are for purposes of obtaining a more precise knowledge and inventory of the Earth's resources, as well as achieving a more profitable use of these resources.

Geodetic studies. Geodetic information from unmanned satellites was first obtained by studying the orbit of *Vanguard 1*, launched Mar. 17, 1958. The study resulted in the determination that the Earth is slightly pear-shaped.

Geodetic studies have been performed using ground-based laser tracking in conjunction with *Explorer 22*, launched Oct. 10, 1964. The Earth's gravitational field was studied by precise Doppler tracking of orbital perturbations of *Explorer 27*, launched Apr. 29, 1961. A worldwide triangulation network was established by optical tracking of the *Pageos 1* satellite, launched on June 24, 1966.

Manned spacecraft have also provided excellent photographic material suitable for detailed geological studies. *Apollo 9*, for example, provided multispectral photographs for geological analysis. Such photographs show definite superiority over color photography for rendering geological structure in areas heavily covered with vegetation. The illustration is a black-and-white rendition of a typical multispectral photograph; it shows the definite contrasts obtainable by this technique.

Other photographs provided for geologists by the manned programs have enabled them to discover many heretofore unmapped geological features; study and remap part of the northern Baja California and Mexico in the vicinity of the Agua Blanca fault; study the tectonics and geology of northern Chihuahua, Mexico, in the vicinity of Palomas; search (unsuccessfully) for the Texas lineament; study regional dune-fracture patterns surrounding the Tibesti massif in Chad and Libya; and study the theory that the Arabian Sea is a sphenochasm. formed by continental drift.

In addition to continuing to use imagery of different types, future geodetic satellites will establish geometric worldwide reference systems and determine fiducial control points to a high degree of accuracy in geocentric coordinates. They will also be used to define more accurately the structure of the Earth's gravitational field.

Meteorology. Satellites have been used as platforms to observe the Earth's weather since *Tiros 1*, launched Apr. 1, 1960, provided the first global cloud-cover pictures.

Cloud-cover pictures of the entire Earth were provided every 24 hr by *Nimbus 1*, launched Aug. 23, 1964. In addition, the satellite provided automatic picture transmission (APT) capability that enabled local readout of weather pictures. It provided high-resolution infrared imagery, which in turn led to both daytime and nighttime cloud-cover information.

A series of other meteorological and applications satellites have been flown both for research and development and on an operational basis. They have provided local readout of daylight cloud cover by the APT television system from polar Sun-synchronous orbits; medium-resolution infrared (MRIR) radiometer for Earth heat balance; high-resolution infrared (HRIR) readout by APT with superimposed orbit data; color coverage of nearly the entire Earth's disk from synchronous altitude; and daily global coverage by advanced vidicon camera system pictures from Sun-synchronous orbit.

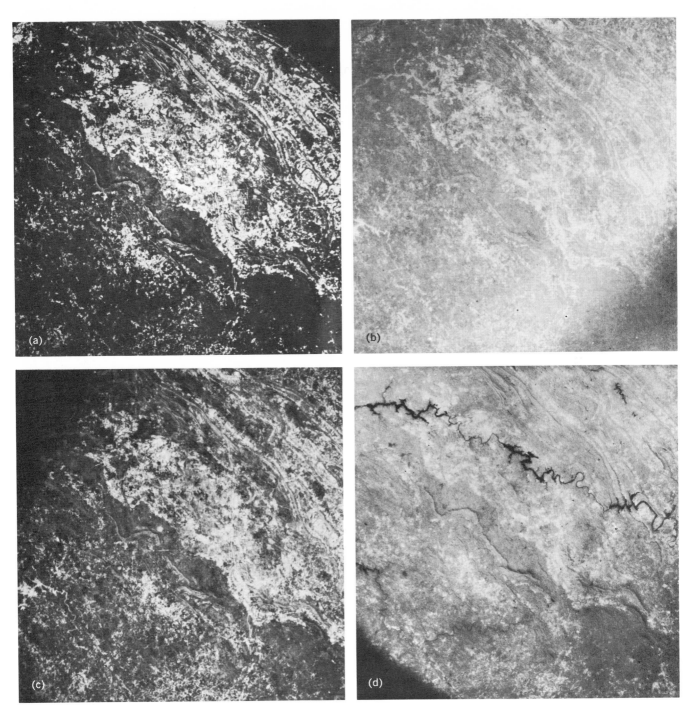

Black-and-white rendition of multispectral photographs of the area east of Birmingham, Ala. (*a*) Color infrared. (*b*) Blue-green spectral band. (*c*) Black-and-white infrared. (*d*) Red spectral band. (*NASA*)

The presently proposed meteorological satellite systems are being configured with the objective of advancing the currently available 1–2-day accurate forecast into a 5–7-day weather forecast of comparable accuracy. Realization of this objective would permit, for example, more efficient routing and scheduling of transportation activities; more dependable scheduling of recreational activities; savings to the farmer by reducing the need for reseeding, by improving his timing for fertilizing or spraying by enabling him to predict heavy rains and by improving his timing on harvesting; and better flood control and water management.

Communications. Communications satellites are used as a means of conveying aural and visual intelligence data and traffic information by relay. They provide a reliable, low-cost method of communicating throughout the world, independent of atmospheric and ionospheric disturbances, and have the capacity to transmit much larger amounts of data (for example, color television) overseas than do undersea cables.

Commercial satellites such as *Intelsat 3*, launched Feb. 6, 1967, are designed to provide 1200 two-way voice circuits, television, and other commercial services from a geostationary orbit.

In addition to their familiar uses, communications satellites can provide the means for timely and efficient gathering of data from a large number of remotely located devices used for such purposes as gathering Earth resources information. They can interrogate sensors and relay the data obtained to central processing stations.

Future communications satellites will provide for multiple-access small terminal (including aircraft) point-to-point services; spacecraft-to-spacecraft-to-ground communications; and broadcasting of monochromatic and color television material either directly to conventional or modified home receivers or to relatively inexpensive community and institutional receivers for distribution of programs to homes and classrooms.

Navigation and traffic control. Satellites are an especially useful tool in aiding navigation and for providing a method to assist in controlling air, land, and sea travel over the oceans and other unpopulated areas.

Traffic control is dependent on both rapid communications and accurate navigation.

The U.S. Navy's use of Transit satellites has demonstrated that precise position fixes can be obtained in all kinds of weather by surface vessels and submarines. The technology already developed in communications satellites is immediately usable in traffic-control systems.

Combined systems will enable rapid and accurate determination of positions of cooperating vehicles on the surface and in the air. Such systems are urgently needed to provide service for air-traffic control over the North Atlantic Ocean, traffic control of surface vessels in confluence areas, and marine and air rescue at sea.

Future navigation systems design should consider a wide spectrum of users, such as the small user (private planes and boats) and the rapid user (supersonic transports). The ground equipment must be inexpensive and simple to operate and maintain. The system should contain provisions for accurately locating and identifying the user for rescue and traffic-control purposes.

Earth resources studies. The Earth is now recognized as one large "spacecraft" whose self-contained environment must be maintained indefinitely by wise management of its food, fiber, water, air, and other natural resource systems. To do so requires nearly real-time knowledge of man's interaction with many of these factors.

Earth resources studies include, in addition to geology and meteorology, hydrology, forestry, agriculture, and oceanography.

Each type of applications satellite previously discussed plays a role in the determination and use of Earth resources information. The proper perspective of Earth resources can be obtained only from the acquisition, cataloging, study, monitoring, and utilization of information obtained periodically on a global scale. Properly instrumented satellites can provide this information on a timely basis.

Space imagery provides information not only on the areas of geology and meteorology previously mentioned but also on vegetation cover types, landforms, shallow submarine forms, generalized land use, and the extent of snow, ice, and glaciers.

In addition to the use of multispectral photographs for geological mapping and for the study of structural geology, other uses have been determined. For example, such techniques are particularly useful in hydrology (the study of the entire water cycle). Applications of hydrology cover not only agriculture and forestry but industrial use, private use, and the weather cycle itself. Applications satellites instrumented for the study of hydrology could reveal snow cover and ice occurrence in rivers and glaciers; nearshore underwater detail in coastal waters, estuaries, large lakes, and major reservoirs; details of saline intrusion, circulation patterns, and variable pollutant distribution in coastal waters, estuaries, and lakes; geomorphology of river basins and changes in shorelines; and land use as regards forestry and other natural vegetation in agricultural and urban areas. *See* HYDROLOGY.

Among the important types of information that must be acquired at frequent intervals in order to study and manage the Earth's resources in the area of agriculture and forestry are snow coverage for water yield predictions; state of air pollution; crop coverage and condition; and forest density, types of trees, and condition and health of the forests.

It has been demonstrated through the use of satellite photography that various kinds of crops and trees reflect light in different degrees. Analysis of this type of photography has enabled limited identification of the types of vegetation observed from spacecraft. In an infrared picture taken by *Apollo 11* of southwestern United States and northwestern Mexico, a difference was apparent between the types of vegetation that exist on either side of the United States – Mexico border below the Salton Sea.

Still in its infancy is the use of satellites as a method of obtaining information for the study of the physical, chemical, and biological character of the ocean, its behavior on a global scale, and its interaction with the atmosphere and the coasts. Limited oceanographic studies have been conducted utilizing data obtained from meteorological satellites, Earth resources technology satellites, and manned satellites. These studies have revealed the great potential of satellites for use in oceanography.

Satellites will provide information on only the top few hundred feet (or about 1%) of the depth of the ocean. Fortuitously, this is the portion of the ocean that is presently most important to man in his fishing, shipping, and coastal activities. The top few hundred feet constitute the photosynthesis zone, which provides the nutritional basis for all of the biological resources of the sea. The major energy exchanges occur across the air-sea interface.

Color photographs from space have shown how man is contaminating the sea. Color photographs of other areas of the oceans have been used to study, on a large scale, subsurface erosion and water currents. Satellite sensors have provided, for example, sea-surface temperature measurements on a global scale. These data are useful to the fishing industry and for locating ocean currents.

The oceanographic parameters that presently appear to have the most significant satellite applications are sea-surface temperature; radar, photographic, and infrared imagery; drift rate of floating objects; location and magnitude of sea ice and icebergs; detection of chlorophyll, bioluminescence, fluorescence, and sea color; sea state; and the dynamic topography of the sea surface.

For background information *see* SATELLITES, APPLICATIONS; SATELLITES, NAVIGATION BY; SATELLITES, TECHNOLOGY in the McGraw-Hill Encyclopedia of Science and Technology.

[ROBERT C. BAUMANN]

Bibliography: J. B. Bird and A. Morrison, Space photography and its geographical applications, *Geogr. Rev.*, 54(4):464–486, October, 1964; J. Bodechtel and H. G. Gierloff, *Weltraumbilder der Erde*, 1969; R. N. Colwell, Remote sensing of national resources, *Sci. Amer.*, 218(1):54–69, January, 1968; E. H. Conghran and R. H. Sullivan, Some considerations affecting satellite data, *Contrib. Fish.*, U.S. Navy Oceanographic Office, July, 1967; P. D. Lowman, Jr., *Apollo 9 Multispectral Photography: Geologic Analysis*, Goddard Space Flight Center Publ. no. X-644-69-423, 1969; P. D. Lowman, Jr., Geologic orbital photography: Experience from the Gemini program, *Photogrammetria*, 24:77–106, 1969; P. D. Lowman, Jr., *Space Panorama*, 1968; H. E. Newell, M. G. Kroshkin, and W. Priester, *Satelliten Erkunden Erde und Mond Herausgeben*, 1969; *Objectives and Goals in Space Science and Applications*, NASA SP-162, 1968; W. J. Powell, C. W. Copeland, and J. A. Drahongel, *Geologic and Hydrologic Research through Space Acquired Data for Alabama: Delineation of Linear Features and Application to Reservoir Engineering Using Apollo 9 Multispectral Photography*, 1970; *Useful Applications of Earth-Oriented Satellites: Summaries of Panel Reports*, National Academy of Science, 1969.

Ship propulsion

During the past 10 years, marine gas turbine engines derived from aircraft jet engines have been developed and employed for the propulsion of increasing numbers of naval and commercial ships. The fact that many inherent characteristics of aircraft jet engines are the same as those desired for certain marine propulsion plants provided the impetus for this development. These inherent characteristics include a high power-to-weight ratio, low space requirements, ease of automation, and rapid start-up.

Fundamentals. When speaking of gas turbine engines for marine use, "engine" is considered to pertain to a complete unit capable of producing shaft horsepower to drive a propeller, usually via a clutch and reduction gear.

The marine gas turbine engines derived from aircraft jet engines consist of two main components: a gas generator to produce high-temperature and high-pressure gases, and an independent or "free" power turbine to convert the thermal energy into mechanical energy. The aircraft jet engine suitably modified is usually used as the gas generator. The principal components of the gas generator are a compressor, the burners and associated combustion chamber or chambers, and a turbine which drives the compressor. The power turbine is specially designed. A typical arrangement is shown schematically in Fig. 1.

Incoming air enters the gas generator at the inlet to the compressor. Inside, the progressive stages compress this air 12–17 times atmospheric pres-

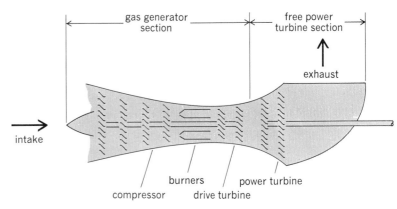

Fig. 1. Schematic of marine gas turbine engine derived from aircraft jet engine.

sure in currently available engines.

The highly compressed air passes into the burner section in the forward part of the combustion chamber or chambers, where the fuel burns with the air. Only about 25% of the air passing through the burner section is actually used for combustion. The remainder serves to cool the burners and turbines to maintain acceptable metal temperatures within the gas generator.

From the burner section the combustion gases and heated excess air pass on to the turbines. Typical turbine inlet temperatures range from 1400 to 1800°F. The turbines absorb power from the expanded high-velocity gases. The mechanical energy thus generated drives the compressor in the forward part of the gas generator.

The gases leaving the turbine at reduced velocity, pressure, and temperature are directed into the free power turbine to produce the torque required to turn the output shaft of the gas turbine engine. Finally, the spent gases are exhausted through ducting.

Engine design. In the development of marine gas turbine engines from aircraft jet engines there are two basic design tasks: First, the aircraft engine must be modified, and second, a suitable power turbine must be developed. These components must then be integrated into a single unit complete with controls and all the auxiliaries necessary for ship propulsion.

The aircraft jet engine which becomes the gas generator is normally chosen on the basis of power output, efficiency, and demonstrated reliability in aircraft service. Engines selected to date have accumulated hundreds of thousands of hours of operation in the air and have been cleared for 3000–6000 hr of operation between overhauls. Nevertheless, important modifications are required to assure equally reliable performance in a marine environment. The specific environmental differences dictating these modifications include the following.

1. The air density at sea level is greater than at the operating altitudes of aircraft. This increases the aerodynamic loads on the engine components. It also increases the rate of fuel consumption, and therefore the combustion equipment must operate at higher heat release rates.

2. The air ingested by the marine engine is always contaminated by salt spray and dust to a greater extent than is the aircraft engine.

3. The marine engine is usually required to burn

Fig. 2. Pratt and Whitney FT4A marine gas turbine engine.

a heavier distillate fuel and one containing more contaminants, such as sulfur and salt water, than is the aircraft engine.

4. The shock loadings the engine must sustain are higher, especially in naval applications, than in aircraft use.

One serious problem created by this environment is hot corrosion; that is, the sulfur in the fuel and the salt in the air and/or salt water in the fuel combine to attack the metal of the hot first-stage vanes and blades of the compressor drive turbine. Gas turbine designers are employing several methods to solve this problem: improved materials, filtering and fuel treatment, and maintaining the metal temperatures below the threshold level at which corrosion becomes significant.

The full extent of the alterations required to adapt for operation in a marine environment depends on the original aircraft jet engine design. This adaptation has generally included the use of improved corrosion-resistant materials, redesigned burners and combustion chambers, and strengthened engine casings in way of the main bearings.

Design of a suitable power turbine presents no

unusual problems. Normally, for smaller engines it is designed to be built into the gas generator, in which case the entire engine can be easily removed for repairs. In the higher ratings the power turbine is usually designed to remain in the ship, and arrangements are made to disconnect the gas generator for repair by replacement.

Essential auxiliaries forming an integral part of a marine gas turbine engine are the fuel-control, lubricating, and engine ventilation systems. Care must be taken during the initial design stages to ensure accessibility and suitable operating conditions for all these systems. The controls must include numerous safety devices which will secure the engine when the operating conditions (such as high exhaust temperature, excessive vibration, or overspeed) might cause serious damage.

Other important design considerations include providing for mounting which will attenuate shock and noise transmission to the hull and minimize misalignment between the engine and gear box, and for local acoustic treatment. A common solution regarding acoustic treatment is to arrange for each engine to have its own closely fitted acoustic enclosure, making the necessary provisions for ease of access and internal ventilation.

An example of the type of marine gas turbine engine discussed here is the Pratt and Whitney FT4A (Fig. 2), which is based on the JT4 engine used extensively to power commercial jet aircraft. The FT4A can produce over 25,000 hp but weighs only 14,200 lb.

Applications. The three basic types of marine propulsion plants used today which derive their energy from fossil fuels are diesel, steam, and gas turbine. The great majority of ships now sailing worldwide are diesel-propelled, with the remainder primarily steam-propelled. Few ships powered by aircraft-type marine gas turbine engines are in operation now. However, the success of early installations and the development programs completed during the past 10 years, as well as the resulting technological advances, have generated increasing interest in these types of power plants.

A number of factors combined to slow the widespread application of aircraft-type marine gas turbine engines after their introduction in the early 1960s. Unlike large diesel and steam power plants, which can burn residual fuels, marine gas turbines

Fig. 3. U.S. Coast Guard's *Hamilton* cutters are powered by two diesels and two Pratt and Whitney FT4A engines.

require more expensive distillate fuels. This requirement compromises the economic advantage to be gained from the weight and space savings of these engines. Another factor that adversely affects gas turbine plant costs is the requirement for special devices (such as reversing gears or reversible pitch propellers) to obtain reverse power from the unidirectional aircraft-type engines. Development costs for marine versions of these engines have also been prohibitive, and the traditionally conservative marine industry has been reluctant to adopt similar engines until reliable operation in a marine environment could be demonstrated.

Therefore, early development was primarily for naval applications. These included combined gas turbine and diesel plants in which the gas turbines were used solely for boost power to achieve maximum speed, and hydrofoil and air-cushion vehicles which required the high power-to-weight ratios only aircraft-type marine gas turbines could offer.

An example of the combined gas turbine and diesel plant is found in the USCG cutter *Hamilton* (Fig. 3), delivered in 1966, the first of 11 similar ships now in operation or on order. Each is powered by two 3500-hp diesel engines during low-speed and endurance cruising operations and by two 20,000-hp Pratt and Whitney FT4A marine gas turbine engines for high-speed runs. Controllable-reversible pitch propellers are employed to maneuver with either the diesels or gas turbines operating.

One of the first large hydrofoil ships, the U.S. Maritime Commission's *HS Denison*, designed in the early 1960s, is powered by a single General Electric LM1500 aircraft-type marine gas turbine of about 14,000 hp. Other large, gas-turbine-powered hydrofoil ships of more recent design are now being operated by the U.S. Navy.

Proven engines are now available in a range from 3000 to 30,000 hp. Advanced designs such as the General Electric LM2500, a 20,000-hp marine gas turbine engine derived from the TF39 aircraft engine, offer fuel rates 20% lower than those of similar "first-generation" engines such as the FT4A and LM1500.

The first large commercial ship designed from inception for aircraft-type marine gas turbine propulsion is a roll-on, roll-off cargo ship chartered to the U.S. Navy's Military Sea Transportation Service. This ship, the *Adm. Wm. M. Callaghan*, has been in operation since late 1967. Although originally powered by two FT4A gas turbines, the ship is now fitted with one FT4A and one of the newer LM2500 engines on a trial basis. Each of the ship's two propellers is driven by one engine via a unique reversing-reduction gear.

Several current shipbuilding programs are indicative of industry's increased interest in these engines and recognition of their advantages. Four 800-ft, 32,000-ton-displacement containerships are now under construction in Germany, each powered by two FT4A marine gas turbine engines. Proposals for the U.S. Navy's new class (DD963) of destroyers include aircraft-type marine gas turbine propulsion plants.

For background information *see* GAS TURBINE; MARINE ENGINE; SHIP POWERING AND STEERING in the McGraw-Hill Encyclopedia of Science and Technology.   [JOHN C. COUCH]

Bibliography: W. A. Brockett et al., *U.S. Navy's Marine Gas Turbines*, American Society of Mechanical Engineers, Zurich, March, 1966; Gas turbine power, *Mar. Eng. Log*, October, 1966; G. A. Hardgrove, *Design Considerations for a Second Generation High Power Marine Gas Turbine*, American Society of Mechanical Engineers, Brussels, May, 1970; W. H. Lindsey, *The Marine Gas Turbine*, Society of Naval Architects and Marine Engineers, New York, June, 1968; E. E. Stoeckly, *Development of the General Electric LM1500 Gas Turbine as a Marine Power Plant*, American Society of Mechanical Engineers, Zurich, March, 1966.

Simulation

Recent developments in the field of simulation include advances in political simulation. Political simulation may be regarded as an experimental technique through which complex political phenomena such as a political campaign or an international relations crisis involving a series of events and a number of nation-state "actors" may be recreated or forecast under quasi-experimental conditions controlled by the simulation director. As such, political simulation techniques provide the advantage of controllability; that is, the circumstances may be altered at will and, especially, the timing of the exercise may be altered to suit the convenience of the researcher or classroom.

Simulation techniques may be utilized for teaching, for managerial decision-making tools, and for research. The simulation approach is not new in the behavioral and managerial sciences and, in fact, has been applied for a number of years in the field of business management. In political science, simulation methodologies have been applied for some years to international relations and foreign policy situations. A particular case in point is the Inter-Nation Simulation, which has been utilized extensively at Northwestern University and several other universities. The latest and perhaps most comprehensive outgrowth of this series of international relations simulation models is the World Politics Simulation (WPS) now in use at the Industrial College of the Armed Forces (ICAF), Washington, D.C., and a few other schools. A teaching version of the original Inter-Nation Simulation is available through Science Research Associates.

In the field of United States government and politics, an election game developed by James S. Coleman has been utilized at the Johns Hopkins University and in the Baltimore high schools. A national political game has been used at Kansas State Teachers College under the direction of Dale Garvey, and a presidential election simulation for instructional purposes has been published by Marvin Weinbaum and Louis Gold.

Most of these simulation exercises are intended to teach the participants particular skills or how to perform particular functions, such as being the head of state of a country, devising viable military strategies, or managing a large business. A major inducement to the utilization of simulation techniques in such situations is that the participant can learn from his mistakes without suffering their real-life consequences. A participant in a business management simulation may, through poorly conceived strategies, bankrupt his company and learn from the experience without actually losing real-world money. The miscalculated foreign policy decisions of the head of a simulated state may lead to war and thus provide the appropriate lessons for the decision maker without incurring the social costs of actual war. The urban political leader may experiment with various strategies to achieve various goals without actually endangering his career at the real-life polls. In short, simulation exercises are useful in teaching skills when the consequences of error in the real-life context are so costly in terms of monetary value or danger to life and property or so time-consuming that they effectively prohibit trial-and-error learning.

Although most simulation model-building efforts in the political science field have been directed toward teaching purposes, there has been some tentative movement into the areas of providing vehicles to assist political decision makers and some efforts at using simulation techniques for research purposes.

International relations. In the simulation of international relations, the major pioneering effort was conducted at Northwestern University under the direction of Harold Guetzkow, and resulted in the Inter-Nation Simulation. This model, which now exists in a number of variants, provides for a number of human players occupying roles such as head of state, economic director, and citizen in a number of hypothetical states (in some runs, how-

ever, the states are modeled after specific real-world nations). In these all-man simulation models, human evaluators, using detailed formulas, calculate the effects of the role players' decisions in each round and feed the results back to the participants for the next round; thus the environment is a dynamic one.

The increasing complexity of this series of models as they evolved over a number of years led to the use of computer facilities (mixed man-computer models) and to the so-called International Processes Simulation work at Northwestern directed in particular by Paul Smoker. The use of the computer enabled the development of more elaborate models, and assisted in the more rapid cycling of the model in play.

The most elaborate of the series of Inter-Nation models is the World Politics Simulation, originally developed by William Coplin and now being used in revised form principally at ICAF. This model provides more detailed and "rich" feedback to the participants than do any of the earlier models.

Other major lines of development in the international simulation field include the Political-Military Exercise (PME), originally developed at The Rand Corporation and now used mainly at the Massachusetts Institute of Technology under Lincoln Bloomfield. It is centered around a detailed scenario of a projected international relations crisis. Role players devise strategies for dealing with the crisis, and the effectiveness of their strategies is evaluated by a control team of knowledgeable political scientists. The exercise does not utilize computers.

The TEMPER international relations model, on the other hand, is an all-computer simulation of international phenomena. The user can "tune" various input parameters but cannot control the programmed interactions contained within the model.

United States politics. In the United States politics area, several teaching models have been devised, as previously mentioned. Additionally, the Woodbury Political Campaign Simulation, a man simulation of an urban mayoralty campaign developed by Bradbury Seasholes, H. Roberts Coward, and Marshall H. Whithed, has been utilized for teaching purposes at a number of colleges and universities, and is now in press with a major publishing house (see illustration). Participants serve in the roles of various campaign actors, such as candidates, pressure groups, and mass media, and the results of their actions are evaluated by a group of political-scientist "umpires." A mixed man-computer model, wherein the computer is used to score the actions taken by the participants, has also been developed under the direction of Whithed, of Temple University. The basic computer algorithms developed in that project are viewed by the developers as being applicable to actual urban electoral situations, and a man-controlled Woodbury-type model has been developed and utilized in a rudimentary way to study electoral processes in an actual urban election.

A computer simulation model of voting behavior was developed by Ithiel de Sola Pool and associates to study certain aspects of the 1960 and 1964 presidential elections, and studies based on this model were made for the Democratic party presi-

Schematic representation of the computer-assisted model of the Woodbury urban mayoralty election simulation system. The voter block reaction (VBR) groups represent ethnic, income, occupational, and religious groupings within the electorate. Gallup polls provide periodic feedback to the role-playing participants.

dential candidate before the election in 1960. The results of the work were enticingly accurate. Eugene Burdick wrote *The 480*, a fiction book on computers, simulation, and the presidential election process; some think that this book was related to the Pool simulation project, especially since the number 480 figures strongly in the Pool simulation study.

Urban planning. In the field of urban analysis, a family of urban-oriented simulation models has been developed. Allan Feldt of Cornell has developed the Community Land Use Game (CLUG), which simulates land-use development patterns in an urban community. A series of more complex urban simulation models has since been developed from this beginning, such as that devised by Richard Duke under the name of Metropolis.

The original CLUG model and many of the subsequent models in this family have been mapboard oriented; that is, a playing board marked in a grid pattern has been provided for the players, who represent private entrepreneurial groups in CLUG. In more complex models the players represent private entrepreneurial groups plus various political and community groups. The players "build" a city in the process of play.

CLUG can be played without computer assistance, but a computer is an absolute necessity in the more complex models. CLUG is provided with a computer-assisted mode by its developers, and a CLUG variant for time-share conversational-mode computer control has been developed under the direction of Whithed.

The basic orientation of these urban models is toward teaching the interrelationships and complexities of urban land-use development. On a wider scope, however, a Washington, D.C., group called Envirometrics has been experimenting with more advanced models such as Region, City I and II, Telecity, and Bigcity. The researchers at Envirometrics hope to develop their models, which contain extensive political and social as well as economic roles, to the point of sophistication where their predictability value and thus their use for decision makers will be high. A. Ray and Duke are also working on the evolution of the Metropolis simulation into what they call M.E.T.R.O., and they see their new model as having research implications as well as educational usages.

Finally, Jay Forrester of the Massachusetts Institute of Technology has developed an Urban Dynamics model of the urban socio-politico-economic environment. This model, which is entirely computer-based, has excited considerable attention in urban planning circles. *See* SOCIAL-SYSTEM DYNAMICS.

Use of time-share computers. Most of the computer-based political simulation models have utilized batch-mode computer input, whereby the decision data were punched on punchcards and taken to the computer, with the results being generated on a computer printer at the computer site and carried back to the simulation participants or simulation director. One significant present trend, however, for both teaching and research simulation models, is to utilize time-share computers with a conversational mode; that is, through the use of telephone lines and remote computer terminals, simulation data can be fed to the computer

in a conversational format (much like the English language) from the simulation site and the results obtained at the same remote terminal in a very short time-span. The immediate feedback is, of course, very desirable from the viewpoint of educational theory, and it also makes research explorations much easier.

For background information *see* COMPUTER, MULTIPLE-ACCESS; GAME THEORY; SIMULATION in the McGraw-Hill Encyclopedia of Science and Technology.   [MARSHALL H. WHITHED]

Bibliography: W. Coplin, *Simulation in the Study of Politics*, 1968; J. Forrester, *Urban Dynamics*, 1969; H. Guetzkow (ed.), *Simulation in Social Science: Readings*, 1962; J. McKenney, *Simulation Gaming for Management Development*, 1967; Ithiel de Sola Pool et al., *Candidates, Issues and Strategies: A Computer Simulation of the 1960 and 1964 Presidential Elections*, 1965; J. R. Raser, *Simulation and Society*, 1969; P. A. Twelker (ed.), *Instructional Simulation Systems: An Annotated Bibliography*, 1969.

Soil chemistry

The public concern over agricultural pollution of the environment has increased the interest of soil scientists in determining where problems of nutrient loss from soils exist. To study this subject more adequately, analytical techniques are being refined and new ones devised to trace the movements of nutrients through the soil. In addition, a number of workers are trying to develop practical means of reducing or eliminating nutrients from groundwater and surface water by using different methods of fertilizer application.

Nutrient mobility. In general, there are two main factors which influence nutrient mobility in soils. The first of these is the affinity of the soil for a particular nutrient, and the second is the net movement of water into and out of the soil.

The affinity of soil for nutrients is variable, but phosphorus tends to be held very tightly, ammonium and potassium are held fairly tightly, and nitrate and sulfate are held weakly by red soils and not at all by other soils. The affinity series for the primary nutrients is summarized by the relationship shown below. It appears obvious that, for a

$$H_2PO_4 > K = NH_4 \gg NO_3^-$$

given amount of water, the portion of nitrate moved out of the soil will be larger than that of any of the other nutrients. In practice, nitrate nitrogen is the only nutrient which is moved out of soil in high concentrations by water. On the other end of the scale, most phosphorus is removed from a soil by erosion, which removes soil as well as the phosphorus.

Evaporation loss. The net water movement through the soil is determined by the total amount which enters the soil through rainfall or irrigation and the water which is lost from the soil by evaporation from the soil surface and by transpiration from plants. For soils on which a grass sod is growing, the total loss of water by evapotranspiration is approximately equivalent to that lost from a U.S. Weather Bureau evaporation pan multiplied by a factor: for example, in the humid part of the United States, the factor is about 0.7. Thus by using Weather Bureau data, an estimate of the excess water available for leaching nutrients can be obtained.

Where soils are left bare, the above relationship does not hold. The reason is that, when a crust or a dry, loose layer forms on the soil surface, evaporation is greatly reduced. Two things favor formation of a crust or a dry, loose layer: a high evaporative potential and a soil with low water conductivity. Because of these factors it is difficult to predict actual water losses from bare soils. For example, in the western United States, evaporative demand often is higher than the water conductivity, and soil moisture loss is much less than that calculated from open-pan evaporation. In more humid regions, the evaporative demand is much lower, and soil water can reach the surface much faster, at least in the more permeable soils. Under these conditions water loss is rather low. At present the evaporation from bare soils cannot be estimated by open-pan evaporation because of the variables mentioned above. It apparently ranges from about 0.15 in the Great Plains to about 0.7 times open-pan evaporation in some permeable soils in humid regions.

The loss of water by evapotranspiration is very complex from a soil on which a short-term row crop such as corn is grown. Evapotranspiration is defined as the total water lost from a soil and from the leaves of plants growing in the soil. When the plants are small, the loss is comparable to that from a bare soil. When the plants approach maturity, the loss by evapotranspiration is even larger than that from a grass sod. In the fall the dry corn stalks may reduce evaporation below that from a bare soil by shielding the soil surface from the wind and Sun.

When a soil is mulched, evaporation is reduced and there is a tendency for water movement, and hence nitrate movement, to increase. Thus nitrate contamination of groundwater tends to be increased by mulching, a practice which can be very beneficial to growing crops.

Phosphorus and nitrogen movement. It appears from the above reasoning that most of the fallowed and row cropland and much of the pasture in the United States have some excess of water available for movement through the soil. The excess water eventually will carry nutrients out of the soil into groundwater and surface water. This natural process has always occurred. The present concern about this process is centered on nitrate and phosphorus losses from the soil, particularly where fertilization has been practiced. As already has been mentioned, phosphorus is tightly held by the soil and is not present in high concentrations in soil water. The pollution from phosphorus involves erosion of soil. Therefore, good soil conservation practices tend to reduce contamination of water by phosphorus. In the case of nitrate, however, the amounts of fertilizer nitrogen added to soils can be expected to increase the concentration in groundwater appreciably. The fact that nitrate has not increased to very high levels in groundwater due to fertilization is mostly because of the great dilution effect of nonfertilized land on the groundwater and surface water. Only a small proportion of the land in most states receives fertilizer, which greatly reduces the "average" nitrogen applied per acre for the United States. In fact, this figure is only 6 lb/acre for the land surface of the United States, whereas average rates of application for corn are 100–150 lb/acre.

Future research on this problem must involve small enclosed watersheds and surveys of well-defined areas so that estimates of water contamination can be made under various agricultural practices. Using this information, one can control the nutrients which will enter water supplies by

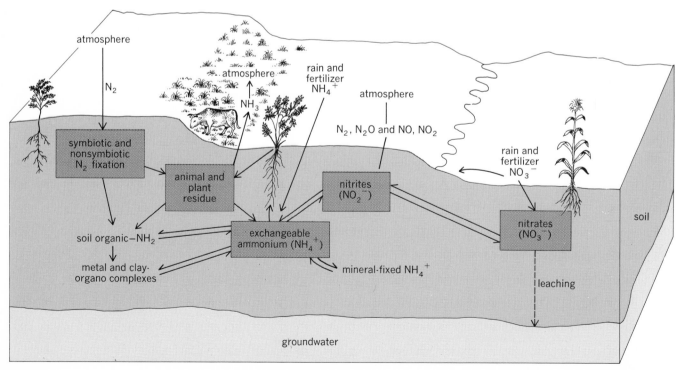

Nitrogen processes occurring in the soil-plant system.

changing fertilizer application rates and using the most appropriate time of application to give a minimum loss of nutrients. [GRANT W. THOMAS]

Analytical techniques. New industrial and agricultural practices may alter and change dynamic nitrogen processes occurring in the soil-plant system. The scientist must be able to trace and provide quantitative descriptions of such dynamic processes (see illustration) as nitrogen fixation, denitrification, nitrification, and plant utilization of inorganic nitrogen if the quality of the environment is to be improved.

Plant production is often limited by available nitrogen, and this is one resource that can be supplied in the form of fertilizers. Fertilizer nitrogen is helping to meet the urgent need of supplying more foodstuffs, especially protein, for the world's exploding population. However, fertilizer nitrogen and animal wastes are being charged with causing eutrophication and pollution problems. It is imperative that what happens to this nitrogen in the environment can be traced and detected.

Nitrogen fixation. The Haber, arc, and cyanamide are three processes widely used to convert molecular nitrogen, N_2, to useful inorganic nitrogen compounds. The inorganic nitrogen compounds are used directly or converted to various salts and used as fertilizer nitrogen. About 20,000,000 tons of fertilizer nitrogen were applied to agricultural lands in 1968. However, nitrogen added to the environment by biological agents probably exceeds by manyfold that added as fertilizers. There are few field estimates of biological N_2 fixation, and little is known concerning the effects of field practices on such fixation.

Organisms such as algae, *Rhizobium, Azotobacter,* and *Clostridium* contain the enzyme nitrogenase, which catalyzes the reduction of N_2 to NH_3 at atmospheric temperatures and pressures. This enzyme also reduces a number of other substrates, such as HCN, RCN, N_2O, and C_2H_2. Acetylene, C_2H_2, is reduced to ethylene, C_2H_4, and can easily be determined by gas chromatography and flame ionization. R. F. W. Hardy and coworkers suggested that as little as 1 $\mu\mu$-mole of ethylene can be detected, providing a sensitivity a thousandfold greater than that possible with tracer N^{15} analysis. With this assay, extensive field studies of nitrogen fixation are possible. However, extensive and detailed comparisons between the acetylene-ethylene and tracer N_2^{15} techniques are needed. The comparison studies between the acetylene-ethylene and N_2^{15} require investigators to sample, transfer, and mix gases. The recently developed technique by L. K. Porter, which uses a high-pressure-cylinder cryosorption pump for mixing and transferring gases, should aid in such studies.

Denitrification. Nitrogen losses from carefully controlled cropping experiments often range from 5 to 25% of the fertilizer applied. R. D. Hauck estimated that this loss of fertilizer costs the world's farmers about $1,000,000,000 per year. Leaching and denitrification are generally cited as major reasons for low recoveries. Nitrate nitrogen from nitrification processes in the soil and applied fertilizers is often leached into groundwaters or carried by field drainage into water systems, where it may promote eutrophication.

Under anaerobic conditions, free oxygen, O_2, is lacking for bacterial metabolism; therefore microbes use the oxygen from the nitrate molecule, releasing molecular nitrogen, N_2, nitrous oxide, N_2O, and small amounts of nitric oxide, NO, and nitrogen dioxide, NO_2. The process leading to the breakdown of nitrate to N_2, N_2O, and so on is called denitrification. This process is being used to avoid nitrate pollution of lakes, streams, and rivers. A greater understanding of this process might also help farmers to reduce losses of fertilizer nitrogen.

Since the main denitrification products (N_2 and N_2O) are gaseous, they must be trapped before they can be detected. Gas lysimeters and gas-sampling wells have been recently used for trapping such products. The gas from these sampling devices can be analyzed by gas chromatography and thermal conductivity detectors using polyaromatic polymer beads for column packing. Tracer $N^{15}O_3$ fertilizer is used to determine whether the nitrogenous gases evolved came from fertilizer. The abundance of N^{15} in gaseous samples is generally analyzed by mass spectrometry. The analysis is complicated, and A. E. Martin and P. J. Ross have pointed out precautions that will minimize error. One of the sources of error in analyzing denitrification gases is that the N_2O is partially decomposed to N_2 when bombarded by electrons in the mass spectrometer. A technique developed recently is the combination of the gas chromatograph with the mass spectrometer. A number of new interfacing devices have been developed which permit the transfer of the chromatographic peak unchanged to the mass spectrometer, where it can be analyzed. Such devices should make it possible to avoid some of the difficulties encountered in analyzing the N_2^{15} and $N_2^{15}O$ in denitrification gases.

Nitrification or mineralization. Another dynamic process occurring in soils is nitrification. Ammonium ions derived from biological decomposition of soil organic matter, from fertilizer ammonium, or from ammonium released from minerals are biologically oxidized to nitrite and then to nitrate ions. This process is extremely important because essentially all the nitrogen available to plants, except those grown in waterlogged conditions, such as rice, is in the nitrate form. Nitrate is extremely water-soluble and, because of this property and as discussed earlier in this article, moves with the water and is found in water runoff and soil percolates. One of the major problems facing modern society is to determine how much nitrate nitrogen is in runoff water and in water percolating from soil into groundwaters. Another problem is to determine whether the amount of nitrate in these waters is increased drastically by nitrogen fertilizer applications or animal feedlot operations.

The classical method for determining nitrate concentrations is the phenoldisulfonic colorimetric method. Using filtered and unfiltered supernatants from soil, R. J. K. Myers and E. A. Paul and A. Oien and A. R. Selmer-Olsen compared the phenoldisulfonic method to a new specific nitrate ion electrode method. The nitrate electrode method gave good correlations to the phenoldisulfonic acid method in the concentration range from 20 to 0.5 μg N/ml. The nitrate electrode operates by developing a potential across a thin layer of water-immiscible ion-exchange resin in response to the activity of the nitrate ion. The potential developed is measured by comparing it to a standard calomel electrode with an expandable pH meter. This method appears to be suitable for rapid routine analysis of soil nitrate.

Tracer $N^{15}O_3$ is used to follow the fate of applied fertilizer nitrate. D. M. Keeney and J. M. Bremmer outlined techniques for the isotope-ratio analysis of nitrate and nitrite ions.

Plant nitrogen metabolism. Many ecologists and environmental biologists believe that preserving the environment is so important that no fertilizer nitrogen should be used. Yet the exploding population will demand more foodstuffs and increasing supplies of protein. Restricting nitrogen fertilizer use would have serious effects on food supplies and would drastically increase food prices.

Supplying nitrogen fertilizer beyond the need of the plant results in many problems; one already mentioned is the deterioration of water supplies. Another problem is that plants assimilate nitrate beyond their ability to metabolize it. Normally a plant reduces nitrate to ammonium ions, which are used to synthesize amides, amino acids, proteins, nucleic acids, and so on. Any excess nitrate accumulates in the plant tissues, causing poor-quality foods and often nitrate toxicity to ruminants feeding on them.

A. S. Baker and R. Smith and J. L. Paul and R. M. Carlson showed that the nitrate electrode technique previously mentioned can be used to rapidly and routinely determine the nitrate content in plant tissues.

Studies of nitrogen metabolism in plants generally require separation and recovery of various nitrogen fractions for isotopic analysis after plants have received either $KN^{15}O_3$ or $(N^{15}H_4)_2SO_4$. A. P. Barker and R. J. Volk developed methods which permit the separation and isotopic analysis of ammonium, amide, amino, and nitrate nitrogen in plant tissue.

For background information *see* NITROGEN CYCLE; SOIL CHEMISTRY in the McGraw-Hill Encyclopedia of Science and Technology.

[LYNN K. PORTER]

Bibliography: A. S. Baker and R. Smith, *J. Agr. Food Chem.*, 17:1284, 1969; A. V. Barker and R. J. Volk, *Anal. Chem.*, 36:439–441, 1964; D. Bennett, *J. Chromatogr.*, 26:482–484, 1967; T. A. Black, W. R. Gardner, and G. W. Thurtell, *Soil Sci. Soc. Amer. Proc.*, 33:655–660, 1969; J. M. Bremner, H. H. Cheng, and A. P. Edwards, in *The Use of Isotopes in Soil Organic Matter Studies*, 1963; J. M. Bremner and D. R. Keeney, *Soil Sci. Soc. Amer. Proc.*, 30:577–582, 1966; E. Bresler and W. D. Kemper, *Soil Sci. Soc. Amer. Proc.*, 34:3–8, 1970; J. R. Burford and R. J. Millington, *Int. Congr. Soil Sci. Trans.*, 9(2):505–511, 1968; R. W. F. Hardy et al., *Plant Physiol.*, 43:1185–1207, 1968; R. D. Hauck, *Int. Congr. Soil Sci. Trans.*, 9(2):475–486, 1968; O. L. Hollis, *Anal. Chem.*, 38:309–315, 1966; F. W. Karasek, *Res. Devel.*, August, 1969; D. R. Keeney and J. M. Bremner, *Soil Sci. Soc. Amer. Proc.*, 30:583–587, 1966; A. E. Martin and P. J. Ross, *Int. Congr. Soil Sci. Trans.*, 9(2):521–529, 1968; S. E. Mustonen and J. L. McGuiness, *Estimating Evapotranspiration in a Humid Region*, Tech. Bull. no. 1389, ARS, USDA, 1968; R. J. K. Myers and E. A. Paul, *Can. J. Soil Sci.*, 48:369–371, 1968; A. Oien and A. R. Selmer-Olsen, *Analyst*, 94:888, 1969; L. N. Overrien, *Soil Sci.*, 106:280–290, 1968; J. L. Paul and R. M. Carlson, *J. Agr. Food Chem.*, 16:766, 1968; L. K. Porter, *Anal. Chem.*, in press; L. K. Porter; *Soil Sci. Soc. Amer. Proc.*, 33:696–702, 1969; P. J. Ross, A. E. Martin, and E. F. Henzell, *Int. Congr. Soil Sci. Trans.*, 9(2): 487–494, 1968; P. J. Ross, A. E. Martin, and E. F. Henzell, *Nature*, 4957:444–447, 1964.

Soil conservation

Recent studies continue to indicate that most pesticides, when applied to soils at recommended dosage rates, exert little effect on soil organisms or soil properties but that some kill or reduce the numbers of nonparasitic soil organisms and may temporarily influence the chemical properties of soil. Generally the effects are not detrimental or may actually improve plant growth, but occasionally they may cause phytotoxicity. The magnitude of these effects depends on dosage, soil type, temperature, moisture, and other factors.

Influence on soil organisms. When applied to crops or to soils at normal field application rates, most herbicides and insecticides have little effect on the soil microbes or may slightly stimulate growth of specific species. Volatile soil fumigants, such as D-D (a mixture of dichloropropane and dichloropropene) and methyl bromide, and many soil fungicides and nematocides, on the other hand, kill numerous nonparasitic microbes, including bacteria, fungi, actinomycetes, yeasts, algae, protozoa, and other soil forms. After the initial kill, numbers of soil bacteria soon reach much higher numbers than had occurred in the original soil (Table 1). In acid soil, fungi may quickly return in greater numbers, whereas in alkaline soils they may return quickly or the numbers may be reduced for a year or longer. Although the total number of species may be greatly reduced, with time, additional species recolonize the soil and total numbers slowly decline to normal. Fumigants and insecticides exert a similar effect on soil insects.

Factors responsible for the increased numbers of certain microbes following a soil treatment which initially kills large numbers of soil organisms are: (1) The pesticide chemical may be utilized as a food source by one or more organisms. (2) The bodies of the organisms killed by the treatment are utilized as a food source by surviving forms or by species that first recolonize the soil. (3) The organisms which survive or first become reestablished can reach very high numbers because the environment is less competitive.

Table 1. Effect of soil fumigation on numbers of bacteria in Yolo sandy loam

Fumigant	Numbers in 1 g of soil after fumigation			
	1 day	10 days	20 days	50+days
None	37	29	23	19
Chloropicrin, 200 lb/acre	4	96	120	61
Chloropicrin, 2000 lb/acre	2	100	153	78

Table 2. Influence of fumigants on extractable manganese in Ramona sandy loam

	Manganese in soil after fumigation, ppm		
Fumigant	1 day	20 days	100 days
None	4	2	5
Chloropicrin	47	46	8
Propylene oxide	22	6	6

Influence on soil chemistry. Pesticides which reach the soil or which are applied directly to the soil may influence soil chemical properties in the following ways: (1) The pesticide chemical or partial decomposition products represent a change in the chemical properties of the soil. (2) Upon decomposition of the pesticide chemical, the constituent elements are released as simple inorganic compounds such as ammonia, phosphate, hydrogen sulfide, sulfate, chloride, and bromide. (3) Pesticidal chemicals which kill an appreciable percentage of the soil organisms often increase the water-soluble salt content of soils. Soluble calcium is especially increased; usually more soluble magnesium, potassium, and phosphorus are noted. (4) Fumigants and fungicides may increase soluble or extractable micronutrient elements, including manganese, copper, and zinc (Table 2). (5) Some fumigants, fungicides, and nematocides kill the relatively sensitive nitrifying bacteria in the soil. These organisms oxidize ammonia to nitrites and nitrates. Until these bacteria become reestablished, relatively more of the available soil nitrogen will be in the form of ammonia. Reestablishment occurs generally within a few weeks to a few months. (6) Pesticides containing the benzene ring may be detoxified and altered with respect to side chains or groups, and may then undergo polymerization reactions with plant and microbial phenolic substances. In this way, they may become constituent units of the beneficial soil humus. *See* SOIL CHEMISTRY.

Increased growth response. The microbiological and chemical effects (side effects) of pesticides in soils generally exert little influence on plant growth. Sometimes, however, growth may be temporarily enhanced or retarded. Increased growth may be related to (1) fertilizer value of nitrogen or phosphorus released during decomposition of specific pesticides; (2) increased availability of soil manganese, phosphorus, and other plant nutrient elements; (3) an auxin-type action of some pesticide chemicals or of their partial decomposition products; (4) recolonization of the soil by microbial species that exert a strong antagonism against plant root parasites; and (5) production of plant growth substances by organisms which become dominant following application of pesticides that have an initial high lethal action on soil microbes.

Reduced plant growth. Occasionally something seems to go wrong following application of a pesticide to soils, and plants may be injured or growth may be retarded for short periods of time. These unexpected results may be caused by several factors:

1. Many pesticide chemicals are toxic to some plant species, and a few, such as the fumigants, are toxic to most plants. If time is not allowed for these chemicals to decompose in the soil or volatilize from the soil, the residual chemical will injure or even kill crop plants. Continuous use of chlorinated hydrocarbon insecticides, such as DDT, which decompose very slowly in the soil, may increase soil levels to a point that growth of sensitive crops, such as cucumbers and potatoes, would be retarded.

2. Simple inorganic substances released during decomposition of certain pesticides may injure sensitive plants. Avocado plants, for example, are highly sensitive to soil chloride. Treatment of the soil with D-D, chloropicrin, or other chloride-containing chemicals may cause or increase chloride injury to this plant. Similarly, several plant species, including onions, carnations, and citrus, are sensitive to bromide. Bromide residues from certain pesticides may temporarily retard growth of these species. Other possibly toxic inorganic decomposition residues of pesticides include arsenic, iodine, copper, and mercury.

3. In greenhouse or ornamental operations, in which fertilization rates are high, the killing of bacteria which oxidize ammonia to nitrates and nitrites may result in the accumulation of toxic concentrations of ammonia from decomposing organic nitrogenous fertilizers.

4. In soils with high manganese levels, fumigation may increase the soluble manganese to toxic levels for a short period of time. Although extractable soil manganese is generally increased, soil fumigation may sometimes cause manganese deficiencies of cauliflower, brussels sprouts, and broccoli.

5. Sometimes treatment of the soil with any pesticide which kills large numbers of microbes will cause a temporary phytotoxicity manifested by reduced absorption of phosphorus, zinc, and copper. The toxicity is quite spotty, and healthy plants may grow next to severely injured ones. The toxicity may last for a few weeks to a year. Young citrus, peach, and certain other tree seedlings are especially sensitive to this toxicity. The toxic condition can be corrected, or partially corrected, by proper fertilization with phosphorus, zinc, and copper.

6. H. D. Dubey recently described a spotty toxicity to sugarcane in soil treated with 26 applications of pesticides during the previous 12 months. In the affected spots, soluble nitrogen and bacterial numbers were low, but very high numbers of a single fungus were present. It was concluded that the repeated applications of fungicides had changed the soil population, reduced decomposition and release of nitrogen from the soil organic matter, and thereby induced a nitrogen deficiency in the soil for sugarcane.

Concluding statement. To provide an overpopulated world with sufficient good-quality food, it will be necessary to continue to use pesticides. The overall evidence indicates that proper use will not make soils sterile or permanently infertile. It is necessary, however, to understand the reactions of pesticides in the soil to use them properly. If this were done, unexpected side effects would be avoided or corrected and it would be possible to replace the occasional pesticide that tends to accumulate in the soil or that contains toxic residues which may accumulate in the soil.

For background information *see* FUNGISTAT AND

FUNGICIDE; PESTICIDE; SOIL BALANCE, MICRO-
BIAL; SOIL CONSERVATION in the McGraw-Hill
Encyclopedia of Science and Technology.

[JAMES P. MARTIN]

Bibliography: H. D. Dubey, Phytopathology, 60:
485–487, 1970; C. A. Edwards, A. R. Thompson,
and K. I. Beynon, Rev. Ecol. Biol. Sol., 2:199–224,
1968; P. C. Kearney and D. D. Kaufman (eds.),
Degradation of Herbicides, 1969; J. P. Martin, in
C. A. I. Goring (ed.), Synthetic Organic Chemicals
in the Soil Environment, in press; C. M. Tu and
W. B. Bollen, Down to Earth, 25:15–18, 1969.

Solid-state physics

The pseudopotential concept has been used recently to describe the interaction of electrons in metals and semiconductors with the atoms which make up the crystalline lattice. Experimental studies of metallic Fermi surfaces have shown that the effects of this interaction are considerably weaker than were initially anticipated. Characterization of these effects by a weak pseudopotential helped explain the success of the traditional free-electron theory of metals and formed the basis for a systematic theory of those properties in metals which were not understandable in the simple theory. The same pseudopotential concept has been used to characterize the interaction in semiconductors and allows a simple parameterization of the electronic structure of semiconductors.

Electrons in metals. It was recognized early in the 20th century that the conduction of electricity and heat in metals could be understood if the electrons were assumed to move freely through metallic crystals. It was expected, however, that the electrons would collide so frequently with the densely packed atoms that the conductivity would be much lower than that observed. The resolution of this dilemma came only with quantum mechanics and the description of electron dynamics in terms of waves. It was then recognized that, if the atoms which make up the metal were arranged in a perfect crystal structure, the scattered wavelets from each of the atoms would interfere in an orderly way such that the propagating electron wave would not be randomized. Thus, only defects in the structure were responsible for reduction in current flow, and the high observed conductivity could be understood.

Although the perfect crystalline lattice was expected not to randomize the electron motion, it was thought to have a large effect on the dynamics of the electrons. The electron wave would be diffracted by a regular array of atoms, just as light is diffracted by a regular array of lines in a spectrometer grid. Because the potential representing the interaction between the electrons and the metallic atoms is very large, this diffraction was expected to be exceedingly strong.

Fermi surfaces. The first quantitative test of this supposition came with the study of Fermi surfaces. The concept of such a surface may be readily developed for the free-electron model. In this model the energy of the electron is assumed to depend only upon the wavelength of the corresponding wave. Each energy state for the electrons is represented by a vector which is parallel to the direction of motion and has a magnitude equal to 2π divided by the wavelength. This is the electron wave number; the greater the length of the wave number vector, the greater the electron energy. In the metal the immense number of electrons present will occupy the lowest energy states, one electron of each of two spin directions occupying each state. Thus if one imagines each electron state to be represented by a vector originating at some point, the states corresponding to those wave vectors lying within a sphere centered at that point will be occupied by electrons; all those outside will be unoccupied. The radius of the sphere is determined by the density of electrons. This surface in wave number space that separates occupied and unoccupied states is called the Fermi surface, and is a sphere in the free-electron model.

The effects of diffraction on the Fermi surface may also be readily understood. If the interaction with the atoms which gives rise to the diffraction is extremely weak, only electrons with wave numbers lying on specific planes, the so-called Bragg planes, will undergo diffraction. For a stronger interaction, electrons with wave numbers near the Bragg plane will also be diffracted, causing a distortion of the Fermi surface which again separates the region of occupied and unoccupied states.

The diagram illustrates the Fermi surface for aluminum. If the electrons indeed constituted a free-electron gas, the Fermi surface would be spherical and the cross section would be a circle. However, interaction with the lattice of metallic atoms gives rise to distortions at the Bragg reflection planes. The size of the distortion is a measure of the strength of the pseudopotential interaction.

There are corresponding modifications of electron dynamics which may be observed experimentally, allowing experimental determination of the true Fermi surface in a real metal. The observed distortion gives a direct measure of the strength of the interaction associated with a given Bragg plane. Since the potential which constitutes the interaction between the electrons and the metallic atoms is large, it was anticipated that the distortions of the Fermi surface would be large.

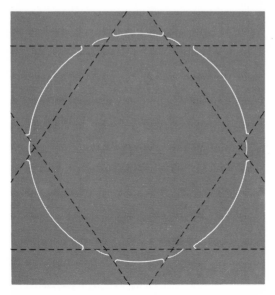

Cross section of the Fermi surface in aluminum. Broken lines represent the Bragg reflection planes.

Experimental studies in the late 1950s demonstrated that, in fact, the distortions of the Fermi surface were quite weak in many metals. The effective interaction giving rise to the weak distortions was called the pseudopotential.

How can a strong interaction have such a weak effect? The phenomenon is analogous to the movement of a car over a rough road. If the car moves sufficiently slowly, it will rise and fall with each bump or may even be brought to rest in a large dip. On the other hand, if the car moves sufficiently rapidly, it will skip from bump to bump, and the ride will be relatively smooth. Electrons at the Fermi surface move very rapidly and are deflected little by the potential dips due to the metallic ions. There are other electrons, frequently called core electrons, which are tightly bound in the atoms. Core electrons have energies far below those of the conduction electrons which have been represented by wave numbers within a Fermi surface. The electrical properties and almost all other properties of metals are determined by these conduction electrons. By focusing attention on the conduction electrons alone, one can represent the interaction with the metallic atoms by a weak pseudopotential.

Perturbation theory. The introduction of the weak pseudopotential did not require any change in the fundamental laws which govern electron dynamics. However, it did provide a convenient concept for describing this behavior, and has allowed the use of simple methods in the calculation of Fermi surface distortions. More important, it provided a basis for relatively simple theories of the entire range of metallic properties. These theories began with the exceedingly simple free-electron model and included the effects of the pseudopotential as a small correction.

The mathematical technique by which this is done, perturbation theory, was possible only after the introduction of the weak pseudopotential, which could be treated as a perturbation. The true potential, being so strong that it can produce the tightly bound core states, could not be used as a perturbation in this theory. The structure of the pseudopotential that is to duplicate the effect of the real potential on the conduction electrons is complicated. However, the simplicity allowed by the use of perturbation theory is decisive and has provided the modern basis for the theory of metals.

One can see the relevance of pseudopotentials to semiconductors by first returning to the Fermi surface. It may be shown that the introduction of a Bragg reflection plane which intersects the Fermi surface always has the effect of reducing the total area of that surface (not counting the flat portions made by the Bragg planes). In some structures, such as those of diamond, silicon, and germanium, the effect of the various Bragg planes is to eliminate the Fermi surface completely. In this circumstance, the surface in wave number space that divides occupied and unoccupied states is made up entirely of Bragg reflection planes. It can then be shown that the lowest energy one-electron state which is unoccupied will have an energy considerably higher, perhaps by an electron volt, than that of the highest occupied state. This is the defining feature of a semiconductor. Such a system will not conduct electricity since, if a field is applied, the electrons will be diffracted just as rapidly as they are accelerated by an applied electric field and no net current will result. Only by disrupting this system by exciting a few electrons to the higher states or by adding impurities which, for example, may put electrons in the higher states can the system be made to conduct.

In a semiconductor, where the Fermi surfaces are eliminated entirely by interaction with the pseudopotential, it is ordinarily not meaningful to treat that pseudopotential as a perturbation. Nonetheless, the concept of a pseudopotential characterizing the strength of the interaction between the electrons and the lattice can greatly simplify the description and calculation of the electronic states in a semiconductor. The pseudopotential concept was, in fact, first introduced in solid-state physics for the theory of semiconductors, and it was only subsequently applied to metals. Because of the introduction of perturbation theory, however, it is in metals where the pseudopotential concept has provided the greatest advance.

For background information *see* BAND THEORY OF SOLIDS; SEMICONDUCTOR in the McGraw-Hill Encyclopedia of Science and Technology.

[WALTER A. HARRISON]

Bibliography: M. L. Cohen and V. Heine, *The Fitting of Pseudopotentials to Experimental Data and Their Subsequent Application*, in press; W. A. Harrison, *Phys. Today*, vol. 22, p. 23, 1969; W. A. Harrison, *Pseudopotentials in the Theory of Metals*, 1965.

Sound-reproducing systems

Recent advances in the field of reproducing sound include, first, a better understanding of the subjective response to reproduced sound and, second, a new technology that makes four-channel systems practical.

Subjective aspects. The psychological and physiological or subjective responses in listening to reproduced sound consist of two parts, namely, the sensational response and the emotional response.

Sensational response. The sensational response is the subjective response involving the sensory performance of the hearing mechanism. The main physical elements involved in the sensational response to the reproduction of sound are frequency range, nonlinear distortion, loudness, spatial effects, and reverberation.

The sensational response R_{sn} involving the nth physical variables is given by Eq. (1), where the

$$R_{sn} = f_n(a_n, b_n, c_n, d_n, e_n) \tag{1}$$

physical variables are as follows: a_n = frequency range n, b_n = nonlinear distortion n, c_n = level of reproduction n, d_n = auditory perspective n, and e_n = reverberation n. Referring to Eq. (1), the highest sensational response will be obtained under the following conditions: full audio-frequency range, low and imperceptible nonlinear distortion, a certain desirable tolerable peak level of sound reproduction depending upon the dimensions and acoustics of the room, true auditory perspective, and an optimum reverberation time.

Subjective responses to the above physical elements are quite definite; consequently, variation in the sensational response from person to person is quite small.

Emotional response. The emotional response is the subjective response involving the arousing of the emotions of fear, anger, joy, grief, surprise, yearning, and so forth. The emotional process is very complex because a very wide variation occurs among listeners in their emotional response to reproduced speech and music. The properties of speech or a musical tone are frequency (pitch), intensity (loudness), growth, duration, decay, portamento, timbre, vibrato, and deviations. The performer employs all the properties of a tone and the arrangement and timing of the tones to create the desired emotional response for the listeners of the reproduced sound. The establishment of the desired emotional response is not amenable to the scientific method and is therefore unlike the case for the sensational response. Individuals involved in the establishment of the emotional response employ, for the most part, empirical and intuitive means.

The emotional response R_{en} involving the nth physical variables is given by Eq. (2), where the

$$R_{en} = f_n(l_n, p_n, r_n, s_n) \qquad (2)$$

physical variables are as follows: l_n = listener n, p_n = program n, r_n = rendition n, and s_n = era n. Equation (2) states that the emotional response depends upon the particular listener, the reproduced program, the rendition, and the era. Under these conditions, there is to be expected a wide variation in emotional response depending upon the variables l, p, r, and s.

In the case of classical music, in which the era extends over several decades, there is a variation

in the emotional response among listeners to a particular program depending upon the rendition. In the case of popular, folk, and rock music, there is a high degree of variation of the emotional response involving the era, the reproduced program, the rendition, and most of all the listener.

Total subjective response. The total subjective response is a combined function of the sensational response and the emotional response, as in Eq. (3),

$$R_{tn} = F(R_{sn}, R_{en}) \qquad (3)$$

where R_{tn} = total subjective response. The sensational response R_{sn} is given by Eq. (1), and the emotional response R_{en} is given by Eq. (2).

If the sound-reproducing system exhibits poor performance characteristics, the sensational response will be deteriorated. Under these conditions, even though the emotional response may be good, the total subjective response will be poor; that is, good subjective response cannot be expected if the physical performance is inadequate.   [HARRY F. OLSON]

Four-channel sound reproduction. Four-channel sound reproduction, viewed as an extension of the widely accepted two-channel stereophonic system, offers the means for heightening the subjective impact of the sound on the listener by employing four rather than two independent sources of sound in the listening area. The four-channel system can produce a variety of psychoacoustic effects, such as a realistic simulation of a concert-hall environment, the sensation of movement of the sound source about the listener, and the complete encircling or envelopment of the listener in the sound field. Recent advances in recording technology and in solid-state electronics now make the commercial offering of four-channel sound systems technically and economically feasible.

Four-channel systems. The first four-channel system considered in this article is a rational extension of the familiar two-channel system. In the latter, two electrically independent signals are fed to two loudspeakers. When the signals are properly balanced, a listener seated in front of the loudspeakers perceives the reproduced sound as being spread out in the region containing the loudspeakers. Moreover, he is usually able to localize certain identifiable segments of the sound, for example, the various instruments in an orchestral ensemble. This perceptual phenomenon is termed auditory perspective. Two-channel stereo has two major shortcomings. One is the diminution of audio perspective when the listener moves away from a location on the central plane between the two loudspeakers. The second is that any reverberation that accompanies the program can be reproduced only along with the direct sound from the same two loudspeakers in front of the listener. This makes realistic simulation of concert-hall listening an impossibility, for in a concert hall the reverberation envelope perceived by the listener consists of a multitude of delayed reflections of the sound arriving from all directions.

These two shortcomings can be corrected to a considerable degree by the four-channel system depicted in Fig. 1. Here a center channel has been added to the usual left and right channels fed to loudspeakers in front of the listeners. It has been found that the use of three separate channels and

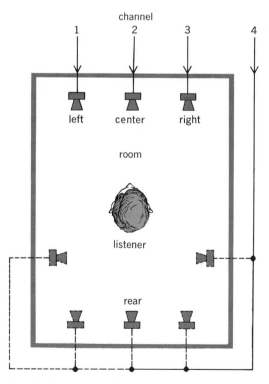

Fig. 1. Arrangement for four-channel simulation of concert-hall conditions. Channels 1, 2, and 3 carry primarily direct sound to give audio perspective. Channel 4 carries reverberation envelope. Simulation may be improved by connecting two or more reverberation loudspeakers in parallel.

loudspeakers provides audio perspective over a larger area than is the case with only two channels. In simulating concert-hall conditions, the program fed to the left, center, and right loudspeakers carries only a small amount of reverberated sound, while the fourth channel, feeding a loudspeaker to the rear of the listener, carries only reverberation. For the most effective simulation, the rear loudspeaker may be augmented by additional loudspeakers connected in parallel with it and situated along the rear and sides of the room. The listeners now perceive the earliest direct sound as coming from the front, while the reverberation envelope comes from the rear and sides of the room. Since the reverberation of the average room in a residence is small compared to that of a concert hall, the acoustics of the listening area do not greatly affect the characteristics of the reproduced sound, and the listeners' impression of being in a much larger room is real. Depending on program requirements, the rear channel may be used for direct program sound for antiphonal or other special effects.

A somewhat different arrangement is depicted in Fig. 2, where the four loudspeakers are in locations designated as left front, left rear, right front, and right rear. Here the direct sound for concert-hall simulation is reproduced by the left front and right front loudspeakers, and the two rear loudspeakers carry independent reverberation envelopes. It is clear that this arrangement is more limited in auditory perspective than that of Fig. 1. This is the price of the increased versatility of the second arrangement. This versatility will be welcomed by some contemporary producers who are less interested in realism and concert-hall simulation than in creating completely new aural sensations. By switching or fading selected components of the sound from one channel to another, "auditory mobiles" can be created wherein the sound source darts about the room. Or, the listener may be instantly transported from a large hall and placed intimately in a group of musicians. Numerous other special effects are possible, limited only by the imagination of the producer.

Other arrangements have been proposed for locating the loudspeakers in the listening room, but they are simply variations of those shown in Figs. 1 and 2.

Four-channel media. The applicability of magnetic-tape recording to a four-channel sound-reproducing system is obvious. The four program channels are recorded on four tracks on the tape and are played back simultaneously with four heads having in-line gaps in a single structure. Four individual preamplifiers and four power amplifiers are required to drive the four loudspeakers. The technology for providing four-channel service by means of this medium is already developed and commercialized. *See* ELECTRONIC MUSIC SYNTHESIS.

Likewise, the technology is available for the transmission of four-channel programs by a frequency-modulated (FM) radio station. Under present standards an FM station fills approximately 106 kHz of its nominal 200 kHz wide channel with the side bands and subcarriers required in stereophonic broadcasting. A third channel can be transmitted in the Subsidiary Communications

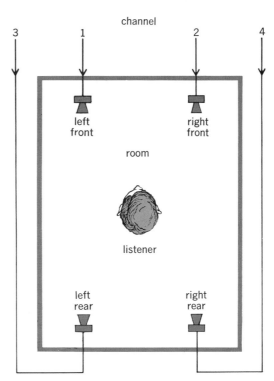

Fig. 2. Versatile four-channel arrangement wherein both front and rear loudspeakers may carry direct sounds for special effects, or rear loudspeakers may be used for the reverberation envelope for concert-hall simulation.

Act (SCA) band, which fills another 44 kHz. A fourth channel can be squeezed in above the SCA band. The SCA channel is a private-subscriber service generally used for the transmission of background music, and preempting this channel for public broadcasting will require modification of transmission standards by the Federal Communications Commission (FCC). It will probably be necessary to reduce the modulation level of the main carrier in order to accommodate four-channel programs without interfering with adjacent FM channels. Several proposals along these lines have been made for distributing the four-channel program power in the FM channel bandwidth, but it remains to be seen which, if any, of them will merit and receive FCC approval.

The problem of providing four-channel sound from phonograph records is considerably more serious, and the technology remains to be developed. The use of two parallel grooves on the record, each groove carrying two channels and being tracked by a separate stylus, is feasible but awkward and difficult. A more acceptable approach, providing it can be made feasible, consists of recording all four channels in the same groove, with the left front and the right front program carried as modulation of the left and right groove walls in the conventional two-channel mode. The other two channels will be added in the form of modulation of an ultrasonic-frequency carrier superposed on the conventional audio modulation of the groove walls. This will require the reproduction of signals extending to frequencies in the vicinity of 40 kHz. Pickups capable of operating over this range have been made. However, tracing distortion and deformation of the groove walls under the pickup stylus produce intermodulation distortion and channel cross talk. Existing data are inadequate for assessing the magnitude of these problems, and more research and development in this area will be required before a practical and economically acceptable four-channel disk system can be offered.

Finally, one must consider other methods that have been proposed for transmitting or recording four-channel programs. These methods involve encoding the four-channel information and arranging it in such a manner that it can be transmitted or recorded in its entirety in the presently existing two-channel stereo systems. In the reproduction electronics, the signals are decoded and distributed to the appropriate four channels and loudspeakers. There has been no public disclosure of the details of the operation of these systems; therefore they cannot be described here and only general statements can be made. Since these approaches involve the inclusion of additional coding and program information in the same channels presently occupied by the two-channel program, nature's inexorable law requires that a price be paid in terms of some sort of degradation of the two-channel program—reduced bandwidth, poorer signal-to-noise ratio, degraded audio perspective, increased distortion, or some combination of these. It may be that certain compromises in degradation can be found that will be judged acceptable by most listeners for most types of program material, in which case one of these encoding systems can be standardized and applied to all media handling four-channel programs. Much work will be required to discover whether such a compromise is possible.

For background information *see* ARCHITECTURAL ACOUSTICS; MUSICAL ACOUSTICS; SOUND-REPRODUCING SYSTEMS in the McGraw-Hill Encyclopedia of Science and Technology.

[J. GUY WOODWARD]

Bibliography: L. Feldman, *Audio,* 78:22–24, January, 1970; R. Miyagawa, T. Nakayama, and T. Miura, Design of reproduced sound by ESP (emotional, sensation and physical) method, *Proceedings of the Sixth International Congress on Acoustics, Tokyo, Japan, Aug. 21–28, 1968,* pp. A129-A132, 1968; B. Whyte, *Audio,* 77:22–28, December, 1969.

Space flight

Failure and near-disaster with the *Apollo 13* mission overshadowed the United States space program in 1970. Subsequent budget cutbacks and scheduled flight slowdowns, although not attributed to the *Apollo 13* failure, did not brighten the picture for 1971 and future space flight programs.

Apollo 13. The launch took place from Kennedy Space Center at 2:13 P.M. Eastern Standard Time on Apr. 11, 1970. The launch vehicle inserted the S-IVB instrument unit and spacecraft combination into an Earth parking orbit of 100.2 naut mi apogee and 98.9 naut mi perigee as planned. During the second-stage boost, because of a slightly slower velocity in one engine, the total burn time was 44 sec longer than predicted, but the vehicle cutoff velocity was within 1.2 ft/sec of the planned one.

After orbital insertion and verification of all systems and a brief video transmission, the second S-

IVB burn was initiated on schedule to implement translunar injection. All maneuvers, safeguards, and procedures were carried out successfully, and at exactly 6:00 hr from launch time, the first burn was started for pointing at lunar target impact. This operation was carried out with sufficient precision to obviate the need for the trim burn scheduled to follow. An unexplained velocity change occurred 13 hr later, which in fact pointed the lunar module–command service module even closer to its target. Consequently, the first scheduled mid-course correction burn was not required.

When all appeared to be going well, at 55:55 hr from launch the crew reported a loud bang and an undervoltage alarm in one of the command service module's main circuits. *Apollo 13* commander James A. Lovell and command service module pilot John L. Swigert, Jr., reported several ensuing problems. The command service module, called *Odyssey,* then lost pressure in its no. 2 oxygen tank, which caused two fuel cells to begin to stop voltage. To sustain power for life support, lunar module pilot Fred W. Haise, Jr., activated fuel cells on board the lunar module. Then the crew observed several repeated firings of the command service module's attitude-control thrusters.

It became apparent that the mission would have to be aborted and, at 61:30 hr after launch, the crew made a mid-course maneuver to effect free-return trajectory around the Moon and back to the Earth. The lunar module had been activated and inertial guidance reference had been transferred from the partially crippled command service module to the lunar module. All command service module systems were turned off, and the world began its anxious wait for the return of the three astronauts.

The remainder of the mission involved two main activities: planning and conducting propulsion maneuvers to return the spacecraft to the Earth, and managing and rationing all consumables to enable the lunar module to support three men and serve as the actual control vehicle for this unexpectedly long period of time. Some difficulty was experienced in sustaining a satisfactory temperature on board the lunar module because the environmental control system had not been designed for the imposed load.

One significant anomaly was observed during the trans-Earth coast. At about 97:14 hr after launch, the crew heard a loud "thump" and observed venting from the lunar module. Later study of this occurrence indicated that this anomaly was unrelated to the problems with the command service module, and that it was probably caused by brief change in the electrical power system's output.

At 105:18 hr into the mission, the crew made a mid-course correction burn to point the vehicle in the direction of its Pacific Ocean target. Later the command module was powered up to test its thermal condition, because this is the vehicle in which the three astronauts would have to reenter the Earth's atmosphere.

Because of the unusual spacecraft configuration (usually only the command service module returns to Earth), new procedures leading up to Earth entry were required. These were verified by simulation on the ground before being used. The procedure which was successfully used involved final

Principal space-flight events for 1970

Launch date	Vehicle	Source	Mission
Jan. 15	Intelsat 3-F-6	United States	Third-generation communications satellite to provide two-way voice circuits
Jan. 23	ITOS 1	United States	Second-generation weather satellite to take both day and night cloud cover photos
Jan. 23	Oscar 5	Australia	Earth-orbiting scientific studies
Feb. 4	SERT-2	United States	Environmental studies
Feb. 11	OHSUMI	Japan	Scientific mission
Feb. 13	Molniya 1 (13th in series)	Soviet Union	Unidentified mission
Mar. 10	DIAL-WIKA	West Germany	Communication
Mar. 10	U.S. Code 016B	France	Unidentified mission
Mar. 17	Meteor 3	Soviet Union	Weather studies
Mar. 18	Cosmos 327	Soviet Union	Unidentified mission
Mar. 20	NATO 1	NATO	Unidentified mission
Apr. 7	Cosmos 330	Soviet Union	Unidentified mission
Apr. 8	Nimbus 4	United States	Weather observation
Apr. 11	Cosmos 332	Soviet Union	Unidentified mission
Apr. 11	Apollo 13	United States	Second manned landing on the Moon
Apr. 23	Intelsat 3-F-7	United States	Two-way voice communication
Apr. 23	Cosmos 334	Soviet Union	Unidentified mission
Apr. 24	U.S. Code 034A, 034B, and 034C	People's Republic of China	Unidentified missions
Apr. 25	Cosmos 336, 337, 338, 339, 340, 341, 342, and 343	Soviet Union	Unidentified missions
Apr. 28	Meteor 4	Soviet Union	Unidentified mission
June 1	Soyuz 9	Soviet Union	Preparation for space station design
June 12	Cosmos 347	Soviet Union	Unidentified mission
June 23	Meteor 5	Soviet Union	Unidentified mission
June 26	Molyniya 1 (14th in series)	Soviet Union	Unidentified mission
June 27	Cosmos 351	Soviet Union	Unidentified mission
July 7	Cosmos 352	Soviet Union	Unidentified mission
July 9	Cosmos 353	Soviet Union	Unidentified mission
July 23	Intelsat 3-F-8	United States	Two-way voice communication
July 28	Cosmos 358	Soviet Union	Unidentified mission
Aug. 19	Skynet-B	United Kingdom	Scientific investigation
Aug. 22	Cosmos 359	Soviet Union	Unidentified mission
Sept. 12	Luna 16	Soviet Union	Unmanned satellite to land in Sea of Fertility on lunar surface, drill for a sample, and return sample to Earth
Sept. 16	Cosmos 362	Soviet Union	Unidentified mission
Sept. 22	Cosmos 364	Soviet Union	Unidentified mission
Sept. 29	Molniya (15th in series)	Soviet Union	Unidentified mission
October	OAO B (2)	United States	Optical astronomy observation from Earth orbit
Nov. 10	Luna 17	Soviet Union	Unmanned satellite to land self-propelled Lunokhod vehicle on Moon
November–December	ITOS A (2)	United States	Weather satellite to photograph cloud cover
November–December	Solrad	United States	Study x-ray radiation from the Sun
November–December	IMP 1	United States	Radiation and geophysical studies from Earth orbit
November–December	SAS A	United States	Study x-ray radiation from all sources in the sky

mid-course correction 137:40 hr after launch, jettison of the service module at 138:02 hr, and power-up of the command module at 141:30 hr; the lunar module was jettisoned after the crew boarded the command module. The command module, just boarded by the crew, was 40,000 ft above the Earth's surface at 15 min before splashdown. The U.S.S. *Iwo Jima* recovered the crew from its Pacific Ocean splashdown target less than an hour after landing.

The work had then just begun. More than 3 months of investigation into the perilous failure of *Apollo 13* was initiated under Edgar Cortright, Director of NASA's Langley Research Center. The study was based on in-flight photographs of the damaged command service module and comparison with an identical module on the Earth, as well as review of the observations made on the mission. The review board concluded that two protective thermostatic switches on an oxygen tank heater assembly were inadequate, and that the tank which failed also had a loose and leaky fill assembly. The switch inadequacy had been overlooked in all six of the previous missions and failure was not expected by anyone. Technicians at the launch site had been unable to boil off liquid oxygen out of the tank prior to refilling because of the faulty fill assembly and therefore applied a voltage which exceeded the capacity of the thermostatic switch. Insulation on wires was damaged and electrical arcing occurred in the flight. The review board published a report exceeding 1000 pages and presenting some 300 findings, determinations, and recommendations for modifications of consumables and emergency equipment.

Alterations of the Apollo flight plans for the future were greatly affected by the *Apollo 13* incident. Two of the remaining missions have been cancelled, and schedules for all missions will slip to facilitate spacecraft modification and budget reduction. The *Apollo 14* mission, scheduled for the first part of 1971, will explore the Fra Mauro region of the Moon, which was the objective of *Apollo 13*. The emphasis of the manned space program will shift toward small Earth-orbiting space stations called Skylabs. These are presently planned for 1972 and 1973. The ultimate 12-man space station would not be launched until the 1976–1978 period. The $2,000,000,000 Skylab program is considered the main effort by which the United States will keep up with Soviet manned space progress.

Soviet program. Though the Soviets lost the race to the Moon, they made great progress in 1970 toward beating the United States in the race to establish an Earth-orbiting space station and were first to return lunar samples to the Earth using an unmanned satellite, *Luna 16*. In June, 1970, cosmonauts Andrian Nikolayev and Vitaly Sevastynov broke the 1965 endurance mark for space travel with their 18-day flight aboard the *Soyuz 9*.

The Soviet program, although aggressive and accomplished in 1970, was not without failures. The program has been beset by engineering problems, rocket failures, and budget cutbacks. The giant launch vehicle, built to boost their 100-ton space station into orbit, exploded on the launch pad. Success of this vehicle would have been a major step past the United States capability, because their thrust capability would have been 10,000,000 lb in comparison to 7,500,000 lb for the United States.

Another part of the Soviet program showing much accomplishment is the unmanned satellite program. The table indicates the principal space flight events for 1970 and the degree to which Soviet launches dominate the picture. Not indicated in this table are more than 25 satellites launched in 1970 that had classified missions related to United States national security, and over 30 related to Soviet national security.

The Cosmos program is ostensibly a scientific one aimed at carrying out environmental, geophysical, and other such studies. The fact that the Cosmos launch schedule practically parallels those of the United States and other nations is, however, not to be overlooked.

United States scientific satellites. The only respect in which the Cosmos program does not parallel the United States scientific program is the fact that the latter has been drastically reduced.

The space science budget has been slashed and an era of second-generation, low-cost, small scientific satellites is being entered. The Orbiting Geophysical Observatory (OGO) and other such large spacecraft will be spaced out over the years ahead to accomplish highly specific goals. The most dramatic of these will be the 1973 Mariner mission past Venus and Mercury, and the 1972–1973 Pioneer missions past Jupiter.

For background information *see* MANNED SPACE-CRAFT in the McGraw-Hill Encyclopedia of Science and Technology.   [JOHN A. KING]

Spacecraft structure

Weight is one of the most important considerations in spacecraft structure design. This is especially true in the case of the Earth-to-orbit shuttle, which is currently receiving much attention among advanced spacecraft designers. This shuttle is nominally a two-stage vehicle consisting of a booster (first stage) and an orbiter (second stage). The design challenge from a structural standpoint is associated with the fact that the same vehicle must operate in several flight modes, each with unique loads, yet be light enough in weight to carry a favorable amount of payload to orbit.

Launch. During launch, both thrusting and aerodynamic loads are encountered. Acceleration loads tend to be most severe near termination of the first stage, when the still fully loaded orbiter stage is being accelerated at the highest rate. Aerodynamic loads are greatest in the transonic region, when dynamic pressure is greatest and when high shears in the wind profile are likely to be encountered. Another important consideration during launch is vehicle stiffness. The vehicle must be sufficiently stiff to avoid coupling of structural dynamic modes with control system or aerodynamic disturbances. Consideration of the structural requirements during the launch phase of flight points out the desirability of using the rocket propellant tanks as the primary structure for axial and bending loads. A tank wall that is sufficiently thick to withstand the hoop stress from internal pressure is adequate for these other loads if the

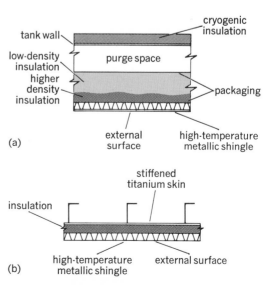

Fig. 1. Metallic thermal protection system of orbiter. (*a*) Fuselage. (*b*) Wing.

skin is properly stiffened.

Entry. The primary concern during entry is dissipation of aerodynamic heating. The weight burden associated with entry may be decreased by minimizing the heating and by proper design of the thermal protection system. From a design standpoint it is important to consider both maximum heating rate and total heating load. In general, the maximum heating in a particular region dictates the maximum skin temperature and thereby the choice and cost of skin material. The total heat load is associated with the total amount of insulation that must be used and is therefore of primary importance from a weight standpoint. Fortunately, there are designs that minimize both heating rates and heating loads.

In general, it is desired to achieve both high drag and high lift during entry. With high drag, the total period of entry is minimized, and thereby the heat load is also minimized. With high lift, entry deceleration occurs at high altitudes, where the air is thinner. This minimizes the heating rate. For most configurations the maximum drag occurs at a 90° angle of attack, and maximum lift occurs near a 55° angle of attack. Thus an angle of attack somewhere between these values should provide a good compromise. Trimming the vehicle at a very high angle of attack during entry has other important natural benefits. Only the lower surfaces of the body, wing, and empennage are directly exposed to the airflow. Other surfaces are exposed to either rarefied flow or completely separated flow, and consequently require either a minimum amount of thermal protection or none at all.

Usually the outer skin of the orbiter reaches an equilibrium temperature, at which point the heat that is radiated from the skin equals the heat transferred to the skin from the hot boundary layer. Temperatures over the majority of the area are typically 2000°F or less. In the case of the booster, which enters the atmosphere at approximately 40% of the velocity of the orbiter, the heating period is too short for equilibrium temperature to be reached. In fact, it may be practical to use aluminum or titanium skin, relying on the heat capacity of the skin to maintain temperatures within the working range of the material.

In the case of the orbiter, heating may be dealt with in a number of ways. A refractory metal skin may be used. Although it may be practical to use the skin as the stressed structure, the more likely case would be a skin made of shingles with insulation covering a cool structure underneath (Fig. 1). A lighter-weight approach is the use of external insulation with sufficient mechanical properties to eliminate the need for a separate outer skin (Fig. 2). Several companies are actively developing promising materials of this type. Where very high heating rates may be encountered, such as wing or tail leading edges or regions of local flow interference, it may be desirable to use replaceable ablative panels rather than to depend on expensive and perhaps undependable materials. This applies to regions where the equilibrium temperature may exceed 2500°F.

Subsonic flight and landing. At the termination of the entry maneuver, it is envisioned that the shuttle will fly subsonically a short distance to its landing field. The approach and landing maneuver

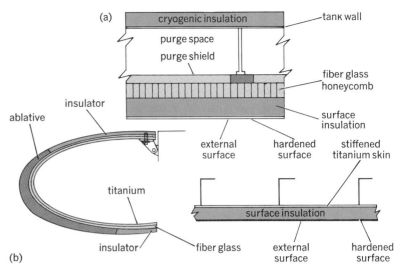

Fig. 2. Nonmetallic thermal protection system of orbiter. (a) Fuselage. (b) Wing.

and the ground facilities will resemble those for conventional airplanes. Both the booster and orbiter will therefore encounter flight and landing loads similar to those associated with large transport aircraft. Although this portion of the shuttle mission may be the briefest, the entire structure of the wings, empennage, and landing gear is provided just for this purpose (Fig. 3). Unswept wings with fairly high aspect ratio and a conventional empennage appear to provide the lightest weight arrangement. With external insulation, a conventional stressed-skin structure is not only suitably lightweight but would appear to be the least expensive approach to wing and tail design.

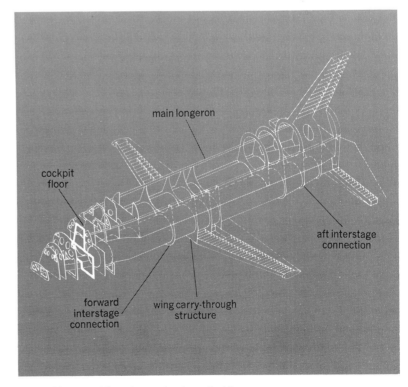

Fig. 3. Diagram of the primary structure of orbiter.

The shuttle vehicle must play the role of booster, entry vehicle, and airplane. Its design and construction will undoubtedly be one of the most interesting engineering projects of the near future.

For background information *see* AERODYNAMIC FORCE; FUSELAGE; SPACECRAFT STRUCTURE in the McGraw-Hill Encyclopedia of Science and Technology.   [MAXIME A. FAGET]

Bibliography: J. S. Butz, Jr., *Air Force Space Dig.*, 52(12):37–45, December, 1969; M. Faget, *Aeronaut. Astronaut.*, 8(1):52–61, January, 1970; R. Lecat, *Space/Aeronaut.*, 53:2, February, 1970.

Space-time

According to the standard view, Einstein's special theory of relativity precludes the possibility of existence of particles that travel faster than light in vacuum. This view was challenged in 1962 by O. M. Bilaniuk, V. K. Deshpande, and E. C. G. Sudarshan, who showed that the existence of superluminal particles is not incompatible with the postulates of special relativity. Since then, and particularly during 1965–1970, a considerable amount of theoretical and experimental work has been devoted to the problem of faster-than-light particles.

Ultimate speed. As early as 1893, J. J. Thompson concluded on the basis of Maxwell equations that infinite amounts of energy would be required to accelerate a charged particle to velocities greater than the velocity of light in vacuum. In 1905 Albert Einstein showed that, irrespective of its charge, an object possessing a nonzero proper (rest) mass could not be accelerated to the velocity of light or beyond. Einstein's mass-energy relationship, Eq. (1), rewritten as Eq. (2), shows that, for

$$E = m_0 c^2 (1 - v^2/c^2)^{-1/2} \qquad (1)$$

$$v = c(1 - m_0^2 c^4/E^2)^{1/2} \qquad (2)$$

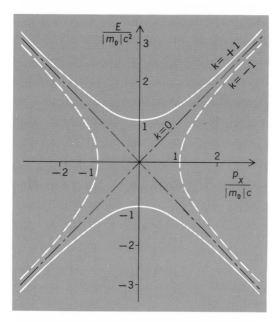

Fig. 1. Graphical representation of the relativistic energy-momentum relationship $(E/|m_0|c^2)^2 - (p_x/|m_0|c)^2 = k$, where k may take values $+1$, 0, and -1. The solid hyperbola, obtained for $k = +1$, corresponds to ordinary, subluminal particles. The dash-dot lines belong to the light cone, $k = 0$. The dashed hyperbola represents faster-than-light particles, for which $k = -1$.

particles having a real proper mass m_0, "velocities greater than that of light . . . have no possibility of existence" (Einstein's words).

Imaginary proper mass. Although it is impossible to accelerate any particle from a subluminal speed to a luminal or superluminal speed, one may conceive of superluminal particles that are created in some as yet unknown process. Such a hypothesis, to be viable, must be consistent with established principles of physics. In particular, the momentum of superluminal particles must remain real, their energy must remain real and positive, and physical processes involving them must not permit communication with the past.

As can be seen from Eq. (2) v can assume values greater than c only if the proper mass of the superluminal particle is imaginary, as in Eq. (3), where

$$m_0 = i m_* \qquad (3)$$

m_* is a real number. The assumption of an imaginary proper mass presents no difficulties in itself, because the proper mass of a superluminal particle is not accessible to measurement. Substitution of $i m_*$ for m_0 in Eq. (2) yields Eq. (4), which

$$v = c(1 + m_*^2 c^4/E^2)^{1/2} \qquad (4)$$

shows that hypothetical particles with imaginary proper mass are confined to velocities greater than c for all values of real energies. The corresponding momenta also remain real throughout.

The relativistic energy-momentum relationship for a particle may be written as Eq. (5). A graphical

$$E^2 - p_x^2 c^2 - p_y^2 c^2 - p_z^2 c^2 = m_0^2 c^4 \qquad (5)$$

representation of this relationship for the case of motion along the x axis is given in Fig. 1. The solid hyperbola corresponds to slower-than-light particles possessing real nonzero proper mass m_0; such particles are called tardons (from the Latin word *tardus*, meaning slow). The two intersecting dash-dot lines belong to the light cone describing light-like particles, or luxons (from the Latin word *lux*, meaning light). The dashed hyperbola corresponds to the case where $m_0^2 c^4$ is negative; that is, m_0 is imaginary. This curve graphically represents the energy-momentum relationship of faster-than-light particles, or tachyons (from the Greek word *tachys*, meaning fast).

Switching principle. An immediate and fundamental objection to the existence of imaginary-mass particles was based on the fact that they can assume negative energies under proper Lorentz transformations. This has been considered to be inadmissible because infinite amounts of (positive) energy could, in principle, be obtained from faster-than-light particles by simply letting them assume an arbitrarily large negative energy. An equally serious objection has been based on a causality argument, proposed in 1917 by Richard Tolman, according to which faster-than-light signals could be used to send signals backward in time and therefore into the past.

By examining these two objections together, the path toward resolution of both has been found. The key to the resolution resides in the circumstance that the sign reversals of energy and of time interval are concurrent. Thus a process in which absorption of a particle (signal) precedes its emission always involves the transmission of negative energy. This permits a reinterpretation of such a process in normal terms: as an emission of a positive-energy particle at an earlier time and its absorption at a later time. In other words, what looks like the absorption of a tachyon in one frame may look like the emission of an antitachyon in another. The proof of the switching principle can be obtained by referring to Fig. 2.

According to reference frame S of Fig. 2a, a tachyon is emitted at $t = 0$, $x = 0$ (event A) and absorbed later at $t = t_1 > 0$, $x = x_1$ (event B). In reference system S', moving with velocity w relative to S, the time sequence of these two events is reversed, $t'_1 < 0$. By comparing the transformations $t'_1 = \gamma(t - wx/c^2)$ and $E' = \gamma(E - wp_x)$, where $\gamma = (1 - w^2/c^2)^{-1/2}$, it can be seen that the sign change of the time interval is always accompanied by a corresponding sign change in energy, as shown in Fig. 2b. The switching of signs takes place when tachyon velocity relative to S becomes $v = c^2/w$.

Although simple causality objections such as Tolman's can be countered by the use of the above switching principle, involved gedanken experiments proposed by David Bohm, Bryce DeWitt, and Roger G. Newton cannot be resolved by means of this principle alone. However, when cosmological boundary conditions are taken into account, in addition to the switching principle, all closed causal chains contrived so far appear to be thwarted.

Experimental searches. Although the problem of causal loops cannot be considered solved definitively, objections to the possibility of existence of superluminal particles arising from causality arguments have not been convincing enough to deter experimenters from looking for these particles. More than half a dozen searches for faster-than-light particles have been carried out since such experiments began in 1963. So far none has produced evidence for the existence of tachyons. However, the negative results of these experiments are not taken as a strong indication that tachyons do not exist, because of the lack of knowledge of interactions in which tachyons may participate. Some of the anticipated properties of tachyons are reviewed below, and their experimental relevance pointed out.

1. The basic property of tachyons is their superluminal velocity in our frame of reference. Time-of-flight measurements would provide the most direct identification of tachyons, if it were known how to produce and detect them; at present, it is not.

2. Since the proper mass of tachyons is imaginary, the square of their proper mass is a real negative number. This property is the central point of several recent experimental tachyon searches based on the so-called missing-mass technique.

3. At a fixed value of momentum, a subluminal particle and a superluminal particle that have the same absolute values for their masses possess different amounts of energy. The first search for faster-than-light particles, carried out at the Nobel Institute of Stockholm, Sweden, during 1963–1966, was based on this property.

4. Charged tachyons, if they exist, would be traveling faster than their electromagnetic field; hence they would very rapidly — over distances of millimeters — "strip" themselves of all their energy by radiating electromagnetic waves according to the Cerenkov process. In a 1968 Princeton experiment, hypothetical tachyons were to travel in an electric field which was to supply them with en-

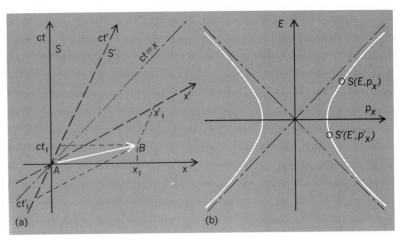

Fig. 2. Illustrations for proving the switching principle. (a) Minkowski diagram showing sign reversal of time interval. (b) Graphical representation of corresponding sign reversal of energy.

ergy at the same rate as they radiated it. Photomultipliers were set up in the hope of detecting this radiation and thus identifying charged tachyons. No evidence for charged tachyons was found.

5. As shown in Fig. 3, extension of the relativistic law of velocity addition, $v = (u + w)/(1 + uw/c^2)$, into the superluminal domain implies that ordinary particles whose velocities are subluminal in our frame of reference ($u = c/2$ in Fig. 3) will travel faster than light relative to hypothetical superluminal reference frames. Conversely, relative to any superluminal reference system, a tachyon (with, for example, $u = 2c$ in our frame of reference) will travel slower than light. Further, it may be seen that the relative velocity of two tachyons is always smaller than c. Conversely, in a hypothetical superluminal frame of reference, an ordinary particle, one which possesses a real positive prop-

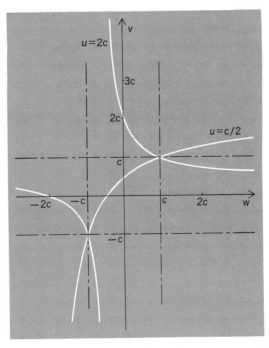

Fig. 3. Graphical representation of the relativistic law of velocity addition extended into the superluminal domain.

er (rest) mass in our frame of reference, would appear to travel faster than light. There seems to be, at least mathematically, a profound symmetry between the subluminal and the superluminal worlds. Consequently, one may be able to deduce tachyon properties by transforming known properties of particles from superluminal to subluminal reference frames. For example, by transforming the field of a hypothetical electric charge at rest in a superluminal system (x',t') into a subluminal reference system (x,t), Leonard Parker obtained a zero-energy magnetic monopole. Only luxon properties remain unaffected by such transformations. Light plays the same role in both the subluminal and the superluminal frames. Thus tachyons should interact at least with lightlike particles. This result led Parker to predict that existence of tachyons would cause the cross section for backward scattering of light on light to be larger than presently predicted by quantum electrodynamics.

For background information *see* CAUSALITY; QUANTUM ELECTRODYNAMICS; RELATIVISTIC MECHANICS; RELATIVITY; SPACE-TIME in the McGraw-Hill Encyclopedia of Science and Technology.   [OLEXA-MYRON BILANIUK]

Bibliography: C. Baltay et al., *Phys. Rev.*, 1:759, 1970; O. M. Bilaniuk and E. C. G. Sudarshan, *Phys. Today*, pp. 43–51, May, 1969; O. M. Bilaniuk and E. C. G. Sudarshan, *Phys. Today*, pp. 50–52, December, 1969; M. B. Davis et al., *Phys. Rev.*, 183:1132, 1969; L. Parker, *Phys. Rev.*, 188:2287, 1969.

Spectroscopy, photoelectron

Electron spectroscopy has been a tool in scientific research for a long time, particularly in the field of nuclear physics. Most recently, however, the application of high-resolution spectroscopy to chemical problems has expanded at such a rate as to make it appear that a new major area of chemical analysis has been created. Its object is to measure the binding energy in atoms and molecules by means of photoelectron emission. Phenomenologically, photoelectron spectroscopy can be divided into two parts: (1) measurement of the inner-shell binding energies as a function of chemical environment and (2) measurement of the outer- or valence-shell binding energies. In the first instance the chemical shifts of the inner-shell binding energies can be correlated with the electron density surrounding each atom or, simply, its oxidation state. This information offers a new dimension to chemical analysis. The second aspect, the determination of the binding energies of the molecular orbitals of the valence shell, offers a fundamental and broad basis on which to evaluate molecular orbital theory. A wide variety of both basic and applied chemical problems have been investigated during 1970, and the pace of interest appears to be ever increasing.

Measurement of binding energies. The basic relationship upon which photoelectron spectroscopy rests is the equation given below. If the incident

$$E_e = h\nu - E_B$$

photon is monochromatic and its energy $h\nu$ is well known and if one measures the kinetic energy of the photoelectron E_e with a high degree of precision, then the binding energy of an electron in a given shell of an atom or molecule E_B will also be known with a high degree of accuracy. To take full advantage of the application to chemical problems, the electron spectrometer should have an energy resolution of at least 0.1%, full width at half maximum, with good transmission. A number of dispersion instruments have been built to meet these requirements using both magnetic and electrostatic focusing. Because of certain practical advantages, particularly lower cost, electrostatic spectrometers seem to be the most appealing, and several versions are now available commercially.

Spectroscopy of inner shell. Measurements of the inner-shell binding energy are important because they reflect changes in the electron density surrounding an atom. Chemical changes affect primarily the valence electrons, but this in turn alters the potential felt by all the inner-shell electrons. For a higher oxidation state the electron density decreases in the valence shell, which results in an increase in the binding energy of all the inner shells of the atom; that is, a photoelectron ejected from an inner shell of an atom in a highly oxidized state feels a positive and therefore retarding potential as it departs, which, from the equation given above, implies a higher binding energy. Photoelectron spectroscopy of the inner shell (PESIS) offers a threefold analysis: (1) qualitative—each element is characterized by the binding energies of its atomic shells; (2) quantitative—the relative intensities of the photoelectrons are proportional to the amount of the different elements present; and (3) interpretive—an evaluation of the electron density surrounding the atom or, in cruder terms, the oxidation state can be obtained.

Figure 1 shows the photoelectron spectrum of acetone. Peaks corresponding to binding energies of approximately 290 and 540 ev correspond to the K binding energies of carbon and oxygen, giving an elemental analysis. The intensities of the peaks are proportional to the amount of the elements present, and thus there is the basis for a quantitative analysis. For carbon there are two peaks of

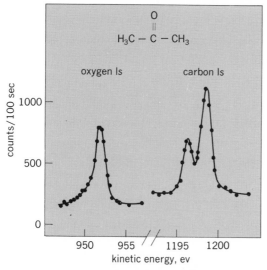

Fig. 1. Photoelectron spectrum of acetone as the result of irradiation with Al *K* x-rays ($h\nu$ = 1487 ev), an example of photoelectron spectroscopy of the inner shell (PESIS).

slightly different binding energies, one twice the size of the other, from which it can be deduced that the carbon is in two oxidation states, the lower state being twice as abundant as the higher. Such an analysis can be carried out on any compound containing elements heavier than helium.

Plots can be made for a given element of experimentally determined binding energy versus some function of electron density. The function may be simply the oxidation number, or it can be derived from electronegativity or from molecular orbital calculations. Once these plots have been established, an unknown molecule may be examined and the electron density evaluated for each of its atoms from the inner-shell binding energies. The elements carbon, nitrogen, and sulfur have been thoroughly studied. In addition, studies have been made on at least 20 other elements from all parts of the periodic table. Only a small sample need be used for study, in practice 1 mg will usually suffice, and as little as one molecular layer has been investigated. The resolution widths of photoelectrons examined in soft x-ray work is normally from 1 to 2 ev, and is limited less by the spectrometer than by the natural width of the x-ray source (usually Al or Mg K_α) and the nature of the material. Although it is possible to determine the position of a single photoelectron peak to 0.1 ev, it would be fair to say that the state of the art has not been reached where subtle differences between a large number of different chemical environments in a complex molecule can be easily resolved.

One of the strengths of PESIS is its wide variety of applications. Solids, condensable vapors, and gases have been studied. Organic and inorganic compounds are equally amenable for investigation. Of the large number of areas that have been recently studied, one might list at random xenon fluoride compounds, inorganic complex ions, components of transfer RNA, enzymes of interest to body fluid analysis, behavior of defoliants in soil, and the nature of back bonding in d-orbitals. Even Moon rocks have been measured. PESIS as used with solids is essentially a surface study since the photoelectrons excited by soft x-rays have enough energy to emerge without degradation only from a depth of the order of 100 Å. Application to surface phenomena has been initiated, with particular attention paid to the behavior of catalysts.

In addition to chemical analysis, PESIS has been most helpful in studying initial excitation processes, since any excitation of a molecule or atom in addition to the energy necessary to create a single inner-shell vacancy will rob the photoelectron of energy equal to that excitation and will be observed as a satellite on the low-energy side of the photoelectron peak. Another phenomenon of substantial interest seen by photoelectron spectroscopy is the splitting of the inner orbital levels due to exchange energy arising from unfilled orbitals or to external electrostatic fields.

Spectroscopy of outer shells. One of the most important ways to characterize the electron structure in molecules is through the determination of the binding energies of the different molecular orbitals, particularly those that make up the valence or outer shells. This effort is greatly helped by the employment of the 21.22-ev He resonance line formed in a gas-discharge tube. This photon is

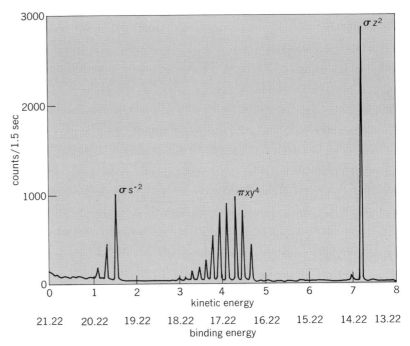

Fig. 2. Photoelectron spectrum of CO as the result of irradiation with He resonance line ($h\nu$ = 21.22 ev), an example of photoelectron spectroscopy of the outer shell (PESOS).

sufficiently energetic to eject electrons from the first few orbitals, and has a natural width of only a few millielectron volts so that one may resolve not only photoelectron peaks representing different electronic states but vibrational structure as well.

Figure 2 shows the photoelectron spectrum of carbon monoxide, CO, using the He resonance line as a photon source. Transitions from three molecular orbitals are observed. The first line in each band is the transition from the ground-state vibrational level in the neutral molecule to the ground-state vibrational level in the singly charged ion or, in other words, the adiabatic ionization potential. The remaining lines are due to transitions to higher vibrational states. This vibrational structure can be most helpful in determining the nature of the molecular orbital. For example, in the case of a diatomic molecule such as CO it can be said that when most of the transitions go to the ground vibrational state, as in the case of the first and third bands in Fig. 2, the orbital is not involved strongly in the binding, whereas in a complex vibrational structure such as the second band the orbital is bonding. Even though the number of possible vibrational degrees of freedom rises sharply as the molecule becomes more complex, the vibrational structure can often be seen quite clearly, since usually only one or two of the vibrational modes are excited in a given electronic transition.

The binding energies, together with the nature of the vibrational structures as measured by photoelectron spectroscopy of the outer shell (PESOS), have provided a broad basis on which to test the validity of molecular orbital theory. About 100 gaseous molecules have been studied. Data from PESOS overlap with those from optical spectrometry, but there are special reasons why photoelectron spectroscopy is particularly powerful or even unique: (1) It can examine the binding energy of all orbitals, no matter how strongly bound; (2) the vibrational states of the singly charged ions can be easily measured; and (3) the spectra are relatively simple to analyze. Photoelectron intensities have been measured as a function of the angle between the incoming photon and outgoing photoelectron. Such measurements can shed additional light on the character of the molecular orbitals, particularly its angular momentum. It has been observed that in some cases the relative population of the final vibrational states may also be angularly dependent.

Although PESOS has so far been used primarily in basic research, it appears to have a great potential for applied problems in gas analysis. In particular, it can be used to assay small amounts of pollutants such as CO, NO, and SO_2 in air.

Expectations for future. The number of fields employing photoelectron spectroscopy is ever increasing. In addition, instrumental improvements are continually being made. From the proper employment of monochromenters, photon sources of widths smaller than those provided for by the natural widths of soft x-rays may be achieved. Plans are also being made to couple new intense sources of synchrotron radiation with a monochrometer in order to obtain photons of reasonably good resolution over a wide energy range. The development of new spectrometer designs and position-sensitive detectors offers the hope that, in the near future, the signal intensity may be increased a thousandfold. Thus, although photoelectron spectroscopy has already demonstrated that it is a powerful tool for chemical analysis, the future looks even brighter.

For background information *see* ELECTRON SPECTROSCOPY; MOLECULAR ORBITAL THEORY; SPECTROSCOPY in the McGraw-Hill Encyclopedia of Science and Technology.

[THOMAS A. CARLSON]

Bibliography: D. Belteridge and A. D. Baker, *Anal. Chem.*, vol. 42, no. 43A, 1970; D. M. Hercules, *Anal. Chem.*, vol. 42, no. 20A, 1970; K. Siegbahn et al., *ESCA Applied to Free Molecules*, 1969; D. W. Turner et al., *High Resolution Molecular Photoelectron Spectroscopy*, 1970.

Sperm cell

Morphological and physiological changes of the mammalian sperm in the female tract are currently under intensive study. These studies relate especially to the mechanisms of sperm passage through barriers in the female tract and surrounding the egg. Sperm passage through the cervical mucus may be aided by proteolytic enzymes and by hyaluronidase, which are known to be present in sperm and seminal plasma. Such cervical passage can be inhibited by antibodies to semen. Concurrently with, or shortly after, cervical passage, sperms begin to undergo capacitation. Subsequently, in the vicinity of the egg they undergo the acrosomal reaction. This apparently results in exposure of acrosomal enzymes including hyaluronidase, a trypsin-like enzyme, possibly a "corona-removing enzyme," a neuriminidase.

These enzymes dissolve a passage between the follicle cells (cumulus oophorus and corona radiata layers) surrounding the egg and digest a path through the zona pellucida. This permits passage of the fertilizing sperm across the perivitelline space to the egg cell membrane, with which the sperm cell membrane fuses.

Capacitation. In order to fertilize eggs, mammalian sperms must first be capacitated by a sojourn in the female reproductive tract. The necessity for capacitation was first demonstrated independently in 1951 by C. R. Austin and M. C. Chang in experiments using rabbits. It was subsequently established in the rat, golden and Chinese hamsters, ferret, pig, and sheep. Some evidence indicates that cow, rhesus monkey, and human sperms must also be capacitated in order to function in fertilization. Finally, amphibian sperms require a capacitationlike exposure to egg jelly before they can fertilize eggs.

In mammals the capacitation phenomenon can be demonstrated experimentally by exposing ovulated eggs directly to sperm either in the test tube or by the technique of intratubal artificial insemination. If the sperms used in such direct insemination are obtained surgically from the male or from fresh semen, they fail to fertilize the eggs. However, if the sperms are obtained from the uterus of a donor female at least 6 hr (for a rabbit) after mating, they do fertilize the ovulated eggs. The need for capacitation can also be inferred from failure of fertilization in animals prematurely ovulated by hormonal treatment and subsequently mated.

The mechanism of capacitation and its specific functions in reproduction are not completely understood, but reasonable explanations are beginning to emerge. Capacitation normally occurs in the uterus but can also occur in the Fallopian tubes. Species specificity is not absolute; for example, hamster sperms become at least partially capacitated in mouse or rat oviductal fluid. The capacitating agent is evidently a large molecule for, although sperm implanted into the rabbit uterus in millipore filters become capacitated, sperms similarly implanted but enclosed in cellophane dialyzing bags do not. The capacitating ability of the female tract is under hormonal control (favored by estrogen and inhibited by progesterone), suggesting that the capacitating agent is a hormonally controlled substance, possibly an enzyme; capacitation may involve more than one agent. Recent reports of capacitation in the test tube by follicular fluid, eosinophils, and heated blood serum provide simplified systems for identifying agents which will cause capacitation and for investigating further the nature of the capacitation change or changes of the sperm.

Capacitation is most readily visualized as a loss or removal of sperm surface material. Mammalian sperms can and do adsorb antigens, such as blood-group substances and "spermatozoan-coating antigen," to their surfaces from seminal plasma. Of greater significance, following exposure to seminal plasma, capacitated sperms become incapable of fertilizing eggs. Such decapacitated sperm will recover the ability to fertilize eggs following a second capacitation in the female tract. Thus capacitation is reversible. The decapacitating factor (DF) in semen is currently under intensive study in the laboratory of W. L. Williams at the University of Georgia. The DF activity of semen is associated with a high-molecular-weight, protein-rich unit. However, digestion with the proteolytic enzyme pronase yields an active fraction of 300–500 molecular weight. DF is found throughout the male reproductive tract. It is reported to inhibit the corona-releasing enzyme of the sperm. It does not inhibit hyaluronidase. Inhibitors of other sperm enzymes that probably function in fertilization are found in seminal plasma. These inhibit the trypsin-like enzyme and a neuraminidase. The trypsin-like enzyme inhibits fertilization by capacitated sperm, and therefore it may be regarded as a second decapacitating substance (DF_2).

The following hypothesis, largely suggested by Williams, accounts for many of the facts of capacitation: During spermatogenesis at least four sperm enzymes essential for fertilization, namely hyaluronidase, corona-removing enzyme, neuraminidase, and trypsin-like enzyme, are deposited in or on the sperm acrosome. In the testis, DF combines with and inhibits the corona-removing enzyme; in the vas deferens, trypsin inhibitor and neuraminidase inhibitor from the male accessory glands inhibit the trypsin-like enzyme and neuraminidase, respectively. In the female, the capacitating agent, or agents, removes these inhibitors from the respective enzymes, which can now function in the subsequent fertilization events detailed below. These fertilization events concern the mechanisms of sperm passage through egg coats. It may be that the sole function of capacitation involves these mechanisms, because J. Toyoda and Chang recently reported on the fertilization of rat denuded eggs (lacking the follicle cell layers and zona) with uncapacitated sperm.

Acrosomal reaction. Subsequent to capacitation and presumably dependent upon it, the fertilizing sperm undergoes the acrosomal reaction. This reaction, or a closely similar one, was first discovered and clearly understood from studies on invertebrate sperms. It consists of fusion of the sperm cell membrane with the subjacent outer acrosomal membrane and rupture of the two membranes at the point or points of fusion. The result of this process is exposure or release of much of the acrosomal contents and establishment of a new membrane arrangement. Under artificial conditions in the hamster, this reaction results in a dramatic loss of much of the original sperm surface in the region of the acrosome. In the test tube this reaction (Fig. 1) approximates events seen in electron micrographs of sectioned hamster sperm interacting with the egg. In the hamster sperm (Fig. 2), many points of fusion between the two membranes are seen producing a vesiculated appearance.

Sperm penetration of egg coats. Normally, the fertilizing sperm approaches the egg in the ampulla of the Fallopian tube. Here the egg in most mammals is surrounded by an outer coat of follicle cells, the cumulus oophorus, and, in some species, an additional inner coat of follicle cells, the corona radiata. The latter covers the innermost coat, the noncellular zona pellucida (Fig. 3). These coats constitute barriers which must be traversed by the sperm prior to fusion with the egg. Penetration through the cumulus coat results primarily from the action of sperm hyaluronidase. Hyaluronidase digests the intercellular cement (mainly hyaluronic acid), permitting passage of the sperm between the cumulus cells. In the rabbit, the corona-removing

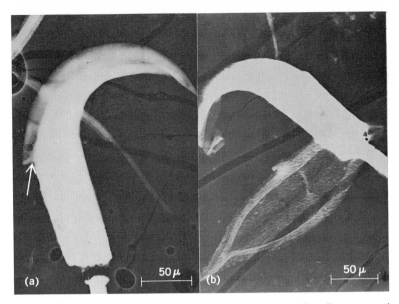

Fig. 1. Epididmyal sperm from a suspension in which only immotile cells were seen to undergo an acrosome reaction. (a) The acrosome swells and begins to detach at the side opposite the curved tip of the acrosomal region (arrow). (b) Detachment progresses across the dorsal and ventral surfaces of the nucleus. The detached acrosome, though swollen, retains its original caplike shape. (*From L. E. Franklin et al., Biol. Rep., in press*)

enzyme similarly provides a path between the corona cells. Finally, the trypsin-like enzyme is primarily responsible for digesting a passage through the zona.

The role of neuraminidase is more obscure. Pretreatment of eggs with neuraminidase renders them unfertilizable by capacitated sperm. This observation suggests that scialic acid (neuraminidase substrate) of the egg performs an essential function in fertilization and that it may be destroyed by sperm neuraminidase in the process.

As indicated above, these enzymes are present largely, if not entirely, in the sperm acrosome and, in fact, can be extracted from isolated acrosomes. Interestingly, the hyaluronidase and trypsin in such extracts appear to be bound together. Release of these enzymes in the vicinity of the egg is

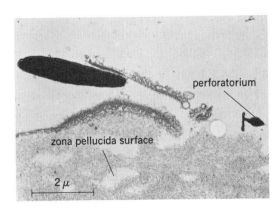

Fig. 2. Transverse section of an epididymal sperm undergoing the "normal" acrosome reaction at the zona pellucida surface. The section passes through the curved perforatorium and through the postacrosomal region of the nucleus. The centrally located acrosomal substance has dispersed. (*From L. E. Franklin et al., Biol. Rep., in press*)

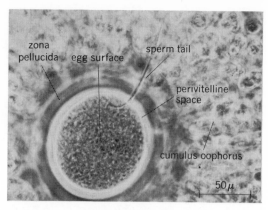

Fig. 3. Recently ovulated Chinese hamster egg surrounded by the cumulus oophorus and zona pellucida. The perivitelline space separates the zona pellucida from the egg surface. A sperm head is attached to the surface of the egg cytoplasm, and the sperm tail extends out into the cumulus. (*From L. Piko, in C. B. Metz and A. Monray, eds., Fertilization, vol. 2, Academic Press, 1969*).

apparently facilitated by, or even dependent upon, the acrosomal reaction. The precise relationship between capacitation and the acrosomal reaction remains to be established. Perhaps removal of the enzyme inhibitors, including DF, at capacitation renders the sperm surface sensitive to an agent, or agents, from the egg which initiates the acrosomal reaction. This agent could be similar to fertilizin, which apparently initiates the acrosomal reaction in sperms of some invertebrates.

For background information *see* FERTILIZATION; SPERM CELL in the McGraw-Hill Encyclopedia of Science and Technology.

[CHARLES B. METZ]

Bibliography: C. R. Austin, *Advan. Biosci.*, 4: 5–11, 1970; L. E. Franklin et al., *Biol. Rep.*, in press; L. Piko, in C. B. Metz and A. Monray (eds.), *Fertilization*, vol. 2, 1969; W. L. Williams et al., *Advan. Biosci.*, 4:61–72, 1970.

Steel manufacture

Innovations in steelmaking technology are continually being sought to reduce the capital outlay for new steelmaking facilities to produce steel at lower costs. One such process which is being given considerable attention is prereduction of high-grade iron-bearing materials, such as pellets, for subsequent direct use in electric steelmaking furnaces. The economics of the conventional coke oven–blast furnace–basic oxygen furnace for steelmaking dictate the need for large-scale units both for the initial facility and expansion, which

require very large capital outlay. Recent trends of higher costs for coking coals and higher scrap prices as experienced in 1969–1970 add further difficulties to attempts at increasing the profits in steel manufacture. The prereduced material–electric furnace processing obviates the need for coking coal, and the availability of prereduced material as a scrap substitute could aid in tempering the effects of upswings in scrap prices. The Midrex process for prereduction of iron ore pellets was put on-stream in 1969 to supply Oregon Steel, a small steel plant, with prereduced pellets. Large tonnage producers such as Stelco and Armco have reported the results of technical evaluations and of economic comparisons of the prereduced material–electric furnace process versus the conventional process.

Prereduction of iron ore. Prereduction is a process for prior preparation of high-grade iron ores for subsequent direct use in steelmaking furnaces. The feed to the prereduction process may be conventional high-grade iron ore pellets or well-screened high-grade lump ore obtained from such areas as South America or Australia. Prereduction increases the iron content of a conventional high-grade iron ore pellet from about 65% to as high as about 93–95% total iron. The prereduced pellet contains about 90% free metallic iron in contrast to the conventional pellet, which contains no iron in the metallic state. With such a high metallic iron content, the prereduced material can be considered to be a scrap substitute but with several advantages over scrap in the steelmaking process, such as (1) stable pricing, which is not expected to be subject to the price fluctuations of scrap; (2) low phosphorous and sulfur contents and high chemical uniformity, both factors are advantageous for close control of the steel composition; (3) very low contents of residual elements such as copper, which may be used to advantage in offsetting high residuals in lower-priced grades of scrap; and (4) suitable physical shape and size, which permit simultaneous charging and refining in the electric furnace, thus reducing heat time and increasing furnace productivity. Since the prereduction process does not affect the gangue constituents (principally silica, SiO_2) of the feed ore, the gangue content of the prereduced product depends only on the composition of the feed ore. Generally, prereduced feed materials having less than about 2–3% SiO_2 are preferred.

Prereduction processes. Several processes for prereduction of high-grade iron-bearing material, such as pellets, have been extensively tested in pilot-plant-scale operations, and several commercial operations are now, or soon will be, on-

Typical heat times and productivities of test furnaces

Test furnace	Scrap	SL	Scrap	SL	Scrap	SL	Scrap	SL
Furnace heat size, ton	22.7		22.0		68.0		145.0	
Transformer output, Mw	10.0		10.5		30.0		56.0	
Heat time, power-on to tap, hr	3.00	1.83	2.17	1.55	2.97	1.95	4.02	2.68
Steel production rate, ton/hr	8.4	13.9	10.2	16.5	25.6	40.1	39.3	59.1
Increase in production rate using SL technique, %	65		43		57		50	

stream. The SL/RN process (Stelco-Lurgi/Republic Steel–National Lead) uses coal as a reductant in a rotary kiln. Coals ranging in rank from anthracite down to lignite have been successfully used, and natural gas may be used as a supplemental fuel. Limestone or dolomite flux is charged to the kiln to keep the sulfur content in the prereduced product at a low level. Conventional indurated pellets, green unfired pellets, or screened high-grade lump ore may be used. As shown in the illustration, a continuous traveling grate is used to preheat ore material being charged to the kiln. Kiln temperature is about 2000°F, and a fuel consumption with bituminous coal of about 780 lb/ton has been reported.

The Midland-Ross Midrex and the Armco processes entail gaseous reduction in shaft furnaces. In each case natural gas is reformed externally to provide the reducing gas mixture. The Midrex process is already on-stream at Oregon Steel with two 200,000 tons/year units. In both processes the prereduced materials are cooled to ambient temperature in a nonoxidizing atmosphere to prevent reoxidation of the product.

Steelmaking. Testing of prereduced materials in electric furnace steelmaking has indicated substantial reductions in heat times and thus increased furnace productivity. At any steel plant a certain amount of domestic (in-house) scrap is generated which must be remelted. Electric furnace shop practice with prereduced materials will undoubtedly include portions of domestic and purchased scrap in the charge. The proportions of each will be dictated by the most favorable economics, taking into consideration purchased scrap availability and price. In times of lower scrap prices the ratio of scrap to prereduced pellets may be 60:40, whereas with higher scrap prices the ratio may be 40:60. Stelco reported the novel practice of continuous charging of prereduced pellets to the electric furnace during the course of the heat which has been named the SL steelmaking process. An initial charge (or charges) of scrap is melted at full furnace power. Once a molten pool is formed, prereduced pellets are charged by way of three downcomers through the furnace roof. Refining of the heat takes place simultaneously during the continuous charging but, because of the favorable chemistry of the prereduced pellets, little refining is required for the metal originating with the prereduced pellets. At the completion of charging, the heat is ready to be tapped. Some comparative data for conventional all-scrap versus SL steelmaking are given in the table.

Outlook. Although increases in the costs of other forms of energy are expected, the relative cost of electrical energy is expected to decrease in future years. The potential of reduced energy cost, together with the new technology of ultrahigh-power electric furnaces, indicates that an increasing share of steel production will be from electric furnaces. The acceptance and growth of prereduced feed for these electric furnaces will depend largely on the price of the prereduced materials. However, some other factors may tend to favor the development of prereduction–electric furnace steelmaking. Since the minimum-size economic units are much smaller than those for conventional steel

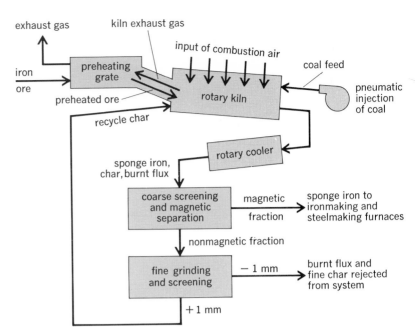

Flow sheet of the SL/RN process.

production, integrated steel plants can be built on a much smaller scale. Likewise, since the economical units for expansion are much smaller, a much closer mating of increased productive capacity with expected increased market is provided. The prereduction–electric furnace process ought to be well suited for developing nations with limited capital, scrap availability, and markets. The most significant indicators of the expected development are the recent commitment to additional construction incorporating the Midrex process at Georgetown Steel and the stated intent by Stelco to use a prereduction process in their future Lake Erie facility.

For background information *see* STEEL MANUFACTURE in the McGraw-Hill Encyclopedia of Science and Technology.   [WALTER NUMMELA]

Bibliography: G. R. Curtis, W. R. Rankin, and C. R. Taylor, *Blast Furnace Steel Plant*, October, 1969; *Eng. Mining J.*, December, 1969; J. P. Gordon, *Blast Furnace Steel Plant*, May, 1970; J. R. Miller, *Eng. Mining J.*, April, 1970.

Sun

For centuries most people, except for a few perceptive scientists, had thought of the space between the Sun and its planets (interplanetary space) as a complete vacuum; however, from effects of solar activity measured on Earth and observations of celestial bodies, some theoreticians had concluded that this space was not completely empty. A new theory on the dynamic processes in interplanetary space was beginning to emerge by 1958. Despite the fact that the material density of solar space is almost a million times less dense than the best vacuum thus far created in a laboratory, the interplanetary medium—although collision-free and fully ionized—was believed to be not completely unlike the atmosphere of Earth. This theory predicted it to be filled with waves and fields, concentrations of energy, winds, and gigantic storms.

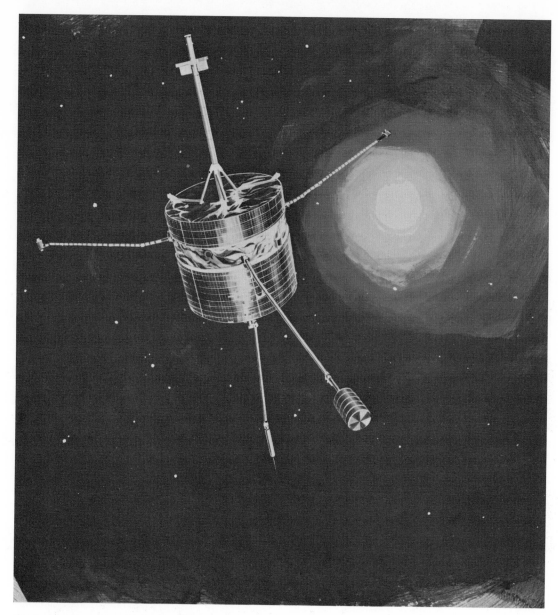

Fig. 1. Pioneer spacecraft developed after 1962.

With theory, Earth-based observations, and the early Pioneer (1–5) spacecraft suggesting that space out from the Sun contained an extremely tenuous fluid of charged, subatomic particles, many scientists believed that the Sun was the driving force on the interplanetary medium and that there must be some interaction between it and the Earth's magnetic field and atmosphere. But most of them openly admitted that the mechanism of the interaction was completely unknown, and sought additional spacecraft instrumented to obtain more precise measurements of interplanetary phenomena. These measurements have been obtained in the last few years.

A new series of Pioneers (6–9) was initiated in November, 1962. The new Pioneer spacecraft (Fig. 1) was developed to provide a capability for returning data on interplanetary particles and fields at great distances from the Earth. Once in interplanetary space, these spacecraft provide simultaneous scientific measurements of the interplanetary environment from locations widely separated relative to the Earth and at distances 0.7–1.2 astronomical units from the Sun (the Earth-Sun distance defining 1.0 astronomical unit). *Pioneers 6, 7, 8,* and *9* have been returning scientific data on a daily basis since their respective launches in 1965, 1966, 1967, and 1968.

Model of interplanetary space. A model of the interplanetary medium developed as a result of data returned from interplanetary flight missions is shown in Fig. 2. The Sun is at the center of this rotating system and revolves on its own axis with about a 27.5-day period. The supersonic plasma, or solar wind, flows radially out from the Sun, transporting with it the solar magnetic field lines. This results in a spiral configuration rotating with the Sun, much like the configuration formed by water drops as they flow outward from a rotating sprinkler head. As the plasma reaches the Earth, the stronger magnetic field of the Earth (magnetosphere) is deflected, and a wake, or geomagnetic tail, is formed behind the Earth, also conforming to the spiral configuration. Solar protons, or cosmic rays, emitted at much higher velocities from active surface regions on the Sun, attempt to flow radially outward but are constrained to propagate in a "corkscrew" manner along the spiral interplanetary magnetic field lines.

Large solar eruptions, or solar flares, explode from the surface of the Sun, disturbing the somewhat stable or quiescent structure of the interplanetary medium and releasing clouds of plasma and energetic particles at even much higher velocities. Many of these disturbances penetrate the Earth's magnetosphere, causing showers of energetic charged particles or cosmic rays, geomagnetic storms, and great fluctuations in the ionosphere.

Forecasting disturbances. While interplanetary research has made tremendous strides since 1958, new operational requirements for accurate forecasts of space disturbances arriving near Earth have been imposed. Adequate forecasting is essential to the successful completion of many aerospace and Earth activities. There are more than 1500 objects orbiting Earth today, and the orbit of each is modified by geomagnetic storms or other manifestations of space disturbances. Fluctuations in the density and altitude of the ionosphere accompanying a geomagnetic storm also disrupt terrestrial communications. Providing an accurate prediction of these storms has been one of the most difficult problems for space-disturbance forecasters. Furthermore, an unexpected arrival of intense showers of solar protons would be dangerous to astronauts on the Moon or exposed in space. These charged particles also degrade or damage solar cells, instruments, and other electronic components used in space vehicles. *See* SPACE FLIGHT.

Since January, 1967, portions of the scientific data telemetered from the Pioneer spacecraft have been used to calculate Pioneer Space Weather Reports. These reports are sent to the Space Disturbance Forecast Center of the Environmental Science Services Administration in Boulder, Colo., NORAD, and other agencies responsible for predicting space disturbances and their effects on Earth. These reports include (1) corotation delay, the expected time in days between the measurement of a disturbance at the spacecraft and its arrival at Earth; (2) solar wind velocity, density, and temperature; (3) the cosmic-ray intensity in several different energy bands; and (4) the general condition of the interplanetary magnetic field.

Normally, Pioneer Space Weather Reports are forwarded to forecasting agencies once each day, but when manned space operations are imminent, hourly reports are transmitted to the Apollo Mission Control Center at Houston, Tex., if warranted by interplanetary activity.

The relative positions of *Pioneers 6, 7, 8,* and *9* to the Sun and the Earth (shown in Fig. 3 for April, 1970) make it possible to monitor activity at different azimuthal locations about the Sun. Thus Pioneer provides forecasters with the capability to investigate, in detail, particle and field emissions from active solar regions days in advance of their impact on Earth. Experience is beginning to show

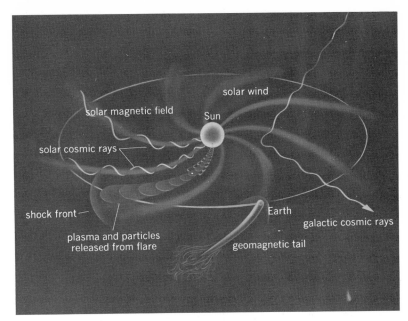

Fig. 2. Solar-system model developed as a result of interplanetary flight missions.

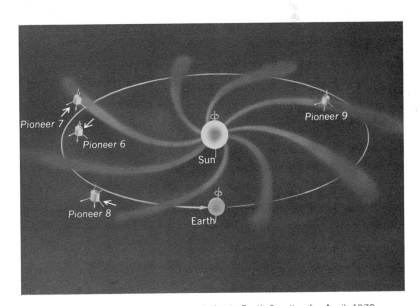

Fig. 3. Pioneer spacecraft positions relative to Earth-Sun line for April, 1970.

that this synoptic view of disturbances in several directions around the Sun can provide a new and improved means for forecasting disturbances.

For background information *see* SOLAR MAGNETIC FIELD; SPACE PROBE; SUN in the McGraw-Hill Encyclopedia of Science and Technology.

[ROBERT R. NUNAMAKER]

Bibliography: H. Friedman, Solar flares, *Astronaut. Aeronaut.*, April, 1969; J. W. King et al., *Solar-Terrestrial Physics*, 1967; R. R. Nunamaker, C. F. Hall, and A. Frosolone, *Solar Weather Monitoring: Pioneer Project*, American Institute of Aeronautics and Astronautics 6th Aerospace Sciences Meeting, New York, N.Y., January, 1968; G. A. Reiff, *The Pioneer Spacecraft Program*, American Astronautical Society–Operations Research Society Joint National Meeting, Denver, Colo., June, 1969.

Switching systems (communications)

Throughout the world there has been intensive study applying electronic technology to the design of switching systems. The switching systems currently serving most of the world's telephone and telex (teletypewriter) stations utilize ingenious electromechanical devices for selection and, in some cases, for storage. With the development of semiconductor logic and bulk electronic storage, there has been considerable recent interest worldwide in applying these techniques to the design of switching systems. The greatest progress in realizing the advantages of electronic switching has been made in the United States, where more than 100 electronic central offices and electronic private branch exchanges (PBXs) are in service.

Since 1958 almost 100 electronic system design experiments have been conducted by manufacturers and administrations throughout the world. The objectives and fields of application of these experiments (or trials, as they are sometimes called) differ from one system design to another. Also, a distinction should be made between systems and offices: Each system design, once placed into production, is installed in many offices.

Variety of networks. Specific system designs may be distinguished by the devices they employ, their organization, and the services they render. The number of such combinations is very large. Competition strongly motivates the variety of approaches. Nevertheless, emerging trends in system designs for application in small-, medium-, and large-size PBXs, local and toll offices, and military systems in countries with high telecommunication development are discernible.

The switching network and the system control are the two principal elements in a modern switching system. These systems are generally known as circuit switching since a path or circuit is established for instantaneous two-way or conversational use. (One-way or message-switching systems generally do not require a switching network.) Frequently systems are named for the type of electromechanical device used in the switching network, but electronic systems are generally named for the class of devices or technique used for the control. Systems which employ electronic devices in the control and electromechanical devices in the switching network are sometimes called quasielectronic systems.

The switching network function using elec-

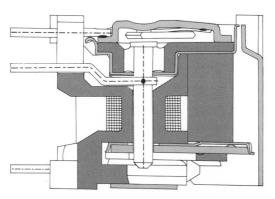

Fig. 1. Reed contact switch sealed in steel.

tronics may be extended from what is known as space division, in which each call-established connection has a specific physical path, to time division. Time-division switching takes advantage of the high-speed capability of electronic devices. The signals from sources being switched are sampled periodically, and the value of the samples from a number of sources in analog or digital form are placed on the same electrical bus or highway in time sequence. The signal samples are removed from the highway by correlating the operation of an electronic device of the destination with that of the source. The source-destination relationship may be changed to meet the needs for switched connections. Analog (pulse-amplitude) time-division switching systems have been developed and are in widespread use in the United States for PBX and military service. In France a network of military offices and in Great Britain a tandem switching office (the Empress exchange in London) use digital (pulse-coded) time-division switching networks in trial operations.

Semiconductor devices have also been used in space-division networks to reduce the physical size of switching networks. Small central offices using discrete *pnpn* triodes have been tried by the N. V. Philips Co. in system ETS no. 3 in the Netherlands and Denmark. These systems also use telephone sets which operate on lower loop currents and employ ringers which respond to an audio tone. PBXs featuring *pnpn* diode cross points are marketed in the United States by the International Telephone and Telegraph Co. as the TE 400 system. Integrated circuit devices are used as cross points in the no. 2750 PABX system developed by the International Business Machine laboratories in France. The network for this system has a separate path for each direction of transmission.

Where semiconductors have not been used for the switching networks, new electromechanical devices have been developed. Generally, these designs take advantage of new materials and automated production. Significant developments are the reed contact switches sealed in glass or steel (Fig. 1) and open magnetic contact switches which are developments of Siemens in West Germany. All these switches are used as individual electromagnetic relay cross points. Smaller crossbar or coordinate switches are also being developed. New coordinate mechanisms generally require more time to develop than do switches made up of individual devices, and therefore systems using them are just beginning to appear.

System controls. To date, the greatest impact of electronics in the design of switching systems has been in the control portions of systems. Not only is less physical space required with electronics but more calls can be served with a single control because of the increased speed. The semiconductor logic circuitry can be compactly mounted on plug-in cards to facilitate system maintenance and repair. The logic for serving calls may be contained in the wiring of these cards, or it may be general-purpose, capable of making decisions on call data based on information supplied by a program or tables stored in random-access high-speed memory.

In the United States the Bell System has introduced into local offices stored programs no. 1 and no. 2 electronic switching systems (ESS) using sealed reed contacts for the switching networks. Similar techniques have also been introduced into the Bell System in the traffic service position system no. 1 for operator-handled calls and have been added to the no. 4A crossbar toll-switching system to provide more flexible and faster translation of the area and office codes.

To provide familiarity with the technology and the operation of stored program-controlled systems, experimental systems have been built, installed, and placed in trial operation by the administrations in England (system TXE 4), Sweden (system A 210), France (system Pericles), and Japan (system DEX-1).

L. M. Ericsson has extended the application of such systems to the toll-switching field with an initial installation of the AKE 13 system in Rotterdam. Earlier versions of this system (AKE 12) were developed for the local and military central office applications. An Americanized version of this system (ETS no. 4) is planned for installation in the United States by North Electric Co. These systems utilize an unusual version of coordinate switch known as the code switch. This matrix switch is designed to respond directly to inputs in one dimension in a binary code.

Second-generation systems using integrated semiconductor logic circuits are under development in France (system E1) and Japan (system DEX-2). Bell Telephone Manufacturing Co. of Belgium is producing systems known as 10C for local, toll, and telex applications.

Electronic bulk memories are important in the implementation of the control portions of switching systems and in time-division switching networks. The storage requirements with respect to size and speed vary widely among the functions they are assigned in the design of switching systems. For programs and data detailing the structure of a particular office, including its lines and trunks (translation data), the stored information is changed relatively infrequently. Various random-access nondestructive read memories have been used for this purpose. For example, in the DEX-1 system a punched metallic card is used for these functions. The data concerning calls in progress have generally been stored in magnetic-core memories. Most newer systems use magnetic cores for all memory functions, with magnetic tape or drums as backup. Nonrandom-access memories of this type, particularly magnetic tapes, are also used to record call billing and traffic data.

Continuity of service. Different design philosophies to ensure continuity of service are being pursued. Earlier systems had duplicate controls that processed calls simultaneously. The results of processing the call data were compared and actions were taken only if the data matched. If they failed to match, a trouble or fault was indicated, requiring a trouble-locating or trouble-indicating routine rather than continuation of call processing.

Another concept is the provision of two or more processors which independently acquire data for different calls, thereby sharing the load. In such systems the expected or engineered loads usually do not exceed the call-carrying capacity of one processor. In event of trouble the entire load may be carried by the remaining processor. However,

Fig. 2. Magnetic ring memory of C1-EAX system.

when both processors are active, such systems are able to handle call-processing overloads, provided other portions of the system are not also overloaded. Multiprocessor systems are also under study to serve large offices where the call demands, even with the high speed of electronics, cannot be processed by a single processor.

Most systems are provided with teleprinter outputs which may be located remotely from the associated switching office. These printers provide information on machine performance and through their keyboards permit changes to be made in the service characteristics of the connected lines and trunks.

Increasing capacity. The data entering a switching office from lines and trunks may be reduced by preprocessing before being sent to the control for the decision-making function. In electronic switching the inputs are usually examined periodically by a scanner. The data obtained may be temporarily stored in circuits associated with the scanner and gathered into a form directly usable by the processor or processors. Units which receive data from a number of sources, generally serially, and prepare it for processing are usually

Fig. 3. General view of TXE 2 system.

known as autonomous scanners.

The data to be sent out from the switching office for transmission over the trunks at slow speed may be taken from the processor or controls at high speed and temporarily stored in circuits which distribute the signals to the trunk circuits or relay sets. Autonomous signal distributors and switching network controllers are finding increased consideration in seeking a balance between highly centralized control and the need for high call-processing-capacity systems. The data-reduction and storage functions of autonomous scanners and signal distributors have been combined successfully in a secondary stored program or wired logic controller, known as a signal processor.

Smaller systems. Generally, outside the United States the need is for smaller systems. Considerable progress is being made in the design and production of systems in the 1000- to 3000-line size range, which use tabular look-up as the action translator or decision-making element in an otherwise wired logic control. The table is "written" by threading wires, which represent input conditions through large magnetic cores, or "rings," which represent output decisions (Fig. 2). After a successful field trial, a production version of a system developed by the Automatic Electric Co. of Canada, known as system C1-EAX, was in service in October, 1970. Another system using a magnetic ring translator, the TXE 2 system (Fig. 3), is in production in Great Britain, with some 50 units in service.

New systems, even trial systems, generally must be designed to work with the existing networks of switching offices. As a result, in most electronic switching systems the supervision of the calls, once a connection is established, has been done by conventional relay trunk circuits. Centralizing of decision logic in stored-program systems permits simpler trunk circuits.

As mentioned above, signaling between offices has generally been accomplished over the trunk assigned to a call. For signaling on international calls, agreement has been reached by a working group of the Consultative Committee International Telephone and Telegraph (CCITT) that one standard form of signaling between countries may use data channels that are separate from the trunks.

This method of intercommunication between toll switching centers is known in the United States as common channel interoffice signaling (CCIS). Generally, adaptation of CCIS is simplified when electronic central controls are employed in these switching systems.

For background information *see* SWITCHING SYSTEMS (COMMUNICATIONS) in the McGraw-Hill Encyclopedia of Science and Technology.

[AMOS E. JOEL, JR.]

Bibliography: *IEEE Conference on International Communications*, Boulder, Colo., 69C20-COM, 1969; *IEEE Conference on International Communications*, Philadelphia, Pa., 68C20-COM, 1968; *Proceedings of International Colloquium on Eelectronic Switching*, Paris, 1966; *Proceedings of the National Electronics Conference*, Chicago, 1965; *Switching Techniques for Telecommunications Networks*, Institute of Electronic Engineers, London, Conf. publ. no. 52, 1969.

Symmetry laws (physics)

The fundamental problem in physics today is to understand the basic laws governing the interactions of the subatomic particles. To accomplish this, the study of the symmetry properties exhibited by nature plays a key role. Investigators cannot expect to understand the details of the basic forces of nature if the general properties of the forces with respect to basic symmetries are not known. Moreover, one's entire concept of the universe, of space and time, is strongly dependent on which symmetry principles are found to be respected by nature. It has long been assumed that the laws of nature were highly symmetric in space, time, and charge. However, from work over the last 15 years, physicists know that processes in nature do distinguish left from right, and world from antiworld, in an absolute sense. Now the problem of considerable interest is the asymmetry to time as well.

Basic symmetries. There are three fundamental symmetries of current interest, denoted by the letters P, C, and T. When inquiries into the symmetry of nature under the operation termed parity P are conducted, physicists are asking whether there is any difference between the world and a mirror image of it. This is equivalent to asking whether nature distinguishes left from right since, in a mirror image, left and right become inverted. One can also ask whether nature is symmetrical with respect to the operation called charge conjugation C, which reverses the sign of all electrically charged particles. In other words, does nature distinguish the world from the antiworld? The third symmetry is time reversal T, and the question is whether the laws of nature are indifferent to the direction of time flow.

Awareness of the importance of these symmetries is due in large part to T. D. Lee and C. N. Yang, who pointed out in 1956 that a paradox resulting from observations of the decay of an elementary particle, the K-meson, would be resolved if parity were violated in the weak interaction responsible for the decay. Subsequent experiments established that this was indeed the case, and hence that nature appeared to distinguish left from right. However, it was subsequently observed that parity P violation was always accompanied by the violation of charge conjugation C and that the combined symmetry CP appeared to be respected by nature. The conservation of CP in physical processes means that left in the world becomes right in the antiworld and that hence nature does not distinguish left from right in an absolute sense, since there was no way to distinguish the world from the antiworld. Nature appeared really quite symmetric after all.

However, a discovery in 1964 by a group of scientists at Princeton University established that this was not the case. The combined CP symmetry was found to be violated in the decay of the neutral K-meson. Thus it was established that nature does distinguish left from right, and world from antiworld, in an absolute sense.

Time reversal invariance. These three basic symmetries are connected formally by a fundamental theorem of physics known as the CPT theorem. This theorem requires that the laws of physics remain unchanged under the combination of all three of these symmetry operations. The proof of the CPT theorem is based on very general principles, and at the present stage of theoretical development it would be difficult to conceive of a basic physical theory that did not satisfy this theorem. A prediction of the theorem is that all elementary particles have antiparticles with opposite charge but with exactly the same mass and lifetime; this prediction has been fulfilled whenever tested. For example, experiment shows that the lifetimes of the pi meson and the anti-pi meson are equal to 1 part in 1000 and that the masses are equal to 5 parts in 10,000.

If the CPT theorem is assumed correct, the observation of CP violation implies that time reversal invariance T is violated. Although it may appear to be contradicted by everyday experience, it has long been a principle of physics that the basic laws of nature do not favor time flow forward or backward. It was thought that only interactions of complicated systems define a direction of time flow. Ordered systems tend to evolve in such a way as to increase their degree of disorder, or randomness, and it is this evolution toward increasing randomness that defines the direction of time. However, observation of CP violation implies that the laws of nature are not necessarily symmetric with respect to the direction of time flow, and has stimulated the search for direct observation of violation of time reversal invariance.

Tests for time reversal invariance. Three fundamental forces are known to be effective between subatomic particles, and searches for violation of time reversal have been made on all three forces. In the domain of the strong interaction force responsible for holding the atomic nucleus together, the tests have involved comparing the cross section for a given nuclear reaction with its inverse process. For example, the reaction $Mg^{24} + d \rightarrow Mg^{25} + p$ has been compared with the time-reversed reaction $Mg^{25} + p \rightarrow Mg^{24} + d$. From experiments on a number of such reactions, it has been established that there is no violation of time reversal in the strong interaction force to an accuracy of 3×10^{-3}.

The familiar electromagnetic interaction became an interesting contender as a source of violation with the realization by Lee and others that there was no good experimental evidence estab-

lishing that this force was invariant under charge conjugation C, although it was known to conserve parity P. Thus it could be the force responsible for CP violation, and hence T violation as well. Accordingly a number of experiments with a variety of methods have been conducted in a search for T violation in the electromagnetic interaction. Measurements on the polarization of gamma rays emitted from aligned nuclei using the Mössbauer effect have shown that the time reversal violation phase is less than 2×10^{-3}. Other tests such as γ-γ correlations in β-decay, electron-deuteron scattering, and electron scattering from polarized protons all have failed to find time reversal violations. Moreover, such a violation would result in the neutron having an electric dipole moment (edm); a series of precise measurements by N. F. Ramsey and collaborators has established that the edm for the neutron is less than 5×10^{-23} e-cm. *See* ATOMIC CONSTANTS.

Although no definite violation has been observed for the electromagnetic interaction, studies on two processes have found small but inconclusive effects. The possibility of C violation in the decay of the eta meson and of T violation in the process $n + p \rightarrow d + \gamma$ have been reported, and new experiments are currently under way to resolve these problems. *See* PARTICLE DETECTOR.

Since the known violations of P and C are observed only in the weak decays of unstable particles, the weak interaction force responsible for these decays is a prime contender for responsibility for T violation as well. Experiments have sought T-violation effects in the β-decay of Ne^{19} as well as in free neutrons and in the decays of the K^0-meson and Λ^0 and Σ hyperons. These experiments are designed to analyze the polarization of decay particles, since T invariance forbids such polarization from having a component in a certain direction. No violation of time reversal invariance has been found to an upper limit of 3×10^{-2} in the T-violating phase. However, since the effects of CP violation are of the order of 2×10^{-3}, it may be that these difficult experiments on the weak interactions must be made an order of magnitude more precise to provide a sensitive test for T violation in this case.

The study of CP violation in K^0 decay continues, the precision having improved in the knowledge of the parameters characterizing the CP-violating decay. Work by R. C. Casella and by J. Ashkin and P. K. Kabir showed how good knowledge of these parameters can establish whether T or CPT violation is also occurring in the decays. Such analysis provided support that T, but not CPT, violation occurs in the CP-violating decays of the K^0-meson. This is the only direct evidence that physicists have of time reversal violation occurring in nature at the present time.

Superweak interaction. Shortly after the discovery of CP violation in K^0 decay, L. Wolfenstein suggested that the source of the violation lay in a new and extremely weak fundamental force which has come to be known as the superweak interaction. Effects of this force would be so small as to be undetectable by experiment, except in the decay of the K^0-meson owing to the special peculiarities of this particle. Thus, in this theory, all observable CP- and T-violating effects would be limited to K^0

decay. The theory makes a set of definite predictions on the values of the parameters characterizing CP-violating K^0 decay, and to within present accuracy experiment agrees with these predictions. The superweak interaction remains a strong contender for the source of CP and time reversal violation in nature.

For background information see ELEMENTARY PARTICLE; SCATTERING EXPERIMENTS, NUCLEAR; SYMMETRY LAWS (PHYSICS) in the McGraw-Hill Encyclopedia of Science and Technology.

[OLIVER E. OVERSETH]

Bibliography: J. Ashkin and P. K. Kabir, Phys. Rev., 1D:868, 1970; R. C. Casella, Phys. Rev. Lett., 22:554, 1969; O. E. Overseth, Sci. Amer., 221(4):89, 1969; K. R. Schubert et al., Phys. Lett., 31B:662, 1970.

Telephone equipment, customer

New telephone equipment giving more services to customers has been developed recently. Some of the new developments have resulted in a dial-tone-first coin service, telephone equipment for people with handicaps, a one-number dialer, and an apartment door answering service.

Dial-tone-first coin service. To make the coin telephone more useful to the customer for normal service and to facilitate the placing of calls to public agencies (fire, police, ambulance) in emergency situations, coin telephone equipment has been redesigned to that the customer can reach these agencies, as well as the telephone company operator, without the deposit of a coin.

This new method of coin service operation is called dial tone first. In this method of operation, the dial tone is received immediately upon lifting the handset (just as in the operation of a normal telephone). The customer, upon hearing the dial tone, is assured of having a properly functioning coin telephone prior to depositing a coin.

The other, perhaps more important, benefit to the user is that in emergency situations he can reach the proper municipal agency (by dialing 911) or telephone operator (by dialing 0) without needing a coin to activate the telephone. He can also, without depositing a coin, communicate with the directory assistance operator (411), repair bureau (611), or business office representative (811).

The dial-tone-first method of coin service operation will also provide improved transmission and will reduce coin-telephone fraud and theft.

Conversion of all coin telephones to this new method of operation will, of necessity, be programmed over 8–10 years because of the magnitude of the work involved in modifying over 1,200,000 coin telephones and corresponding circuitry in central offices.   [PAUL LUBLINER]

Impaired motion, sight, or hearing. There is standard mass-produced telephone equipment for people with common handicaps such as a minor hearing impairment. For more unusual handicaps, standard equipment for the nonhandicapped can be modified, or locally engineered equipment can be supplied.

Motion. A 2-year research program at the Institute of Rehabilitation Medicine of New York University showed that almost anyone can be provided with a telephone that he can use, regardless of the severity of his motion handicap. Rather surprisingly, it was found that a great many handicapped people can be served with standard sets intended for the nonhandicapped. In some cases, however, these sets must be manipulated in an unconventional manner, or used for purposes other than originally intended.

Most other motion-handicapped individuals can be served with slightly modified standard equipment. Custom-built equipment appears to be needed in only a few of the most extreme cases.

The findings of the research program have been published in two books, one for rehabilitation physicians and occupational therapists and the other for the telephone industry.

Basically, telephone installations for the motion handicapped have two special features: (1) a control switch activated by touch or minimal pressure, and (2) hands-free operation. Hands-free operation may involve use of a headset, a handset on an adjustable arm, or a Speakerphone.

The TOUCH-TONE dial meets the needs of most motion-handicapped individuals. Those unable to operate it with a finger or a prosthesis can usually dial with a mouth stick. In extreme cases an automatic one-number dialer can be employed to dial "operator," or manual service may be provided (the operator answers whenever the control switch is activated).

Sight. A modification has been devised to enable the blind to operate the six-button telephones found in most offices. The modified six-button set provides an audible counterpart for the light signal from each illuminated button.

Hearing. A volume-control handset has been developed for use with public telephones. A slide switch in the handle of the handset enables the user to select either of two levels of amplification. When the switch is released, it returns automatically to the normal, or no-amplification, position.

A tone ringer is available as an auxiliary signal. The sound output has been "shaped" so that the ringer sounds much louder than the regular telephone bell to a person who is hard of hearing but not appreciably louder to the listener with normal hearing.

The new CODE-COM set for the totally deaf is simple and inexpensive. A set is required only at the deaf person's end of the telephone connection. This set has both visual and tactile outputs, and thus it can be used by the deaf-blind as well as by the sighted deaf. A sending key and oscillator are provided for those deaf people who are without speech. The CODE-COM unit is associated with an ordinary telephone, and the dial on this telephone·is used for outgoing calls. The deaf person with speech can talk into the telephone's handset in the normal manner.

Operation of the CODE-COM set is very simple. Any incoming sound signal causes a light to flash on and a small touch pad to vibrate. A caller can produce these effects by whistling, humming, making a buzzing sound, or merely saying "dit" and "dah" into the mouthpiece of his own telephone. The CODE-COM set also produces visual and tactile representations of call progress signals (dial tone, busy signal, ringing signal, and so on).

As the name implies, the CODE-COM set enables the deaf person to communicate by means of a code. The most elementary code is a yes-no one.

The deaf person asks a question and his caller hums once for "yes," twice for "no," or three times for "I don't know." Information is drawn from the caller by restricting questions to those which can be answered by a simple affirmative or negative.

The next simplest communication scheme employs a homemade code book with numbered messages. The deaf person distributes copies among his friends. A third and more advanced method of code communication presumes that both the deaf person and his friends are reasonably proficient in a code such as Morse code. A message of any degree of complexity can be transmitted. Highly motivated and highly skilled individuals can communicate by a fourth method, the Phillips code. The Phillips code is a kind of shorthand Morse code that permits much higher speeds of information transfer.

At the present time, the CODE-COM set is the only piece of telephone equipment for the deaf which does a complete signal-processing job, converting both call progress signals and incoming coded messages into visual or tactile signals. Other telephone devices for the deaf require an associated CODE-COM set or its equivalent to provide the call progress indications.

[GALE M. SMITH]

One-number dialer. As its name implies, the one-number dialer is an automatic device which dials a single preset number. The number can be of any length up to 14 digits, and is selected when the dialer is installed by adjusting a series of slider switches. Dialing begins when a single contact switch is closed, either manually by a person pushing a button or automatically by a device that detects the dial tone.

Two versions of the one-number dialer are made: one simulates the dc pulsing of a rotary dial, and the other dials by TOUCH-TONE signaling. Both dialers give private-line communication between two points over the telephone exhange network instead of using a much more costly direct line. Typical applications include providing a convenient way for travelers at a train station to contact a taxicab dispatcher; servicing an unmanned reservation desk at an airport for a downtown hotel; and giving doctors at a clinic quick access to a pharmacy's prescription counter.

The one-number dialer is about as large as a cigar box and is connected to a telephone set. It is usually mounted out of sight under a counter or in a closet nearby, but it could be located as far away as the central office switching center. The dialing mechanism is powered through a small transformer connected to an ac power outlet.

Slider positions that can be used for programming the dial are U (for unused) and digits 1 through 9 and 0. In addition to these positions, the TOUCH-TONE model has slider positions for * and # signals. When the number to be dialed has less than 14 digits, the U position of the slider is used to make the dialer skip the unnecessary digit positions. Another feature permits the dialer to be set to stop after the first or third digit to wait for a second dial tone. Automatically dialing a complete number takes the dc pulse dialer about 20 sec, and the TOUCH-TONE dialer only about 2 sec.

[JOHN P. ELLIS]

Apartment door answering service. The apartment door answering service (ADAS) is a new lobby-to-apartment communication service which enables the apartment dweller to admit visitors after they identify themselves through a locked lobby door. The system utilizes equipment in the central office of the telephone company to associate this service with the apartment tenant's regular telephone line. Thus every telephone or extension becomes a means for admitting visitors through the locked lobby door.

The lobby equipment consists of a specially designed decorative TOUCH-TONE telephone set on which the visitor dials a three-digit code associated with the tenant he wishes to visit. The call is completed over the tenant's regular telephone line, and his instrument rings with a distinctive ringing code that indicates a lobby call. After conversing with the visitor, the tenant can admit him by dialing the digit 4. This activates the lobby door release mechanism (provided by the building owner) to open the door. Only the tenant called can open the door, and, if the building has more than one entrance, only the particular lobby door associated with the lobby from which the call was originated will be released.

If the tenant's line is busy with a regular exchange call when the lobby call is made, the tenant hears a distinctive "call-waiting" beep tone superimposed on the conversation. He may put his regular telephone call on hold and transfer to the lobby call by flashing the switch hook. Upon completion of his lobby conversation and by flashing the switch hook again, he is reconnected to his original call. In the event that he forgets his original call and inadvertently hangs up, his line will be rerung for the regular telephone call. If the tenant is busy with a lobby call when a regular-exchange caller is trying to reach him, a similar "call-waiting" tone is applied, and the apartment tenant can proceed in a similar manner. Therefore ADAS will in no way affect the tenant's regular telephone service.

The ADAS central office equipment is compatible with all types and vintages of telephone switching equipment used in the Bell System.

For background information *see* TELEPHONE; TELEPHONE EQUIPMENT, CUSTOMER; TELEPHONE SIGNALING, SPECIAL-SUBSCRIBER in the McGraw-Hill Encyclopedia of Science and Technology.

[WILLIAM R. FLOOD]

Bibliography: AT&T, *Telephone Services for the Motion Handicapped*, 1968; G. M. Smith, *Volta Rev.*, 67:7, 1965; R. A. Sullivan, F. H. Frieden, and J. Cordery, *Telephone Services for the Handicapped*, Institute of Rehabilitation Medicine, New York University Medical Center, 1968.

Telephone service

As of Jan. 1, 1969, there were 237,900,000 telephones in the world, with the rate increasing at about 6% a year. The United States is the leader, with 109,256,000 telephones, or about 45.9% of the world's total. Japan is second, with 20,525,000, and the United Kingdom is third, with 12,910,000. The Soviet Union has an estimated 9,900,000 telephones. Other totals include West Germany, 11,248,000; Canada, 8,820,000; Italy, 7,752,000; France, 7,503,000; and Sweden, 4,110,000.

North America and Europe have a total of 82%

of the world's telephones, whereas Asia, the most heavily populated continent, has 11.6%. There are 6.8 telephones for every 100 people in the world. In the United States, there are 54 telephones for every 100 people.

A study of the world's telephones shows some interesting contrasts. Formosa (Taiwan) has 280,-000, but all of mainland China has 244,000. East Germany has 1,896,000 telephones, compared with West Germany's more than 11,000,000. Of the Latin American nations, three have more than a million telephones: Argentina, 1,599,000; Brazil, 1,560,000; and Mexico, 1,174,000. Pitcairn Island in Oceania has 15 telephones. The area of Bhutan in Asia and Canton Island and the Tokelau Islands in Oceania are among the relatively few places that have none.

New York City has 5,723,000 telephones, more than any other city in the world; Tokyo is second, with 3,641,000; and London is third, with 3,199,000.

Number of conversations. Overseas telephone calls are increasing at an annual rate of 25%. Telephone users in the United States can now reach 209 countries and areas, and 96.4% of the world's telephones. In 1969 many remote places were brought within reach by telephone. They included the Admiralty, Ellice, Gilbert, and Chatham islands; Saipan, New Ireland; Nauru; and Bougainville—all in the South Pacific. Islands in the Atlantic that are now within reach by telephone are São Tomé and Principe, situated off the coast of Africa.

People in the United States hold more telephone conversations than do any other people on Earth. In 1969 the average telephone user in the United States held 701 telephone conversations. Canadians were second, with an average of 692.9 telephone conversations each. Icelanders were third, with an average of 632 each. (Iceland has only 66,-000 telephones, fewer than the city of Covington.) Bahamians averaged 507 telephone conversations each. They have only 36,000 telephones, but they are in fifth place among the world's telephone talkers. The Swedes are ahead of them, with an average of 624 conversations each. The people of the tiny nation of Liechtenstein, with 8000 telephones, averaged 279.6 conversations each. India, with its masses of people and 1,500,000 telephones, averaged 3.1 conversations for each person.

Monaco has the most telephones for its people, 60.93 per 100. Upper Volta in Africa has the fewest telephones per 100 people: 0.03 (discounting those countries that have none).

Overseas service. Overseas telephone calls are transmitted by undersea cable, communications satellite, and high-frequency and over-the-horizon radio, either directly or in combination. Cables are laid on the ocean's floor by special cable ships, and now link the major continents and most principal islands. Today, most offshore cable is buried beneath the continental shelf to prevent accidental cutting by fishing trawlers. See COMMUNICATION CABLES.

A typical satellite call is placed in this manner: The signal is transmitted over regular telephone land facilities from the point of origin to a ground station. From there it is transmitted to a satellite and bounced to a distant ground station in another country or on a different continent. The signal is then fed back into land facilities and directed to the telephone being called. See COMMUNICATIONS SATELLITE.

The principal nations in the world are working with the International Satellite Consortium (INTELSAT) to expand the global satellite system for transmission of all types of messages, including telephone, radio, and television.

Ownership. Telephone systems in most countries, particularly in Europe, Asia, and Africa, are government-owned. In South America, telephone ownership is partly government-owned and partly privately owned. Argentina's telephone system, the South American continent's largest, is government-owned. The next largest system, Brazil's, is mostly government-owned. Telephones in the United States, Japan, and Italy are, for the most part, privately owned. Those of the United Kingdom, the Soviet Union, France, and India are government-owned.

In the United States, the Bell System maintains 85% of the telephones, as well as the vast nationwide switching network for long-distance telecommunications. The rest of the telephones are maintained by about 1000 independent telephone companies, and the Federal government operates 93,000 telephones. More than 99% of the telephones in the United States have been converted to dial or push-button calling. Most long-distance calls can be dialed direct without operator assistance. Electronic switching, high-capacity cable and microwave facilities, and see-as-you-talk Picturephone service are being developed and installed in the United States. See SWITCHING SYSTEMS (COMMUNICATIONS); TELEPHONE EQUIPMENT, CUSTOMER.

For background information see TELEPHONE SERVICE in the McGraw-Hill Encyclopedia of Science and Technology.   [MARCO GILLIAM]

Bibliography: *The World's Telephones*, published annually by AT&T.

Temperature measurement

The temperature scale in common usage in scientific and industrial laboratories is called the International Practical Temperature Scale (IPTS). It is designed to conform as closely as is practical to the thermodynamic temperature scale (on which experimentally determined values are known as International Practical Kelvin Temperatures), but is expressed in terms of the Celsius (previously known as centigrade) scale. Originally established by international agreement in 1927 as the International Temperature Scale, the International Practical Temperature Scale has been updated periodically to reflect improvements in the knowledge of the thermodynamic temperature scale. The resulting scales are designated according to the year in which significant changes have been internationally agreed upon. Thus the scale agreed upon in 1948 was designated IPTS-48, and was in use until October, 1968, when it was superseded by IPTS-68. The differences between the upper portions of IPTS-48 and IPTS-68 are substantial in terms of the more accurate work in common practice, and are the subject of the discussion that follows.

Definition of temperature unit. Among the significant developments leading to IPTS-68 was a basic change in the manner of definition of the

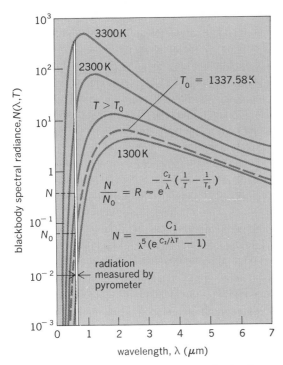

Fig. 1. Planck's blackbody radiation distribution function, in which area under the curve represents energy being radiated from a blackbody surface.

unit of temperature on the thermodynamic temperature scale, implemented in 1954, such that the fundamental fixed points of 0 and 100°C were replaced by a single fixed point, the triple-point temperature of water, to which was assigned the temperature 273.16°K exactly (reproducible at present to within approximately 0.0001°K). Using constant-volume gas thermometry in such a manner as to simulate ideal gas law conditions, highly reproducible temperatures (such as the freezing points of various pure metals) were determined relative to the triple point of water; these temperatures are denoted the International Practical Kelvin Temperatures. In 1967 the degree notation on the thermodynamic temperature scale was changed such that the unit of temperature is the kelvin, instead of the degree kelvin; that is, the triple point of water is 273.16 kelvins, or 273.16K. Using gas thermometry as described above, the temperature of the freezing point of pure gold (commonly called the gold point) was determined to be 1337.58K ± 0.2K, about 1.4K higher than the previously accepted value. The change was attributed to improvements in the gas thermometry instrumentation and measurement methods.

IPTS below gold point. Below the gold point, the IPTS is formed by specifying instruments and techniques for interpolating between the fixed points whose thermodynamic temperatures are determined by gas thermometry. The interpolation instruments include platinum resistance thermometers and platinum versus platinum/10% rhodium thermocouples; they are used in accordance with specified equations and tables to generate a continuous temperature scale (IPTS-68) that agrees closely with the thermodynamic temperature scale (to within the experimental uncertainties of the

International Practical Kelvin Temperatures of the fixed points). A significant change was the extension of IPTS-68 down to the triple point of normal hydrogen (13.81K), whereas IPTS-48 extended down only to the boiling point of oxygen (90.188K on IPTS-68).

IPTS above gold point. The higher the temperature to which a sensor is exposed, the less reproducible the sensor calibration tends to be as a function of temperature and time, since at high temperatures the physical properties of the sensor begin to undergo irreversible change. Ultimately, as temperature is increased, contact thermometry (where the sensor is at the temperature being measured) becomes impractical because of calibration instability, and it becomes necessary to infer the temperature by measuring some property of the electromagnetic radiation emitted by the heated object. Although there are several methods whereby the temperature of an object may be so inferred, one method (ordinarily referred to as optical pyrometry) stands out as being clearly more accurate than other methods, when applied under prescribed conditions; it is used to define the IPTS at temperatures above the gold point and for standardization purposes. In this method, the radiance (at some specified wavelength, usually 650 nm) of a blackbody is measured in terms of its ratio to the radiance (at the same wavelength) of a blackbody at the gold point. The radiance of a blackbody is exactly described by the well-known Planck's radiation distribution as represented in Fig. 1, where the area under the curve represents radiant energy.

Radiant energy measured by an optical pyrometer is from a narrow spectral band, as seen in Fig. 1, conventionally at 650 nm. By measuring the ratio R of the radiant energy $N\Delta\lambda$ (in the band of width $\Delta\lambda$) at some elevated temperature $T > T_0$ to the radiant energy at the temperature $T_0 = 1337.58$K (the gold point), the temperature T may be calculated, assuming that C_2, the second radiation constant, and λ are known. The blackbody thus calibrated may serve as a standard with which to calibrate the pyrometer.

Radiance measurement in terms of temperature has been accomplished for many years with an optical pyrometer of the disappearing-filament type (Fig. 2). As indicated, a red filter and the response

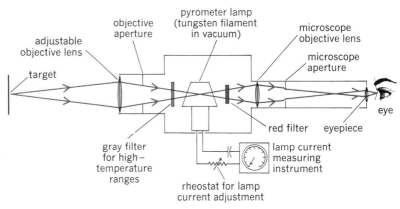

Fig. 2. Diagram of a conventional disappearing-filament optical pyrometer, which is a low-power telescope with a small tungsten lamp filament in the focal plane of the objective lens.

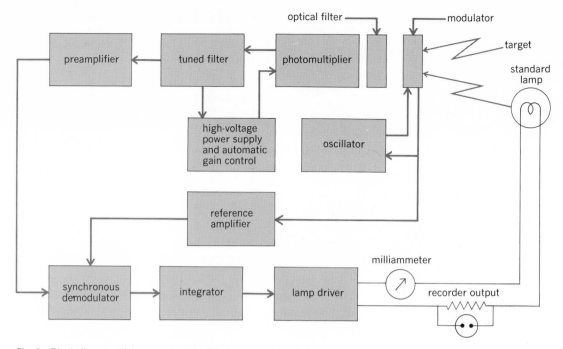

Fig. 3. Block diagram of the components of the automatic optical pyrometer.

of the eye combine to limit the spectral band being passed, so that the eye sees a red image of the target on which is superimposed the glowing lamp filament. The lamp current is adjusted until the brightness of the filament matches the brightness of the target, at which time the filament image blends into the target image and disappears. The pyrometer lamp current is calibrated as a function of temperature when a calibrated blackbody (as in Fig. 1) is used as a target.

In 1956 precise photoelectric pyrometry was introduced in the Soviet Union, followed closely by similar instruments in the national standardizing laboratories in the United States and Australia. These were essentially disappearing-filament pyrometers in which the eye was replaced by a photomultiplier tube. Self-balancing commercial versions of the type shown in Fig. 3 have made their appearance in more recent years. These are not as precise as those used in the national standardizing laboratories but are substantially more precise than the conventional disappearing-filament type; a detailed error analysis of this type of instrument is available in the literature.

As indicated in Fig. 3, the modulator alternately passes radiation from the target and then from the pyrometer standard lamp at some frequency such as 90 or 400 Hz. If the photomultiplier receives unequal amounts of radiant energy from the two sources, its response is a square-wave signal, the phase of which determines whether the lamp current is too high or too low. This signal is filtered and synchronously demodulated, and then integrated. The integrated signal controls the lamp current, driving it up or down to achieve a zero-amplitude square wave from the photomultiplier, at which time the lamp current is said to be in null balance. The pyrometer is calibrated in essentially the same manner as is the conventional disappearing-filament optical pyrometer.

The equation relating the two absolute tempera-

tures of the blackbody radiances (Fig. 1) measured by an optical pyrometer is known exactly, and is given to a close approximation by Eq. (1), where λ

$$\frac{1}{T} = \frac{1}{T_0} - \frac{\lambda}{C_2} \ln R \qquad (1)$$

is the mean effective wavelength of the spectral band in which the radiance ratio R is measured, T_0 is the absolute temperature of the gold point 1337.58K, and T is the temperature of the blackbody at some higher temperature. Equation (2)

$$C_2 = \frac{hc}{k} = 1.4388 \text{ cm K} \qquad (2)$$

defines C_2. Here h is Planck's constant, c is the speed of light, and k is the Boltzmann constant.

Uncertainties in IPTS-68. From Eq. (1) it can be seen that errors in temperatures T above the gold point depend upon errors in T_0, λ, C_2, and R, each of which has undergone substantial reduction in recent years. The magnitude of the previous error in T_0, about 1.4K, and the presently accepted value of its error, 0.2K, together with the cause for that improvement, have been discussed above. Improvement in the determination of both λ and R result from the introduction of electronic photodetectors, which are now used in place of the human eye to sense the radiant energy measured by the optical pyrometer.

The value of λ is obtained by a weighted averaging process involving numerical integration with respect to λ across the spectral band of radiation sensed by the photodetector; the uncertainty in the averaged value of λ thus obtained is determined in part by the width of the spectral band being detected. The very high resolution of small changes in radiant power attendant to the use of photodetectors permits the use of a narrower spectral bandwidth than is possible with the human eye, with a consequent reduction in uncertainty in the averaged value of λ. This is possible through the use of thin-film interference filters which, because of recent technological advances, can now be made of very good quality. The numerical integration also involves the use of the relative response curve of the photodetector as a function of wavelength; this curve is more accurately measured by a photodetector than by the human eye. The net effect of the reduced spectral bandwidth and the more accurately measured relative response curve is to reduce the uncertainty in the mean effective wavelength λ to slightly less than 1A. *See* DIFFRACTION GRATING.

The value of R, the measured radiance ratio, was previously determined through the use of sectored disks, using the human eye to determine equality of two radiances. This technique has been replaced by the use of an optical beam conjoiner, a device which, in its simplest form, permits two radiances to be made equal to within the resolution of the photodetector and then permits the two radiances to be summed, yielding a radiance exactly double that of the initial radiance (within the resolution of the detector). By a repeated process of doubling, the gold point radiance may thus be accurately multiplied to a much higher value, corresponding to a much higher temperature through an accurately determined large multiple value of R.

The value of C_2, as shown in Eq. (2), is determined by the values of three physical constants: h, c, and k. The best values and uncertainties of these three constants are determined as part of an elaborate least-squares adjustment process involving many other physical constants, all of which are interrelated through the various equations of physics, and many of which have been accurately determined experimentally. The value of C_2 is thus subject to gradual improvement through a very indirect process involving the results of a wide range of experimental physics. The value presently accepted as part of the definition of IPTS-68 is $C_2 = 1.4388$ cm K, whereas the value for IPTS-48 was $C_2 = 1.438$ cm°K.

Practical problems. The principal difficulty in the accurate maintenance of IPTS-68 above the gold point presently lies not so much in realizing the scale accurately as in preserving the initial primary calibration of a pyrometer. The major source of calibration change is in the current-radiance relationship of the pyrometer lamp filament. The basic physical phenomena involved in the lamp calibration instability are not well understood. However, recent research has led to sufficient improvement that the best available lamp-manufacturing technology can provide lamps adequate for virtually all present-day industrial and scientific applications.

The maximum achievable accuracy can rarely be obtained in common laboratory applications because of inability to simulate blackbody conditions or to apply sufficiently accurate corrections to account for nonblackbody conditions. Of the various measurement methods that have been tried in an attempt to alleviate this limitation, none has proved to be generally satisfactory. Of these methods, such as ratio pyrometry, mirrored chambers to simulate blackbody radiation, and the use of polarization techniques, each has limitations that restrict its proper realm of application more or less severely, for the present time at least. The greatest need in the field of applied high-temperature measurements is for a much more general solution to this problem than that which presently exists.

For background information *see* PHYSICAL MEASUREMENT; TEMPERATURE; TEMPERATURE MEASUREMENT in the McGraw-Hill Encyclopedia of Science and Technology.   [GENE D. NUTTER]

Bibliography: The International Practical Temperature Scale of 1968 (Foreword by C. R. Barber), *Metrologia*, vol. 5, no. 2, April, 1969; R. D. Lee, *Metrologia*, vol. 2, no. 4, October, 1966; G. D. Nutter, *I.S.A. Pap.*, no. 16, Dec. 2, 1966; T. P. Murray, *Rev. Sci. Instrum.*, vol. 38, no. 6, June, 1967.

Toxoplasmosis

Discovery of the sexual and oocyst stages of the protozoan *Toxoplasma* has helped to explain its transmission to herbivorous animals and man. It had previously been known that toxoplasmosis can be acquired by eating raw or undercooked meat which may contain *Toxoplasma* cysts. Recent findings indicate that when cats so acquire infection, a series of multiplicative stages are initiated in their intestinal epithelium (Figs. 1 and 2); these events lead to the formation of male and female gametes and the production of oocysts, which are shed in the feces. The oocysts develop infectivity in 1–5 days by dividing into two sporocysts, each of which develops four sporozoites. The mature oocyst is highly resistant to environmental influences and can survive in moist soil for weeks or months, serving as a potential source of infection (Fig. 3).

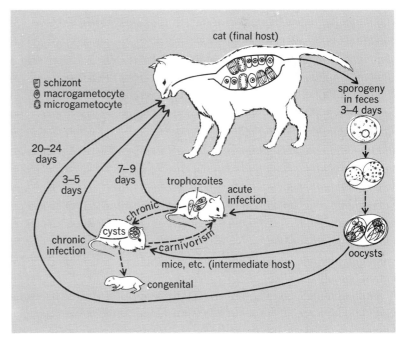

Fig. 1. Postulated life cycle of *Toxoplasma* showing means of transmission by oocysts from cat feces and by ingestion of trophozoites or cysts from intermediate hosts. Days indicated represent the intervals from ingestion of *Toxoplasma* by cats to the reexcretion of oocysts. Mice are shown to represent the many mammals and birds that can serve as intermediary hosts and among which infection may spread, usually by carnivorism. Congenital transmission may take place during chronic infection, as in mice, or during acute infection, as in humans. (*J. K. Frenkel, J. P. Dubey, and N. L. Miller, Toxoplasma gondii in cats: Fecal stages identified as coccidian oocysts, Science, 167:893–896, Feb. 6, 1970*)

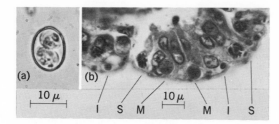

Fig. 2. Multiplicative forms. (*a*) Oocyst containing two sporocysts and four sporozoites. (*b*) Schizonts (S), macrogametes (M), and immature form (I) in epithelium of small intestine of a kitten that had been fed cysts of *Toxoplasma* in mouse brain 8 days earlier.

Search for manner of infection. *Toxoplasma* has been known since 1908, when it was discovered almost simultaneously by A. Splendore in a laboratory rabbit in Brazil and by C. Nicolle and L. Manceaux in gondis, small wild rodents used as laboratory animals in the Pasteur Institute of Tunis, Tunisia. This discovery started the search for the mechanism by which these animals became infected. In 1917 E. Chatton and G. Blanc established that the gondis were not infected in their native habitat, the foothills and mountains of southern Tunisia. However, many gondis developed toxoplasmosis and died after they had been in the laboratory for several weeks, especially in the rainy winter months. Transmission by an insect was suspected. For this reason these French workers and, in the 1950s, L. Jacobs and J. K. Frenkel studied many species of bloodsucking flies, mosquitoes, fleas, ticks, mites, and bugs, but without finding a vector.

Carnivorism. In 1954 D. Weinman considered transmission in pigs and rodents that are carnivorous. In 1960 the frequent finding of *Toxoplasma* cysts in the skeletal muscle of sheep, pigs, and occasionally cattle led L. Jacobs, J. S. Remington, and M. L. Melton to biologically compare the "resting" cyst with the proliferative forms of *Toxoplasma*. They found the cyst organisms to be resistant to gastric juices and to be highly infectious orally, whereas the proliferative forms were rarely infectious by mouth. G. Desmonts suspected that ingestion of undercooked or raw meat might be an important mode of infection in France. He proved this in 1965 by showing that the acquisition rate of infection could be increased by additional amounts of undercooked meat. These observations could explain the occurrence ot toxoplasmosis in carnivores; however, the high incidence of infection in some herbivores such as sheep and rabbits remained a mystery.

Fecal transmission. W. M. Hutchison conceived the idea that *Toxoplasma* might be transmitted in the feces of animals. In 1965 he showed that after cats ate *Toxoplasma*-infected mice, their feces became infectious. Stored feces retained infectivity for several months. Since *Toxoplasma* is ordinarily a very fragile organism, Hutchison suspected and seemingly proved that *Toxoplasma* was contained within the eggs of a roundworm, the cat ascarid (*Toxocara cati*), with which these cats were also infected. That a protozoan organism might be transmitted in a worm egg stimulated several investigators to reexamine the characteristics of *Toxoplasma* in cat feces. Some workers supported the role of the nematode in transmission. Other investigators separated worm eggs and

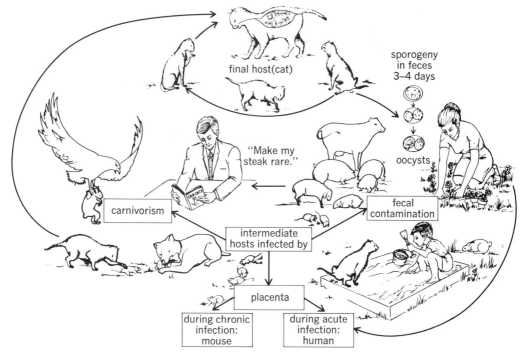

Fig. 3. Postulated transmission of toxoplasmosis. Oocysts are excreted in the feces of cats and, after sporulation, are infectious to a great variety of hosts by fecal contamination (right). Carnivorism is the other important means of transmission (left). Cats probably become infected more often by carnivorism than by the ingestion of oocysts. Transplacental transmission is indicated below. (*From R. A. Marcial-Rojas, ed., The Protozoan and Helminthic Diseases of Man, Williams and Wilkins, 1971*)

larvae from *Toxoplasma* by washing and sieving, and also recovered infectious feces from worm-free cats.

Because many small structures are found in cat feces, no single test could furnish conclusive evidence. It became necessary, therefore, to accumulate evidence for or against their identicalness with *Toxoplasma* by scrutinizing each of the candidate forms in a series of mutually independent tests.

Fecal stages as coccidian oocysts. As a candidate form oocysts were compared with *Toxoplasma* infectivity by J. P. Dubey, N. L. Miller, and Frenkel. Oxygenation was necessary both for sporulation of the oocyst and for attaining infectivity, as was a temperature of 20–30°C (68–80°F). The oocyst size was identical with that of the infectious particle, as shown by filtration through a graded series of filters. The two entities coincided quantitatively, in appearance and disappearance from cat feces, in mobility characteristics in an electric field, and in their densities. Antigenic similarities were found between conventional *Toxoplasma*, the oocyst, and some of the new stages. Infection of newborn kittens with *Toxoplasma* cysts resulted in excretion of oocysts and in the finding of multiplicative and sexual stages in the gut; all these stages were absent in uninoculated littermate kittens. Feeding the three different stages of *Toxoplasma* to cats resulted in different incubation periods for the appearance of oocysts and for *Toxoplasma* infectivity, suggesting a cyclic relationship: Infection appears 3–5 days after feeding of cysts, 7–9 days after feeding of proliferative forms, and 20–24 days after feeding of infectious cat feces. Finally, it was shown that a given number of oocysts increased fourfold in infectivity after some of the enclosed sporozoites had been liberated. Harley G. Sheffield and Marjorie L. Melton observed that the oöcyst became infectious to tissue cultures only after liberation of the sporozoites, and that the sporozoites had the general ultrastructure of *Toxoplasma*. J. P. Overdulve microisolated some oocysts and showed them to be infectious, whereas fluid without such oocysts was not infectious. Because all these tests indicated identicalness in quantity or quality between oocysts and *Toxoplasma* infectivity, the term *Toxoplasma* oocyst can be used with assurance.

Comparison with other coccidia. *Toxoplasma* oocysts resemble the oocysts of the coccidium *Isospora bigemina*, which was described by C. W. Stiles and by A. Railliet and A. Lucet in 1891. It is not known whether this rarely observed form is the long unrecognized component of the *Toxoplasma* cycle or whether *I. bigemina* exists as a separate entity, similar to the other cat coccidia *I. felis* and *I. rivolta*. The identification of an oocyst stage and of the formation of many male gametes in the life cycle of *Toxoplasma* indicates a closer affinity to the Eimeriidae (Coccidia) than to the Haemosporidiida. Coccidia generally live in one host, usually only in the gut, and sometimes only in its epithelium. The thrust of phylogenetic development to parasitize other tissues is shown by *Toxoplasma* and the two related organisms *Schellackia* from reptiles and *Lankesterella* from birds; in addition, *Toxoplasma* also infects numerous mammals and birds as intermediary hosts in which only the asexual, proliferative cycle and the resting cyst have

been observed. The sexual cycle has been found only in cats, the final host. It is likely that other related mammals may serve as final hosts; also, because cats bury their feces, transport hosts (earthworms, flies, beetles) may be involved in the dissemination of oocysts.

Epidemiology. The proportion of people who become infected from meat or by oocysts from domestic cats is unknown. Cultural factors determine how much meat is eaten raw, or undercooked, whether cats are kept as pets, and the frequency of contact with soil possibly contaminated by cat feces. Limiting climatic factors include freezing, drying, and heating above 50°C (122°F), all of which kill the sporozoites in the oocysts.

Prevention. Ingestion of raw or undercooked meat should be avoided. Cats that hunt, soil contaminated by their feces, and their litter boxes may intermittently represent sources of infection. Cats that are fed on a canned, dried, or frozen food have little opportunity to become infected. Disease rarely develops from infection in adult humans; however, pregnant women should avoid exposure since the unborn fetus is especially vulnerable to toxoplasmosis.

Toxoplasmosis in cats. Almost all cats without antibody become infected after eating cysts, and excrete oocysts for 1–2 weeks. Adult cats occasionally become sick from lesions in lungs, liver, lymph nodes, brain, and the inner eye. Young cats sometimes die from penumonia and enteritis. A recently infected cat generally does not reexcrete *Toxoplasma* following a second feeding. A study showed that about 45% of adult cats in Kansas City have *Toxoplasma* antibody, and since half of these excrete oocysts after cyst feeding, apparently immunity is not absolute and a cat may be infectious several times during its lifetime.

For background information *see* COCCIDIA; TOXOPLASMOSIS in the McGraw-Hill Encyclopedia of Science and Technology.   [J. K. FRENKEL]

Bibliography: J. P. Dubey, N. L. Miller, and J. K. Frenkel, *J. Exper. Med.*, 132:636–662, 1970; J. K. Frenkel, Toxoplasmosis, in R. A. Marcial-Rojas (ed.), *The Protozoan and Helminthic Diseases of Man*, 1971; J. K. Frenkel, J. P. Dubey, and N. L. Miller, *Science*, 167:893–896, Feb. 6, 1970; L. Jacobs, *Advan. Parasitol.*, 4:1–45, 1967.

Transfer cell, plant

The architecture of plant tissues allows the living protoplasts of virtually all cells to exchange dissolved substances directly with their surroundings, whether these surroundings be nonliving spaces within the plant or the medium external to the plant body. Probably all cells engage in, and to some extent depend on, this type of transfer, but a newly recognized type of plant cell, the transfer cell, is considered to be especially well adapted.

Structural features. The structural feature which immediately sets the transfer cell apart from all other cells is the irregular ingrowths of material which protrude into the protoplast from the inner faces of its walls (Figs. 1 and 2). These ingrowths are apparently composed of materials similar to those of the parent cell wall, each ingrowth possessing a skeletal framework of microfibrils and a matrix of polysaccharides, possibly of acidic character. There are usually many ingrowths to each cell; in some instances they may be small, discrete, and papillate; in others, long and filiform; and in yet others, branched or even anastomosed into a labyrinth. The crevices between the ingrowths are penetrated by living cytoplasm and, in consequence, the surface area of the boundary membrane of the protoplast is unusually extensive. It is presumed that this enhanced surface area, together with other cytoplasmic features (Fig. 1), endows the transfer cell with an especially high capacity for absorption of solutes from, or for their secretion to, the nonliving region outside the plasma membrane.

Evolution and functions. In evolutionary terms the first appearance of transfer cells was probably related to the nutrition of very young embryos (sporophytes) while still in contact with, and partially dependent for nutrient on, their parent gametophytic generation. Cells of either or both generations along the line of contact may be specialized as transfer cells, and it is probable that those of the host (gametophyte) secrete nutrients which are in turn absorbed by those of the sporophyte. Once evolved, this nutritive apparatus seems to have persisted with remarkable stability, being encountered in mosses, liverworts, hornworts, horsetails, and ferns, and in comparable locations in flowering plants. In flowering plants, transfer cells are found where the embryo sac (a gametophyte) comes into contact with its host (a sporophyte) or, later, where the embryo absorbs from nutritive tissues such as the endosperm.

The appearance of the conducting channels of the plant, xylem and phloem, were key evolutionary events, and in many species transfer cells may be found located strategically, it would seem, for the loading and unloading of these channels. They occur alongside the xylem of stems of ferns and conifers and are especially well displayed in the nodes of the shoots of flowering plants (Fig. 2). In all these situations they may absorb solutes from the ascending transpiration stream. There are also instances when xylem-associated transfer cells occur in belowground parts of a plant (such as in the root of the water fern *Azolla* and in the leguminous root nodule); here they may well secrete solutes into the xylem. Transfer cells are commonly found in association with sieve cells or sieve tubes of the phloem. Stem nodes and the minor veins of leaves are the two most usual anatomical sites for phloem transfer cells; in the former they occur in horsetails and in many flowering plants, and in the latter they are best developed in herbaceous dicotyledons. At these sites there is evidence that their function is the collection and concentration of solutes from the nonliving spaces of the tissues. Some of these solutes may leak from living cells, and others may originate from the xylem elements. Once concentrated by the transfer cells, the solutes are transferred via cytoplasmic intercellular connections to the neighboring sieve elements for export to other regions of the plant. Phloem transfer cells almost certainly have key functions in the export of solutes (including products of photosynthesis) and in the exchange of solutes between the xylem and phloem.

In addition to their occurrence in the internal locations mentioned above, transfer cells may also be associated with the external surfaces of plants, very frequently, for example, in nectaries and in secreting glands of insectivores, hemiparasites,

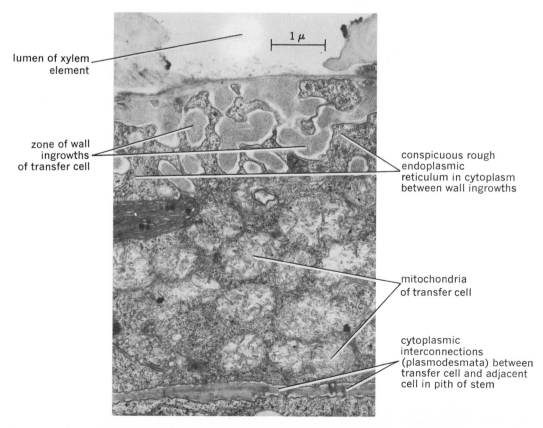

lumen of xylem element

zone of wall ingrowths of transfer cell

conspicuous rough endoplasmic reticulum in cytoplasm between wall ingrowths

mitochondria of transfer cell

cytoplasmic interconnections (plasmodesmata) between transfer cell and adjacent cell in pith of stem

Fig. 1. Electron micrograph of part of a xylem-associated transfer cell in a stem node of *Galium aparine*.

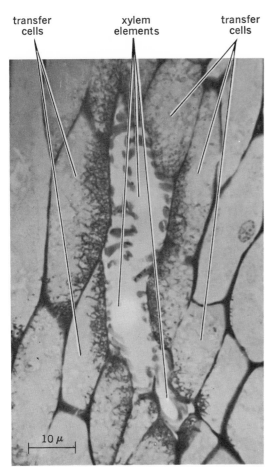

transfer
cells

xylem
elements

transfer
cells

10 μ

Fig. 2. Light micrograph of transfer cells surrounding
the xylem elements of a departing leaf trace in a stem
node of *Ammobium alatum*. Wall ingrowths of transfer
cells line walls closest to the xylem.

and halophytes, less frequently on the absorptive
surfaces of the submerged leaves of water plants.
Significantly, they do not appear to be present on
the absorptive surfaces of roots; probably these
surfaces are already of vast area and can perform
with adequate efficiency without further amplification at the ultrastructural level.

Animal counterpart. The counterpart of transfer
cells in animal tissues are cells which bear microvilli. In both instances the enlarged plasma membrane may be regarded simply as a device which
accentuates the capacity of the cells for solute
transport, and which has evolved on many independent occasions and in a wide diversity of anatomical situations in response to physiological
selection pressures generated by the requirement
of handling intense fluxes of an equally wide diversity of solutes.

For background information *see* CELL MEMBRANES; CELL WALLS (PLANT); PHLOEM; REPRODUCTION (PLANT).

[BRIAN E. S. GUNNING; JOHN S. PATE]
Bibliography: B. E. S. Gunning and J. S. Pate,
Protoplasma, 68:107–133, 1969; J. S. Pate and
B. E. S. Gunning, *Protoplasma*, 68:134–156, 1969.

Transformation, cell

Since the discovery that human diploid fibroblast
cells could be adapted to grow in tissue culture,
there has been extensive investigation of the nor-

mal growth properties of human cells and of the
effects of various oncogenic viruses on these properties. Of the known tumor viruses, the RNA-containing avian and murine sarcoma viruses and the
DNA-containing adenovirus 12 and simian virus 40
(SV40) have been shown to induce morphologic
alteration of human cells in culture. This article is
concerned primarily with human cell transformation by SV40 and its isolated deoxyribonucleic acid
(DNA).

SV40 was first discovered as a contaminant of
monkey cells in tissue culture, where it normally
kills the cells as a result of lytic infection. It has
since been shown to produce tumors in several
rodent species and induce morphologic alteration
of the same species in cell culture. In mouse or
hamster cells the virus is unable to reproduce, and
infected cells are either cured or transformed.
H. Shein and J. F. Enders, with H. Koprowski and
coworkers, first showed that SV40 could induce
cellular changes in human fibroblasts, thus providing one of the first laboratory demonstrations of
the potential oncogenicity of a virus for human
cells. In human cells the virus is able to grow to a
limited extent so that, following infection, some
cells may be killed. Of the others, many spontaneously recover, but a small fraction go on to be
permanently transformed. By this process, called
transformation, SV40 induces persistent changes
in cellular growth characteristics in a manner
which is genetically transmitted from cell to
daughter cell.

SV40-transformed human cells. Transformation leads to a number of striking alterations. Some
of these acquired properties are very similar to the
properties that can be demonstrated in apparently
spontaneously occurring tumors, though there is
no evidence that SV40 itself is actually responsible
for any class of human neoplasia. The release of
transformed cells from normal growth restraints
results in their ability to form multiple cell layers
in culture. This permits the densely growing SV40-
transformed human colonies to be easily distinguished from the monolayer of surrounding normal
cells and provides the basis for the quantitative
assay of SV40 transformation. In animal model
systems the loss of contact inhibition induced by
SV40 is clearly associated with the development of
tumorgenic potential by the cells. At least some
spontaneously occurring human tumors that have
been studied in tissue culture also show a lack of
contact inhibition of cell division.

Another cellular alteration associated with SV40
transformation of human cells is the production of
nuclear abnormalities such as chromosomal
breaks, fragments, and dicentric chromosomes.
This alteration may be partly responsible for the
large number of bizarre giant cells and degenerating cells seen in SV40-transformed human cultures. Similar cellular abnormalities can be detected in tumor cells from naturally occurring
neoplasms.

Normal human fibroblasts have a finite life-span
in tissue culture of 30–100 generations, whereas
cells derived from human tumors can be established as permanent cell lines in culture. SV40
transformation of normal human cells also results
in the acquisition of this property of indefinite
growth capacity.

Recent evidence strongly suggests that SV40

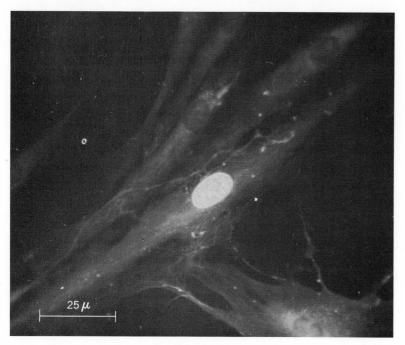

Fig. 1. SV40 T antigen localized to the nucleus of an infected human cell by the fluorescent antibody method. Several negative cells can also be seen in field.

becomes permanently associated with the host-cell genome since it can be recovered from transformants after many cell generations in the absence of any infectious virus production. Other evidence from DNA-hybridization experiments indicates that the viral genome is probably integrated within the DNA of the host cell. In fact, the host-cell genome may contain several copies of the viral DNA.

SV40-induced antigens. New antigens appear in human cells infected with SV40. The tumor (T) antigen is not part of the virus itself but neverthe-

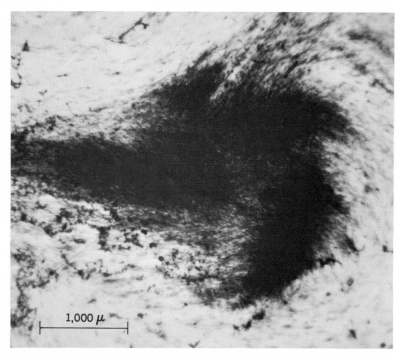

Fig. 2. An SV40 DNA–transformed human cell colony shown growing against a background of normal human cells.

less is consistently found in SV40-infected cells and in SV40-induced tumors. It can be visualized entirely in the nucleus by a fluorescent-antibody staining procedure. Since T antigen persists in all SV40-transformed cells, it is thought to be important in the process of transformation.

Figure 1 shows an example of an SV40 T antigen positive human fibroblast cell stained by the fluorescent-antibody method. The fluorescein-conjugated T antibody is concentrated in the nucleus of the positive cell, whereas the nuclei of the surrounding negative cells have failed to take up any stain.

SV40 can replicate in human cells to a limited extent, but cannot do so in cells of species such as the hamster or mouse. Virion (V) antigen, which is thought to be virus capsid proteins, can be detected in the nuclei of recently infected human cells. A small fraction of SV40-transformed human cells continues to synthesize V antigen, even many cell generations after the initial transforming event.

Some alterations induced by SV40 are localized to the cell surface. In model systems, SV40-transformed cells elicit a strong immune response when injected into the animal because of the presence of virus-specified transplantation antigens. Recent evidence indicates that SV40 also induces conformational changes in the cell surface, thereby exposing cellular components which normally are masked. The functional role of each newly acquired antigen in the process of transformation has yet to be clarified.

SV40 DNA. The DNA of SV40 exists in the form of a double-stranded twisted circle with a molecular weight of about 2×10^6 to 3×10^6 daltons. Purification of viral DNA requires the preparation of a large amount of virus grown by infection of green monkey kidney cells. The virus is then concentrated and purified by density gradient centrifugation in cesium chloride. After extraction with sodium dodecylsulfate to remove the DNA from its protein coat, the viral DNA is purified by equilibrium density centrifugation in cesium chloride with ethidium bromide. In this step the double-stranded twisted circular DNA (DNA I) can be separated from the nicked circular (DNA II) and linear DNA forms.

Isolated SV40 DNA was first shown by P. Gerber to be infectious for monkey cells. Other investigators later reported that SV40 DNA can induce the formation of T antigen in hamster cells and can produce morphologic transformation of bovine embryo cells. This was particularly interesting because the latter cells were highly resistant to infection by the whole virus. In each of these studies, nonphysiological conditions, including the use of hypertonic solutions, were necessary in order to get the cells to take up enough DNA to produce a detectable effect.

More recently it has been shown that the infectivity of SV40 DNA can be enhanced by more than a thousandfold with the use of diethylaminoethyl-dextran (DEAE-D). DEAE-D may act either by enhancing DNA entry into the cell or by inhibiting degradative enzymes which might normally inactivate it. The use of DEAE-D thus allows relatively efficient infection with DNA under physiological conditions, so that cellular growth is not greatly impaired. Healthy cells are critically important to transformation assays since the cells can only be recognized as transformed several cell divisions after infection. Any procedure which will damage the cell sufficiently to inhibit its reproductive capacity cannot be used; the most nearly physiological methods are the most preferable.

Human cell transformation by SV40 DNA. Transformed colonies can be observed in SV40 DNA–infected human cultures 16 to 25 days after infection. The tissue culture properties of the transformed colonies are identical to those produced by whole virus. The cells are epithelioid, show pronounced loss of contact inhibition of cell division, and readily form multiple cell layers. They also show persistent nuclear abnormalities and many giant cells. An example of an SV40 DNA–transformed human cell colony is shown in Fig. 2. The cells have been fixed with formalin and stained with hemotoxylin. Note the densely staining transformed cells against the background of normal human cells.

The transforming efficiency of DNA is far greater per infectious unit than that of whole virus. This is the case for unfractionated DNA as well as purified DNA I. As few as 200 infectious units of SV40 DNA I have been found sufficient to produce one transformed human cell colony. With intact SV40 small plaque virus, on the other hand, approximately 2.5×10^5 infectious units are required to produce a transformed colony. Viral DNA is, therefore, in the order of a thousandfold more efficient per infectious unit than whole virus. The inefficiency of whole virus must be due to the inefficiency in human cell infection in an early step, for example, at the level of adsorption, penetration, or uncoating. Once either the virus or the viral DNA reaches the stage of inducing SV40 T antigen, a comparable but low probability exists that the cell will develop into a transformed colony.

The most recent studies indicate that DNA II, the single-nicked circular form, is also capable of transforming human cells. Linear single-stranded fragments of DNA I with up to one-third of the molecular weight of the whole molecule can be produced by mechanical shearing. These fragments are inactive both in inducing T antigen and in transforming human cells.

Colonies of SV40 DNA–transformed human cells that have been isolated have all been found to contain T antigen positive cells. By cocultivation with monkey cells in the presence of ultraviolet-inactivated Sendai virus, a sensitive method for recovery of infectious virus from whole virus–transformed cells, infectious virus has been recovered from every one of a large number of DNA transformants tested. Therefore, as with transformation by intact virus, the entire viral genome is generally present and recoverable from DNA-transformed cells.

Transformation and T antigen induction appear, at least over a limited range, to be linearly related to the number of SV40 DNA molecules added. This suggests that a single molecule of DNA is capable of inducing T antigen and also of permanently transforming human cells. At high concentrations of DNA (above 2 μg/ml), there appears to be a plateau for transformation. The reasons for this plateau are not clear, but it may be caused by toxicity of the DNA or by a requirement for the infected cell to be at a particular phase of the

growth cycle at the time of DNA infection. Such a plateau, however, is not observed for whole virus, where, with enough virus, essentially all of the cells will express T antigen.

The fact that SV40 DNA can transform human cells demonstrates that infectious nucleic acid can enter and permanently alter the genetic makeup of a normal human cell. Whether SV40 has an integration function similar to that of bacteriophages λ and P 22 is not yet known. If the ability of the viral DNA to integrate could be separated from its transforming ability, it can be speculated that the "integrating" DNA might be used as a carrier to insert specific genetic information into human cells.

For background information *see* BACTERIO-PHAGE; CULTURE, TISSUE; DEOXYRIBONUCLEIC ACID (DNA); TUMOR VIRUSES in the McGraw-Hill Encyclopedia of Science and Technology.

[STUART A. AARONSON]

Bibliography: S. A. Aaronson and G. J. Todaro, *Science*, 166:390, 1969; S. A. Aaronson and G. J. Todaro, *Virology*, 36:254, 1968; P. H. Black, *Ann. Rev. Microbiol.*, 22:391, 1968; R. Dulbecco, *Science*, 166:962, 1970.

Tritium

Tritium oxide is an isotopic form of water and thus is difficult to separate from ordinary water by simple chemical means. As a result, much of the tritium produced in nuclear processes is released to the environment as water. Concern over the incorporation to tritium in hydrologic and biologic cycles has led to several recent reviews on its environmental cycling. Most of the predicted behavior of tritium is based on existing information regarding the cycling of water, though this information has been supplemented by more recent observations of the cycling of tritium produced during atmospheric testing of nuclear weapons. It is hoped that such reviews will provide the impetus for additional studies of the environmental behavior of tritium.

Tritium is the heaviest isotope of the element hydrogen and the only one which is radioactive. It is produced in nature by a number of processes, including bombardment of nitrogen in the stratosphere by cosmic rays. It is also produced by different types of nuclear activities, both purposefully for further use and as a by-product. Tritium is used as a fuel for thermonuclear reactions, as a radioactive tracer for water and organic compounds, and as a material for luminous dials.

Global cycling. Conditions in the environment favor the conversion of tritium gas to tritiated water; thus it behaves generally, though not exclusively, like ordinary water during its global cycling. Tritium can also become an integral part of any chemical compound containing hydrogen atoms, including the organic compounds that make up living tissue. Since water and organic materials are basic to life, there is considerable interest in the fate of tritium released to the environment from its various sources.

During the years following the large-scale atmospheric testing of hydrogen bombs, it was found that tritium introduced into the atmosphere tended to distribute in about the same manner as did naturally produced tritium, with tritiated water being the predominant species (see table). Natural triti-

Natural tritium distribution in the atmosphere*

Atmospheric component	Tritium distribution, %
Hydrosphere	~ 90
Troposphere	
Water vapor	0.1
Molecular hydrogen	0.02–0.2
Methane	$\leq$ 0.04
Stratosphere	
Water vapor	~ 10
Molecular hydrogen	0.004–0.007

*From F. Begemann, *Earth Science and Meteorites*, North Holland Publishing Co., 1963.

um and tritium produced from atmospheric testing of nuclear weapons are introduced into the stratosphere, where vertical mixing processes are slow. Tritium remains in the stratosphere for a mean residence time of 1–10 years before it is mixed into the troposphere. In the troposphere the flushing action of rainfall causes removal of tritium as water vapor with a mean residence time of 20–40 days. During its stay in the stratosphere, tritium is rather uniformly mixed over the entire hemisphere into which it has been introduced. Since mixing between the stratosphere and the troposphere occurs preferentially in the mid-latitudes, most of the tritium introduced into the northern stratosphere is deposited on the Earth's surface in the zone between 30 and 50° north latitude.

Practically all continental deposition of tritium is caused by precipitation, although transfer can occur by vapor exchange between the air and the Earth's surface. Vapor exchange is especially important over water surfaces, such as the ocean, and may account for about two-thirds of the tritium deposition into oceans. Since mixing occurs when water vapor exchanges with the ocean surface, re-evaporated water has a much lower concentration of tritium than the originally deposited water vapor; thus a net deposition of tritium occurs. Over land surfaces there is a smaller amount of water for dilution and mixing is not so rapid. In this case the reevaporated water has a tritium content nearly equal to that of the exchanging vapor; therefore the net deposition of tritium is smaller. Since the same mixing processes occur with rainfall, the tritium content of precipitation over continental areas is generally higher than that over oceans.

The annual rainfall over continental areas of the world is about 0.66 m per year. Of this amount, 37% is lost by immediate runoff and underground flow of water into surface streams. The remaining 63% is reintroduced into the atmosphere by evaporation and transpiration. Water that percolates into the ground may remain there for times ranging from a few days to tens of thousands of years. Tritiated water that reaches deep underground reservoirs is essentially removed from circulation if the residence time exceeds a few decades.

Tritiated water can be deposited directly onto vegetative cover and become incorporated into the biological cycle, or it can enter the biological cycle from the groundwater in the root feeding zone. When the tritium concentration in groundwater of the root feeding zone is maintained at a relatively constant level for an entire growing season, the organic tissue reaches and maintains approximately the same tritium-hydrogen ratio as does its substrate. Food obtained from such locations can be expected to have about the same tritium-hydrogen ratio in their organic components as in their fluids. Hence, there has been considerable interest, from the view of the effect of radiation on health, in describing the quantitative ecological cycling of tritium.

Stimulation of natural gas reservoirs. Two experiments were conducted for the purpose of determining the feasibility of stimulation of natural gas production by use of nuclear detonations, Project Gasbuggy and Project Rulison. Gas is just now being recovered for analysis from the Rulison experiment, but it was found in the Gasbuggy experiment that tritium produced in the device and by activation of lithium in the soil during the detonation became distributed in essentially the same manner as ordinary hydrogen. About 95% of the tritium formed tritiated water. However, a portion of the tritium became incorporated in the hydrocarbons, predominantly methane, which constitutes the natural gas product. The primary use of natural gas is for combustion as fuel; during this process the hydrocarbons are oxidized to carbon dioxide and water vapor. Thus utilization of natural gas from nuclearly stimulated gas reservoirs constitutes another potential source of release of tritiated water to the environment. It is doubtful that release of tritium from this source would ever reach the level of natural tritium production of 4,000,000–8,000,000 curies per year, but local situations may require observation.

Tritium release reactors. Several reactions lead to the production of tritium in reactors—ternary fission and activation of boron and lithium are the most important reactions in light water reactors while activation of deuterium is the predominant reaction in heavy water reactors. The total production of tritium in light water reactors has been estimated to range from 50 to 90 curies per day for the generation of 1000 megawatts of electricity. The major fraction of the tritium produced in light water reactors probably arises from ternary fission within the fuel element. Most of the tritium produced inside the fuel element is contained until the fuel is processed for recovery of unburned fissile materials at a chemical processing plant. At the Connecticut Yankee pressurized water reactor (PWR), 16% of the fission product tritium was estimated to have penetrated the fuel element cladding; this contributed about two-thirds of the total release of tritium from the power station with the remainder coming primarily from activation of boron and lithium. Releases of tritium from boiling water reactors (BWRs) have been significantly lower than those from PWRs for the production of equivalent amounts of electric power.

Projected production of tritium from the production of electricity using nuclear reactors suggests that by about 1990 this rate of production will about equal the natural rate of tritium production of 4 to 8 megacuries per year and that by the year 2000 the accumulated tritium from the nuclear power industry will be about 100 megacuries or in the same range as that accumulated from natural production. However, throughout this time period the actual levels of tritium in the environment should be decreasing because the major contributor to present tritium levels was atmospheric testing of thermonuclear weapons which released an estimated 1700 megacuries of tritium to the environment.

For background information *see* ATMOSPHERIC CHEMISTRY; HYDROLOGY; TRITIUM in the McGraw-Hill Encyclopedia of Science and Technology.   [DONALD G. JACOBS]

Bibliography: F. Begemann, *Earth Science and Meteorites*, 1963; *Environmental Effects of Producing Electric Power*, Hearings before the Joint Committee on Atomic Energy, 91st Congress of the United States, Oct. 28–31, Nov. 5–7, 1969; A. G. Evans, *Health Phys.*, 16(1):57–63, January, 1969; A. Holzer, Gasbuggy in perspective, *Proceedings of the ANS Topical Meeting: Engineering with Nuclear Explosives*, Las Vegas, Nev., January, 1970, CONF-700101, 1:62, 1970; D. G. Jacobs, *Sources of Tritium and Its Behavior Upon Release to the Environment*, USAEC Critical Review Series, 1968; W. F. Libby, *J. Geophys. Res.*, 68:4485–4494, August, 1963; J. E. Mountain and J. H. Leonard, Tritium production in a pressurized water reactor, *16th Annual Meeting of the American Nuclear Society*, Los Angeles, Calif., June, 1970; K. Rankama and T. G. Sahama, *Geochemistry*, 1949; C. E. Weaver, E. D. Harward, and H. T. Peterson, Jr., Tritium in the environment from nuclear power plants, *Environmental Effects of Producing Electric Power*, Hearings before the Joint Committee on Atomic Energy, 91st Congress of the United States, August, 1969.

Turbojet

Rapid growth of civil air transportation has resulted in increased frequency of aircraft takeoffs and landings and the use of jet aircraft at a large number of cities. As a result, more people are in the neighborhood of jet airports and are exposed to annoying noise levels more frequently. Community response has resulted in a substantial effort by government and industry to reduce the noise of near-airport operations of the civil air fleet. Achievement of this goal will improve the existing airport neighborhood environment and ease the problem of site selection of future airports. Recently, a high-bypass turbojet engine has been developed to alleviate the problem of annoying noise levels near jet airports.

Bypass engine design. The first civil jet-powered transports were powered with pure turbojet engines. Noise of the exhaust jet mixing with the atmosphere was a major concern accompanying the introduction of these aircraft into commercial service. The development of the bypass engine permitted use of lower exhaust velocities of the gas from the engine than those from the pure turbojet while maintaining the required thrust levels. Continued development of the bypass engine has resulted in the availability of several engines with high bypass ratios. These high-bypass-ratio engines have better fuel economy and also result in less jet-mixing noise than earlier engines. These engines handle very large quantities of air and obtain the required thrust by exhausting these large quantities of air at relatively low velocity. The reduced jet velocity is inherently quieter because the lower jet velocity more than compensates for the increased gas flow.

Thus the development of the high-bypass-ratio engine has resulted in a reduction of jet-mixing noise. This component of the engine noise is a broad-band noise and is distributed generally across the acoustic frequency range up to approximately 1000 hertz (Hz). This noise is at a maximum in the rear quadrant of the engine and thus affects a ground observer after the aircraft has passed by. It is more pronounced during takeoff than on landing.

The first engine of this type to enter service was the General Electric TF-39 engine for the Lockheed C-5 military transport. The Pratt and Whitney Division of United Aircraft followed closely with the JT9D for the Boeing 747 long-range civil transport. Rolls-Royce is developing the RB-211 for the Lockheed L-1011 medium-range transport, and General Electric is supplying the CF-6 engine, a derivative of the TF-39, for the McDonnell-Douglas DC-10 medium-range transport.

Fan noise. The high-bypass-ratio engine has a large-diameter fan on the front of the engine. This fan is a significant source of noise that has become the dominant source as bypass ratio increases and jet-mixing noise decreases. In the past several years there has been considerable emphasis in government and industry on research on fan noise.

The fan consists of a set of rotating blades with airfoil cross sections followed by a set of stationary vanes. Kinetic energy is imparted to the air by the rotating blades. The swirling air exits from the rotating blade cascade and enters the stationary blade row. The swirl kinetic energy is converted to static pressure rise in the passage through the stationary cascade. Downstream of the stationary cascade, the air has attained a pressure higher than that at the inlet to the fan. The pressurized air is then exhausted through a convergent nozzle to produce thrust.

The airflow over the rotating and stationary blades is unsteady. The resulting pressure fluctuations propagate as acoustic waves out of the engine and are transmitted to the observer as noise. The acoustic energy emanating from the blades covers a wide frequency range, with the major part of the energy in the frequency range close to the frequency of rotation of the rotating blades. This frequency is referred to as the blade passage frequency and is usually in the range of 1000 to 3000 Hz.

In contrast to jet noise, which is generated outside the engine as the exhaust jet mixes with the surrounding atmosphere, fan noise is generated within the engine housing. Thus this noise can be reduced by lining the inside surfaces of the engine and the nacelle with sound-absorbing material. The technology of engine and nacelle acoustic treatment has advanced considerably in the last several years. Flight tests of 707 and DC-8 aircraft have shown that fan noise can be decreased 10–15 perceived noise decibels by the use of such linings. The perceived noise decibel is a unit of measurement of the human annoyance caused by noise exposure.

Additional reductions in noise output can be made by appropriate design of the fan itself. Proper selection of rotational speed, spacing between rotating and stationary blades, and the number of rotating and stationary blades can result in significant reductions in the amount of noise generated. These concepts are being used in a pro-

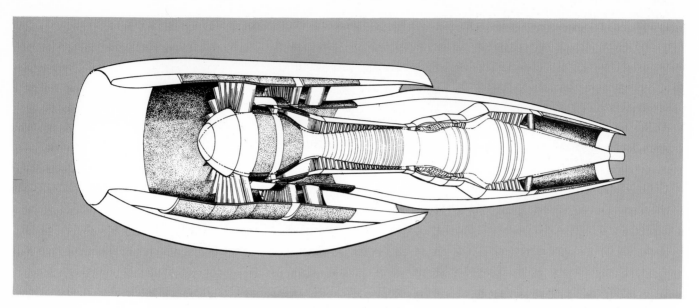

Diagram of the chief components of the NASA experimental quiet engine. Sound-absorbing materials are indicated by the shaded sections. Engine is being built by the General Electric Co.

gram known as the Quiet Engine Program to build and test experimental engines with low noise performance as the prime requirement. A low-noise engine is being fabricated and tested by the General Electric Co. under NASA contract (see illustration). After the engine is delivered to NASA in late 1972, it will be fitted with an acoustically treated nacelle to form a low-noise propulsion system. This propulsion system, suitable for subsonic flight, will be ground-tested for noise performance in 1973.

Noise pollution in the neighborhood of jet airports has resulted in the Federal Aviation Administration imposing a noise performance requirement for aircraft certification. This requirement specifies maximum noise levels at several reference locations for new aircraft during takeoff and landing operations. This rule formalizes the trend of recent years for low-noise operation to be a prime requirement for aircraft propulsion systems.

Supersonic aircraft. This discussion has been concerned with propulsion for subsonic aircraft. The noise problems of supersonic aircraft are sonic boom and jet-mixing noise. The boom is determined by the shape of the aircraft, flight speed, and altitude. It affects observers under the flight path during supersonic cruise. The jet-mixing noise is a concern for airport neighborhoods during takeoffs and landing operations. It is not practical to use the high-bypass-ratio engine to reduce the jet-mixing noise of a supersonic aircraft. Considerable effort is being expended on development of jet noise suppressors for the turbojet engine being developed for the supersonic transport.

For background information *see* NOISE CONTROL; SUPERSONIC FLIGHT; TURBOJET in the McGraw-Hill Encyclopedia of Science and Technology.

[JAMES J. KRAMER]

Bibliography: Federal Aviation Agency, Noise standards, pt. 36: Aircraft type certification, *Fed. Regist.*, Nov. 25 and 29, 1969; *Progress of NASA Research Relating to Noise Alleviation of Large Subsonic Aircraft*, NASA SP-189, 1968.

Venus

In the past few years, the astronomer's concept of the planet Venus has changed radically; the former "near-twin of Earth" is now known to have a surface temperature of about 750°K (1350°F), a carbon dioxide atmosphere with a pressure 100 times that at the Earth's surface, and clouds to a height of 57 km. Though much of the new information came from probes by the Soviet Union and the United States to the planet, essential contributions have been made by Earth-based radar and radio studies and by spectroscopy at large optical telescopes. These methods have revealed a very slow rotation (a Venus day lasts 117 Earth days) in the direction opposite to that of Earth, have accurately measured the diameter of the planet (12,106 km), and have found traces of CO, HCl, and HF. Water vapor has been observed too, but the amount is still very much in doubt. There could be a few percent of inert gas such as nitrogen.

Surface temperature. The high surface temperature was indicated by the first measurements, in 1956, of the shortwave radio emission from Venus. The result was so astonishing, however, that many scientists sought other explanations for the radio brightness. In October, 1967, the entry probe *Venera 4* reached a region with a pressure of 20 atm and a temperature of 544°K. For a short time it was assumed that these results represented surface conditions, but the complementary measurements from *Mariner 5* the following day strongly suggested that *Venera 4* did not reach the surface seen by radar from the Earth. In May, 1969, these inferences were confirmed by *Venera 5* and *6*. Not only did these probes penetrate somewhat deeper before expiring but their radars showed the presence of a surface at still greater depths.

Atmosphere and clouds. Almost as strange as the high surface temperature is the total opacity of the atmosphere and clouds in the infrared and visible; the infrared brightness temperature is a nearly uniform 207°K on both day and night sides, with only a slight cooling toward the poles. If enough visible solar radiation could reach the surface, the high temperature could perhaps be explained by the "greenhouse effect," since outgoing infrared is certainly blocked. However, there is reason to doubt that solar radiation can penetrate; even if it did, heat transfer by convection might upset the equilibrium. Indeed, some convection or circulation is clearly necessary to support the clouds. A more convincing model suggests that solar heat is deposited in the region of the cloud tops, driving a large-scale circulation of the atmosphere which carries the necessary small amount of heat downward.

Despite much research effort, the nature of the clouds themselves remains in doubt. It is possible that there are several cloud layers of different composition, but nearly all the available observations necessarily pertain to the visible surface. There is strong evidence that this surface is very hazy, like a mist or fog, and unlike most terrestrial clouds. Ice, which naturally comes first to mind, seems to be ruled out because the expected infrared absorption lines of water vapor are far too weak. Perhaps there is a layer of water cloud hidden from view; all the Venera probes observed water vapor at a point about 15 km below the visible cloud tops. (Actually, any cloud drops would be rather strong hydrochloric acid, because HCl vapor is present.) The substance that best matches the reflection spectrum of the planet is ferrous chloride dihydrate, $FeCl_2 \cdot 2H_2O$. The presence of this substance in the atmosphere is, however, difficult to explain; it would have to be classed as a dust, raised by either strong winds or volcanic eruptions, and it is strange that other, commoner dusts seem to be absent. Another popular suggestion is ammonium chloride, NH_4Cl; and other chlorides may be present in lower layers, hidden from view. The problem of cloud composition and distribution will probably not be settled until another probe mission permits detailed measurements right in the clouds.

The existence of such a huge CO_2 atmosphere seems strange at first, but in fact Earth's atmosphere has probably contained almost as much, a bit at a time, in the past. On Earth, CO_2 dissolves in liquid water, and is converted into limestone and the shells of marine organisms. Estimates of the total amount present on Earth in this form are about half as much as has been found in the Venus atmosphere. Without liquid water, the CO_2 should remain a gas; the anomaly is that most of the water that would be expected on Venus is missing. Even by the highest estimates, Venus has only 1/500 as much water (vapor) as does the Earth (mostly in the oceans). Perhaps this difference reflects the original compositions of the two planets, or perhaps Venus has lost most of its water. It is likely that water vapor on Venus is far more subject to dissociation by sunlight than on Earth, because the atmosphere does not absorb ultraviolet as strongly. After dissociation of the water vapor, the hydrogen would escape from the planet because of its large thermal velocity, and the oxygen would react with the rocks.

Corona. The presence of a hydrogen cloud, or corona, around Venus was revealed by both *Venera 4* and *Mariner 5*. Detection was by means of the far-ultraviolet light, the Lyman-alpha radiation,

which is strongly scattered by hydrogen. The Mariner data show an anomalous variation with distance from the planet, which is best explained if heavy hydrogen, or deuterium, is more abundant than light hydrogen is. (Less probable explanations have been based on molecular hydrogen.) The thermal escape of deuterium is much slower than that of hydrogen, and the observations thus suggest that much of the original deuterium has been retained, while most of the light hydrogen has been lost.

Measurements of the ionosphere were made by two radio-propagation experiments on *Mariner 5*. On the day side, a peak, thought to be of CO_2^+, was observed at a height of 140 km. A long tail, presumably of light ions, extends to about 500 km, where it suddenly disappears. This sharp cutoff, or plasmapause, is thought to be due to the pressure of the solar wind. On the night side, the main layer is much weaker and the tail extends about 5000 km back from the surface.

For background information *see* VENUS in the McGraw-Hill Encyclopedia of Science and Technology.

[DONALD M. HUNTEN]

Bibliography: D. M. Hunten and R. M. Goody, *Science*, 165:1317–1323, 1969; Proceedings of the 2d Arizona Conference on Planetary Atmospheres, *J. Atmos. Sci.*, 25:533–671, 1968.

Vision

An aspect of visual perception, eidetic imagery, has been clarified recently and found to be not as rare as was once thought. Eidetic imagery has been defined as a visual image of a previously viewed picture. The image is located in front of the eyes and persists for several minutes. It has been described as the most vivid of any kind of visual imagery, comparable to an afterimage though persisting for a much longer time. An eidetic perceiver can scan the stimulus continuously during inspection without any interference with the production of an eidetic image. In fact, scanning is necessary to generate an image of the entire figure. Such a perceiver can also move his eyes over the image after the stimulus is no longer present without the image moving or being destroyed.

Eidetic images have been distinguished from memory by the perceiver's report that the visual image persists after the stimulus has been removed, and by his behavior which indicates that he is indeed looking at such an image. There have also been attempts to use the amount of detail as a criterion for distinguishing eidetic images from memory, but this is not a consistent criterion. Eidetic images have been distinguished from afterimages by their persistence; by the ease with which they are evoked, even from low-contrast stimuli; by their positive representations of color; by the independence of visual fixation (afterimages require fixation to form); and by the lack of interference from eye movements made during the report of the image.

Prevalence. An extensive history spanning more than 50 years can be briefly summarized. Percentages of children said to possess some form of eidetic imagery range from 30 to 90%, depending upon the age and the population studied, with a rough average of all studies around 50%. Nearly

The nonverbal test for eidetic imagery. The upper and middle portions of the figure represent the individual stimuli used to elicit images. The bottom portion represents the composite picture when the two stimuli are superimposed. (From J. Leask, R. N. Haber, and R. B. Haber, Eidetic imagery in children: II. Longitudinal and experimental results, Psychonomic Monogr. Suppl., 3(3):46, 1969)

every investigator has reported that eidetic imagery is common and that eidetic children can be easily found among any population. Different investigators have reported different peak ages or a negative correlation with age, and others have pointed to puberty or shortly before as the age of greatest

prevalence. All investigators report zero or nearly zero frequencies among adults. It is not necessary, however, to read this literature closely to see the many serious methodological errors of omission and commission. There has been relatively little recent work, although extensive work by R. N. Haber and several others in the last 5 years has shown that eidetic imagery can be reliably studied and that it must be accounted for by current perceptual theory.

Nearly all of the recent studies have shown that the percentage of children of elementary school age who possess eidetic imagery is less than 10%. The only exception to that conclusion has been one study using institutionalized retarded children which found that more than 20% were eidetic; that high percentage was due primarily to the inclusion of brain-damaged children in the sample. In one recent study in central Africa 20% of the subjects were found to be eidetic. These percentages were lower than the ones from the older literature, undoubtedly because of the better differentiation in these studies between eidetic imagery and after-imagery or vivid memory.

In each of the experiments or samples reported in the last 5 years, there has been no correlation found with age. Generally, experimenters have had difficulty finding eidetic children under age 6 or 7, but this seems to be primarily due to the difficulties of testing. With only one exception, adults have not been tested in the United States. In the only longitudinal study carried out recently, Haber found that there seemed to be no substantial loss of eidetic imagery over a 5-year period.

Relationship with memory. There is strong indication that eidetic images are not used as a device for strengthening memory. Quite the contrary: If a child wishes to remember a scene, then he behaves as an adult would, making some conscious effort to retain the information, either by naming each part and rehearsing the names or by some other related strategy. In such cases, eidetic children do not report eidetic images of the stimulus. Only when their perception is passive, without any conscious attempt at remembering the picture, does an eidetic image develop. It appears as if eidetic imagery and normal memory for pictures are independent and perhaps even antagonistic mechanisms for preserving and maintaining information.

Most children can control their eidetic images, either by conscious effort to prevent them or by exaggerated eye blinks. No children tested seemed to be inundated with such images or unhappy with them, although most children reported that sometimes unwanted images developed.

The duration, vividness, and content of eidetic imagery vary greatly among eidetic subjects. Using as a criterion the child's report that he is seeing an image in front of his eyes, the duration of that image lasts from 20 sec to several hours or days, with several cases reported in which the image can be reevoked upon command weeks or months later. Most of the eidetic subjects tested could quite easily maintain images for a minute or more. The content of the images depended explicitly on the part of the stimulus scanned. If the child did not explicitly look at a particular part, then that part was absent in his image, even though he might be aware of that part from having quickly glanced at it or from peripheral vision and memory. Thus eidetic images are in no sense like a photographic memory, from which the child can report details from merely a brief glance.

Eidetic images appear to fade with time, and often the various parts of the image fade at different rates. It is a typical and perhaps an almost universal report that, although the initial image may have been relatively complete, later one part of the image disappears, although another part is still seen. This process continues until all of the parts are gone. Eidetic images of linguistic material are generally very poor. Children do not have complete images of print. This is probably because they read the page rather than look at the characters passively.

Several recent experiments have reported attempts to isolate memory factors entirely from the measurement of eidetic imagery. Haber reported a composite picture study in which a child was first shown one picture (top portion of the illustration) and told to scan it. The picture was then removed, presumably leaving the child with an eidetic image. A second picture (middle portion of the figure) was then presented. Had the two pictures been superimposed, they would have formed a composite picture (bottom of the figure) that could not have been predicted from either part alone. Children with very good eidetic imagery have no difficulty in identifying the composite. In the most extreme version of this test, reported by C. Strolmaier and J. Merritt, a pair of stereo random dot patterns was presented to the subject, one to each eye. If the patterns had been presented simultaneously to a normal subject, stereoscopic fusion of them would have permitted the viewer to see a figure in depth stand out from the dots, even though it would be impossible to see such a figure when viewing just one member of the pair. Strolmaier and Merritt reported that one eidetic subject could see the pattern in depth even when there was a separation of days, or even weeks, between the presentation of one member of the stereo pair to one eye and the other member to the other eye. In this sense, then, an image of the random dot pattern was maintained, or was recoverable, during this period of time.

Summary. A long-known area in visual perception has again been shown to be important. The proportion of the population that is eidetic is probably much smaller than had originally been thought. Though very few adults have been tested, the phenomenon is, nevertheless, not a rarity. In the samples reported none of the experimenters has been able to find correlates with eidetic imagery that have held up in any other experiment. The children do not seem to be abnormal or atypical; they do not seem to suffer from having eidetic imagery in terms of other cognitive skills.

For background information *see* MEMORY; PERCEPTION; VISION in the McGraw-Hill Encyclopedia of Science and Technology.

[RALPH NORMAN HABER]

Bibliography: R. N. Haber and R. B. Haber, Eidetic imagery: I. Frequency, *Percept. Mot. Skills*, 19:131–138, 1964; J. Leask, R. N. Haber, and R. B. Haber, Eidetic imagery in children: II. Longitudinal and experimental studies, *Psychonomic Monogr. Suppl.*, 3(3):25–48, 1969; C. F. Stromeyer, III, and J. Psotka, The detailed texture of eidetic images, *Nature*, 225:346–349, 1970.

Volcano

On a worldwide scale, the location of active volcanic belts is explained by the sea-floor-spreading hypothesis. On the scale of an individual volcano, small earthquakes and ground surface movements occur between and during eruptions. These geophysical investigations are leading to a better understanding of the roots of volcanoes and the mechanics of eruptions. Of the world's 543 volcanoes

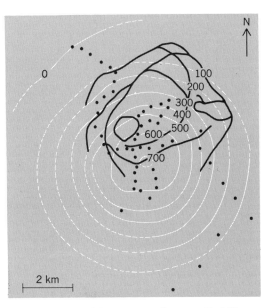

Fig. 1. Uplift in millimeters near the summit of Kilauea Volcano, Hawaii, from January, 1966, to October, 1967. The heavy lines are the outlines of the summit craters and calderas of Kilauea. (*R. S. Fiske and W. T. Kinoshita, Inflation of Kilauea Volcano prior to its 1967–1968 eruption, Science, 165:341–349, 1969*)

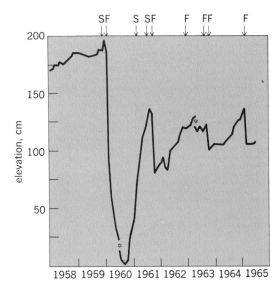

Fig. 2. Elevation changes of the summit of Kilauea Volcano, Hawaii, from 1958 to 1965. Asterisk represents leveling adjustment; S, summit eruption; and F, flank eruption. (*R. W. Decker, D. P. Hill, and T. L. Wright, Deformation measurements on Kilauea Volcano, Hawaii, Bull. Volcanol., 29:721–732, 1966*)

with known recorded activity, 90 have erupted during 1966 to early 1970.

Location of volcanic belts. More than 80% of known active volcanoes occur on the rim of the Pacific Ocean, "the ring of fire," and in the belt from the Mediterranean Sea to Indonesia. This zone of island arcs and mountain ranges is also the location of major earthquakes and young folded and faulted mountains. According to the sea-floor-spreading hypothesis, large plates of the Earth's crust are slowly moving relative to one another, like giant slabs of floating ice. In the Pacific and Mediterranean mountain belts, the plates apparently override one another (Japan) or slide side by side (San Andreas Fault in California). The volcanoes are mainly along the overriding margins. Another belt of volcanoes is located along the oceanic ridges, where the ocean floors are apparently spreading and new volcanic crust is being formed to replace the void of the separating plates (Iceland). Only 15% of the known active volcanoes are along these trends, but many more active, but unknown, volcanoes probably occur along the deep submarine portions of the oceanic ridges.

The fractures formed along the edges of the moving crustal plates provide the channels through which the melted rock ascends to the surface. Molten rock may form a few percent of the common mantle material of the Earth at depths of 50–200 mi, or it may be generated by special processes along the separating and overriding crustal plates. Volcanoes of the continental margins are explosive, and erupt mostly lava fragments of variable composition, whereas volcanoes of the oceanic ridges generally erupt basaltic lava flows in relatively quiet eruptions. *See* EARTH, HEAT FLOW IN.

Volcano roots. The deep source of magma (melted rock) is a zone 50–200 mi below the Earth's surface where temperatures approach or reach the 1000–1200°C melting temperature of rock. The magma is less dense than the surrounding unmelted rock and either rises in a blob which melts and forces its way upward or, more likely, rises in a plumbing system of fractures formed by the breaking edges of large moving crustal plates. As magma nears the surface, but still at depths of a few miles, it forms a magma chamber, a complexly shaped body several thousand feet across. The main evidence for this storage chamber comes from small changes in the elevation of the summit area of the volcano (Fig. 1). Kilauea Volcano in Hawaii slowly inflates during periods of several months to a few years between eruptions, and rapidly collapses in a few days or weeks during an eruption (Fig. 2). The shape of the bulging and sagging area is controlled by the depth to the magma chamber; an increase in volume at 4 mi depth will affect a much larger area than will an increase in volume at 2 mi depth. In Hawaii, a 1- to 3-mi depth is indicated by several eruptions since 1959.

Small earthquakes, many of which are "felt" only by sensitive seismographs, also provide data on the roots of a volcano. Most earthquakes are caused by the underground breaking of rocks which have been strained beyond their bending limit. Because there are many forces at work on a volcano, the strains are complex. Some relate to the magma pressure changes which cause the surface of the volcano to swell and sag. Others are caused by the growing load of the volcano on the Earth's crust. Still others are probably related to the rapid temperature changes in the rocks surrounding the magma chamber. As a result of these differing stresses and strains, earthquakes related to active volcanoes belong to several families. The underground location and seismic "signatures" of different earthquakes can be deciphered only after information is accumulated for many years. At Kilauea, the earthquake data indicate a deep source of magma 30–40 mi down and a fracture system which connects the source to the shallow magma chamber. Fracture zones which radiate out into the sides of Kilauea and allow magma to escape in flank eruptions several miles from the summit are also outlined by the earthquake locations (Fig. 3). *See* EARTHQUAKE.

The slow uplift between eruptions and the rapid collapse during eruptions are analogous to an electrical or mechanical device that slowly stores energy or mass to a certain threshold and then suddenly discharges. An example of such an electrical device is the winking neon bulb on a highway danger signal. A battery supplies current through a resistor which is stored on a capacitor. The voltage builds up on the capacitor toward the limiting voltage of the battery. A neon bulb across

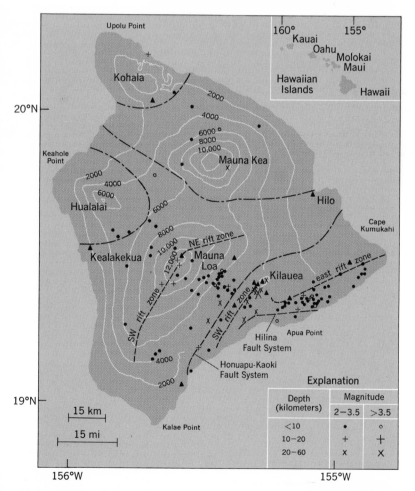

Fig. 3. Locations of earthquakes on Hawaii during the first quarter of 1968. Dot-and-dash lines are boundaries of volcanic systems. Contour interval is 2000 ft and datum is mean sea level. (*R. Y. Koyanagi, Hawaiian seismic events during 1968, USGS Prof. Pap., no. 650-D, 1969*)

Volcanic eruptions since 1966

Location and volcano	Year of eruption	Location and volcano	Year of eruption
Italy		Mariana Islands	
Stromboli	1966–1970	unnamed*	1966
Etna	1966–1970	North Pagan	1966
Ethiopia		Farallon de Pajaros*	1967, 1969
Erta ale	1968	Japan	
Tanzania		Sin-Iwo-zima	1968
Ol Doinyo Lengai	1966	Myojinsho*	1970
Congo		O-sima	1966–1968
Nyamuragira	1967	Okinawa-tori-sima	1967, 1968
Nyiragongo	1966–1970	Suwanose-zima	1967, 1968
Indian Ocean, Réunion Is.		Kutinoerabu-zima	1966, 1968
Piton de la Fournaise	1966	Sakura-zima	1966–1969
Indonesia		Aso	1966, 1970
Marapi	1966–1968	Me-Akan	1966
Talang	1967, 1968	Kurile Islands	
Kerintji	1966–1968	Ebeko	1967
Tangkuban Prahu	1967, 1969	Kamchatka	
Slamet	1967	Karymsky	1967
Merapi	1967–1969	Bezymianny	1966–1970
Kelut	1966, 1967	Kliuchevskoi	1966
Semeru	1967–1969	Alaska	
Batur	1966, 1968	Kiska	1969
Rindjani	1966	near Amlia*	1966, 1967
Sangeang Api	1966	Trident	1966–1969
Ebulobo	1969	Redoubt	1966–1968
Ija	1969	Hawaii	
Keli Mutu	1968	Kilauea	1967–1970
Rokatenda	1966	Guatemala	
Lewotobi Lakilaki	1968	Santiaguito	1966–1970
Nila	1968	Fuego	1966, 1967
Soputan	1966–1968	Pacaya	1966–1970
Lokon-Empung	1966, 1969	El Salvador	
Api Siau	1966, 1967	Izalco	1966
Banua Wuhu*	1968	San Miguel	1970
Awu	1966, 1968	Nicaragua	
New Zealand		Telica	1969
Ruapehu	1966–1969	Cerro Negro	1968–1970
Ngauruhoe	1968	Masaya	1970
White Island	1966, 1968	Costa Rica	
Rumble IV*	1966	Rincon de la Vieja	1966, 1967, 1969
Rumble III*	1966	Arenal	1968–1970
Tonga Islands		Poas	1969
Metis Shoal*	1968	Ecuador	
unnamed*	1970	Sangay	1966, 1967
Santa Cruz Islands		Galapagos Islands	
Matthew	1966	Fernandina	1968
Yasour	1966–1970	Cerro Azul	1969
Lopevi	1968	Peru	
Ambrym	1967–1970	Ubinas	1969
Tinakula	1966	Chile	
Solomon Islands		Lascar	1968
Kovachi*	1966–1970	Austral Islands	
Bagana	1966	Macdonald*	1967
New Britain Islands		West Indies	
Ulawun	1970	Kick-'em-Jenny*	1966
Philippines		Iceland	
Canlaon	1969	Hekla	1970
Mayon	1968	Surtsey	1966, 1967
Taal	1966, 1967, 1969	Antarctica	
Didicas	1969	Deception Island	1967, 1969

*Submarine eruption.

the capacitor with a firing voltage less than the battery voltage flashes on and allows the rapid discharge of the capacitor. The voltage drops below the sustaining level of the neon bulb and the bulb winks out, and the buildup cycle begins again. The time lag between flashes depends on the value of the battery, resistor, capacitor, and neon bulb. In a hypothetical volcanic system, the battery is analogous to the rising force of the magma. Any increase in this force shortens the repose time between eruptions. The electrical resistor is similar to the flow resistance in the fracture plumbing of the volcano between the deep source and the shallow magma chamber. An increase in resistance lengthens the repose time. The capacitor is analogous to the shallow magma chamber, and an increase in capacity lengthens repose time. The neon bulb is similar to a potential fracture between the magma chamber and the surface. Rupture occurs when the swelling reaches some critical level, allows magma to escape, and then seals itself by the cooling of the lava in the fracture. The strain needed to cause the next fracture is probably not exactly the same as that required for the previous one.

The eruption "cycle" of the volcano is therefore erratic in time and difficult to predict unless specific values can be assigned to all the variables. The more explosive volcanoes often have a longer repose time between eruptions, and gathering the needed geophysical data is very long-term and often dangerous work. The electrical analog model may not be applicable. At this time it appears that the magma source of the Pacific rim volcanoes is deeper than that of the ocean ridge volcanoes. The explosive volcanoes seem to have shallow magma chambers, but they are less well defined than in Hawaii. Their magma is generally very viscous, and its breakthrough to the surface is accompanied by an increase in shallow earthquakes. Mapping the earthquake pattern with seismographs is the best danger alert technique presently available.

Eruptions since 1966. About 40 volcanoes are in eruption on a yearly basis. Half of these are persistently active over several years, and the others erupt briefly after long repose periods. The table lists the world's active volcanoes in eruption between January, 1966, and May, 1970. The most interesting eruption during this interval occurred at Fernandina in the Galapagos Islands. Although the island volcano is uninhabited and the eruption was not closely observed, the explosions and earthquake swarm of June 11–23, 1968, were related to a major collapse of the caldera of Fernandina. The nearly 3-mi² caldera floor dropped as much as 1000 ft below its pre-1968 elevation.

For background information *see* MARINE GEOLOGY; SEISMOLOGY; VOLCANO in the McGraw-Hill Encyclopedia of Science and Technology.

[ROBERT W. DECKER]

Bibliography: R. S. Fiske and W. T. Konoshita, *Science,* 165:341–349, 1969; D. Herbert and F. Bardossi, *Kilauea: Case History of an Eruption,* 1968; R. Y. Koyanagi, *USGS Prof. Pap.,* no. 600-C, 1968; Smithsonian Center for Short-Lived Phenomena, *Event Notifications,* 1968–1970.

Water pollution

The use of commercial nitrogen fertilizer has greatly increased during the past 20 years. Also, there has been a tremendous increase in the number of livestock fed for slaughter. A high percentage of the livestock are fed today in concentrated feeding operations involving 25,000–100,000 animals. These changes in agricultural practices have increased production efficiency and greatly enhanced food quality. At the same time, however, questions have arisen regarding the effects of such practices on groundwater supplies. Much of the concern has centered around the possible buildup of the nitrate level, since water containing more than 45 parts per million nitrate is considered by the U.S. Public Health Service as unsafe for drinking. Nitrogen is also involved, along with other required nutrients, in the growth of algae in surface waters.

Nitrogen fertilizers. Evaluating the extent of water pollution caused by nitrogen fertilizers is very complex because fertilizer is only one contributor. In certain parts of the United States, natural geologic deposits of nitrates contribute a large percent of the nitrate leached to groundwater formations. Soil organic matter, animal wastes, and plant residues also contribute. Decomposition of soil organic matter is a major contributor of nitrate, especially during the first few years of cultivation. Nitrate generally does not accumulate under native grasslands because it is used by the plants as it is formed. Cultivated lands normally produce crops for only a portion of the year, and the cultivation stimulates decomposition of organic matter. Nitrate produced while plants are not growing may leach below the root zone and accumulate in the soil or move into groundwater supplies.

Nitrogen fertilizer can increase the amount of nitrate available for leaching, but this is generally true only when fields are fertilized at the wrong time of year or with excessive amounts. Both practices reflect poor management systems that should be corrected. Leaching is most likely to occur in the fall and winter, when precipitation may exceed evaporation and transpiration. Many fertilized fields are contributing less nitrate to water resources than they did several years ago because the total nitrate available from the soil organic matter and fertilizer is less than that which the soil organic matter alone furnished during the first few years of cultivation.

Feedlots. Presently, the large, mechanized commercial feedlot accounts for many of the livestock fed for slaughter. In the past, animals were fed in small units, and wastes were considered an asset. Today, the concentration of many animals in one unit creates enormous waste-management problems. A cattle feedlot of 50,000 capacity covers approximately 200 acres and produces approximately 450,000 tons of wet manure and urine annually. This manure contains about 15,000 tons of dry mineral matter, 60,000 tons of dry organic matter, and 2800 tons of nitrogen. Pollution of groundwater can occur directly beneath the pens, beneath basins used to impound runoff, and beneath cropland treated with wastes or runoff from the feedlots.

Effect on water supplies. Increased nitrate concentrations found in Missouri water supplies were attributed to feedlots and not to fertilizer use on farms. In a Colorado study profile samples were obtained from the surface to bedrock or to the water table. Nitrate content of the profiles was influenced by land use. Nitrate contents in 20 ft of profile under cattle feedlots ranged from none to over 5000 lb per acre, the average for 47 feedlots being 1436 lb. For other kinds of land use the average values were: virgin grassland, 90 lb; unfertilized wheat-fallow land, 261 lb; and irrigated land not in alfalfa, 506 lb. Nitrate gradients in the profile showed accumulations at the soil surface, with concentrations decreasing with depth. Nitrate content in the water at the water table surface and at the 20-ft depth in the soil showed little difference for land-use treatment. The question is what happened to the nitrate? There was evidence that water was moving through the profile and that some nitrate was moving with it. The decrease in nitrate with depth was attributed to denitrification, a process whereby microorganisms reduce nitrate to nitrogen gas. The greatest decrease in nitrate concentration occurred under feedlots where an abundant supply of carbonaceous material was present in the soil water to serve as an energy source for the microorganisms. Denitrification, therefore, appears to be significant in determining the fate of nitrate moving through soil profiles, particularly those under barnyards and feedlots.

Even if nitrate concentrations remain low, other

pollutants can degrade the quality of groundwater. The Colorado study revealed that several wells were abandoned near feedlots, even for livestock use, because of poor water quality. The table shows analyses of water samples taken from core holes in the Colorado study. The largest differences between water samples collected under feedlots and adjacent irrigated fields were in the concentration of ammonium and organic carbon, a measure of soluble organic matter.

Feedlots and the use of fertilizer have affected some water supplies. However, detailed studies show that even when fairly high rates of commercial fertilizers are used, well-managed operations do not contribute to water pollution. Commercial fertilizer nitrogen poses no additional threat to water supplies unless it is applied in excess or at the wrong time of year. Management also is the key in preventing large livestock feeding operations from polluting water resources. Therefore, feedlot operators should give special consideration to general site location, including soils, and to runoff control capabilities. *See* SOIL CHEMISTRY.

[B. A. STEWART]

Effect on aquatic vegetation. Foul-smelling beaches, dead fish, scums of stringy decaying plant material, and green-colored water are commonly brought about by additions to surface water of domestic sewage (treated and untreated), animal wastes, agricultural fertilizers, and nitrogen- and phosphorus-containing industrial wastes. Although these perturbations of nature are often observed, few investigations have examined the qualitative and quantitative effects of specific nutrients on aquatic vegetation under controlled or semicontrolled conditions. This diverse assemblage of aquatic vegetation (Fig. 1) comprises algae and aquatic vascular plants, which together provide the energy to support all life in the water.

Algae. Although algae are present in either microscopic or macroscopic forms in all natural waters, excess growths can occur: Blooms of algae are discolorations of the water brought about by the presence of millions of microscopic algae in each drop of water; pond scums are usually clumps of macroscopic algae in filaments floating at the shorelines of ponds and lakes. Algae are not differentiated into roots, stems, and leaves, have no internal circulatory system, take in nutrients through each cell surface, and, in all cases, depend upon the water which surrounds them for their nutrients.

Aquatic vascular plants. Aquatic vascular plants have many modes of existence in water and respond differently to similar nutrient enrichment.

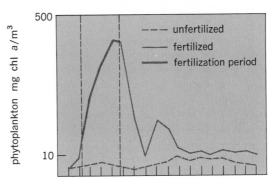

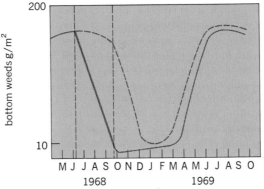

Fig. 1. Response to eutrophication.

Emergent plants (plants rooted in mud at the shoreline) bear the closest resemblance to terrestrial plants in growth habit and respond to similar fertilization practices. Several emergent plants provide crops of tremendous economic significance, and standardized fertilizer procedures have been developed for them. These include lowland rice, a crop which provides food for millions of people all over the world, and *Phragmites communis*, the major source of cellulose in Romania. Floating plants with their roots suspended in the water take their nutrients from the water. The normal existence of these plants is in hydroponics-type cultures, a cultural technique sometimes utilized for growing commercial plants. Submerged plants have evolved leaves with a high surface-to-volume ratio and relatively little waxy cuticle, a condition which facilitates nutrient uptake through the leaves. Since these plants often have well-developed root systems, they assimilate nutrients both from the soil and from the water.

Unpolluted vegetative cycle. The annual vegetative cycle in unpolluted lakes is as follows: Microscopic algae normally reach bloom proportions in

Analysis of water samples taken from core holes

Location	Depth to water table, ft	Nitrate nitrogen, ppm	Ammonium nitrogen, ppm	Organic carbon, ppm
Feedlot	30	8.6	5.1	130
Irrigated field	30	0.1	0.0	18
Feedlot	1.5	18	5.7	130
Irrigated field	10	31	0.1	12
Feedlot	12	21	5.8	90
Irrigated field	10	8.5	0.0	9
Feedlot	35	1.1	38	170
Irrigated field	35	18	0.4	26

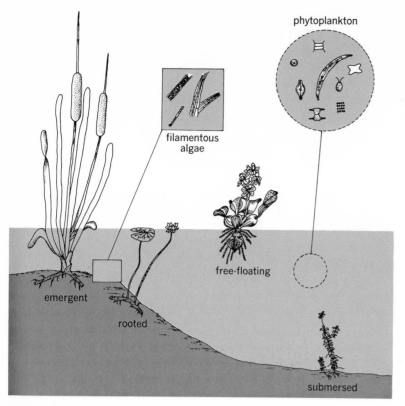

phytoplankton

filamentous
algae

free-floating

emergent

rooted

submersed

Fig. 2. Aquatic primary producers. Emergent plant: *Typha* (cattail); rooted plant: *Nymphaea* (water lily); free-floating plant: *Eichhornia* (water hyacinth); and submerged plant: *Elodea*.

the spring and deplete plant nutrients from surface waters for the remainder of the summer; filamentous algae grow at the shorelines in the spring and then stop growing and turn brown or yellow in midsummer; and large aquatic plants begin to grow in shallow water in April or May and reach peak standing crops in late summer.

Cycle in polluted lakes. Plant nutrients in surface waters are taken up by aquatic vegetation capable of absorbing nutrients from water, including microscopic and macroscopic algae, floating plants, and submerged plants. In nutrient-rich water, emergent plants can develop roots which protrude from the mud and extend into the water. Because they possess a greater surface-to-volume ratio than do the larger aquatic plants, algae are more effective in removing nutrients from the water. If the dissolved nutrients are present in large quantities, few kinds of algae occur and blooms of single species of algae may develop and cause the water to appear green or brown.

Submerged aquatic plants have evolved the capacity to grow in subdued light, but floating plants and algae in the water above them modify the light reaching the bottom. The light intensity is reduced, and the red and blue portions of the light spectrum, which are essential for photosynthesis, are selectively removed. When the nutrient supply to the water is large and continuous, algal blooms are sustained and the bottom plants die because of the reduction in light. After construction of ponds or lakes, surface scums of filamentous algae often develop as a consequence of nutrients being leached from the soil by the water. Ultimately, these excess nutrients will be taken up from the

water by the algae, the algae will settle to the bottom, and, if no new nutrients are added, no future explosive plant growths will occur.

Nitrogen and phosphorus effect on plants. Studies of algae in greenhouses and experimental ponds at Cornell University have verified the proposition that the more nitrogen and phosphorus added to an aquatic ecosystem, the denser the algal growths. In these laboratory studies, natural populations of algae increased in abundance at nitrogen and phosphorus nutrient levels up to and above those found in lakes or ponds.

Of four common submerged plants studied in replicated ponds, *Elodea canadensis* and *Potamogeton crispus* survived periods of heavy nitrogen and phosphorus fertilization and showed increased growth after fertilization had been curtailed, whereas *Myriophyllum exalbescens* and *Chara* were eliminated by the same treatments. When algae are removed from aquatic systems, as in laboratory studies by G. Gerloff and P. Krumholz carried out in 1965, aquatic plants studied responded to nitrogen and phosphorus fertilization by showing increased growth. However, in 1968 R. Wetzel and D. McGregor showed growth inhibition of another species of macrophyte at very low phosphorus concentrations. It is obvious that some submerged aquatic plants are stimulated by nitrogen and phosphorus nutrient additions, whereas the growth of others is inhibited.

If nitrogen and phosphorus were no longer added to lakes, the water quality would rapidly recover. In the summer, following the cessation of nitrogen and phosphorus nutrient additions, algal blooms would be much less frequent and less intense, and quantities of submerged aquatic plants would soon return to their prefertilization levels (Fig. 2). Initially, only the nitrogen and phosphorus nutrient-pollution-tolerant species of aquatic plants would grow, but eventually other submerged plants would return.

For background information *see* ALGAE; EUTHROPHICATION, CULTURAL; FRESH-WATER ECOSYSTEM; WATER MICROBIOLOGY; WATER POLLUTION.
[HUGH F. MULLIGAN]

Bibliography: G. Gerloff and P. Krumholz, *Limnol. Oceanog.*, 10:529–537, 1965; H. F. Mulligan, *Eutrophication: Causes, Consequences, Correctives*, National Academy of Sciences, 1969; H. F. Mulligan, *2d Annual Cornell Agricultural Waste Symposium*, Rochester, N.Y., 1970; H. F. Mulligan and A. Baranowski, *Verh. Intern. Ver. Limnol.*, 17:802–810, 1969; G. E. Smith, in *Agriculture and the Quality of Our Environment*, AAAS Publ. no. 85, 1967; B. A. Stewart et al., *Environmental Science and Technology*, 1:736–739, 1967; P. R. Stout and R. C. Burau, in *Agriculture and the Quality of Our Environment*, AAAS Publ. no. 85, 1967; R. Wetzel and D. McGregor, *Amer. Midland Nat.*, 80:52–54, 1968.

Water purification

Recent studies have shown the benefits of wastewater (sewage) effluent purification by the use of soil filtration. Conventionally treated sewage effluent can be put on land, where the nitrogen, phosphorus, organic carbon, and other substances in the effluent can be utilized by a crop or decomposed in the soil. This practice is ecologically more desirable than discharging such waste into streams, lakes, or other surface water where these substances will be pollution hazards. Moreover, as the portion of the waste water not used by the plants moves deeper into the soil, most of the organic and inorganic pollutants are removed, leaving reclaimed water. This water flows naturally underground to surface water, or it is collected by wells, drains, or other facilities for reuse (see illustration). In a number of cases, reclamation and reuse of the water are the main objectives of land application, using soil filtration as a treatment process. Settled effluents from chicken-processing plants, canneries, and so on can also be treated in this way. *See* HYDROLOGY.

Application of effluent. The waste water can be applied to the soil with sprinkler systems, furrows, borders, or basins, depending on soil and topography. Soils with a high salt content or soils that may release undesirable elements should be avoided. If the infiltration rate of the soil is low, relatively large areas will be needed, and it will probably be economical to combine the waste disposal fields with agricultural utilization of the land. Much less land is needed for soils with high infiltration rates. In this case it may be desirable to use a certain area for disposal or reclamation of waste water only, with no agricultural utilization of the land. Loading rates vary from a few centimeters to a few meters per week, depending on the permeability of the soil and whether the land is used for crop production.

Infiltration periods must be alternated with dry-up periods to allow oxygen to enter the soil and to restore the infiltration rate, which usually decreases during water application because of clogging of the surface soil. The desired lengths of the infiltration and dry-up periods depend on the oxygen demand of the waste water, the permeability of the soil, and the depth of the groundwater. Application schedules commonly in use range from a few hours of infiltration each day to infiltration and dry-up periods of several weeks each.

As the effluent moves through the soil, the biodegradable material is metabolized by the soil microorganisms. Pathogenic and coliform bacteria, being in a hostile and competitive environment, generally do not survive after a few meters of travel. Viruses are adsorbed to the soil particles. Phosphates precipitate in the soil as calcium phosphate complexes (apatites), or they are fixed by iron and aluminum oxides. These oxides also immobilize boron. Fluorides may be adsorbed by clay minerals, or they may be precipitated as calcium fluoride or as fluorapatites. Metallic ions are adsorbed on the clay minerals, although chelating effects by the organic fraction may cause a certain mobility of the heavy metals. Nitrogen removal processes are (1) uptake by plant roots, (2) fixation in microbial tissue, (3) denitrification (which forms free nitrogen gas that escapes to the atmosphere), (4) volatilization of ammonia, (5) adsorption of ammonium on clay and organic matter, and (6) fixation of ammonia by organic matter.

Unless the total nitrogen load on a disposal field is not much higher than that which the crop can absorb, significant nitrogen removal rates can be obtained only by careful scheduling of the infiltration and dry-up periods. In the infiltration

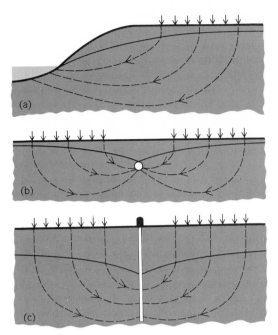

Schematic of waste-water renovation. Infiltration areas (arrows) with reclaimed water (*a*) draining to a stream or lake, (*b*) collected by drains, and (*c*) pumped by wells. Streamlines are shown as dashed lines, the groundwater table as a solid line.

basins west of Phoenix, Ariz., inundation and dry-up periods of 2–4 days each gave no nitrogen removal, but inundation and dry-up periods of 10–20 days yielded about 80% removal of the nitrogen after the passage of reclaimed water with a high nitrate concentration. This "nitrate peak" is caused by nitrate formed during a dry-up by oxidation of ammonium in the unsaturated portion of the soil and the subsequent flushing out of this nitrate by the water entering the soil with a new inundation period. Nitrogen removal decreased after more than 20 days of inundation and completely ceased after 100 days of continuous inundation.

Most of the end products of the waste-water renovation processes in the soil are water-soluble or gaseous. Phosphorus is almost the only element that is fixed or precipitated in quantity. The useful life of the system may thus be eventually limited, but only after decades or even centuries. Periodically, the infiltration fields may be taken out of the system to grow crops. This would utilize nutrients that may have accumulated in the soil and at the same time rejuvenate the land for waste-water reclamation.

System design. The design of a land treatment system for renovating waste water should generally be based on (1) avoiding excessively high groundwater mounds beneath the infiltration areas, (2) locating the wells or drains for collecting the reclaimed water far enough away from the infiltration areas to obtain sufficient time and distance of underground travel, and (3) avoiding lateral movement of the reclaimed water into the aquifer outside the system of infiltration areas and wells or drains. The last is important if the quality of the reclaimed water is inferior to that of the native groundwater and if contamination should be minimized. The desired time and distance of underground travel are of the order of at least several weeks and at least 50–100 m, respectively, depending on water quality and soil materials.

Soil filtration is essentially an advanced treatment process for waste water, which under favorable hydrogeologic conditions yields reclaimed water at considerably less cost than equivalent in-plant tertiary treatment. In view of the current interest in keeping waste water out of surface waters, and of the inability of conventional primary and secondary treatment to remove significant amounts of nitrogen and phosphorus and all organic carbon, land treatment of sewage effluent and similar liquid waste may well play an increasing role in the efforts to control water pollution and to reuse waste water.

For background information see HYDROLOGY; SEWAGE; WATER CONSERVATION in the McGraw-Hill Encyclopedia of Science and Technology.

[HERMAN BOUWER]

Bibliography: H. Bouwer, J. Sanit., Eng. Div., Amer. Soc. Civil Eng., 96SA1:59, 1970; H. Bouwer, Proceedings of the Ground Water Recharge Conference, Water Research Association, Medmenham, England, 1970; P. H. McGauhey and R. B. Krone, SERL Rep., no. 67-11, University of California, Berkeley, 1967; J. C. Romero, Ground Water, 8:37, 1970.

Wood products

The first fuel used by primitive man was undoubtedly wood, and its metamorphosis through the ages has supplied much of the fossil fuels of today. Although coal and oil are now in plentiful supply, the rapid increase in their consumption will sooner or later deplete the reserves and it will again be necessary to revert to wood as a source of fuel. Concurrently, a rapid increase in population is depleting the food supply, and resource must eventually be made to that portion of the total plant material which currently is unused.

The annual yield of plant matter is about 15×10^{10} to 20×10^{10} tons. The current food consumption is estimated at 5×10^8 tons, a mere 0.3 of 1% of the plant substance produced. Thus a vast amount of material is available for exploitation. However, the technology for its economic utilization is lacking and must be developed. Roughly one-half of all plant substance is cellulose, an insoluble polymer of glucose. Cellulose itself cannot be used as food by most animals, ruminants being the exception. But glucose is a sugar that nearly all living organisms can use. The problem, then, is to convert cellulose into glucose by depolymerization.

Abundant as cellulose is as a primary product of nature, it is the waste cellulose, the unwanted by-product or the no longer useful end item, that offers the most immediate promise. Cellulose is a major component of (1) agricultural wastes (straw, stubble, leaves and stalks of many plants, rice and other hulls, peanut, almond, and other shells, corn cobs, bagasse, and so on); (2) food-processing wastes (fruit peels, pulp, pomace, vegetable trimmings, and so on); (3) wood wastes (brush, chips, bark, sawdust, paper-mill fines, and the like); and (4) municipal wastes (40–60% of solid wastes chiefly as garbage and waste paper). The disposal of these wastes by composting, land fill, or burning is expensive and greatly increases the incentive to use cellulose as a novel energy source. The availability of cellulosic wastes ensures abundant cheap substrates for any process that is developed.

Although cellulose has long been recognized as a potential source of food, the financial support required for investigating its conversion has never been great. The wealthy nations were not short of food but lacked vision; the poor nations were hungry but lacked money. Now that pollution is a problem of the wealthy nations, funding for the disposal of waste cellulose may be expected. Since waste disposal is recognized as being costly, conversion of cellulose to glucose need not show a direct profit. It is necessary only that conversion to glucose be less costly than other means available for the disposal of cellulose. Future thinking must be directed less to making a profit and more toward minimizing the cost to government of disposing of cellulosic wastes and toward the obligation of all men to see that their less affluent fellow beings do not go hungry.

The conversion of cellulose to glucose is a hydrolytic reaction that can be catalyzed by acid or by enzymes (cellulases) produced by microorganisms capable of digesting cellulose.

Acid hydrolysis of cellulose. Many attempts have been made to put acid hydrolysis on an economically feasible basis. Wood wastes have been used extensively, since they are available in large quantities. However, they have the disadvantage of containing lignins, phenols, and polysaccharides other than cellulose. Softwoods are preferable to hardwoods because of their greater glucose content.

The first commercial application of the acid hydrolysis method was in 1913 at Georgetown, S.C., where a plant was built to hydrolyze southern pine mill waste in a rotary steam-heated digester with 2% sulfuric acid at 175°C. The dilute sugar solutions produced (25% yield) were fermented to

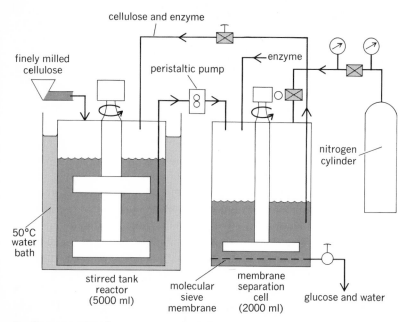

Continuous system for cellulose saccharification using enzymes. (From T. K. Ghose and J. A. Kostick, Model for continuous enzymatic saccharification of cellulose, Biotechnol. Bioeng., vol. 12, in press)

5000 gal of ethyl alchol per day. This plant and a second one at Fullerton, La., operated until 1923, when a decline in the price of sugar-rich black-strap molasses (which could also be used for alcohol production) made them unprofitable.

In the Scholler process, which was used during World War II, wood was pretreated with 1% HCl and then digested with 0.5% sulfuric acid at 130–190°C for 18–24 hr in stationary digesters. Sugar yield was 40–50%. In the Bergius process concentrated HCl was used in special acid-resistant equipment. It gave cleaner syrup but was expensive because of the need to recover the acid.

Although none of these processes has been a commercial success since World War II, it should be noted that the methods do work and that they can be used again when the economy of the times requires them.

Enzymatic hydrolysis of cellulose. Until recently, enzymatic hydrolysis of cellulose was too slow and too inefficient to compete either with acid hydrolysis or with the successful conversion of starch to glucose by fungal enzyme. Current developments have changed the outlook. The important factors are the importance of pretreatment of the cellulose and the development of much more active enzyme preparations. Fine grinding of cellulose greatly increases its susceptibility to hydrolysis by enzymes. When cellulose pulp is heated and milled, a product is formed from which high-density suspensions can be prepared. This is important to processes in which solutions of high glucose content are desired, since the cost of evaporating more dilute solutions can be a major economic factor.

The enzyme system from the fungus *Trichoderma viride* has great promise. This preparation can completely degrade even the most resistant forms of cellulose, such as crystalline cellulose. In laboratory experiments, treatment of heated and milled cellulose with a concentrated *T. viride* cellulase has given a glucose syrup of 30% concentration in 4 days. On a larger scale, 14% glucose solutions were obtained in 50 hr.

A continuous system has been conceived in which (1) the enzyme and cellulose are retained in the reaction vessel, (2) the products are constantly being removed, and (3) water, makeup enzyme, and solid cellulose are constantly added to maintain the steady-state condition (see illustration). Finely milled cellulose is hydrolyzed by enzyme in a reactor. The reaction mixture is transferred by means of a peristaltic pump to an ultrafiltration cell. Glucose and water are forced through the molecular sieve membrane by pressure from a nitrogen cylinder. Cellulose and enzyme are retained by the membrane and are returned to the reactor, where additional cellulose and water may be added. The ability to reuse the enzyme is an important advance resulting from this concept. Preliminary cost estimates based on a 100 ton/day plant are encouraging.

Once the conversion of cellulose to glucose is commercially successful, the production of derived foods and fuels will gain momentum. The technology for conversion of glucose to protein and fat (by yeasts) and of glucose to ethanol and methane is already well advanced. It is inevitable that these processes will become interwoven with the diverse problems of pollution, feeding, and fuel development. Only the question of when these developments will take place remains.

For background information *see* FUNGI; INDUSTRIAL MICROBIOLOGY; WOOD PRODUCTS in the McGraw-Hill Encyclopedia of Science and Technology.　　　　　　　　　　[ELWYN T. REESE]

Bibliography: W. D. Gray, *Advan. Chem. Ser.*, no. 57, p. 261, 1966; G. Hajny and E. Reese (eds.), *Cellulases and Their Applications, Advan. Chem. Ser.*, no. 95, 1969.

X-ray astronomy

Although x-ray astronomy is very young, it is already well established as an important branch of astronomy, capable of yielding extremely significant information regarding the solution of

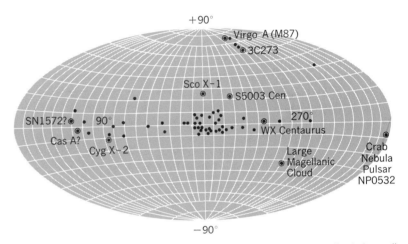

Fig. 1. Distribution of the cosmic x-ray sources plotted in the galactic (l^{11}, b^{11}) coordinate system. Most of the sources lie close to the equator and are in the Milky Way Galaxy. Two sources, Virgo A and the Large Magellanic Cloud, have been positively identified with other galaxies. Encircled sources have been optically identified.

Fig. 2. X-ray telescope to study solar x-ray phenomena from Skylab. The x-rays enter the narrow annulus in the front and are focused by glancing-incidence fused-silica mirrors. (*NASA*)

Cosmic x-ray sources known as of April, 1970

Constellation	Source	Right ascension (1950.0)	Declination, degrees	l^{II}, degrees	b^{II}, degrees	Flux, photons/cm² 2–10 kev	Flux, photons/cm² 20–35 kev	Remarks
Aquila	Aq1 X-1	19h12m	0	036	−05	0.13		
	GX+48.7	19h18m	+13.9	049	00±15	0.19 (2–5 kev)		
Ara	Ara X-1	17h5m	−45.9	340	−03			
	GX341-6	17h16m	−48.7	341	−06	2.8 (1.5–6.5)		
	GX350-2	16h52m	−46.2	340	−03	0.3 (4–8)		
	Lockheed 3	16h49m	−45.3	341	−01	0.1 (4–8)		Possibly Ara X-1
Cassiopeia	Cas X-1	23h21m	58.5	112	−02	0.3	<0.10	Cas A?
	Cas B?	0h23m	63.9					Unconfirmed
Centaurus	Cen X-1	14h28m	−63	314	−03	0.17		
	Cen X-2	13h24m	−62	307	+02	Variable (nova)	0.06	WX Centaurus
	Cen X-3	11h23m	−59	292	+02	0.7		
	Cen X-4	14h56m	−32.3	332	+23	Variable (nova)		S5003
	Cen A	13h22m	−42.8			0.039		Cen A
Cepheus	Cep X-1	0h15m	66	120	+04	0.3	<0.01	SN1572?
	Cep X-2	22h42m	62	109	+03	0.18		
	Cep X-3	23h54m	72.9	119	+11	0.15		
Cetus		02h00m	−06	164	−60	1.5 (1.5–6)		
Cygnus	Cyg X-1	19h53m	34.5					
		19h57m	34.5					
		19h53m	34.6	071	+03	1.1	0.086–0.21	Variable
		19h58m	35.1					
		19h56m34s	35.6					
	Cyg X-2	21h43m	38.8	087	−11	0.8	<0.06	Optically identified 15th magnitude spectroscopic binary
	Cyg X-3	20h30m52s	40.9	080	+0.7	0.4	<0.02	
	Cyg X-4	21h9m11s	38°33′	082.9	−6.4	0.3		Cyg A?
Dorado		05h20m	−69	280	−33	0.2 (1.5–10.5)	<0.02 (20–52)	Large Magellanic Cloud. Source extended to 12° in direction of source
Lacerta	Lac X-1	22h34m	+53.8	92 to 107	−7 to −20	0.17		
Leo	Leo X-1	9h35m	+8.6	226	+41	0.2		
Lupus	Lup X-1	15h2m	−52			0.2	<0.026	
		14h59m	−52	323	+05	1.6	<0.03	
Lyra	Lyr X-1	18h18m	+36	063	+21	0.17		
Norma	Nor X-1	16h24m	−51	334	−02	0.8	0.019	Lockheed 1?
	Nor X-2	15h38m	−57	324	−02	0.4	0.03	
Ophiuchus	Oph X-1	17h32m	−20.7	006	+06	1.3	<0.0020	SN 1604
	Oph X-2	17h14m	−23.5	001	+08	0.5		
	GX9+9	17h30m	−16.5	009	+09	0.41 0.25 (1.5–6)		
Sagittarius	Sgr X-1	17h55m	−29.2	001	−03	1.6	0.15 (20–40)	
	Sgr X-2	18h10m	−17.1	013	−01			
		18h11m	−17.2					
	Sgr X-3	17h56m	−21.6	008	+01			
		18h03m	−20.7					
	Sgr X-4	18h26m	−31	003	−09	0.6		
	Sgr X-5	18h02m	−24.9					
Sagittarius, Scorpius	GX3+1	17h43.4m	−26.8	002.5	+01.3	0.68	0.18	
	GX5−1	17h58.6m	−25	005.2	−01.1	1.4	<0.02	
	GX9+1	17h59.7m	−20.5	009	+01	2.8		
	GX17+2	18h12.7m	−13°48′	016.6	+01.5	2.0		
	GX9+9	17h30.2m	−16°36′	008.9	+08.8			
	GX354-5	17h47m	−36	354	−05	0.11 (2–5)	0.25	
	GX(+16.7)	18h09m	−12.5	017	+04 ±03			
	GX(+13.5)	17h27m	−19.1	013	−04 ±08			
	GX(+9.1)	18h02m	−21	009	+00	0.95		
	GX(+5.2)	17h58m	−25	005	−01	0.77		
	GX(+2.6)	17h50m	−27	003	00 ±05			
	GX(−2.5)	17h25m	−31.5	357	00 ±10	0.40		
	GX(−5.6)	17h30m	−33.5	354	00 ±10			
	GX(−10.7)	17h10m	−37.8	349	00 ±05			
	GX(−12.9)	16h12m	−29.8	347	>+10	0.33		
	GX(−14.1)	17h0m	−40.5	346	00 ±10	0.1		
	GX13+1	18h10m	−17.1	013.3	+00.7	1.5		

Cosmic x-ray sources known as of April, 1970 (cont.)

Constellation	Source	Right ascension (1950.0)	Declination, degrees	l^{11}, degrees	b^{11}, degrees	Flux, photons/cm² 2 – 10 kev	Flux, photons/cm² 20 – 35 kev	Remarks
Scorpius	Sco X-1	$16^h17.07^m$	−15.53	359	+24	18.7 – 67	0.066 – 0.26	Optically identified with 12th magnitude irregular variable
	Sco X-2	17^h8^m	−36.4	350	+01	1.1		
	Sco X-3	17^h23^m	−44.3	345	−06	1.1		
	Sco X-4	16^h25^m	−40	342	+06	0.8	<0.016 (27 – 77)	
	Sco X-5	17^h37^m	−40.4	350	−05	0.7		
	Sco X-6	18^h7^m	−35.6	357	−08	0.3		
	Lockheed 5	16^h51^m	−40.8	344	+02	0.11 (2 – 5)		
	Lockheed 6	17^h09^m	−37.7	349	+01	2.6		
	Lockheed 7	16^h42^m	−31.9	350	+09	0.15 (4 – 8)		
	Lockheed 8	17^h01^m	−30.5	354	+07	0.15 (4 – 8)		
	Lockheed 9	17^h36^m	−35.2	354	−02			
	Lockheed 10	17^h45^m	−33.7	356	−03	0.15 (4 – 8)	0.085	
	Lockheed 12	17^h49^m	−31.5	359	−03	>0.35 (4 – 8)		
	Lockheed 14	17^h49^m	−28.2	001	−01	0.40 (1.5 – 6)		
	Lockheed 16	18^h10^m	−27.6	004	−05	<0.6 (4 – 8)		
	Lockheed 17	18^h01^m	−25.6	005	−01			
	Lockheed 18	18^h07^m	−22.3	009	−01	0.3 (4 – 8)		
	Lockheed 19	18^h02^m	−21	009	+00	1.28		
	Lockheed 20	18^h15^m	−18	013	−01	0.43		
Serpens	Ser X-1	18^h45^m	05.3	037	+02	0.7	<0.045	
	Ser X-2	18^h10^m	−12.9	016	+00			
	Lockheed 21	18^h16^m	−14.8	016	+00	1.6		
	GX+36.3	18^h23^m	+07	036	09±08	0.15 (2 – 5)	<0.04	
Taurus	Tau X-1	$5^h31^m30^s$	21°59.1′	185	−06	2.7	0.10	Pulsar NP0532
Vela	Vel X-1	8^h52^m	−44	263	+03	0.5		
	GX263+3	8^h57^m	−41.3	263	+03	<0.4		
Virgo	Vir X-1	12^h28^m	12.7	284	+75	0.2		Virgo A (M-87)
		12^h30^m	+04.8	291	+67			
		12^h27^m	+02.4	290	+65	0.02 (0.044)		3C273
		12^h31^m	+07.0	290	+69			
		12^h32^m	+10.1	289	+72			
Vulpecula	Vul X-1	20^h38^m	29	071	−08	0.15		

cosmological problems. Numerous galactic and extragalactic x-ray sources have already been observed and found to have totally unexpected properties. As of April, 1970, at least 60 discrete celestial objects, most of which are galactic in origin, have been found to be emitting measurable fluxes of x-rays in the 1 – 500-kev region (see table).

Some of these sources have been identified with starlike objects, and others with supernova remnants; but all radiate enormous amounts of power in the x-ray region. The vast majority of the sources discovered lie within the Milky Way Galaxy and radiate approximately $10^{30} – 10^{31}$ watts each (Fig. 1). This is comparable to the power output in the visible light region of a star of high intrinsic brightness and is 10^{14} times the x-ray power output of the quiet Sun. Permeating all space there is an isotropic background of diffuse x-ray emissions. This cosmic x-ray background may be of extragalactic origin and could have profound cosmological implications. One source has been reliably identified with another galaxy, Virgo A (M-87). This is an astonishing result, since any extragalactic object observed with detectors currently available must radiate more power in the x-ray region than a normal galaxy emits in the visible light region (>10^{37} watts). Two x-ray sources have appeared suddenly in the constellation Centaurus, grown to great brilliance, and then decayed, thus exhibiting a nova behavior. Another significant development of x-ray astronomy was the recent discovery of an x-ray pulsar in the Crab Nebula. Observations of this object have given great insight into the nature of cosmic rays and the long-sought "neutron stars."

INSTRUMENTATION

Several types of instrumentation have been developed to study x-ray sources. These include x-ray telescopes, spectrometers, and polarimeters.

X-ray telescope. The glancing-incidence x-ray telescope is one of the most important developments in instrumentation for x-ray astronomy. X-ray telescopes using confocal paraboloidal-hyperboloidal mirrors have been fabricated and flown on Aerobee rockets to obtain photographs of the Sun in the soft x-ray region. Two glancing-incidence x-ray telescopes with a spatial resolution of a few seconds of arc will soon be orbited by NASA in Skylab to study solar x-ray phenomena (Fig. 2). Although glancing-incidence x-ray telescopes have been used primarily for solar studies, they permit measurements which differ qualitatively from those obtainable with mechanical collimators, and will certainly play an important role in future investigations of cosmic sources. The high spatial resolution inherent in glancing-incidence telescopes makes it possible to examine the structure of extended sources and to precisely establish the source position. The large collecting area obtainable permits the use of small, well-shielded detectors which minimize background problems and produce a superior signal-to-noise ratio. The high x-ray fluxes produced by focusing x-ray telescopes facilitate spectrometry and polarimetry.

Spectrometry. Information regarding the spectra of cosmic x-ray sources has thus far been obtained with low-resolution nondispersive techniques (that is, proportional counters). Focusing x-ray telescopes now make it possible to study cosmic sources with high spectral resolution devices, such as Bragg crystal spectrometers, which have already provided valuable data on the solar x-ray spectrum. The Bragg crystal spectrometer works best in the study of line emission. The observation of line emission from highly stripped atoms, such as Fe XXV or Fe XXVI, would be proof of a thermal mechanism of x-ray emission in the source.

The "slitless" spectrometer, which consists of a transmission diffraction grating placed in front of an x-ray telescope, may also prove to be a valuable tool for cosmic x-ray investigations. This device permits an entire spectrum to be recorded simultaneously and should be of great value in studies of emission lines and absorption edges from the interstellar medium.

Polarimetry. The amount of polarization of the emitted x-rays provides a valuable clue to the emission mechanism. Polarized radiation strongly indicates that the emission arises from a synchrotron mechanism or a bremsstrahlung source with highly nonisotropic exciting electrons. An x-ray polarimeter, based on the assymetry of Thompson scattering induced by a polarized flux, has been built by the Columbia Radiation Laboratory. The device consists of lithium hydride blocks as scatterers, surrounded by banks of proportional counters.

OBSERVATIONAL RESULTS

Significant observations have been obtained in studies of Sco X-1, x-ray novae, x-ray galaxies, the diffuse x-ray background in space, and x-ray pulsars.

Sco X-1. The first cosmic x-ray source was discovered in the constellation Scorpius and was designated Sco X-1. It has been identified with a blue starlike variable of the twelfth magnitude. Simultaneous observations of the x-ray and visible light emissions of this object have shown the emissions to vary sporadically, generally increasing and decreasing together. The object becomes bluer as the x-ray emission increases. In visible light its magnitude varies from 12.2 to 13.2, often as rapidly as 0.3 magnitudes per hour, and it flickers near maximum brightness. The flickers frequently appear in triplets approximately 10 min apart and are roughly constant in duration and intensity. Radio observations of Sco X-1 in 1968 showed it to vary by more than 3 stellar magnitudes in 2 hr. Although related short-term x-ray variations have not been confirmed, there is evidence for long-term variability. The x-ray emission of Sco X-1 has been observed to change in intensity by a factor of two in a month. In 1968 a flare which consisted of comparatively soft x-rays with energies less than 4 kev rose to a maximum of about four times the average intensity of Sco X-1 in 10 min and then decreased in the following 20 min. The x-ray emission of Sco X-1 has been examined for pulsations of frequencies from 50 msec to several seconds with only negative results. An Aerobee rocket equipped with a Thompson scattering x-ray polarimeter was flown in July, 1968, to observe the degree of polarization of x-rays from Sco X-1. A null result was obtained, which indicates that the degree of polarization must be less than 5%.

X-ray novae. Two remarkable x-ray objects have suddenly appeared, grown to great brilliance, and then decayed, thus exhibiting a behavior very similar to that observed in a typical nova outburst. In April, 1967, three Skylark rockets launched from Woomera, South Australia, discovered a source in the constellation Centaurus (Cen X-2) which was not detectable in an October, 1965, sky survey. On Apr. 10, 1967, Cen X-2 was more intense in the 2–5 kev range than Sco X-1 (the brightest x-ray

source), which was observed on the same flight.

Six weeks later the source was found to have decreased in brightness by a factor of six. By September, 1967, it was undetectable, which indicated that its intensity was at least 100 times less than at maximum. Cen X-2 reappeared in the 2–20 kev range in November, 1968, and may give recurring x-ray outbursts in a manner similar to those of the well-known recurrent novae seen in visible light. Recently, Cen X-2 has been identified with the variable star WX Centaurus, which has an emission line spectrum similar to that of Sco X-1. Another x-ray nova (Cen X-4) appeared on the Centaurus-Lupus boundary between July 6 and 9, 1969. After its sudden appearance, Cen X-4 was observed repeatedly in the 3- to 12-kev region by instruments on two Vela satellites and a rocket experiment. Cen X-4 rapidly brightened to twice the intensity of Sco X-1 and then declined in a fluctuating manner to slightly less than half its maximum by mid-August. At this point, the rate of decline increased, and by Sept. 24, 1969, the source was no longer visible. The time variation in the intensity of x-rays from Cen X-4 shows an astounding similarity to the visible light emission of novae of the DQ Herculis type. It is interesting that the light curves of these novae reach a deep minimum and then increase to a secondary maximum followed by a long decay. This secondary maximum has apparently occurred in Cen X-2 and should be observed if it occurs in Cen X-4.

X-ray galaxies. Much effort has been devoted recently to the search for x-rays from the strongest radio galaxies (Cygnus A, Virgo A (M-87), and Centaurus A) and quasars. The observations have revealed that Virgo A is emitting about 100 times as much energy in the x-ray region as in the radio region; whereas any x-ray emission from Cygnus A must be less than 1/10 of its radio output. The spectrum of Virgo A has been observed in the radio, infrared, visible, x-ray, and gamma-ray regions. Over this vast range the spectrum is consistent with that produced by synchrotron emission from relativistic electrons.

The detection of x-rays from Centaurus A has been reported by the University of California, Berkeley, group. Their results indicate that Centaurus A radiates approximately twice as much power in the x-ray (1–12 A) region as in the radio (0.3 mm to 30 m) region.

The Naval Research Laboratory group reported a flux of soft x-rays from the quasar 3C273, believed to be 2×10^9 light-years away. This astonishing result has been confirmed by the University of California, Berkeley, group. The power output of 3C273 in the soft x-ray region must be of the order of 10^{39} watts.

The nearest galaxy, the Large Magellanic Cloud, has been identified as an extended source of low-energy x-rays (1.5–10.5 kev). The x-ray emission extends over 12° and is consistent with the visible character of the cloud. The observed flux indicates that the relative population of x-ray stars in the Large Magellanic Cloud is approximately the same as the Milky Way Galaxy.

Diffuse background. An isotropic diffuse x-ray background exists which appears to permeate all space. Its spectrum above 1 kev suggests a nonthermal origin. This emission possibly results from inverse Compton scattering of relativistic electrons with cosmic blackbody photons in extremely distant radio sources (red shifts of the order of five). However, evidence obtained in 1969 revealed an excess of radiation in the 1.4- to 18-kev region in the galactic plane in the Vela-Carina region. One possible mechanism for the galactic origin of the diffuse x-ray emissions arises from collisions of 2- to 20-mev cosmic-ray protons with ambient electrons in the spiral arms of the galaxy.

At low energies (<1 kev), where galactic absorption plays an important role, the diffuse background is more intense at high galactic latitudes. This strongly suggests an extragalactic origin. This portion of the background possibly results from free-free emissions in an intergalactic gas of density ~10 particles/m^3, with a temperature of 3×10^5 to 8×10^5°K. The existence of an intergalactic gas of this density is of great importance to a number of cosmological models which require the universe to have a density of this order. Observations of the diffuse background at low energies indicate that no more than 2% of the intergalactic gas can be composed of the heavier elements. *See* INTERSTELLAR MATTER.

X-ray pulsars. One of the most remarkable developments of x-ray astronomy was the Mar. 13, 1969, discovery of an x-ray pulsar in the Crab Nebula. Pulsars, celestial objects which emit vast amounts of power in brief intense bursts with extremely precise periodicity, were first observed in the radio region. The Crab pulsar (designated NP 0532) has also been observed in the visible light, x-ray, and gamma-ray regions, and the pulses in the various spectral regions coincide not only in period but also in time of origin. The periods of the pulsars increase with time. The increase is observed to be the greatest in pulsars with the shortest periods. The Crab pulsar has the shortest period (33 msec) and the greatest increase rate (40 nsec/day). This behavior suggests that pulsars are rapidly rotating, highly magnetic neutron stars. A neutron star would come into existence when the interior of an old massive star collapsed, packing the mass of the Sun into a small sphere of ≈ 10-km diameter (density ≈ 10^{14} gm/cm^3). To conserve angular momentum, the object must spin faster as its diameter decreases. T. Gold has calculated that at the moment of formation, in A.D. 1054, the Crab pulsar must have been spinning at 50 rps. The period of the pulsations indicates that it is now spinning at 30 rps. The power output of the Crab pulsar (10^{31} watts) agrees well with the rotational energy available, based on the theoretical mass and moment of inertia of the neutron star and the observed slowdown rate. Furthermore, the prodigious rate of energy dissipation of NP 0532 is equal to, and may account for, the total radiant energy output of the vast Crab Nebula at all frequencies.

Most of the x-ray emission of the Crab Nebula may result from relativistic particles emitted by the pulsar interacting with intense magnetic fields associated with the wisp structures. In visible light, the wisps ripple outward from the region of the pulsar every few months at expansion velocities 20–25 times greater than those of the Nebula. Following the abnormal increase in period (spin-up) of NP 0532 which occurred on Sept. 28, 1969, the wisps have exhibited definite changes in shape

and brightness, which indicates an intimate interrelationship between wisp behavior and energy releases from the pulsar. The theoretician G. Burbidge recently suggested that all discrete galactic x-ray sources may be rotating neutron stars embedded in small dense gas clouds. The rotational energy available is sufficient to maintain the x-ray source, and the highly directional pulsations would be seen only from stars in the proper orientations. Furthermore, the energy and plasma contributions from a number of pulsars may account for the cosmic rays which continuously bombard the Earth, thereby solving a decades old mystery.

For background information *see* PULSAR; ROCKET ASTRONOMY; X-RAY ASTRONOMY; X-RAY STAR; X-RAY TELESCOPE in the McGraw-Hill Encyclopedia of Science and Technology.

[RICHARD B. HOOVER]

Bibliography: K. A. Pounds, *Sci. J.*, vol. 6, no. 4, April, 1970; *Proc. Roy. Soc. London Ser. A*, 313: 299–402, 1969; *Pulsating Stars*, Nature reprints, vols. 1 and 2, 1968, 1969; S. Tilson, *IEEE Spectrum*, vol. 7, no. 2, February, 1970.

Xylem

Two approaches to the study of xylem differentiation have been particularly fruitful in recent years. The first of these has been the investigation of factors necessary for the initiation of xylem differentiation, and the second approach has been to examine the structural and biochemical changes that accompany the formation of xylem elements. There is good evidence that cytokinins are necessary for xylem differentiation, in addition to auxin and sucrose. Also, the evidence strongly suggests that a cell must divide before one or both of its daughter cells can be stimulated to differentiate by hormonal and other environmental factors. At the structural level, permanganate has been shown to act as an electron-dense stain for

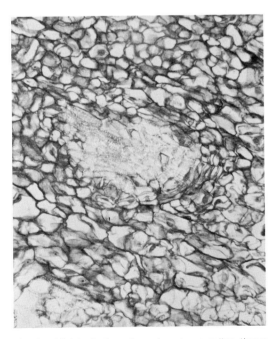

Fig. 1. Histological section of soybean callus tissue showing a nest of tracheary elements surrounded by a cambium-like layer of dividing cells.

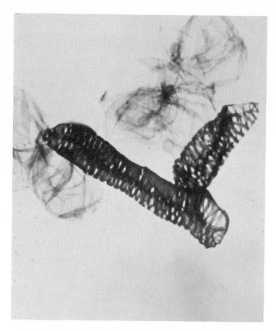

Fig. 2. Two tracheary elements with well-defined perforation plates obtained from a maceration of soybean callus tissue.

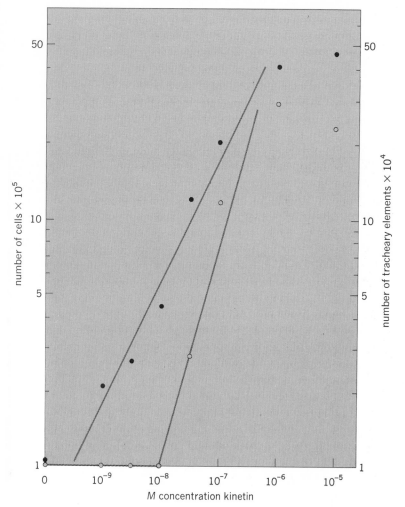

Fig. 3. The effect of different kinetin levels upon cell production and xylem differentiation in soybean callus tissue. Closed circles represent total cells; open circles, tracheary elements.

the lignin component of the cell wall of tracheary elements. By means of permanganate staining, lignification in *Coleus* was found to begin in the middle lamella and primary cell wall as soon as secondary cell wall deposition began in stem tissues. Associated with these structural changes was the activation or synthesis of the key enzyme in the biosynthetic pathway of the lignin monomers, phenylalanine ammonia lyase.

Role of auxin. It has been known since 1908 that wounding higher plant tissues triggers a chain of events that leads to wound xylem differentiation, if the trauma is sufficiently great to sever or disrupt a vascular strand. As a result of the work of William P. Jacobs and coworkers, it is known that auxin acts physiologically as the principal factor limiting, or controlling, the appearance of wound xylem elements. Young expanding leaves were shown to be the main sources of auxin for stem tissues. In recent years aphids have been used to tap the phloem selectively and to obtain its translocated materials for analysis. Through the use of such techniques, auxin has been shown to be translocated from leaves into the stem by means of the sieve tubes. Thus it appears quite likely that the most significant event brought about by wounding is the flow of auxin from the severed sieve tubes into the surrounding parenchymatous tissues.

Since the now classic work of Jacobs demonstrating the central role of auxin in the control of xylogenesis, attention has turned to the definition of chemical factors which may be required for xylem differentiation, even though they may be present in abundance in the intact or wounded plant and thus not actually limiting for differentiation. These more recent studies have been conducted with excised plant parts in axenic culture and with cultured callus tissues. The use of such materials makes it possible to manipulate the hormonal and nutritional milieu of cells so that the significance of a given substance to xylem differentiation may be evaluated.

Role of cytokinin. If the cotyledons of soybeans are surface-sterilized and placed in culture on a chemically defined medium containing inorganic ions, sucrose, vitamins, an auxin, and a cytokinin, a callus tissue is formed within a few weeks. Although this callus tissue will continue to grow indefinitely, if subcultured to fresh medium periodically, cell proliferation is absolutely dependent upon the presence of cytokinin and, to a lesser extent, upon the presence of auxin in the medium. If the tissue is transferred to a medium lacking cytokinin, growth stops completely within a few days. When the soybean callus tissue is grown in the dark with 5×10^{-7} M kinetin, 8–12% of the cells differentiate as xylem elements. Since these tracheary elements usually form perforation plates, most of them appear to be modified vessel elements (Figs. 1 and 2).

Because the soybean callus tissue could not produce its own cytokinin, it was possible to determine whether cytokinin played a role in xylem differentiation, independent of its role in cell proliferation. When the callus was grown on media containing various concentrations of cytokinin, increasing levels of cytokinin proved to result both in greater cell proliferation and additional xylem differentiation (Fig. 3). However, cell proliferation was stimulated over a much broader range of cytokinin concentration than was xylem differentiation. As a result, increasing levels of cytokinin actually raised the percentage of cells which differentiated as xylem elements. Furthermore, low levels of cytokinin (below 10^{-8} M) were completely inadequate for the initiation of xylem differentiation, although some cell proliferation continued. Because of this differential sensitivity to cytokinin, a tracheary element – free population of cells was produced when the callus was subcultured several times on a medium containing 10^{-8} M kinetin. Xylem differentiation resumed when the tracheid-free tissue was placed on a medium containing a higher level of cytokinin.

Additional evidence that a cytokinin is necessary for xylem differentiation came from the study of the behavior in culture of pea root segments. A 1-mm thick segment of a seedling pea root taken from the region 10–11 mm from the root tip can be placed in culture on a complex but chemically defined medium. When the medium contained auxin but no cytokinin, the diploid cells of the pericycle entered into division and their descendants ultimately produced a callus tissue. Xylem elements were not formed from the pericycle derivatives or from other tissues of the pea root segment under these conditions. By isolating the root segments on a medium containing cytokinin in addition to auxin, the cortical cells were stimulated to divide. When these cortical cells entered into mitosis, beginning on the third day of culture, they were found to be polyploid. After one or two divisions, many of the cortical derivatives were found to differentiate as tracheary elements. After 10 days in culture, almost 16% of the total number of new cells differentiated as xylem elements. As was the case with the soybean callus, increasing levels of cytokinin stimulated both cortical cell proliferation and tracheid differentiation. The formation of xylem elements, however, was much more strongly promoted. Increasing the kinetin level from 0.1 to 1.0 ppm brought about a fivefold increase in cell production and a fiftyfold increase in tracheary element formation.

Role of cell division. During normal plant development, xylem differentiation tends to occur in regions of prolonged, persistent, or recurrent cell division activity. In primary vascular development, the procambium exhibits continued mitotic activity after cell division has largely stopped in adjacent tissues. As already pointed out, tracheary element formation was preceded and accompanied by active cell division in the cultured pea root segments and the soybean callus tissue. Both cell division activity and xylem differentiation were hormonally regulated in these systems. The question of what role cell division plays in the process of xylem differentiation remains. Of course, cell division will always be necessary for xylem differentiation in the sense that it is the source of additional cells that may be induced to differentiate. For example, in woody species, the production of new tracheary elements is dependent upon continued mitotic activity of the cells of the vascular cambium. In addition to this relationship, can an existing mitotically quiescent parenchyma cell be induced to form a tracheary element without first stimulating that cell to divide?

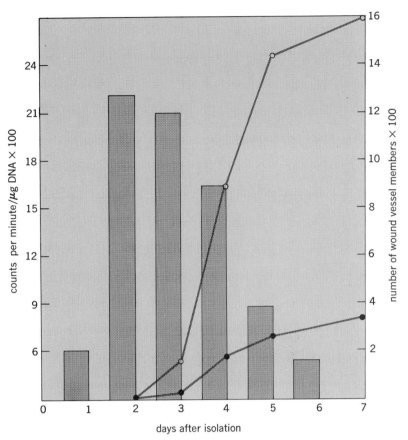

Fig. 4. The time course of xylem differentiation and DNA synthesis in cultured *Coleus* stem segments. Vertical bars represent H³-thymidine incorporation into DNA; closed circles represent number of wound vessel members without auxin in the medium; and open circles represent number of wound vessel members with 50 µg per liter of auxin.

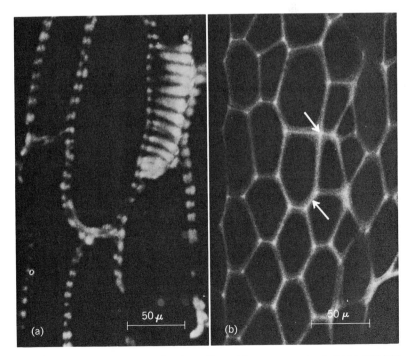

Fig. 5. Light micrographs showing the pattern of fluorescence due to lignin. (a) Light areas represent regions of lignin deposition in the secondary thickenings of differentiating wound xylem elements in *Coleus*. (b) Pitted xylem elements from the secondary xylem of *Coleus* showing fluorescence due to lignin throughout the primary and secondary walls.

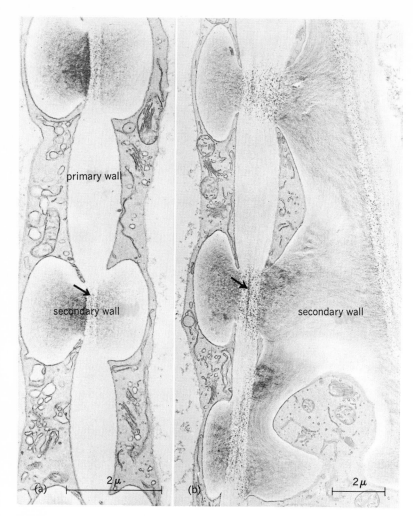

primary wall

secondary wall

secondary wall

(a)   2 μ   (b)   2 μ

Fig. 6(a, b). Electron micrographs of permanganate-fixed differentiating xylem elements. Granules in the primary wall are more concentrated in the vicinity of the middle lamella.

With regard to wound xylem differentiation this question has been answered negatively. When 2-mm-thick segments of the second internode of *Coleus* were placed in axenic culture on a simple medium consisting of 2% sucrose, wound xylem elements were formed in the time course shown in Fig. 4. No wound vessel members were observed until the third day of culture, whereas almost 50% of the wound vessel members appeared on the fourth day of culture. The addition of 50 μg per liter of auxin to the medium increased the number of xylem elements almost fourfold, but auxin at this concentration did not alter the time course of xylem element formation. Upon examining the time course of deoxyribonucleic acid (DNA) synthesis by measuring the rate of H³-thymidine incorporation into this macromolecule, DNA synthesis was found to peak on the second day of culture, 2 days prior to the maximum period of xylem differentiation. An examination of Feulgen-stained squashes for mitotic activity demonstrated that the daily rate of H³-thymidine incorporation into DNA reflected cell division activity.

If cell division were not necessary for xylem differentiation, it could be expected that xylem differentiation would occur normally in the *Coleus* stem segments under culture conditions that pre-

vented cell division. This, however, did not prove to be the case. Inhibitors of DNA synthesis, such as 5-fluoro-deoxyuridine and mitomycin c, as well as inhibitors of cytokinesis such as colchicine, were found to block wound xylem differentiation at concentrations that inhibited cell division. Furthermore, these inhibitors of cell division were largely without effect if given after the peak period of cell division but before most of the xylem elements had begun to form a secondary wall. Thus it would appear that cells are receptive to hormonal and other environmental stimuli that initiate xylem differentiation only during some phase of the mitotic cycle.

Structural studies. A number of techniques have been developed to examine the pattern of lignin deposition in xylem elements. These techniques include a variety of histochemical stains as well as fluorescence microscopy. The latter method takes advantage of the fact that lignin absorbs light in the region of 260–280 nm and emits light in the region of 460–480 nm. Fluorescent micrographs of two kinds of xylem elements are shown in Fig. 5. Tracheary elements with pitted secondary walls exhibit a characteristic pattern of lignin deposition in which the primary wall and middle lamella are more heavily lignified than the layers of the secondary wall (Fig. 5b). With reticulate wound xylem elements, a different pattern of lignification is observed (Fig. 5a). In such xylem elements lignin is confined exclusively to the secondary walls and to the regions of the primary wall upon which the secondary wall is built. Areas of primary wall between the secondary thickenings are not lignified.

Although fluorescence microscopy has been a valuable tool in the study of lignification, its usefulness is limited by the resolution of the light microscope. Recently it was found that permanganate acts as a highly specific electron-dense stain for the lignin component of the cell wall. Thus lignification may be studied with the very high resolution afforded by the electron microscope. An examination of the time course of lignification in the reticulate wound xylem elements of *Coleus*, as seen in the electron microscope after permanganate staining, has revealed that lignification begins as soon as secondary wall deposition is perceptible in the differentiating xylem element. The earliest sites of lignin deposits are the middle lamella and the primary wall underlying the developing secondary thickening. At this time lignin is visible as an electron-dense granulation in the KMnO₄-fixed material. Lignin extends into the growing secondary thickening, but in the secondary wall it is seen as fine, short striations rather than granules (Fig. 6). As the secondary thickening develops, lignin continues to be deposited, progressing from the base of the thickening toward its periphery (Fig. 6a).

Lignin is a complex polymer which is constructed by joining together hundreds of smaller molecules which are all chemically related. They are phenylpropane derivatives and possess an aromatic ring which has a 3-carbon side chain with a double bond. The immediate source of these compounds is the amino acid phenylalanine. Phenylalanine is deaminated by the enzyme phenylalanine ammonia lyase (PAL) to produce cinnamic acid. Cinnamic acid is further modified enzymatically, resulting in the production of the various lignin monomers. Do differentiating xylem elements produce their own lignin precursors or are these phenylpropane derivatives synthesized elsewhere in the plant and transported to the differentiating xylem element? This question was investigated recently by assaying for PAL activity during xylem differentiation. In cultured *Coleus* stem segments a threefold increase in PAL activity was observed as lignin began to be deposited in the differentiating wound vessel members. Thus it would appear likely that the differentiating xylem elements produce their own lignin precursors.

For background information *see* PLANT ANATOMY; PLANT TISSUE SYSTEMS; XYLEM in the McGraw-Hill Encyclopedia of Science and Technology.

[DONALD E. FOSKET]

Bibliography: D. E. Fosket, *Plant Physiol.*, in press; D. E. Fosket and J. G. Torrey, *Plant Physiol.*, 44:871–880, 1969; P. K. Hepler, D. E. Fosket, and E. H. Newcomb, *Amer. J. Bot.*, 51:85–96, 1970; P. H. Rubery and D. E. Fosket, *Planta*, 87:54–62, 1969; J. G. Torrey and D. E. Fosket, *Amer. J. Bot.*, in press.

List of contributors

A

Aaronson, Dr. Stuart A. *Viral Carcinogenesis Branch, National Cancer Institute, Bethesda, Md.* TRANSFORMATION, CELL.

Alper, Dr. Chester A. *The Blood Grouping Laboratory, Boston, Mass.* COMPLEMENT, SERUM.

B

Bach, Dr. Fritz H. *Department of Medical Genetics, University of Wisconsin.* LEUKEMIA.

Bahcall, Dr. John N. *California Institute of Technology.* QUASAR.

Barham, Dr. E. G. *Naval Undersea Research and Development Center, San Diego, Calif.* ELECTRIC ORGAN (BIOLOGY).

Barrett, Dr. Alan H. *Department of Physics, Massachusetts Institute of Technology.* INTERSTELLAR MATTER.

Battista, Dr. O. A. *FMC Corporation, Princeton, N.J.* POLYMER.

Baumann, Dr. Robert C. *Goddard Space Flight Center.* SATELLITES, APPLICATIONS.

Becker, Dr. Yechiel. *Department of Virology, The Hebrew University—Hadassah Medical School, Jerusalem, Israel.* ANTIBIOTIC.

Belter, Dr. Walter G. *Environmental and Sanitary Engineering Branch, Division of Reactor Technology and Development, U.S. Atomic Energy Commission, Washington, D.C.* RADIOACTIVE WASTE DISPOSAL.

Beroza, Dr. Morton. *Agricultural Research Services, U.S. Department of Agriculture, Beltsville, Md.* OPTICAL FIBERS.

Bessis, Dr. M. *Institut de Pathologie Cellulaire, Hôpital de Bicêtre, Paris, France.* CYTOLOGY.

Bilaniuk, Dr. Olexa-Myron. *Department of Physics, Swarthmore College.* SPACE-TIME.

Blair, Dr. G. J. *Department of Soils, University of Guelph, Ontario, Canada.* PLANT, MINERAL RELATIONS OF (coauthored).

Bobo, Dr. Roy A. *Roy Bobo Engineering, Houston, Tex.* OIL AND GAS WELL DRILLING.

Bouwer, Dr. Herman. *U.S. Water Conservation Laboratory, Phoenix, Ariz.* WATER PURIFICATION.

Bowers, Dr. Cyril Y. *Department of Medicine, Tulane University School of Medicine.* HORMONE (in part).

Boyle, Dr. John. *Hope Radium Institute, Manchester, England.* DEOXYRIBONUCLEIC ACID (DNA).

Braun, Dr. Werner. *Institute of Microbiology, Rutgers State University.* ANTIBODY.

Brewer, George E. *Ford Motor Company, Dearborn, Mich.* ELECTRODEPOSITION.

Briskman, Dr. Robert D. *Communications Satellite Corporation, Washington, D.C.* COMMUNICATIONS SATELLITE.

Bromley, Prof. D. Allan. *Nuclear Structure Laboratory, Yale University.* NUCLEAR STRUCTURE.

Brown, Dr. Glenn H. *Liquid Crystal Institute, Kent State University.* LIQUID CRYSTAL.

Brown, Dr. Joel E. *Massachusetts Institute of Technology.* EYE (INVERTEBRATE).

Brown, Dr. Robert M. *Brookhaven National Laboratory.* ATMOSPHERE.

Brownell, Dr. Gordon L. *Massachusetts Institute of Technology.* RADIATION BIOLOGY.

Bruce, Albert W. *Pacific Gas and Electric Company, San Francisco, Calif.* ENERGY SOURCES.

Buddenhagen, Dr. Ivan W. *Department of Plant Pathology, University of Hawaii.* PLANT DISEASE.

Bühler, Dr. F. *Physikalisches Institut, Universität Bern, Switzerland.* MOON (in part).

C

Caddell, Dr. Joan L. *Anemia and Malnutrition Research Center, Chiang Mai Medical College and St. Louis University School of Medicine, Thailand.* NUTRITION.

Cadle, Dr. Richard D. *Chemistry Department, National Center For Atmospheric Research, Boulder, Colo.* ATMOSPHERIC PARTICLES.

Carlson, Dr. Thomas A. *Oak Ridge National Laboratory.* SPECTROSCOPY, PHOTOELECTRON.

Chapman, Dr. J. A. *Rheumatism Research Center, University of Manchester, England.* COLLAGEN.

Clayton, Prof. Donald D. *Department of Space Science, Rice University.* ELEMENTS AND NUCLIDES, ORIGIN OF.

Collins, Dr. J. H. *Electrical Engineering Department, University of Edinburgh, Scotland.* FILTER, ELECTRIC.

Couch, J. C. *Litton Systems, Inc., Pascagoula, Miss.* SHIP PROPULSION.

Couch, Dr. Richard B. *Naval Architecture Research Office, The University of Michigan.* HYDRODYNAMICS (coauthored).

D

Dalziel, Dr. Charles F. *Department of Electrical Engineering, University of California, Berkeley.* ELECTRIC SHOCK PROTECTION.

Dansgaard, Dr. W. *H. C. Oersted Institute, University of Copenhagen, Denmark.* PALEOCLIMATOLOGY.

Davies, Dr. D. Roy. *Department of Applied Genetics, John Innes Institute, Norwich, England.* CULTURE, TISSUE.

Davis, Dr. Robert J. *Smithsonian Institution Astrophysical Observatory, Cambridge, Mass.* ASTRONOMICAL SPECTROSCOPY.

Decker, Prof. Robert W. *Department of Geology, Dartmouth College.* VOLCANO.

Delerno, Joseph. *Offshore Raydist, Inc., New Orleans, La.* DIRECTION-FINDING EQUIPMENT.

Donaldson, Dr. Virginia H. *Shriners Hospitals for Crippled Children, Burns Institute, Cincinnati, Ohio.* BLOOD.

Dowds, John P. *Adamana, Limited, Oklahoma City, Okla.* OIL FIELD MODEL.

Dwyer, Dr. O. E. *Department of Applied Sciences, Brookhaven National Laboratory.* HEAT TRANSFER.

E

Edgar, Dr. Terry. *Scripps Institution of Oceanography, La Jolla, California.* DRILLING, DEEP-SEA (coauthored).

Edgington, Prof. Lloyd V. *Department of Botany, University of Guelph, Ontario, Canada.* PLANT DISEASE CONTROL.

Ellis, John P. *AT&T, New York, N.Y.* TELEPHONE EQUIPMENT, CUSTOMER (in part).

F

Faget, Dr. Maxime A. *NASA Manned Spacecraft Center, Houston, Tex.* SPACECRAFT STRUCTURE.

Falk, Dr. Richard H. *Department of Botany, University of California, Davis.* PLANT ANATOMY.

Farley, Dr. F. F. *Shell Development Company, Emeryville, Calif.* GASOLINE; PETROLEUM PRODUCTS.

Fell, Dame Honor B. *Director, Strangeways Research Laboratory, Cambridge, England.* ORGAN CULTURE.

Fingerman, Dr. Milton. *Department of Biology, Tulane University, New Orleans, La.* CHROMATOPHORE.

Flood, William R. *AT&T, New York, N.Y.* TELEPHONE EQUIPMENT, CUSTOMER (in part).

Forrester, Prof. Jay. *Alfred P. Sloan School of Management, Massachusetts Institute of Technology.* SOCIAL-SYSTEM DYNAMICS.

Fosket, Dr. Donald. *The Biological Laboratories, Harvard University.* XYLEM.

Freidman, Prof. Gerald M. *Department of Geology, Rensselaer Polytechnic Institute, Troy, N.Y.* PROSPECTING, PETROLEUM.

Frenkel, Dr. J. K. *Department of Pathology and Oncology, University of Kansas Medical Center.* TOXOPLASMOSIS.

Furth, Prof. Harold P. *Plasma Physics Laboratory, Princeton University.* PLASMA PHYSICS (in part).

G

Gaffney, Dr. Eugene S. *Department of Vertebrate Paleontology, American Museum of Natural History, New York, N.Y.* CHELONIA.

Gazzaniga, Dr. Michael. *Department of Psychology, New York University.* PSYCHOLOGY, PHYSIOLOGICAL AND EXPERIMENTAL.

Gentry, Dr. R. Cecil. *National Hurricane Research Laboratory, Miami, Fla.* HURRICANE.

Ghiorso, Dr. Albert. *Lawrence Radiation Laboratory, University of California, Berkeley.* ELEMENT 105.

Gilliam, Marco. *AT&T, New York, N.Y.* TELEPHONE SERVICE.

Goldman, Dr. Morton I. *NUS Corporation, Rockville, Md.* NUCLEAR POWER.

Granhall, Dr. Ulf. *Department of Microbiology, Agricultural College, Uppsala, Sweden.* ALGAE.

Grauer, William K. *AT&T, New York, N.Y.* COMMUNICATIONS CABLES.

Grinspoon, Dr. Lester. *Department of Psychiatry, Harvard Medical School.* MARIJUANA.

Gunning, Dr. Brian E. S. *Department of Botany, The Queen's University, Northern Ireland.* TRANSFER CELL, PLANT (coauthored).

H

Haber, Dr. R. N. *Department of Psychology, University of Rochester.* VISION.

Hamilton, Dr. Robert M. *U.S. Geological Survey, Menlo Park, Calif.* EARTHQUAKE (in part).

Haneda, Dr. Yata. *City Museum, Yokosuka, Japan.* BIOLUMINESCENCE (coauthored).

Harrison, Prof. George R. *Department of Physics, Massachusetts Institute of Technology.* DIFFRACTION GRATING.

Harrison, Prof. Walter. *W. W. Hansen Laboratories of Physics, Stanford University.* SOLID-STATE PHYSICS.

Heindl, Dr. L. A. *U.S. National Committee for the International Hydrological Decade.* HYDROLOGY (in part).

Hirshfield, Prof. J. L. *Yale University.* PLASMA PHYSICS (in part).

Hochachka, Dr. Peter. *Department of Zoology, University of British Columbia, Vancouver, Canada.* ISOZYMES.

Hofstadter, Dr. Robert. *Department of Physics, Stanford University.* PARTICLE DETECTOR (coauthored).

Holdeman, Dr. L. V. *Virginia Polytechnic Institute, Blacksburg, Va.* BACTERIA, TAXONOMY OF (coauthored).

Hoover, Richard B. *Astrionics Laboratory, Marshall Space Flight Center.* X-RAY ASTRONOMY.

Horai, Dr. Ki-iti. *Department of Earth and Planetary Science, Massachusetts Institute of Technology.* EARTH, HEAT FLOW IN.

Howatson, Dr. Allan F. *Department of Medical Biophysics, University of Toronto, Ontario, Canada.* MEASLES.

Howell, Dr. F. Clark. *Department of Anthropology, University of Chicago.* HOMINIDAE (in part).

Hughes, Dr. E. B. *Department of Physics, Stanford University.* PARTICLE DETECTOR (coauthored).

Hunten, Dr. Donald M. *Planetary Science Division, Kitt Peak National Observatory.* VENUS.

I

Irving, Dr. Lawrence. *Institute of Arctic Biology, University of Alaska.* ARCTIC BIOLOGY.

Isenhour, Dr. Thomas L. *Department of Chemistry, University of North Carolina, Chapel Hill.* CHELATION.

Ishiguro, Dr. Edward E. *Department of Microbiology, University of Illinois.* ACTINOMYCETALES.

Ishizaka, Dr. Kimishige. *The Johns Hopkins University, School of Medicine at the Good Samaritan Hospital.* HYPERSENSITIVITY.

J

Jacobs, Dr. Donald G. *Health Physics Division, Oak Ridge National Laboratory.* TRITIUM.

Jensen, Dr. Marvin R. *Agricultural Research Service, U.S. Department of Agriculture.* PLANT, WATER RELATION OF.

Joel, Amos E., Jr. *Bell Telephone Laboratories, Holmdel, N.J.* SWITCHING SYSTEMS (COMMUNICATIONS).

Jones, Prof. Lawrence W. *Department of Physics, University of Michigan.* HADRON.

Josephson, Dr. Robert K. *Case-Western Reserve.* NERVOUS SYSTEM (INVERTEBRATE).

K

van de Kamp, Prof. Peter. *Sproul Observatory, Swarthmore College.* BARNARD'S STAR.

Kaplan, Dr. Donald R. *Department of Botany, University of California, Berkeley.* LEAF (BOTANY).

Kasha, Dr. Henry. *Brookhaven National Laboratory.* QUARK.

Kasten, Dr. Paul R. *Oak Ridge National Laboratory.* REACTOR, NUCLEAR (in part).

Kent, Dr. Earle L. *C. G. Conn Ltd., Elkhart, Ind.* MUSICAL INSTRUMENTS.

King, John A. *Space/Aeronautics Magazine.* SPACE FLIGHT.

Kittridge, Dr. James S. *Division of Neurosciences, City of Hope National Medical Center, Duarte, Calif.* PHEROMONES.

Klein, Dr. Cornelis. *Department of Geological Sciences, Harvard University.* MOON (in part).

Koehler, Dr. James S. *Department of Physics, University of Illinois.* CHANNELING IN SOLIDS.

Korringa, Prof. P. *Netherlands Institute for Fishery Investigations.* MARICULTURE.

Kramer, James J. *Lewis Research Center, Cleveland, Ohio.* TURBOJET.

Krinsley, Prof. David. *Department of Geology, Queens College, Flushing, N.Y.* MICROSCOPE, ELECTRON.

Kryter, Dr. K. D. *Sensory Sciences Research Center, Stanford Research Institute, Menlo Park, Calif.* NOISE CONTROL.

Kuo, Dr. John. *Columbia University.* EARTH TIDES (co-authored).

L

Laetsch, Dr. W. M. *Department of Botany, University of California, Berkeley.* PHOTOSYNTHESIS.

Lamparter, William C. *Staley/Graphics, Division A. E. Staley Manufacturing Company, Columbus, Ohio.* PRINTING IN COLOR.

Latham, Dr. Gary V. *Lamont-Doherty Geological Observatory, Columbia University.* MOON (in part).

Lederman, Prof. Leon M. *Department of Physics, Columbia University.* MUON.

Li, Dr. Choh Hao. *Hormone Research Laboratory, University of California, San Francisco.* HORMONE (in part).

Lounasmaa, Prof. O. V. *Department of Physics, Technical University of Helsinki, Otaniema, Finland.* ABSOLUTE ZERO.

Low, Dr. Frank J. *Lunar Planetary Observatory, University of Arizona.* GALAXY, INFRARED.

Lubin, Dr. Moshe. *Department of Mechanical and Aerospace Sciences, University of Rochester.* PLASMA PHYSICS (in part).

Lubliner, Paul. *AT&T, New York, N.Y.* TELEPHONE EQUIPMENT, CUSTOMER (in part).

M

McCarthy, Dr. Eugene D. *Universidad Catolica de Valparaiso, Chile.* ORIGIN OF LIFE.

Machattie, Dr. Lorne A. *Department of Medical Cell Biology, University of Toronto, Ontario, Canada.* GENE.

McMillan, Dr. J. A. *Argonne National Laboratory.* ELECTRON PARAMAGNETIC RESONANCE (EPR).

MacPherson, Dr. H. G. *Oak Ridge National Laboratory.* REACTOR, NUCLEAR (in part).

Mamay, Dr. Sergius H. *U.S. Geological Survey, National Museum, Washington, D.C.* CYCADALES.

Manuali, Bernard. *Centre National d'Etudes Spatiales, Bretigny-sur-Orge, France.* NAVIGATION.

Marrus, Dr. Richard. *Department of Physics, University of California, Berkeley.* ATOMIC BEAMS.

Martell, Dr. Edward A. *National Center for Atmospheric Research, Boulder, Colo.* AIR SAMPLERS, ROCKET-BORNE.

Martin, Dr. James P. *College of Biological and Agricultural Sciences, University of California, Riverside.* SOIL CONSERVATION.

Meiman, Dr. James R. *Department of Watershed Sciences, Colorado State University.* HYDROLOGY (in part).

Metz, Dr. Charles. *Institute of Molecular Evolution, University of Miami.* SPERM CELL.

Miles, Dr. Roger S. *Department of Palaeontology, British Museum (Natural History), London, England.* PALEOZOIC FISH.

Miller, Dr. J. F. A. P. *The Walter and Eliza Hall Institute for Medical Research, Royal Melbourne Hospital, Victoria, Australia.* IMMUNOLOGY.

Miller, Dr. M. H. *Department of Soil Science, University of Guelph, Ontario, Canada.* PLANT, MINERAL RELATIONS OF (coauthored).

Moore, Dr. W. E. C. *Virginia Polytechnic Institute, Blacksburg, Va.* BACTERIA, TAXONOMY OF (coauthored).

Moss, J. L. *College of Engineering, The University of Michigan.* HYDRODYNAMICS (coauthored).

Mulligan, Dr. Hugh F. *Department of Agronomy, Cornell University.* WATER POLLUTION (in part).

N

Narahashi, Dr. Toshio. *Duke University Medical Center.* BIOPOTENTIALS AND ELECTROPHYSIOLOGY.

Nesbit, E. A. *Bell Telephone Laboratories, Murray Hill, N.J.* MAGNET.

Norrby, Dr. Elring. *Department of Virus Research, Karolinska Institutet, Stockholm, Sweden.* ADENOVIRUSES.

Nummela, Walter. *Cleveland Cliffs Iron Company, Cleveland, Ohio.* STEEL MANUFACTURE.

Nunamaker, Dr. Robert R. *Ames Research Center, Moffet Field, Calif.* SUN.

Nutter, Gene D. *Instrumentation Systems Center, University of Wisconsin.* TEMPERATURE MEASUREMENT.

O

O'Brien, Thomas. *PRB Electronic, Inc., Syosset, Long Island, N.Y.* CIRCUIT (ELECTRONICS).

Olmsted, Leonard M. *Electrical World Magazine, McGraw-Hill Publications, New York, N.Y.* ELECTRICAL UTILITY INDUSTRY.

Olson, Dr. Harry F. *RCA Laboratories, Princeton, N.J.* ELECTRONIC MUSIC SYNTHESIS; SOUND-REPRODUCING SYSTEMS (in part).

Ostriker, Dr. J. P. *W. K. Kellogg Radiation Laboratory, California Institute of Technology.* PULSAR.

Overseth, Prof. Oliver E. *Department of Physics, The University of Michigan.* SYMMETRY LAWS (PHYSICS).

P

Pakiser, Dr. Louis C. *U.S. Geological Survey, National Center for Earthquake Research, Menlo Park, Calif.* EARTHQUAKE (in part).

Pate, Prof. John S. *Botany Department, The Queen's University, Northern Ireland.* TRANSFER CELL, PLANT (coauthored).

Peterson, Dr. M. N. A. *Scripps Institution of Oceanography.* DRILLING, DEEP-SEA (coauthored).

van Poollen, Dr. H. K. *H. K. van Poollen and Associates, Littleton, Colo.* PETROLEUM SECONDARY RECOVERY.

Porter, Dr. Lynn K. *Agricultural Research Service, U.S. Department of Agriculture.* SOIL CHEMISTRY (in part).

Posner, Dr. Michael. *Department of Psychology, University of Oregon.* PERCEPTION.

R

Reese, Dr. Elwyn T. *U.S. Army Natick Laboratories, Natick, Mass.* WOOD PRODUCTS.

Renick, Dr. Robert C. *The Mitre Corporation, Bedford, Mass.* NAVIGATION SYSTEMS, ELECTRONIC.

Reynolds, Dr. John J. *Strangeways Research Laboratory, Cambridge, England.* CALCITONIN.

Riggs, Dr. Robert F. *Sperry Marine Systems Division, Division of Sperry Rand Corp., Charlottesville, Va.* NAVIGATION INSTRUMENTS.

Risebrough, Dr. Robert. *Department of Nutritional Sciences, University of California, Berkeley.* PESTICIDE.

Rom, Frank. *Lewis Research Center, Cleveland, Ohio.* NUCLEAR AIRCRAFT.

S

Sandler, Dr. Harold. *Ames Research Center, Moffett Field, Calif.* BIOTELEMETRY.

Schmidt, Dr. Jean Marie. *Department of Botany and Microbiology, Arizona State University.* CAULOBACTERACEAE.

Schultes, Dr. Richard Evans. *Botanical Museum, Harvard University.* PSYCHOPHARMACOLOGICAL DRUGS (in part).

Schwartz, Dr. Mischa. *Department of Electrical Engineering, Polytechnic Institute of Brooklyn.* COMMUNICATIONS, ELECTRICAL.

Schweitzer, Richard. *Citizens and Southern National Bank, Atlanta, Ga.* INDUSTRIAL ENGINEERING.

Shaw, Milton. *U.S. Atomic Energy Commission, Washington, D.C.* NUCLEAR FUELS.

Shen-Miller, Dr. Jane. *Division of Biological and Medical Research, Argonne National Laboratory.* PLANT MOVEMENTS.

Shier, Dr. W. Thomas. *The Salk Institute for Biological Studies, San Diego, Calif.* BACTERIAL GENETICS.

Silberman, Dr. Edward. *Civil Engineering Department, University of Minnesota.* HYDROLOGY (in part).

Simons, Dr. Elwyn L. *Peabody Museum, Yale University.* PRIMATES, FOSSIL.

Smith, Gale M. *AT&T, New York, N.Y.* TELEPHONE EQUIPMENT, CUSTOMER (in part).

Smith, Prof. Gustavus E. *Department of Electrical Engineering, Ohio University.* DOPPLER RADAR.

Smythies, Dr. John R. *Department of Psychiatry, University of Edinburgh, Scotland.* DRUG ADDICTION.

Sperling, Dr. Albert R. *Bureau of Narcotics and Dangerous Drugs, U.S. Department of Justice.* PSYCHOPHARMACOLOGICAL DRUGS (in part).

Stewart, Dr. B. A. *Soil and Water Conservation Research Division, Southwestern Great Plains Research Center, Bushland, Tex.* WATER POLLUTION (in part).

Stow, Dr. C. D. *Department of Physics, University of Auckland, New Zealand.* LIGHTNING.

Swider, Dr. William. *Air Force Cambridge Research Laboratories, Hanscom Field, Bedford, Mass.* ECLIPSE, ASTRONOMICAL; IONOSPHERE.

T

Tattersall, Dr. Ian. *Peabody Museum, Yale University.* HOMINIDAE (in part).

Taylor, Dr. Barry. *National Bureau of Standards, Washington, D.C.* ATOMIC CONSTANTS.

Thomas, Dr. Grant W. *Department of Agronomy, University of Kentucky.* SOIL CHEMISTRY (in part).

Tichauer, Dr. Erwin R. *Institute of Rehabilitation Medicine, New York University Medical Center.* HUMAN-FACTORS ENGINEERING.

Tochner, Dr. Max. *Silicone Products Department, General Electric Company, Waterford, N.Y.* COMPUTER-ASSISTED ANALYTICAL CHEMISTRY.

Tsuji, Dr. Frederick I. *Department of Biophysics and Microbiology, University of Pittsburgh.* BIOLUMINESCENCE (coauthored).

U

Uttal, Dr. William R. *Mental Health Research Institute, The University of Michigan.* NEUROPHYSIOLOGY.

V

Valentine, Dr. James W. *Department of Geology, University of California, Davis.* ECOSYSTEM.

W

Waller, Dr. Thomas R. *Department of Paleobiology, Smithsonian Institution, Washington, D.C.* ANIMAL EVOLUTION.

Wasserman, Dr. Aaron O. *Department of Biology, The City College of the City University of New York.* POLYPLOIDY.

Whithed, Prof. Marshall H. *Department of Political Science, Rensselaer Polytechnic Institute, Troy, N.Y.* SIMULATION.

Willens, Dr. Ronald H. *Bell Telephone Laboratories, Murray Hill, N.J.* BAND THEORY OF SOLIDS.

Williams, Dr. Theodore J. *Purdue Laboratory for Applied Industrial Control, Purdue University.* PROCESS CONTROL.

Withbroe, Dr. George L. *Astronomy Department, Harvard University.* OBSERVATORY, ASTRONOMICAL.

Wood, Dr. Francis A. *Department of Plant Pathology, Pennsylvania State University.* AIR POLLUTION.

Woodward, Dr. J. G. *RCA Laboratories, Princeton, N.J.* SOUND-REPRODUCING SYSTEMS (in part).

Z

Zetler, Dr. Bernard D. *Environmental Science and Services Agency, U.S. Department of Commerce, Miami, Fla.* EARTH TIDES (coauthored).

Zuman, Dr. Petr. *Department of Chemistry, Clarkson College of Technology, Pottsdam, N.Y.* POLAROGRAPHIC ANALYSIS.

Zurbrick, John R. *Avco Corporation, Lowell, Mass.* NON-DESTRUCTIVE TESTING.

McGRAW-HILL YEARBOOK OF SCIENCE AND TECHNOLOGY

Index

Index